GEOTECHNICAL SPECIAL PUBLICATION NO. 187

CONTEMPORARY TOPICS IN GROUND MODIFICATION, PROBLEM SOILS, AND GEO-SUPPORT

SELECTED PAPERS FROM THE 2009 INTERNATIONAL FOUNDATION CONGRESS AND EQUIPMENT EXPO

March 15–19, 2009
Orlando, Florida

SPONSORED BY
The Geo-Institute of the American Society of Civil Engineers

International Association of Foundation Drilling (ADSC)

Pile Driving Contractors Association (PDCA)

EDITED BY
Magued Iskander, Ph.D., P.E.
Debra F. Laefer, Ph.D.
Mohamad H. Hussein, P.E.

Published by the American Society of Civil Engineers

Library of Congress Cataloging-in-Publication Data.

International Foundation Congress and Equipment Expo (2009 : Orlando, Fla.)
Contemporary topics in ground modification, problem soils, and geo-support : selected papers from the 2009 International Foundation Congress and Equipment Expo, March 15-19, 2009, Orlando, Florida / sponsored by the Geo-Institute of the American Society of Civil Engineers [and] International Association of Foundation Drilling (ADSC [and] Pile Driving Contractors Association (PDCA) ; edited by Magued Iskander, Debra F. Laefer, Mohamad H. Hussein.
p. cm. – (Geotechnical special publication ; no. 187)
Includes bibliographical references and index.
ISBN 978-0-7844-1023-3
1. Soil stabilization--Congresses. I. Iskander, Magued. II. Laefer, Debra F. III. Hussein, Mohamad H. IV. American Society of Civil Engineers. Geo-Institute. V. Association of Drilled Shaft Contractors (U.S.) VI. Pile Driving Contractors Association (U.S.) VII. Title.

TA749.I587 2009
624.1'51363--dc22 2009004035

American Society of Civil Engineers
1801 Alexander Bell Drive
Reston, Virginia, 20191-4400

www.pubs.asce.org

ISBN 978-0-7844-1023-3
Manufactured in the United States of America.

Geotechnical Special Publications

1 *Terzaghi Lectures*
2 *Geotechnical Aspects of Stiff and Hard Clays*
3 *Landslide Dams: Processes, Risk, and Mitigation*
7 *Timber Bulkheads*
9 *Foundations & Excavations in Decomposed Rock of the Piedmont Province*
11 *Dynamic Response of Pile Foundations-Experiment, Analysis and Observation*
14 *Geotechnical Aspects of Karst Terrains*
15 *Measured Performance Shallow Foundations*
16 *Special Topics in Foundations*
17 *Soil Properties Evaluation from Centrifugal Models*
18 *Geosynthetics for Soil Improvement*
19 *Mine Induced Subsidence: Effects on Engineered Structures*
21 *Hydraulic Fill Structures*
22 *Foundation Engineering*
23 *Predicted and Observed Axial Behavior of Piles*
24 *Resilient Moduli of Soils: Laboratory Conditions*
25 *Design and Performance of Earth Retaining Structures*
27 *Geotechnical Engineering Congress*
28 *Detection of and Construction at the Soil/Rock Interface*
29 *Recent Advances in Instrumentation, Data Acquisition and Testing in Soil Dynamics*
32 *Embankment of Dams-James L. Sherard Contributions*
33 *Excavation and Support for the Urban Infrastructure*
34 *Piles Under Dynamic Loads*
35 *Geotechnical Practice in Dam Rehabilitation*
37 *Advances in Site Characterization: Data Acquisition, Data Management and Data Interpretation*
39 *Unsaturated Soils*
40 *Vertical and Horizontal Deformations of Foundations and Embankments*
41 *Predicted and Measured Behavior of Five Spread Footings on Sand*
42 *Serviceability of Earth Retaining Structures*
43 *Fracture Mechanics Applied to Geotechnical Engineering*
44 *Ground Failures Under Seismic Conditions*
45 *In Situ Deep Soil Improvement*
46 *Geoenvironment 2000*
47 *Geo-Environmental Issues Facing the Americas*
48 *Soil Suction Applications in Geotechnical Engineering*
49 *Soil Improvement for Earthquake Hazard Mitigation*
50 *Foundation Upgrading and Repair for Infrastructure Improvement*
51 *Performance of Deep Foundations Under Seismic Loading*
52 *Landslides Under Static and Dynamic Conditions-Analysis, Monitoring, and Mitigation*
53 *Landfill Closures-Environmental Protection and Land Recovery*
54 *Earthquake Design and Performance of Solid Waste Landfills*
55 *Earthquake-Induced Movements and Seismic Remediation of Existing Foundations and Abutments*
56 *Static and Dynamic Properties of Gravelly Soils*
57 *Verification of Geotechnical Grouting*
58 *Uncertainty in the Geologic Environment*
59 *Engineered Contaminated Soils and Interaction of Soil Geomembranes*
60 *Analysis and Design of Retaining Structures Against Earthquakes*
61 *Measuring and Modeling Time Dependent Soil Behavior*
62 *Case Histories of Geophysics Applied to Civil Engineering and Public Policy*
63 *Design with Residual Materials: Geotechnical and Construction Considerations*
64 *Observation and Modeling in Numerical Analysis and Model Tests in Dynamic Soil-Structure Interaction Problems*
65 *Dredging and Management of Dredged Material*
66 *Grouting: Compaction, Remediation and Testing*
67 *Spatial Analysis in Soil Dynamics and Earthquake Engineering*
68 *Unsaturated Soil Engineering Practice*
69 *Ground Improvement, Ground Reinforcement, Ground Treatment: Developments 1987-1997*
70 *Seismic Analysis and Design for Soil-Pile-Structure Interactions*
71 *In Situ Remediation of the Geoenvironment*
72 *Degradation of Natural Building Stone*
73 *Innovative Design and Construction for Foundations and Substructures Subject to Freezing and Frost*

136 *Innovations in Grouting and Soil Improvement*
137 *Legal and Liability Issues in Geotechnical Engineering*
138 *Site Characterization and Modeling*
139 *Calibration of Constitutive Models*
140 *Slopes and Retaining Structures under Seismic and Static Conditions*
141 *International Perspectives on Soil Reinforcement Applications*
142 *Waste Containment and Remediation*
143 *Geomechanics: Testing, Modeling, and Simulation*
144 *Sinkholes and the Engineering and Environmental Impacts of Karst*
145 *Seismic Performance and Simulation of Pile Foundations in Liquefied and Laterally Spreading Ground*
146 *Asphalt Concrete: Simulation, Modeling and Experimental Characterization*
147 *Unsaturated Soils 2006*
148 *Advances in Unsaturated Soil, Seepage, and Environmental Geotechnics*
149 *Site and Geomaterial Characterization*
150 *Soil and Rock Behavior and Modeling*
151 *Advances in Earth Structures: Research to Practice*
152 *Ground Modification and Seismic Mitigation*
153 *Foundation Analysis and Design: Innovative Methods*
154 *Pavement Mechanics and Performance*
155 *Underground Construction and Ground Movement*
156 *Geomechanics II: Testing, Modeling, and Simulation*
157 *Computer Applications in Geotechnical Engineering*
158 *Contemporary Issues in Deep Foundations*
159 *Case Studies in Earth Retaining Structures*
160 *Dynamic Response and Soil Properties*
161 *Embankments, Dams, and Slopes: Lessons from the New Orleans Levee Failures and Other Issues*
162 *Problematic Soils and Rocks and In Situ Characterization*
163 *Geoenvironmental Engineering*
164 *Innovative Applications of Geophysics in Civil Engineering*
165 *Geosynthetics in Reinforcement and Hydraulic Applications*
166 *Educational Activities in Geotechnical Engineering*
167 *Geotechnics of Soil Erosion*
168 *Grouting for Ground Improvement: Innovative Concepts and Applications*
169 *Soil and Material Inputs for Mechanistic-Empirical Pavement Design*
170 *Probabilistic Applications in Geotechnical Engineering*
171 *Advances in Shallow Foundations*
172 *Soil Improvement*
173 *Advances in Measurement and Modeling of Soil Behavior*
174 *Designing Our Underground Space*
175 *Field Measurements in Geomechanics 2007*
176 *Analysis of Asphalt Pavement Materials and Systems: Emerging Methods*
177 *GeoCongress 2008: Geotechnics of Waste Management and Remediation*
178 *GeoCongress 2008: Geosustainability and Geohazard Mitigation*
179 *GeoCongress 2008: Characterization, Monitoring, and Modeling of GeoSystems*
180 *From Research to Practice in Geotechnical Engineering*
181 *Geotechnical Earthquake Engineering and Soil Dynamics IV*
182 *Pavements and Materials: Characterization, Modeling, and Simulation*
183 *Sinkholes and the Engineering and Environmental Impacts of Karst*
184 *Pavements and Materials: Modeling, Testing, and Performance*
185 *Contemporary Topics in Deep Foundations*
186 *Contemporary Topics in In Situ Testing, Analysis, and Reliability of Foundations*

Preface

This Geotechnical Special Publication (GSP), along with its two companion volumes contain selected papers from the International Foundations Congress and Equipment Expo (IFCEE09) held in Orlando, Florida on March 15–19, 2009. This unprecedented Congress combined the 2009 annual meetings of the Geo-Institute of ASCE, the International Association of Foundation Drilling (ADSC), and the Pile Driving Contractors Association (PDCA). IFCEE09 provided an international forum to discuss technological advances, case histories, and present challenges related to foundations, ground support, and earth retention. The congress was heavily attended by a wide range of geo-professionals including engineers, contractors, academicians, equipment manufacturers, geo-technologists, researchers, and suppliers of tools and services.

An international call for papers and a rigorous peer review process yielded 242 accepted technical papers that were presented in 44 sessions, in addition to invited keynote presentations and 12 panel discussions. These contributions are organized by theme in three GSPs, as follows:

- GSP **185**: Contemporary Topics in Deep Foundations
- GSP **186**: Contemporary Topics in In Situ Testing, Analysis, and Reliability of Foundations
- GSP **187**: Contemporary Topics in Ground Modification, Problem Soils, and Geo-Support

The editors would like to express their appreciation for having been provided the opportunity to be a part of this Congress' organization, and hope that these proceedings will be of use to the geotechnical engineering community, for many decades to come.

The Editors

Magued Iskander, PhD, PE
Polytechnic Institute of New York University
New York, NY, USA
Iskander@poly.edu

Debra F. Laefer, PhD
University College Dublin, Ireland
debra.laefer@ucd.ie

Mohamad H. Hussein, PE
GRL Engineers, Inc.
Orlando, Florida, USA
MHGRL@pile.com

Acknowledgments

The success of this Congress is credited to the authors of the technical papers for their valuable contributions to the geotechnical literature. The commendable quality of the Congress' technical content and these proceedings is also the result of the efforts of the hundreds of reviewers who conducted over 960 reviews. We thank the reviewers for generously contributing their valuable time and expertise, in many cases, under time pressures and demanding deadlines.

A new ASCE paper management system was utilized for the first time for the papers submission and management of IFCEE09. We thank Patrick Driscoll of Mira Digital publishing system for his excellent customer service, and also thank all involved for their patience and understanding during the difficult phase in process.

Acknowledgment is also due to members of the Geo-Institute's Deep Foundations Committee (Mohamad Hussein, Chairman) who for over three years of preparations contributed thousands of hours of volunteer effort, and to members of the technical committees of the ADSC and PDCA who provided valuable and much needed support to assure the success of this Congress. In particular, the following individuals deserve special acknowledgment and recognition for their extraordinary efforts in making this first-of-its-kind Congress a reality and a resounding success: Dr. Dan Brown, Mr. William (Billy) Camp III, Dr. Brian Anderson, Ms. Carol Bowers, Mr. Scot Litke, Mr. Steve Hall, Dr. Paul Mayne, Mr. Allen Cadden, Mr. Michael Moore, Dr. Sarah Gassman, Dr. Sam Paikowsky, and Ms. Jan Hall. To all of you, and the many others who are too numerous to name, THANK YOU.

Finally, we wish to gratefully acknowledge the support we received from our employers Polytechnic Institute of New York University, University College Dublin, and GRL Engineers, Inc.

Review Criteria for Papers Published in This Geotechnical Special Publication

Following standards of practice of the Geo-Institute of The American Society of Civil Engineers, each papers published in this Geotechnical Special Publication was peer reviewed by at least two anonymous, qualified, technical reviewers (and in many cases three or four reviewers) and selected for publication by the proceedings editors. An advanced document management software was utilized in order to assure anonymity and maintain uniformity of standards. As such, the papers contained in these proceedings are eligible for discussion in the ASCE Journal of Geotechnical and Geoenvironmental Engineering and for ASCE awards.

REVIEW BOARD

The following individuals served as Track and Session Chairs. They coordinated and managed the peer review process of the papers included for publication in this Geotechnical Special Publication and its companion volumes. The quality of theses world-class proceedings is a testament to their professionalism and dedication. Listed in alphabetical order, by first name, they are:

Ahmed Elgamal, PhD, University of California, San Diego
Alan Lutenegger, PhD, PE, University of Massachusetts at Amherst
Alan Macnab, PEng, Condon-Johnson & Associates
Allen W. Cadden, PE, Schnabel Engineering
Amit Bhasin, PhD, The University of Texas at Austin
Amy Cerato, PhD, PE, University of Oklahoma
Anand J. Puppala, PhD, PE, The University of Texas at Arlington
Andrea Welker, PhD, PE, Villanova University
C. Hsein Juang, PhD, PE, Clemson University
Charles Schwartz, PhD, University of Maryland
Dan Brown, PhD, PE, Dan Brown & Associates
David White, PhD, Iowa State University
Debra F. Laefer, PhD, University College Dublin
Edward L. Hajduk, DEng, PE, The Citadel
Elizabeth (Liz) Smith, PE, GE, Terracon Consultants
Emin Kutay, PhD, PE, Michigan State University
George Filz, PhD, PE, Virginia Polytechnic Institute and State University
George K. Burke, PE, Hayward Baker, Inc.
J. Brian Anderson, PhD, PE, University of North Carolina at Charlotte
Jesus Gomez, PE, PhD, Schnabel Engineering
John Delphia, PE, Texas Department of Transportation
Kok Kwang (KK) Phoon, PhD, PEng, National University of Singapore
L. Sebastian Bryson, PhD, PE, University of Kentucky
Lance Roberts, PhD, PE, South Dakota School of Mines and Technology
Larry Olson, PE, Olson Engineering
Laureano R. Hoyos, Ph.D., P.E., The University of Texas at Arlington
Limin Zhang, PhD, Hong Kong University of Science and Technology
Luis Garcia, PE, Geo Cim, Inc.
Magued Iskander, PhD, PE, Polytechnic Institute of New York University
Miguel Pando, PhD, PEng, University of Puerto Rico at Mayagüez
Mike Justason, PEng, BERMINGHAM Foundation Solutions
Mohamed A. Gabr, PhD, PE, North Carolina State University
Mohamed Hussein, PE, GRL Engineers, Inc.
Monica Prezzi, PhD, Purdue University
Mourad Zeghal, PhD, Rensselaer Polytechnic Institute
Muhannad T. Suleiman, PhD, Lafayette College

Nick Hudyma, PhD, PE, University of North Florida
Paul Bullock, PhD, PE, Tolunay-Wong Engineers, Inc.
Paul Mayne, PhD, PE, Georgia Institute of Technology
Phillip Ooi, PhD, PE, University of Hawaii at Manoa
Rami El-Sherbiny, PhD, PE, Geosyntec Consultants
Raymond J. Castelli, PE, Parsons Brinckerhoff
Rick Deschamps, PhD, PE, Nicholson Construction Company
Robert A. Carnevale, DBM Contractors, Inc.
Robert Mokwa, PhD, PE, Montana State University
Robert Y. Liang, PhD, PE, University of Akron
Samuel Paikowsky, ScD, University of Massachusetts Lowell
Sanjeev Malahotra, PE, Parsons Brinckerhoff
Sarah Gassman, PhD, PE, University of South Carolina
Silas Nichols, PE, Federal Highway Administration
Susan Burns, PhD, PE, Georgia Institute of Technology
Thomas J. Tuozzolo, PE, Moretrench American Corporation
Timothy C. Siegel, PE, Berkel & Company Contractors
W. Robert Thompson, III, PE, Dan Brown & Associates
William M. Camp, PE, S&ME, Inc.
Xiong (Bill) Yu, PhD, PE, Case Western Reserve University

LIST OF REVIEWERS

The editors are grateful to the following 358 individuals for working tirelessly under a tight schedule in order to peer review the papers included for publication in this Geotechnical Special Publication and its companion volumes. As a group the following individuals performed over 960 peer reviews, and their service to the geotechnical community is gratefully appreciated.

Aaron Budge, Minnesota State Univ. at Mankato
Abdul-Hamid Soubra, Univ. of Nantes
Ahmed Elgamal, Univ. of California, San Diego
Ahmet Aydilek, Univ. of Maryland
Al Griffith, Aramco
Alaa Ashmawy, American Univ. in Dubai
Alex Smith, Subsurface Construction
Alexander McGillivray, Berkel & Company Contractors, Inc.
Ali Tolooiyan, Univ. College, Dublin
Allen Jones, South Dakota State Univ.
Americo Fernandez, Synchropile, Inc.
Amit Bhasin, Univ. of Texas At Austin
Amr Ragy, Treviicos
Amy Cerato, Univ. of Oklahoma
Anand Puppala, Univ. of Texas At Arlington
Anastasia Santoso, National Univ. of Singapore
An-Bin Huang, National Chiao Tung University
Andrea Welker, Villanova Univ.
Andrew Graettinger, Univ. of Alabama
Anil Misra, The Univ. of Kansas
Anirban De, Manhattan College
Antonio Carraro, Colorado State Univ.

Antonio Marinucci, Univ. of Texas At Austin
António Viana Da Fonseca, Univ. of Porto
Art Hoffmann, Gannett Fleming, Inc.
Ashish Juneja, Indian Inst. of Technology, Bombay
Aspasia Zerva, Drexel Univ.
Assaf Klar, Technion - Israel Inst. of Technology
Ayhan Gurbuz, Atilim Univ.
Aykut Erkal, Univ. College Dublin
Ayoade Adeyefa, Worleyparsons Group
G L Sivakumar Babu, Indian Inst. of Science, Bangalore
Ba'te Ba'te, Georgia Institute of Technology
Beena Sukumaran, Rowan Univ.
Ben Cote, North Carolina State Univ.
Ben Possiel, Subsurface Construction
Bengt Fellenius, Independent Consultant
Bernard Hertlein, STS-AECOM
Bill Starke, Deep Foundations Contractors Inc.
Bin Zhang, Case Western Reserve Univ.
Botao Lin, Univ. of Oklahoma
Brent Robinson, North Carolina State Univ.
Brent Rosenblad, Univ. of Missouri-Columbia
Brian Anderson, Univ. of North Carolina at Charlotte
Brian Jasperse, Geocon
Burcin Avar, Scott-Wilson UK
Chad Freed, Widener Univ.
Chadi El Mohtar, Univ. of Texas At Austin
Chamil (Hiroshan) Hettiarachchi, Lawrence Technological University
Charles Ng, Hong Kong Univ. of Science & Technology
Charles Schwartz, Univ. of Maryland
Chianan Liu, National Chi Nan University
Ching Kuo, PSI
Ching Tsai, Louisiana DOTD
Chris Nickel, Louisiana DOTD
Christina Vulova, Tensar
Christine Weber, MWH Americas, Inc.
Christopher Clarke, ODOT
Chu Ho, Arup
Chuck Dowding, Northwestern Univ.
Chung Song, Univ. of Mississippi
Claudia Zapata, Arizona State University
Conrad Felice, Lachel Felice & Associates
Craig Calabria, Geosystems
Cumaraswamy Vipulanandan, Univ. of Houston
Curt Basnett, CH2M Hill
Curt Christensen, Parsons Brinckerhoff
Daehyeon Kim, Chosun University
Dan Cadenhead, Anderson Drilling
Darrell Wilder, Schnabel Engineering
David Arellano, The Univ. of Memphis
David Berti, Berti Lindquist Consulting Engineers
David Cotton, Kleinfelder
David Crapps, David K Crapps, Inc.
David Druss, Bechtel
David Salter, Terracosta Consulting Group
David Weatherby, Schnabel Foundation Company
David White, Iowa State Univ.
Dean Hardister, North Carolina DOT
Debra F. Laefer, Univ. College Dublin
Dennis Boehm, Hayward Baker, Inc.
Dennis Grubb, Schnabel Engineering
Dennis Hiltunen, Univ. of Florida
Dimitrios Zekkos, Univ. of Michigan

Dipanjan Basu, Univ. of Connecticut
Dominic Parmantier, Condon-Johnson & Associates, Inc.
Donald Bruce, Geosystems, L.P.
Donald Gerken, Bechtel, Inc. San Fransisco, Ca
Dongwook Kim, Purdue Univ.
Donovan Ledford, URS Corporation
Doug Green, Ohio Univ.
Edward Glynn, Villanova University
Edward Hajduk, The Citadel
Elizabeth Dwyre, Parsons Brinckerhoff
Elizabeth Smith, Terracon Consultants
Eric Brandner, South Dakota School of Mines & Technology
Eric Burke, Wilbur Smith Associates
Eric Drumm, Univ. of Tennessee, Knoxville
Eric Weaver, Federal Highway Administration
Ernian Pan, Univ. of Akron
Eyad Masad, Texas A&M University
Frank Pepe, Parsons Brinckerhoff
Frank Townsend, Univ. of Florida
Fred Kulhawy, Cornell University
Frederick Rhyner, Mueser Rutledge Consulting Engineers
Garland Likins, Pile Dynamics, Inc.
Gary LeCoq, Louisiana DOTD
Gary Norris, Univ. of Nevada, Reno
Gary Taylor, Parsons Brinckerhoff
George Burke, Hayward Baker Inc
George Burrough, Condon-Johnson & Assoc., Inc.
George Filz, Virginia Polytechnic Institute and State Univ.
George Goble, Consulting Engineer
Gerald Miller, Univ. of Oklahoma
Gerald Verbeek, VMS
Gray Mullins, Univ. of South Florida
Greg Hebeler, Golder Associates
Gregory Baecher, Univ. of Maryland
Gregory Canivan, S&ME, Inc.
Guillermo Zavala, Ardaman & Associates
Guney Olgun, Virginia Polytechnic Institute and State Univ.
Gunjin Yun, Univ. of Akron
Gyaneswor Pokharel, Mountain State University
Halil Ceylan, Iowa State University
Hani Titi, Univ. of Wisconsin - Milwaukee
Hanping Hong, Univ. of Western Ontario
Hoe Ling, Columbia Univ.
Hong Bo Zhou, Shanghai Jianke Project Management Co. Ltd
Hongwei Huang, Tongji Univ.
Ioannis Zevgolis, Purdue Univ.
Ivan Au, City Univ. of Hong Kong
Jamal Nusairat, El Robinson
James Blackburn, Hayward Baker Inc
James Langer, Gannet Fleming Inc.
James Nevels, Jr., Retired ODOT
James Niehoff, Professional Services Industries, Inc.
James Parkes, Parsons Brinckerhoff
James Pegues, Southern Company
James Sheahan, HDR Inc.
Jan Hartman, Parsons Brinckerhoff
Jason Dejong, Univ. of California, Davis
Jean Benoit, Univ. New Hampshire
Jeff Bean, Layne Geoconstruction
Jeffry Evans, Bucknell University
Jerry Dimaggio, FHWA
Jesus Gomez, Schnabel Engineering
Jianye Ching, National Taiwan University
Jie Han, Univ. of Kansas
Jie Zhang, Hong Kong Univ. of Science & Technology
Jie Zhang, New Mexico State University
Jiliang Li, Univ. of Akron
Jim Collin, The Collin Group

Jin Hung Hwang, National Central University
Jinyuan Liu, Ryerson Univ.
Joan Manuel Larrahondo-Cruz, Georgia Institute of Technology
Joan O'brien, Parsons Brinckerhoff
Jodie Nixon, Subsurface Construction
Joe Wang, Parsons Brinckerhoff
Joel Sprague, TRI
John Bowders, Univ. of Missouri-Columbia
John Byrne, Ground Support Pllc
John Christian, Consulting Engineer
John Davie, Bechtel
John Delphia, Texas Department of Transportation
John Horne, Parsons Brinckerhoff
John Jenkins, Hanson Professional Services, Inc.
John Lessley, S&ME, Inc.
John Magnavita, Langan Engineering
John McCartney, Univ. of Colorado at Boulder
John Quaranta, West Virginia University
John Schmertmann, John H. Schmertmann Inc.
John Turner, Univ. of Wyoming
John Wilson, McMaster University
John Wolosick, Hayward Baker
Jorge Zornberg, Univ. of Texas at-Austin
Joseph Caliendo, Utah State University
Joseph Wartman, Drexel University
Joshua Adams, WPC Inc.
Juan Bernal, Univ. of Puerto Rico
K.K. Muraleetharan, Univ. of Oklahoma
Kallol Sett, Univ. of California, Davis
Ke Fan, Parsons Brinckerhoff
Ke Yang, CH2M Hill
Kenneth Bell, Bechtel Power Corp.
Kerstin Lesny, Univ. of Duisburg-Essen
Kevin Foye, CTI & Associates, Inc
Khaled Sobhan, Florida Atlantic Univ.
Kimberly Warren, UNC Charlotte
K. J. Kim, NCDOT
Kok-Kwang Phoon, National Univ. of Singapore
Kord Wissmann, Geopier Foundation Company, Inc.
Kwang Ro, Parsons Brinckerhoff
Kyle Rollins, Brigham Young University
Kyungseob Cha, Daewoo E&C
Lance Roberts, South Dakota School of Mines & Technology
Larry Olson, Olson Engineering
Laureano Hoyos, Univ. of Texas At Arlington
Laurence R. Bentley, Univ. of Calgary
Les Chernauskas, GTR Inc.
Liang Cui, Univ. College Dublin
Limin Zhang, Hong Kong Univ. of Science & Technology
Lindsey Bryson, Univ. of Kentucky
Lucian Bogdan, Dywidag Systems, USA
Luis Garcia, Geo Cim, Inc.
Lyndi Blackburn, Alabama Department of Transportation
M. Emin Kutay, Michigan State University
Madira Madhav, JNT University
Magued Iskander, Polytechnic Institute of New York University
Mahongo Dithinde, Univ. of Botswana
Malay Hajra, PSI, Inc.
Manuel Ochoa, Fugro
Marco Boscardin, Boscardin Consulting Engineers
Marco Uzielli, Georisk Engineering S.R.L.
Mark Dunscomb, Schnabel Engineering

Mark Jaksa, Univ. of Adelaide
Mark Thompson, CH2M Hill
Mark Woodward, USACE
Mary Ellen Bruce, Geotechnica, S.A., Inc.
Matthew Janes, Isherwood Associates
Matthew Miller, Terracon, Inc.
Mehmet Tumay, Louisiana State University
Michael Justason, Bermingham Foundation Solutions
Michael Walker, GEI Consultants, Inc.
Michael Woodward, Mactec Engineering & Consulting, Inc.
Michel (Mike) Gambin
Miguel Pando, Univ. of Puerto Rico
Miguel Ruiz, National Univ. of Cordoba
Mike Adams, FHWA
Mike Lewis, Bechtel Corporation
Mike Muchard, Applied Foundation Testing
Mike Wysockey, Thatcher Foundations
Milind Khire, Michigan State Univ.
Ming Xiao, California State Univ. Fresno
Mo Nasim, Schnabel Engineering
Mohamed Ashour, Univ. of Alabama
Mohamed Gabr, North Carolina State Univ.
Mohamed Mekkawy, Iowa State Univ.
Mohamed Nofal, Renewable Energy Systems
Monica Prezzi, Purdue Univ.
Muhannad Suleiman, Lafayette College
Munwar Basha, Indian Inst. of Science, Bangalore
Murad Abu-Farsakh, Louisiana State Univ.
Muthucumarasamy Yogendrakumar, Golder Associates
Nadarajah Sivaneswaran, FHWA
Nadir Ansari, Isherwood Associates
Naji Khoury, Temple Univ.
Naresh Samtani, NCS Consultants
Nasser Massoudi, Bechtel
Nelson Gibson, FHWA
Nick Hudyma, Univ. of North Florida
Ning Lu, Colorado School of Mines
Norman Dennis, Univ. of Arkansas
Nortey Yeboah, Georgia Institute of Technology
Paul Axtell, Dan Brown & Associates
Paul Cosentino, Florida Institute of Technology
Paul Doherty, Univ. College Dublin
Paul Mayne, Georgia Institute of Technology
Paul Passe, PSI
Paulo Pinto, Univ. of Coimbra, Portugal
Pete Cali, USACE
Phil King, Synchropile, Inc.
Phillip Ooi, Univ. of Hawaii
Priyantha Jayawickrama, Texas Tech Univ.
Qian Zhao, Georgia Institute of Technology
Rafael Castro, Jacobs Associates
Rami El-Sherbiny, Geosyntec Consultants
Ray Franz, Hayward Baker Inc.
Raymond Castelli, Parsons Brinckerhoff
Renpeng Chen, Zhejiang Univ.
Richard Meininger, FHWA
Richard Reid, SDSU
Rick Deschamps, Nicholson Construction
Robert Carnevale, DBM Contractors, Inc.
Robert Holtz, Univ. of Washington
Robert Liang, Univ. of Akron
Robert Mokwa, Montana State Univ.
Robert Parsons, Univ. of Kansas

Roger Failmezger, In-Situ Soil Testing LLC
Roger Gobin, Parsons Brinckerhoff
Ron Xia, Mcclymont & Rak Engineers
Ross Cutts, Georgia Institute of Technology
Ross McGillivray, Ardaman & Associates
Roy Borden, North Carolina State Univ.
Roy Olson, The Univ. of Texas at Austin
Rudolph Frizzi, Langan Engineering
Russell Green, Virginia Polytechnic Institute and State Univ.
Sahadat Hossain, Univ. of Texas at Arlington
Sal Italiano, SPI Consulting Engineers
Salah Sadek, American Univ. of Beirut
Sam Frydman, Israel Institute of Technology
Samer Dessouky, The Univ. of Texas at San Antonio
Samuel Khor, Ansys
Samuel Paikowsky, Univ. of Massachusetts Lowell
Sandra Houston, Arizona State Univ.
Sangchul Bang, South Dakota School of Mines & Technology
Sanjeev Malhotra, Parsons Brinckerhoff
Sanjoy Chakraborty, Wilbur Smith Associates
San-Shyan Lin, Taiwan Ocean Univ.
Sarah Gassman, Univ. of South Carolina
Scott Jacobs, ECS Carolinas LLp
Scott Mackiewicz, Kleinfelder
Scott Raschke, Schnabel Engineering
Scott Wasman, Univ. of Florida
Seyit Ceribasi, Univ. College Dublin
Shadi Najjar, American Univ. of Beirut
Shailendra Amatya, Univ. of Massachusetts Lowell
Shawn Kelley, Geodesign Inc
Sherif Elfass, Univ. of Nevada, Reno
Silas Nichols, FHWA
Sireesh Saride, Univ. of Texas at Arlington
Soheil Nazarian, Univ. of Texas, El Paso
Sridhar Chandran, Polytechnic Institute of New York University
Stefan Romanoschi, Univ. of Texas At Arlington
Steven Levorson, Terracon Consultants, Inc.
Storer Boone, Golder Associates
Susan Burns, Georgia Institute of Technology
Syam Nair, Texas A&M University
T. Evans, North Carolina State Univ.
T. Wu, Ohio State Univ.
Tae Ju, Subsurface Construction Company
Tanner Blackburn, Hayward Baker Inc.
Tanya Kunberger, Florida Gulf Coast Univ.
Tara Hutchinson, Univ. of California, San Diego
Terence Holman, Moretrench American Corporation
Terry Micnhimer, Tolunay-Wong Engineering
Thomas Bird, Williams Form Engineering
Thomas Cooling, URS Corporation
Thomas Lee, Parsons Brinckerhoff
Thomas Tuozzolo, Moretrench American Corporation
Thomas Weaver, Univ. of Idaho
Thomas Zimmie, Rensselaer Polytechnic Institute
Tim Myers, Layne Geoconstruction

Timothy Siegel, Berkel & Company Contractors
Tom Armour, DBM Contractors
Tom Gurtowski, Shannon & Wilson
Tom Richards, Nicholson Construction
Usama El Shamy, Southern Methodist Univ.
Van Komurka, Wagner Komurka Geotechnical Group, Inc.
Vishnu Diyaljee, Gaea Engineering Ltd
W. Nesmith, Berkel & Company Contractors
W. Robert Thompson, Dan Brown & Associates
Weslie Ross, Fundaciones, S. A.
William Camp, S&ME, Inc.
William Isenhower, Ensoft, Inc.
William Kaggwa, Univ. of Adelaide
William Kennard, Geosolutions, LLC
William Likos, Univ. of Missouri - Columbia
William McCuskey, Ardaman & Associates Inc.
Wing Heung, Parsons Brinckerhoff
Wolfgang Fritz, NCS Consultants
Wout Broere, Delft Univ. of Technology
Xiong Yu, Case Western Reserve Univ.
Yan Liu, Case Western Reserve Univ.
Yao Xu, Hong Kong Univ. of Science & Technology
Yoichi Watabe, Port & Airport Research Institute
Yu Wang, City Univ. of Hong Kong
Zhen Liu, Case Western Reserve Univ.
Zoran Batchko, Parsons Brinckerhoff

Contents

Earth Retaining Structures

Geosynthetics

Indexes

Field Monitoring of Active Zone and Cracked Retaining Wall

C. Vipulanandan[1], Kalaiarasi Vembu[2], N. Sivaruban[3] and Ö. Bilgin[4]

[1]Chairman, Professor and Director of Center for Innovative Grouting Materials and Technology (CIGMAT), and Director of Texas hurricane Center for Innovative Technology, Department of Civil and Environmental Engineering, University of Houston, Houston, Texas 77204-4003; cvipulanandan@uh.edu
[2 & 3]Graduate Students, Center for Innovative Grouting Materials and Technology (CIGMAT),, University of Houston, Houston, Texas 77204-4003; kvembu@mail.uh.edu
[4]Assisstant Professor, Department of Civil and Environmental Engineering, University of Houston, Houston, Texas 77204-4003; obilgin@uh.edu

ABSTRACT: In this study, the causes of distress (crack movement) in the face panels of a 244 m (800 ft) long and 4.6 m (15 ft) high retaining wall was investigated by instrumenting the active zone adjacent to the wall. The foundation of the retaining wall was placed in a CH soil few feet below the surface and the ground water table was about 7 feet below ground. Total of three boreholes were instrumented to determine the vertical and horizontal ground movements in the active zone and to determine the variation of matric soil suction adjacent to the wall. The instrumentation included tensiometers, extensometers and an inclinometer. Two additional boreholes were instrumented with extensometers and pore-water transducers to determine the consolidation settlement. The crack opening and closing in the retaining wall panels were measured regularly.

Based on the measurements over period of nine months, greatest vertical and horizontal displacements were observed in the top five feet of the ground. The tensiometers measured the changes in the matric suction pressures during this period. The pressures varied from 77 kPa to 10 kPa in the dry and wet weather conditions. Highest vertical settlement and swelling in the active zone were -0.51 cm (-0.2 in.) and 2.03 cm (0.8 in.) respectively and were in phase with suction pressure measurements. The lateral movement in the top five feet, during the monitoring period was about 1.02 cm (0.4 in.) (moving away from the wall). The consolidation settlement was about 0.25 cm (0.1 in.), much smaller than what is being observed in the active zone. Hence the movements in the active zone have had the greatest influence on the crack movement in the face panels of the retaining wall.

INTRODUCTION

Active zone is the portion of the ground that experiences seasonal changes in moisture content. The depth of active zone is influenced by the soil type, vegetation, water table, drainage and climate. The measurement of soil suction can be used to measure the depth of active zone (depth of constant suction). Aubeny et al., (2007) measured the depth of the moisture active zone y_{ma}, based on moisture diffusion coefficient α and the frequency of suction variations n by the general expression

$$y_{ma} = C_{ma}\sqrt{\alpha / n} \quad (1)$$

Hence the depth of the active zone is very much influenced by the moisture diffusion coefficient. Variation of matric suction influences the shear strength capacity of soil deposits in the unsaturated zone (active zone). The time history of suction at the ground surface could be approximated as a sinusoidal function (Mitchell, 1979).

$$u(0,t) = U_e + U_0\cos(2\pi tn) \quad (2)$$

where n = frequency of the suction cycles; U_e = invariant component of suction; U_0 = amplitude of the suction variations at the ground surface. The magnitude of U_0 is estimated from observed seasonal variations at the ground surface. Fredlund and Rohardjo (1993) modified the Mohr-Coulomb failure criterion (Eq. 3) for unsaturated soil by incorporating modified effective stress expression, based on the independent stress state variable approach.

$$\tau_f = c' + [(\sigma - u_a) \tan \varphi' + (u_a - u_w)] \tan \varphi^b \quad (3)$$

where c'= effective cohesion intercept; φ'= effective angle of internal friction; $(\sigma - u_a)$ = net normal stress; $(u_a - u_w)$ = matric suction and φ^b = rate of increase in shear strength with respect to matric suction. Similarly, the matric suction will influence the volume changes (swelling and shrinkage) in the soils near and below the retaining wall.

Therefore studying the behavior of the soil in the active zone based on the matric suction pressure gives better understanding of the soil behavior. Nichol et al., (2008) studied the long term measurements of in situ matric suction for an unsaturated mine waste rock piles. In this study, the variation of matric suction in the active with time and climatic change was studied. The lateral movement in the ground was measured using an inclinometer.

OBJECTIVES

The overall objective of this study was to investigate the causes of distress in the retaining wall panels due to near surface ground conditions. The specific objectives are as follows:

1. Quantify the vertical and horizontal movements in the ground (active zone and consolidation) near the retaining wall over period of time.
2. Determine the changes in the suction pressure in the ground over period of time.
3. Measure the changes in the retaining wall panel crack widths with time

SITE CONDITIONS

The 244 m (800 ft.) long retaining wall about 4.6 m (15 ft.) high, in front of an embankment, has been in service for over fourteen years. Inspection of the site indicated highly distressed panels with numerous cracks, Three borings were drilled in 2007 (Fig. 1) to measure the settlement and pore pressure changes in the soft soil layer (Fig. 2) Figure 2 shows the typical soil profile below the retaining wall. The top 1-7.5 m was CH clay soil and below it was CL soil. The moisture content varied from 19-60%. The liquid limit varied from 30-75%. The undrained shear strength obtained from the unconfined compression test varied between 13.7 and 120.5 kPa in the top 14 m of CH clay.

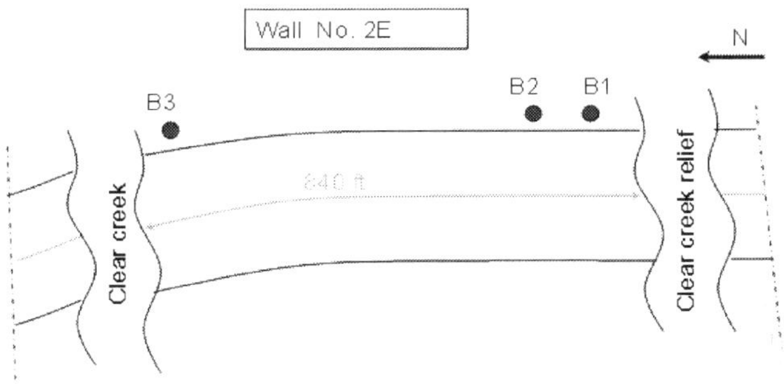

Figure 1 Schematic View of the Location of Boreholes

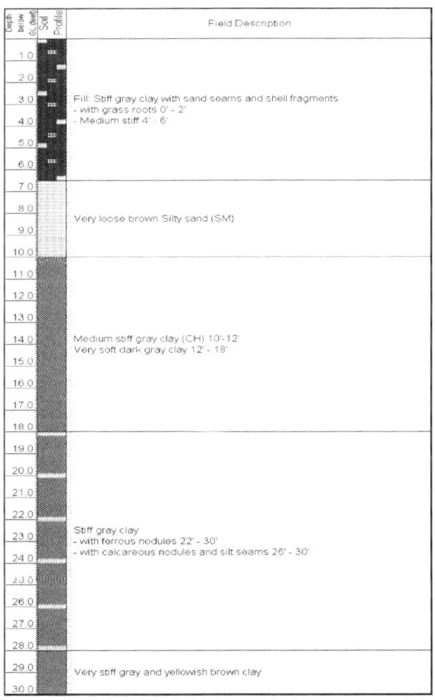

Figure 2 Soil Profile

The retaining wall had noticeable failures due to settlement and Fig. 3 shows one of the cracked panels in the retaining wall. Figure 4 shows the measured changes in the cracked wall and it varied from 0.2 mm (opening) to -0.4 mm. The change in the length of the crack opening varied with time and it followed the similar trend with the settlement in the active zone.

Figure 3 Cracked Panel in the Retaining Wall

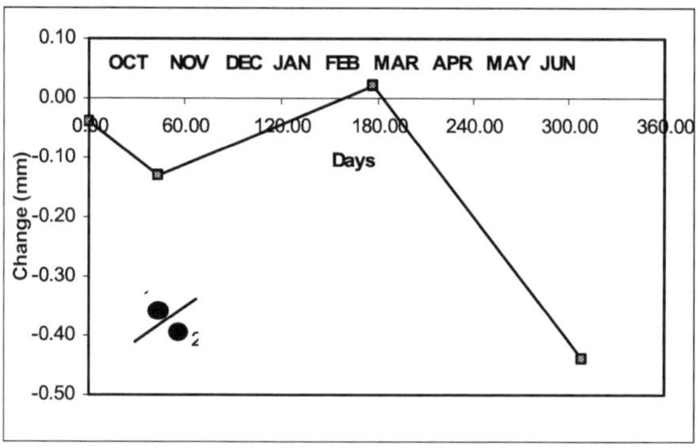

Figure 4 Changes in Crack Opening inthe Wall

FIELD INSTRUMENTATION

Two boreholes (B1 & B3) were instrumented with piezometers to measure the pore water pressure and extensometers to measure the total settlement. Borehole (B2) was instrumented with an inclinometer to measure the lateral deformation. In addition two boreholes of 1.5 m depth were drilled and instrumented with tensiometer and extensometer Fig. 5). Direct measurement of matric suction in a borehole can be obtained by using a tensiometer. A tensiometer consists of a tube with a porous ceramic tip on the bottom, a vacuum gauge near the top and a sealing cap. When tensiometer was filled with water and inserted into the soil and thewater can move into and out of the tensiometer through the connecting pores at the tip. As the soil dries and water moves out of the tensiometer, it creates a vacuum inside the tensiometer which is indicated on the gauge. When the vacuum created is equals the 'Soil Suction', and the water stops flowing out of the tensiometer. If the soil dries further, additional water moves out until a higher vacuum level is reached. Because water can move back and forth through the pores in the porous ceramic tip, the gauge reading is always in balance with the soil suction. The extensometer was monitored with time to measure settlement of the ground based on the reference plate. Therefore the differential settlement within the 1.5 m active zone was identified using the difference in the reading. Figure 4 shows the schematic view of the instruments used.

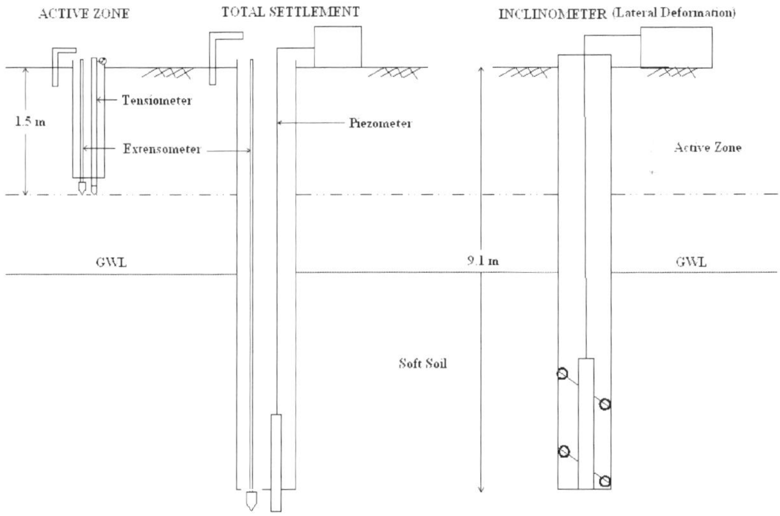

Figure 5 Schematic View of the Instruments

RESULTS AND DISCUSSION

(a) Active Zone

Tensiometer was used to measure the soil suction within the active zone. Two tensiometers with extensometer were installed to a depth of 1.5 m (5 ft) to measure the matric suction and the settlement in the active zone. Figure 6 shows the suction pressure measured. Figure 7 shows the settlement measured within the active zone.

When soil shrinks due to dry weather suction pressure will increase, the ground will settle and the extensometer reading will be negative. During wet conditions suction pressure will decrease, the ground will swell and the extensometer reading will be positive. The maximum suction measured during dry weather and wet weather conditions were 77 kPa and 10 kPa. The corresponding vertical settlement and swelling in the soil measured by the extensometer was -0.51 cm (-0.2 in.) (settlement) and 2.03 cm (0.8 in.) (ground heave). In order to understand the drying and wetting phenomena in the soil, therainfall and temperature data is shown in the Fig. 7. The maximum average rainfall occurred during the months of January to March with average monthly precipitation from 75 mm to 166 mm. This is reflected in the reduced suction pressure and with swelling of the ground due

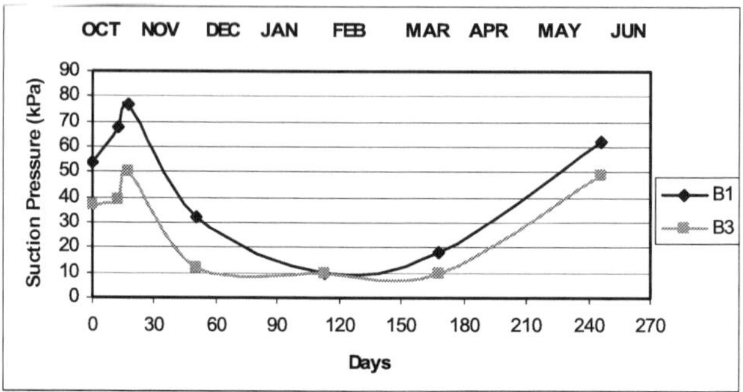

Figure 6 Change in Suction Pressure with Time

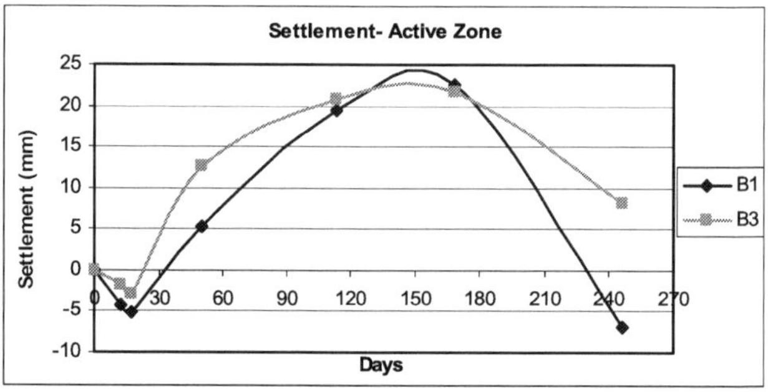

Figure 7 Variation of Settlement in Active Zone

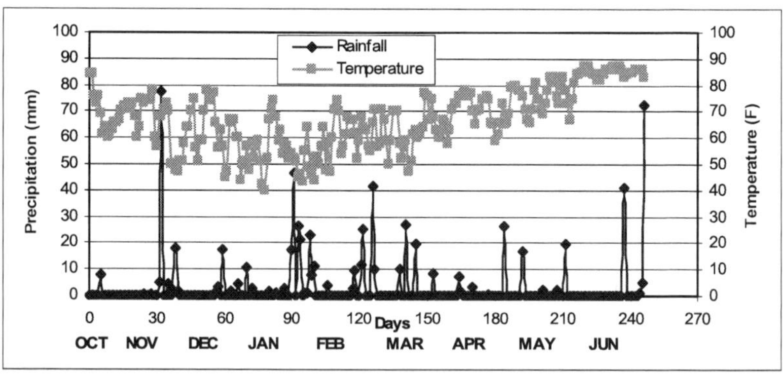

Figure 8 Rainfall and Temperature for the Houston Area (www.weather.gov)

to increased moisture content. Maximum temperature was recorded for the months of May and June with temperatures of 79.5 °F and 84.7 °F respectively. High temperature results in reduced moisture in the ground, higher suction pressure and settlement (Figs. 6 and 7).

It is of interest to note that the crack opening and closing (Fig. 4) trend in the retaining wall had similar trend to the suction pressures and ground settlement and heave and as shown in Figs. 6 and 7 respectively. When there is ground settlement, the crack tends to close and when there is ground heave the crack tends to open.

(b) Consolidation Settlement

Total settlement of the embankment was calculated from the extensometer installed to a depth of 9.14 m (30 ft). Consolidation settlement of the soil was calculated from the total settlement measured from the extensometer and from the active zone settlement. It has been found that the consolidation settlement was low as compared to the active zone settlement. Figure 9 shows the consolidation settlement measured. The vertical settlement measured varied from 0.04 in. to 0.16 in., lower than what was measured in the active zone.

(c) Lateral Deformation

Lateral deformation of the soil was measured using the inclinometer. The inclinometer casing was installed to a depth of 9.14 m (30 ft). The inclinometer survey was taken every 0.6 m (2 ft). Subsequent surveys of the inclinometer casing, when compared with the original survey, will reveal any changes of inclination of the casing and locations at which these changes are taking place. Figures 10 show the lateral movement along the borehole B2. From the figure it had been found that the soil moved laterally away from the wall by more than 6 mm (0.25 in.) in the top 1.5 m (5 ft) and reduces to less than 2.5 mm below 2.4 m (8 ft), which shows that there was more lateral movement within the active zone.

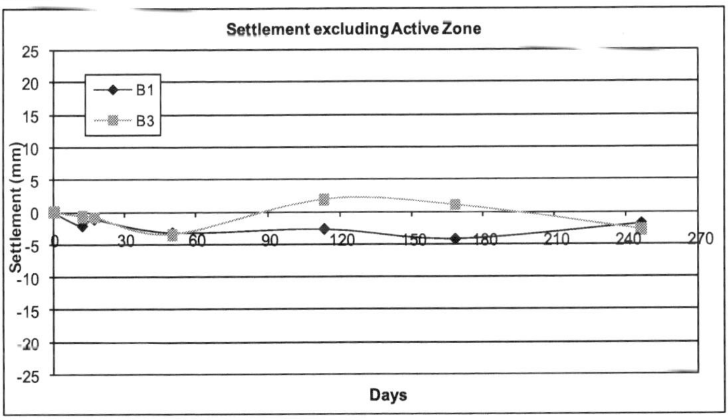

Figure 9 Variation of Consolidation Settlement with Time

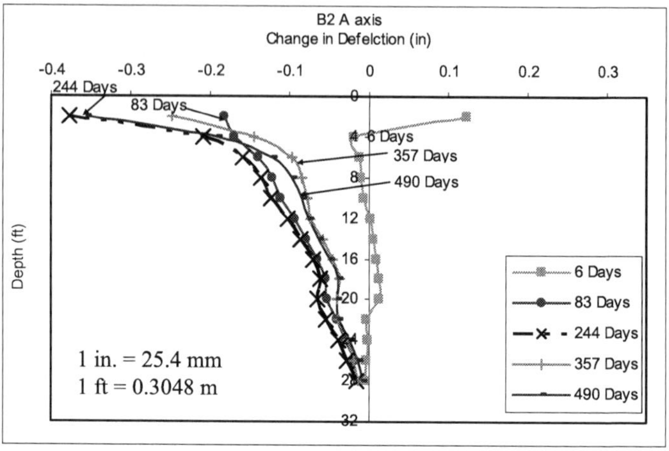

Figure 10 Lateral Deformation Measured in Borehole B2

CONCLUSIONS

Based on the field measurements over period of over nine months, following can be concluded:

1. There was swelling and settlement in the active zone due to changes in the ground moisture content. The maximum and minimum suction pressures were 77 kPa and 10 kPa respectively. Maximum and minimum ground movement in the active zone was 0.8 in (20 mm swelling) and -0.2 in. (-5 mm) respectively.
2. Largest lateral movement was observed in the active zone.
3. Crack opening and closing in the retaining wall panel was directly affected by the active zone movements.

Consolidation settlement was much smaller than what was observed in the active zone. Hence the active zone contributes more toward the distress in the wall.

ACKNOWLEDGEMENT

This study was supported by the Center for Innovative Grouting Materials and Technology (CIGMAT) with funding from the Texas Department of Transportation.

REFERENCES

Aubeny, C. P., and Lytton, R. L. (2007). "Moisture Diffusion in Shallow Clay Masses." *J. Geotech geoenviron. Engrg.*, 133(10), 1241-1248.

Fredlund, D. G., and Rahardjo, H. (1993). *Soil Mechanics for Unsaturated Soils*, Wiley, New York.

Mitchell, P. W. (1979). "The Structural Analysis of Footings on Expansive Soils." *Research Rep. No. 1.* K. W. G. Smith and Associates Pty. Ltd., Newton, South Australia.

Nichol, C., Rowlett, D. K., and Barbour, S. L. (2008). "A New Standpipe Lysimeter Design for the measurement of Soil Matric Suction. " *Vadose Zone J.*, 7, 919-929.

Characteristics of ground expanded by Pulsed Power for Anchorage

Tae-Hoon Kim, Kyung-Seob Cha, and Seon-Ju Kim

Daewoo Institute of Construction Technology, Daewoo E&C, Researcher

Gyeonggi-Do, S. Korea

Buhm-Soo Chang

Korea Infrastructure Safety & Technology Corporation, Ph.d

Gyeonggi-Do, S. Korea

Abstract

Anchor capacity is influenced by ground condition such as shear strength. In general, it has been known that the capacity in soft soil is several times lower that in rock. In addition, it also influenced by the configuration and size of the anchor zone as well. Four types of anchors have been introduced generally according to mechanism of stress transfer and how to be installed. Among them is a bell or under reamed anchor, which can increase resistance. However, the effect of enlargement of fixed zone is controversial. Since such an enlargement is formed by cut mechanically, there always exists so-called smear zone of ground which gives rise to the reduction of shear strength. This paper introduces PDT(Pulse Discharge Technology) anchoring method. The PDT anchoring method is to install an anchor using electrical power so called Pulse power. Since this technique can make not only a bore hole to be expanded but also the ground to be improved by compaction, the anchor installed by PDT is expected to develop shaft and end bearing capacity efficiently. In this study, as preliminary study, we introduce characteristics of ground expansion as well as improvement. In order for quantitative evaluation of the relation between pulse power and ground expansion a series of laboratory tests are conducted. The tests are carried out on sand with different relative densities and various number of pulse discharge. As a result, it is concluded that the expansion and soil improvement appeared to increase with an increase of the number of discharge and with a decrease of relative density.

Key words : Pulse Power, Expansion, Improvement, Pulse discharge, Relative density, Anchors, Laboratory testing

Introduction

Ground anchor which was introduced in 1970s for the first time is one of the most popular ground improvement methods. It has been frequently adopted in constructions such as, subway, skyscraper, underground parking space as well as large scale of excavation in urban area because of its stability and efficient site usage. However, such a ground anchorage has several shortcomings. Firstly, it requires to lengthen of its free zone up to bed rock and to narrow the space between anchors for soft soil condition. It gives rise to an increase of total construction cost inevitably. Secondly, since the ground anchor goes deeper in order to secure proper resistance it is apt to intrude the boundary especially in urban area. On the effort of overcoming, a few alternative methods such as, load distribution compression type of anchor, and JSP and pressure grouting using packer have been developed and used. However, the former may induce local failure since the length of each tendon is different one another in order to mobilize load distribution effect, which requires alternative tensioning method or equipment to prevent load concentration. Meanwhile, it may be hard for the latter to expect the ground strength improvement because of uncertainty of infiltration into the ground. For these reasons, an alternative method is required to be developed. PDT(Pulse Discharge Technology) is to compress electrical energy. By means of electrical impact, the ground is compacted, and therefore the strength can be improved. In this study, the relationship between pulsed power and the ground is investigated. To do so, a series of laboratory tests are carried out.

Pulsed Power

Pulsed power is the physical value that indicates the energy change per unit time (dE/dt). The power depends on how fast the energy is released. In other words, if a unit of energy is discharged in 1 sec, a watt of power will be generated. However, if the energy is discharged in 1 μs (10^{-6}sec), the power will be 1MW(10^{6}Watt) as shown in Fig. 1.

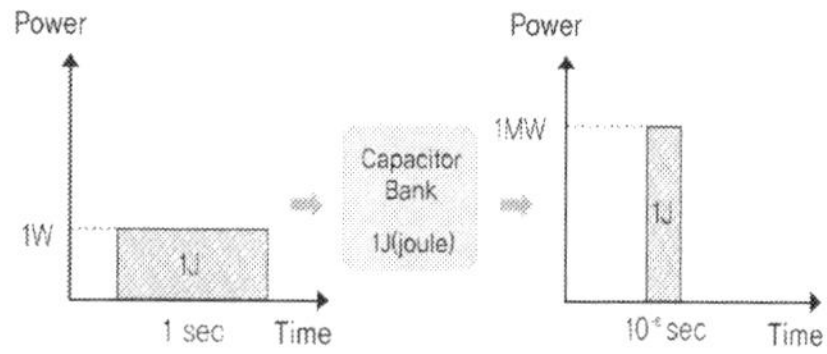

Fig. 1. Pulse power

Shock wave

Shock wave can be generated by the pulse power in a fluidal medium. When a high voltage generates between two electrodes inserted in the medium, the fluidal material between electrodes may heat up. Then the material will be going to expand outward. Meanwhile the material outside the electrodes remains cold and will restrain the heat-up material from expanding. Because of this opposite characteristics between expansion and restraint, a high pressure will be generated at the boundary. Finally this pressure is changed into a different type of energy so called shock wave and propagates through the medium until meets a different type of medium. Once the shock wave reaches the different type of medium, which herein represents a soil layer, it pushes the soil media. Finally the soil layer gets expanded. Fig. 2 shows how the electrical energy changes to the shock wave and potential energy by stage.

Fig. 2. Energy Transformation

The shock wave generated through the several energy transformations as shown earlier propagates to the ground and make it expanded and recompacted. Fig. 3 shows a sketch of concept of expansion by the shock wave induced by the pulse power.

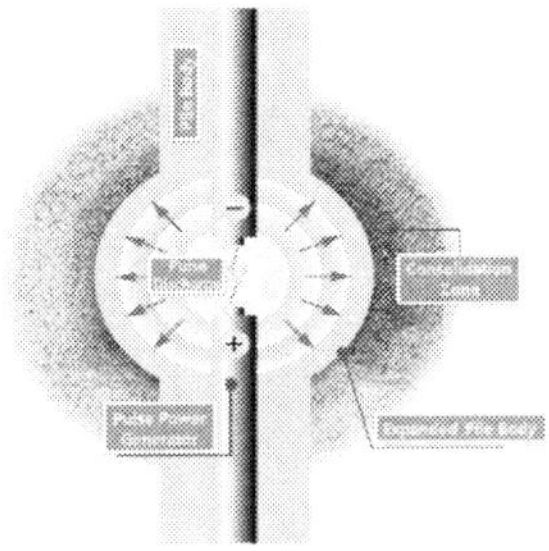

Fig. 3. Conceptual view of ground expansion and an example

COEFFICIENT OF EXPANSION (E_C)

In order to calculate the bearing capacity of a pile, the areas of shaft and tip of the pile are to be determined. In general, most piles have constant shape and diameter along the longitudinal direction, which means the areas are constant. However, for the piles installed using the pulsed power, the shape varies and therefore the area is different one another depending on the pulsed power. Therefore the most important thing for the calculation of bearing capacity is how to take into account the varied area. In this section, the coefficient of expansion is introduced. The details of derivation of the coefficient of expansion (E_c) are as follows. First, the shape of expansion in a hole is assumed to be sphere as shown in Fig.4, for the shock wave is expected to propagate in radial direction.

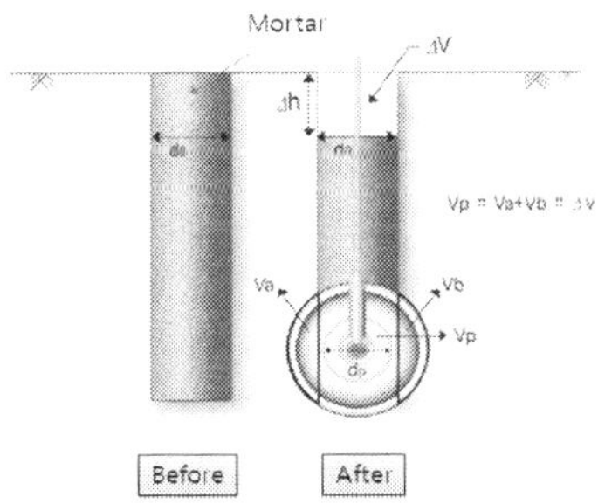

Fig. 4 The assumed shape of a hole expansion by pulse discharge

In the figure, d_0 represents the initial diameter of the hole, and Vp represents the volume of a space occurring instantly at impulse which is assumed to be sphere. If it is assumed that there is no loss of mortar during impulse, then it could be said that there occurs volume change, Δv, as much as the mortar is falling down because of the hole expansion.

The volumetric change by pulse discharge, Vp, can be calculated with easy on the basis of material mechanics.

In the meantime, the coefficient of expansion (E_C) can be defined as the ratio of the final to the initial diameter as following

$$E_c = \frac{d_f}{d_0}$$

EXPERIMENT

PDT Equipment

The pulse discharge equipment consists of main switch, capacitor bank, coaxial power cable and load impact cell as shown in Fig. 5. The process of pulsed power generation is as follows. At first, the commercial electricity is stored and compacted in a capacitor bank. Then the stored electricity is discharged in a moment using a special switching system for high electricity. The momentary discharged electricity generates a high electrical power, so called pulsed power. Finally this pulsed power of high electricity is translated to the ground through the load impact cell.

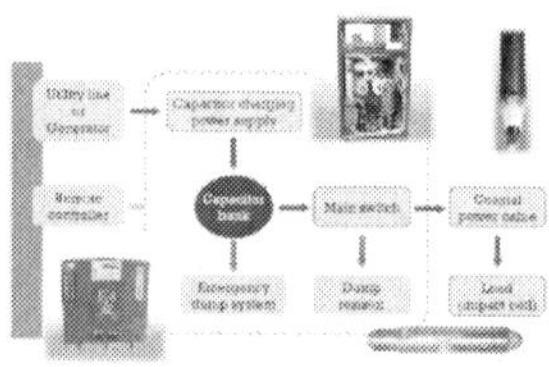

Fig. 5 Component of Pulse Discharge Equipment

Container

In order to investigate the expansion characteristics a series of experiments were conducted in a model container at a laboratory. In general, the space between anchors is dependent on designed force, diameter of an anchor, length of fixed zone and so on. According to previous studies, when the ratio of space to diameter of anchors is same or greater than 6, a similar pullout capacity would be expected for single and group anchors. Similar result was introduced from the experiments on Kanto loam layer in Japan. The results showed that the deformation characteristics of anchor group are almost equal to those of single anchor if the ratio of embeded length to diameter of an anchor is larger than 20, or the space is larger than 1.0m. Besides Japanese geotechnical society recommend the space between anchors to be 10D and the minimum to be 1.5m. In addition, the effecting range of pulse power according to previous research is about 1.2m. On the basis of the above references the model container used in this study has dimensions 1.5m diameter and 1.0m height so as to take into account anchorage space in practice. A schematic view of the container is shown in Fig. 6.

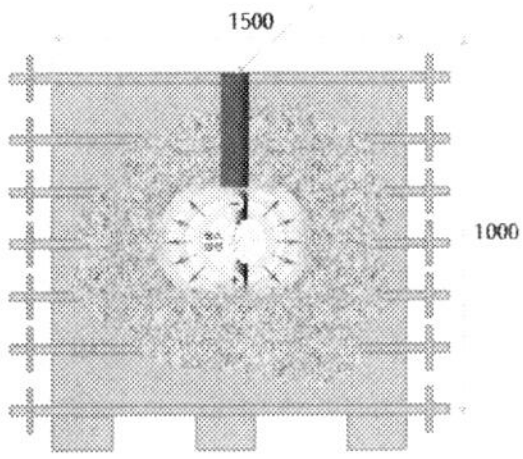

Fig. 6 Container

Model Ground Preparation

In the tests model grounds were prepared using Jumunjin sands. In general, there are two methods to make sand grounds broadly. One is raining technique, and another is compaction. In this study, the latter was selected since the former takes too much time because of the large container. A ground is prepared by compaction dividing into several layers. Compaction is carried by means of a vibrator and a static compacter. During compaction, a hollow pipe which is for grout injection is installed in the middle of the ground. Having completed the ground preparation, dynamic cone penetration test is carried out in order to verify the ground uniformity. Fig.7 shows the procedure of ground preparation

(a) Sand raining (b) Compaction (c) Pipe installation (d) compaction (e) completion

Laboratory Tests

The target ground condition to improve is medium dense sandy soils. Therefore two different soil conditions like relative density 40%, and 60% were selected as shown in Table 1. Since the goal of this study is to figure out the expansion characteristics of the ground we also conduct experiments on each soil condition with different number of impulse as shown in Table. 1

Table 1. Matrix of Experiment

Material	Relative Density	No. of Impulse	sets
Sand (Jumunjin sand)	40%	3	1
		5	2
		10	1
		15	1
		20	1
		40	1
	60%	5	1
		20	1
		40	1

Once the ground is prepared cement paste, of which W/C is 50%, is poured into the hollow pipe up to the top. Then the pipe is pulled up about 15cm in order for later impulse. After injection, the pipe is sealed perfectly with rubber valve so that no energy loss occurs during impulse. Then energy is discharged as designed. After impulse the falling depth of the grout was measured.

Discussion

Fig. 8 shows the shape of anchor bodies at different number of impulse for the relative density of 40%. The shape seems to be rather cylinder than sphere, and the size becomes bigger with an increase of the number of impulse.

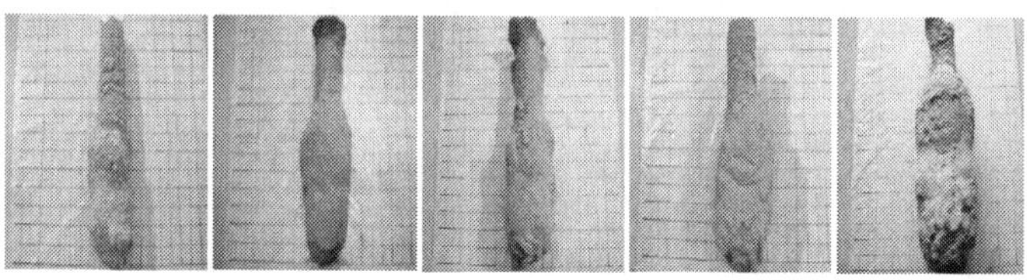

5 10 15 20 40

(a) Dr =40%

5 40

(b) Dr= 60%

Fig. 5 Shape of grouted anchor body

Fig. 9 presents falling depth of the grout with respect to the number of impulse. According to the figure, the relationship between the coefficient of expansion and discharging number seems to be linear in the zone of the number of discharge less than 20. When the number is over 20 the slope appears to gradually decrease, and the degree of decrease seems to be more noticeable for stiff soils.

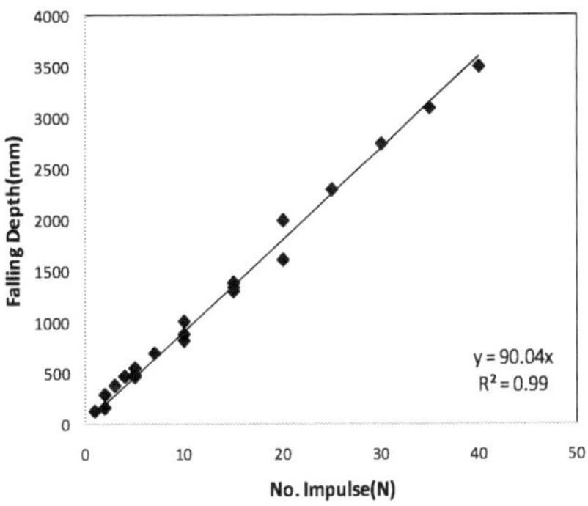

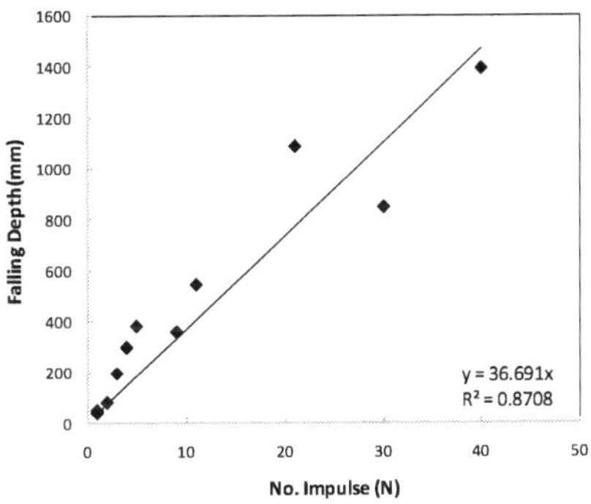

(a)Dr= 40% (b)Dr=60%

Fig. 9 Falling depth of grout vs. No. impulse

From the result, it may be said that the effect of the number of pulse discharge on ground expansion is dependent of ground condition, and remains constant at the end. The average falling depth of grout appears to be 91cm for 40% and 36 cm for 60%. Fig. 10 represents the coefficient of expansion with respect to the number of impulse on the semi-log scale. In case of relative density of 40%, the coefficients appear to be 2.37 at 5 times impulse and 3.95 at 40 times impulse. On the other hand, it represents for 60% of relative density from 2.13 to 3.01. From these observation, it is clear that the more the impulse the larger the expansion coefficient. Meanwhile, it seems that the prediction estimated from the theory is much larger by 15 to 32% that those obtained from measurement.

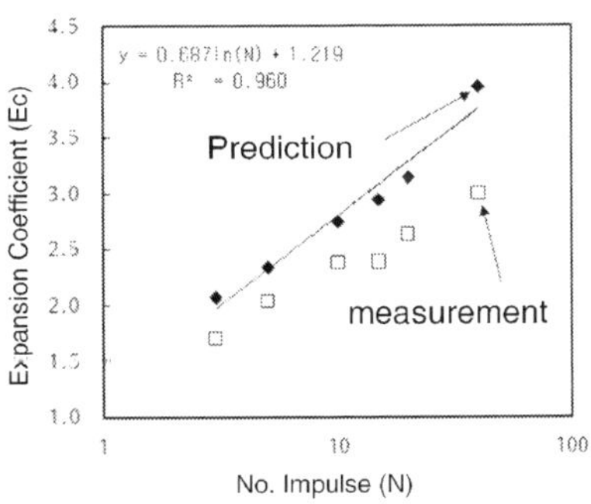

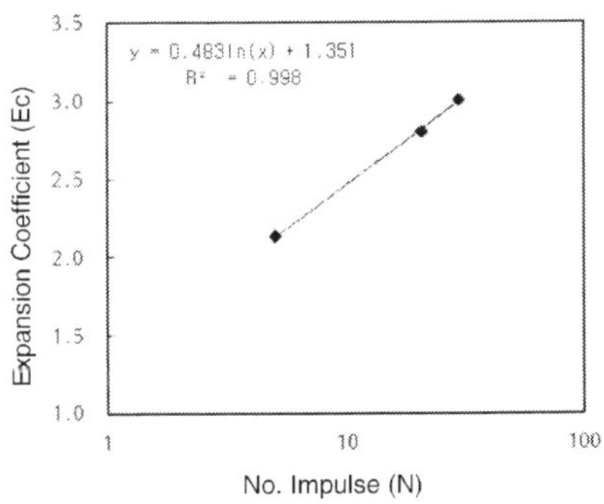

(a)Dr= 40% (b)Dr=60%

Fig. 10 Coefficient of Expansion vs. No. impulse

CONCLUSION

In this study, a series of laboratory tests were performed in order to investigate the effects of pulse power on the ground improvement for anchorage. As a result, the following conclusions could be derived.

1) The shape seems to be rather cylinder than sphere, and the size becomes bigger with an increase of the number of impulse.

2) the effect of the number of pulse discharge on ground expansion is dependent of ground condition

3) The coefficient of Expansion increases with an increase of the number of pulse discharges.

ACKNOWLEDGMENT

This research was supported by a grant (R&D 06C12) funded by Ministry of Construction & Transportation of Korean government

REFERENCE

Hanna, T.H., and M.W. Leonard.(1969), " Some Design and Const. Considerations on the Use of Anchorages and Tiebacks," Piling Committee, Inst. Civ. Eng., Jan. 1969, London.

Kim, T.H., Chai, S.G. and Park, J.M.(2005), "A Study of Applicability of PDT pile", Proceedings of KSCE Annual Conference, pp.4309~4312

Littlejohn, G. S. (1973). "Ground Anchors today-a foreword", Ground Engineering, Vol.6, pp. 20-23.

Lo, K.Y., SHANG, J.Q.(1994), "Effects of intervening media on dielectrophoretic strengthening of soft clays", Canadian Geotechnical journal, Vol.34, pp. 607~613

Lo, K.Y., SHANG, J.Q.,INCULET, I.I.(1994), "Electrical strengthening of clays by dielectrophoresis", Canadian Geotechnical journal, Vol.34, pp. 192~203

PETROS, P.X.(1990),"Ground Anchors And Anchored Structures", JOHN WILLEY & SONS INC.

SHANG, J.Q., Dunlap, W.A.(1996), "Improvement of soft clays by high voltage electrokinetics", Journal of Geotechnical Engineering, pp. 274~280

Dyna Force-An Elasto-Magnetic Sensor for Force Measuring of Ground Anchors

Lucian Bogdan, P.E., Member ASCE

DSI (DYWIDAG-Systems International USA, Inc.), 2154 South Street, Long Beach, CA 90805
PH (562) 531-6161, FAX (562) 531-3266, email: lucian.bogdan@dsiamerica.com

ABSTRACT

Dyna Force sensor uses the elasto magnetic technology (EM) and non destructive testing (NDT) to direct measure stress level in prestressed strands or bars installed on ground anchors or post tensioning tendons.

Dyna Force sensor can be used to rapidly read the change in load during proof and performance tests and compare with the pressure gauge and during the life on the anchor to monitor existing lock-off load. It can be used also to determine the load along the bonded length of the anchor allowing the contractor to increase or decrease this length.

The paper presents Dyna Force operating principle and some applications. In conclusion, the elasto magnetic technology applied through Dyna Force is a new and easy to use approach to monitor anchor load during proof test and service life of the structure.

OPERATING TECHNOLOGY PRINCIPLE[1]

On many occasions during construction and service life of the structure, it is crucial to know the load in post tensioning tendons and anchors. Although there are many methods to measure the tendon force, most of them are cumbersome, expensive and the accuracy differs depending on the method used.

The sensor combines NDT and EM methods and measures directly the internal stress in steel tendons. Dyna Force is manufactured based on the theory that the permeability of steel to a magnetic field changes with the stress level in the steel. By measuring the change in a magnetic field, the magnitude of the stress in the steel element can be obtained.
The Dyna Force sensor is essentially a solenoid composed of a primary coil and a secondary coil (sensing coil), which are insulated from each other and work together

to formalize the elastic magnetic characterization of the material. (Figure 1).

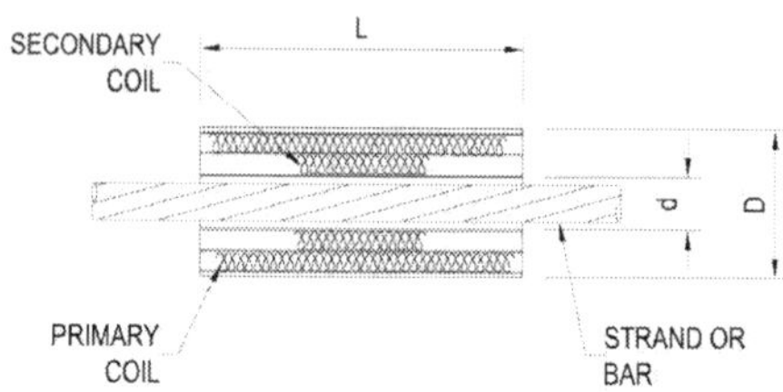

Figure 1

Pulsed current is passing through the primary coil and the sensing coil picks up the induced electromotive force that is directly proportional to the change rate of the applied magnetic flux and the relative permeability of the prestressed steel. The input received by the secondary coil is output to the power stress reading devise. The sensor is designed as a self-contained unit which can be applied onto the strand or bar anchor whose properties are to be measured.

The primary and secondary coils must be wrapped around a non-metallic core that does not exhibit ferromagnetic properties. Plastics and other polymers are ideal materials for the core, which acts as an "insulating" layer between the specimen and the sensor coils and protects the coils from being damaged. Because of the large number of turns in the sensor, there are usually many layers of windings. These layers must be insulated from each other in order to control the heat generated by the current pulse.

ACCURACY AND DURABILITY

The EM sensory technology can reliably monitor actual stress in tensioned steel bars and cables based on the fact that the permeability of ferromagnetic material is a function of magnetic history and applied stress and ambient temperature. Laboratory and field test results shown that Dyna Force sensor monitoring accuracy is within +/-1%.

The sensor durability depends of: copper wire, potting compound, steel cover and connecting cable. Copper wire has a very high durability. When sheathed with special insulations, copper wire may sustain for hundreds of years. Potting compound also protects the copper wire. The shell steel cover is zinc plated and sometimes embedded in grout or concrete. The weakest point of the sensor is the connecting cable so contractor should be careful during sensor and lead wire installation

The sensor and measuring unit has temperature compensating function. Variation in ambient temperature when sensor is placed very close to the anchorage is compensated automatically.

The measuring unit is not critical since the measuring conditions are precisely defined and the damaged measuring unit parts can be easily replaced. The technology will continue to grow but the physical and the measuring steel principle will remain the same. The minimum service life of the Dyna Force sensor is expected to be as long as the structure where is installed.

SENSOR ASSEMBLY AND CALIBRATION

Dyna Force does not alter the characteristics of the tendon other than its magnetization and the calibration process eliminates the impact of these variables. The program placed inside measuring unit, takes into consideration the temperature change effect automatically. Dyna Force sensor must be encased by an external steel shell and sealed in order to protect the coils inside from moisture, corrosion and other elements presented in the field. A lead wire will connect sensor to a portable Power Stress (read-out) unit. This unit will create the magnetic field and then measure the residual value and convert it to a direct force reading. No other devices are necessary. The sensor is installed over strand or bar during construction and a zero reading is taken before applying post tensioning force.

Simple and Robust Dyna Force Sensor

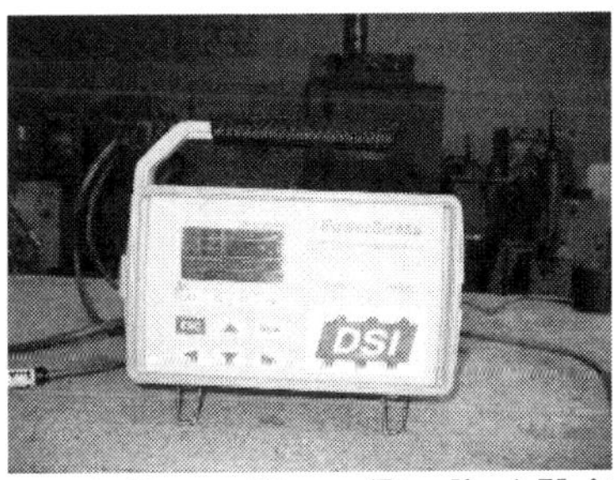

Accurate Power Stress (Reading) Unit

Every sensor is calibrated inside lab before is sent to the field. Items used are stressing jack, test block, a computer, load cell, power stress unit that will storage calibration data for each unit and sample of strand or bar used in the field. Pressure load cell is placed between plates and lead wire is hooked to the power stress unit. The results show that reading on the power stress unit is matching readings of the load cell and jack pressure gauge.

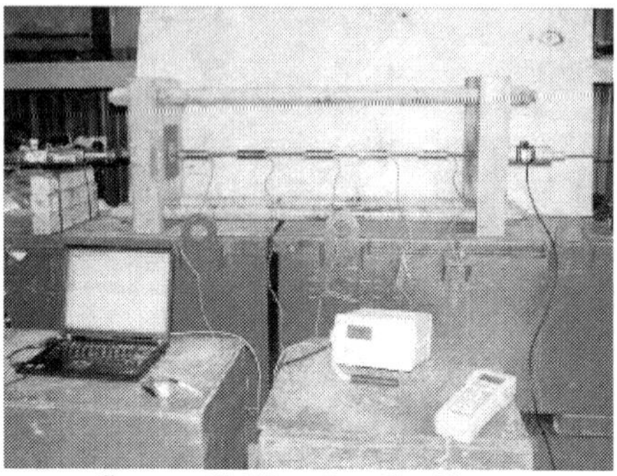

Sensor Calibration Set-Up

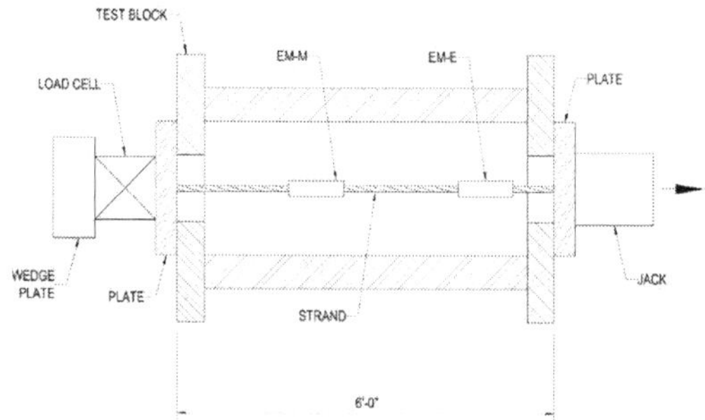

Test Block Set-Up

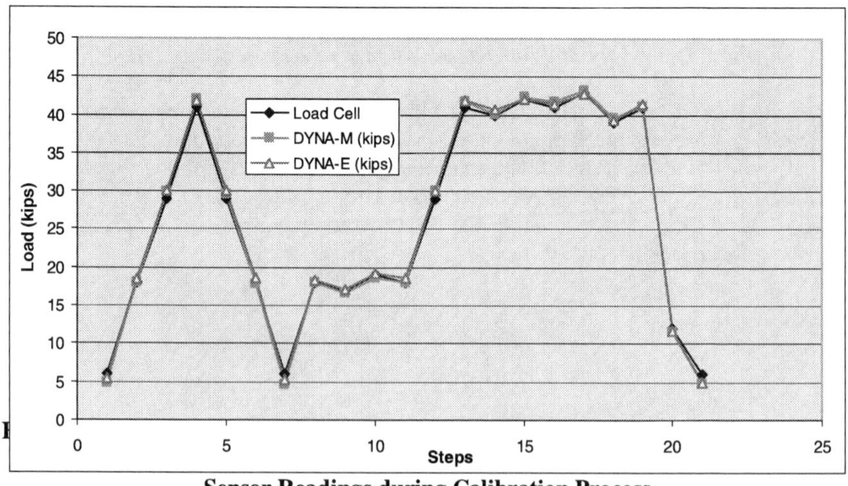

Sensor Readings during Calibration Process

Retrofitting of Boundary Dam, Saskatchewan Province, Canada

Retrofitting of the Boundary Dam, Saskatchewan Province in Canada, involved placement of post tensioning strand rock anchors placed in spillway chute and stilling basin to enhance slope stability and resist uplift forces during eventual high flow. A test program was executed prior to determine length and capacity of the production anchors. Test anchors used ten 9/15.2mm (0.6 in) uncoated strands (ASTM A 416) with single corrosion protection. Design load (DL) per anchor was 910 KN and ultimate load 2400 KN. Soil was a mixed of mudstone and siltstone and anchor bond length was pressure grouted and post grouted. Drilled hole was 197mm diameter.

Two of the test anchors .with intentionally longer bond length (18m), were equipped with three Dyna Force sensors placed on the bonded length of the anchors. Sensors were installed from the top of the bonded length at 0.2m, 6.2m and 12.2m. Anchors unbonded length was 20m and 24m.

The main purpose of the sensors was to determine how the stress is distributed throughout the bond zone in this soil and whether there is any potential benefit to lengthening the bonded length behind the limits recommended in the specifications.

Readings during testing to 80% of ultimate (1920 KN) shown that sensor at the top of the bond zone matched tensioning force, middle sensor indicated significant load reduction and bottom sensor indicated zero loads.

The results of the sensor measurements confirmed that entire 1900 KN load was transferred to the soil in less than first half of the bonded length and drilled length can be reduced.

The shear strength of the existing siltstone and mudstone can be mobilized to sustain the entire test load provided that post grouting method is carried out and the hole size and drilling method for production anchors will be consistent with those employed during testing program.

Anchor Installation

Sensor Placement in the Bonded Length

Patton Creek Retaining Wall Shopping Center, Birmingham, AL

Border of the large Patton Creek shopping center is retained by more than 250 m (820 ft) long crib retaining wall that started to show signs of slow moving. Retrofitting with long, strand anchors that stabilized the slope was the right approach. Hundreds of 9/15.2mm (0.6 in) permanent strand anchors were installed and anchored to the crib wall with large precast concrete pads. Nine Dyna Force sensors were installed in the unbonded length of nine different anchor locations and vary wall heights (7m to 14m). Load was monitored during proof test, at the load transfer, after four hours and after 28 days and graph shows below readings.

Sensor readings during proof test at 80% of ultimate of 209 KN/strand, matched stressing equipment pressure gauge and readings after 28 days vary but where in expected design load range.

Anchor Stressing Operation

Anchors Placed Through Precast Pads

Reading of the Power Stress Unit

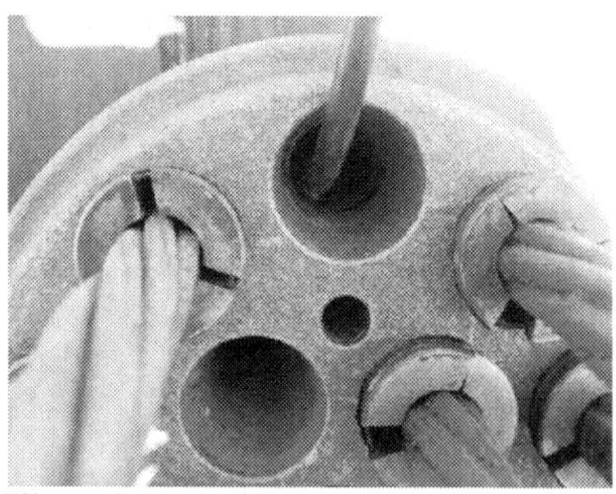

View of the Anchorage and Lead Wire

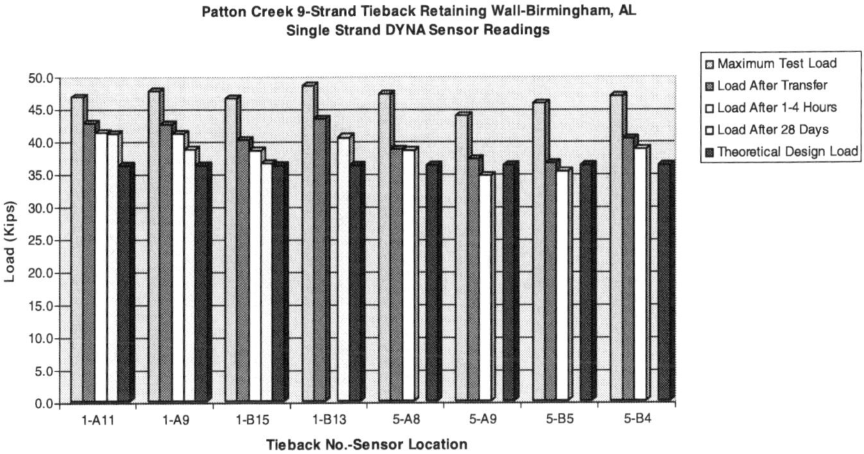

Cable Stay Waldow Bridge, Maine, USA

Dyna Force sensors have been used in more than twelve bridges and other superstructure applications in USA and oversee.

Two hundred forty sensors were installed on the stays of the Waldo Cable Stay Bridge. Each cable had six sensors and the cable forces at each stage of construction were monitored using these sensors.

Periodic lift-off operations were made on the strands with sensors and a good correlation was observed between them. The bridge is open to the traffic and forces in any cable are monitored anytime without any disruption to traffic.

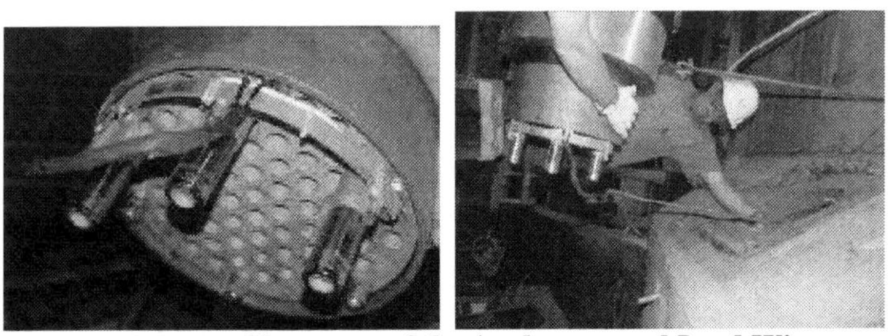

Placement of the Anchorage and Lead Wire

View of the Bridge

Other Applications

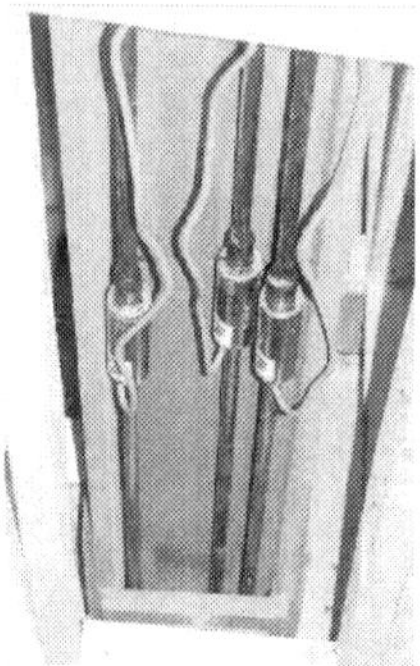

Over Plastic Sheathed Strand in New Building

Over Strand Tendon for Bridge Girder Retrofit

CONCLUSIONS

1. The Dyna Force sensor allows reliable load monitoring of post tensioning anchors and tendons during proof or performance test, lift-off test, during service life and for anchor bonded length evaluation.
2. Sensor can be applied over uncoated or epoxy coated strand or bar or greased and extruded HDPE sheathed steel
3. It is a robust system with free maintenance providing quick and easy installation and readings. One man can do the job
4. Long-term durability is ensured by zinc coated steel cover, polyurethane insulation and sometimes by the grout or concrete surrounding the sensor.
5. High accuracy of +/-1% is normally achieved
6. Ambient temperature influence is corrected by sensor calibration and does not influence the service life readings

References

[1]Yang Zhao PhD, (2006). "EM Technology and Its Application", Paper presented as advertising for his company Intelligent Instrument System, Inc.

Load Development in Soil Nails from a Strain-Gauge Instrumented Wall

Terence P. Holman[1], Ph.D., P.E., M.ASCE and Thomas J. Tuozzolo[2], P.E., M.ASCE

[1]Senior Engineer, Moretrench, 100 Stickle Avenue, Rockaway, NJ 07866; tholman@mtac.com
[2]Vice President, Moretrench, 100 Stickle Avenue, Rockaway, NJ 07866; ttuozzolo@mtac.com

ABSTRACT: This paper contains analyses of an instrumented temporary soil nail wall in which strain-gauged nails were installed and monitored during and after construction. Three test sections were installed consisting of a total of 43 strain gauges on 10 soil nails and monitored for a total of nearly 6 months. The maximum stresses measured in the monitoring were significantly less than those predicted by the limit equilibrium design process used to analyze the walls. Loads continued to increase in the nails until the end of the monitoring program without any observable decrease in stress-accumulation rate. These observed mechanisms have far-reaching impacts on the design of temporary and permanent soil nail walls and should be further investigated with additional instrumentation.

INTRODUCTION

Soil nailing is commonly employed in the northeastern United States for temporary and permanent earth retention where stable geologic deposits exist. Current analysis and design methods focus on assessing the global factor of safety as a function of nail spacing, length, bond stresses, and limit equilibrium. The prediction of nail loads by currently used analytical software programs is based on limit equilibrium and the distribution set in accordance with assumed patterns. The magnitude and development of nail load distribution patterns is actually a function of construction sequence, time, and wall movement; factors not explicitly considered in current analysis methods.

This paper contains analyses of an instrumented temporary soil nail wall in which strain-gauged nails were installed and monitored during and after construction. The project involved a 6.4 to 7.6 m deep, 883 m^2 excavation in the Borough of Manhattan, New York City to allow construction of a large electrical vault. This project is believed to be the first soil nail wall in Manhattan and was located on a university campus near to the East River. The site was very limited in size and access and the construction schedule was aggressive, which when matched with the natural silty sand soils and underlying competent bedrock, made soil nailing an appropriate technology. Three test sections of three to four instrumented soil nails each were installed as part of the construction process. The monitoring program for this project lasted for the duration of soil nail wall construction, about 1 to 1.5 months, and then for an additional 4 months, resulting in a total time of about 5.5 months. The instrumentation data accumulated on this project has provided much insight into the development and distribution of loads in the soil nails supporting these walls and the adequacy of existing analysis methods.

PROJECT DESCRIPTION AND WALL DESIGN

The Transformer Vault project required that a deep excavation be made on the edge of an active university campus and adjacent to active New York City streets. The required excavation footprint was 54.9 m long by 12.5 m wide. Based on the topography of the project site, the depth of the excavation ranged from 6.1 to 7.6 m below adjacent ground surface, which was situated between Elev. 7.9 and 9 m (Manhattan Datum). Planned subgrade elevation was Elev. 2.0. A temporary excavation support system was required for this project to allow for unobstructed construction of the Transformer Vault.

The specialty geotechnical contractor determined that a temporary soil nail wall was the most appropriate excavation support system for the project site, geometric constraints, and subsurface conditions. A test pit was excavated to observe the stand-up time of the fills and natural soils. The subsurface conditions at the project site consisted of man-made granular fills underlain by stable natural silty sands, decomposed rock, and intact rock. No static groundwater was encountered in the borings and test pit. Figure 1 depicts the typical subsurface conditions assumed in the initial feasibility assessment and design. The surface of the decomposed rock and intact rock was inconsistent along the length of the site.

The design of the soil nailing was conducted by a consultant retained by the specialty contractor. Force equilibrium, allowable stress-based design procedures were employed using the public domain computer program SNAILZ (Caltrans, 1991) to analyze the system. The assumed unit weights and angles of internal friction for the Fill and Silty Sand were 18 kN/m^3 and 19.6 kN/m^3, and 30° and 38°, respectively. The design nail spacing ranged from 1.22 to 1.83 m vertically and was 1.83 m horizontally. Nail lengths of No. 25 Grade 520 reinforcing bar ranged from 5.8 m in the upper tiers to 2.7 m at the base of the excavation. The consultant's target design factor of safety was 1.35 for the completed excavation. Based on these input parameters, the calculated factors of safety for the final condition were 1.43 to 1.51.

INSTRUMENTATION INSTALLATION AND DATA COLLECTION

Two unique features of the soil nail wall design and construction made this project a good candidate for a detailed instrumentation program. First, the excavation was to remain open for a significant period of time allowing for both short and medium-term performance data to be collected. Second, the excavation geometry was expected to yield two-dimensional plane strain behavior as well as three-dimensional behavior based on criteria presented by Finno et al. (2007). The specialty contractor decided to employ a significant quantity of strain-gauged soil nails to investigate the system performance and collect detailed time and geometry-related data.

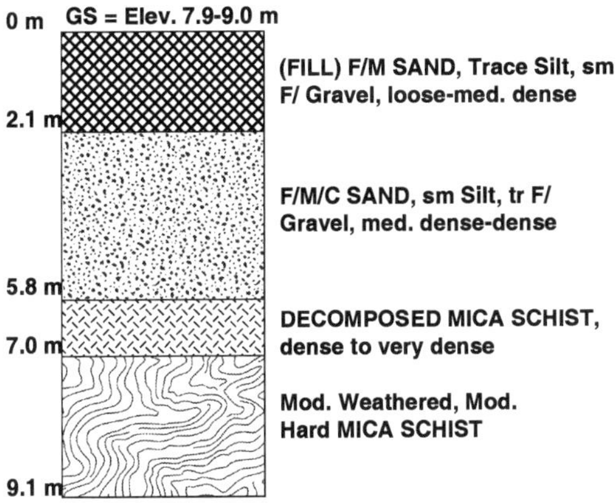

FIG. 1. Typical subsurface profile

Three columns of instrumented soil nails were installed in the southern portion of the excavation at the locations shown in Figure 2. Test section IN-1 was located at approximately the midpoint of the excavation near the highest portion of the wall, where the planned height H_e was about 7.0 m ft. Test section IN-2 was situated at about the midpoint of the short wall where H_e was 6.1 m. The third instrumented column was at location IN-3, adjacent to an excavation corner with the same cut height as IN-2. The nails were instrumented with three to five spot-weldable strain gauges. Figures 3 and 4 depict cross sections of test sections IN-1 and IN-2/IN-3 with nail level designations, lengths, and vertical spacings. The nails were installed over a period of about 40 to 42 days beginning on August 24, 2007 as the excavation was carried sequentially downward. The time-history of the excavation is shown in Figure 5 and indicates that each test section in the southern portion of the excavation was constructed similarly.

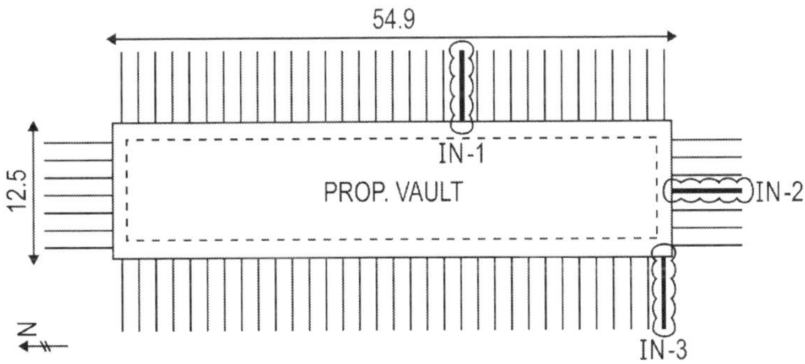

FIG. 2. Plan view of excavation geometry and test sections (all dim. in meters)

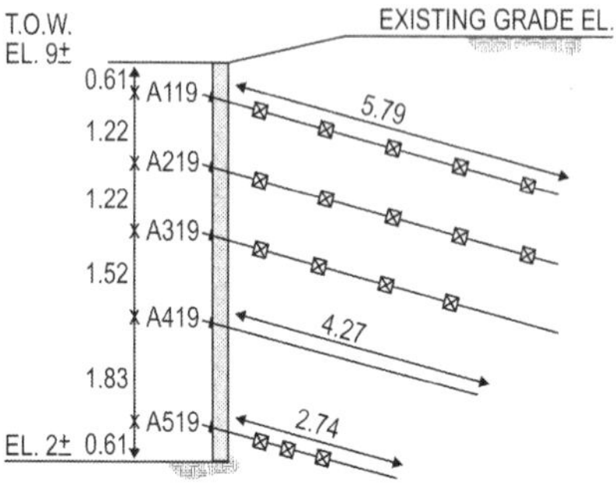

FIG. 3. Test section IN-1 (all dim. in meters)

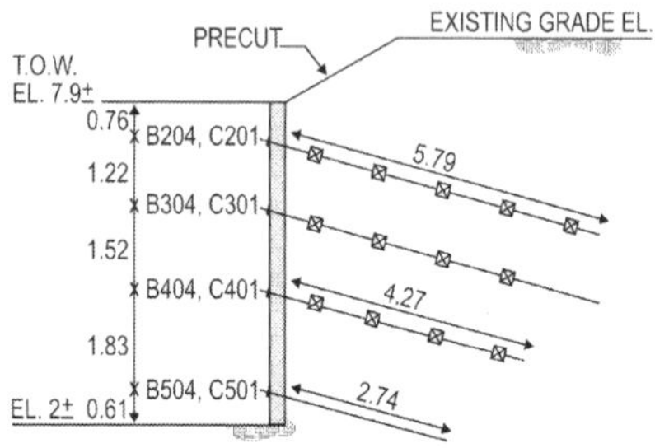

FIG. 4. Test sections IN-2 and IN-3 (all dim. in meters)

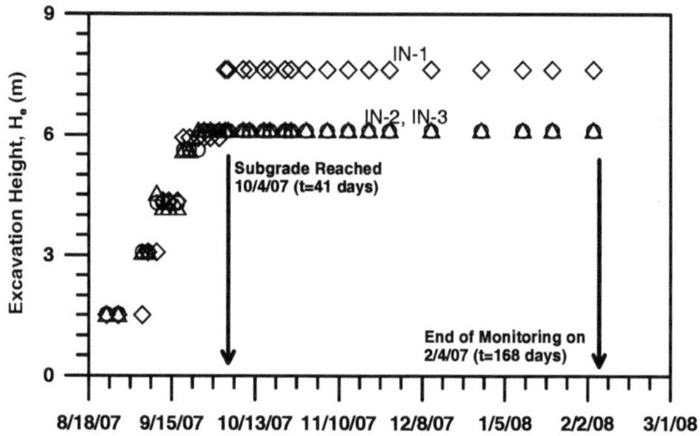

FIG. 5. Time history of Transformer Vault excavation

The specialty contractor monitored the strains in the gauges throughout construction. Readings were typically taken every two days during the excavation process, twice weekly for nearly 1.5 months after reaching subgrade on October 4, 2007, and weekly up to February 4, 2008, a total of 5.5 months later. At this point, the foundation contractor was in the process of backfilling the completed vault structure and damaged the instrument cables. Excavation heights were measured along with the nail strains. The overall excavation height was closer to 7.6 m at IN-1 because of less precutting along the boundaries and overexcavation into the subgrade during rock removal. No deformation monitoring of the soil nail wall was performed.

ANALYSIS AND IMPLICATIONS FOR DESIGN OF SOIL NAIL WALLS

The strain gauge monitoring information obtained from this project furnished valuable data that impacts the design and performance of soil nail walls, both temporary and permanent. The measured versus design-predicted nail stresses and nail stress/load-time effects are the areas to which the strain gauge data is applied. The distribution pattern of the maximum nail loads was similar to that documented in Byrne et al. (1996) and CLOUTERRE (1993), but is not discussed here.

Comparison of Design versus Field Nail Stresses

Mobilized maximum nail loads and stresses should be compared to those predicted by limit equilibrium analyses used to predict the global factor of safety and assess the wall design. For this project, the maximum stresses predicted by the design software SNAILZ are part of the program output, but their location is not specifically known along the nail length. Maximum stresses calculated from the strain gauge measurements are from known locations and for two points in time at maximum excavation height. Table 1 presents a comparison of these predicted and measured data.

The ratios between predicted and measured nail stresses are very significant, with overall mean values approaching 2.1 and a standard deviation greater than 75% of the mean. The design stress farthest from the field-measured value occurred at nail A519 in test section IN-1, where the ratio was 6.8. Tuozzolo (2003 reported similar results for a soil nailed excavation of similar height with SNAILZ-calculated stresses far in excess of measured stresses. The results for section IN-3 also reinforce the three-dimensional nature of this excavation in that the ratio in computed to measured stresses is 2.0, 1.25 times greater than for IN-1, an excavation section more characteristically two-dimensional or plane strain. These variations in stresses are thought to be rooted in the simplified two-dimensional force equilibrium design procedure to describe a three-dimensional, kinematic, soil-structure interaction problem where geometric effects influence excavation behavior. The soil behavior on the inside of the reinforced mass is not linear and the development of stresses in the soil nailed structure is a function of shear and volumetric strains (i.e. a constitutive relationship). The software cannot adequately predict realistic stresses as a result.

Table 1. Comparison of predicted and measured nail stresses

Test Section No.	Nail No.	Maximum Tensile Stresses (MPa) SNAILZ (1)	Measured 10/4/2007 (2)	Measured 2/4/2008 (3)	Ratio (1)/(3)	Ave. Ratio
IN-1	A119	64.9	46.3	76.1	0.85	2.50
	A219	114.5	65.3	116.4	0.98	
	A319	197.7	73.5	145.4	1.36	
	A519	334.8	27.7	49.2	6.81	
IN-2	B204	104.2	27.8	61.9	1.68	1.63
	B304	181.1	66.7	99.0	1.83	
	B404	122.0	31.0	87.8	1.39	
IN-3	C201	104.2	39.4	43.8	2.38	2.03
	C301	181.1	25.0	90.7	2.00	
	C401	122.0	25.7	70.9	1.72	
				Mean	**2.10**	
				St. Dev.	**1.72**	
				COV (%)	**82%**	

Nail Stress/Load-Time Effects

The loads in a soil nailed wall develop and evolve as a function of time, and are not necessarily constant at any point during or after a construction process. For the test sections installed at this project, nail loads and stresses continued to increase over the entire time after reaching subgrade and full excavation depth to the end of monitoring 127 days later. Figure 6 demonstrates the time-related development of loads and changes in the tension load distribution. From this figure, one can observe that T_{MAX} nearly doubles from 33.3 kN to 59.3 kN in 4 months. This trend was seen in all of the nail strain gauge data. The actual rate of load or stress development is of great interest to designers and analysts because it has implications for long term versus short term loads and serviceability of these structures. Plots of maximum tensile nail stresses versus time are included in Figure 7 for all three test sections.

Zones of constant slope are present within the data set for each instrumented nail, particularly after t=56 days. This monitoring program cannot be considered long-term because it only involves data collection for a period of time less than 2 log cycles long and the curves are likely to tail off to some asymptotic value. However, the data trends are strong: there is some accumulation of additional load in each soil nail as a function of time, even in the short term, after reaching the final subgrade level. This behavior is important to consider. Loads will increase in soil nails under service loadings after construction and in the long-term could result in lower factors of safety for nail structural capacity, particularly for permanent walls. Furthermore, if nail loading is assumed concurrent with soil mass deformation, then wall deflections and rotations are also likely to increase with time.

Linear regression lines were fit through each data set to determine the linear stress accumulation rate. The extracted data is summarized in Table 2. The average rate was 0.285 MPa/day and the standard deviation was approximately half of that value, indicating significant scatter in the data. The bulk of the stress accumulation in the soil nails is due to additional long-term strains in the granular soils reinforced by the nails that result from the unloading stress-path of the wall construction. Nail stresses result from soil mass strains as the wall is constructed and afterwards. It is these creep-related strains that result in slowly accumulating additional loads in the passive nail elements. For the data obtained in this project, the creep rates were somewhat constant and expected to reduce with time beyond the monitoring period. More limited amounts of the stress increases could be a result of vibrations due to mechanical rock removal at subgrade.

CONCLUSIONS

The strain-gauged test sections employed for this project provided data related to the development of loads in soil nails. The mean difference between measured and design-predicted maximum nail stresses is nearly 40%, a substantial figure suggesting that current limit equilibrium design methods may not be rigorous and sufficient to describe the internal behavior of the soil nailed structure. Significant conservatism may exist in the design of soil nails based on typical computer software output. Continued load generation was observed in the instrumented nails well after reaching excavation subgrade and is thought to be a result of continued unloading stress-strain behavior in the silty sand. Stress accumulation rates were estimated from the linear portions of the curves for a period over 100 days.

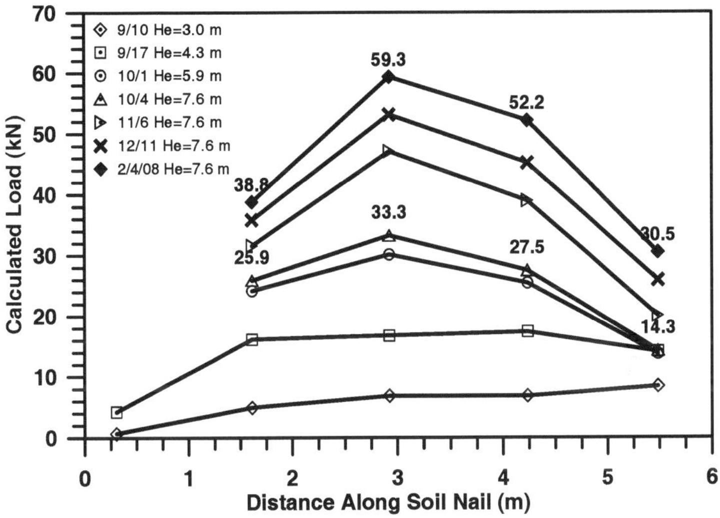

FIG. 6. Time-related load development in nail A219 (IN-1)

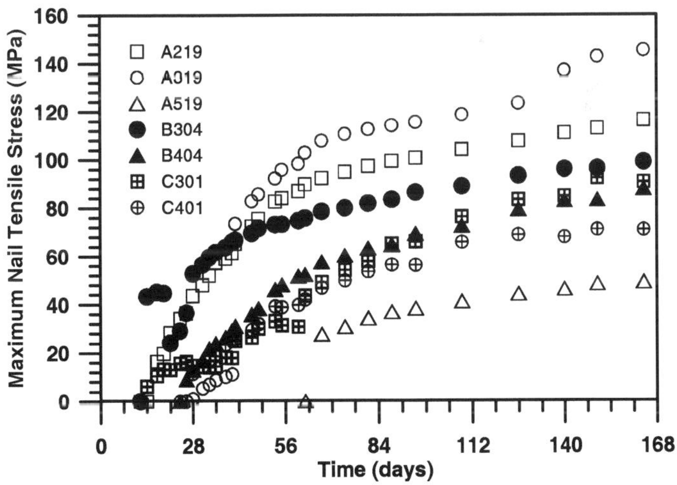

FIG. 7. Tension stress increases in soil nails versus time

Table 2. Summary of stress accumulation rates for soil nails

Test Section No.	Nail No.	Calc. Stress Accumulation Rate (MPa/day)
IN-1	A119	0.131
	A219	0.262
	A319	0.414
	A519	0.317
IN-2	B204	0.310
	B304	0.234
	B404	0.345
IN-3	C201	0.021
	C301	0.524
	C401	0.296
	Mean:	0.285
	St. Dev.	0.140

ACKNOWLEDGEMENTS

The authors would like to thank their engineering and field staff for their attention to care in constructing and monitoring the test sections; specifically Michael McNicholas, Jason VanDuzer, and Craig Anderson. Morris-Flood Associates of Hampton, New Jersey designed the temporary soil nail wall.

REFERENCES

Byrne, R.J., Cotton, D., Porterfield, J., Wolschlag, C., and Ueblacker, G. (1998). *Manual for Design and Construction Monitoring of Soil Nail Walls*, Report FHWA-SA-96-69R, Federal Highway Administration, Washington, D.C.

CALTRANS (1991). "A User's Manual for the SNAIL Program, Version 2.02 – Updated PC Version," California Department of Transportation, Division of New Technology, Material and Research, Office of Geotechnical Engineering, Sacramento, California.

CLOUTERRE (1993). *Recommandations CLOUTERRE 1991 - Soil Nailing Recommendations 1991*, English Translation, Presses de l'Ecole Nationale des Ponts et Chaussées, Paris, France.

Finno, R.J., Blackburn, J.T., and Roboski, J.F (2007). "Three-dimensional effects for supported excavations in clay." *Journal of Geotechnical and Geoenvironmental Engineering*, ASCE, Vol. 133, No. 1, 30-36.

Tuozzolo, T.J. (2003). "Soil Nailing: Where, When, and Why – A Practical Guide." *Proceedings of the PennDOT/Central Pennsylvania Geotechnical Conference*, ASCE, Hershey, 26 pp.

Repeated Lift-off Testing of Single Bore Multiple Anchors for Dam Retaining Wall over a 5-Year Period

Mary Ellen C. Bruce[1], M. ASCE, P.E., Jesús Gómez[2], M. ASCE, Ph.D., P.E., and Robert P. Traylor[3], A.M. ASCE

[1]President, geotechnica, s.a., Inc., P.O. Box 178, Venetia, PA, 15367, geotechnica@comcast.net
[2]Schnabel Engineering, 510 E. Gay St., West Chester, PA, jgomez@schnabel-eng.com
[3]Traylor LLC, 3525 Ellicott Mills Drive, Unit C, Ellicott City, MD, bobt@traylorgroupllc.com

ABSTRACT: The right downstream retaining wall of Hodenpyl Dam in Michigan had undergone significant lateral displacement due to global instability of the in situ soils. The remediation design called for installation of additional anchors bonded within a high plasticity, stiff clay layer. To increase the unit bond strength and reduce the potential for creep observed in other anchors installed at the dam, the contractor selected the Single Bore Multiple Anchor (SBMA) system with intensive, selective post grouting. The SBMA anchors have been lift-off tested four times since their installation. This paper describes the design and construction of the SBMA anchors, testing of a sacrificial SBMA anchor, lift-off testing procedures and results, and conclusions.

INTRODUCTION

Hodenpyl Dam, owned and operated by Consumers Electric Company of Cadillac, MI, is located in Wexford and Manistee Counties, Michigan, USA, and is one of a series of dams along the Manistee River. The facility was constructed between 1923 and 1925 and consists of an earthen embankment running roughly in the north-south direction. The powerhouse, located on the southern end of the embankment, discharges water to a tailrace channel excavated in natural soils. Cantilever retaining walls of varying height were originally constructed on the east and west sides of the tailrace channel. The right (east) downstream wall is approximately 76 m (250 feet) long and 9 to 15 m (30 to 50 feet) high (Figure 1).

After significant displacements of the wall were first noticed, a repair attempt was conducted in 1996. The repair consisted of conventional strand anchors bonded into the high plasticity clays located behind the wall backfill. In addition, steel sheeting was installed bearing on the wall footing and attached to the wall face. The sheeting was intended to provide structural reinforcement and was not driven into the ground. Subsequent monitoring of the wall revealed that the wall has continued to displace laterally, and that additional stabilization measures were required. The monitoring data collected show that the wall is rotating around a vertical axis located at the upstream end of the shorter height section seen in Figure 1.

The Engineer specified a total of 13 restressable, regroutable anchors to improve the wall stability. The remediation was designed to provide immediate improvement of the wall stability while controlling construction costs. The Engineers also specified a detailed monitoring/action plan, the results of which were to be used to determine the need for subsequent treatments. Full details of the investigation and analysis are discussed by Gómez et al. (2004).

Figure 1. View of the tailrace channel and right downstream retaining wall. Note the sheeting and tiebacks installed in 1996.

BOND ZONE MATERIALS

The soils behind the retaining wall comprise a layer of poorly graded sand, underlain by an interval of overconsolidated clay (bond zone materials). The clay typically contains a 10- to 30-percent fraction of fine to coarse sand (passing #4 sieve and retained in #200 sieve) and traces of gravel-size particles (retained in #4 sieve), plasticity index ranging between 32 and 50 percent and a liquid limit of 50 to 72 percent. Sporadic lenses and/or seams of sand were also detected at some locations within the clay mass. The plasticity of the clay materials increases with depth. The 2002 tiebacks were bonded to these clays. The clay stratum is underlain by a confined aquifer consisting of fine to coarse sands. The presence of the aquifer induces upward flow within the stiff clay and above-hydrostatic pore pressures around the tiebacks.

During design of the 2002 remediation, the Engineer realized that the available space for bonding the tiebacks was limited. The existing tiebacks were located near the top of the wall and inclined at angles steeper than 25 degrees with horizontal (see Figure 2). The Engineer also recognized that the new tiebacks should not penetrate the underlying aquifer. The available space for the new tiebacks did not allow for bond zone lengths longer than 12 m (40 ft). Finally, the number of tiebacks was also limited due to the configuration of the reinforcing sheeting installed in 1996.

Because of the limitation on the number and maximum bond length of the tiebacks, there was a concern over the bond strength capacity of the tiebacks. Also, because the tiebacks were to be bonded to stiff, high plasticity clay, the potential for creep of the tiebacks was considered during design. The available proof testing data of the 1996 tiebacks showed that creep was observed during at least one of the tests.

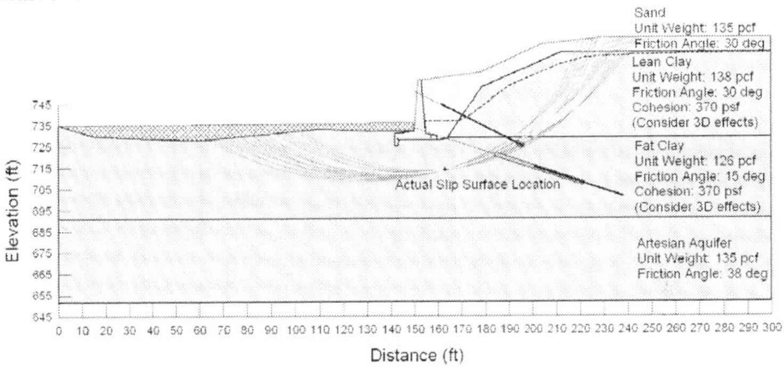

Figure 2. Typical section through the wall showing the 1996 (upper) tiebacks, the 2002 (lower) SBMAs, and the geotechnical conditions. 1 pcf = 0.157 kN/m^3; 1 psf = 0.048 kN/m^2

SINGLE BORE MULTIPLE ANCHOR SYSTEM

Researchers who have investigated grout/ground load transfer conclude that the distribution of stress along the fixed anchor is non-uniform due to general incompatibility between elastic moduli of the anchor tendon, anchor grout, and the ground (Littlejohn and Bruce, 1977; Barley, 2000). The top portion of Figure 3 depicts the "progressive debonding" that occurs in a traditional tieback due to this incompatibility. The area under the bond stress distribution line represents the ultimate load in the anchor. As shown, the load does not increase uniformly with increasing length. In an SBMA, load transfer from the strand to the grout takes place over a series of short lengths at staggered intervals along the borehole, thereby reducing or eliminating the occurrence of progressive debonding and substantially increasing the efficiency of the overall anchor. Full details on the SBMA system are discussed by Barley and Windsor (2000).

A typical SBMA (Figure 4) consists of several unit anchors (single or double strand) with varying lengths installed in a borehole [nominally 100 to 200 mm (4- to 8-inch) diameter] such that their respective load transfer lengths are located at predetermined positions within the total fixed length.

SACRIFICIAL ANCHOR INSTALLATION AND TESTING

The Engineer required the installation and testing of a sacrificial SBMA anchor prior to construction of the production anchors. This anchor was installed using identical construction methods and materials (see next section for construction details) and bonded in the same soils as those proposed during production with the following exceptions a) the sacrificial anchor was installed vertically, as opposed to being inclined at an angle of 20° below horizontal; b) three 3-m-long (10-foot) unit encapsulations were installed as opposed to four to avoid penetration into the sand aquifer; and c) the upper two unit anchors contained two strands in each encapsulation to allow load testing to very high grout/ground bond stress.

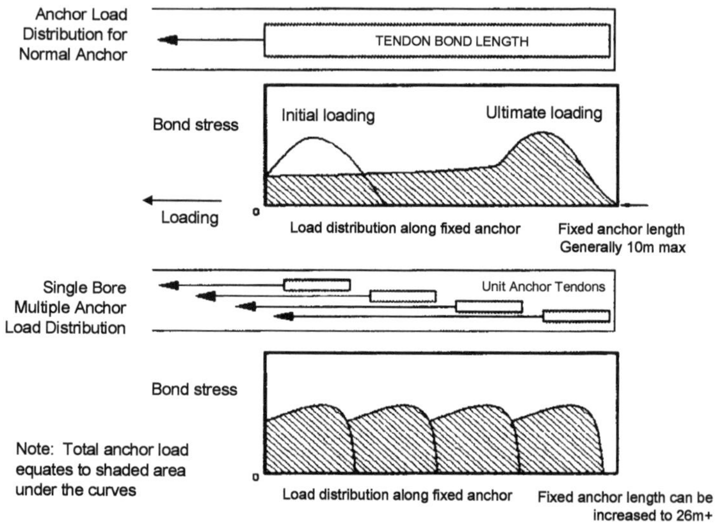

Figure 3. Comparison of load distribution along a conventional anchor and an SBMA (Barley, 2000).

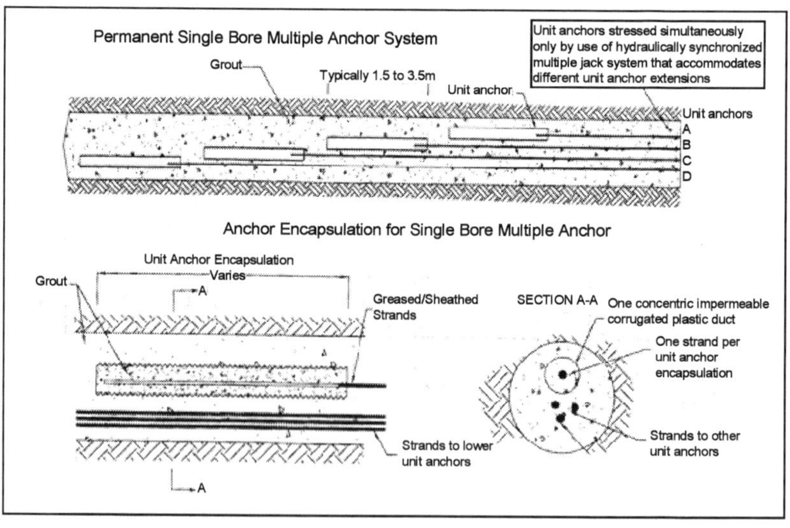

Figure 4. Elevations and cross section of a typical 4-unit SBMA.

The jack arrangement for a three-unit SBMA includes three hydraulic rams that are synchronized by coupling to the same hydraulic powerpack, so that the same load is applied simultaneously to each unit anchor. The jacking arrangement is shown in Figure 5. A primary gauge and a reference gauge were calibrated with one of the jacks. The ram extensions were recorded during performance testing using a stiff steel rule, and during creep testing by using a vernier caliper. Extensions were corrected for movement of the reaction pad, which was measured by dial gauges mounted on an independent reference beam.

Two tests were conducted on the sacrificial anchor: 1) Performance/Extended Creep Tests (PTI, 1996), and 2) a high bond capacity test. Since creep was a concern on this project, the load-hold period was extended overnight (total load-hold time 810 minutes) at maximum test load. The linearity of the load/movement plots indicated that virtually no progressive debonding occurred into the unit fixed anchor. The maximum test load for the high bond capacity test was limited by the steel capacity, and so an ultimate bond stress was not measured. The average mobilized bond stress was 239 kPa (5.0 kip/ft^2) at the maximum test load with no creep or failure.

Figure 5. Sacrificial test anchor stressing jack.

DETAILS OF ANCHOR CONSTRUCTION

Anchor holes were drilled using rotary duplex with water flush and nominally 175 mm (7-inch) casing with an approximate 25 mm (1-inch) overcut from the casing shoe teeth, resulting in an nominally 200 to 216 mm (8- to 8½-inch) hole diameter. Casing was advanced to the target tip elevation for each anchor.

The specified working load for each anchor was 600 kN (135 kips). Anchor geometry and design requirements dictated a minimum free length of 15 m (50 feet); a minimum bond length of 12 m (40 feet); a minimum drill hole diameter of

127 mm (5 inches); and an anchor inclination of 20° (or less). Class I corrosion protection (PTI, 2004) was required for the anchors. A summary of "unit" anchor lengths and SBMA geometry is provided in Table 1.

Table 1. Summary of unit anchor lengths and SBMA geometry.

UNIT ANCHOR	FREE LENGTH	FIXED LENGTH	TOTAL TENDON LENGTH*
A	16.7 m (55 ft)	3.5 m (11.5 ft)	21.8 m (71.5 ft)
B	20.2 m (66.5 ft)	3.1 m (10 ft)	24.8 m (81.5 ft)
C	23.3 m (76.5 ft)	3.1 m (10 ft)	27.9 m (91.5 ft)
D	26.3 m (86.5 ft)	3.1 m (10 ft)	31.0 m (101.5 ft)

*Total tendon length = 1.5 m (5-foot) tail + free length + fixed length

All anchors were subjected to two phases of post grouting. Two performance and extended creep tests were conducted on production anchors. All other production anchors were proof-tested. The maximum test load was 805 kN (180 kip), or 201 kN (44.9 kips) per strand (see Bruce et al. 2004).

All unit anchors were locked off after testing at 152 kN (34 kips) using a mono-strand jack. The strands were then trimmed to approximately 203 mm (8 inches) above the gripping wedges. The trumpets were grouted, and the steel caps were filled with grease and installed as designed.

LIFT-OFF TESTING PROCEDURES

Lift-off tests were conducted in April 2004, April 2005, May 2007 and June 2008. The steel anchor head caps were removed and the grease stripped from the strands. A mono-strand jack was used to lift-off each of the four strands of the SBMA individually. Lift-off was defined as the load at which the gauge pressure increase hesitated during loading and at which a 0.5-mm-thick (0.020-inch) feeler gauge could first be fitted below the barrel.

During the tests, it was observed that the strands and wedges did not show evidence of corrosion as a consequence of removing the greased-filled cap in each of the lift-off events. Indentations of the stressing wedges on the strands occurred during the prior lift-off events were visible, although they did not pose significant concerns.

LIFT-OFF TESTING RESULTS

Lift-off loads have increased consistently. The 2008 lift-off loads ranged from 135 to 181 kN (30.44 to 40.64 kips) per strand. The highest load measured was 116% of the generalized lock off load, which is equivalent to 67% GUTS. The lift-off loads measured are shown graphically in Figure 6. The lift-off load for each anchor (total of four unit anchors) increases in the downstream direction along the wall. This trend correlates with the pattern of wall displacement measured, which indicates that the wall is rotating around a vertical axis located near upstream tiebacks.

Figure 7 shows a comparison between the increase in tieback load measured during each lift-off event, and the theoretical increase in tieback load calculated from the measured wall displacements at each tieback location. The 45-degree line shown corresponds to full agreement between the theoretical and measured load increases. The data in the figure show that there is good agreement between the data and the 45-degree line, thus suggesting that the tiebacks have not undergone yielding or creep since installation.

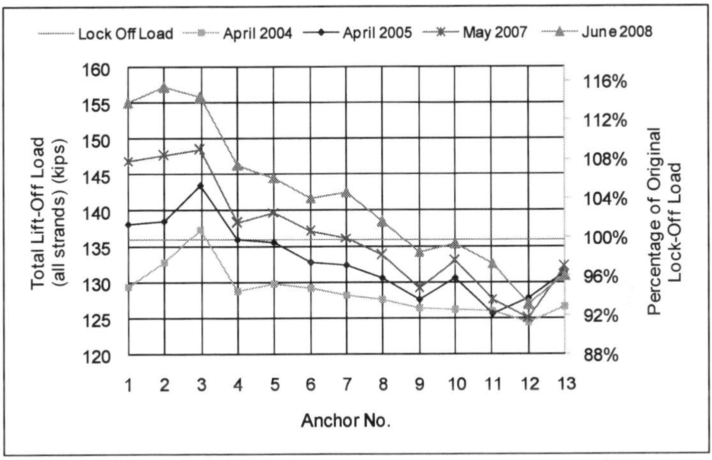

Figure 6. Lift-off loads (total of all strands). 1 kip = 4.448 kN

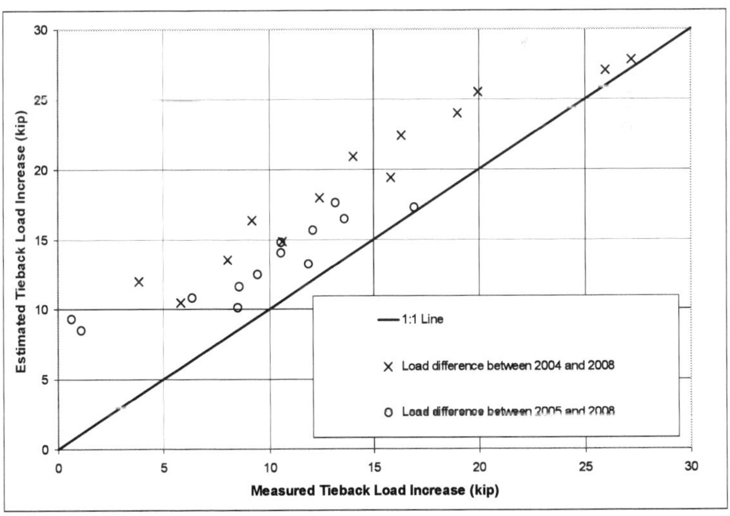

Figure 7. Comparison of estimated anchor load (based on wall movement) and anchor load measured during lift-off testing (2007). 1 kip = 4.448 kN

CONCLUSIONS

Based on the results obtained during load testing and lift-off testing of the SBMA elements, we may conclude the following:

- The sacrificial test showed that the SBMA installed in high plasticity clay may develop significant capacity.
- The sacrificial and production tests showed that creep in the high plasticity clay was insignificant, even though design bond stresses were relatively high.
- Comparison between the lift-off test data and the wall monitoring data shows that the production tiebacks have not undergone significant creep during the 6 years that have elapsed since their installation in 2002.
- Repeated lift-off testing of the tiebacks did not appear to have a detrimental effect on the tiebacks.
- The SBMA technology is a valuable tool in projects on which it is necessary to install large-capacity tiebacks in high-plasticity clay.

ACKNOWLEDGEMENTS

We thank Consumers Energy Company and especially Mr. Stuart Johnson for kindly allowing the publication of the information contained in this paper. Additional thanks must also go to Gerace Construction Company for their cooperation during the construction and lift-off testing phases. Tony Barley (SBMA, LLC) and Donald A. Bruce (Geosystems, L.P.) provided important insight during design, construction and testing.

REFERENCES

Barley, A.D. (2000). "Trial soil nails for tunnel face support in London Clay and the detected influence of tendon stiffness and bond length on load transfer." Proceedings of the Third International Geosystems, London, UK, June.

Barley, A.D., and Windsor, C.R. (2000). "Recent advances in ground anchor and ground reinforcement technology with reference to the development of the art," GeoEng 2000, International Conference on Geotechnical and Geological Engineering, Melbourne, November 12-19, pp, 1084-1094.

Bruce, M.E., R.P. Traylor, A.D. Barley, D.A. Bruce and J. Gómez. (2004). "Post Grouted Single Bore Multiple Anchors at Hodenpyl Dam, Michigan," *GeoSupport 2004 Drilled Shafts, Micropiling, Deep Mixing, Remedial Methods, and Specialty Foundation Systems*, ADSC and ASCE, GSP No. 124, Orlando, FL, January 29-31, 361-373.

Gómez, J., M.E. Bruce, D.A. Bruce, D.L. Basinger and A.W. Cadden. (2004). "Stabilization of Retaining Wall along Tailrace Channel of Hodenpyl Dam," *GeoSupport 2004 Drilled Shafts, Micropiling, Deep Mixing, Remedial Methods, and Specialty Foundation Systems*, , ADSC and ADSC, GSP No. 124, Orlando, FL, January 29-31, 346-360.

Littlejohn, G.S. and Bruce, D.A. (1977). "Rock Anchors - State of the Art." Foundation Publications, Essex, England, 50 p. (Previously published in *Ground Engineering* in 5 parts, 1975-1976.)

Post Tensioning Institute (PTI) (1996). "Recommendations for prestressed rock and soil anchors." *Post Tensioning Manual*. Fourth Edition. Phoenix, Arizona. 41 p.

Measurement and Simulation of Ground Deformation Due to Pneumatic Caisson Construction

Peng Fang-Le[1,2] ,Wang Hai-Lin[1,2], Guo Jian-Gang[1,2], Xu Zhen-Liang[3]

([1] Department of Geotechnical Engineering, Tongji University, Shanghai 200092; [2] China Key Laboratory of Geotechnical and Underground Engineering of Ministry of Education, Tongji University, Shanghai 200092, China; [3] Shanghai Urban Construction Design & Research Institute, Shanghai 200125, China)

ABSTRACT

Combined with one shaft construction employing pneumatic caisson in Shanghai, in-situ measurement of ground deformation was conducted. Based on the particular construction process of pneumatic caisson, one kinematic numerical analysis method was proposed for evaluating its influence on the surrounding strata. Through comparisons between calculated results and measurements, the proposed mechanical model and numerical analysis method were verified.

Introduction

As one of the safest and most effective ground excavation methods, the pneumatic caisson is frequently used. There are a number of pneumatic caisson cases both large and small, especially in Japan (Ohuchi et al. 2002). Until recently, this method was applied for the first time in Shanghai for a shaft construction and achieved great success. These engineering practices have indicated that the pneumatic caisson method is of considerable merits in deep foundation excavation. However, with its wider application in underground engineering, it is urgent and meaningful to carry out research on its effects on the surrounding strata and adjacent structures during the process of caisson sinking. In Japan there are relatively mature experiences in design and construction of pneumatic caisson. Some Japanese scholars (e.g. Umeda et al. 2006; Inoue et al. 2006) have conducted a lot of research on the pneumatic caisson. They proposed a simple mechanical model to predict the deformation of the surrounding ground. However, in order to simplify the analysis, the model was built based on plane strain assumption, the soil stress-strain relationship was assumed linear elastic and they did not take the caisson structure into account. In China, research on the pneumatic caisson is not yet initiated because of its rare application, but many scholars have done much work on the open caisson and achieved some progress. They proposed a dynamic mechanical model in view of rigid motion of multiple-degrees-of-freedom system (Wang et al. 2006). However, this study is focused on the control of the caisson, aiming to predict the relation between the sink depth and the volume of excavation, and it did not touch on the subject of predicting of ground deformation during caisson sinking.

Therefore, this paper aims to propose one kinematic mechanical model to describe the interaction between the caisson and its surrounding ground, providing some non-linear numerical simulation techniques, and verifying the proposed model and numerical analysis method.

Outline of shaft construction and its in-situ measurement

The pneumatic caisson is manufactured wholly or partially (at least the working chamber and part of the vertical walls) at ground level. Material is excavated in the working chamber, the groundwater being kept out with compressed air. Due to the constant shear failure below the cutting edge, the caisson sinks into the ground slowly under its own weight, as shown in Fig.1 (a).

This method has been applied successfully in one shaft construction of Shanghai. The shaft is used for ventilation in the shield tunnel of Metro Line 7. It is one fully-embedded four-storied reinforced caisson structure with plane dimension of 25.24m×15.60m and a total sinking depth 29.06m. Two partition walls are installed in the longitude and transverse direction, and thus several compartments form. The partition walls can strengthen the global rigidity of the structure, and moreover, they make it possible to adjust the caisson sinking attitude during sinking by filling some of the compartments with water or other weights. The pressure of the compressed air within the working chamber prevents the ground water from flowing in and cushions the caisson for smoother sinking. Most importantly, it can control the ground water more effectively, minimizing the disturbance in the surrounding ground.

Soils encountered at the site consist of clayey silt, muddy clay, sandy silt and silty clay from surface downward. The silt clay layer is flow-plastic and thixotropic, and has low bearing capacity. Below that are the layers of clayey silt and sandy silt, which are characterized by high water permeability and possibility of sand flowing and piping erosion under the action of flowing water. Thus, the geological condition is complex and it brings some challenges for caisson sinking.

The caisson structure is fabricated in sections of six. Before sinking, one shallow excavation is made for the treatment of shallow ground. When the first three sections are made, the caisson sinks by 6.00m. Afterwards, the caisson sinks in three stages after each section is made. Depths of sinking at each stage are 4.20m, 8.80m, and 10.06m, respectively. To obtain related data, some measurement points were laid out around the caisson at regular intervals to investigate the horizontal displacements and surface settlement, deep settlement, pore water pressure and the ground water table. The plan view of in-situ measurement is shown in Fig.1 (b).

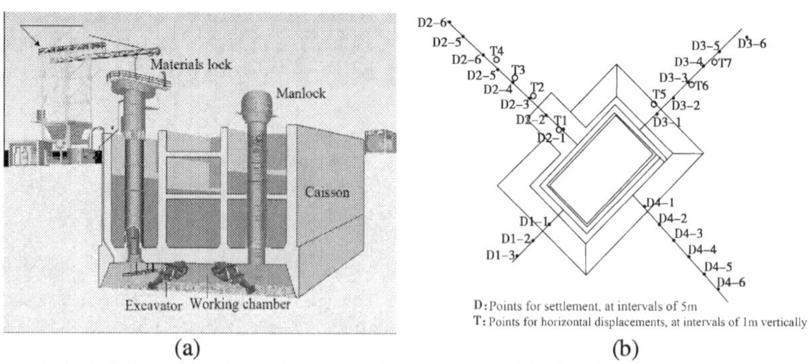

(a) (b)

Fig.1 Schematic view of pneumatic caisson and its in-situ measurement

Mechanical model and numerical methods of full construction process

Mechanical model for the whole process of sinking. For simplification, this study assumes that the rigidity of the caisson is much larger than that of the soil and the caisson structure shows little deformation. Previous studies (e.g., Umeda et al. 2006; Inoue et al. 2006) show that the accuracy of soil deformation prediction is acceptable when the horizontal displacements and inclinations of caisson during sinking are ignored. Moreover, this study aims to predict the ground deformation due to pneumatic caisson construction. Therefore, the strata are modeled without taking into account of the computation of the caisson structure, and thus all related force are directly applied to the soil. Assuming that the caisson maintains steady during sinking, the whole construction process can be divided into five typical load cases, as shown in Fig.2 (a). The step (0-0) is the state of initial ground stress distribution and the step (0-1) denotes the first fabrication of caisson structure before sinking. The self-weight of caisson (W_1) is balanced by the resultant reaction force (R_l) acting on the cutting edge. The r_l in step (0-0) is the vertical distributive force per unit area that results in R_l. Their relationship can be expressed as follows:

$$W_1=R_l\ (=\Sigma r_l) \tag{1}$$

The step (1-0) shows that the caisson has sunk to the depth of h_l when the working chamber is filled with compressed air. Assume that f_1 is the skin friction force per unit area resulting in F_l, h_l the depth of sinking, γ_w the unit density, A the bottom area of the caisson, and P_1 the resultant force acting on the caisson due to compressed air. When the ground water level agrees with the ground surface, $P_1=\gamma_w h_l$ is obtained. During this stage, the relationship of these forces is shown as follows:

$$W_1=P_1+F_l+R_2\ (=\gamma_w h_l A+\Sigma f_1+\Sigma r_2) \tag{2}$$

In the same way, two similar formulations can be obtained as Eqs. (3) and (4), according to step (1-1) and step (2-0), respectively.

$$W_2=P_1+F_2+R_3\ (=\gamma_w A h_l+\Sigma f_2+\Sigma r_3) \tag{3}$$

$$W_2=P_2+F_3+R_4\ (=\gamma_w A h_2+\Sigma f_3+\Sigma r_4) \tag{4}$$

Where W_2 is the updated caisson self-weight, F_2, F_3 are the actual resultant skin friction, and R_3, R_4 are the resultant cutting edge reaction force corresponding to each load case.

Based on the above analyses, the forces acting on the strata are illustrated in Fig.2 (b). The force acting on the soil beneath the caisson points upwards, because of the unloading due to excavation. In addition, the stage from step (c) (1-0) to (d) step (1-1) is the extension of caisson, in which the pressure of the compressed air remains almost the same, but the force on the bearing ground and skin friction force will increase due to the extension of the caisson structure.

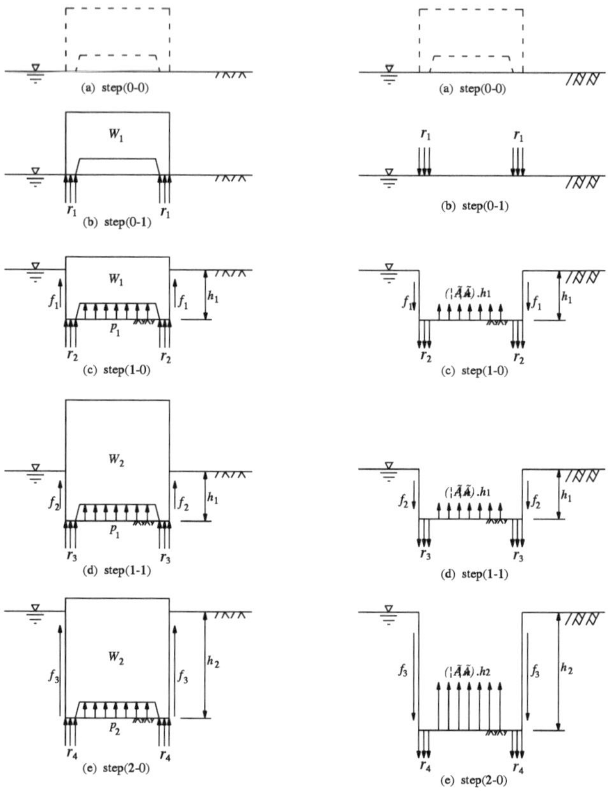

(a) (b)

Fig.2 Mechanical analysis of pneumatic caisson and surrounding strata

Implementation in the FEM codes. In the present study, the finite element computer program developed by the Technical University of Delft (Plaxis B.V. 1998) is adopted to model the whole construction process of pneumatic caisson. For simplification, the plane strain assumption was adopted and a model grid of transverse section was established as shown in Fig.3. In addition, the symmetries of the analyzed problem were also taken into consideration.

In this numerical analysis, three types of constraints were applied on the model as follows: (1) The symmetry property was applied on the left side of the model, which allowed no horizontal displacements; (2) The soil around the caisson wall was constrained in the horizontal direction but was free in the vertical direction; (3) The top side was unconstrained; horizontal displacement was allowed for the bottom side and vertical displacement for the right side of the model. Based on the construction scheme, the sinking process was modeled by activating corresponding loads in each load case, in which element birth/death techniques were used.

To model the soil behavior more accurately, the advanced elasto-plastic constitutive model Hardening-Soil (HS) model was adopted which is uniquely incorporated in Plaxis codes. This model has the ability to simulate the advanced behavior of many types of soil including both soft and hard soil (Schanz et al. 1999), and particularly, it can reflect the stiffness difference of soil for primary loading and un-/reloading, which is better accorded with the actual excavation process. The material parameters used in the analysis are summarized in Table 1.

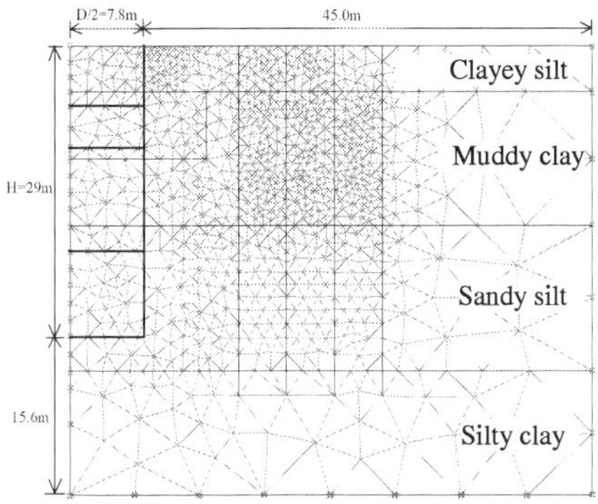

Fig.3 FEM mesh in the present study

Table 1. Physical properties used in FEM analysis

Layers of soil	Submerged unit weight γ' (kN/m³)	Cohesive strength c (kN/m²)	Internal friction angle ϕ (°)	Poisson's ratio ν	Dilatancy angle ψ (°)	E_{50}^{ref} (MN/m²)	E_{ur}^{ref} (MN/m²)
Clayey silt	9.3	20	19.5	0.2	0	17.4	69.4
Muddy clay	8.2	15	11	0.2	0	10	40
Sandy silt	9.2	6	30	0.2	0	24	96
Silty clay	9.0	16	19.5	0.2	0	10.3	41

Notes: E_{50}^{ref} is the reference stiffness modulus and E_{ur}^{ref} is the reference Young' modulus for un-/reloading.

Results and Discussions

The shaft construction took 11 months and achieved rather satisfactory results. In this section, the calculated results and in-situ measurements at some representative measurement points are presented and compared in terms of the following two aspects.

Surface settlements. It can be seen from Fig.4 that, the results of numerical calculation seem to agree with the measurements rather well. In these graphs, the x-coordinate is described as the ratio of distance from the center of the caisson (D) to the depth of total sinking (H) for normalization, and h is the depth by which the caisson has sunk. The final surface settlement curve obtained from calculation behaviors as a parabola, and with the increase of D, the settlement decreases rapidly. The results also indicate that, with the process of caisson sinking, the settlement is increasing and the scope of surface settlement involved is extending gradually. When the shaft caisson is sunk to the designed level, the total calculated maximum settlement around the caisson is 78.8mm.

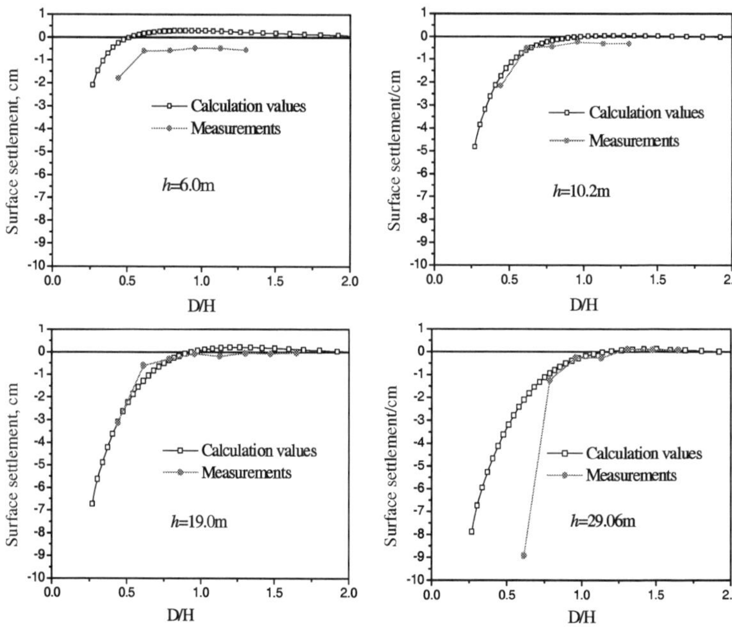

Fig.4 Calculated values and measurements of surface settlement in D2 direction

Deep horizontal displacements. Herein, only the point of T1, 5m distant from the caisson, is illustrated in Fig.5. It can be seen that, both the calculated values and measurements are relatively small and vary with the depth of soil. As the caisson sinking progresses, the deep horizontal displacements are increasing gradually. As to the distribution pattern of the horizontal displacements curves, it is highly dependent on depth of caisson sinking. Fig.5 shows that, with the caisson sinking, soils above some critical level (assumed as H_0) tend to move towards the caisson and the displacement reaches the maximum at the surface, whereas soils below H_0 tend to move away from the caisson and the displacement increases until the maximum then decrease to zero gradually.

Discussion. With respect to surface settlement, the calculated values deviate from the measurements at the first stage of caisson sinking, as shown in Fig.4. This phenomenon is attributed to the fact that, the numerical model does not take into account certain factors, such as the shallow excavation and backfill, and instability of caisson at initial sinking. It can be verified that the calculated values and measurements agree with each other in the following process of sinking. The high depth of lowering at the last sinking stage and a lack of appropriate construction control measures, those can explain why the calculated values shows relatively large deviation from measured ones.

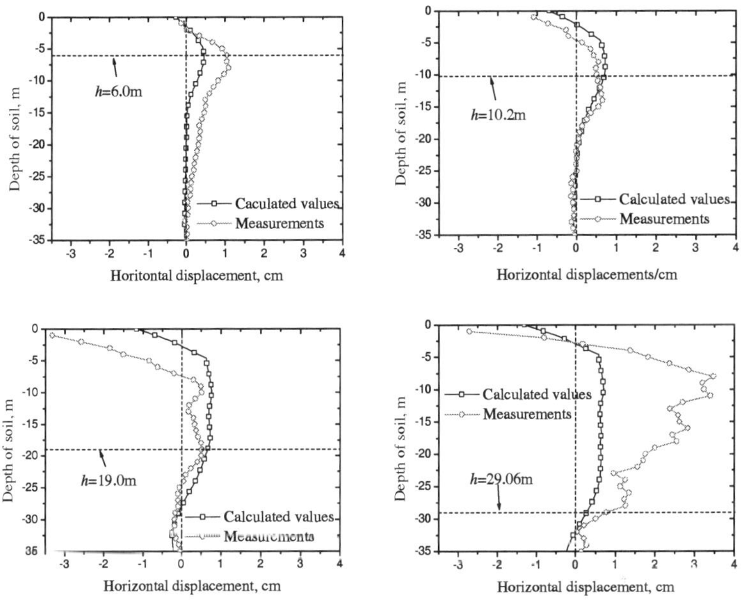

Fig.5 Calculated values and measurements of horizontal movements of T1

The patterns of deep horizontal displacements for calculated values and measurements are approximately similar but small deviation still exists. In the process of caisson sinking, the surrounding ground is subjected to complex and even some unexpected actions, and the cutting and extruding of caisson cutting edge will always disturb the soil repeatedly. Besides, the adjustment of caisson attitude, procedure of soil excavation, and some auxiliary techniques such as mud coating to assist the caisson sinking indeed influence the distribution of the horizontal displacements around the caisson. Outside the wall of caisson, no bench for friction reducing is designed for the shaft caisson. Some gap between the caisson and surrounding soil occurs during the adjustment of caisson attitude. Therefore, the soil above H_0 moves towards the caisson due to self weight of soil and surcharge around the caisson. The cutting edge' extruding soil out of caisson also explains why the soil below H_0 moves away from the caisson.

Conclusions

(1) The proposed numerical analysis could simulate well the whole construction process of pneumatic caisson, and it has taken the pressure of the compressed air, skin friction and the cutting edge reaction force into account; besides, unloading effect of soil under the caisson is also incorporated.

(2) Hardening-soil model was employed to conduct non-linear analysis in the proposed numerical calculation. Numerical analysis results were in good agreement with the measured ones, which verified the proposed mechanics model and numerical analysis method. It is shown that, the environmental effects of pneumatic caisson construction are highly related to the depth of caisson sinking. The influence zone on the ground surface is approximately one time the total depth of caisson sinking. With the increase of distance away from the caisson, the surface settlement decreases more rapidly compared to other deep excavation method.

(3) In the above study, the horizontal displacement and inclination of caisson and the three dimensional effect of structure are not taken into account. The further researches need to be conducted.

Acknowledgements

This work was supported by funds from Grant 2006BAJ27B02-02 from National Key Technology R&D Program as well as Shuguang Project (No.05SG25), NSFC (No.50679056) and NCET (06-0378).

References

Inoue, T., Fujii, N., and Ohuchi, M. (2006). "Numerical simulation of ground deformation due to pneumatic caisson tunnel." *Journal of tunnel engineering*, Vol.16, 155-164 (in Japanese).

Ohuchi, M., Peng, F.L., Sun, D.X., Zhu, H.H. and Liao, S.M. (2003). "New Pneumatic Caisson". *Geotechnical Engineer*, 15(1), 19-22 (in Chinese).

Plaxis B.V. (1998). "PLAXIS Professional version 7.2." *Geotechnical Finite Element Code for Soil and Rock Analyses* (CD-Rom), Delft, Netherlands.

Schanz, T., Vermeer P.A., and Bonnier P.G. (1999). "The Hardening Soil model–formulation and verification." *Beyond 2000 in Computational Geotechnics –10 Years of PLAXIS International, Balkema, Rotterdam*, 281–296.

Umeda, N., Fujii, N., and Inoue, T. (2006). "Numerical Modeling of Pneumatic Caisson Method." *Journal of Applied Mechanics*, Vol.9, 603-612 (in Japanese).

WANG, H.X., and WANG, D.Y. (2003). "Mechanics Model Study in Large Extra Caissons Construction." *CHINESE QUARTERLY OF MECHANICS*, Vol.24, No.1, 68-74 (in Chinese).

Temperature Effect on Tieback Loads

Nasser Massoudi [1], M. ASCE, P.E.

[1] Principal Engineer, Bechtel Power, 5275 Westview Dr., Frederick, MD, 21703; nmassoud@bechtel.com

ABSTRACT: More than 50 tiebacks were installed into a structural concrete slurry wall supporting a deep excavation. The tiebacks consisted of 7-wire strands and ranged in length from about 50-80 ft (15.2-24.4 m). Three of the tiebacks were fitted with load cells and were monitored for a period of 14 months. Results indicate that the tieback loads fluctuated over this period by as high as about 10 percent from the initial lock off load. The variation in load is attributed to large fluctuations in temperature, ranging from about 25-100 deg F (-4 to 38 deg C) over the monitoring period. A maximum load loss of about 1 kip (4.5 kN) for every 10 deg F (5.6 deg C) rise in temperature was recorded. This can be significant for excavations with restricted movement requirement.

INTRODUCTION

Temperature fluctuations are known to affect performance of excavation support systems. Hashash et al. (2003) describes the results of a 100-ft (30.5-m) deep section of an excavation in Boston supported by a concrete slurry wall, presenting variations in strut loads with temperature, difficulties with load interpretations, and separating earth loads from thermal loads in struts. Duchesne (2003) reports on the effects of frost heave on performance of several soil nail walls, indicating high tensile loads in nails during the colder months; in one case the nail loads doubling during the colder months. Boone and Crawford (2000) documents the effects of temperature on a 66-ft (20.1-m) deep braced excavation in glacial soils, supported by soldier piles and lagging with pipes for struts, establishing correlations between temperature and strut loads. Morgenstern and Sego (1981) reports on effects of frost on an instrumented 23-ft (7-m) high tieback wall in fissured clays of Canada where loads as high as 120% of the design loads were measured in anchors during winter periods. Tieback walls are also reported to have failed as caused by freezing temperatures (Sandegren et al., 1972).

This paper documents a case history of tiebacks installed for a deep excavation supported by jet grouted soils and a reinforced concrete slurry wall. It presents measurements as well as correlations among temperature, tieback load, and movements.

BACKGROUND

Renovation of the Virginia State Capitol Building in Richmond required a deep excavation about 40 ft (12.2 m) below the ground surface. The excavation was designed to be supported by a permanent concrete slurry wall, with soils between the slurry wall and the Capitol building improved using jet grouting and compensation grouting. The excavation support system is depicted in Figure 1.

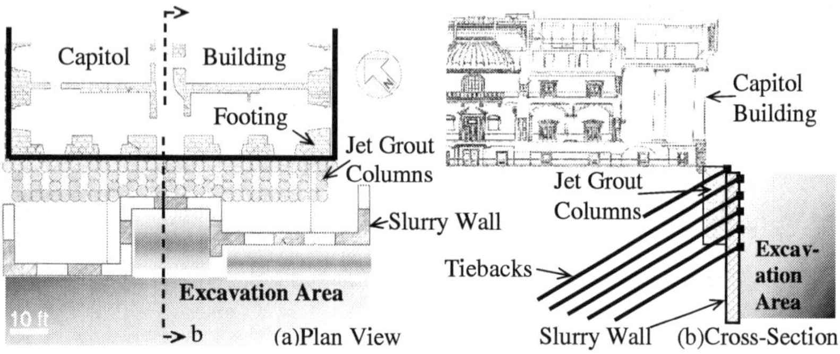

FIG. 1. Excavation Support System at the Virginia State Capitol Building.

The site soils consist primarily of marine deposits. Typical subsurface stratigraphy consists of approximately 2 ft (0.6 m) of medium dense sandy fill, 20 ft (6.1 m) of loose-very dense clayey sand with cobbles and boulders (average Standard Penetration Test N-value of 40 blows/ft), 70 ft (21.3 m) of firm-hard clay and silt of Miocene Age (average N-value of 15 blows/ft with average undrained shear strength of 2,000 psf (95 kPa)), 50 ft (15.2 m) of very dense sand and gravel (SPT values typically greater than 100 blows/ft), and granite bedrock at a depth of about 140 ft (42.7 m) with average unconfined compressive strength of 15,000 psi (103 MPa). Perched groundwater level is about 20 ft (6.1 m) below the surface. Only soils of the upper 80 ft (24.4 m) are of geotechnical significance for the project, i.e., the loose-very dense clayey sand and the firm-hard clay and silt, as these soils have the greatest influence upon the stability of the excavation as well as serving as the bond soils for the tiebacks. The Miocene-age firm-hard clay and silt soils are known to exhibit sensitive properties.

TIEBACK DESIGN AND INSTALLATION

The excavation was tiedback using 6 rows of temporary tiebacks. The upper row of tiebacks (Row A) was drilled directly into jet grouted soils. The remaining 5 rows (Rows B through F) were installed through the concrete slurry wall. The tieback locations are shown in Figure 2, as well as the locations of the three tiebacks which were fitted with load cells. Details of the instrumented tiebacks are given in Table 1. As noted earlier, the tiebacks were only temporary; once the structure was completed all tiebacks were de-tensioned.

The tiebacks consisted of three to six 7-wire strands. All tiebacks were post-grouted, pre-stressed, and tested. Tiebacks in the jet grout columns were stressed to 70 kips (311 kN) each, although one tieback was locked off at slightly higher than its design load. Tiebacks in the slurry wall were stressed to loads varying from 60-160 kips (267-712 kN). For tieback installation, excavation of the soils in front of the completed slurry wall was limited to about 2 ft (0.6 m) below the tieback levels. All tiebacks were drilled with a Klemm 806 drill rig, using a 5.5-inch (140-mm) diameter sacrificial point bit, anticipated to result in a bond diameter of about 6 inches (150 mm). The drilling was carried out using water for fluid. Rotary drilling was performed using casing with a tricone roller bit until the casing was seated in the Miocene clay/silt. Thereafter, drilling was performed using the tricone bit to completion. Once the hole was completed, grout was pumped through the inner rods from the bottom of the hole until fresh grout reached the ground surface. The tendons were then inserted and the casings removed. Additional grout was pumped into the hole as needed.

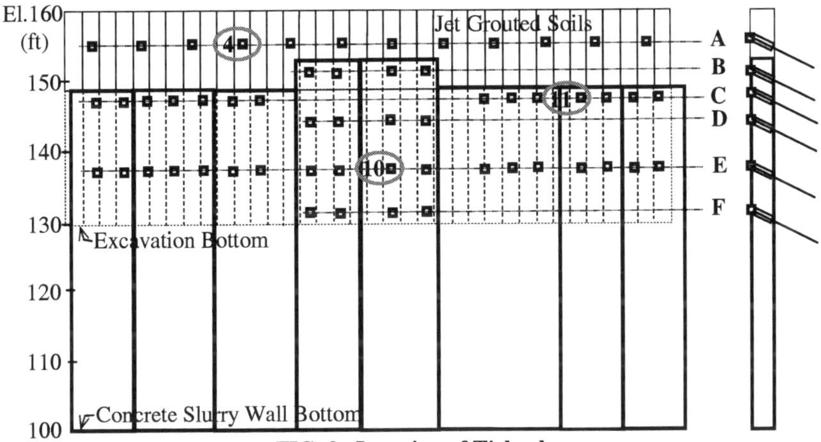

FIG. 2. Location of Tiebacks.

TABLE 1. Details of Tiebacks Fitted with Load Cells.

Instrumented Tieback	A4	C11	E10
Total length, ft (m)	63 (19.2)	94 (28.7)	76 (23.2)
Bond length, ft (m)	25 (7.6)	33 (10.1)	40 (12.2)
No. of strands	3	5	6
Allowable bond stress, Kip/ft (kN/m)	2.7 (39)	4 (58)	4 (58)
Design load, Kip (kN)	71 (316)	134 (596)	159 (707)
Date installed and grouted	Jan-12-2005	Feb-2-2005	Feb-23-2005
Bonded portion installed in	Clayey Sand	Clay/Silt	Clay/Silt
Anchor head installed in	Jet grouted soil	Slurry wall	Slurry wall
Date pressure grouted	Jan-13-2005	Feb-4-2005	Feb-24-2005
Date post-grouted	Jan-14-2005	Feb-5-2005	Feb-25-2005
Date proof-tested	Jan-18-2005	Feb-5-2005	Feb-28-2005
Max. test load, Kip (kN)	110 (489)	179 (796)	212 (943)
Est. free length movement, in.(mm)	2.7 (69)	4.2 (107)	2.5 (64)
Lock-off load, Kip (kN)	86 (383)	134 (596)	159 (707)

The tendons were pre-assembled off-site, individually marked, and shipped coiled. They were fabricated of 58K Polystrand®, 7-wire, and cut to the required length with the required number of strands. Assembly of centralizers and placement of post-grout tubes and spacers were performed on site. Post-grout tubes were PVC with an external diameter of 1.25 inches (32 mm) and with grout ports at 1-ft (0.3-m) center-to-center. Centralizers were placed at 5-ft (1.5-m) intervals along the bond portion of the tendons.

The grout W/C ratio was in the range of 0.5-0.67. Grout for pressure grouting of tendons, however, had W/C of 0.67. Type 3 cement was used for achieving early strength so that tendons could be stressed within 1-2 days of their installation. Laboratory testing of unconfined compressive strength on grout cube samples are shown in Figure 3. Very high compressive strengths were achieved early; the average strength values approached 8,000 psi (55 MPa) at 28 days.

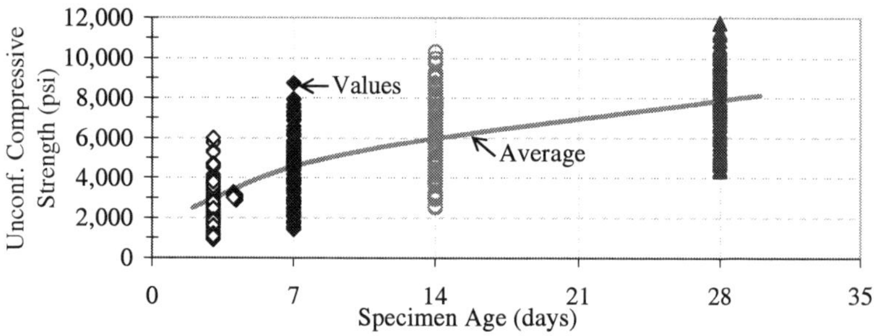

FIG. 3. Unconfined Compressive Strength of Grout for Tiebacks.

Allowable bond load used for design of the tiebacks was 2.7 kips/lineal ft (39 kN/m) for level A tiebacks (bonded in the clayey sand) and 4 kips/lineal ft (58 kN/m) for all others which were bonded in the Miocene age clay/silt.

All tiebacks were post-grouted within 1-2 days of their installation. Post-grouting included pressure-grouting every third grout port based on construction considerations. The distance between every third port is 3 ft (0.9 m). The criterion for post-grouting was injecting 50 gallons (190 liters) of grout at a pressure of 300 psi (2.1 MPa). If the 300-psi pressure was not achieved, the tieback was re-grouted the next day, at the same ports, until the 300-psi pressure was realized. In a majority of cases, the desired pressure was achieved at the second post-grouting period. Few tiebacks were re-grouted for the third time. This aggressive post-grouting practice was significant in improving the tiebacks bond zones, and reflected positively in tiebacks performance during load testing as discussed later. At project completion, about 12 tons of cement had been used for grouting the tiebacks compared to an initial cement budget of 6 tons.

TIEBACK LOAD TESTING

All tiebacks were load-tested to establish their installed capacity. Two verification load tests (to 150% and 200% of design load) and one performance load test (to 150% of design load) were performed. The maximum test load was 300 kips (1,334 kN). All tiebacks were proof-tested to 133% of their design load, and locked-off at 100% of their design load. The proof-test results for the three instrumented tiebacks are shown in Figure 4. Refer to Table 1 for the details on instrumented tiebacks.

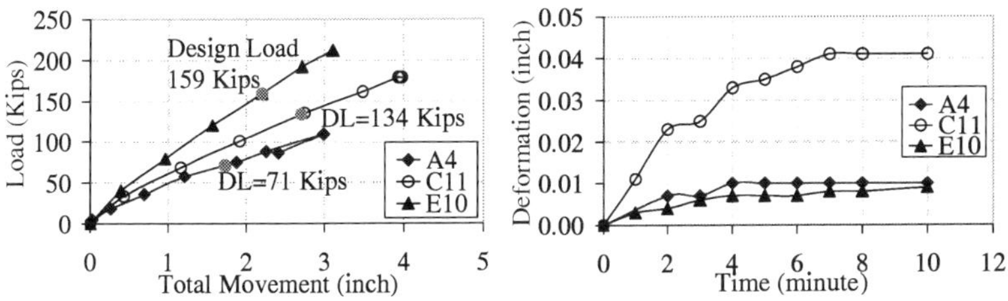

FIG. 4. Proof Load Test Results for Instrumented Tiebacks.

INSTRUMENTATION MONITORING

Movement requirements for the Capitol Building foundations were extraordinarily limiting, established at 0.25 inch (6 mm). This necessitated the availability of information to evaluate the performance of the excavation on a continual basis, and to enable rapid response to developing trends. Hence, a real-time system of geotechnical instruments was used for monitoring the building and the excavation. They consisted of three total station theodolites with about 80 optical prisms, two in-place inclinometers (and one manual as back-up), temperature sensors, data acquisition system, and Internet connection. As previously noted, three of the tiebacks were also fitted with load cells. The load cells were vibrating wire type, model VH-1000 and VH-1500, manufactured by RocTest, with accuracy of 1.1 and 1.7 kips (5 and 8 kN), respectively. Monitoring instruments served as a critical component of the verification and quality assurance.

RESULTS FROM INSTRUMENTED TIEBACKS

Load and ambient temperature data were collected during the period from January 2005 through March 2006 for the three instrumented tiebacks until the tiebacks were de-tensioned. Load histories for the three tiebacks are shown in Figure 5.

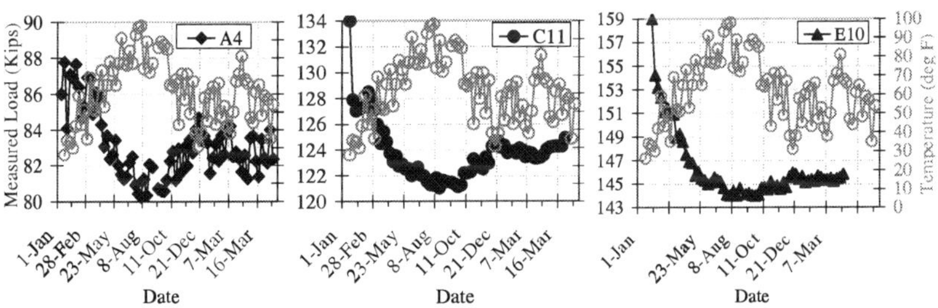

FIG. 5. Load and Temperature History for Instrumented Tiebacks.

The trend clearly shows a reduction in tieback load with rise in temperature and a reversal in load with reversal in temperature. Tieback A4 showed the highest percentage of load fluctuation than the other two tiebacks, which may be related to its lower lock-off load. Also, starting in December 2005, heaters were placed in front of the slurry wall, where the other two tiebacks reside, and the area was covered with plastic sheeting. A few measurements of ambient air in the heated area indicated temperatures in mid 60's degree F (16° C). Regardless, the maximum recorded change in tieback loads for a maximum temperature change of 72 degree F (22° C) is shown in Table 2.

TABLE 2. Maximum Change in Tieback Load for 72°F Temperature Change.

Instrumented Tieback	A4	C11	E10
Lock-off Load, Kip (kN)	86 (383)	134 (596)	159 (707)
Max. Measured Drop in Load, Kip (kN)	7.5 (33)	6.9 (31)	10.2 (45)
Max. Measured Drop in Load (%)	9	5	7

A possible mechanism that leads to such behavior would be due to changes in that portion of tendons near the ground surface or the excavation face as impacted by the temperature fluctuation. Tendon expansion is expected to result in load loss. Conversely, tendon contraction is expected to recover the load or place additional load into the tendon as the anchor head is firmly supported on the concrete slurry wall, or on the steel waler system, which limits the contraction beyond that imposed during the locking off period. Another mechanism is described by Clough and Duncan (1972) where a concrete wall moves away from the excavation due to temperature increase. This mechanism can result in a loss of load in tiebacks.

The change in load versus change in temperature for all three tiebacks is shown in Figure 6. The load and temperature relationship may be established as

$$\Delta P = m\,\Delta T + n \qquad (1)$$

where, ΔP= load change in kips, ΔT= temperature change in degree F, with n and m being regression coefficients. The latter is a thermal load coefficient representing the stiffness of the excavation support system for small deformations (Boone and Crawford, 2000). A value of 4.817 kips/°F is reported by Hashash et al. (2003) for an excavation support system consisting of a 3-ft thick concrete slurry wall supported by cross-lot bracing with steel struts jacked to about 50% of their design load. From Figure 6, the m values range from 0.0302 to 0.0721 kips/°F, indicative of a very stiff excavation support system. Such large system stiffness is consistent with the project excavation support for its several redundant and complementary components such as those offered by the rigidity of the slurry wall, its geometry in plan and elevation, jet grouting and compensation grouting improvements, relatively stiff site soils, early capping of slurry wall panels with a reinforced concrete slab to engage the panels, and tieback lock-off loads at high (100%) design values. It is noted that the lock-off loads for these tiebacks

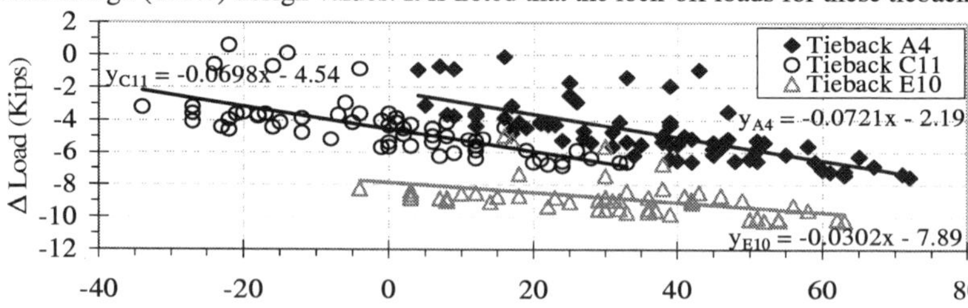

FIG. 6. Change in Measured Load vs. Temperature.

were also examined relative variations in coefficient *m*, however, they were found to have practically no effect on this coefficient.

The instrumentation included monitoring deflection along the entire depth of the slurry wall as well as monitoring of building movement. The slurry wall movements were recorded via in-place inclinometers in the slurry wall whereas the building movements were by theodolites. Measured lateral movements versus temperature are shown in Figure 7. Although there is wide scatter in data from the inclinometers, the results indicate that relationships exist between temperature fluctuation and foundation (or slurry wall) movement such that with increasing temperature, foundations (or the slurry wall) move away from the excavation. These results corroborate with a scenario described by Clough and Duncan (1972) where increase in temperature of a concrete wall resulted in the wall moving away from the excavation. The lateral movement and temperature relationship may be established as

$$\Delta Y = a\, \Delta T + b \tag{2}$$

where, ΔY= change in lateral movement in mm, ΔT= change in temperature in degree F, with *b* and *a* being regression coefficients. The latter is a thermal coefficient which is noted as 0.0241 to 0.0301 mm/°F for the concrete slurry wall and 0.0037 mm/°F for the jet grouted soils, indicating that the jet grouted soils experienced less thermal expansion than the concrete slurry wall by a factor of 7 to 8. For the 3-ft (0.9-m) thick reinforced concrete slurry wall, its coefficient of thermal expansion is back-calculated to be about 3E-5/°F. For the 3-ft (0.9-m) thick jet grouted columns, the coefficient of thermal expansion is back-calculated to be about 4E-6/°F. These values approach the thermal coefficient of normal concrete which is in the 3E-6 to 8E-6/°F range.

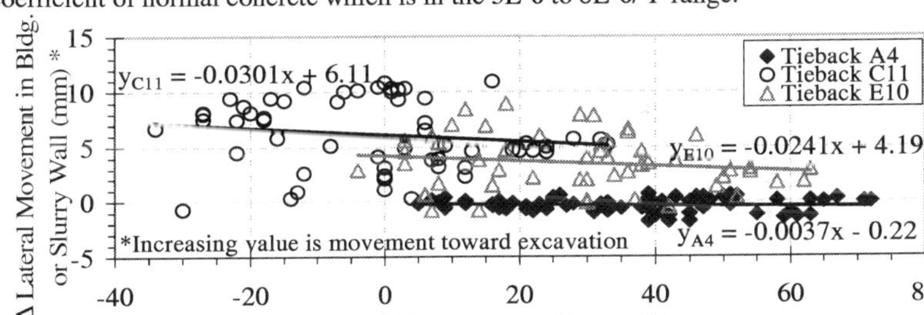

FIG. 7. Change in Measured Lateral Movement vs. Temperature.

Measured lateral movements are also compared with measured tieback loads, as shown in Figure 8. Results indicate a trend between fluctuation in tieback load and movement. The load and movement relationship may be established as

$$\Delta P = x\, \Delta Y + y \tag{3}$$

where, ΔP and ΔY are as previously defined, with *y* and *x* being regression coefficients. Coefficient *x* is a system stiffness (compliance) value, 0.0609 and 0.2656 kip/mm for the concrete slurry wall and 1.0418 kip/mm for the jet grouted soils. The greatest loss in tieback load is in the upper tieback anchored in jet grouted soils, or about 1.0 kip (4.5 kN) loss in load for every 1 mm of lateral movement, compared to less than 0.1 kip for every 1 mm of lateral movement in the lowest tieback row connected to the slurry wall.

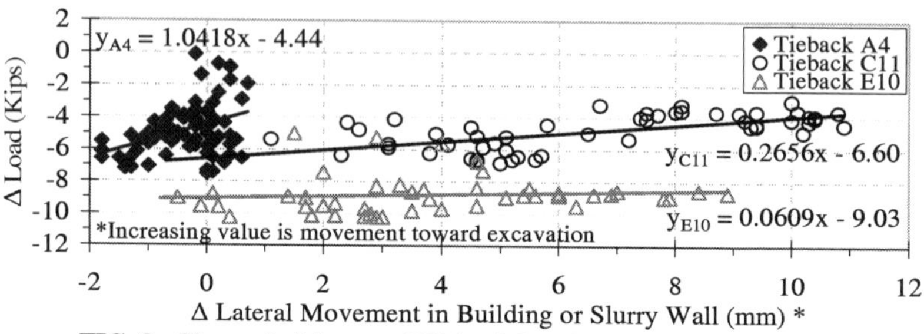

FIG. 8. Change in Measured Tieback Load vs. Lateral Movement.

CONCLUSIONS

Elevated or sub-zero temperatures can highly impact tieback loads, resulting in undesirable movements or even failure. Results from three instrumented tiebacks for a deep excavation supported by jet grouted soils and a concrete slurry wall indicate a relationship among tieback load, temperature, and lateral movement. Although the magnitudes of measured movements or load fluctuations were not large, such changes may be significant for satisfactory performance of excavation support systems near sensitive structures as large fluctuations in load could manifest large movements. Consideration of temperature effect on load, and their monitoring, should be a part of any excavation support system with restrictive movement criterion.

ACKNOWLEDGEMENT

The cooperation of the Virginia Department of General Services is appreciated. The author gratefully acknowledges the encouragement and support of Mr Richard Sliwoski.

REFERENCES

Boone, S.J. and Crawford, A.M. (2000). "Braced Excavations: Temperature, Elastic Modulus, and Strut Loads," *ASCE Journal of Geotechnical and Geoenvironmental Engineering*, Vol. 126, No. 10, pp. 870-881.

Clough, G.W. and Duncan, J.M. (1972). "Temperature Effects on Behavior of Port Allen Locks," *Proceedings, ASCE Specialty Conference on Performance of Earth and Earth-Supported Structure*, Vol. 1, Part 2, pp. 1467-1479.

Duchesne, S.M. (2003). "Effects of Frost Heave on a Soil Nail Wall in Brunswick, Maine," Master of Science Thesis, University of Maine.

Hashash, Y.M.A., Marulanda, C., Kershaw, K.A., Cording, E.J., Druss, D.L., Bobrow, D.J., and Das, P.K. (2003). "Temperature Correction and Strut Loads in Central Artery Excavations," *ASCE Journal of Geotechnical and Geoenvironmental Engineering*, Vol. 129, No. 6, pp. 495-505.

Morgenstern, N.R. and Sego, D.D. (1981). "Performance of Temporary Tiebacks Under Winter Conditions," *Canadian Geotechnical Journal*, Vol. 18, pp. 566-572.

Sandegren, E., Sahlstrom, P.O., and Stille, H. (1972). "Behavior of Anchored Sheet Pile Exposed to Frost Action," *Proceedings, 5th European Conference on Soil Mechanics and Foundation Engineering*, Madrid, Vol. 1, pp. 285-293.

STEEL INNOVATION: CHANGING THE ECONOMICS OF BELOW GRADE FOUNDATIONS

Dean Abbondanza,
Skyline Steel LLC, Engineering Group, Pittsburgh, PA, USA

Below grade foundation design and construction can be a complicated, multi phase process. Conventional means and methods have additional issues that at grade foundations do not. The cost factors and systems are dependent on soil conditions and vary by region, but the process remains fairly standard. Install temporary shoring, brace as excavation proceeds to final elevation, build your foundations, construct a permanent wall, waterproof, and work your way up out of the ground. These multiple contracting phases require form work, curing time, and possible offset from the property line. Subterranean structures offer shelter from the elements, minimal temperature variance, and maximize use of the property. Those benefits are overshadowed by the costly variables of building a geostructure. It is no surprise that underground construction can be more than twice the cost of above grade. For the past three years, domestic owners, consultants and contractors have begun to realize the benefits of below grade permanent sheet pile walls for parking and basement structures.

CONVENTIONAL FOUNDATION CONSTRUCTION

Soldier beam and lagging is the predominant temporary shoring method. Soil nailing, soil/cement mixing columns, insitu secant piles, temporary sheeting, and slurry diaphragm walls are other methods considered where deeper excavation, high water table, and adjacent structures are present. The common denominator with all these methods is; they usually require a permanent wall component be constructed within.

Material Perception

Sheet piles are well known for bulkheads, docks, seawalls, and cofferdam applications. They perform well in a range of soil and water environments. Beyond marine structures, the industry perception is "temporary support of excavation", but steel sheet piling is capable of much more. Sheet piles are independent geostructural foundation elements with the potential to carry lateral, transverse, and axial forces in permanent loading applications.

For the last 30 years, European sheet pile producers have promoted sheet pile walls for a variety of permanent applications. From grade separation walls for roadways, to bridge abutments and underground car park foundations, this unconventional approach to sheet pile design has made a significant impact on European construction practices, none greater than underground parking garages.

SHEET PILE PRODUCTION AND EVOLUTION

Hot rolled steel sheet piling is produced in three general shapes by less than fifteen steel mills throughout the world. It is considered one of the most difficult structural shapes to roll. The intricate interlock connection and differential web and flange thicknesses pose a variety of rolling and pass design issues. The slightest oversight can bring a mill rolling to a stand still, resulting in higher scrap yields.

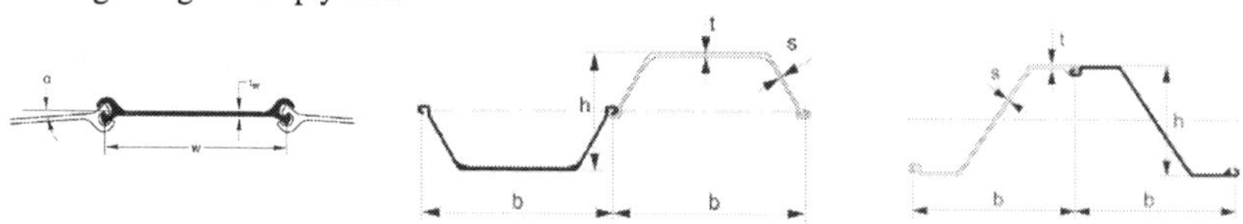

FLAT WEB **U - PILE** **Z - PILE**

Hot Rolled Sheet Pile Profiles
FIG 1

Flat web piles were the first hot rolled pile produced. Designed for membrane structures, they derive strength from interlock tension. These sections are predominantly specified on circular cell structures, and offer no value for support of excavation or bending applications. The U section evolved from the flat web shape, giving birth to pile retention walls. Over time, designers realized that U piles may not achieve full design strength due to the interlock location through the neutral axis. Research would validate these claims proving that the interlocks tend to rotate about the centroid, thus are susceptible to oblique bending under surface loading. Even with crimping and welding, the section modulus and moment of inertia are significantly less than calculated. Based on these findings, Euro codes applied reduction factors to both Section Modulus (0.7 – 1.0) and Moment of Inertia (0.8 – 1.0) properties for U pile sections. Among the current hot rolled sheet pile manufacturers in North America, none produce U shapes. The domestic mills have committed to the Z sections. Design engineers prefer the excellent strength-to-weight properties, and contractors like the profile for driving ease and efficiency.

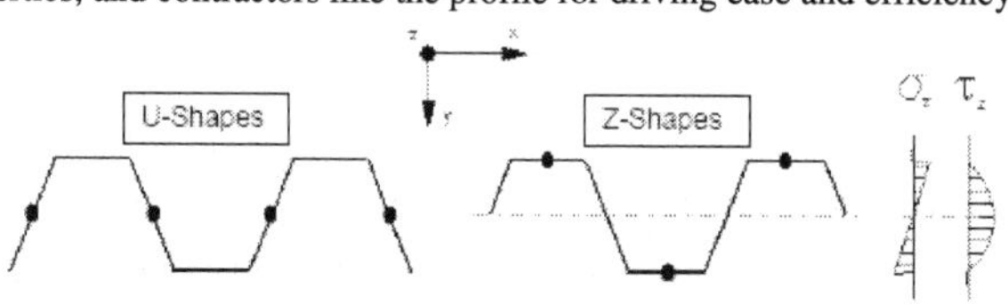

Location of Interlocks and Stress Distribution
FIG 2

Note: Cold formed sheet piles have a rolled "hook & grip" interlock that does not provide the same level of permeability as the hot rolled connections. Therefore, hot rolled piles are preferred for this permanent application.

STRUCTURAL ANALYSIS

Design and construction of permanent sheet pile parking garage and basement structures demands a thorough knowledge of all forces acting between the wall and soil mass. Fixed and free earth methods of analysis are standard for temporary shored retaining structures. Pressure diagrams are often required by local building code authorities for shoring design, but this is only a static snapshot of post construction loading conditions. The excavation sequence introduces dynamic loads into the wall structure. Analyzing the magnitude and location of these loads at different phases of excavation is helpful to simulating true site conditions. Although the forces are temporary, they can be significant. Therefore, an elasto-plastic modeling could be applied to check if the unique dynamic construction loads will affect final design requirements.

Axial Loading

Steel sheet pile walls are independent deep foundation systems capable of carrying forces in multiple planes. Their primary design function is to resist lateral loads. They also have a substantial surface area for bearing perimeter vertical loads in low to mid-rise structures. Sheet piles have a corrugated Z shape that gives high surface to soil contact per square foot of wall. Vertical capacity can only be achieved if the soils offer enough cohesive value, or bearing capacity. More often the axial loading is limited by the geo-mechanics, and not the piles material strength. To avoid eccentric forces on the sheets, the perimeter loads are distributed uniformly to the sheet pile wall by casting the tops of the piles into a grade beam. In contrast, conventional cast in place concrete walls have to be founded on perimeter footings, and usually do not carry any vertical loads from the above structure.

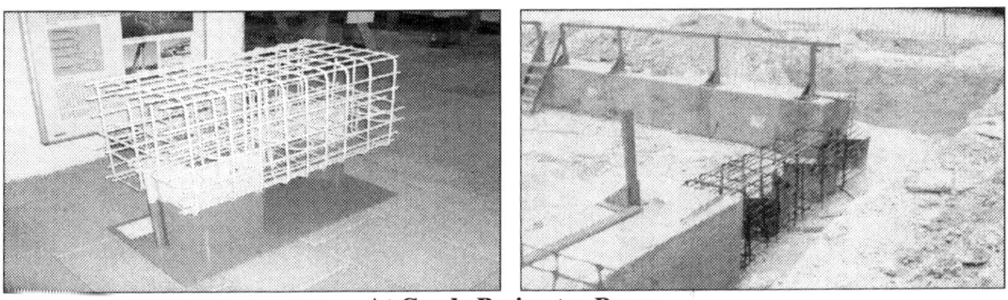

At Grade Perimeter Beam
FIG 3

Drivability

The structural requirement for a retaining wall is driven by application loads, but there are other design and construction factors that help to determine the sheet pile section. "Drivability" is equally important for permanent sheet pile walls. This is a foundation structure. Vertical tolerance and driving stress are important to consider when selecting a wall section. The bending moment and deflection limits may be achieved by a smaller section, but a heavier shape may be needed to achieve installation of the piles without damage. The goal is to balance design efficiency with constructability.

Advancements in hammer technology have broadened pile driving applications. Variable high frequency hammers and hydraulic press equipment with driving assistance attachments for auguring and water jetting, are recapturing pile projects that might have been lost to difficult soils or environmental noise and vibration.

Hydraulic Press Pile Hammers
FIG 4

Durability

Steel corrosion is an electro-chemical process where exposed material degrades over time. The service life of a steel section is reached when the material has degraded to the point of 2/3 the yield strength, or in the case of grade 50, 3.0 mm thickness. The durability of steel pile structures has been documented in numerous marine conditions, and in a comprehensive range of soil chemistries. It is from this empirical data, we can extrapolate the corrosion rates for sheet pile in this permanent application.

Loss of Thickness (mm)	5 Years	25 Years	50 Years	75 Years	100 Years
Undisturbed Natural Soils	0.00	0.30	0.60	0.90	1.20
Polluted Natural Soils and Industrial Grounds	0.15	0.75	1.50	2.25	3.00
Aggressive Natural Soils (Swamp, Marsh, Peat)	0.20	1.00	1.75	2.50	3.25
Non-Compacted and Non-Aggressive Fills	0.18	0.70	1.20	1.70	2.20
Non-Compacted and Aggressive Fills (Ashes Slag)	0.50	2.00	3.25	4.50	5.75

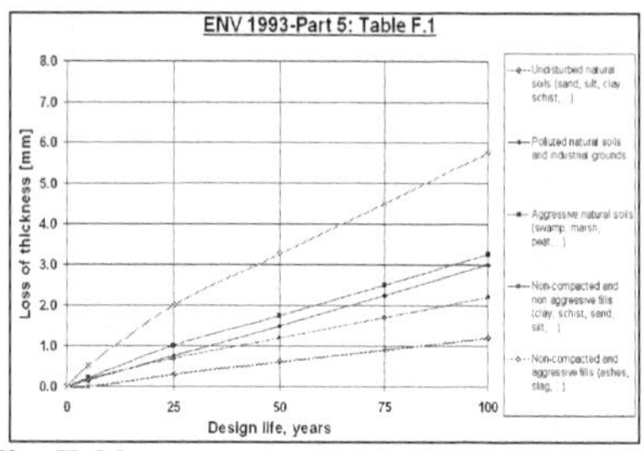

Steel Durability Tables
FIG 5

The backside is the only portion of the wall in direct contact with soil. The exposed side is primed and painted for a dry, climate controlled environment. These are robust steel profiles with a minimum wall thickness of 9.50 mm (0.375 in), ranging up to 20.00 mm (0.787 in). With additional design consideration for driving stresses and fire safety engineering, the added material thickness serves as a sacrificial "belt and suspenders" measure for a relatively non aggressive environment. Conservative design life expectancies start at seventy five and exceed more than one hundred fifty years. Durability software is incorporated in addition to corrosion calculations to verify service life expectancy.

WATERPROOFING

Vertical Walls

Steel is an impervious material; the only lateral pathway for groundwater or vapor along the sheet pile wall is through the vertical joint. In relation to the total wall area, the interlock makes up less than 1%. With more than 99% of the wall impervious at installation, we have identified and minimized the pathway potential. For building owners, this is of great value. This sheet pile section is manufactured with a Larrsen interlock. It has a four point contact connection with conductivity of 6.98E-09 m/s in soil. Although this is very low permeability, a non structural seal weld is specified down the exposed interlock after excavation. This eliminates virtually any pathway potential along the vertical wall providing a waterproofing system designed for the life of the structure.

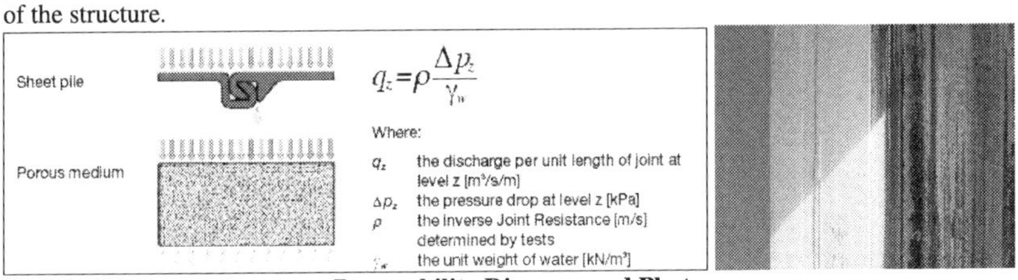

Permeability Diagram and Photo
FIG 6

Waterstop: Base Slab

As with conventional structures, the base slab waterstop issues are really no different for the permanent sheet pile application. Based on the geotechnical engineers' advice, a barrier membrane or under drain system is specified for waterproofing the underside of the base slab. To ensure a tight bond at the base of the sheets, the reinforcement steel is welded to the sheet pile around the perimeter. This connection resists shrinkage when the concrete cures at the slab/wall connection. A hydrophilic stripping is added to the perimeter of the sheet pile at the top and bottom elevation of the slab/pile interface.

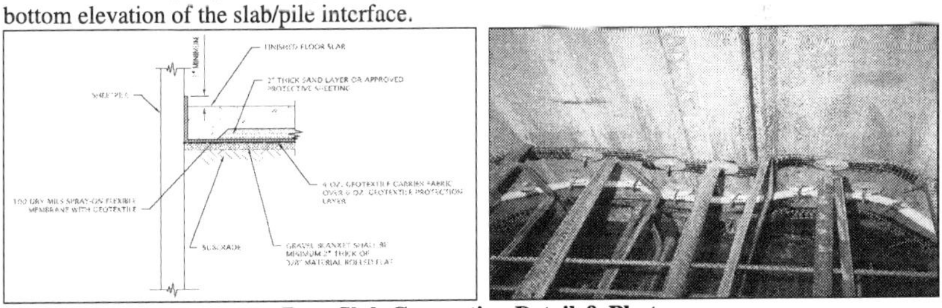

Base Slab Connection Detail & Photo
FIG 7

CONNECTION DETAILS

Depending on the floor system, a structural connection between the slab and sheets may or may not be required. Whether precast, post tensioned, or cast in place, there are no limitations or increased cost associated with the sheet pile permanent wall system. As previously mentioned, the sheets can carry axial load. A variety of simply supported and structural connection details are available for any flooring system.

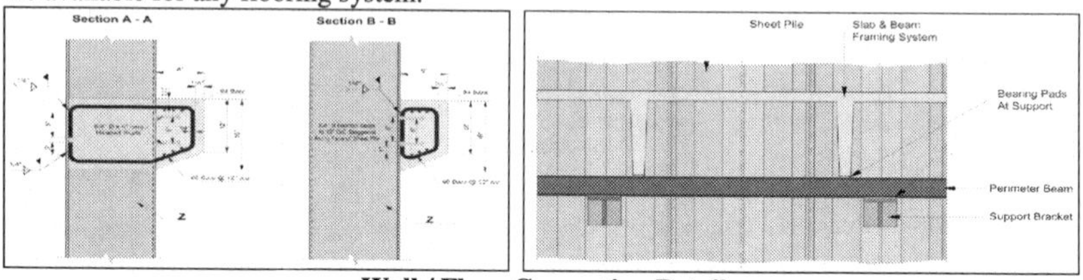

Wall / Floor Connection Details
FIG 8

FOOTPRINT

The permanent sheet pile design is a temporary shoring system and permanent wall structure in one. It requires no offset for waterproofing or dampproofing on the backside. No other below grade foundation system offers a more efficient layout. The shallow profile potentially could add thousands of square feet around the perimeter of each parking or basement level.

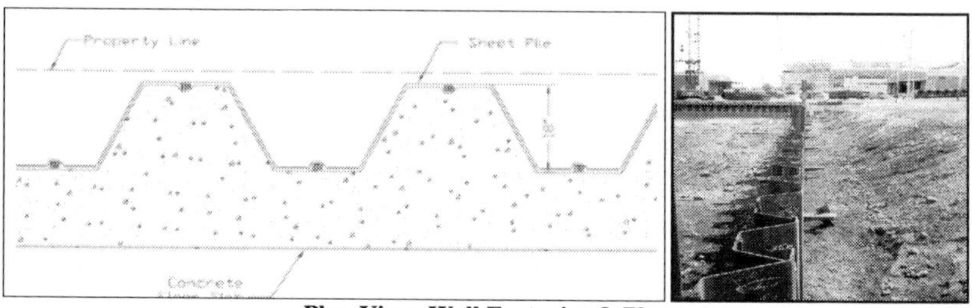

Plan View: Wall Footprint & Photo
FIG 9

CONSTRUCTION METHODS

Open Pit Excavation

Constructing a below grade structure with conventional shoring methods requires temporary support of excavation. For conventional concrete designs, the permanent wall structure is cast and the waterproofing is installed on the backside. The sheet pile design is both temporary shoring, and permanent wall. The material and time needed to construct a reinforced concrete wall element and vertical waterproofing system is eliminated. In comparison, this offers some material costs savings, but more importantly, dramatically reduces the foundation schedule.

Open Pit Excavation
FIG 10

Hybrid Top Down Excavation

For any traditional project, the foundation is the critical path. The above grade superstructure cannot begin taking shape until the foundation is structurally complete. The hybrid top down method allows the subsurface work to proceed at it's own pace, while the superstructure is constructed concurrently, removing it from the critical path. The three sided excavation makes for slower removal of dirt, but the financial impact is offset as the above grade structure commences well ahead of conventional schedules. In addition, the floor slabs serve as the bracing system during excavation. This eliminates the design & installation of anchors or tiebacks for multiple subterranean levels.

Building Information Modeling (BIM) is a fairly new 5D rendering tool that is gaining acceptance with designers and General Contractors. It uses real material and labor cost data to simulate a virtual construction model. This tool is very effective at demonstrating critical path issues and schedules with a project in a virtual world. A hybrid top down project model has been rendered to demonstrate the features and benefits of the permanent sheet pile wall application.

Hybrid Top Down Excavation
FIG 11

FIRE CODE REQUIREMENTS

In the United States, local building authorities are responsible for interpreting and enforcing fire life safety guidelines for building construction in their jurisdiction. National codes are typically used as the benchmark, but the local AHJ (authority having jurisdiction) has the ability to modify or interpret for their construction preferences. This makes good sense when you look at the varying terrain, weather conditions and seismic zones throughout the US. Wall vicinity, adjacent property, ventilation and vertical loading all play a part in the code interpretation. No UL/UBC codes were in place for permanent sheet pile foundation walls.

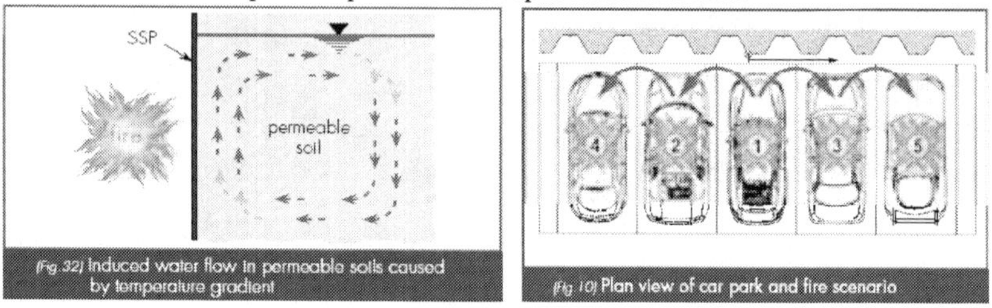

Fire Test Concept
FIG 12

Fire ratings and code interpretation vary by local jurisdiction throughout the US. The initiative was taken to employ an international fire consultant to perform a code interpretation. Recommendations were made to have an independent approved laboratory fire test sheet pile. The test was based on ASTM E-119 building standards. Parameters modeled combined lateral & vertical loads on the sheet pile section with a moist soil mass behind the sheets. The results demonstrated that the loaded sheet pile sections with a minimum wall thickness of 9.50 mm (0.375 in) achieved a 4 hour fire rating by ASTM test standards. Building authorities in the states of California, Florida, Minnesota, and Oregon, have already accepted this test report for approval on multiple projects.

AESTHETICS

Sheet piles have a unique profile that cast in place concrete walls do not. These walls require no more additional maintenance than conventional systems. The corrugated section provides a surface that can house utilities, and allows the architect to incorporate the building form and function into the basement with lighting. Color options are limitless.

Aesthetics of Steel Sheet Pile Wall
FIG 13

SUSTAINABILITY

Hot rolled steel sheet piles are produced from recycled content. They qualify for both 4.1 & 4.2 LEED credits on green building construction projects. Sustainability aims to improve occupant well being, environmental performance, and economic return for innovative structures.

DAMAGE APPROXIMATION METHOD FOR EXCAVATION-INDUCED DAMAGE TO ADJACENT BUILDINGS

Michael J. Kotheimer[1] and L. Sebastian Bryson[2]

[1] Associate II, Wiss, Janney, Elstner Associates, Inc., 2751 Prosperity Avenue, Suite 450, Fairfax, VA 22031, USA, Mkotheimer@wje.com

[2]Assistant Professor, Dept. of Civil Engineering, 161 Raymond Bldg, University of Kentucky, Lexington, KY 40506, USA, Phone 859.257.3247, FAX 859.257.4404, bryson@engr.uky.edu

ABSTRACT: A new damage approximation method is proposed that directly relates ground distortions resulting from deep excavations to building damage. Several factors were considered for the method to accurately predict building damage in varying situations including; horizontal strains, critical strain criteria, wall dimensions, and wall flexibility. Previous methods of damage approximation have related ground distortions to varying degrees of general building damage by using simplified beam bending theory. A new damage approximation method is proposed that attempts to improve upon current methods relating ground distortions to specific building damage in terms of previously defined damage criteria. The proposed method quantifies building damages in terms of strain capacities and crack widths. The proposed damage approximation method was compared to measured crack gauge data and was shown to perform quite well.

INTRODUCTION

In metropolitan areas around the world, infrastructure construction and renovation projects often require deep excavations. The prospect of deep excavations being close to existing structures is almost a certainty. In light of this reality, the importance of predicting and minimizing the resulting excavation-related building damage is of major concern to contractors, owners, and the public. A concise yet reliable method is required to approximate damage.

A new damage approximation method is proposed that directly relates ground distortions resulting from deep excavations to building damage. The method predicts building damage as a function of excavation-induced horizontal strains, critical strain criteria, and infill wall dimensions and flexibility. The metric used to quantify building damage is crack width. This metric has been used with some success by other researchers to describe building damage resulting from self-weight settlement (Dulacska, 1992) and adjacent excavations (Boone, 1996; Son and Cording, 2005).

BEAM BENDING APPROXIMATIONS

The proposed damage approximation method idealizes building wall sections as elastic beams. This elastic beam idealization is based on equations given by Burland and Wroth (1975) to approximate the response of a building settling due to its own self-weight. These equations were later extended by Boscardin and Cording (1989) to describe the response of a building adjacent to a deep excavation. The Boscardin and

Cording (1989) equations explicitly consider the effects of horizontal strains associated with excavation-induced movements. These equations describe a simply supported beam with a central point load and consider the effects due to both shear and bending deformations. However, only the equations for shear-related deformations are used for the derivations in this study. This is because the initial excavation-induced cracking in adjacent buildings tends to result from shear deformations (Boscardin and Cording, 1989).

For framed buildings and reinforced load bearing walls, Boscardin and Cording (1989) presented the following equation that relates deflection ratio (Δ/L) to horizontal strain for pure shear conditions:

$$\frac{\Delta}{L}=\left(0.064\frac{L^2}{H^2}+1\right)\left(\frac{\varepsilon_{crit}-\varepsilon_h\cos^2\theta_{max}}{2\cos\theta_{max}\sin\theta_{max}}\right) \tag{1}$$

where Δ= the relative deflection of the beam; L and H = the length and height of the beam, respectively; ε_{crit}= the critical strain of the beam (i.e. strain level at which cracking occurs); ε_h= the applied horizontal strain; and θ_{max}= the maximum inclination of the beam from the horizontal. For conditions of pure shear, the maximum inclination of the beam is 45 degrees.

The maximum change in slope along the beam, or the angular distortion (β), is related to the deflection ratio for the case of a beam with a central point load by the following equation.

$$\beta=\frac{3\Delta}{L}\left(\frac{1+4\left(\frac{E}{G}\right)\left(\frac{H^2}{L^2}\right)}{1+6\left(\frac{E}{G}\right)\left(\frac{H^2}{L^2}\right)}\right) \tag{2}$$

where E = the modulus of elasticity and G = the modulus of rigidity. It is noted that the deflection ratio is common to both equations. Thus, by substituting Equation 1 into Equation 2, a relationship between the critical strain and the angular distortion can be developed. From the combined expressions, it is now possible to estimate the critical angular distortion (β_{crit}) at which cracking will begin, given a particular critical strain. Assuming an isotropic material (i.e. $\nu=0.3$, $E/G=2.6$) the expression for any length-to-height ratio (L/H) ratio is given as

$$\beta_{crit}=\frac{3}{125}\left(5\frac{L^2}{H^2}+52\right)\left[\frac{8\frac{L^2}{H^2}+125}{5\frac{L^2}{H^2}+78}\right]\left(\varepsilon_{crit}-\frac{\varepsilon_h}{2}\right) \tag{3}$$

METHOD FORMULATION

The next step in developing the proposed damage approximation method is to relate the angular distortion to the estimated crack width. Both Burland and Wroth (1975) and Boscardin and Cording (1989) concluded that for structures with relatively low stiffness in shear or a significant degree of tensile restraint (i.e. framed buildings and reinforced load bearing walls), the limiting factor would be cracking due to diagonal tensile strain occurring at the neutral axis. Following this conclusion, Dulacska (1992) suggested that crack width of interior infill wall panels could be estimated by relating building distortions and diagonal strain.

Figure 1 presents the deformations of a simple frame and an approximation that utilizes common mechanics of materials relationships. The simple frame (Figure 1a) shows the angular distortion resulting from differential movement. In the figure, δ = differential movement, β = angular distortion, l_d = original length of the diagonal, Δl_d = change in the length of the diagonal, and θ = angle of rotation. The mechanics of materials approximation (Figure 1b) illustrates the deformation of an element in plane strain due to shear strain. In the figure, dx = the elemental length along the x-axis, dy = the elemental length along the y-axis, ds = the elemental length along the diagonal, γ_{xy} = the shear strain in the xy-plane, and $\gamma_{xy} dx \cos\theta$ = the increase in the diagonal length.

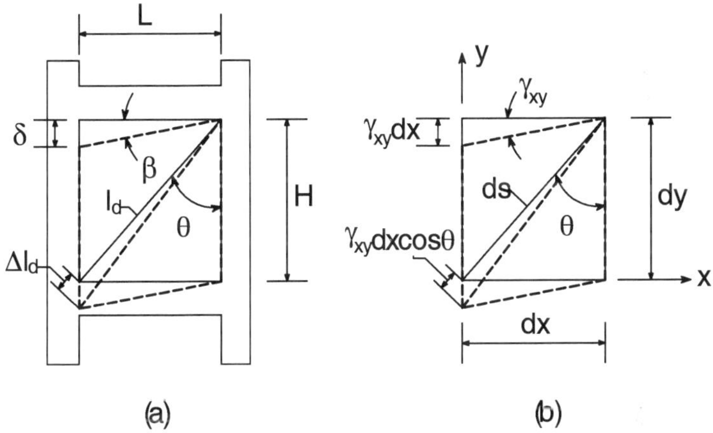

Figure 1. Deformation of a simple frame: (a) actual simple frame; (b) mechanics of materials approximation.

The implication from the work done by Dulacska (1992) is that increasing crack width in an infill wall panel is proportional to the increase in length along the diagonal. An expression for increasing crack width can be developed by substituting the horizontal length of the wall panel (L) for the horizontal elemental length (dx), the vertical height of the wall panel (H) for the vertical elemental length (dy), and the angular distortion (β) for the shear strain (γ_{xy}). The progression of these substitutions is given as:

$$\Delta l_d = \gamma_{xy} dx \cos\theta = \beta L \cos\theta = \beta L\left(\frac{H}{\sqrt{H^2 + L^2}}\right) \quad (4)$$

The implicit assumption of Equation 4 is that a crack forms at the immediate onset of angular distortion. However, fundamental to the assumptions put forth in this study is that most civil engineering materials do not experience cracking until some critical distortion is exceeded. Thus, the excavation-induced distortion must be reduced by the critical distortion to realistically reflect crack growth. Also, because of the rigid body response of the building to the excavation-related deformations, the building distortions can differ from the measured ground distortions. Dulacska (1992) suggested that stiffer buildings are less affected by the ground distortions. He presented a flexibility factor (η) to account for the stiffness of a building or wall section when representing distortion. This factor is dependent upon the number of openings within a wall section and ranges from 0.5 for a solid or stiff wall section to 1.0 for a highly punctured or flexible wall section. Utilizing the flexibility factor and the critical distortion, the normalized crack width is given as

$$\frac{\Delta l_c}{L} = \eta(\beta_{meas} - \beta_{crit})\left(\frac{H}{\sqrt{H^2 + L^2}}\right) = \beta_{build}\mu \tag{5}$$

where β_{meas} = the measured or anticipated ground distortion; β_{build} = the building distortion = $\eta(\beta_{meas} - \beta_{crit}) \geq 0$; and μ = the geometric factor = $\left(H/\sqrt{H^2 + L^2}\right)$and is dependent upon the L/H ratio. It is noted that this simplified relationship assumes that only one crack develops at a characteristic place within the wall element resulting in the widest cumulative crack width estimate.

DEVELOPMENT AND USE OF DESIGN CHARTS

Design charts were developed to aid the user in making preliminary estimates of input variables required to predicting damage to interior wall panels of structures adjacent to deep excavations. The charts are based on the relationships given in the previous section.

Figure 2 presents an aid for making preliminary estimates of critical strain.

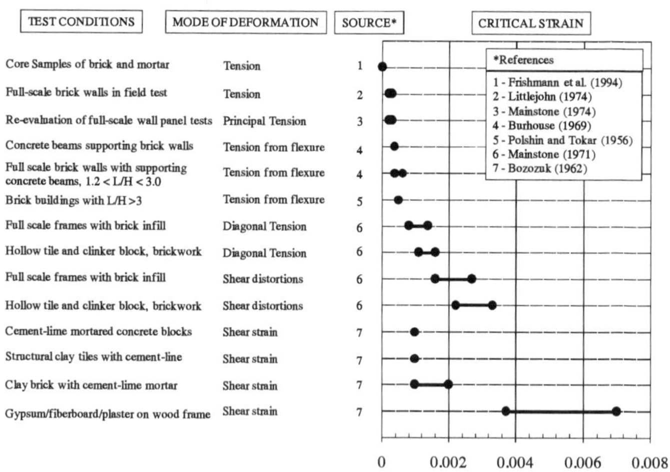

Figure 2. Critical Strain Criteria (data from Boone, 1996).

The critical strain criteria presented in Figure 2 was originally presented in tabular form in Boone (1996). The data were reorganized in terms of the mode of deformation and put into graphical form to give a more logical presentation. It is observed that a trend emerges in the reorganized data. The critical strain values for the test conditions subjected to direct tension are the lowest critical strain values in the figure. The critical strain values increase as the mode of deformation changes from direct tension, to tension from flexure, to diagonal tension, and finally to shear. This trend implies that a relationship may be developed between modes of deformation and rotation of principal planes. However, the investigation of such a relationship was outside the scope of work for this study. As is, the user will determine an approximate critical strain value, or range of values, for the test conditions and deformation modes that best represent the existing or anticipated situation. It is also noted that if horizontal strain is anticipated, but will not be measured, Figure 2 can be used to provide estimates of maximum permissible horizontal strain.

An estimate of critical angular distortion (i.e. angular distortion at which the onset of cracking will occur) can be determined by inserting the estimated critical strain into Equation 3. Figure 3 is a graphical representation of the Equation 3 and allows the user to visualize the relationship between critical angular distortion and critical strain.

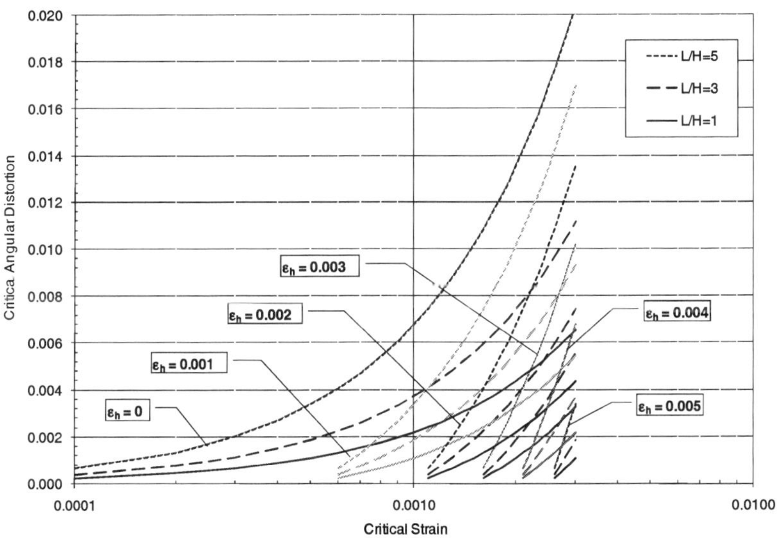

Figure 3. Graphical relationship between critical angular distortion and critical strain.

The figure shows variation of two important variables; the horizontal strain and the length-to-height ratio of the wall panel. The horizontal strain ranges from 0 to 0.005 and is given in increments of 0.001. This variation is displayed on the chart by the six separate groupings of lines. Three length-to-height ratios are provided in the design chart:

L/H = 1, 3, and 5, and each ratio is displayed on the chart with a different line type. The relative clarity of the design chart allows for interpolation of intermediate values. To determine the critical angular distortion the user will input the appropriate critical strain value along the x-axis and take a vertical line to the corresponding values of horizontal strain and the L/H ratio. Next the user will draw a horizontal line from the intersection of the vertical line and the corresponding L/H ratio line to the y-axis to determine the corresponding value of critical angular distortion. This value of critical angular distortion will be used in Equation 5 to determine the crack width and ultimately will be used to estimate the corresponding building damage severity.

VERIFICATION OF THE PROPOSED METHOD

The proposed crack width approximation methodology was compared to the crack gauge data reported by Bryson (2002) for a building adjacent to the Chicago Avenue and State Street Excavation. The crack gauge was located in the south wall of the third floor faculty room. *[The reader is referred to Finno et al., 2002; Finno and Bryson, 2002; and Bryson, 2002 for additional information on the Chicago Avenue and State Street Excavation and the response of the adjacent building].* Figure 4 shows the cumulative crack width (i.e. sum of the all the incremental changes in crack width) predicted from Equation 5 compared to the cumulative resultant crack gauge data. The predicted crack width estimate assumed $\varepsilon_{crit} = 0.0022, \varepsilon_h = 0$; L/H = 0.95, $\eta = 1$, and $\mu = 0.724$. These values were chosen based on the type and dimensions of the building and the assume mode of deformation. Utilizing Equation 3, these values correspond to $\beta_{crit} = 0.0047$.

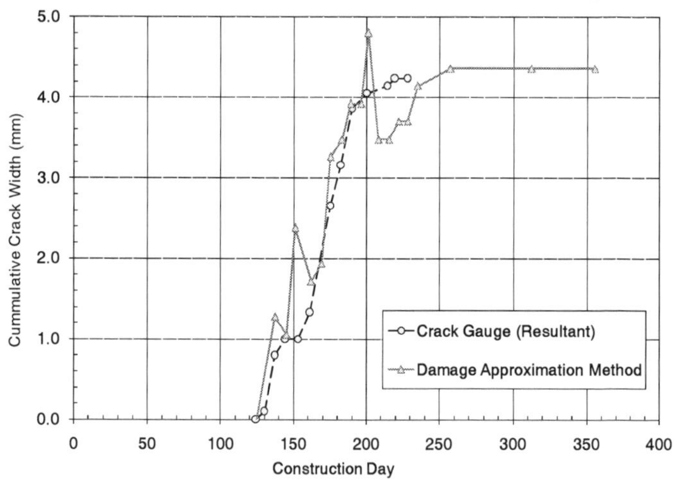

Figure 4. Cumulative crack width approximation.

The figure shows that the approximation method is quite accurate despite minor fluctuations. The final predicted crack width from the approximation method was roughly 4.4 mm. This compares to the final measured resultant crack width of 4.2 mm. Thus, crack width approximation method was deemed to be valid.

CASE HISTORY DATA

Further verification of the damage approximation methodology was performed by developing a damage severity chart similar to those presented by Boscardin and Cording (1989) and Boone (1996). Damage approximation predictions were calculated using specific case history data taken from Boone (1996). The steps for developing the severity charts included: (i) estimating the critical strain criteria for each case using descriptions of their individual test conditions; (ii) estimating the critical angular distortion using Equation 3; and (iii) using the critical angular distortions, the measured angular distortions and the wall section dimensions to estimate the resulting crack widths for each case history using the crack width approximation method. The predicted crack widths resulting from the proposed damage approximation method were then compared to the actual crack widths recorded in the field. This comparison is presented in Figure 5.

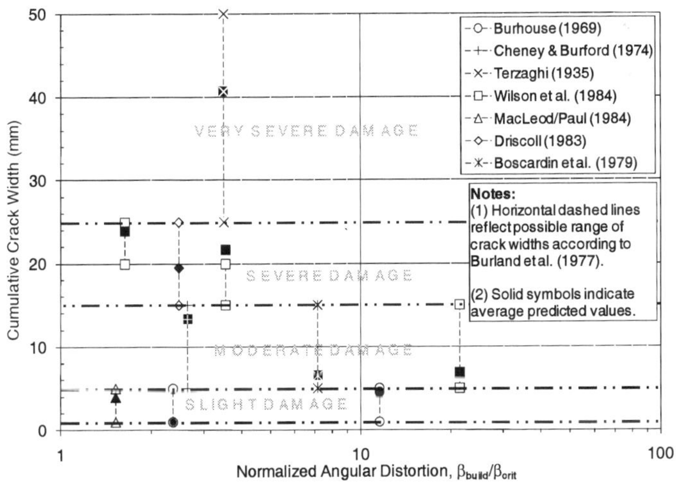

Figure 5. Damage severity chart.

In Figure 5, normalized angular distortion ($\beta_{build}/\beta_{crit}$) is plotted against cumulative crack width. The solid symbols indicate the average predicted value of the cumulative crack widths according to the new damage approximation method. The horizontal dashed lines indicate borders between classification zones of visible damage presented by Burland et al. (1977). The vertical dashed lines signify the range of crack width given by the case history damage reports. For cases in which only the damage severity was given, the range of possible crack widths extend between the upper and lower ends of the classification zones.

Figure 5 shows that all average predicted values of cumulative crack widths fall within the classification zones given by the case history damage reports. Thus, the new damage approximation method compares extremely well with the case history data.

CONCLUSIONS

The damage approximation method proposed herein directly relates ground distortions resulting from deep excavations to specific building damage in terms of previously defined damage criteria. The new methodology utilizes several important variables to approximate building damage. These factors include: critical strain criteria, horizontal strain, infill wall panel dimensions, excavation-induced ground distortions, and wall flexibility. The verification studies showed the efficacy of the new methodology in that predicted results compared well with reported case history data. In conclusion, the new methodology is shown to be a viable means to predict excavation-related damage to adjacent buildings.

ACKNOWLEDGEMENTS

The material presented in this paper is based upon work supported by the National Science Foundation under grant No. CMS 06-50911, under program director Dr. R. Fragaszy. This financial support is greatly appreciated.

REFERENCES

Boone, S.J. (1996). "Ground-Movement-Related Building Damage," *Journal of Geotechnical Engineering*, ASCE, Vol. 122, No. 11, pp. 886-896.

Boscardin, M.D. and Cording, E.J. (1989). "Building Response to Excavation-Induced Settlement," *Journal of Geotechnical Engineering*, ASCE, Vol. 115, No. 1, pp. 1-21.

Bryson, L. S. (2002). "Performance of a Stiff Excavation Support System in Soft Clay and the Response of an Adjacent Building." *PhD Thesis*, Northwestern University, Evanston, IL.

Burland, J. B., and Wroth, C. P., (1975). "Settlement of Buildings and Associated Damage," *Proceedings of a Conference on Settlement of Structures*, Cambridge, pp. 611-654.

Burland, J. B., Broms, B. B., and DeMello, V. F. (1977). "Behavior of foundations and structures." State of the Art Report, *Proceedings of the 9th International Conference on Soil Mechanics and Foundation Engineering*, Tokyo, Vol. 2, 495–546.

Dulacska, E. (1992). "Soil Settlement Effects on Buildings," *Developments in Geotechnical Engineering*, No. 69, Elsevier, New York, pp. 447.

Finno, R. J. and Bryson, L. S. (2002). "Response of Buildings Adjacent to Stiff Excavation Support Systems in Soft Clay," *Journal of Performance of Constructed Facilities*, ASCE, Vol. 16, No. 1, pp. 10-20.

Finno, R. J., Bryson, L. S., and Calvello, M. (2002) "Performance of a Stiff Excavation Support System in Soft Clay," *Journal of Geotechnical and Geoenvironmental Engineering*, ASCE, Vol. 128, No. 8, pp. 660-671.

Son, M. and Cording, E. J. (2005). "Estimation of Building Damage Due to Excavation-Induced Ground Movements," *Journal of Geotechnical and Geoenvironmental Engineering*, ASCE, Vol. 131, No. 2, pp. 162-177.

Geotechnical Challenges on the BART Fremont Central Park Subway Project

Mitchell L. Fong[1], MASCE, PE, GE and Thomas S. Lee[2], MASCE, PE, GE

[1]Professional Associate, PB Americas, 303 Second St, Suite 700N, San Francisco, CA 94107
[2]Principal Professional Associate, PB Americas, 303 Second St, Suite 700N, San Francisco, CA 94107

ABSTRACT: The Bay Area Rapid Transit (BART) District will be extending BART service south along an 8.7-km (5.4-mile) -long corridor in the city of Fremont, California as part of the Warm Springs Extension (WSX). The WSX project will link the existing BART system to the proposed BART to San Jose extension. A segment of WSX is located in Fremont Central Park (referred to as the Fremont Central Park Subway Project). This segment is considered the most geologically and geomorphologically complicated portion of the project. The project consists of (1) an underground subway box being constructed not only in the middle of the largest park in Fremont and crossing under a lake, but also subject to high hydrostatic water pressure from the underlying aquifer and (2) a fill embankment over the Hayward Fault.

This paper presents the geotechnical challenges that have to be addressed both in the design stage and during construction.

INTRODUCTION

The Bay Area Rapid Transit (BART) is a heavy rail electric train system that has provided the Bay Area congestion relief for over 30 years. BART plans to extend its service south along an 8.7-km (5.4-mile) -long corridor in Fremont to the Warm Springs District, California as part of the Warm Springs Extension (WSX). The WSX project has been designated for Design-Build, however due to limited available funding, a segment referred to as the Fremont Central Park Subway project (Figure 1) was separated out and will be constructed under a traditional Design-Bid-Build procurement.

The 2.3-km (1.4-mile) -long Fremont Central Park project is dominated by two significant geologic features, the Hayward Fault and the Above Hayward Fault (AHF) aquifer. The alignment crosses the seismically active Hayward Fault twice, once in Tule Pond and the other at Washington Blvd. Extensive previous fault investigations have pinpointed the location of the Hayward fault in Tule Pond and an embankment wide enough to accommodate the anticipated seismic displacement is planned. The other crossing is at grade and within the Design-Build portion of the project. The second geologic feature, the AHF aquifer, provides a significant source of water for the local water district and is also used for water storage. The underground subway penetrates this aquifer and is up to 11 m (36 ft) below the groundwater table.

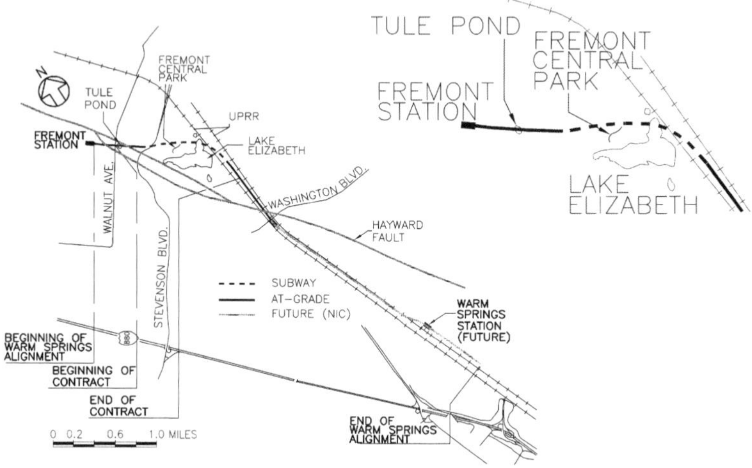

FIG. 1. BART WSX & Fremont Central Park Subway Alignment

GEOLOGICAL, SEISMOLOGICAL, AND HYDROGEOLOGIC SETTING

The San Francisco Bay Area is located in the California Coast Range Physiographic Province which is characterized by northwest-southeast trending valleys and ridges. These features are controlled by folds and faults that resulted from the collision of the Farallon and North American plates and subsequent strike-slip faulting along the San Andreas fault zone. The Bay Area experienced uplift and faulting in several episodes during late Tertiary time (about 25 to 2 million years ago) that produced a series of northwest-trending valleys and mountain ranges, including the Berkeley Hills, the San Francisco Peninsula, and the intervening San Francisco Bay.

The active faults of most significance to the project are the Hayward fault, the San Andreas Fault, and the Calaveras Fault. These faults have caused severe ground shaking at the site and have the potential to do so in the future. The BART tracks will cross the Hayward fault in Tule Pond as well as further south at Washington Blvd. Historically, the most significant earthquake near the WSX alignment occurred in 1868 along the southern section of the Hayward Fault and there is about a 60% chance that the next magnitude 6.8 or greater earthquake in the Bay Area will occur in the next 30 years.

The Fremont Central Park Subway project is located within the Niles Cone Subbasin of the Santa Clara Valley Groundwater Basin. The Nile Cone Subbasin is further divided into the Above Hayward Fault (AHF) aquifer and the Below Hayward Fault (BHF) aquifer. The WSX subway is located within the AHF aquifer where groundwater elevations have varied historically from 15.3 m (50 ft) above mean sea level (MSL) to 3.1 m (10 ft) above MSL.

The geotechnical investigations consisted of mud rotary borings with field sampling and testing including Standard Penetration Test (SPT), Cone Penetration Test (CPT),

trenching, monitoring wells, and pump tests. General subsurface conditions (Figure 2) as inferred by the geotechnical investigations for the project site include:
(1) Fill – Roadway or railroad embankments, varies from silty clay to track ballast,
(2) Basin Deposits – Holocene localized marsh deposits of soft to firm clay, organic clay, and loose to medium dense sands and silts,
(3) Young Alluvium – Holocene age alluvial fan deposits and Holocene alluvial deposits, primarily fine-grained, and
(4) Old Alluvium – Late Pleistocene or early Holocene alluvial deposits, primary coarse-grained within the unconfined aquifer.

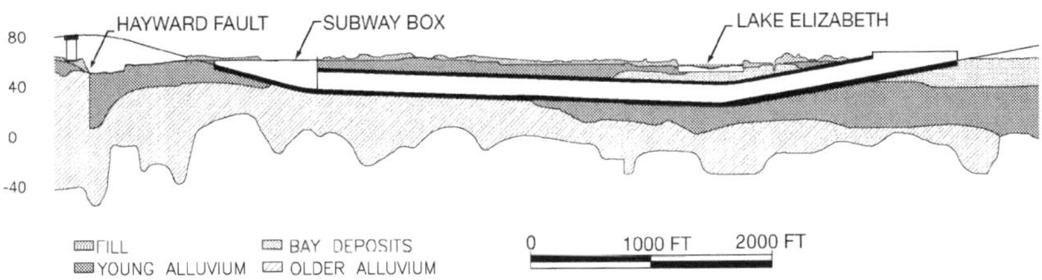

FIG. 2. Inferred Geologic Profile

SITE HISTORIC DEVELOPMENTS

Two significant historic developments have taken place along the proposed alignment. The first one is the formation of Tule Pond due to a depression between the two strands of the Hayward Fault. Tule Pond is presently used by the Alameda County Flood Control District as a retention pond for storm runoff. The second is Lake Elizabeth which was a natural lagoon called Stiver Lagoon before it was developed into an artificial lake. Stiver Lagoon was a wet-season pond ringed with marshes. The eastern part of the lagoon was possibly filled in the late 19th century to build railroad tracks. In the 1960's and 1970's, Fremont Central Park and Lake Elizabeth were developed by filling parts of the lagoon and constructing the artificial lake in the basin of the natural pond.

GEOTECHNICAL CHALLENGES

Geotechnical challenges confronting the Fremont Central Park Subway Project include the following:

- Design of a structure to support the BART trackway located directly over the Hayward Fault in Tule Pond;
- Design of a fill embankment over a large diameter drainage pipes in a compressible marshy soil over the Tule Pond;
- Selection of a temporary cutoff wall system to resist a high groundwater table at 0.6 to 1.5 m (2 to 5 ft) below ground surface; and

- Design of the subway box to counteract uplift hydrostatic pressures in an existing aquifer, which serves as the main water source for the municipal areas.

Support of the BART trackway over the Hayward Fault

The active Hayward Fault crosses Tule Pond and is capable of generating an earthquake large enough to result in surface rupture that may damage the BART trackway crossing the fault trace. William Lettis & Associates (WLA) 2008 report estimated the total surface displacement around 1.3 m (4.2 ft) with a range of possible displacements of 0.9 to 2.0 m (2.9 to 6.4 ft) and vertical displacements of less than 0.4 m (1.3 ft) after a magnitude 7.0 earthquake on the Hayward Fault. Any structure supporting this section of the alignment over the Hayward Fault must be able to accommodate this large lateral offset and vertical slip. The BART performance criterion states that the structure is to survive a magnitude 7.2 earthquake and be capable of being operational within 72 hours. A fill embankment was selected after weighing the pros and cons of various alternatives including conventional elevated structures. A much wider width at the top of the embankment was selected to accommodate the large deformations during and after a major earthquake and would provide the space needed to realign the damaged rails, therefore minimizing down time and service impacts.

Fill embankment over Tule Pond

The WSX alignment crosses the western trace of the Hayward Fault at the northwestern edge of Tule Pond near Walnut Avenue (Figure 3). The alignment at this location consists of a maximum 9.1 m (30-ft) high fill embankment which extends through the Tule Pond - a seasonal pond which is usually filled with water in the winter and dry in the summer. The pond is about 2.4 to 3.7 m (8 to 12 ft) below the adjacent roadway and is currently covered with shrubs, weeds, trees, and construction debris. Below the fill embankment, a culvert consisting of seven 1.8-m (72-in) diameter drainage pipes is proposed to hydraulically connect Tule Pond to the replacement pond, however the pipes will not be used for flow.

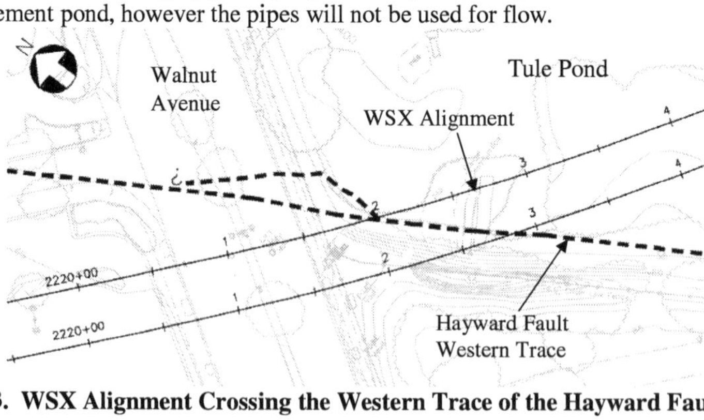

FIG. 3. WSX Alignment Crossing the Western Trace of the Hayward Fault

Subsurface conditions at Tule Pond are as follows (in descending order): 0.3 to 0.6 m (1 to 2 ft) of fill, 6.1 to 13.7 m (20 to 45 ft) of Young Alluvium with pockets of Basin Deposits underlain by Older Alluvium. The Older Alluvium is composed typically of dense to very dense sands and gravels. Potentially liquefiable deposits were identified both in the Basin Deposits and the Older Alluvium, but they are erratic, discontinuous, and of limited thickness and extent. Therefore, these potentially liquefiable materials are judged not to pose any detrimental impact to the seismic stability of the proposed embankment.

The 9.1-m (30-ft) high fill embankment over Tule Pond is estimated to induce consolidation settlements ranging from 15 to 25 cm (6 to 10 in). These settlements will have a detrimental impact to the seven 1.8-m (72-in) diameter drainage pipes that cross over Tule Pond. Conventional mitigation measures to mitigate the settlements were considered during the design phase. They included (1) surcharging the area and installing wick drains to expedite consolidation settlements and (2) sub-excavating and replacing the soft to firm Basin Deposits with engineered fill. Due to overall schedule constraints and seasonal access limitations in Tule Pond that are critical to completing this portion of the project, the above-mentioned mitigation measures were ruled out, and alternative settlement mitigation measures were considered. They are:

- column supported, geosynthetic reinforced embankment,
- lightweight fill over the drainage pipe area, or
- cambering the center part of the pipe to compensate for unequal settlements under the weight of the embankment.

Both the column supported, geosynthetic reinforced embankment and lightweight fill alternatives over the drainage pipe area are very expensive and subsequently discarded. Cambering the center part of the pipes to compensate for unequal settlements under the weight of the embankment is considered the most practical and economical alternative. The cambering will prevent the drainage pipes from sagging in the middle portion as the foundation consolidates with time. To ensure more even differential settlements, the existing subgrade will be sub-excavated to a depth of 1.5 m (5 ft), have a geosynthetic reinforcement layer placed at the bottom of the excavation, and then backfilled with compacted engineered fill.

Temporary cutoff wall for subway box excavation below grade

In order to limit the surface disruption to Fremont Central Park, a 1.9-km (1.2-mile) -long subway box will be built using cut-and-cover techniques through the Fremont Central Park. Along the cut-and-cover subway alignment, there is a significant variation in the composition, consistency, and engineering properties of the subsurface soils encountered due mainly to several historic developments that took place along the alignment as state previously.

Subsurface soils in the cut-and-cover segment of the alignment consist of Fill, Basin Deposits, Young Alluvium, and Older Alluvium. The bottom of the subway box at its deepest section just south of Lake Elizabeth is about 11.6 m (38 ft) below the existing ground surface. To minimize dewatering of the AHF aquifer during excavation for the subway box, watertight cutoff walls and a grout plug at the base of the subway box excavation will be required to limit groundwater inflow and resist large hydrostatic uplift forces.

The types of watertight cutoff walls considered for construction of the subway box included cement deep soil mixing (CDSM) walls, diaphragm walls (slurry walls), secant pile walls, and ground freezing. The pros and cons of these four options are shown in Table 1 below. Based on evaluation of different methods and their applicability to our project, it was concluded that a combination of CDSM walls and a grout plug is the most practical option to collectively cut off the water inflow into the subway box excavation (Figure 4). The impact of the cutoff walls to the water supply aquifer is considered minimal due to the shallow depth of the walls – ±18 m (60 feet).

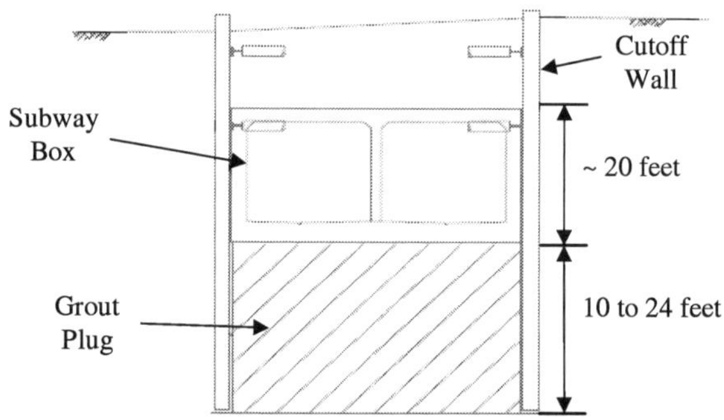

FIG. 4. Cutoff Walls and Grout Plug Proposed for Construction of the Cut-and-Cover Subway Box Alignment

Due to a high differential hydrostatic head between the bottom excavation of the subway box and the design groundwater table, upward seepage through the excavation bottom will occur and substantial uplift pressures will be developed at the base of the excavation upon removal of overburden soils. These uplift pressures must be resisted by the counter weight of the clayey soil (sometimes called soil plug) located above the aquifer. If the weight of the soil plug is not heavy enough to counteract the uplift forces, the soil will lose strength and eventually become "quick" and begin to boil. In this case, a grout plug will be needed to resist the uplift forces.

Grout plug to counteract uplift hydrostatic pressures

For the excavation of the proposed subway box structure, stability analyses of the excavation bottom were conducted to determine base rupture (blow-ins) caused by water pressure acting on the bottom plug. A factor of safety of 1.5 was selected against bottom heave. The side friction of the grout plug was assumed to be mobilized at the interface between the cutoff wall and the plug. This factor of safety was chosen to account for soil/grout variability (in terms of consistency and shear resistance) within the Young Alluvium. Analyses indicated that failure along the grout wall interface is the controlling failure mode. A lower factor of safety of 1.2

TABLE 1. Pros and Cons of Various Cutoff Wall Alternatives

Wall Construction Option	Relevant Characteristics	
	Pros	**Cons**
Diaphragm Walls	Common technology	High mobilization cost
	Local experience in Bay Area	Possible different equipment for walls and base plug
	Vertical alignment easier to control, depending upon equipment used	Risk of side wall collapse during construction
	Can be used as a single structural wall	Risk of bentonite slurry loss in high permeability ground
	Fewer joints between structural elements	Requires large work area for equipment
	Well suited to planned depths	Underground utilities conflicts problematic
	Boulder obstructions easier to excavate	Spoils contaminated with bentonite slurry
	Provides a smooth wall for waterproofing installation	
Secant Piles	Common technology	Moderate mobilization cost
	Local experience in Bay Area	Frequent joints
	Well suited to planned depths	Vertical alignment more difficult
		Different equipment for walls and base plug
		Risk of slurry loss in high permeability ground
		Vertical misalignment leads to open "windows" between adjacent columns
		Underground utilities conflicts problematic
		Produces a rough irregular surface for waterproofing installation
CDSM Walls	Same equipment for both walls and base plug, if CDSM used for plug	High mobilization cost
	Local experience in Bay Area	Relatively new technology
	Good overlap between adjacent column elements	Risk of heterogeneous mix
	Generally considered to have low unit construction cost	Requires large work area for equipment
	Well suited to planned depths	Underground utilities conflicts problematic
	Full soil replacement not required	Boulder obstructions potentially problematic
	In-situ quality verifiable	
	Provides a relatively smooth wall for waterproofing installation	
Ground Freezing	Could use the same equipment as for a frozen soil plug.	High mobilization costs
	Applicable to a variety of soil types.	Linear facility requires multiple freeze plants
	Can maintain a continuous freeze as long as desired	Cost to maintain frozen ground for long periods of time
	Does not require the injection of any chemicals or cements into the ground with a potential to contaminate the ground water as a drinking water source.	Exposed frozen wall during construction may require insulation to prevent thawing and degradation during the process of excavation and box construction. Requires substantial cost of energy to maintain frozen ground.
		Limited previous applications in Bay Area
		Requires instrumentation temperature monitoring for ice wall closure

may also be considered if the perimeter shear is ignored. Design groundwater tables of El. 15.2 m (50 ft) in the northern subway section and El. 15.9 m (52 ft) in the southern subway section were used for conditions during construction. This groundwater table accounts for higher groundwater levels during wet seasons. The calculated plug thickness ranged from 3.1 to 7.3 m (10 to 24 ft). After excavation to the planned invert, a minimum 10-cm (4-in) mud slab will be poured over the foundation subgrade to protect a firm working surface. This slab will also help prevent loosening and boiling of the subgrade. For long term stability, the weight of the backfill soil and subway box is adequate to resist uplift hydrostatic forces from the 500-year flood event without consideration of the grout plug. In other words, the grout plug is only needed for construction as the completed structure and backfill have sufficient weight to resist buoyancy.

CONCLUSIONS

- Two major geologic features at the Fremont Central Park project site have a significant impact on the geotechnical design of the project, namely the Hayward Fault and the AHF aquifer. The Hayward Fault crosses the alignment at Tule Pond and the underground subway box structure penetrates and sits in the main water bearing formation of the AHF aquifer.
- A fill embankment is considered to be most practical option to support the BART trackway over the Tule Pond which crosses the Hayward Fault. The fill embankment would not only minimize damage to the BART trackway from a major seismic event on the Hayward Fault, but also allow relatively quick repairs to damaged tracks in case of a fault rupture.
- Drainage pipes below the fill embankment can be cambered to compensate for the consolidation settlements induced by the loading of the fill embankment.
- The subway box will be constructed using cut-and-cover techniques. A water tight temporary excavation support system such as CDSM walls will have to be used to minimize the impact to the AHF aquifer and limit dewatering during construction of the subway box. A grout plug will be used to seal off the bottom of the temporary excavation and balance high hydrostatic pressures.

ACKNOWLEDGEMENTS

The authors are thankful to BART and PB Americas, Inc for the support of publishing this paper. The conclusions of this paper reflect the views and opinions of the authors, and do not necessarily reflect official views and opinions of BART and PB Americas, Inc.

REFERENCES

William Lettis & Associates, Inc. (2008). Geologic Evaluation of Hayward Fault Crossing near Walnut Avenue, BART WSX, Final Report, October 24.

Instrumented Sheet Pile Wall Load Test to Indirectly Measure Earth Pressure

Richard E. Burrage[1], Student. M. ASCE, J. Brian Anderson[2], M. ASCE, and Vincent O. Ogunro[3], M. ASCE

[1]Graduate Student, University of North Carolina at Charlotte, Department of Civil and Environmental Engineering, 9201 University City Boulevard, Charlotte, NC USA, rburrage@uncc.edu
[2]Associate Professor, University of North Carolina at Charlotte, Department of Civil and Environmental Engineering, 9201 University City Boulevard, Charlotte, NC USA, jbanders@uncc.edu
[3]Associate Professor, University of North Carolina at Charlotte, Department of Civil and Environmental Engineering, 9201 University City Boulevard, Charlotte, NC USA, vogunro@uncc.edu

ABSTRACT: Two parallel retaining walls were constructed from 10.4 m embedded PZ22 sheet piles in order to indirectly measure the amount of earth pressure induced during excavation. Selected sheets were instrumented in an effort to measure bending moment and deflection, whereby the induced earth pressure could be discerned. Four sheets were instrumented, each with 16 strain gages (eight gages per flange) to detect bending moment. Steel tubing was attached to four additional piles for use with an inclinometer to measure horizontal movement.

The soil between the walls was excavated in five equal lifts over a ten day period until the total depth of excavation was 6.1 meters. Strain gages were continuously monitored while inclinometer readings were taken at the completion of each excavation lift. The qualitative and quantitative results of the load tests showed no mobilization of lateral earth pressure on the walls with an excavation depth of 6.1 meters.

INTRODUCTION

Earth pressure is typically calculated using Rankine stresses or Coulomb forces. Both methods rely on the Mohr-Coulomb criteria to define the amount of vertical overburden stress that is translated to horizontal pressure. Naturally, the greater the shearing strength, the less pressure transmitted.

This research project concerns the determination of earth pressure in Piedmont residual soils (PRS). Residual soils often retain cohesion-like shearing strength that can result in a reduction of the calculated earth pressures; in practice however, this cohesion is often neglected for conservatism. Measuring earth pressure insitu is difficult for textbook soils, and even more so for PRS. To measure the lateral stress in place, a device would need to be inserted into the soil profile with minimal soil disturbance.

In order to determine the earth pressure in PRS, it was proposed to instrument a full scale retaining structure. Sheet piles were selected because they could be instrumented, then vibrated or driven into place with minimal soil disturbance. Once installed, the soil between the walls could then be excavated, allowing the sheets to translate and bend due to the earth pressure pushing on them. The earth pressure could then be back-calculated based on the amount of deflection in the sheet piles.

An experiment consisting of two parallel cantilevered sheet pile walls was devised to measure the insitu lateral earth pressure of the PRS. This paper describes the experimental apparatus including sheet piles, instrumentation, and a data acquisition system. In addition, the resulting structural measurements recorded during a staged 6.1m excavation are presented.

LITERATURE REVIEW

Researchers such as Peck (1969) and Clough and O'Rourke (1990) have investigated retaining wall pressure and earth movements due to excavations. Other case studies of worldwide experience are also presented by Long (2001). In addition, studies seeking to determine lateral earth pressures have been presented by researchers like Fang et al. (1994) who utilized a movable model retaining wall to study passive pressures. The same moveable retaining wall was presented by Fang et al. (1997) to study the effect of sloping backfill on earth pressures. Similarly, Georgiadis and Anagnostopoulos (1998) used a model cantilever sheet pile wall in a tank to study surcharge effects. Both studies dealt with carefully placed sand in a controlled environment.

Full scale tests have been rare. A case was discovered where a full scale sheet pile supported excavation was instrumented. Kort (2002) presents a study carried out on an anchored sheet pile wall in Rotterdam, Holland. The soils in this study were predominantly weak clay and peat, and the objectives of the study were oblique bending and plastic hinging of sheet piles. The report includes extensive documentation of the field instrumentation plan.

INSTRUMENTATION PLAN AND LOAD TEST

In order to determine the appropriate instrumentation for sheet piles in PRS, the finite element program Plaxis (Brinkgreve and Vermeer, 1998) was used to determine the potential earth pressure, shear, and bending in the wall, as well as the displacements that the wall would undergo during loading. The results of the initial study showed that the optimum size sheet pile was the PZ22 because it would allow a measurable amount of deflection, without sacrificing the safety of the excavation. Using the same principles, the model was used to design the overall test system; the sheet piles were designed to be 10.7 m in length and driven to an embedment of 10.4 m. The walls were designed to be a minimum of 12.2 m apart. The maximum safe excavation depth, as permitted by the funding agency, between the walls would be 6.1 m, leaving the sheet piles embedded 4.3 m.

Many factors contributed to the instrumentation plan, most notably budget and survivability. There was approximately $25,000 available to purchase and install instruments. Therefore, a system consisting of vibrating wire (VW) strain gages and a slope inclinometer was proposed. The plan view of instrumented piles is depicted in Figure 1. VW strain gages have been widely used in the testing of driven piles in axial and lateral load (Bullock et al. 2005, Rieke and Crowser 1987, and Paikowski and Hajduk 2004). Furthermore, VW strain gages use analog frequency signals which, in comparison standard strain gages, are not affected by lead distance or electrical noise. These gages are reusable, durable, believed to have very good long term stability, and have low power requirements. The gages have a 3000 microstrain limit which was sufficient to measure the expected pile deflection during testing. Theoretical maximum allowable pile deflection after each excavation ranged from 12 mm at 1.5 m excavation to 190 mm at 6 m excavation. Four sheet piles were instrumented with 16 gages each for a total of 64 strain gages. Gages were installed in pairs at 1.22 m intervals along the sheet piles, and were protected from installation damage by a steel angle cover. The strain gages were continuously monitored using dataloggers. The relative bending of the sheets with respect to the placement of the strain gages is shown in Figure 2.

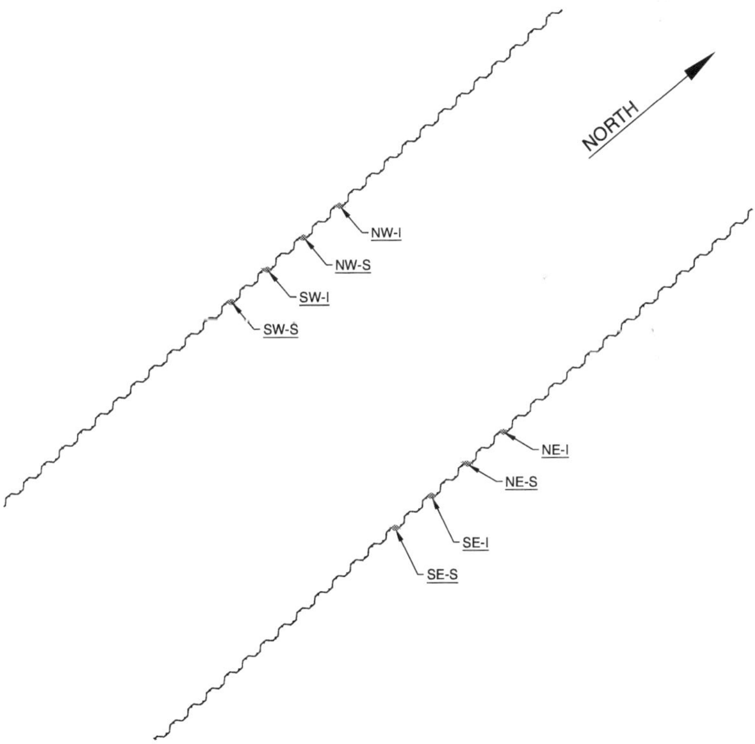

FIG 1. Site plan showing location of instrumented sheet piles (S = strain gages, I = inclinometer).

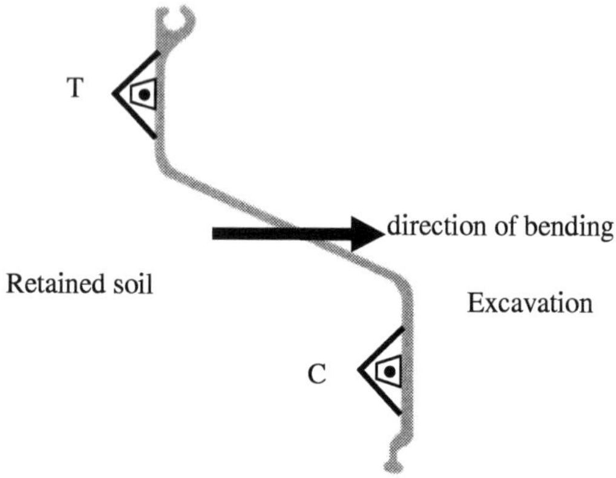

FIG 2. Diagram of strain gage attachment with respect to pile bending.

In order to provide some redundancy in the data acquisition system, the slope inclinometer was used (in case the strain gages were damaged during installation). Steel tubing was welded to the non excavation side of four sheet piles and used as an inclinometer casing. Unlike typical slope inclinometer tests, the axes of measurement using square tubing are skewed at 45° from direction of wall movement. Figure 3 shows how the measurements were rotated in the data reduction equations to match the offset angle. A single probe inclinometer with data collector was used to perform the measurements.

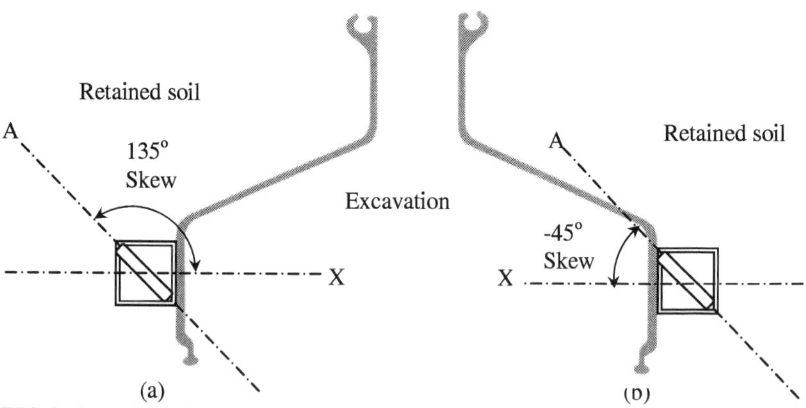

FIG 3. Rotation of measurements to the axis of wall bending for the (a) west side and (b) east side.

Since it was not desirable to operate heavy machinery directly behind the walls, the soil between the walls was to be removed using a hydraulic excavator operating between the walls; thus, operators and support personnel would be inside of the active excavation. Based on finite element models, two limit states were defined for monitoring the wall. First, the maximum allowable bending moment was determined to be 192.6 kN*m for PZ22sheet piles formed from 345 MPa steel. Varying the soil properties within a reasonable range, the maximum top of sheet pile deflection, with a stable soil mass, was in the range of 150 to 225 mm.

RESULTING FIELD INSTALLATION AND EXCAVATION

The sheet piles were installed beginning September 12, 2005. As mentioned previously, the sheets were to be driven 10.4 m leaving 0.3 m of exposure. In the northwest corner of the site, this was possible. However, the PRS provided much higher resistance to driving than predicted by the initial tests. As a result, many of the piles were significantly under-driven. The as-built sheet pile walls at the Statesville site were 36.9 m long consisting of 66 sheets per side. The strain gage sheets on the west wall were installed at 17.9 m and 22.4 m from the north end, and the inclinometer sheets were installed at 2.2 m from the strain gage sheets. Four of the topmost gages on pile NE-S were installed before gages could be properly set, resulting in out of range strain measurements, that were of no use in the analysis.

The soil between the sheet pile walls was excavated in five 1.2 m lifts over a period of ten days between October 17 and October 27, 2005. After each excavation step, inclinometer readings were immediately taken. Subsequently, strain gage readings were downloaded from the compact flash cards attached to the dataloggers.

Bending moments were calculated from the strain by converting the strain difference to curvature then multiplying by the sheet pile stiffness EI. Plots of the evolution of the moment with depth of excavation for all piles are shown in Figure 4. A clear progression of intensity and depth of maximum moment with excavation depth can be seen, particularly in the piles on the west wall where the piles were driven more uniformly. In any case, the calculated bending moments were much lower than the allowable bending moment of 192.6 kN*m.

Similar plots of the sheet pile deflections are shown in Figure 5. The deflections are shown for the sheet piles with the ground surface as reference depth. Even though one of the sheets was driven short of the specified 0.3 m exposure, the total deflections of the sheets fell below the 150 mm threshold specified at the project outset. The eastern piles displaced significantly more than those on the western side, due to a combination of being under driven as well as the stress relaxation. Gaps were observed between the sheet piles and the retained soil as deep as 14 feet.

SUMMARY

Two instrumented sheet pile walls were installed 12.2 m apart to determine the earth pressure in Piedmont residual soils, experimentally. Strain gages and slope inclinometer casings allowed the measurement of bending moments and deflections, respectively. Soil between the two walls was excavated to a depth of 6.1 m, allowing the earth pressure to mobilize behind the sheet piles, and in-turn causing the sheet piles to deflect. Strain gages were continuously monitored, and the results were used to calculate bending moments, while deflections were read daily using a slope inclinometer probe. The calculated bending moments, measured deflections, and field observations supported that there was no appreciable movement of the soil mass during the load test which suggests no mobilization of lateral earth pressure.

REFERENCES

Brinkgreve, R.B.J. and Vermeer, P.A., (1998). *Plaxis version 7.12: users guide,* A.A. Balkema, Rotterdam.

Bullock, P.J., Schmertmann, J.H., McVay, M.C., and Townsend, F.C., (2005). "Side shear setup. I: test piles driven in Florida." *Journal of Geotechnical and Geoenvironmental Engineering*, ASCE, Vol. 131, No. 3, pp. 292-300.

Clough, G.W., and O'Rourke, T.D. (1990). "Construction induced movements of insitu walls." *Design and Performance of Earth Retaining Structures*, GSP No. 25, ASCE, New York, NY, pp. 439-470.

Fang, Y., Chen, T., and Wu, B., (1994). "Passive earth pressures with various wall movements." *Journal of Geotechnical Engineering*, Vol. 120(8), p. 1307 - 1323.

Fang, Y., Chen, J., and Chen, C., (1997). "Earth pressures with sloping backfill." *Journal of Geotechnical Engineering*, Vol. 123(3), p. 250 - 259.

Georgiadis, M. and Anagnostopoulos, C., (1998). "Lateral pressure on sheet pile walls due to strip load." *Journal of Geotechnical & Geoenvironmental Engineering*, Vol. 124(1), pp. 95 - 98.

Kort, D. A. (2002). *Steel Sheet Pile Walls in Soft Soil* . Delft University Press, Delft, The Netherlands.

Long, M. (2001). "Database for retaining wall and ground movements due to deep excavations." *Journal of Geotechnical and Geoenvironmental Engineering*, Vol. 127, No. 3, ASCE, March, pp. 203-224.

Peck, R.B. (1969). "Deep excavations and tunneling in soft ground." *Proceedings of the 7th International Conference of Soil Mechanics and Foundation Engineering*. Mexico City, pp. 225-290.

Paikowski, S.G. and Hajduk, E.L., (2004). "Design and construction of three instrumented test piles to examine time dependent pile capacity gain." *Geotechnical Testing Journal*, ASTM, Vol. 27, Issue 6, pp.

Rieke, R.D. and Crowser, J.C., (1987). "Interpretation of a pile load test considering residual stresses." *Journal of Geotechnical Engineering*, ASCE, Vol. 113, No. 4, pp. 320-334.

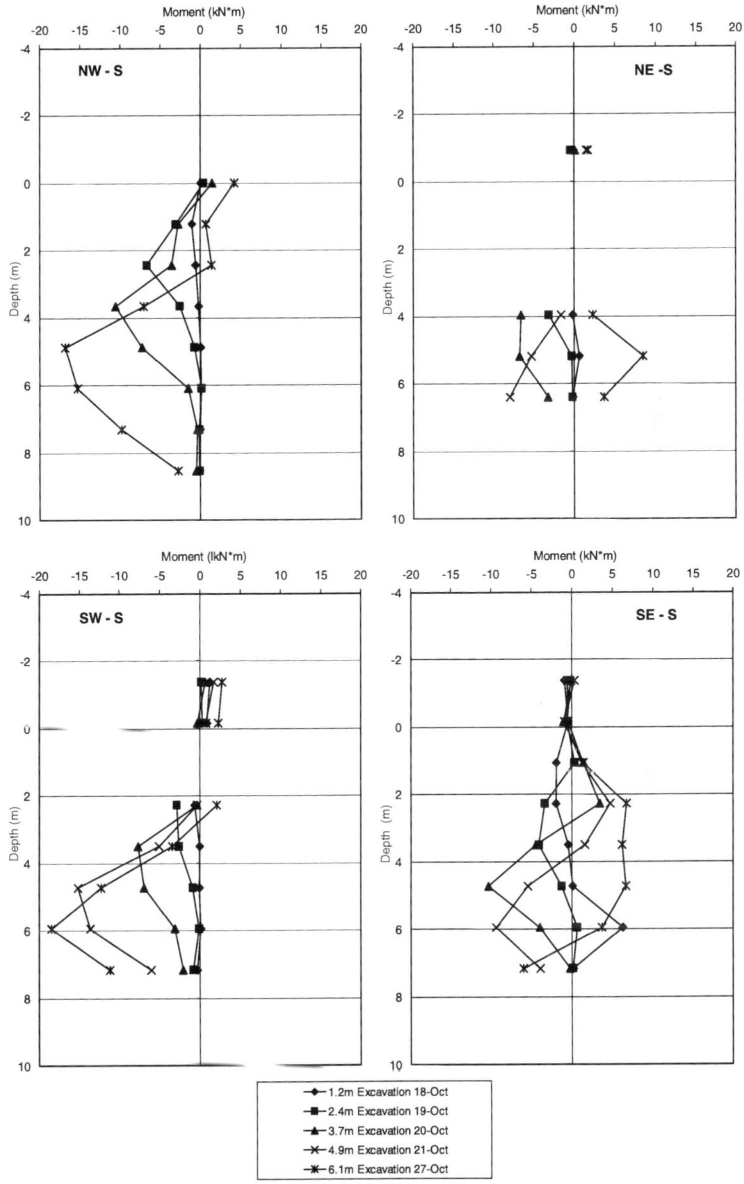

FIG 4. Measured bending moments per pile and excavation step.

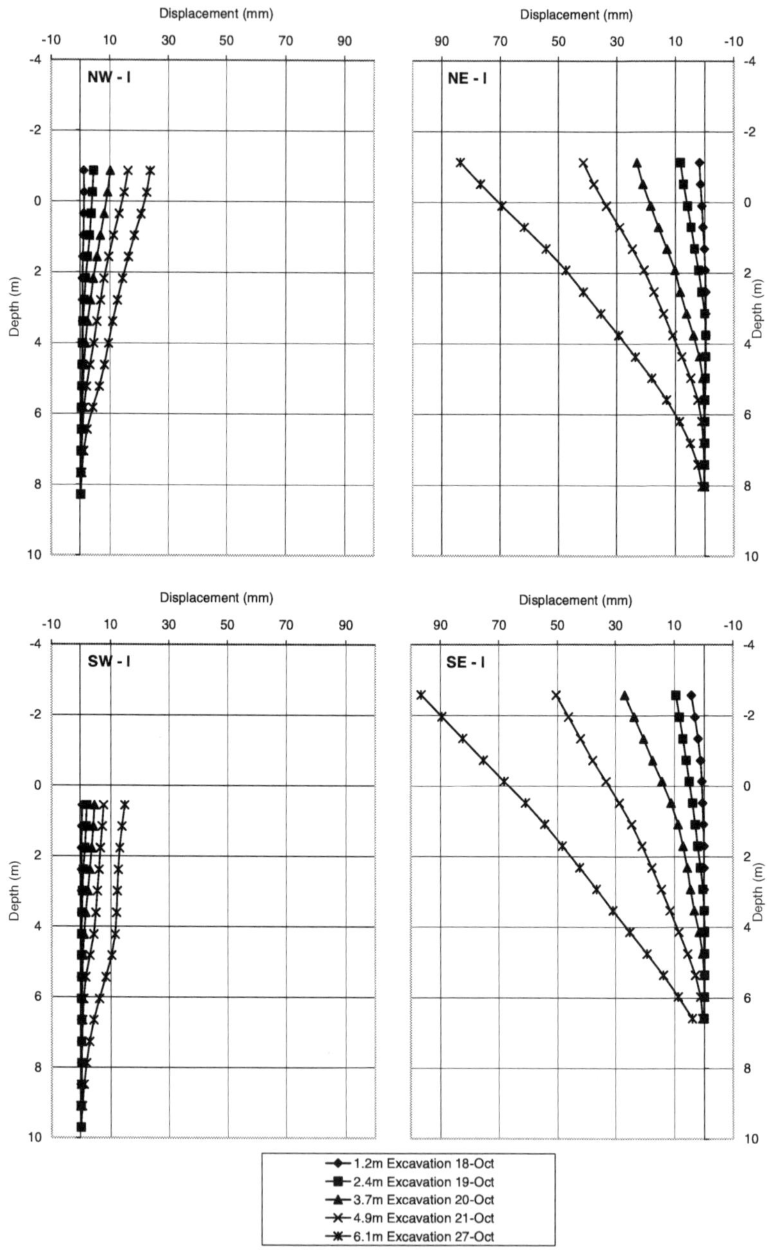

FIG 5. Measured lateral displacements per pile and excavation step.

Third Dimension Simplifies Earth Retention at the Calumet Plant

Gregory A. Terri, P.E., A.M. ASCE

Project Manager, Hayward Baker Inc., 1350 West Lake Street, Roselle, Illinois, USA
gaterri@haywardbaker.com

ABSTRACT: The Hydraulic Improvements at the Calumet Water Reclamation Plant project in Chicago, Illinois required excavation of over 76,000 cubic meters (100,000 cubic yards) of soil. To accommodate the excavation, an earth retention system (ERS) was designed and installed within the existing dense network of active utilities at the site. The potential for conflicts between the ERS and the utilities was very high. The use of three dimensional modeling, considerable pre-planning and design was required to ensure that a cost effective, stable and constructible earth retention system could be installed that met the project requirements yet did not disrupt the existing facility. This paper covers the development of the design, installation and performance of the earth retention system.

INTRODUCTION

Set to open in April 2010, the improvements at the Calumet Water Reclamation Plant project will include construction of a new raw wastewater pumping station, screen building, diversion chambers and associated structures. The new pump station for the facility will have a design capacity of 2,270 megaliters per day (600 million gallons per day), treating wastewater from an approximately 775 square kilometers (300 square mile area) of metro Chicago.

In order to construct such a facility, excavations as deep as 19 meters (62 feet) below the existing grade were required within the existing operating plant. The new facility was to be integrated into the existing utilities including two eight-foot diameter TARP discharge pipes from the deep tunnel system, two nine-foot diameter South Park/Blue Island sewage interceptor lines, arched triple barrel influent interceptor and effluent tunnels, a seven foot diameter Harvey interceptor line and a 13.2 kV electric service encased in a two foot by three foot concrete duct bank. The existing features at the Calumet site are depicted in Figures 1 and 2.

Figure 1: Aerial view of the site – view south

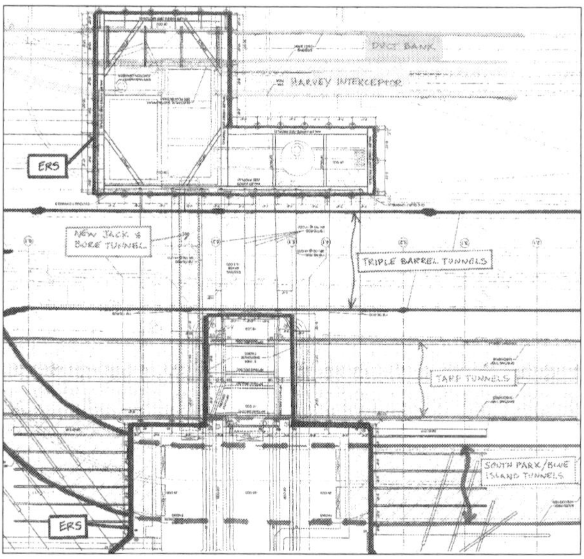

Figure 2: Detailed Plan View at south end of site

DESIGN REQUIREMENTS

Earth retention was required around the perimeter of the new structures to accommodate the deep excavations and to protect the existing utilities. The deepest excavation at the center of the site was 19 m (62 ft), with excavations along the ERS as deep as 13.5 m (44 ft) below grade. The total wall footage around the perimeter was approximately 425 linear meters (1,400 linear feet). Figure 3 presents the required depths of excavation for the new structures. The configuration of the new facility resulted in eight re-entrant corners of the earth retention system as it jogged around the proposed structures. The active utilities, re-entrant corners and new jack and bore tunneling resulted in a significant potential for conflicts between the earth retention system components and existing underground structures.

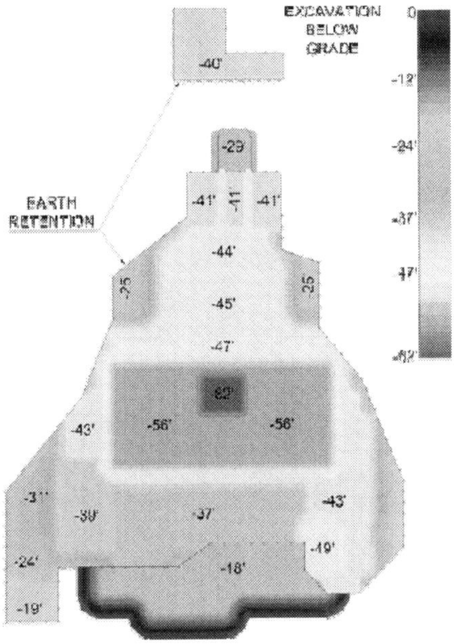

Figure 3: Excavation depths at the site (feet)

The specifications for the work consisted of a strict ground movement criteria to ensure that the existing utilities were not impacted by the deep excavations. Lateral movement of the earth retention system was limited to 0.5% of the cut height. Maximum vertical displacement of the ground surface behind the ERS was required to be less than 38 millimeters (1.5 inches). In order to meet these criteria, the stiffness of the ERS was increased in the areas of the deepest cut and where utilities were present.

GROUND CONDITIONS AND PRESSURES

The soil conditions at the site consist of silty clay underlain by dolomitic limestone at about 18 m (60 ft) below grade. The clay strengths are summarized in Table 1.

Depth, m (ft)	Cohesion, kPa (psf)	Depth, m (ft)	Cohesion, kPa (psf)
0–6 (0–19)	115 (2,350)	10–12.5 (32–41)	180 (3,750)
6–8 (19–25)	160 (3,300)	12.5–14 (41–47)	260 (5,500)
8–10 (25–32)	200 (4,250)	14–18 (47–60)	320 (6,750)

Table 1: Clay Strengths at the Site

Design earth pressure diagrams for the tied-back and internally braced soldier beam and lagging systems were developed based on apparent pressure envelopes for stiff to hard clays. The maximum pressure for the envelope was approximately $\sigma = 0.25*(\gamma H+q)$, as determined by re-distributing the Rankine pressures to a trapezoidal shape. The 0.25 factor falls within the range of published values for the clay soil types encountered. The trapezoidal shape was a function of the upper and lower level tieback elevations and is schematically shown in Figure 4.

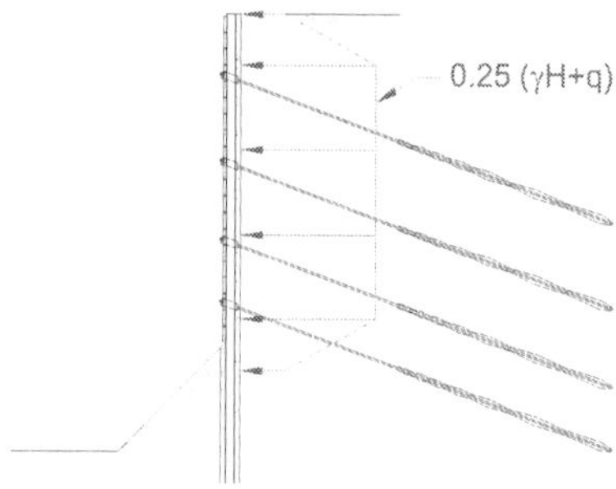

Figure 4: Earth Pressure Diagram

DESIGN METHODOLOGY

Based on the subsurface conditions at the site, existing utilities and configuration of the structures, a combined tied-back soldier beams and lagging, soil nails and shotcrete, and internally braced beams and lagging was selected for earth retention. The stiff cohesive soils were conducive to drilling and setting soldier beams, allowed for setting lagging with minimal exposure to soil loss, and provided adequate bond material for the grouted tiebacks. The soldier beam system allowed for flexibility of the alignment of the retention and for the system to jog over existing utilities. The soil nail and shotcrete provided earth retention above the large diameter South Park/Blue Island pipes which precluded the embedment of the ERS. The internally braced earth retention was more cost effective in the confined Harvey structure and TARP areas, and where the bracing members could be conveniently removed as the structures were installed.

3D MODELING

The geotechnical contractor used AutoCAD to create three dimensional models to avoid conflicts between the active utilities and the externally braced system including tiebacks and soil nails. The eight re-entrant corners presented further conflict possibilities between ERS components which were managed using the 3D modeling.

As the apparent pressure diagram is a function of the tieback load and elevation, many iterations and recalculations were required before settling on an optimal and constructible system. Iterations were required to make certain the soldier beams, tiebacks and soil nails were appropriately designed and that conflicts were not encountered. Examples of the 3D models are presented in Figures 5 and 6.

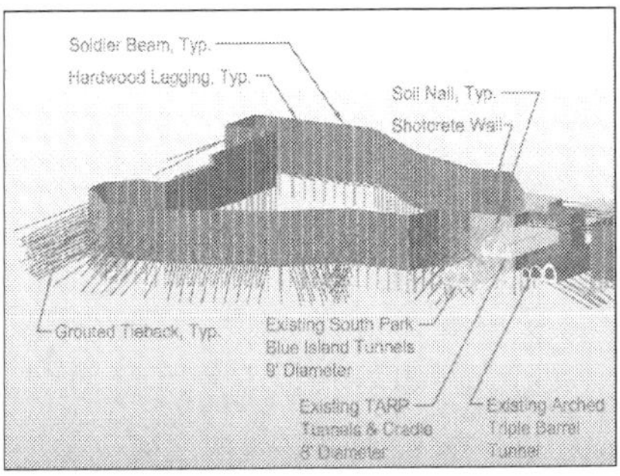

Figure 5: 3D Modeling – Entire Site

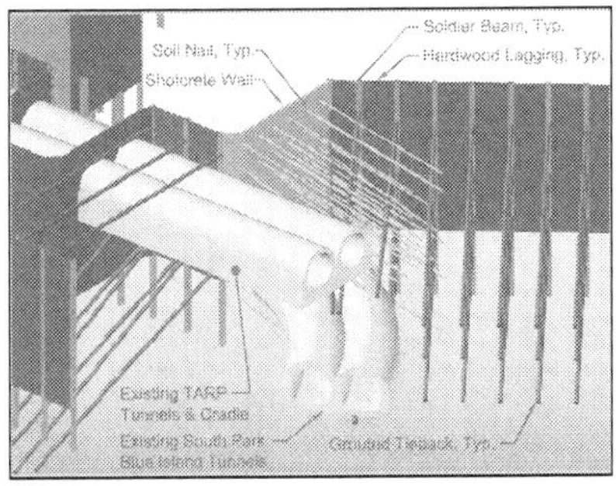

Figure 6: 3D Modeling Detail

The high tieback loading on the ERS dictated that the most economical solution to externally brace the soldier beams was to insert pre-fabricated pipe sleeves through the beams. The cut heights required as many as four pipe sleeves per soldier beam. Due to the variation in pipe sleeve angles and elevations of the sleeves along the beams, nearly every soldier beam in the vicinity of the existing utilities was unique. There were 36 different types of beams of the total 175 installed for the project. Communication between the modeling, designing, fabricating and install stages was critical in ensuring that all beams were installed in the proper locations.

The variation in elevations, angle of the tiebacks, and cut heights at almost every soldier beam along the length of the earth retention system, resulted in over 45 different types of grouted tiebacks to optimize and simplify installation. The tie-backs ranged from 15 to 27 m (50 to 90 ft) long with capacities ranging from 310 kN (70 kips) to 630 kN (142 kips). Figure 7 presents a view of the excavation, installed soldier beams and lagging, and tiebacks.

Due to the intersecting tiebacks, installation of complex grids of soil nails was required to support the earth above the South Park/Blue Island Tunnels. A total of 75 soil nails were installed to support wall heights, as high as 12 m (40 ft). An active electric duct bank that feeds the existing plant was also underpinned to accommodate excavations 10 m (32 ft) below the utility. In total, over 4,200 square meters (45,000 square feet) of exposed earth retention was installed. Figures 7 and 8 present views of the installed beams and lagging and soil nails and shotcrete.

Figure 7: View north of the soldier beam and lagging ERS;

Figure 8: View south of the soil nails and shotcrete and soldier beam and lagging ERS near the South Park/Blue Island tunnel.

INSTALLATION

The stiff clay soils were conducive to open-hole drilling of the soldier beams. The beams were installed with a European style track mounted fixed mast piling rig. The maximum drill length for the beams was 18 m (60 feet). The tiebacks and soil nails were also open-hole drilled to lengths of up to 27 m (90 feet). Figures 9 and 10 show the rigs used for drilling of the beams and tiebacks/soil nails.

Figure 9: Beam installation.

Figure 10: Tieback installation.

QUALITY CONTROL AND ASSURANCE

Another critical phase of this project was testing of the installed components of the ERS. Performance tests were executed on 5% of the installed grouted tiebacks, in accordance with the Post Tensioning Institute (2004). The balance of the tiebacks were proof tested. Verification tests were performed on about 10% of the soil nails installed. Five percent of the balance of the soil nails were proof tested. All soil nails were tested in accordance with the Federal Highway Administration (2003). Cylinders were cast daily for the grout used for the tiebacks, soil nails and shotcrete.

The test results confirmed that the tiebacks, nails and grout installed met the parameters required for the design. An example of the tieback test results is presented in Figure 11.

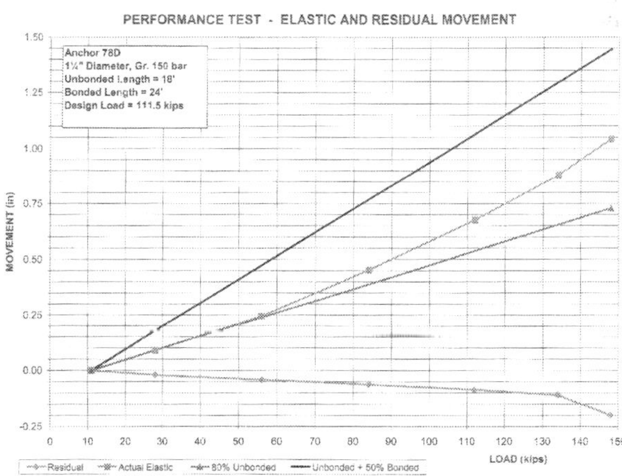

Figure 10: Performance Test results for a grouted tieback.

Settlement monitoring of the utilities was also performed throughout the duration of the project. Settlement points were placed on various utilities within the active zone of the ERS. The results of the monitoring indicated no recordable settlement of any utilities. Lateral movements of the ERS were not recorded during the excavation.

CONCLUSIONS

For deep excavations within dense networks of utilities, three dimensional modeling and extensive pre-planning and design is essential in developing a builable and optimized earth retention system. For the work at the Calumet Water Reclamation Plant, the models were the starting point for creating an efficient system at the site. The flexibility of the models allowed for quick and thorough reviews of determining the most cost effective ERS while also simplifying the field operations. Where the potential for conflicts exist to the degree encountered at the Calumet site, 3D modeling is the only solution.

ACKNOWLEDGEMENTS

The author wishes to acknowledge the professional contributions to this project by Mr. Benjamin Strahan, CAD designer at Hayward Baker, Mr. Mark Hampton and Mr. Don Trexler, field engineer and superintendent for the Calumet project, respectively.

REFERENCES

Lazarte, Carlos A., Elias, Victor, Espinoza, R. David, and Sabatini, P.J. (2003). "Geotechnical Engineering Circular No. 7: Soil Nail Walls". *Publication No. FHWA-IF-03-017 Federal Highway Administration.*

Post Tensioning Institute (2004). *Recommendations for Prestressed Rock and Soil Anchors.* Post Tensioning Institute, Phoenix, AZ.

Sabatini, P.J., Pass, D.G and Bachus, R.C. (1999). "Geotechnical Engineering Circular No. 4: Ground Anchors and Anchored Systems". *Publication No. FHWA-IF-99-015 Federal Highway Administration.*

INNOVATIVE EARTH RETENTION, UNDERPINNING AND WATER CONTROL TECHNIQUES FOR A MEGA-DEVELOPMENT IN IN SALT LAKE CITY, UTAH

Tom Hurley, M. ASCE, Project Manager Nicholson Construction Company, Salt Lake City UT 84116
Mike Walker, P.E., Vice President GEI Consultants, Woburn MA 01801

ABSTRACT

The City Creek Center Block 76 Project includes the demolition and redevelopment of an entire city block for mixed use retail, office and residential properties directly adjacent to Historic Temple Square in Salt Lake City, Utah. Prior to construction, the Block 76 Project footprint was completely occupied by an existing shopping mall, subsurface parking structure and a twenty two story office building. Demolition of these structures, including the implosion of the office tower, was required to make way for the new construction on the project. However an adjacent light rail system and four existing buildings between 9 and 23 stories remained at the corners of the project.

Excavation was required to 65ft (19.8m) below street grade and by as much as 45ft (13.7m) directly below adjacent buildings supported by shallow mat and strip type footings. Movements of the earth retention system and adjacent buildings were monitored in real time using automated equipment which complimented a suite of traditional geotechnical instrumentation.

Due to the high profile nature of this project and considering deep excavations through granular soils adjacent to high rise structures on shallow foundations, careful consideration was applied to evaluate the proposed support system. Anchored diaphragm walls were employed in the areas immediately adjacent to the existing buildings greater than 10 stories. This is believed to be the first anchored diaphragm wall for earth retention in Salt Lake City, Utah.

Soil nail wall systems were utilized at areas of the site not supported by diaphragm walls employed. The location of soil nail walls was generally adjacent to streets where larger wall movements were acceptable to the project team. Soil nail walls were also employed directly below existing structures, but were enhanced by an underpinning system consisting of Nicholson Pin Piles with an integral cap beam to limit anticipated movements to 1" (25mm).

Jet grouting was employed as an alternate for water cutoff in lieu of dewatering to mitigate water inflow along the North and East sides of project. Prior to commencing production jet grouting, a field test program was instituted to optimize injection parameters and to confirm jet grout element geometric and mechanical properties. The test program involved the construction of test elements followed by a 3 day cure time. Jet grout test elements were exposed by excavation and geometric properties obtained by direct measurement. The jet grout wall consisted of overlapping secant columns and was constructed along two alignments, providing a connection between diaphragm walls adjacent to buildings along three corners of the project.

This paper will provide an overall description of the Block 76 project and discuss the approach for the selection of the earth retention systems employed on the site. In addition, the location and reasoning behind the use of underpinning and jet grouting will be discussed. A detailed account of the installation of the earth retention, underpinning and water cutoff systems will be provided. Soil conditions encountered at the site will also be discussed. Predicted versus actual earth retention wall movements will be presented. In closing a brief summary will be provided with regards to the 24" (600mm) diameter, 450tn capacity augercast piles currently in place for the new structure that is taking shape.

PROJECT OVERVIEW

City Creek Center is a US$1 billion project that will transform the downtown area of Salt Lake City, Utah. The project is spread over 3 city blocks encompassing 20 acres (81,000 Sq M) and involves the demolition and redevelopment for mixed use retail, office and residential properties directly adjacent to Historic Temple Square, (Figure 1).

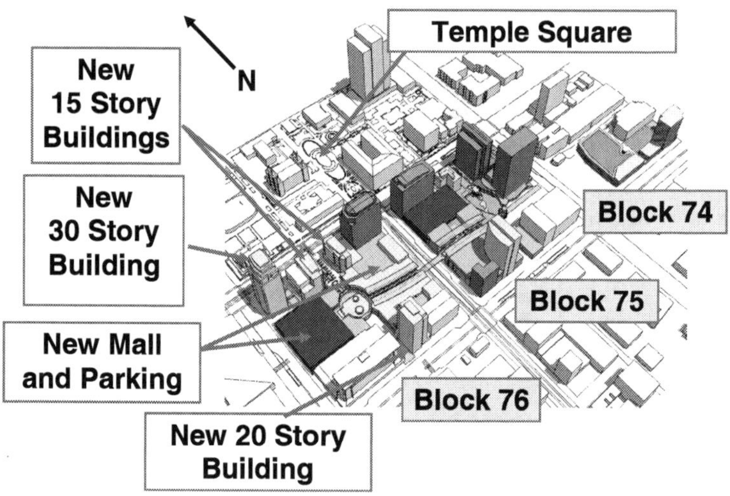

FIGURE 1 – City Creek Center Project

Prior to construction, the Block 76 Project footprint was completely occupied by an existing shopping mall, subsurface parking structure and office buildings. Demolition of these structures, including the implosion of the office tower, was required to make way for the new construction on the project. However an adjacent light rail system and four existing buildings between 9 and 23 stories remained at the corners of the project (Fig 2).

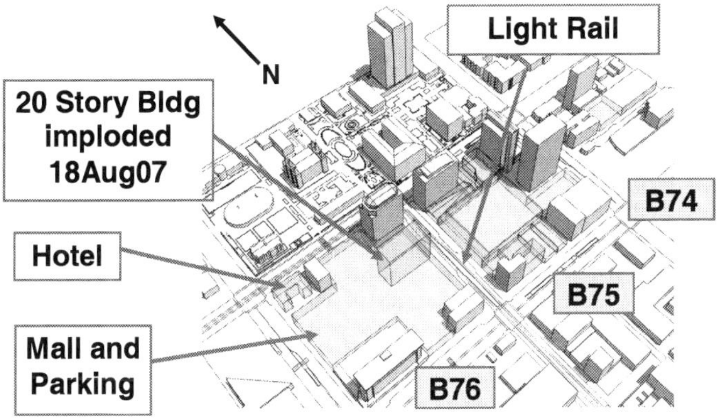

FIGURE 2 – Block 76 Demolition

DESIGN CONSIDERATIONS AND SOIL CONDITIONS

Excavation was required up to 65ft (19.8m) below street grade and by as much as 45ft (13.7m) directly below adjacent buildings supported by shallow foundations. Controlling movement of the earth retention system and adjacent buildings was of critically important.

The soil stratigraphy consists of inter layered alluvial and lacustrine deposits. The top layers are medium to very dense sand and gravel over medium stiff clays and silts under which is a very dense layer of sand and gravel. The clay and lower gravel layer are deepest at the southwest corner of the site and rise towards the northeast corner of the site. The depth to existing groundwater is approximately 40-54 ft (12-16m) below grade with the highest groundwater at the northeast corner of the site. (Figure 3).

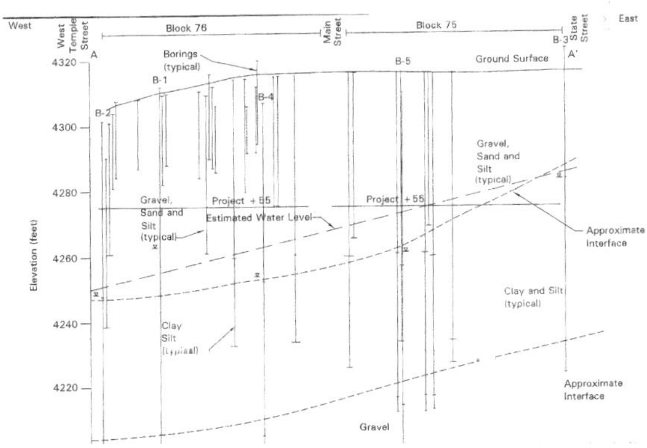

FIGURE 3 – Block 76 Subsurface Profile from Geotechnical Report

Due to the high profile nature of this project and taking into consideration the requirement of deep excavations through granular soils and the adjacent high rise structures on shallow foundations, careful consideration was applied to evaluate the proposed support system.

Anchored diaphragm walls were employed in the areas immediately adjacent to the existing buildings greater than 9 stories. The diaphragm wall design incorporated 3 to 4 levels of 6 strand tie back anchors to control movement. A compromise was reached with the Owner in regards to diaphragm wall thickness and encroachment on precious retail space which led to the design of a 24" (600mm) reinforced concrete thick wall versus the 30" (800mm) wall originally envisioned.

Soil nail wall systems were utilized at areas of the site not supported by diaphragm walls employed. The location of soil nail walls was generally adjacent to streets where larger wall movements, up to 2" (50mm), were acceptable to the project team. Because the soil nail walls were not permanent, self drilling hollow bars were utilized in the design. Shotcrete facing was employed with wire mesh reinforcement.

Soil nail walls were also employed directly below existing structures, but were enhanced by an underpinning system consisting of micropiles with an integral cap beam to limit anticipated settlement to 1" (25mm). The underpinning system was designed to transfer the building loads to an elevation below the earth retention system. This allowed some of the building surcharge to be eliminated from the soil nail wall design. The micropile system consisted of both self drilling hollow bars and Nicholson cased Pin-Piles and were bonded in the lower dense gravel layer, generally 100 feet (30m) below working grade. Load was transferred from the existing structure through a doweled connection to the reinforced concrete cap beam into the micropiles and down to the bond zone in the lower gravel.

As seen in Figure 3 Subsurface Profile, the excavation to project elevation 36ft would be below the existing ground water at the north east corner of the site. The soil nail wall requires a stable face for drilling and shotcreting, therefore a groundwater control system needed to be implemented. A dewatering system was evaluated and a proposal was reviewed by the General Contractor. The main concerns with the dewatering systems were two fold. 1) The dewatering system was to be located in a zone of fine material with low hydraulic conductivity, requiring many small diameter wells at a very close spacing and 2) the potential for removal of fines by the dewatering process could lead to settlement. Ultimately, jet grouting was employed as an alternate to dewatering to mitigate water inflow along the North and East sides of project (Figure 4). A series of overlapping columns was designed to create a cutoff wall in between diaphragm wall segments along the north east part of the site.

The entire earth retention system and all existing adjacent structures were monitored with Sol Data's Cyclops real time monitoring system. This system was a design requisite and an overall risk management tool for the project team. The Cyclops system utilizes robotic total stations that cycle through prism targets mounted on the earth retention systems and adjacent structures and transmits data through wireless connection to Sol Data where the proprietary software reduces and reports the data to project participants in real time. Complementing the Cyclops system was an array of inclinometers behind soil nail walls and directly in the diaphragm walls, observation wells, and surface settlement points.

PRODUCTION WORK

Soil Nail Walls: The project began with a demolition phase that was supported by soil nail wall construction. As subsurface demolition occurred, the soil nail walls were sequentially installed as earth retention to support further demolition below grade.

The soil nail walls were constructed using an excavator mounted, top drive rotary percussion drill. The excavator mount provided flexibility with drill setup and tracking over variable terrain. In addition, the drill mast has 360 degrees of movement in 2 directions making setup and installation at corner locations easier. As demolition areas were completed, a more productive pace was realized and soil nail wall construction was largely governed by the excavation progress. Over 5,000 soil nails were installed totaling over 150,000 lf (45,720 m) of drilling and over 104,000sf (9,660 sq m) of shotcrete facing.

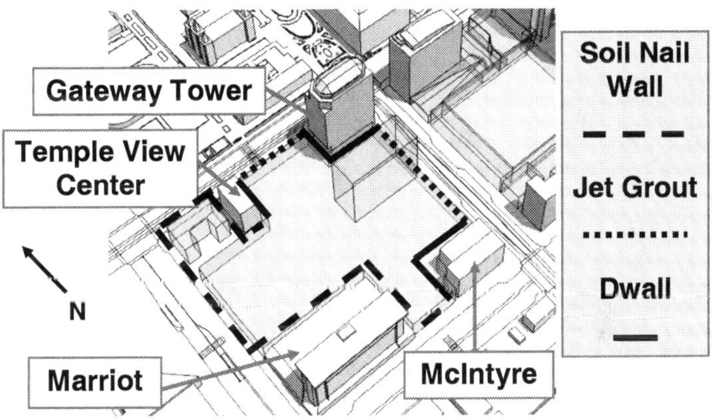

FIGURE 4 – Block 76 Earth Support Technique Key Plan

During the initial soil nail sequence, underpinning activities were underway along the Marriott Hotel. (Figure 4 shows the earth support technique key plan). The use of hollow bar micropiles at this location was driven by the drill proximity to the existing building. Only a one foot clearance was available from the existing building to the backside of the new construction. To mitigate potential ground loss under adjacent footings, the drill rig also installed inclined grouted bars to aid the first 3 lifts of soil nailing. Micropiles were installed to a depth of 102 ft (31 m) below grade and were bonded into the lower dense gravel formation. A total of 103 micropiles and 850 linear feet (259 m) of cap beams were installed on the project at 3 locations; Marriott Hotel, Temple View Center, and at the Main Street Garage.

Jet Grouting: As discussed above, jet grouting was selected as an alternate to dewatering along South Temple and Main Streets. Prior to commencing production jet grouting, a jet grout field test program was instituted to optimize injection parameters and to confirm jet grout element geometric and mechanical properties. The test program involved the construction of 4 sets of 2 test columns followed by a 3 day cure time. The injection times were varied for each set of test columns, while all other parameters remained constant. The grout mix included a small amount of bentonite (2% by weight) to stabilize the grout mix and provide additional water control properties. The target column diameter was 4ft (1.2 m) using the single fluid technique. The test columns were then exposed by excavation and geometric properties obtained by direct measurement. All test sets achieved 4ft (1.2m) column diameters or greater and production work commenced.

Two jet grout walls were constructed with a single row of overlapping secant columns and connected the diaphragm wall segments between the Temple View Center Building and the Gateway Tower on South Temple Street and between the Gateway Tower and the McIntyre Building on Main Street. By connecting to the Diaphragm walls, the northeast quadrant of the site was essentially cutoff from major water inflows.

The jet grout columns were installed so the bottom of column was 5-10ft (1.5–3m) below the base of the excavation and the top of the column was installed to 5ft (1.5m) above the known high water elevation determined from the observation well readings. The average drill length for each column was approximately 50 ft (15m) and the jet length was 30 ft (9m). 236 columns were installed in a 2 month time frame to complete the water control activities enabling completion of soil nail walls below the ground water table. Figure 5 shows the exposed jet grout wall and completed soil nails.

FIGURE 5 – Block 76 Jet Grout Wall South Temple Street

Diaphragm Wall: D-Walls were constructed adjacent to the Temple View Center building (9 stories), the Gateway Tower (23 Stories) and the McIntyre and Crandall Buildings (9 Stories). Each building is on shallow foundations, and excavation was required to approximately 40ft (12m) below the foundations.

Each Building had its own challenges. The D-Wall adjacent to the Temple View Center needed to be installed within 1ft (300mm) of the existing building facade and required removal of a portion of the existing isolated spread footings that protruded into the D-Wall alignment. Prior to any footing removal, a micropile underpinning system was installed to replace the capacity of the existing footings that were to be removed. After the micropiles were installed, excavation for footing removal revealed a larger footing than was expected and required a wire saw to remove. D-Wall Panels in this area were installed as "single bites" meaning open D-Wall excavation trenches were limited to the length of the bucket, which was 9ft (2.75m). This technique was employed because the TVC building was supported by independent spread footings. The Project team did not want to induce any potential bearing failure by having open trenches greater than the length of the bucket.

Along the Gateway Tower alignment, an existing building overhang was present approximately 86ft (26m) above our work platform elevation. This required careful consideration of crane requirements and rigging to ensure the reinforcing cages, when placed into the excavated panel, would fit under the overhang without the added costs of cage splicing. The actual placement of cages required the crane boom tip to be within 3ft (914mm) of the underside of the overhang. As a precautionary measure, the crane boom was fitted with a proximity sensor that warned the operator when he was too close to the building. The Gateway Tower was founded on an 8ft thick concrete mat, enabling the D-Wall to be constructed with traditional 22ft (6.7m) long "three bite" panels.

No as-built drawings were in existence for the McIntyre and Crandall Buildings. These building were constructed in the 1900's and are some of the oldest in Salt Lake City. The Project team expected surprises, but only one surfaced which required adjusting the D-Wall alignment north to miss the existing McIntyre footing.

In total 43 panels were installed to an average depth of 60ft (18m). Over 45,000sf (4,180sqm) of D-Wall was installed in two months, ahead of the project schedule eliminating a remobilization because all D-Walls were complete prior to the implosion of the Key Bank Tower, a significant project milestone. The D-wall for Block 76 is believed to be the first diaphragm wall for earth retention in Salt Lake City, Utah.

MONITORING OF EARTH RETNTION SYSTEM DURING EXCATATION

Nicholson's overall scope included the monitoring, collection, and reporting of geotechnical instrumentation on the project. The Sol Data Cyclops system provided the real time monitoring of the shoring systems and building movements. Traditional inclinometers, observation wells and surface settlement points were reported on a weekly basis. Evaluation of earth retention system and building movements were done routinely by our Engineer, GEI Consultants, and Nicholson's project and home office staff.

Movement of the soil nail walls was expected to follow the typical pattern of rotation about the toe, similar to a gravity or cantilevered retaining wall. Maximum movements were expected at the top of the walls, with magnitudes in the range of 1½ to 2 inches for the deepest cuts of 65 feet (19.8m). Observed movements were typically much less than expected generally by more than 50%. We attribute the good performance of the soil nail walls to the predominance of the upper gravel stratum and its slightly cemented nature which was conservatively neglected in the design. In isolated areas where the upper gravel stratum was disturbed, the observed deformations were closer in magnitude to the anticipated values.

Movement of the diaphragm walls was expected to follow a pattern of increasing deflection with depth, corresponding to greater driving loads and the weaker clay and silt strata near the bottom of the cut. Maximum movements were expected between mid-height and the toe of the walls, with magnitudes of approximately 1 inch for the deepest cuts of 40 feet (12.2m). Observed movements were typically in line with expectations, although additional movement control measures were required in some portions of the work. Subgrade struts and supplemental anchors were used to control diaphragm wall deflections adjacent to the Gateway Tower, where the weaker clay & silt strata was found at a higher elevation than anticipated.

CLOSING

In total, Nichoson's earth retention scope included 45,000 sf (4,180 sq m) of Diaphragm Wall, 104,000 sf (9,660 sq m) of Soil Nail Wall, Pin Pile underpinning of 3 buildings, and jet grouting for water control where soil nails walls were below the ground water table. Sol Data's Cyclops monitoring system provided real time data in a web based format for quick evaluation of earth retention systems and building deformations. Traditional geotechnical instrumentation, such as inclinometers, validated the Cyclops data and provided wall movement at incremental depths for subsequent wall behavior analyses. The scope and magnitude of this project is unprecedented in Salt Lake City. The amount of earth retention for deep excavation accomplished in around and below existing buildings was truly unique and full of challenges. Figure 6 depicts a wide angle view of the project to give the reader some prospective.

FIGURE 6 – Wide angle view of Block 76 Site

Secant Pile Wall Design and Construction in Manhattan, New York

Cem Altuntas[1], M. ASCE, P.E., Deo Persaud[2], A. M. ASCE, Alan R. Poeppel[3], M. ASCE, P.E.

[1] Assistant Project Manager, Langan Engineering and Environmental Services, P.C., 360 West 31st Street, 8th Floor, New York, NY 10001; caltuntas@langan.com
[2] Senior Staff Engineer, Langan Engineering and Environmental Services, P.C., 360 West 31st Street, 8th Floor, New York, NY 10001; dpersaud@langan.com
[3] Senior Associate, Langan Engineering and Environmental Services, P.C., 360 West 31st Street, 8th Floor, New York, NY 10001; apoeppel@langan.com

ABSTRACT: Below-grade excavation for basement construction of real estate developments is a risky and challenging undertaking in Manhattan. New development sites are frequently adjacent to old buildings (with shallow foundations) that require underpinning to the excavation levels. Presence of poor surficial soil conditions and urban fill warrant the extension of underpinning to underlying deeper satisfactory bearing materials or possibly bedrock. Furthermore, groundwater is generally shallow, dictating excavations below the groundwater table. Secant pile walls, a stiff excavation support system primarily consisting of overlapping drilled concrete piles, are becoming more common in Manhattan because this support system can eliminate the need for underpinning and dewatering.

This paper presents the state-of-the-practice design and construction of a secant pile wall for a major real estate development in Lower Manhattan. At this site, the presence of a 1.5-meter wide rear alley between the adjacent buildings and the property line precluded the use of conventional underpinning. As an alternative, a secant pile wall was constructed to provide an excavation support system for the basement construction and to eliminate underpinning of adjacent buildings. The secant pile wall inherently created a dry excavation that avoided costly dewatering operations. The secant pile wall also served as the permanent foundation wall of the new building. The 58-meter long wall was completed in less than five months.

INTRODUCTION

This paper presents the design and successful construction of a secant pile wall for a real estate development located in the Greenwich Village section of Manhattan, New York. The project site is located on the city block bordered by Spring Street on the north, Canal Street on the south, Greenwich Street on the west and Hudson Street on the east. The site area is about 836 square meters with a generally flat surface grade at about Elevation 2.1 Manhattan Borough Datum (MBD), which is 0.84 meters above the U.S. Coast and Geodetic Datum (Mean Sea Level at Sandy Hook, NJ 1929). The southern part of the sidewalk fronting the site is within the 100-year flood level and the northern part of the sidewalk is within the 500-year flood level.

The project is an 11-story residential building with one cellar level. The cellar houses mechanical and electrical rooms, a swimming pool and a parking garage. The building is constructed of cast-in-place concrete slabs with reinforced concrete columns and shear walls. Column loads (dead plus live service loads) ranged from 445 kilonewton (kN) to 3,100 kN.

The development site is adjacent to buildings on the north, south and west, and Renwick Street to the east. A brick retaining wall and a rear alley of an adjacent building abut the project site on the west. The adjacent residential buildings are brick, 6- to 10-stories with single cellar levels.

Langan Engineering and Environmental Services, P.C. was the geotechnical design engineer and Hayward Baker was the contractor for the secant pile wall construction.

GEOLOGY

The site is just inboard of the 17th century Manhattan shore line, according to the Sanitary and Topographical Map (Viele 1865). The southern half of the site was marsh and the northern half of the site was meadow. Marsh deposits in the vicinity of the site generally consisted of dark brown peat and organic clayey silt with varying amounts of sand.

The marsh deposits are underlain by glacial deposits. The New York metropolitan area was covered by the great continental glaciers during the Pleistocene epoch. During glaciations, ice transported and redeposited a layer of glacial till which generally consisted of unsorted mixture of sand, gravel and clay with zones of cobble and boulders. These materials vary widely in their texture, plasticity and engineering properties.

The project site is underlain by the Manhattan Formation, which predominantly consists of dark gray, medium to coarse-grained muscovite-biotite (mica) schist with localized concentrations of garnet and intrusions of coarse-grained granitic pegmatite.

SUBSURFACE INVESTIGATION

The subsurface investigation consisted of nine test borings that were advanced using rotary drilling techniques with steel casing and drilling mud used for soil support. Several of the borings extended into bedrock, which was cored with an NX-sized double-tube core barrel with a diamond cutting bit. Several completed borings were converted into observation wells to determine the groundwater levels. A boring and site location plan is shown in Figure 1.

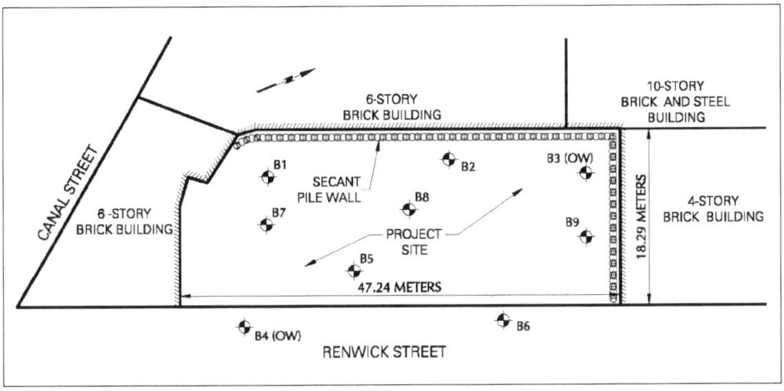

FIG. 1. Boring and Site Location Plan

Subsurface soil units consisted of uncontrolled fill and organic materials extending 6 meters below the grade, 4.5 to 6 meters of soft to stiff, grey to brown silt with varying amounts of sand and clay, followed by 12 to 18 meters of sand and gravel, and finally bedrock. The bedrock was characterized as dark grey, slightly weathered mica schist. Joints were fresh to slightly weathered. The depth to bedrock ranged from 25 to 37 meters below the grade. A representative subsurface profile is presented in Figure 2. The depth to groundwater ranged from 2.3 to 2.6 meters below the existing grade.

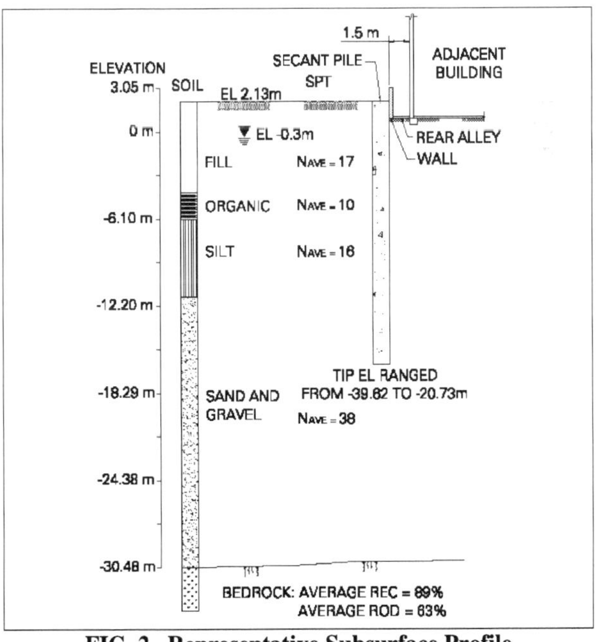

FIG. 2. Representative Subsurface Profile

SECANT PILE WALL ANALYSES AND DESIGN

Excavations extending seven meters below grade in some areas for the basement construction required an excavation support system along Renwick Street and potential underpinning of the adjacent buildings.

Test pits were completed to identify the foundation configuration of adjacent buildings and to determine underpinning requirements for constructing the new building's basement.

- On the north, a four-story brick building was supported on a rubble-foundation wall bearing on strip footing.

- On the northwest, a 10-story brick and steel renovated building supported on drilled auger cast-in-place piles did not require underpinning.

- On the south, a six-story brick building's foundation level extended to the new building's foundation level; therefore this building did not require underpinning.

- On the west, foundation conditions for 6-story buildings could not be determined because of a 1.5-meter wide rear alley. This alley precluded the use of an underpinning system such as contiguous concrete piers or bracket piles.

Since the 1.5-meter wide rear alley precluded an underpinning system such as contiguous concrete piers or bracket piles for the western building, the project team considered a secant pile wall. The secant pile wall, as a stiff and relatively impervious support system, eliminated the need for underpinning and dewatering. The secant pile wall was extended along the project site's north boundary to eliminate the need to underpin the adjacent northern building. The excavation support system along Renwick Street consisted of soldier piles with steel sheeting to minimize water inflow to the site and to minimize the dewatering effort during foundation excavation.

A secant pile wall is essentially a line of overlapping drilled concrete piles. Piles that are drilled first and then filled with concrete are called primary piles. Piles that are drilled in between the primary piles are called secondary piles and these are generally filled with steel reinforcement. The overlapping of the piles creates a continuous, stiff and water-tight wall.

Design Analyses

A commercially available finite-element software program, RISA-2D, was used for the design analyses. The secant pile was modeled as a continuous beam with rigid supports. Rankine active soil pressures, water pressure, surcharge from the adjacent buildings and dynamic soil loadings (represented with an equivalent static loading) were considered. A line load of 60 kN/meter/story (2/tons/foot/story) was assumed for adjacent buildings. Since the secant pile wall served as the permanent foundation wall, the service column loads supported by the secant pile were also included in the loading schedule. A portion of the secant pile wall along the northern boundary supported the shear wall.

Two levels of lateral support were considered. The design analyses were performed for each stage of excavation. The secant pile wall section consisted of 76.2-centimeter (30 inches) diameter piles overlapping for 15.2 centimeters (6 inches) and a W18x175 steel section at every secondary pile (Figure 3). The concrete for primary and secondary piles was designed for a 28-day compressive strength of 41 MPa. Shear studs were added at the top and bottom of the core beam to act against the uplift loads.

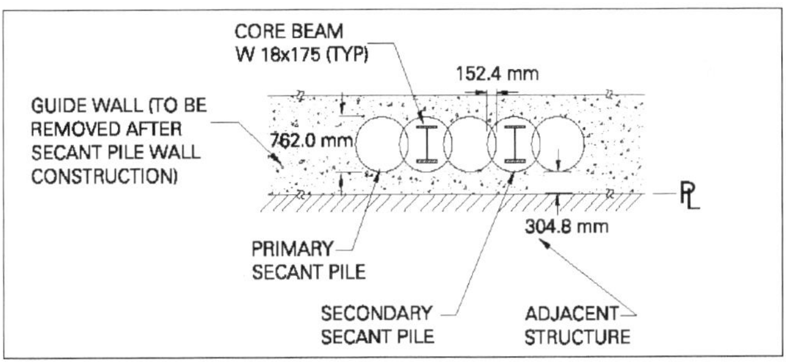

FIG. 3. Plan View of Secant Pile Wall

The axial capacity design of the secant pile wall to support the column loads was determined based on design methodologies recommended by FHWA (1999) and skin resistances known from local experience. Pile lengths ranged from 13.7 to 21.3 meters with the longer piles located under the shear wall. Piles were required to penetrate a minimum of 1.5 meters into the sand and gravel stratum.

Permanent Water Control

The cellar slab and below-grade walls were required to be fully waterproofed to 0.3 meters (1 foot) above the 100-year flood level. A membrane-type waterproofing, which is the commonly used waterproofing method in the New York City area, was recommended. Both bentonite waterproofing or negative side crystalline waterproofing are not generally used in the city for basement waterproofing below the water table.

Since the secant pile wall would be installed at the site boundary, installing the membrane waterproofing to the outside face (positive side) of the secant pile wall was not possible. Instead, a 20.3-centimeter (8-inch) thick interior foundation wall was designed to confine the waterproofing membrane. This interior wall also created a smooth finish for the basement.

With the addition of a 20.3-centimeter (8-inch) thick interior wall, the basement extended 1.27 meters within the property line which resulted in loss of space within the basement. The loss of space was a concern to the owner as property values in Manhattan are among the highest in the country.

SECANT PILE WALL CONSTRUCTION

Secant pile wall construction began with the construction of a cast-in-place reinforced concrete guide wall. The purpose of the guide wall was to maintain horizontal and vertical alignment of each pile. The guide wall, which was constructed approximately 30.5 centimeters from the adjacent property line, had a plan width of 1.4 meters and a depth of 1.2 meters.

Sequence of Pile Installation

The secant pile wall was constructed by drilling primary and secondary piles in the sequence shown in the Figure 4. Phase I primary piles (A1, A2 and A3) were first drilled and concreted. After the concrete in Phase I piles had set, Phase II primary piles (B1 and B2) were drilled and concreted. This was followed by drilling the secondary piles (C1, C2, C3 and C4) before the concrete at primary piles achieved full strength. The secondary piles cut into and overlapped the adjacent primary piles. Steel core beams were installed in all secondary piles and secondary piles were filled with concrete.

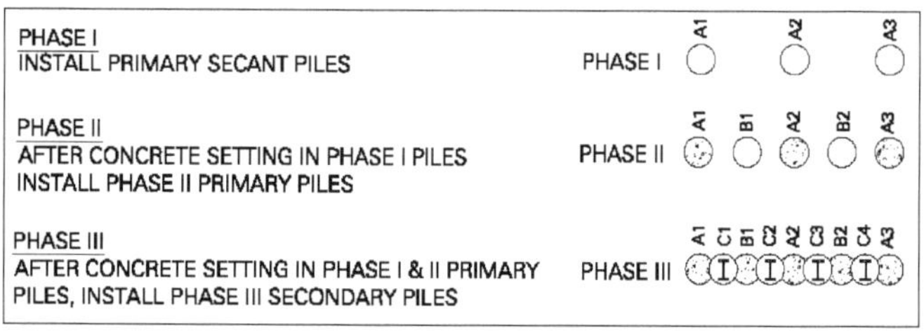

FIG. 4. Construction sequence of secant piles

Drilling of the Piles

Secant piles were drilled with a Bauer BG 24H double rotary drive drill rig with a Kelly bar extension. The double rotary system consisted of two independent motor drives, an upper rotary drive connected to an auger, and a lower rotary drive connected to the casing. The two drives rotated opposite one another and permitted removal of the auger and soil in a continuous pass while leaving a cased hole.

The 76.2-centimeter outside-diameter steel casing (with a wall thickness of 3.8 centimeters) was rotated and pushed into the ground while soil from within the casing was simultaneously removed with a soil auger or a soil bucket. The casing provided borehole stability. The borehole was advanced in one continuous process by adding

3-meter or 4-meter sections to the required termination depths. Pile termination depths ranged from 10.7 to 19.8 meters below the groundwater table; therefore, the casing was kept filled with water during drilling to maintain a positive head. The casing was usually advanced about 0.6 meters beyond the termination depth of the secant pile to form a plug at the bottom of the casing. This plug was formed to maintain stability of the pile base. The bottom of the pile was cleaned with a soil bucket before concreting.

Installation of the Core Beam

When a secondary secant pile was cased and drilled to the design pile-tip elevation, the steel core-beam was installed into the casing using a heavy-duty crane. Steel rods were welded to the core beam to serve as centralizers.

The contractor initially attempted to install the core beam after concreting the borehole. However, several feet of the core beam did not sink in the concrete under its own weight. Therefore, the contractor used a vibratory hammer to lower the core beams in place. However, this practice was abandoned after completion of the first three piles in order to prevent potential adverse effects of vibration to adjacent buildings. The holes were concreted after the placement of the core beam.

Concreting of the Borehole

With the core beam installed to the design pile-tip elevation, a continuous water-tight 0.25 meter diameter tremie pipe formed by attaching 3-meter pipe sections and a top-mounted hopper was inserted into the space between the steel core-beam and the casing's internal wall. The tremie pipe was suspended about 0.3 to 0.4 meters above the pile-tip elevation and a plug was placed in the tremie pipe. The plug was used as a stopper to prevent soil and water from getting into the tremie pipe during the concrete pour. Concrete was then poured continuously into the hopper from a 1.5-cubic-meter bucket. The discharge end of the tremie pipe was kept submerged in the fresh concrete and a head of concrete was always maintained in the tremie pipe above the groundwater level. The casing sections were withdrawn as the concrete pour progressed.

Settlement and Vibration Monitoring Program

The adjacent buildings were monitored for vertical and horizontal movements throughout the secant pile wall construction. The monitoring program consisted of optical survey readings of monitoring points installed on the facades of the adjacent buildings. In addition, crack gauges were installed on several cracks at the adjacent buildings. Settlements up to 3.0 cm and horizontal movements up to 2.0 cm were observed for the northern adjacent building. The movements at the western adjacent buildings were smaller than 1.5 cm. The higher movements at the northern building were attributed to the fact that this building was a more fragile building with rubble wall foundation and was within 30 cm of the secant pile wall construction. The western buildings were more robust and were separated from the secant pile wall by the 1.5-meter wide rear alley.

Vibration monitoring was performed during the course of secant pile wall construction. Multiseis type seismographs were placed in the basement of adjacent buildings. The vibration levels were well below the city limit of peak particle velocity of 1.3 centimeters per second (0.5 inches per second) for construction activities adjacent to historic structures (Esrig and Ciancia, 1981).

CONCLUSIONS

1. The 58-meter long secant pile wall was completed in less than five months. A total of 96 secant piles with a total length of 1,303 meters were drilled. The secant pile served as a cutoff wall, a soil retention system, and the permanent foundation wall, eliminating the need for continuous site dewatering and underpinning of adjacent buildings.

2. The 76.2-centimeter diameter secant pile wall installed 30.5 centimeters away from the property line and the 20.3-centimeter thick interior wall resulted in loss of space within the basement. The loss of space was a concern to the owner as property values in Manhattan are among the highest in the country.

3. Settlements up to 3.0 cm and horizontal movements up to 2.0 cm were observed for the northern adjacent building. The movements at the western adjacent buildings were smaller than 1.5 cm. The higher movements at the northern building were attributed to the fact that this building was a more fragile building with rubble wall foundation and was within 30 cm of the secant pile wall construction. The western buildings were more robust and separated from the secant pile wall by the 1.5-meter wide rear alley.

4. The design team concluded that the secant pile wall construction even drilled with casings can significantly impact adjacent structures. The use of bentonite slurry as a drilling agent to minimize base heave of the borehole will reduce impact to adjacent structures. However, use of bentonite slurry will result in smaller skin resistance warranting longer piles.

REFERENCES

Esrig, M. I. and Ciancia A. J. (1981). "The Avoidance of Damage to Historic Structures Resulting from Adjacent Construction" *ASCE Preprint 81-052*, New York, NY, May 11th-15th.

Federal Highway Administration (1999). "Drilled Shafts: Construction Procedures and Design Methods." *Publication No. FHWA-IF-99-025*

Viele, E. L. (1865). "Sanitary and topographical map of the city of New York."

SOIL NAIL WALLS SUPPORTING STUB ABUTMENTS ON SHALLOW FOUNDATIONS – A CASE STUDY

Ragui Wilson-Fahmy[1], P.E., Ph.D., Member, ASCE, Geo-Institute; Kuangyu Yang[2], P.E., Member, ASCE, Geo-Institute; Kwang Ro[3], P.E., Ph.D., Member, ASCE, Geo-Institute; Sherif Hanna[4], Associate Member, ASCE

ABSTRACT: The Route 47 Improvements Project in Cumberland County, New Jersey involved widening Route 47 at the interchange with Route 55. The north and south bounds of Route 55 pass over Route 47 through two bridges supported on stub abutments. Three meter high soil nail walls were constructed in front of the abutments in order to accommodate the widening at the Route 55 bridge structures. The abutments of the Route 55 bridges are founded on shallow foundations and rest on top of a 1.0V:1.5H slope. Design of the soil nail walls was conducted using a limiting equilibrium approach. The finite element method was also used to estimate the deformations of the bridge abutments under the various stages of construction of the wall. Based on the finite element analysis, it was concluded that installing a system of tiebacks penetrating the abutment stem and tying it to a stable soil zone far from the back of the abutment would considerably reduce the amount of abutment deformation. The soil nail walls were instrumented to assess the stability of the system during and after construction. The instrumentation consisted of strain gages attached to selected soil nails for measuring the distribution of tension force along the nails and tiltmeters and optical survey points mounted at selected locations on the wall and abutment to detect any rotation and movement. The instrumentation data indicated that the system was stable and that the abutment and wall deformation were minimal and within acceptable limits. Design, construction and performance of the soil nail walls are discussed in this paper.

---

[1] Technical Leader, Geotechnical Department, PB Americas, Inc., 506 Carnegie Center Blvd., Princeton, NJ 08540, fahmy@pbworld.com

[2] Manager, Geotechnical Unit, New Jersey Department of Transportation, 1035 Parkway Avenue, Trenton, NJ 08625, Kuangyu.Yang@dot.state.nj.us

[3] Department Head, Geotechnical Department, PB Americas, Inc., 506 Carnegie Center Blvd., Princeton, NJ 08540, ro@pbworld.com

[4] Senior Geotechnical Engineer, PB Americas, Inc., One Penn Plaza, NY, Ny 08540, hannas@pbworld.com

INTRODUCTION

The Route 47 Improvements Project located in Cumberland County, New Jersey involved widening Route 47 at the interchange with Route 55. The north and south bounds of Route 55 pass over Route 47 through two bridges supported on stub abutments, Figure 1. In order to accommodate the widening at the Route 55 bridge structures, two three meter high retaining walls were required in front of the abutments on both sides of Route 47. A soil nail approach was selected as it was considered the least disruptive method to construct the walls. The abutments of the Route 55 bridges are founded on shallow foundations and rest on the top of a 1.0V:1.5H slope which makes them susceptible to movement during construction of the soil nail walls. This in addition to the fact that the edge of the abutment footing would be at 1.2 meter behind the wall and that the maximum bearing pressure at the toe of the abutment was over 260 kN/m^2 (5000 psf) necessitated special considerations in the design and construction of the walls.

Design of the soil nail walls was based on limiting equilibrium techniques. However, the finite element method was used to estimate the deformations of the bridge abutments at the various stages of construction. Based on this analysis, it was concluded that installing a system of tiebacks penetrating the abutment stem would considerably reduce the amount of abutment deformation. Accordingly, tiebacks were installed prior to construction of the soil nail walls.

The soil nail walls were instrumented to assess the stability of the system during and after construction. The instrumentation consisted of strain gages attached to selected soil nails and tiltmeters and optical survey points mounted at selected locations on the walls and abutments.

Details of the limit equilibrium and finite element analyses will be presented in the paper. Comparison between measured and predicted stresses and deformation will also be made.

SUBSURFACE CONDITIONS

Five borings were drilled for each of the two soil nail walls from the top of the slopes supporting the bridge abutments, Figure 1. The depth of borings varied between 12.5 to 18.5 meters. The soil to be retained by the soil nail walls consists in general of coarse to fine sand with a fine content of 7 to 17 percent. The underlying soil can be described as fine to fine to coarse sand with varying amounts of silt. Standard penetration tests (SPT) indicate in general that the SPT-N values exceed 40 from the top to the bottom of borings with an average value of 56 indicating dense to very dense soil conditions. Water level readings indicated that the groundwater table is below the bottom of the finished ground surface in front of the soil nail walls by about 5 meters.

DESIGN METHOD

Design of the soil nail walls was based on the slip surface limit equilibrium technique and applied using the computer program GOLDNAIL developed by Golder Associates. The design concepts and methodology adopted by GOLDNAIL can be found in the FHWA publication No. FHWA-SA-96-069 (1992) and GOLDNAIL Reference Manual (1996).

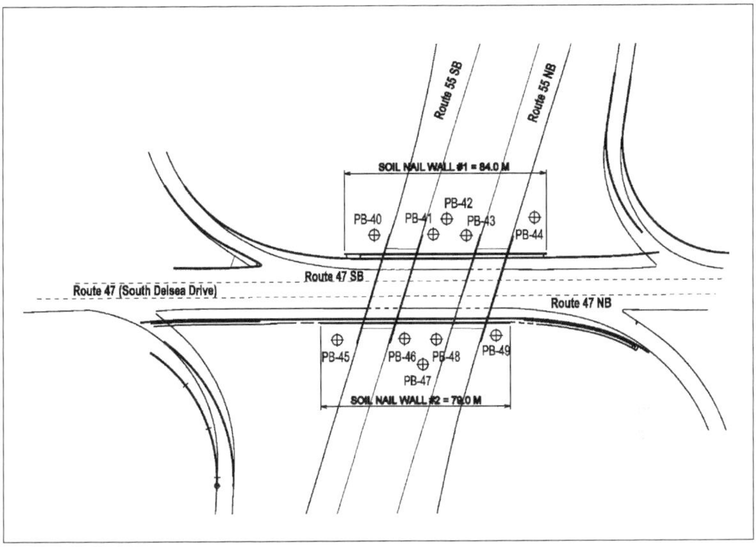

FIG. 1 LOCATION OF SOIL NAIL WALLS AND BORING LOCATIONS

GOLDNAIL allows consideration of factored strengths when the Load and Resistance Factor Design (LRFD) method is selected and safety factors when the Allowable Stress Design (ASD) is chosen. The program supports both analysis and design modes. In the design mode, the program provides the minimum required nail lengths and nail strengths for a given nail pattern. In the analysis mode, the program calculates either the resistance/load ratio (LRFD) or the global factor of safety (ASD) depending on which method is used. The nail service loads can be estimated using the ASD method by setting the soil strength factors of safety equal to unity.

SOIL AND NAIL DESIGN PARAMETERS

As already mentioned, the soil behind the walls was found to be in a dense to very dense state. The soil friction angle was assumed to be in the range of 38 to 40 degrees and the unit weight was assumed equal to 22 kN/m^3 for wall design. The nominal nail head strength was determined to be 170 kN. The selected tendon was 26 mm diameter threaded bar of a nominal tensile strength of 567 kN.

LOADING CONDITIONS

The soil nail wall was designed according to the LRFD method. Bearing pressures and horizontal forces acting at the level of the abutment footings are summarized in Table 1. A surcharge loading of 12 kN/m^2 was applied behind the abutments to simulate traffic loading.

Table 1: Soil nail Wall Abutment Design Loads

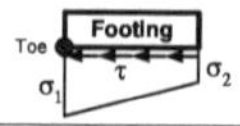	**Vertical (σ_1) @ Toe KN/m^2**	**Vertical (σ_2) @ Heel KN/m^2**	**Horizontal (τ) KN/m^2**
Service I	260	50	40
Strength I	355	75	50
Strength II	360	110	50
Strength III	250	40	40
Strength V	330	65	50
Extreme Event I	325	20	55

SOIL NAIL WALL CONFIGURATION

Various computer runs were performed for both permanent and temporary construction phases in order to obtain the optimum nail configuration. Three rows of nails spaced at 1.0 meter vertically and 1.2 meters horizontally were required to support the retained soil and the overlying abutment. Between and outside the abutment areas the horizontal spacing was increased to 1.5 meters. The nail inclination was selected at 15 degrees to the horizontal direction. The required nail length was 10 meters for the top two rows of nails and 5 meters for the bottom row. Slightly longer nails were actually used in the wall construction. Figure 2 shows the nail configuration in an elevation view while Figure 3 shows a cross-sectional view.

WALL DEFORMATION

Several requirements were included in the contract drawings and project specfications to limit the wall deformation and to minimize the risk of wall instability which included the following: (i) Three tiebacks to be installed through each abutment stem prior to construction of any portion of the soil nail wall, (ii) Pretensioning the top nail rows under the abutments to a lock-off load equal to 25 percent of the design service load, and (iii) Soil excavation, shotcreting and installation of nails to be performed longitudinally in segments not exceeding 15 feet in length. Excavation of any segment would be allowed only after the nail grout and shotcrete have reached the specified 3-day compressive strength. Pretensioning nails was previously recommended by Tuozzolo (1997) for supporting a sensitive structure. A PVC bond breaker was used over a length of 0.5 meters for production nails subjected to lock-off load and one meter for proof test nails to avoid overstressing the shotcrete facing during load application.

The effect of tiebacks on wall deformations was investigated using the finite element program PLAXIS (1996). Two analyses were performed, one without and one with tiebacks. The analyses were carried out in stages according to the following sequence: (i) Construction of the embankment which supports the stub abutment, (ii) Construction of the stub abutment, (iii) Backfilling behind the abutment, (iv) Applying the bridge forces acting at the abutment seat, (v) Applying the prestressing forces to the abutment tiebacks and (vi) Constructing the soil nail wall in stages as shown in Figure 3 until the full excavation is complete. The elements representing the tiebacks and soil nails were activated during the analysis at the relevant construction stages.

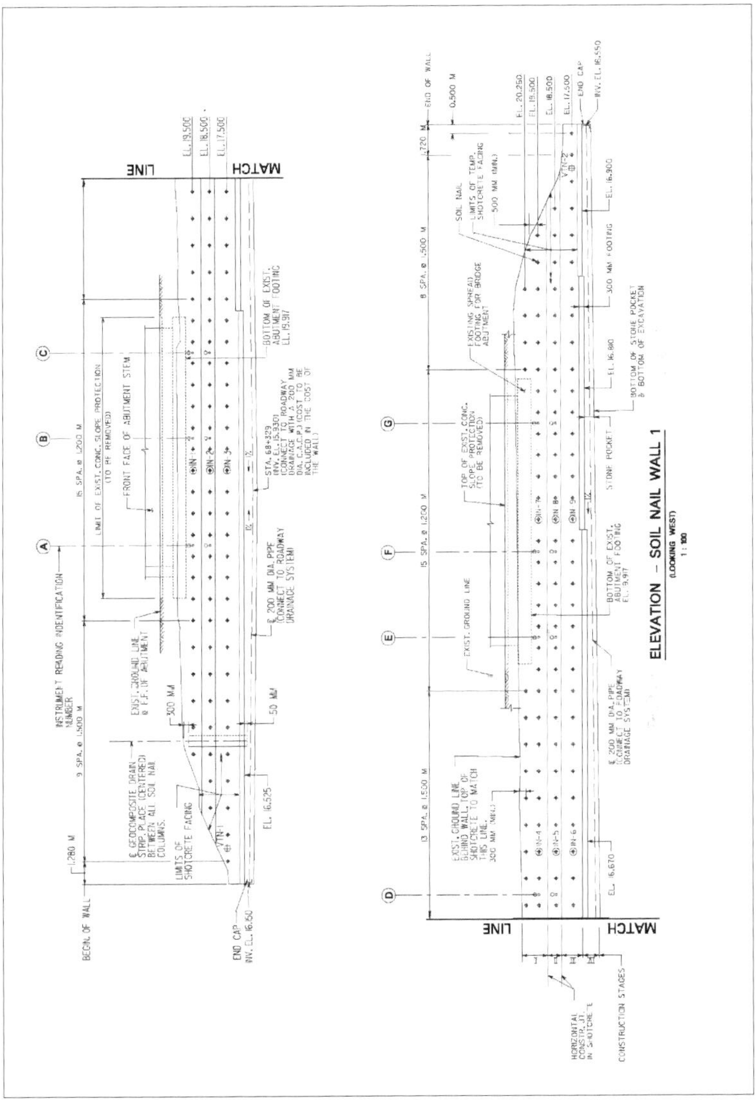

FIG. 2 ELEVATION VIEW OF SOIL NAIL WALL 1

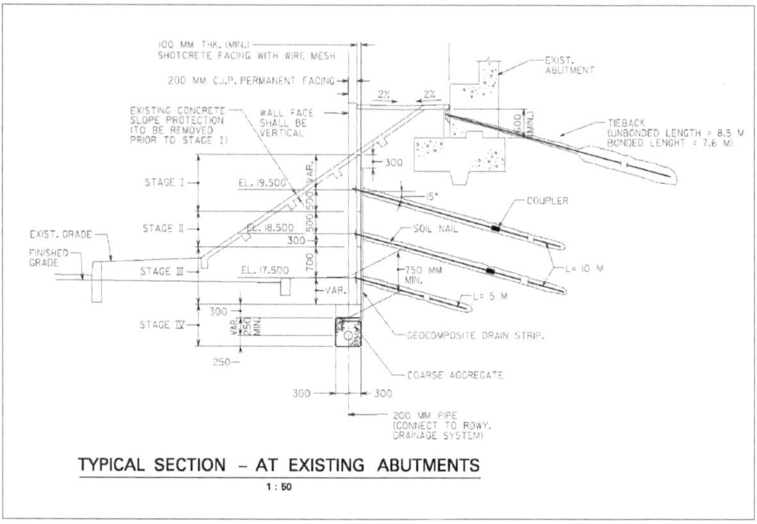

Tieback Tendon:Three 7-wire strands encapsulated in a 50 mm corrugated HDPE tubing and installed in a 150 mm grouted hole. Service and lock-off loads: 335 and 310 kN respectively.

FIG. 3 CROSS-SECTIONAL VIEW AND CONSTRUCTION STAGES

All the system components with the exception of the soil elements were modeled assuming linear elastic response. The soil behavior was simulated assuming a Hardening Soil Model. The parameters describing the soil model were obtained from published data for soils similar in gradation and relative density to the soil used in the construction of the embankment. Predicted vertical and lateral displacements at the abutment seat are given in Table 2. These displacements correspond to Steps (v) and (vi). The results indicate a substantial reduction of lateral and vertical displacements if tiebacks are installed. Analyses were also conducted to investigate the sensitivity of the results to the assumed soil strength. It was found that relatively small variations in soil strength resulted in relatively large variations in the predicted deformations for the case without tiebacks and negligible effect for the case with tiebacks. This indicates that a more stable system can be constructed by adding tiebacks. The axial force distribution in the three rows of nails expressed as force per unit wall length is given in Figure 4 for the case with and without tiebacks. The figure also provides the service loads predicted by GOLDNAIL assuming no tiebacks. It can be seen that the maximum loads predicted by PLAXIS and GOLDNAIL are comparable for the case without tiebacks and the predicted axial forces with tiebacks are substantially smaller than without tiebacks.

Table 2 – Predicted Abutment Displacements With and Without Tiebacks

Case	Lateral Displacement	Vertical Displacement
No Tiebacks	18 mm (0.71 in)	7 mm (0.28 in)
With Tiebacks	3 mm (0.12 in)	4 mm (0.16 in)

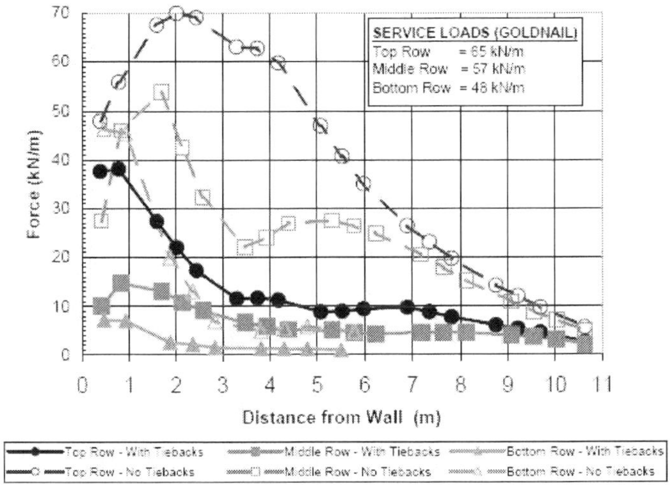

FIG.4 PREDICTED AXIAL NAIL FORCES WITH AND WITHOUT TIEBACKS

INSTRUMENTATION

At the middle section of each abutment, soil nails were instrumented with strain gages. Optical survey points and tiltmeters were also installed at different locations along the wall. Since periodic instrumentation monitoring did not indicate any signs of instability such as excessive wall movement or high nail forces, wall construction proceeded without interruption.

Figure 5 shows the predicted axial force distribution as well as the axial forces obtained by back analysis of the strain gage data for the soil nail walls at the four abutments. It can be seen that although the predicted and measured data are comparable in magnitude, the forces based on strain gauge data are generally higher than those predicted. However, by comparison with Figure 4, nail forces estimated based on strain gauge data are considerably lower than those predicted using both the finite element and limit equilibrium methods for the case with no tiebacks. It may be noted that in calculating the nail forces from strain gage data, grout cracking was assumed to occur and hence no tensile force was considered to be taken by the grout. Following the methodology developed by Banerjee et al (1998) to account for grout contribution, nail forces less than about 8.7 kN/m in Figure 5 could be somewhat underestimated due to the fact that grout cracking may not occur at low stress levels.

Displacements of the optical survey points could not be accurately determined by the surveying operation due to the fact that the displacements were small and the readings were erratic fluctuating in opposing directions from one reading to the other. The tiltmeter readings showed a wall inclination varying between +0.05 and -0.05 degrees. For a 3 meter high wall and assuming an average rotation of 0.05 degrees, this translates into a maximum movement of about 2.6 millimeters which was within anticipated values (see Table 2).

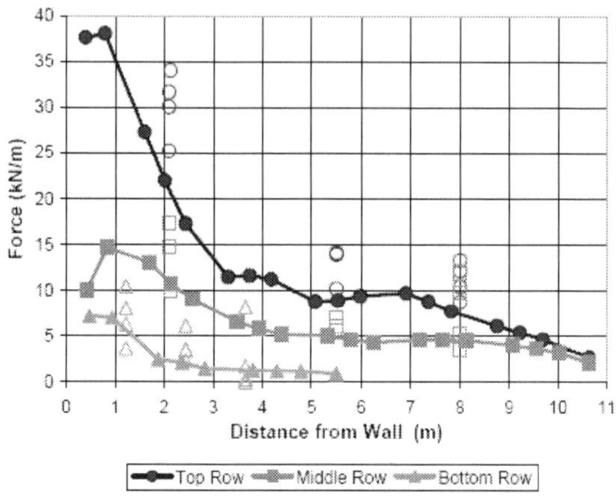

FIG. 5 VARIATION OF PREDICTED AND BACK-CALCULATED AXIAL NAIL FORCES (Hollow circles, squares and triangles are based on measurements)

CONCLUSIONS

- Limit equilibrium design methods provide a safe design of soil nail walls
- Finite element method provides a useful tool for deformation and stress analysis
- The use of tiebacks reduces the magnitude of nail forces and deformations according to the finite element analysis
- Instrumentation provides a useful tool for monitoring complex conditions and should be analyzed periodically during wall construction to detect any signs of instability.

REFERENCES

Federal Highway Administration (FHWA) (1996), "Manual for Design & Construction Monitoring of Soil Nail Walls", *Publication No. FHWA-SA-96-069.*

GOLDNAIL (1996), "A Stability Analysis Computer Program for Soil Nail Wall Design, Reference Manual, Version 3.11, *prepared by Golder Associates.*

Plaxis (1998), "Finite Element Code for Soil and Rock Analyses", *University of Delft.*

Banerjee, S., Finney, A., Wentworth, T. and Bahiradhan, M. (1998), "Evaluation of Design Methodologies for Soil Nail Walls, Volume 1", Report No. WA-RD-371.1, *Washington State Transportation Center (TRAC), University of Washington*, Seattle, Washington.

Tuozzolo, T.J. (1997), "Stabilizing the Stacks", *Civil Engineering*.

SOIL NAILING EARTH RETENTION ON THE PARK LAFAYETTE DEVELOPMENT IN MILWAUKEE, WISCONSIN

Eric W. Bahner, P.E., M.ASCE[1] and Mike Christensen[2]

Abstract: The Park Lafayette development currently under construction in Milwaukee, Wisconsin consists of two 20-story towers supported on a combination of spread footings and drilled shafts. The development includes four levels of underground parking, requiring a 14m (45 ft) deep excavation immediately adjacent to city streets and Milwaukee County property. The presence of the adjacent County property required an unprecedented level of pre-construction planning by the Developer, Construction Manager Hunzinger Construction Company, and Specialty Contractor Edward E. Gillen Company. The small site footprint required that the earth retention system be installed such that the retention system would also serve as the backside form for the foundation walls. The 13 corners in the foundation footprint complicated this necessity. Gillen and Hunzinger determined that soil nailing would be the most cost effective method to retain the excavation. To avoid conflicts with new foundation walls, the excavation footprint was oversized 150 mm (6 inches) on each side to allow for irregularities in the excavation and shotcrete face and lateral movement of the earth retention system. Construction of the soil nail wall proceeded through the winter of 2006 to 2007 with minimal delay. Optical survey monitoring by the construction manager indicated that the 150 mm (6-inch) wall offset was sufficient to allow foundation wall construction to proceed with modifications to neither the retention system nor the foundation walls.

INTRODUCTION

The latest addition to the City of Milwaukee's skyline is the Park Lafayette condominium development on the city's east side. Dense urban development and severely limited/restricted parking in the area required that the Developer incorporate substantial underground parking into the design to accommodate the needs of the businesses and private residents. To address the earth retention needs for this project, the authors' companies worked with the Developer to secure the necessary permits/easements to allow construction, and then designed an earth retention system to safely and cost effectively retain the excavation. The principal author and his firm successfully designed and constructed a temporary soil nail wall to retain the excavation. To minimize excavation backfilling costs and expedite construction of the underground parking levels, the soil nail wall was designed to incorporate the

[1] Chief Engineer, Edward E. Gillen Company, 218 West Becher Street, Milwaukee, WI, USA, 53207, Tel: 414-769-3120, Email: eric.bahner@gillenco.com

[2] Sr. Project Manager, Hunzinger Construction Company, 21100 Enterprise Avenue, Brookfield, WI, USA, 53045, Tel: (262)-797-0797, Email: mchristensen@hunzinger.com

construction of the perimeter drainage system as the excavation was advanced, and the new foundation walls were formed and poured against the shotcrete facing of the soil nail walls.

PROJECT DESCRIPTION

Work on the project proceeded in the late summer/fall of 2006 upon the successful completion of negotiations with the Milwaukee County Parks Department [Parks Department]. The site is located at the northeast corner of North Prospect Avenue and East Lafayette Place.

The site is triangular shaped and encompasses slightly more than 1 acre. The west leg of the site is bounded by major east side arterial North Prospect Avenue, and the south leg of the triangle, by East Lafayette Place. The hypotenuse of the triangle runs parallel to land owned by the Parks Department. The site was relatively flat and sloped down to the North. Beyond the east property line, the terrain slopes down sharply to a former railroad grade that now serves as a bike trail.

The development features two 20-story towers and four levels of underground parking extending to depths of 12 to 14 m (40 to 45 ft). Refer to Figures 1 and 2. The majority of the development is supported by a combination of spread and combined footings at the base of the parking structure excavation, while the south tower area is unexcavated and supported on belled caissons.

SUBSURFACE CONDITIONS

The subsurface profile consists of approximately 2 to 4 m (8 to 10 ft) of miscellaneous urban fill underlain by stiff to hard clay. Very dense/hard till deposits (hardpan) are present below the clay beginning at depths of 12 m (40 ft) and greater.

Occasional zones of buried rubble containing perched water were noted on the south end of the site. Occasional boulders were noted in the profile at the clay/hardpan interface.

Groundwater was observed at depths of 8 to 18 m (26 to over 60 ft) below original site grades up to 14 days after completion of the subsurface exploration for the project.

RETENTION SYSTEM DESIGN

The design process began with the development of preliminary drawings and an extremely complicated negotiation process with the Milwaukee County Parks Department to secure a temporary easement to allow soil nails to encroach onto Milwaukee County parklands. Permit drawings were completed in April 2006 and approval was received from the Parks Department later that summer.

Based on our experience, the very stiff to hard clays and hard/extremely dense hardpan are well suited to soil nailing, and typically demonstrate very little lateral movement while in service. The soil nail walls were designed using a Braced Cut approach as described by Nicholson (1986) and Bahner (1994, 2005), and checked using the CALTRANS

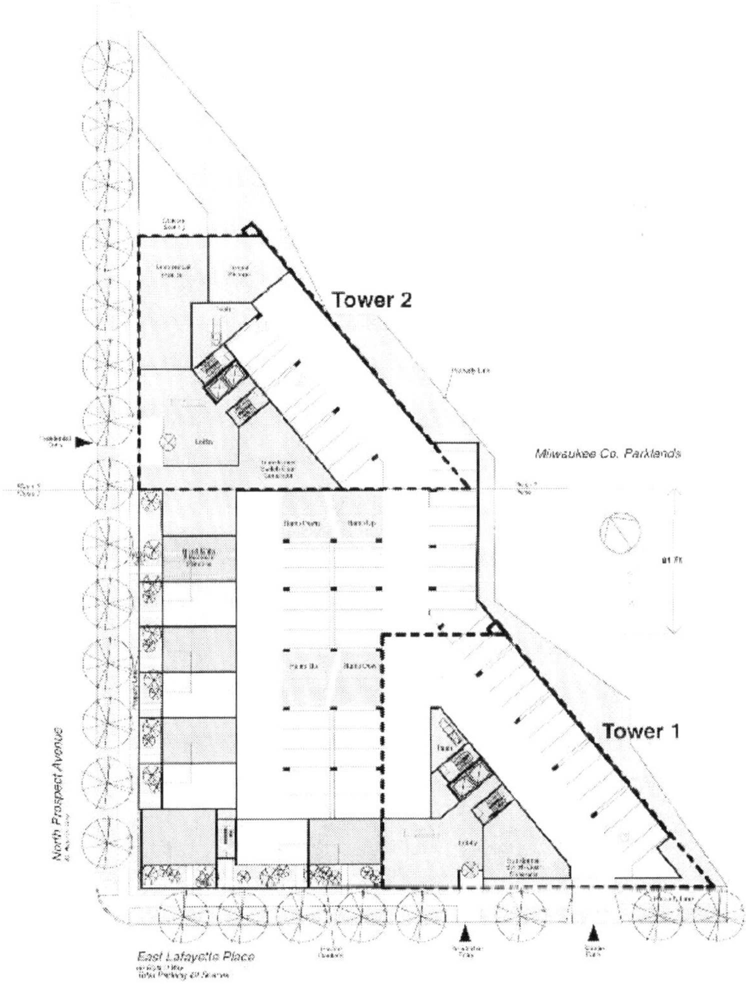

Figure 1
Development Layout

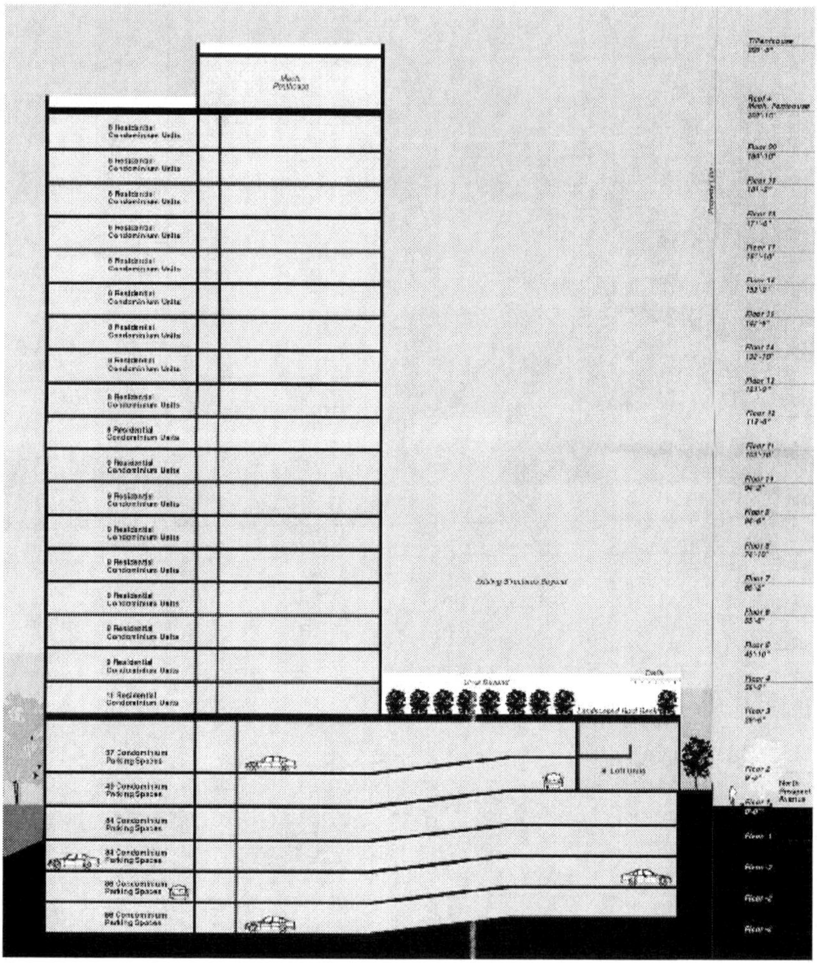

Figure 2
Typical Building Section

computer program SNAILZ. Both approaches provided similar results. The nails consisted of 200 mm (8 in) diameter drilled shafts with grouted in place steel bar tendons. The nails were installed in a 1.5 m (5 ft) vertical by 1.8 m (6 ft) horizontal grid pattern as the excavation was advanced. Nail lengths were determined based on the expected soil pullout resistance, which was confirmed during construction, and estimated soil and surcharge loads. In general, the nail lengths typically ranged between 0.6 and 0.8 times the total height of the excavation. Many examples of typical wall cross-sections and shotcrete facing details are presented in the literature. A typical soil nail wall cross section is provided in Figure 3 below.

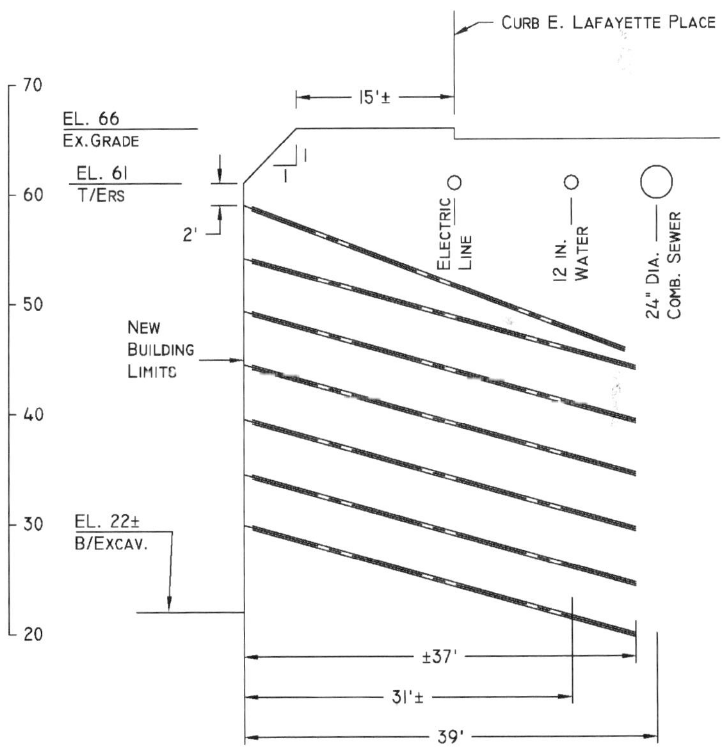

Figure 3
Typical Soil Nail Wall Cross-Section

Where sufficient space was available, the upper 1.5 to 3 m (5 to 10 ft) of the excavation was open cut at a 1H:1V slope to minimize the quantity of earth retention necessary. These slopes were protected with reinforced visqueen and wire mesh anchored to the slope. The visqueen is used to prevent seepage behind the shotcrete facing and promote runoff into the excavation. Based on the experience of the principal author, proper protection of open cut slopes above soil nail walls is critical to the overall long-term performance of the earth retention system. Consequently, the principal author and his company [the Specialty Contractor (SC)] requires that slope protection be designed and installed by their own field crews, or that this protection be installed to the specifications of the SC.

To reduce the costs of backfilling between the retention system and the new basement walls, the Construction Manager required that the earth retention system be positioned such that the front face of the retention system was flush with the back face of the new foundation wall. As a result, the retention system was constructed at, or very near, the property line, and included 13 corners. The excavation was oversized 150 mm (±6 in) to account for irregularities in the excavated face of the cut, irregularities in the shotcrete face and potential lateral deflection of the soil nail wall. Perimeter drainage was provided by tacking geocomposite drain board strips onto the excavation face between the soil nail columns before shotcrete application. The drain board strips were extended to the excavation base and connected to the sub-floor drainage system via bleeder pipes extending through the new foundation walls. A waterproofing membrane was applied to the face of the shotcrete before constructing the foundation walls. The one-sided forming system used to construct the new foundation walls was self-supporting and did not rely on the earth retention system for lateral support.

CONSTRUCTION PROCEDURES

As excavation proceeded, open cut slopes were systematically protected with visqueen and wire mesh as described above. Once excavation reached earth retention grade, excavation and construction of the soil nail walls proceeded in 1.5 m (5 ft) vertical lifts to excavation base grade. Many examples of a typical installation sequence are presented in the literature. To prevent overbreak of the excavated face, which can add to shotcreting costs, each successive lift was rough-cut and later fine trimmed after the nails were installed. Depending upon ground conditions, the nail holes were drilled with hollow stem augers and/or duplex drilling methods. Nail tendons were placed in the completed holes and gravity grouted by tremie methods. The nail design was confirmed by verification testing at the start of the work, and by proof testing 3 to 5 percent of the production nails. The shotcrete facing was constructed after nail installation was complete for each lift.

Since the majority of the work proceeded during the winter of 2006/2007 in Wisconsin, cold weather protection was essential to the quality of the completed wall. Insulating blankets were used to protect newly excavated lifts of the excavation, and newly placed shotcrete during initial curing. Work was stopped in the event of rainy or snowy weather, and when the temperature was -7° C (20° F) and falling to avoid problems with the quality of the constructed wall and the construction equipment. On many occasions, drilling was completed during a late evening/early morning shift so shotcrete work could be completed during warmer, daytime hours.

CONSTRUCTION MONITORING/WALL PERFORMANCE

During the course of mass excavation and retention system construction, the Construction Manager retained a surveyor to monitor the excavation using basic, optical survey methods. Survey measurements were taken at every shotcrete level as the excavation proceeded to base grade. The goal of the monitoring program was to confirm that the excavation and temporary retention system did not encroach into the footprint of the new building foundation walls. There was little budget to allow more elaborate monitoring of the retention system.

Though very simple, the monitoring program satisfactorily demonstrated that the excavation footprint was sufficiently oversized to allow earth retention construction and one-sided forming of the foundation walls. Although the field monitoring program did not provide a description of lateral movement of the excavation over time, it effectively conveyed information regarding the main concern of the Construction Manager: Was the excavation and earth retention system encroaching into the basement wall footprint?

CONCLUSIONS

From work on this project, and past experience with the design and construction of soil nail walls in similar ground, the Specialty Contractor concludes the following:

- A successful soil nailing project requires teamwork with a knowledgeable excavating contractor. Installation of the perimeter drainage system as the excavation proceeded and one-side forming of the foundation walls required a level of control not typically required for mass excavation.

- Accommodating the building perimeter drainage system in the soil nail wall design, and one-side forming the perimeter foundation walls can be effective in reducing costs associated with conventional drainage system installation and excavation backfilling, and compressing the construction schedule.

- With favorable subsurface conditions and proper care, soil nail construction can proceed in winter conditions.

ACKNOWLEDGEMENTS

The authors acknowledge the following parties involved on this project:

Renaissant Development Group

PappaGeorge/Haymes, Ltd., the Project Architect

Bowman, Barrett and Associates, the Project Structural Engineer

The field crews of Edward E. Gillen Company and Hunzinger Construction Company

Schneider Excavating

REFERENCES

Nicholson, Peter J. (1986). "Insitu Ground Reinforcement Techniques" presented at the International Conference on Deep Foundations, Beijing, China, September 1986. (Meeting Reprint).

Bahner, Eric W. (1994). "Soil Nailing in Madison Wisconsin", Proceedings—Woodward Clyde Shareholder's Conference. Phoenix, Arizona, October 1994.

Bahner, Eric W. (2005). "Soil Nailing: A Local Perspective". *Geotechnical Applications for Transportation Infrastructure*, Geotechnical Practice Publication No. 3, ASCE, pp. 28-39.

STEMMERS RUN STEEL SHEET PILE COFFERDAM AND TRESTLE ACROSS THE BACK RIVER

Jennifer Peirce Brandt[1], P.E., John J. Peirce[2], M. ASCE, P.E., and Frank M. Vibbert[3], M. ASCE, P.E.

[1]Vice President, Peirce Engineering, Inc., Phoenixville, PA 19460; Jennifer@PeirceEngineering.com
[2]Vice President, Peirce Engineering, Inc., Phoenixville, PA 19460; JPeirce@PeirceEngineering.com
[3] Senior Consultant, Peirce Engineering, Inc., Phoenixville, PA 19460; Frank@PeirceEngineering.com

ABSTRACT Project documents and permits for construction of the Stemmers Run Relief Wastewater Force Main across the Back River in Baltimore County, Maryland specified a minimal width, sheet pile cofferdam and pile supported, work trestle approximately 518 m long. The contractor combined the impractically narrow cofferdam and trestle structures to minimize wetland disturbance and maximize access. The resulting structure was a braced, sheet pile cofferdam with timber deck mats as the work trestle for excavation, pipe installation, and backfill. Construction of this cofferdam/trestle progressed from the north shoreline, toward the south shoreline, to the two-thirds point of the structure. Then, working northward, the contractor excavated, installed the 1372 mm diameter pipe and support bents and backfilled the trench in increments of approximately 18.3 m. As the pipe installation progressed toward the north, the sheet piling and timber deck were removed. As sufficient materials were removed, the contractor began the process anew from the south shoreline using the reclaimed materials. Deep, very soft silts and organic clays and an unbalanced, hydrostatic head complicated the project. Concurrent pile driving, excavating, and backfilling in successive cofferdam sections required large backhoes and crawler cranes which controlled the design of the trestle and supporting sheet piles.

BACKGROUND

As part of a court injunction and consent decree between United States Environmental Protection Agency (EPA) and the State of Maryland for violation of the Clean Water Act, a design was completed by Rummel, Klepper, and Kahl, LLP (RKK) of Baltimore, Maryland for construction of approximately 4023 m of 1372 mm diameter force main running southwest from the proposed Stemmers Run Relief Wastewater Pump Station at Golden Ring Road to an existing force main junction chamber at the Back River Wastewater Treatment Facility on Eastern Avenue in the County of Baltimore, Maryland. In August 2005, the contract was awarded to American Infrastructure-Maryland (AIMD), a division of American Infrastructure, which is headquartered in Worcester, Pennsylvania. This project was valued at approximately $26.4 million dollars. AIMD began work in October 2005.

The work being discussed in this paper is the construction of a pile supported, 1372 mm diameter, sanitary sewer force main extending approximately 518 m across the Back River, an estuary to the Chesapeake Bay (Figure 1). This river crossing is east of I-695 over the Back River, just north of the I-695 interchange with Route 150, Eastern Avenue. The construction cost for this river crossing was approximately $7.9 million dollars. The force main was constructed using cut and cover methods to a depth 4.88 m and 5.64 m below mean high water.

FIG. 1. Aerial view of the river crossing site, east of I-695 bridge.

The original RKK design concept for the project was to excavate, lay, and backfill the pipe between two parallel rows of steel sheet piling (SSP). The work would take place from a parallel, pile supported, timber deck, work trestle about 3.66 m wide.

AIMD decided to investigate the elimination of the separate work trestle. Their proposed concept was to perform the entire river crossing work from a timber deck spanning between and supported by SSP. AIMD approached Peirce Engineering, Inc. (PE Inc.) of Phoenixville, Pennsylvania to investigate their concept and provide design calculations and drawings. AIMD's alternate work concept required a permit revision by the U.S. Army Corps of Engineers. The concept proved feasible. PE Inc.'s design submission was approved as the method of construction.

Advantages of the alternate cofferdam/trestle were: 1) it eliminated one of the structures needed to cross the river; 2) it allowed for an 8.53 m wide work platform rather than the 3.66 m wide platform; 3) it allowed excavation, pipe placement, and backfill to be performed directly in front of the equipment and in line with the cofferdam and pipeline rather than from the side; 4) it improved visibility, safety, and productivity; and 5) minimized wetland disturbance along the estuary.

DESIGN

The design for the cofferdam used the Shoring Suite Plus, Version 7.3, computer program from CivilTech Software. Lateral earth pressures were calculated using CivilTech's EPres software based on input of the expected soil properties. These calculated earth pressures were then imported to CivilTech's Shoring software to compute the required SSP section and the wale and brace loads.

The design was for a single tier, braced system with a triangular earth pressure distribution due to the very soft, retained soils (Figure 2). The required depth of SSP toe embedment was calculated by the CivilTech software. The upper level brace at Elevation +1.22 m was not considered in the SSP design but was incorporated into the deck construction to maintain deck alignment and provide deck stiffness.

Ground conditions through which the pipeline was to be constructed consisted of a layer of dark gray, very soft, silt/clay mixture. The silt/clay layer was overlain in some areas by a mass of organic material, often referred to as meadow mat. This weight-of-hammer material extended to as much as 9.14 m deep. Below the very soft silt/clay were layers of medium dense to dense sand and gravel, 1.52 to 3.05 m thick, (N = 32 to 50 blows per 305 mm) over very stiff silty clay (N = 50 blows per 305 mm to 100 blows per 127 mm). During installation, the SSP would typically sink under its own weight for 4.57 to 7.62 m and would then be vibrated an additional 3.05 to 4.57 m with an ICE Model 22-30 vibratory pile hammer. The SSP was then driven to its depth with an ICE Model I-19 diesel hammer with a rated energy of 58 605 N · m.

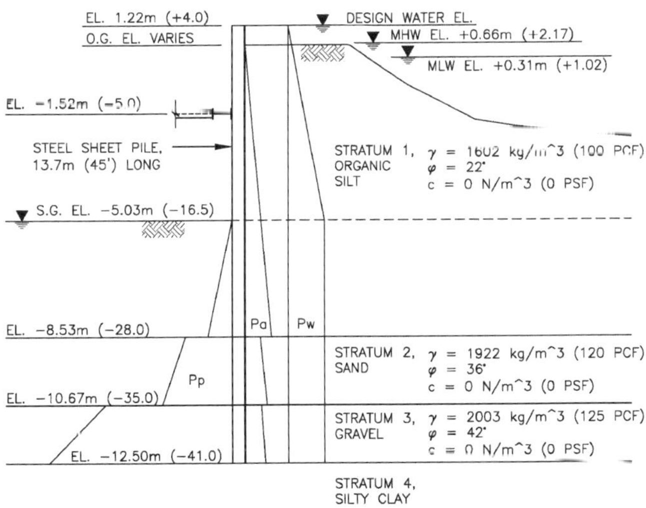

FIG. 2. Generalized earth and water pressure diagram.

The normal river depth varied from 0 to 914 mm. The water level was tidal with Mean High Water (MHW) at Elevation +0.661 m and Mean Low Water (MLW) at Elevation +0.311 m. The top of the SSP was at Elevation +1.83 m with regularly spaced openings at Elevation +1.22 to limit the design hydrostatic pressure on the SSP during a high water event.

CONSTRUCTION PHASING

Starting from the north shore, AIMD installed two-thirds of the proposed cofferdam. As construction proceeded and the middle third of the pipe installation was completed, its SSP was removed for use at the opposite shoreline. An initial starter cofferdam was left-in-place for making the final connection. As work was being completed in the second-third of the cofferdam at the north shore, the reclaimed sheets were then installed starting at the south shore and proceeding toward the starter cofferdam. This sequence allowed AIMD to work both sides of the estuary simultaneously. The schedule benefits of working in thirds far outweighed the cost of purchasing only half of the SSP and doing the work in halves.

COFFERDAM AND TRESTLE CONSTRUCTION

The cofferdam utilized hot-rolled, ASTM A572, Grade 345, F_y = 345 MPa, Z-section, steel sheet piling, 13.7 m in length, with a required section modulus of 698 cm^3 per meter of wall. The two rows of parallel SSP were spaced at 4.72 m which corresponded with an even number of sheet piles which formed the bulkheads between cofferdam segments. As the SSP was being installed, W530X150 cap beams were installed with their webs placed horizontally onto and capped the top of the SSP walls. The cap beams were then welded to the SSP. Sitting on the cap beam webs were HP360X132 deck beams which supported the 1.52 m by 305 mm by 8.53 m long timber deck mats which served as the trestle along the top of the cofferdam.

HP250X62 braces were installed across the excavation between the W530X150 cap beams to maintain deck beam alignment and provide top support for the SSP. These braces were centered at 5.18 to 6.71 m for setting of the 6.1 m long pipe sections.

As the SSP installation progressed southward across the river, the deck mats were installed and attached to the deck beams with welded, threaded studs. Plates and nuts, countersunk below the top of the mats, held the mats in place. This bolted connection allowed easy mat removal for pipe installation and allowed equipment to concurrently work both ends of any segment. The work deck was designed for AASHTO HS20-44 truck loading, a three-axle truck having a 106.8 kN axle design load for timber structures. Vehicles were restricted from portions of the timber deck which cantilevered beyond the SSP and deck beams. The 1.9 m long, cantilevered portions of the deck were designed for a construction load of 14 370 Pa.

Crane loads were supported directly over the SSP walls which resisted these vertical loads in friction only. Additional, load resistance due to end bearing was conservatively ignored. The ability of the SSP to support the crane and deck loads was field verified by counting driving blows per 305 mm of SSP penetration and by applying dynamic pile driving formulae. Continuous timber curbs were installed longitudinally along the deck as guides to maintain the cranes, trucks, and other equipment in their proper position (Figures 3).

FIG. 3. Phase 1 construction photo.

A 136 t capacity, Manitowoc Model 555 crawler crane sitting on the timber deck was used to drive the SSP, set braces, and place deck mats. The SSP was set into position and partially driven with the ICE Model 22-30 vibratory hammer. The ICE Model I-19 diesel hammer was then used to drive to the sheets to their final length.

FIG. 4. North side, initial cofferdam (at right) at the two-thirds point.

Roughly at the two-thirds point of the cofferdam, additional steel sheet piling was driven in the EW direction to form a cofferdam bulkhead which served as the starting point for the pipe installation (Figure 4). Approximately 6.09 m away, north toward the shoreline, a second SSP bulkhead was installed and, at an additional 18.3 m, the first segment bulkhead was installed. Excavation in the starter cofferdam, between the first two bulkheads was initiated using a CAT 345 hydraulic excavator. These initial excavation spoils where trucked from the site.

When the excavation reached 3.05 m deep, a W610X241 wale was installed against each SSP wall. Next, W250X101 cross braces were installed at 5.18 to 6.71 m on center (Figure 5). Excavation then continued to Elevation -5.03 m. To vertically support each 6.09 m pipe section, one double HP310X79 pile bent was installed.

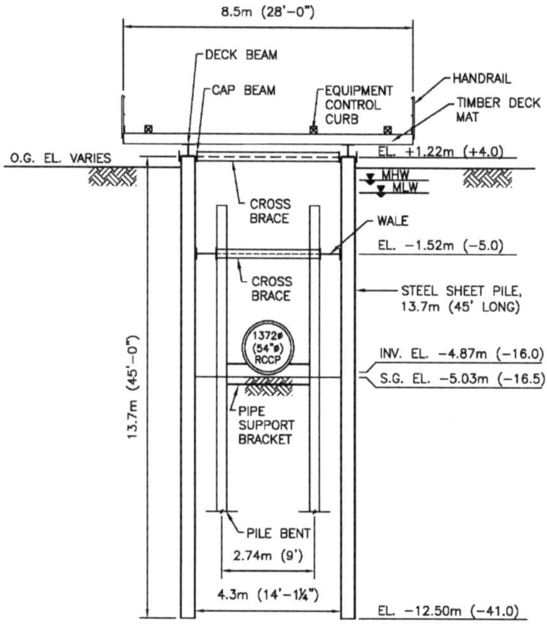

FIG. 5. Typical cofferdam section.

PIPE INSTALLATION

The initial pipe section was installed and it extended into the 6.09 m long starter cofferdam from the first 18.3 m long northern bulkhead. A reinforced concrete collar was poured around the pipe and against the second SSP bulkhead which then remained in place for the duration of the pipe installation. The far end of the first pipe section was then plugged. The pipe plug provided a seal to keep water out of the first 366 m long cofferdam as the pipe was installed toward the north shore. The second SSP bulkhead with the concrete collar provided a seal to keep water out of the remaining 152.4 m long cofferdam as the pipe was installed toward the south shore.

Three lengths of pipe were placed within the first 18.3 m long cofferdam segment and 6.09 m long starter segment. This placed the northernmost end of the initial pipe run approximately 2.44 m from the northern bulkhead of the first 18.3 m long cofferdam segment (Figure 6). Each 6.09 m long pipe section weighed 11 340 kg and was supported by a single, steel, pipe cradle attached to the HP310X79 bearing pile bent. The bearing piles were driven and pipe was placed using the Manitowoc 555 crawler crane sitting on the trestle deck mats.

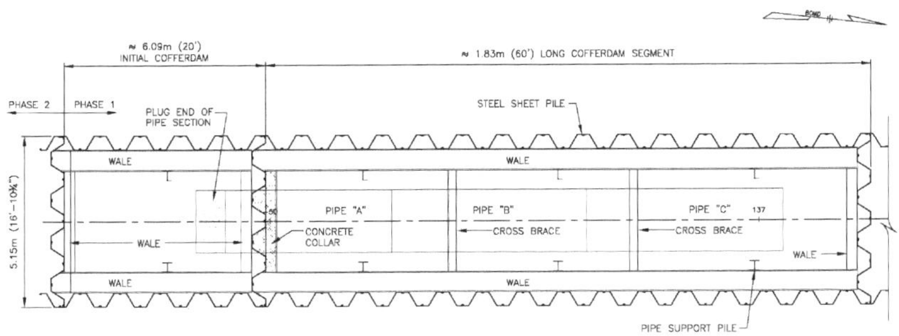

FIG. 6. Initial cofferdam staging detail.

Using the pipe bent bearing piles as soldier beams, the area just south of the northern bulkhead was sealed around and over the pipe with 102 nominal thickness, timber lagging boards. After the timber lagging and a pipe braces were installed at the bent near the 18.3 m bulkhead, the installed pipe sections were backfilled using material being excavated from the next, consecutive, 18.3 m, cofferdam segment.

This leapfrogging process of using the excavated material from one 18.3 m long cofferdam segment to immediately backfill the previous 18.3 m long segment eliminated the need to haul, stockpile, and then return soft, saturated, excavation spoils from each cofferdam segment. This procedure was repeated and progressed northward for one 18.3 m long cofferdam segment and 3 pipe sections at a time until one-third of the required pipe was installed and backfilled. At that time, the first third SSP was removed and trucked to the south shore of the river.

FIG. 7. North side steel sheet pile removal for reuse on south side.

While pipe installation continued in the second third at the north shore, work for the remaining third, approximately 152.4 m long phase then began from the south shore and proceeded north toward the initial 6.09 m bulkhead and the first-laid pipe section in the middle of the Back River (Figure 7). The SSP, upper bracing, and decking were installed using a 181 t capacity, Manitowoc Model 777 crawler crane. Excavation, lower wale and brace installation, pipe placement, and backfill then progressed southward toward the south shoreline. The SSP, deck beams, deck mats, wales, and braces were reused for the second phase cofferdam and trestle and, at completion of the river crossing, were reclaimed for use on other projects.

CONCLUSION

AIMD proposed and implemented an alternate method of access and construction for a 518 m crossing of the Back River estuary in Baltimore County, Maryland and for the installation of a 1372 mm diameter force main as part of a 4023 m project for Stemmers Run Relief Wastewater Force Main project. This alternate eliminated a separate, parallel work trestle for the entire length of the estuary crossing and made use of the proposed cofferdam for the support of the timber work deck. This plan minimized disturbance in this important estuary of the Chesapeake Bay. As well, it improved access to the work by providing a wider work deck which improved the contractor's visibility, productivity, and job safety. Cofferdam, trestle, and the installation of the 1372 mm diameter force main river crossing were completed in March of 2008. The final phase of the project is in progress including the non-tidal wetland restoration and tidal wetland mitigation.

REFERENCES

AMERICAN ASSOCIATION OF STATE HIGHWAY AND TRANSPORTAION OFFICIALS (1994). "Standard Specifications of Highway Bridges, Interim Specifications, Bridges, Division 1 – Design." Washington D.C., Figure 3.7.7A, pg. 22.

BOWLES, J. E. (1988). "Foundation Analysis and Design Forth Edition." McGraw-Hill Book Company, New York, pp. 791 - 795

NAVAL FACILITIES ENGINEERING COMMAND (1982). "Foundations and Earth Structures, NAVFAC Design Manual 7.2." Department of the Navy, pg. 7.2-196

PEIRCE, J. J. (2006). "Design Calculations. Temporary Sheet Pile Cofferdam and Trestle for Baltimore County", Department of Public Works, Stemmers Run Relief Wastewater Force Main, Peirce Engineering, Inc., 31 p.

RUMMEL, KLEPPER AND KAHL, LLP (2005). "Subsurface Exploration and Geotechnical Evaluation." Stemmers Run Relief Wastewater Force Main. Baltimore County Department of Public Works, 21 p.

Analysis of Anchored Sheet Pile Wall Deformations

Ömer Bilgin[1], P.E., M.ASCE and M. Bahadir Erten[2]

[1] Assistant Professor, Dept. of Civil and Environmental Engineering, University of Houston, Houston, TX 77204-4003. E-mail: obilgin@uh.edu

[2] Graduate Research Assistant, Dept. of Civil, Architectural, and Environmental Engineering, University of Texas, Austin, TX 78712

ABSTRACT: Sheet pile walls are one of the oldest earth retention systems utilized in civil engineering projects. The conventional methods used in the design of sheet pile walls are based on the limit equilibrium approach using active and passive earth pressures. These methods, based on force and moment equilibrium, do not consider wall deformations, which are important for serviceability considerations. For varying soil conditions and wall heights, effects of anchor location, sheet pile stiffness, anchor stiffness, and number of anchors on wall and soil deformations were studied. The free earth support method was used to calculate penetration depths of single-anchored walls for varying soil conditions and wall heights. Using these calculated penetration depths a parametric study was performed to investigate the effect of parameters considered on wall and soil deformations. Analyses were performed using the finite element method. The analysis results show that while having multiple anchor levels is the most efficient way to reduce wall and soil deformations, using pile profiles larger than the one required by the structural design can also be very effective. This paper presents the results and findings of the parametric study performed.

INTRODUCTION

Sheet pile walls are used for various purposes; such as excavation support system, cofferdams, cut-off walls under dams, slope stabilization, waterfront structures, and floodwalls. Although there are several other materials (such as timber, reinforced concrete, and plastics) used for sheet piles, steel sheet piles are the most common in retaining walls. The sheet pile walls can be either cantilever or anchored depending on the wall height. While relatively shorter sheet pile walls can be cantilever, higher walls require anchors. The selection of wall type, either cantilever or anchored, is based on the function of wall, the characteristics of foundation soils, and the proximity of wall to existing structures (ASCE 1996).

The movement of in-situ walls is an important issue because of the development of new technologies in this field and increase in litigation associated with damages caused by the movements of these walls to adjacent structures (Clough and O'Rourke 1990). The anchored sheet pile walls, being considered as flexible walls, experience more movements compared to diaphragm walls, such as slurry walls.

The conventional methods used in the design of anchored sheet pile walls are based on active and passive earth pressures that are concerned with the failure condition based on the Mohr-Coulomb failure criterion. These design methods do not specifically consider wall displacements. A factor of safety is applied to the passive pressures during the determination of lateral earth pressures (NAVFAC 1986). The safety factors applied to passive pressures are used to take into account for uncertainties in soil conditions, method of stability analysis, loading conditions, as well as to restraint the soil movements to an acceptable level (Potts and Fourie 1984).

The objective of this study was to investigate the deformations of anchored sheet pile walls, designed using conventional methods, and then to investigate the effect of different parameters in reducing the lateral wall movements and soil settlements behind the wall. Parametric study was performed for varying anchor depth, wall bending stiffness, anchor stiffness, and number of anchor levels. The finite element method was used to perform the analyses.

NUMERICAL MODELING AND ANALYSES

Geometry and Material Properties

A schematic of a typical wall and soil section analyzed is shown in FIG. 1. Three type of soil conditions and three different wall heights were used in the parametric study. The walls studied were 6, 9, and 12 m high. Three different soil types considered were dense (D), medium dense (M), and loose (L) sands. The cases were identified with a number followed by a letter code. The number refers to the wall height and one-letter code, given in the parentheses above, indicates the soil type. In each case uniform soil conditions at the site was assumed. The properties selected for the soil types are given in TABLE 1. The groundwater elevation was assumed to be at the anchor level on both sides of the wall.

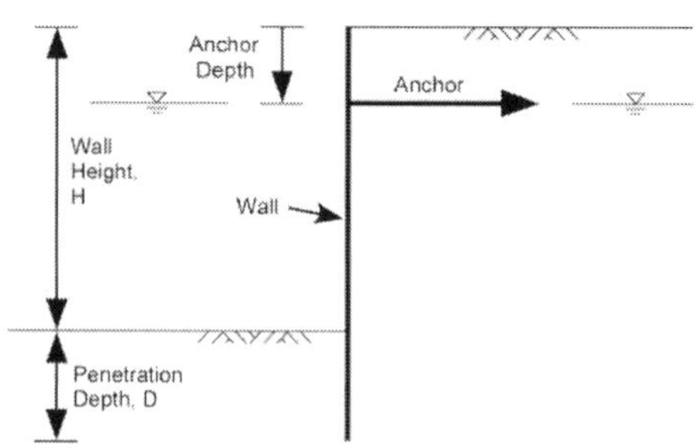

FIG. 1. Typical soil and wall profile used in parametric study

TABLE 1. Soil Properties

Property	Dense sand (D)	Medium dense sand (M)	Loose sand (L)
Saturated unit weight	20 (kN/m^3)	18 (kN/m^3)	16 (kN/m^3)
Unsaturated unit weight	18 (kN/m^3)	17 (kN/m^3)	16 (kN/m^3)
Cohesion	0.3 (kPa)	0.3 (kPa)	0.3 (kPa)
Friction angle	42 (°)	36 (°)	30 (°)
Dilatancy angle	12 (°)	6 (°)	0
Modulus of elasticity	75,000 (kPa)	35,000 (kPa)	15,000 (kPa)
R_{int}	0.63	0.65	0.67

NOTE: R_{int} = Soil/wall interface strength = $\tan\delta/\tan\phi$ (δ=interface and ϕ=soil friction angle)

By using the free earth support method the anchored sheet pile walls were designed for the three wall height and three soil type combinations. The depth of wall penetration calculated and pile section selected were used in numerical modeling during the parametric study. The calculated anchor forces were also used to determine the anchor stiffness and to study the effect of increased anchor stiffness on wall and soil deformations. The anchor force calculated by using the free earth support method was divided by the anchor's allowable stress, assumed as 138 MPa, to obtain the anchor area needed. The design anchor stiffnesses, *EA*, used in the numerical analyses were obtained by multiplying the anchor area by the elastic modulus of steel. The variables and their ranges considered in the parametric study are given in TABLE 2.

Finite Element Software and Constitutive Model

Finite element analyses were performed using Plaxis finite element code (Brinkgreve et al. 2006). The finite element modeling comprised two-dimensional plane strain analysis. The soil layers and sheet pile walls were modeled using 6-node triangular elements and 3-node plate elements, respectively. A finer mesh was used around the wall because of the stress concentration. The excavation was simulated by removing soil in lifts. The complete excavation was performed in eight steps, i.e. the thickness of the soil lifts was equal to one-eight of the total wall height. This resulted in 0.75, 1.125, and 1.50 m soil lift thicknesses for 6, 9, and 12-m-high walls, respectively. The anchor was installed when excavation reached the anchor level. Because of the cohesionless soils, the analyses were performed considering fully drained conditions.

TABLE 2. Variables and Ranges Used in Parametric Study

Variable	Range Considered
Anchor depth	0, 0.125*H*, 0.25*H*, 0375*H*
Wall (section profile) stiffness	PZ22, PZ27, PZ35, PZ40
Anchor stiffness	(1, 5, and 10) × Design stiffness
Number of anchor levels	1 and 2

The Mohr-Coulomb constitutive model, a linearly elastic and perfectly plastic model, was used to model soil behavior. In the elastic perfectly plastic models, the material behavior is elastic within a region of stress space, below the yield surface, where recoverable deformations occur. When the stress states reach the boundary of elastic region, i.e. yield surface, the material starts to deform under constant stress and ideal plastic deformations occur. The total strain that accompany the stress change is divided into elastic and plastic strain components.

RESULTS AND ANALYSIS

Effect of Anchor Location

First set of analyses were performed to investigate the effect of anchor depth, measured from the top of wall, on wall and soil deformations. The effect of anchor location on deformations at the end of excavation is shown in FIG. 2. The results from five different parametric study sets are given in the figure, to be able to present the effect of soil type using the same wall height as well as the effect of wall height using the same soil type. FIG. 2(a) shows the maximum horizontal wall deformations, normalized by the wall height, and FIG. 2(b) shows maximum ground settlements behind the wall. The grounds settlements are also normalized by the wall height.

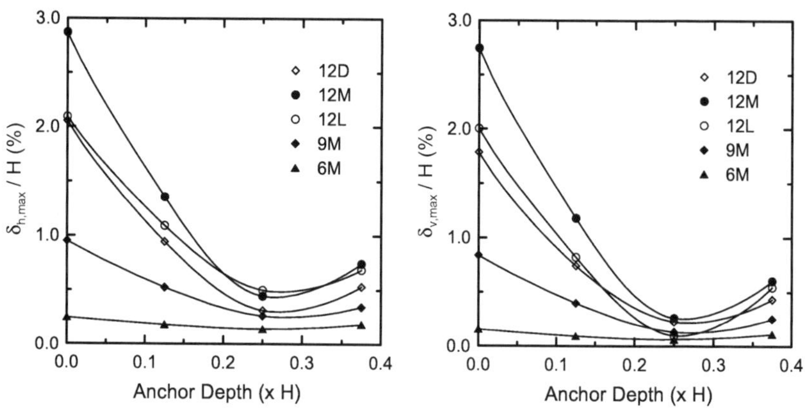

(a) Maximum wall displacement (b) Maximum soil settlement

FIG. 2. Effect of anchor location on wall and soil movements

The results given in FIG. 2 show that as the anchor depth from the top of wall increases to a certain depth, the maximum horizontal wall deformations and maximum settlements behind the wall decrease. When the anchor depth is between $0.25H$ and $0.27H$ the maximum deformations reach a minimum value. The maximum deformations start to increase when the anchor depth is more than $0.27H$.

FIG. 2 also shows that the effect of anchor location on maximum wall and soil deformations are more significant for relatively higher walls. For example, while the anchor location can change the maximum wall deformations almost four times for relatively higher walls (12D cases), the change can be as low as 25 percent (6M cases) for walls with lower heights.

Once the investigation of anchor location showed that the anchor depth of 0.25*H* from the top of wall results in the lowest deformations, the anchor level was fixed at this depth during the investigation of other parameters.

Effect of Sheet Pile Section

The effect of wall stiffness on maximum horizontal wall deflection and maximum soil settlement behind the wall are shown in FIG. 3(a) and (b), respectively. This parametric study was performed by using a pile section determined from the structural design of walls using conventional design and then using larger pile sections than needed by the design. The results show that as the pile size increases both wall and soil deformations decrease, as expected. Using larger pile sections has more effect on reducing the deformations when the walls are relatively higher. While using larger sections can reduce the wall deformations by 50 percent [FIG. 3(a)], the decrease in soil settlements behind the wall can be as much as 80 percent [FIG. 3(b)].

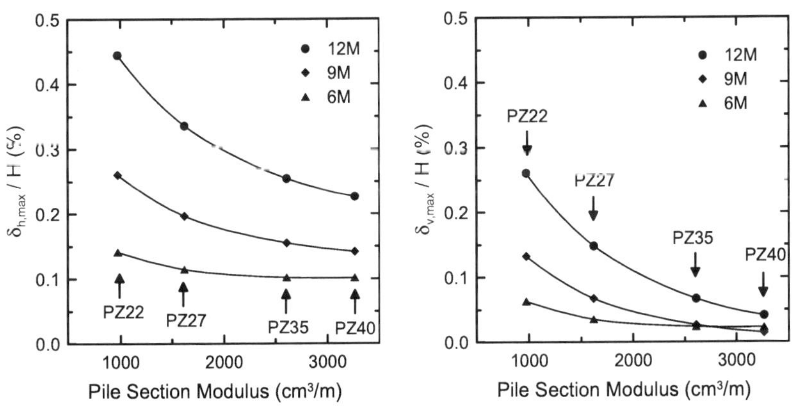

(a) Maximum wall displacement (b) Maximum soil settlement

FIG. 3. Effect of sheet pile section on wall and soil movements

Effect of Anchor Stiffness

The effect of anchor stiffness on wall deformations for varying wall heights and soil types is shown in FIG. 4. The anchor stiffness is normalized by the anchor stiffness obtained from the anchor design for each soil and wall height combinations. The stiffness values in the parametric study were increased five and ten times, i.e. increased anchor's cross-sectional area. The analyses results show that as the anchor stiffness increases maximum horizontal wall deformations decrease. However, the reduction in deformations is stabilized very rapidly. An increase in the anchor stiffness after a certain value has a minimal effect on wall deformations. The maximum effect is observed when the walls have lower heights and about 50 percent decrease can occur in maximum wall deformations.

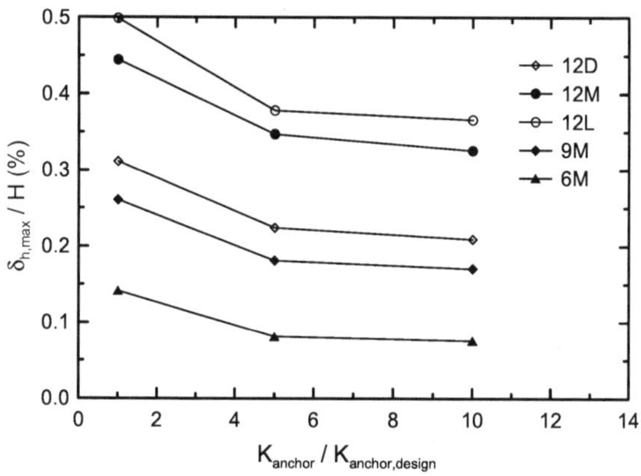

FIG. 4. Effect of anchor stiffness on wall displacements

Effect of Number of Anchor Levels

The effect of number of anchor levels on wall deformations was investigated by introducing second anchor level on the wall and soil combinations studied. The second anchor level was located at the mid-height between the first anchor level and the bottom of excavation, i.e. $0.625H$ below from the top of wall. The effect of having a second level of anchors on the maximum horizontal wall deformations is shown in FIG. 5. Adding the second level of anchors reduces the wall deformations, as expected. The reduction is more significant for higher walls and when the walls are installed in loose sands. Maximum wall deformations are reduced more than 65 percent when the second level of anchor is installed for the 12L case. The reduction in wall deformations was 40 percent for the 6M case. Although having two levels of anchors is not common for walls with 6 m height, the analyses with two layers of anchors for 6 m walls were also performed for completeness of the parametric study.

Comparisons with Other Investigations

Based on the observations from case studies, Goldberg et al. (1976) showed that in-situ wall movements and soil settlements behind the wall are usually less than 0.5

percent of wall height, *H*. In another study, based on case studies and field observations, Clough and O'Rourke (1990) concluded that the horizontal movements tend to average about 0.2 percent of the wall height and the vertical movements tend to average about 0.15 percent of the wall height. The observations made by Goldberg et al. (1976) and Clough and O'Rourke (1990) along with the maximum lateral wall movements and maximum soil settlements behind the wall obtained from the numerical analyses performed during this study are shown in FIG. 6.

FIG. 6 shows that the results obtained from numerical analyses during this study agree well with the observations from the case studies reported by others. In all the cases analyzed, the maximum wall and soil deformations were less than 0.5 percent of *H*, which was the observation made by Goldberg et al. (1976). The maximum horizontal wall deformation trend of 0.2 percent of *H* given by Clough and O'Rourke (1990) also agrees reasonably well with the data of this study. An exponential trend line obtained from this study results in slightly lower wall deformations for walls less than 10 m and slightly higher for walls more than 10 m high. The maximum soil settlement trend obtained from this study using an exponential trend line results in lower settlements than the observational data. However, it should be noted that the purpose of this study was to investigate the parameters to reduce the wall and soil deformations and FIG. 6 shows the results of all the cases analyzed. Therefore, most of the cases were much stiffer than the actual design cases and having wall displacements and soil settlements lower than the field observations is reasonable.

CONCLUSIONS

The anchored sheet pile wall deformations and soil settlements behind the wall were studied for varying wall heights and soil conditions. The effect of several parameters on the wall and soil movements was investigated. The parameters considered were; anchor depth, pile stiffness, anchor stiffness, and number of anchor levels. Based on the variables and the ranges considered, the main findings from this research are as follows:

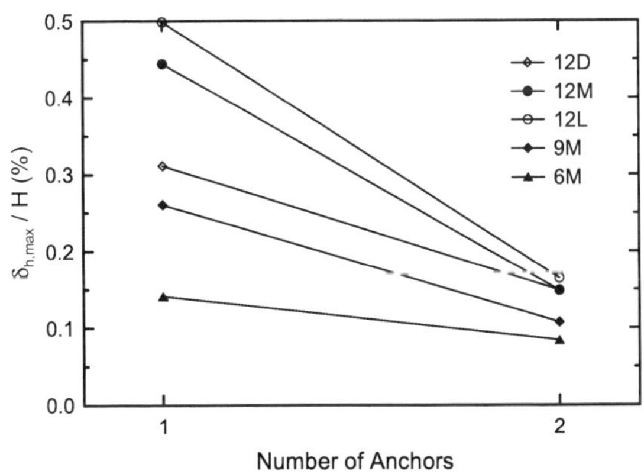

FIG. 5. Effect of number of anchors on wall displacements

- The lateral wall movements and soil settlements behind the wall are minimum when the anchor is installed at approximately $0.25H$ depth from the top of wall;
- Using pile profiles larger than required by wall's structural design can decrease both wall and soil displacements significantly. By using larger pile profiles, 50 percent reduction in lateral wall movements and soil settlements were observed;
- Although has minimal effect on soil settlements, increased anchor stiffness, i.e. using larger anchor sections, can help to reduce wall movements; and
- By using the second level of anchors wall movements can be reduced more than 65 percent for relatively high walls.

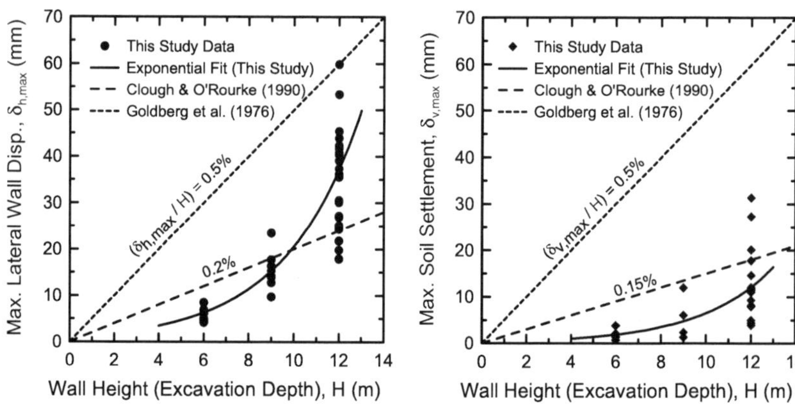

(a) Maximum wall displacement (b) Maximum soil settlement

FIG. 6. Comparison of wall and soil movements with observed data by others

REFERENCES

ASCE (1996). *Design of sheet pile walls*. Technical engineering and design guides as adapted from the US Army Corps of Engineers, No. 15.

Brinkgreve, R.B.J. et al. (ed.) (2006). *Plaxis 2D-Version 8 users manual*. Rotterdam: A.A. Balkema.

Clough, G.W. and O'Rourke. T.D. (1990). "Construction induced movements of insitu walls." *Proceedings, ASCE Specialty Conference on Design and Performance of Earth Retaining Structures*, Ithaca, New York, 439-470.

Goldberg, D.T., Jaworski, W.E., and Gordon, M.D. (1976). "Lateral support systems and underpinning." *Report FHWA-RD-75-128*, Vol. 1 Federal Highway Administration, Washington, D.C.

Potts, D.M. and Fourie, A.B. (1984). "The behavior of a propped retaining wall: results of a numerical experiment." *Geotechnique*, 34(3), 383-404.

United States Navy, Naval Facilities Engineering Command (NAVFAC), (1986). *Design Manual, 7.02, Foundations and Earth Structures*. Alexandria, Virginia.

Anchored Sheet Pile Walls Constructed on Sloping Ground

Ömer Bilgin[1], P.E., M.ASCE and M. Bahadir Erten[2]

[1] Assistant Professor, Dept. of Civil and Environmental Engineering, University of Houston, Houston, TX 77204-4003. E-mail: obilgin@uh.edu
[2] Graduate Research Assistant, Dept. of Civil, Architectural, and Environmental Engineering, University of Texas, Austin, TX 78712

ABSTRACT: Conventional methods used in the design of anchored sheet pile walls are based on the limit equilibrium approach and they do not consider processes involved during construction. The sheet pile walls constructed on slopes may require both cut and fill operations. Varying amounts of cut and fill sections cause different loading and unloading of soils around the wall resulting in different wall behavior. The effects of simultaneous cut and fill operations for varying slope angles and cut to fill ratios on wall behavior were investigated using numerical methods. Sheet piling with a single anchor was considered. Wall deformations, bending moments, and anchor forces were investigated for the conditions studied. The results of the study show that the wall deformations and bending moments are significantly affected by the amount of cut and fill that occur during the construction of anchored sheet pile walls. This paper presents the results and findings of the parametric study performed for the sheet pile walls constructed on slopes, where both cut and fill operations are required during construction.

INTRODUCTION

The anchored sheet pile walls are used as either permanent or temporary lateral earth support system in various civil engineering projects. In some cases, these walls are needed to be installed on slopes. These situations mostly arise in transportation projects where one highway can cross through many geologically and topographically changing conditions. Sloping ground conditions may also occur when the sheet pile walls are installed as waterfront structures.

Depending on the location of a wall on a slope section, ground surface elevations may require either cut behind the wall or filling in front of the wall. Some cases may require both cutting and filling. If the wall is located at the top of the slope then only cut is required in front of the wall as shown in FIG. 1(a). On the other hand, the wall located at the toe of the slope requires only backfilling [FIG. 1(c)]. When the wall is located somewhere along the slope then both cutting in the front and filling behind the wall are needed as shown in FIG. 1(b). The amount of cut and fill depends on the location of the wall along the slope, e.g. a wall closer to the top of the slope requires more cut than fill while a wall closer to the toe of the slope needs more fill than cut quantities.

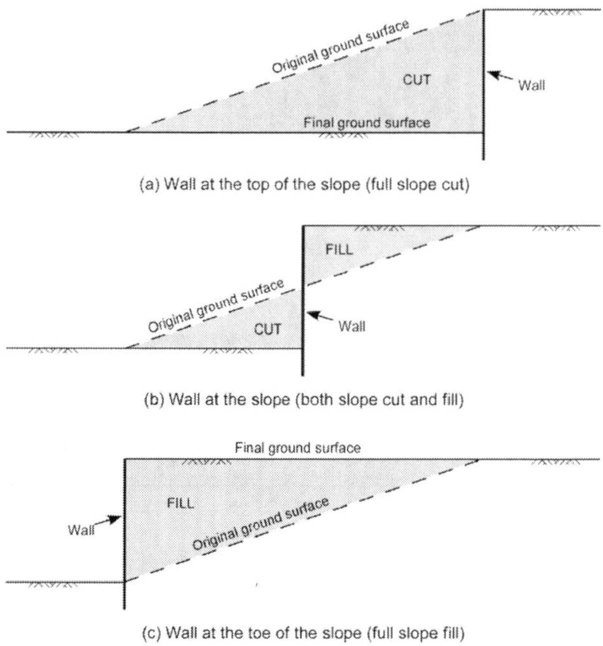

FIG. 1. Wall locations on a sloping ground

Most of the time location of a wall along a slope will be determined by the construction activities and structures planned at the site. However, availability of a suitable borrow material to be used for fill sections can also be a determining factor for the wall location. For example, if the fill material needs to be hauled long distances then having the wall at the toe of the slope may be very expensive. If the natural soils are appropriate to be used as backfill material, then placing the wall in the middle of the slope and using the excavated material from the front as backfill material behind the wall help in reducing construction costs.

Depending on the cut and fill quantities, anchored sheet pile walls at the slopes are expected to behave differently because of the stress paths that are generated in surrounding soils during cut (unloading) and filling (loading) operations. The free earth support method, conventional design method commonly used in anchored sheet pile wall design, does not consider whether the walls are constructed by cutting, filling, or both. Numerical analyses, on the other hand, allow modeling of the construction processes. The finite element method has been used for the analysis of anchored or braced retaining structures by Bjerrum et al. (1972), Potts and Fourie (1984 and 1985), Briaud and Lim (1999), Dawkins (2001), and Krabbenhoft et al. (2005).

The objective of this study was to investigate the behavior of anchored sheet pile walls constructed on slopes. The finite element method was used to perform analyses. The behavior of walls were investigated for varying slopes and cut/fill quantities.

MODELING

Geometry and Material Properties

A 12-m-high slope was selected for the parametric study. Slope ratios (horizontal to vertical, H:V) of 5H:1V, 3H:1V, and 2H:1V were investigated. The anchor location was fixed at 3 m (25% of wall height) below the top of wall. A schematic of a typical wall and slope section analyzed along with the parameters considered is shown in FIG. 2 and the cases analyzed during the parametric study are given in TABLE 1.

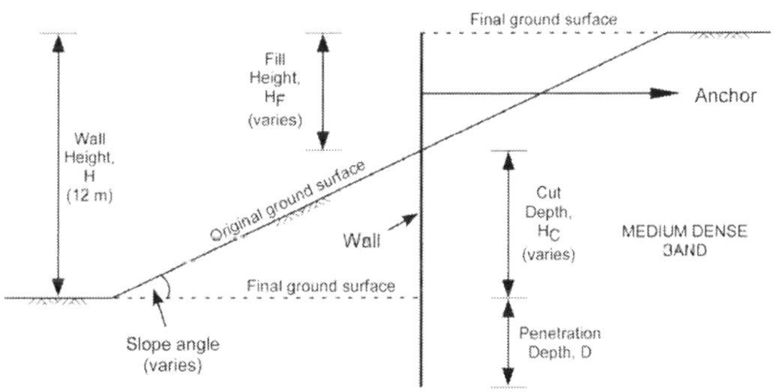

FIG. 2. Schematic of wall and slope section analyzed

The depth of wall penetration, *D*, used in numerical modeling and analyses was 2.61 m, which was calculated using the free earth support method. The pile section was also selected based on the conventional design calculations. The pile section selected was PZ35. The soils at the site were selected as dry medium dense sand. The soil and wall material properties used in the conventional design calculations and in the finite element analyses are given in TABLE 2. Although sand was used in this study, because of the numerical calculation complications that may occur when performing some of the options in finite element analysis software used (Brinkgreve et al. 2006), small soil cohesion value of 0.3 kPa was used in the numerical analyses.

The anchor was installed as soon as the anchor location on the wall is available, i.e. at the beginning of construction in fill cases and when excavation reached the anchor level in excavation cases. The anchor force calculated from the free earth support method was divided by the anchor's allowable stress, assumed as 138 MPa, to obtain the anchor area needed. The design anchor stiffness, EA, of 2.5×10^5 kN/m used in the numerical analyses was obtained by multiplying the anchor area by the elastic modulus of steel.

TABLE 1. Cases Used in Parametric Study

Group Name	Case Name	Cut Depth H_C, (m)	Fill Height H_F (m)	Percent of Cut Height	Percent of Fill Height	Slope (Hor:Ver)
Group A	A51	12	0	100	0	5H:1V
	A31					3H:1V
	A21					2H:1V
Group B	B51	9	3	75	25	5H:1V
	B31					3H:1V
	B21					2H:1V
Group C	C51	6	6	50	50	5H:1V
	C31					3H:1V
	C21					2H:1V
Group D	D51	3	9	25	75	5H:1V
	D31					3H:1V
	D21					2H:1V
Group E	E51	0	12	0	100	5H:1V
	E31					3H:1V
	E21					2H:1V

TABLE 2. Soil and Wall Properties

Property	Value	Unit
Soil:		
Unit weight, γ	17	kN/m^3
Friction angle, ϕ	36	degree
Dilatancy angle, ψ	6	degree
Modulus of elasticity, E_{50}	35,000	kPa
Poisson's ratio, ν	0.28	
Wall:		
Axial stiffness, EA	4.36×10^6	kN/m
Bending stiffness, EI	9.86×10^4	kNm^2/m
Soil/Wall interface strength[a], R_{int}	0.65	

NOTE: [a]$R_{int} = \tan\delta/\tan\phi$ (δ=interface and ϕ=soil friction angle)

Finite Element Software and Constitutive Model

The finite element analyses were performed by using Plaxis finite element code (Brinkgreve et al. 2006). The finite element modeling comprised two-dimensional plane strain analysis. Soil layers and sheet pile walls were modeled using 6-node triangular elements and 3-node plate elements, respectively. A typical finite element model mesh consisted of, on average, 580 soil elements and 34 plate elements. A finer mesh was used around the wall because of the stress concentration. The construction process was modeled by removing soil in layers for cut and by adding soil layers for fill situation. The thickness of each lift was 1.5 m.

The Mohr-Coulomb constitutive model, a linearly elastic and perfectly plastic model, was used to model soil behavior. The Mohr-Coulomb model has been commonly used in the finite element modeling of retaining walls (Potts and Fourie 1984 and 1985; Krabbenhoft et al. 2005; and Tan and Paikowsky 2008). The yield function, f, is defined by the Mohr-Coulomb criterion using soil stresses (normal stress, σ, and shear stress, τ) and effective strength parameters (cohesion, c, and friction angle, ϕ) as

$$f = |\tau| - \sigma \tan\phi - c \tag{1}$$

and when the yield function is equal to zero (f = 0), plastic yielding occurs. The elastic soil behavior below the yield surface is characterized by Hooke's law to relate the stresses to strains.

RESULTS AND ANALYSIS

Wall Deformations

Wall deformed shapes for Group A, C, and E cases of varying slopes are shown in FIG. 3. Group A represents the walls located at the top of the slope (100% cut cases) while Group E represents the walls located at the toe of the slope (100% fill cases). Group C cases are the cases where the wall is located at the mid-height of the slope therefore requiring same cut and fill quantities (50% of wall height requires cut and the remaining 50% requires fill).

The results show that the cut to fill ratio has more effect on wall deformation behavior than the slope angle. The top of wall in cut cases moves towards the excavation, however, wall top deformations decrease as the amount of cut decreases and the fill increases (moving from Group A towards Group E in FIG. 3). Eventually the top of the wall starts moving backwards towards the soil. This is because of the installation of an anchor at the beginning of construction in fill cases compared to the installation when excavation reaches the anchor level in cut cases. In fill cases the wall deformations are restricted at the anchor level from the beginning of construction. In cut cases, however, the wall can move as a cantilever until the excavation reaches the anchor level and the anchor is installed.

The results in FIG. 3 also show that the effect of slope angle on wall deformations increases as the fill height increases. When the wall is located at the top of the slope and only cut is required, changing the slope angle between 5H:1V and 2H:1V does not affect the wall deformations. This is due to the soil providing enough confinement in front of the wall within the slope range studied. As the wall location moves towards the toe, the slope starts to effect wall movements and the deformations increase slightly as the slope angle gets smaller. The reason for this is that the smaller slope angles require more soil volume and loads behind the wall. The movements of foundation soil because of the heavier loads can be seen from the pile tip movements of Group E cases (FIG. 3).

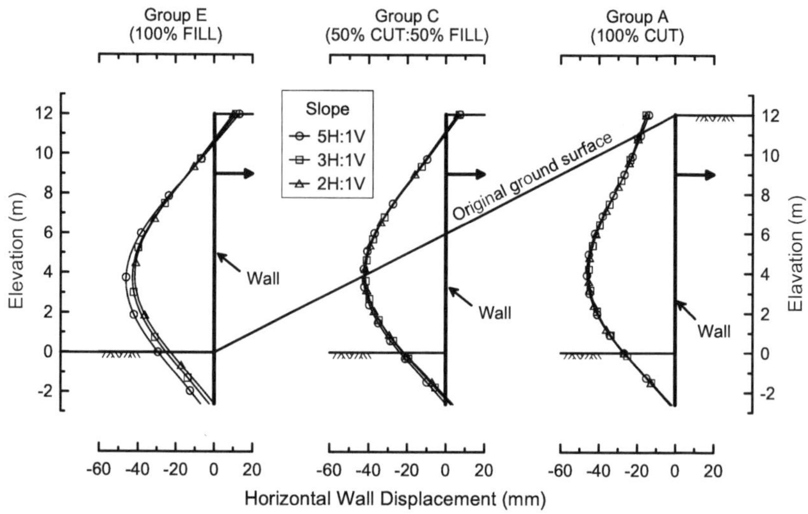

FIG. 3. Horizontal deformations of walls located along the slope

Wall Bending Moments

The bending moments along the wall height is shown in FIG. 4 for varying cut/fill height percentages and varying slopes. The results show that the location of the wall along the slope, which affects the cut and fill ratios, has more effect on the bending moments compared to the slope angle. FIG. 4(b) shows that the change in slope only slightly affects the wall bending moments. On the other hand, as the amount of cut decreases and the fill amount increases, i.e. as the wall location moves from the top of the slope towards the toe of the slope, the maximum moments increase [FIG. 4(a)]. For the cases studied the increase in bending moment was approximately 14%. In addition, increased fill amounts result in moment reversal close to the tip of the wall. The results also show much higher moments, as much as three times, at the anchor level in cut cases compared to fill cases [FIG. 4(a)].

Anchor Forces

The anchor forces obtained from the parametric study results are shown in FIG. 5. FIG. 5(a) shows the effect of slope angle on the anchor forces for Group A, C, and E cases. It can be seen in the figure that the slope angle has a minimal effect on the anchor force for a given cut and fill ratio. However, the location of the wall along the slope, and therefore the cut and fill ratios, effects the anchor forces significantly as shown in both FIG. 5(a) and (b). As the cut amount decreases and the fill amount increases, the anchor forces drop significantly. There is approximately 30% difference between only cut and only fill cases. The reason of the decrease in anchor forces is related to the wall deformations. Since the anchor is installed at the beginning of the fill process, the movements are restricted at the anchor level. The stresses developed by the backfill soil cause wall tip to move outwards and rotate the wall about the anchor location. This rotation also causes wall top to move backwards as it was discussed previously and shown in FIG. 3.

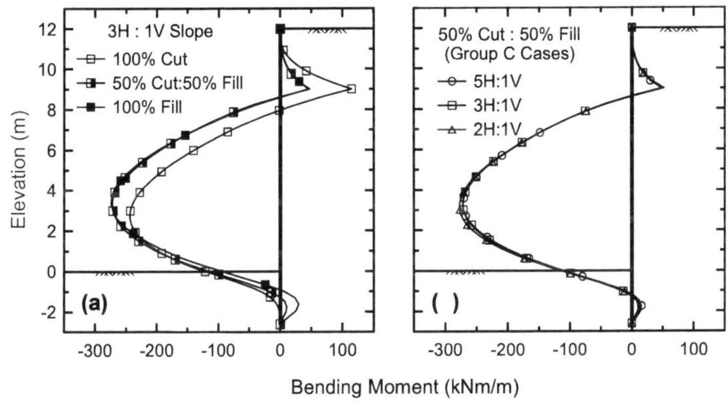

FIG. 4. Effect of (a) cut & fill quantities and (b) slope on wall moment diagram

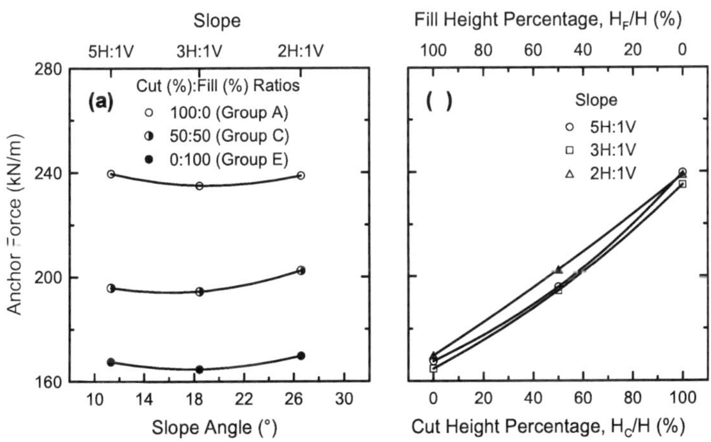

FIG. 5. Effect of (a) slope and (b) cut & fill quantities on anchor force

CONCLUSIONS

The parametric study using finite element method was performed to investigate the behavior of anchored sheet pile walls constructed on slopes. The impact of wall location along the slope, which affects the cut and fill amounts, and slope angle on wall deformations, bending moments, and anchor forces were investigated. The study was performed for fixed soil and wall properties. The study results show that for the cases studied the slope angle, varying between 5H:1V to 2H:1V, has a very minimal effect on wall behavior.

The study results also show that the location of anchored sheet pile wall along the slope, on the other hand, has a significant effect on wall behavior. Based on the cases and parameters studied, as the location of wall moves from the top of the slope towards the toe, i.e. decreasing cut and increasing fill quantities:

- The top of wall starts moving backwards towards the soil due to the wall rotation about the anchor location;
- Maximum wall bending moments increase approximately 14 percent, when the wall moves from the top of the slope to the tip of the slope;
- Wall bending moments at the anchor level decrease almost 65 percent when the wall moves from the top of the slope to the middle of the slope; and
- Anchor forces decrease significantly, approximately 30 percent when the wall moves from the top of the slope to the tip of the slope.

REFERENCES

Bjerrum, L., Clausen, C.J.F., and Duncan, J.M. (1972). "Earth pressures on flexible structures-A state-of-the-art report." *Proceeding, 5th European Conference on Soil Mechanics and Foundation Engineering*, Vol. 2, Madrid.

Briaud, J.L. and Lim, Y. (1999). "Tieback walls in sand numerical simulation." *Journal of Geotechnical and Geoenvironmental Engineering*, 125(2), 101-109.

Brinkgreve, R.B.J. et al. (ed.) (2006). *Plaxis 2D-Version 8 users manual.* Rotterdam: A.A. Balkema.

Dawkins, W.P. (2001). "Investigation of wall friction, surcharge loads, and moment reduction curves for anchored sheet pile walls." *US Army Corps of Engineers, ERDC/ITL TR-01-4.*

Krabbenhoft, K., Damkilde, L., and Krabbenhoft, S. (2005). "Ultimate limit state design of sheet pile walls by finite elements and nonlinear programming." *Computers and Structures*, 83(4-5), 383-393.

Potts, D.M. and Fourie, A.B. (1984). "The behavior of a propped retaining wall: results of a numerical experiment." *Geotechnique*, 34(3), 383-404.

Potts, D.M. and Fourie, A.B. (1985). "The effect of wall stiffness on the behavior of a propped retaining wall." *Geotechnique*, 35(3), 347-352.

Tan, Y. and Paikowsky, S.G. (2008). "Performance of sheet pile wall in peat." *Journal of Geotechnical and Geoenvironmental Engineering*, 134(4), 445-458.

Case Study: Cement-Bentonite Pre-Trenching and Cutter Soil Mixing (CSM) for Temporary Shoring and Groundwater Cutoff

Dominic M. Parmantier[1], M. ASCE, P.E; Rowland F P Stow[2], P.Eng.; R. John Byrne[3], P.E.

[1]Project Manager, Condon-Johnson & Associates, Inc., 651 Strander Blvd., Suite 110, Tukwila, WA 98188, dparmantier@condon-johnson.com
[2] Project Manager, Condon-Johnson & Associates, Inc., 651 Strander Blvd., Suite 110, Tukwila, WA 98188, rstow@condon-johnson.com
[3] Partner, Ground Support PLLC, 2475 152nd Avenue NE, Redmond, WA 98052, johnb@groundsupport.com.

ABSTRACT: A multi-story office building in Seattle, Washington with four levels of underground parking was sited adjacent to the existing Alaska Way Viaduct in an area which had previously been occupied by turn of the century sawmills before being developed for commercial uses in more recent history. The soils at the site consist of approximately 8.5 m (29 ft) of sand and wood debris fill over estuary deposits and glacial soils. The ground water regime at the site consists of an upper and lower aquifer separated by an aquitard. The piezometric heads of the aquifers were 2.7 m (9 ft) and 2.4 m (8 ft), respectively, below existing grade.

The proposed excavation was 13.1 m (43 ft) deep with a requirement that the shoring system extend an additional 7.6 m (15 ft) below the bottom of excavation to limit seepage from the upper aquifer into the excavation. Although initially tendered for construction with an owner designed secant wall or freeze wall system, the successful shoring contractor offered a design-build cutter soil mixed (CSM) shoring system which included pre-trenching the wall alignment under a cement-bentonite (C-B) slurry to remove the identified wood debris prior to installing the shoring system. This paper addresses the stability analysis and excavation of the C-B trench for removal of obstructions along with the design, construction, and performance of the CSM shoring system and associated depressurization of the deep aquifer.

INTRODUCTION

The 505 1st Avenue site is two blocks from the current water front in Seattle, Washington. The site is situated on reclaimed ground which consists of approximately 8.5 m (29-ft) of fill which was placed in Elliot Bay in the early 1900's. The presence of a high water table and permeable fill, which is susceptible to consolidation from dewatering, have previously limited the depths of excavations in this area to one level below grade. The 505 1st Ave building is the first structure in the Sodo District of Seattle to tackle the challenges associated with a deep excavation below the water table for the construction of multiple levels of below grade parking.

FIG. 1. Proposed 505 1st Ave Building Location and Completed Excavation

The 505 1st Ave building site is a triangular parcel located between two city streets and two existing buildings. An on-ramp to the Alaska Way Viaduct, which was damaged during the 2001 Nisqually Earthquake, is also located directly to the west of the project site. A three story historic building with no available foundation information is south of the building site. An eight story building founded on driven timber piles is north of the building site and abuts the excavation. Looking north, Figure 1 shows the completed excavation and adjacent structures.

Soil and Groundwater Conditions

A total of nine borings have been advanced at the site to assess the soil profile and properties. Four deep wells and numerous shallow monitoring wells have been installed to assess the ground water regime in the area. The stratigraphy at the site consists of 2 to 3 m (6.5 to 10 ft) of sand fill over another 5.2 to 7.4 m (17 to 24 ft) of sand and wood debris fill . SPT blow counts ranged from 2 to refusal in the sand and wood debris fill. Below the fill and debris is a thin layer of sandy-silt which was deposited by the waters of Elliot Bay prior to the land reclamation in the area.

Beneath these estuary deposits, there is a very dense glacially overridden till which is classified as a sandy-silt with some sand lenses. SPT blow counts in the till were typically over 50 blows for 150 mm (6 in). The borings also indicated the presence of a substantial glacial outwash layer of clean gravelly-sands below the sandy-silt till in some of the borings. When the outwash was encountered, it occurred at depths of 20 to 26 m (66 to 85 ft) and extended the full depth of the 30.5 m (100 ft) borings.

The ground water conditions at the site consist of an upper and lower aquifer which are separated by the sandy-silt till layer. The upper aquifer exists in the sand and wood fill material and has a piezometric head at a depth of 2.7 m (9 ft). Slug testing indicated a permeability of 10^{-2} to 10^{-3} cm/sec in the fill. The lower aquifer exists in the glacial outwash sand and has a piezometric head at a depth of 2.4 m (8 ft). Pump tests indicated a permeability of 8 to 9 x 10^{-3} cm/sec in the outwash sand. Because of the potential for dewatering of the fill material to result in settlement of adjacent buildings, roads, and utilities and increased down drag on the piling for the adjacent Alaska Way Viaduct, the water level in the upper aquifer could not be lowered. Two pump tests verified the ability of the glacial till to act as an aquitard and prevent draw down of the upper aquifer during pumping of the lower aquifer. Based on the pump tests, the extraction rate required to draw down the deep aquifer to prevent base heave in the excavation was estimated to be 75 to 150 lpm (20 to 40 gpm).

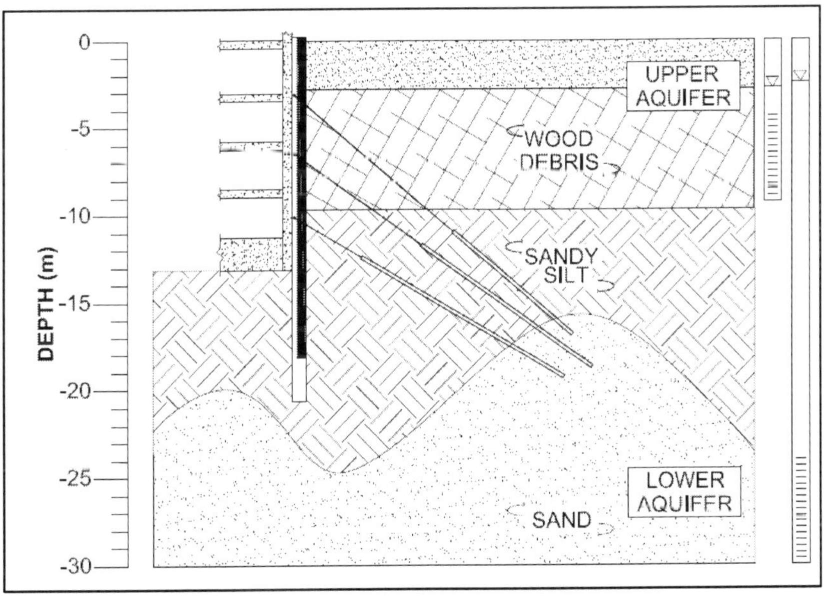

FIG. 2. Idealized Soil Profile and Shoring System

Original Shoring Design

Once the soil and ground water conditions were determined, the owner's design team established the lateral earth pressures and cutoff requirements based on the building size and site restraints. The excavation was planned for a depth of 13.1 m (43-ft) below grade. The excavation support system would have to incorporate underpinning of the adjacent eight story building, and it would have to isolate the excavation from the upper aquifer to prevent lowering of the water level in the loose fill material. The deep aquifer would need to be depressurized so that the excavation could be made safely without bottom heave of the glacial till on which the mat foundation for the new building will be founded. Based on seepage analyses, the owner's design team stipulated that the shoring extend 7.6 m (15-ft) below the final excavation depth of 13.1 m (25-ft) to limit in flow from the upper aquifer under the cutoff system. Preliminary designs for a secant pile system and a freeze wall system with four rows of tiebacks were provided for pricing by shoring contractors.

CONTRACTOR DESIGNED EXCAVATION SYSTEM ALTERNATIVE

The successful shoring contractor provided an alternate design-build excavation support system which relied on pre-trenching through the wood debris to remove known obstructions in the fill before installing a 20.7 m (68-ft) deep soil mixed cutoff with three rows of tiebacks to support the excavation. The deep soil mix wall system would be installed using Cutter Soil Mixing (CSM) technology. After evaluating the potential cost and schedule risks of the three competing systems, the owner selected the design build CSM system for two reasons. First, the proposed C-B slurry pre-trenching eliminated impacts to the shoring installation process by removing the known debris in the fill. Second, the real-time verticality controls of the CSM equipment provide an assurance of panel overlap and a continuous water cutoff which neither soil freezing nor secant piles are able to provide.

Cement-Bentonite (C-B) Slurry Pre-Trenching

Although the true extent of the wood debris would not be known until mass excavation began, it was never considered feasible to construct a structural soil-cement wall through a soil profile with significant amounts of wood debris. The pre-trenching was intended to remove the obstructions and wood debris in the CSM wall alignment to maximum a depth of 10.4 m (34-ft) so subsequent CSM could achieve a homogeneous soil-cement product.

C-B slurry was chosen over conventional bentonite slurry for the pre-trenching for two reasons. First, the C-B slurry is self hardening to a compressive strength in the range of 0.1 to 0.3 Mpa (20 to 50 psi) so it was not necessary to backfill the trench for stability reasons. However, sand was pushed into the open end of the slurry trench to provide a backfill more conducive to achieving consistent soil-cement strengths with the follow on CSM installation. Second, the C-B slurry increases trench stability. With a greater density than bentonite slurry, the C-B slurry generates a larger stabilizing pressure on the trench walls. With a higher viscosity compared to bentonite slurry, the C-B slurry reduces the risk of slurry loss into voids along the trench walls and maintains more spoil material in suspension, thereby increasing the density of the slurry in the trench during excavation.

While stability analyses are seldom done for slurry wall excavations, the City of Seattle Department of Transportation (SDOT) required the submission of a stability analysis for any C-B slurry pre-trenching adjacent to city streets. Utilizing the soil parameters outlined in the geotechnical report for a loose sand fill and the C-B slurry properties recorded in earlier bench testing, a factor of safety against trench failure was calculated using the closed form solution proposed by Filz et. al. (2002). Table 1 outlines the parameters used to calculate the factor of safety. Since strength parameters for the wood debris were not available, the sensitivity of the factor of safety to decreases in the strength and unit weight was assessed prior to beginning the pre-trenching. In general, the reduction in the factor of safety with a decrease in strength tended to be offset by the decrease in unit weight expected with wood debris.

Table 1 – Slurry Trench Stability Parameters and Resulting Factors of Safety

Trench Depth	Slurry Depth	GW Depth	Surcharge Loading	Slurry Density	Soil Total Unit Wt.	Phi Angle	F.S.
m (ft)	m (ft)	m (ft)	kN/m^2 (psf)	Kg/m^3 (pcf)	Kg/m^3 (psf)	(deg)	
3 (10)	0.3 (1)	2.1 (7)	4.8 (100)	1185 (74)	1826 (114)	30	1.30
6.1 (20)	0.3 (1)	2.1 (7)	4.8 (100)	1185 (74)	1826 (114)	30	1.25
9.1 (30)	0.3 (1)	2.1 (7)	4.8 (100)	1185 (74)	1826 (114)	30	1.15

During the initial pre-trenching, it became obvious that the wood debris below the upper sand fill consisted almost entirely of a matrix of mill ends and slab wood with interstitial sawdust and the occasional timber pile. There was essentially no sand in the wood debris. While the amount of wood was more substantial than initially anticipated, the interlacing of the wood debris had the effect of reinforcing and stabilizing the sidewalls of the pre-trench excavation.

The contractor regularly sampled and measured the density of the C-B slurry from the active trench. This testing verified that the fresh slurry, which had a density of 1126 Kg/ m^3 (70 pcf), was suspending sufficient soil particles from the excavation to achieve the minimum in trench density of 1185 Kg/m^3 (74 pcf) used in the stability analysis. There were no significant slurry loses and no observed sidewall stability problems. Survey monitoring points located approximately 3 m (10 ft) from the center of the trench indicated 5 to 18 mm (0.2 to 0.6 in) of lateral movement. One monitoring point, located on the curb adjacent to a sidewalk damaged by the weight of the excavator, indicated a lateral movement of 31 mm (1.2 in). The pre-trenching successfully removed and replaced the wood debris to a maximum depth of 10.4-m (34 ft) without any unacceptable movements of adjacent structures or utilities.

Cutter Soil Mixing (CSM) Excavation Support System

CSM Method

The CSM equipment was developed by Soletanche-Bachy and Bauer utilizing the cutter head technology from hydro fraises, which are used to excavate diaphragm walls. The CSM equipment has two hydraulically driven motors mounted horizontally at the bottom of a single element kelly bar as shown in Figure 3. As the cutters rotate, the soil is broken down and mixed with the slurry being injected just above the cutter wheels. Depending on soil conditions and depth, bentonite slurry or cement slurry is injected during penetration. When the CSM reaches the maximum depth, cement slurry is injected as cutter head is retracted in a series of steps. The rate of slurry injection and the withdraw speed of the cutter head are pre-determined to insure that each 0.5 m (1.6 ft) step of the panel receives the target amount of cement and the minimum number of wheel rotations for adequate mixing. Once the panel is completed, any steel reinforcing is then inserted into the wet soil-cement mixture.

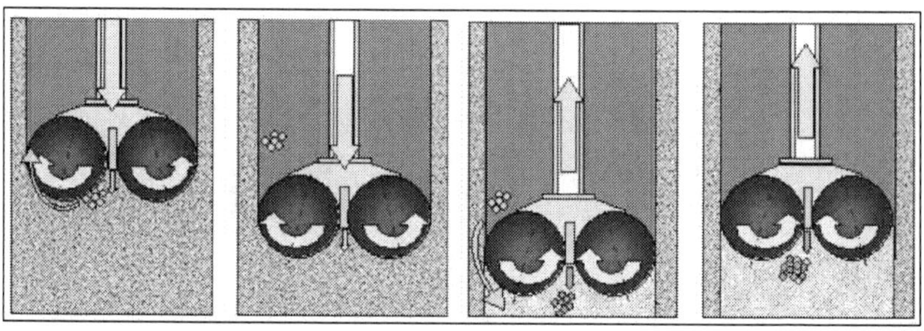

FIG. 3. Continuous Soil Mixing (CSM) Equipment and Process

Although the CSM method of soil mixing is able to treat a wider range of soils than conventional multi-axis soil mixing, the CSM's greatest advantage for cutoff applications is its verticality monitoring and control system. The operator of the CSM has a display of the location of the cutter head in the ground relative to the plan location of the panel. The operator can adjust the left and right location of the cutter head by changing the rotation speed and direction of one or both of the cutter wheels. The fore and aft location of the head is changed by manipulating the boom angle.

CSM Excavation Support Design

With the depth of the CSM excavation support system stipulated to be 20.7 m (68 ft) to limit seepage, the shoring contractor's design had to determine the pile sizes and spacing, tieback locations, and compressive strength of the soil-cement. The CSM unit used for this project creates a panel with plan dimensions of 0.8 m x 2.8 m (2.6 ft x 9.2 ft). By utilizing a primary-secondary sequence of panel installation and inserting one pile in the primary panels and two piles in the secondary panels as shown in Figure 4, the resulting pile spacing is 1.7 m (5.5 ft) on center.

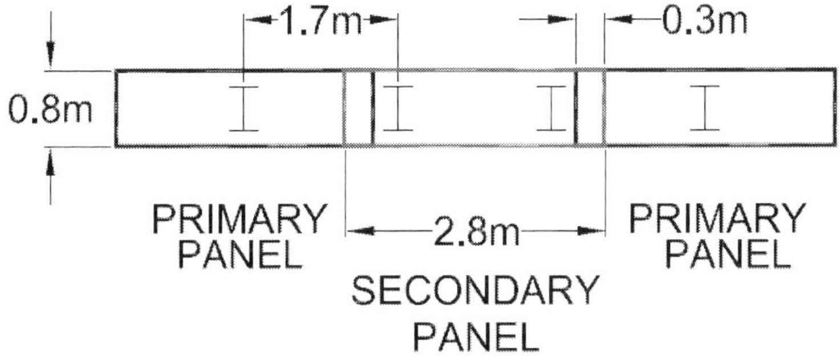

FIG. 4. Typical CSM Panel and Soldier Pile Plan View

Because of the dominance of water pressure loading versus earth pressure loading, the individual tieback design loads are controlled by the intermediate stages of construction, not the completed excavation. The total anchor design loads are approximately 25 percent higher than would be inferred from the design earth pressure diagram. The anchors are installed at steep 30° to 40° angles to develop their bond in the underlying till, resulting in axial loads of over 1300 kN (300 kip) on each soldier pile. Punching shear stability of the pile within the soil-cement panel requires consideration of skin friction around the pile perimeter both above and below the base of excavation and end bearing of the pile within the soil-cement panel. The soil-cement panel acts as watertight lagging and transfers the water and lateral earth pressure loads to both flanges of the soldier piles by arching. The calculated compressive arch loading within the soil-cement panel is 0.4 Mpa (60 psi). The specified strength of 1.4 Mpa (200 psi) represents a factor of safety greater than 3.

CSM Construction

After the pre-trenching operation had removed the wood debris, the CSM began. During penetration of the cutter, a bentonite water slurry with a SG of 1.03 was injected. During retraction of the cutter, a cement, water slurry with a SG of 1.68 was injected. In addition to the real time monitoring and control of grout flow, grout density, cutter wheel rotations, and depth, wet soil-cement samples were cast into 75 mm x 150 mm cylinders for compression testing.

The effectiveness of the CSM system is judged by its ability to safely support the adjacent ground with minimal movement and to isolate the excavation from the ground water. Monitoring of the system indicated that the only appreciable lateral movement was out of the excavation. When the first row of tiebacks was locked off, the tops of the soldier beams moved over 25 mm (1 in) out of the excavation. Even with the subsequent excavation phases, very little of this outward movement has been reversed. There were no leaks in the CSM wall system although some weeping and dampness occurred, especially, in the outside corners where the combined movements of the panels out of the excavation tended to open the corner joints.

Dewatering

Since CSM excavation support and water cutoff extends 7.6 m (15 ft) below the bottom of the excavation in the till layer, the water in the upper aquifer is prevented from flowing into the excavation. However, the water trapped inside the cutoff had to be removed to permit excavation. A series of six shallow wells were installed inside the perimeter of the excavation to a depth of 19.8 m (65-ft) and screened full length to remove the upper aquifer water trapped by the CSM cutoff. They were decommissioned when the bottom of the excavation was reached.

With the excavation extending to a depth of 13.1 m (43 ft) and the lower aquifer having a piezometric head at a depth of 2.4 m (9 ft), the lower aquifer had to be depressurized to prevent bottom heave in the completed excavation. During the installation of the upper two rows of tiebacks, the piezometric head in the lower aquifer was lowered in stages to avoid the potential for large upward gradients (between the lower aquifer and the open drill casing) and large downward gradients (between the upper and lower aquifers in a recently installed tieback hole) that could displace the fresh anchor grout. For constructability reasons, the piezometric head of the lower aquifer was lowered to a minimum depth of 14.6 m (48 ft) by the time the bottom of the excavation was reached. Three 30.5 m (100 ft) deep wells are currently pumping 75 lpm (20 gpm) from the lower aquifer to maintain depressurization until the exterior building cladding is installed.

CONCLUSIONS

The use of C-B pre-trenching to remove the wood debris to depths of 10.4 m (34 ft) in an urban environment permitted the subsequent installation of a CSM excavation support and water cutoff system. By removing the wood debris and replacing it with sand and C-B slurry, a soil mix wall with a minimum strength of 1.4 Mpa (200 psi) was successfully installed to a depth of 20.7 m (68 ft) without the potential for delays and increased costs associated with the wood debris and timber piles located in the fill material. The CSM system with its real-time verticality monitoring and controls on the cutter head was used successfully for the first time to construct a tied back excavation support and water cutoff system in the Western United States.

ACKNOWLEDGMENTS

The authors appreciate advice received from Prof George M. Filz, PhD, P.E. of Virginia Polytechnic Institute and State University. Hart Crowser is the owner's Geotechnical Engineer and Geotechnical Special Inspector for this project.

REFERENCES

Filz, G.M., Adams, T., and Davidson, R.R. (2004). "Stability of Long Trenches in Sand Supported by Bentonite-Water Slurry." *J. Geotechnical & Geoenv. Engr.*, Vol. 130 (9): 915-921.

Failure of a Pile Retaining Wall

Vishnu Diyaljee, Ph.D., F.ASCE, P.Eng

Managing Director, GAEA Engineering Ltd, 212, 9054-51 Avenue, Edmonton, Alberta, T6E 5X4
diy001@gaeaeng.com

ABSTRACT: This paper addresses the failure of a pile retaining wall constructed as a remedial measure to arrest instability of a highway embankment resulting from movement of its sideslope which was contiguous with the valley slope of a nearby creek. A drainage blanket was incorporated within the distressed section of the highway in 1983 to mitigate the pavement distress that was occurring. This remedial measure proved to be effective over a two-year period after which pavement distress re-occurred in the summer of 1990 following a period of increased rainfall. Further geotechnical investigation in 1990 resulted in the construction of a cantilever pile wall to isolate the highway from the perceived slide activity of the highway sideslope. Failure of this pile wall occurred in the summer of 1993 about 3 years after its construction, eventually requiring realignment of the highway. Failure to fully understand the site conditions from the inception of the geotechnical investigation played a major role in the short-lived performance of the remedial measures.

INTRODUCTION

A 50 m section of Secondary Highway (SH) 507:02 between Pincher Creek and Beaver Coal Mines in Southern Alberta, Canada was experiencing pavement distress in the form of cracking and settlement along the eastbound lane for a number of years (Fig.1). Based on visual inspections, the pavement distress was determined to be an indirect consequence of slide activity occurring beyond the sideslope on the south side of the highway.

In 1983 and 1990, this section of highway was the subject of geotechnical investigations and the implementation of remedial measures aimed at mitigating periodic pavement distress. A drilled shaft cantilever pile retaining wall was constructed in May 1990 to isolate the highway prism from a larger slide activity encroaching the highway. However, this pile wall failed in the summer of 1993 by tilting as a result of the loss of soil support behind the walls. This paper addresses the site conditions, investigations, remedial measures and the pile wall failure and concludes with lessons learned from the pile wall failure.

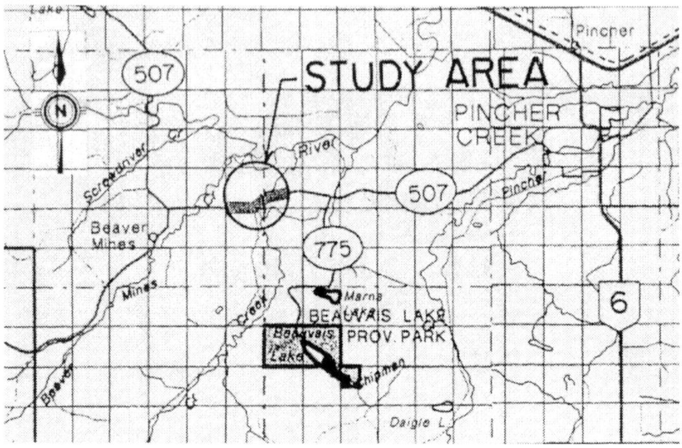

FIG.1. Project Location

SITE CONDITIONS AND HISTORY

The SH 507:02 alignment is along an east-west route through the hilly terrain prevalent in the Rocky Mountain foothills around Pincher Creek, Alberta. At the location of the pavement distress, the highway follows a ridge between two valley sections. There are two creeks on either side of the ridge - Mill Creek on the south and a no-name creek on the north. The pavement cracking and settlement was on the eastbound lane in proximity to the north valley slope of Mill Creek, the crest/top of bank of which is only about 60 m south of the centreline of the highway. Mill Creek is on a bend in this area, and its flow is eroding the valley face at its toe thus leading indirectly to slumping of the material from the higher elevations of the valley face.

The approximate elevation difference between the highway and the Mill Creek flood plain is about 40 metres. From a review of the 1950 air photos, it was observed that the entire Mill Creek valley at this location was plagued with landslide problems in the past.

From historical records, pavement distress along this section of highway dated back to 1973 when the existing roadway was upgraded. Apparently, a spring source, approximately 100 m west of the distressed area, was covered up at the time and pit run gravel used to bridge the location.

GEOTECHNICAL INVESTIGATIONS

1983 Investigation

A limited geotechnical investigation, consisting of drilling a few test holes, was undertaken in March 1983 within the highway right-of-way. The stratigraphy in the distressed area from testhole drilling consisted of 6 m of silty sand followed by 6 m of silty clay overlying clay shale and sandstone. The top 2 m of soil was found to be well above standard Proctor optimum moisture for the material. As a result, the installation of a granular blanket to a depth of about 3 m within the affected area and daylighting on the sideslope was recommended (Fig.2). This measure was implemented in 1985. The purpose of the granular blanket was mainly to facilitate the removal of moisture from the roadbed, which was thought at the time to be responsible for movement of the sideslope and hence pavement distress occurring within the traveled highway.

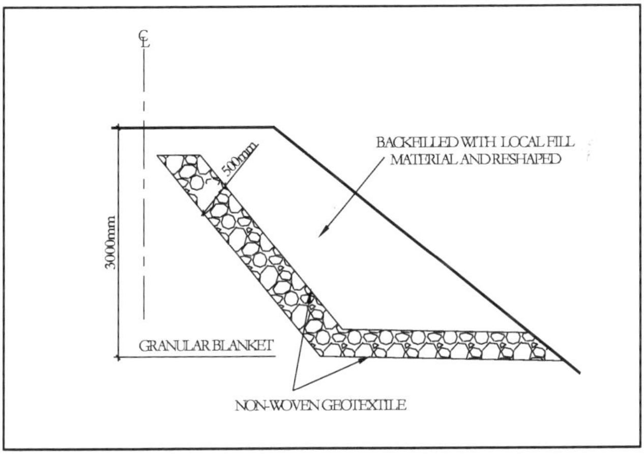

FIG.2. Drainage Blanket

1990 Investigation

The implementation of the drainage blanket remedial measure resulted in a decrease in pavement distress until around 1990 after which pavement distress resumed at an increasing rate. This renewed activity after a five (5) year period was felt to have been promoted by increased rainfall than in previous years. It is well known in Alberta that prolonged periods of low rainfall or drought conditions for several years are often followed by periods of heavy rainfall.

During the first week of May 1990, the distressed pavement section was reported to have settled about 1 m. As a result of this significant distress and the need to maintain traffic along this highway immediate attention was required. A detailed site review was undertaken on May 16. This was followed by testpitting and testhole drilling. The testhole drilling was undertaken with a Texoma Drill Rig normally used for the construction of bored concrete piles. This drill was used in the investigation instead of the conventional soil testing drill to determine whether there would be any problems in constructing bored piles should this alternative be chosen for stabilizing the roadway.

Subsequent to the site visit, on May 29, the distressed section was reported to be worsening. The fence along the south side of the highway right-of-way moved 2 m and the distressed section of highway settled considerably at its centreline. As a result only one lane of the highway was trafficable. This renewed activity coincided with continuous heavy rainfall.

During the first week in June drilling with the Texoma rig was undertaken to a depth of 12 m where hard clay shale was encountered and drilling terminated after penetrating this layer about 1 m. The drilled hole stood up without any caving of its sides indicating that bored piles could be constructed without the need for casing of the pile holes.

1990 Remedial Measures

In view of the urgency of the situation and the observed site conditions, two remedial options were considered - toe berm construction and the construction of a pile retaining wall. The toe berm option was ruled out due to the proximity of the steep valley face to the highway, river erosion causing steepening of the valley face, and the fragile nature of the south slope to the valley crest. Other forms of stabilization were not considered practical since little disturbance of the buffer zone was advisable because existing trees and vegetation were assisting in maintaining the stability of the area between the crest and highway right-of -way.

In deciding on the pile retaining wall option, several assumptions/judgments were made. These included the opinion that while the creek was actively eroding the banks, there was a sufficient buffer zone between the banks and roadway and it would take some time before this event would create a global instability of this area. Also, the presence of very hard material and sandstone and clay shale inclusions would create a restraint against the river erosion forces.

The fast tracked pile retaining wall required thirty, 760 mm diameter drilled shaft piles installed at 1.5 m centres to span across the length of the slide area. Each pile was reinforced with a 14 m long H-pile (W410x60) and nominal helical reinforcement, with the top 3 m consisting of H-pile and timber lagging to retain the roadway fill (Fig.3). The pile design was provided by the Contractor and installation of this pile retaining wall was implemented in July 1990.

The side slope portion on either outer side of the pile wall system was also reconstructed with granular material backfill to help free drainage to the sideslope.

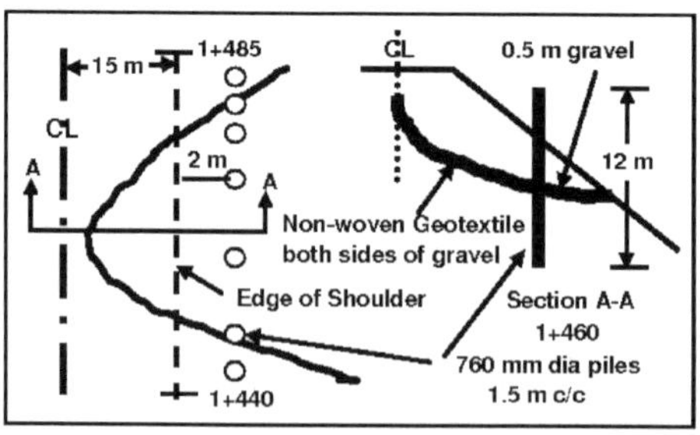

FIG.3. Plan and X-Section Showing Pile Layout

A subsurface trench drain with a perforated pipe was also installed in the ditch of the westbound lane as an added precaution to minimize the flow of subsurface water from the higher ground existing on the north side of the distressed section. The perforated pipe was daylighted in the treed area on the south side away from the slide area.

SITE STATUS IN 1990-1991

Following the implementation of the pile retaining wall, a site inspection was undertaken on October 29, 1990. This inspection showed no cracks in the highway surface. The sideslope where the piles were installed was neatly trimmed and showed no signs of cracking and bulging. During this inspection, the installation of three (3) slope indicators was considered to monitor the performance of the retaining wall and surrounding area. However, installation of these slope indicators was not undertaken. Although the pile wall system was effective in preventing further settlement of the eastbound lane, the south sideslope slumped once again in the summer of 1991 following heavy rainfall in the general area.

An inspection of the site was undertaken on July 1991 following the reported slumping. This inspection showed that the retaining wall was still performing well since no cracks could be observed along the highway within the length of the pile wall. However, it was noted that the sideslope outside of the pile wall had slumped exposing the backface of the pile wall (Fig.4).

FIG.4. Sideslope Slump

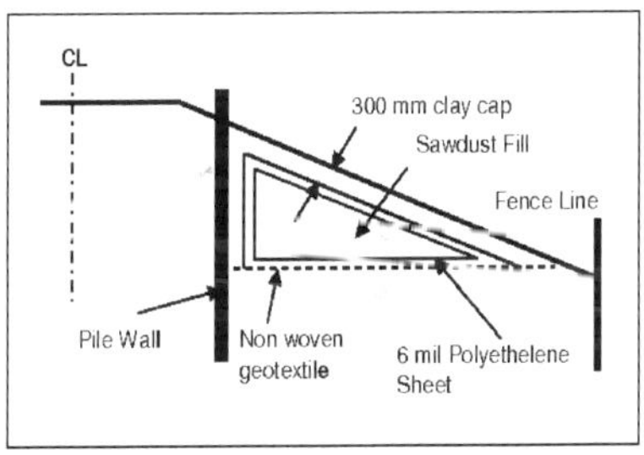

FIG.5. Sawdust Fill in Sideslope Repair

This failure was attributed to river erosion and the distinct possibility of seepage within the highway grade since instability of the distressed section was noted to be coincident with periods of heavy rainfall. This opinion was confirmed during the 1990 investigation prior to the recommendation of the pile wall as the preferred stabilization measure. There was also some evidence of cracking towards the highway shoulder outside of the limits of the pile wall.

Based on the site observations, it was decided to restore the failed sideslope using sawdust lightweight fill to minimize the load on the sideslope in the hope that it would minimize further slumping of the sideslope. Whether or not the pile wall should be extended was to be considered following restoration of the sideslope.

Figure 5 shows a schematic outline of remedial measures implemented. A thin layer of top soil was used overtop of the sawdust to encourage vegetation growth.

1993 Failure

The slide activity increased at a fast rate in the month of July 1993 due to heavy rains resulting in mass movement of the soil towards the Mill Creek valley slope. As a result of loss of lateral support, many of the piles experienced bending away from the road, the lateral movement being small in the centre of the slide area but more towards the outer limits of the slide area. The roadway subgrade soil also slowly moved through the openings in between the concrete piles below the elevation of the timber lagging resulting indirectly in a 0.6 metre drop in the road grade. Typical photographs of tilted piles and slide area are shown in Figures 6 through 8.

FIG.6. Tilted Piles (1993)

FIG.7. Displaced Tilted Piles (1994)

FIG.8. Overview of Slide Area

Two slope indicators (SI 1A and SI 1B) were installed within the slide scarp area in August 1993 to monitor the depth of slide movement. These SIs were located over the length of the pile wall parallel to the south shoulder of the highway, outside of the pile wall but within the highway right-of-way. Their monitoring indicated a slip plane at 6 m below the ground surface at the location of the SI 1A.

A cross-section of the highway within the distressed section, Figure 9 shows the location of the SI 1A, relative to centreline of the roadway, inferred plane of movement and soil profile. The depth of movement corresponded with the interface of the silty clay and gravelly sand layers. The slide plane was at about a depth of 11 m in relation to the finished roadway grade. In relation to the length of pile installed, the slide plane removed most of the pile fixity, which resulted in the tilting of the piles.

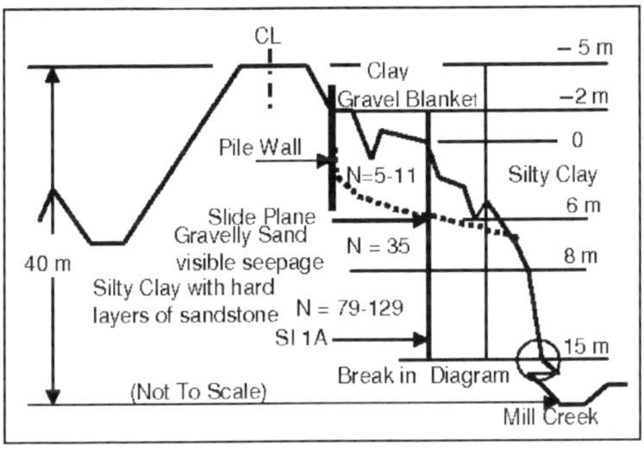

FIG.9. Cross-section of Highway at Slide

REMEDIAL MEASURES

A review of possible remedial options indicated that it would be quite expensive to restore the verticality of the piles and to implement subsurface trench drainage measures within the slide area between the piles and the top of the valley face. Even then, it was felt that the long term effectiveness of the subsurface drainage measures in stabilizing the slide was debatable, since the toe erosion in the creek was very active.

FIG.10. Bank Slumping at Bend in Mill Creek

In the final analysis, it was decided to shift the alignment of the highway by 20 m north of the slide area filling in the creek on the north side. The highway grade line was also lowered such that the new grade would be close to the elevation of the plane of movement. The principle behind the lowering of the grade line was to remove as much soil with moisture content wetter than optimum. Excavation of additional material from the south backslope which is likely being influenced by the slide activity was also recommended. The horizontal alignment was adjusted for a length of about 2 km for consistency with the ground features that existed beyond the slide area.

Since the no name creek on the north side would be covered up with the new grade, the creek alignment was also shifted further north and would flow partly in open cut and partly through a long 600 mm diameter culvert.

DISCUSSION AND CONCLUSIONS

Based on observations from 1983 to 1993, groundwater seepage and toe erosion appeared to be primary contributory factors to the recurring roadway distress since distress only occurred during periods of heavy rainfall. These factors may have acted singly or in combination.

The gravel blanket constructed in 1983 at the 3 m depth to remove groundwater seepage to ensure stable site conditions was not entirely successful since the source of groundwater was at around 11 m below the roadway grade at centreline. It is unlikely that the sawdust fill could have become saturated and contributed to increasing the weight of the sideslope leading to the 1993 failure since the sawdust was encapsulated with polyethylene and capped with clay.

The failure of the pile wall resulted in a reflection of whether the pile wall system might not have failed if a tie-back system were designed and installed in 1991 or in 1993. Of course, the tie-back system might not have solved the problem permanently since the failure plane was deeper than the depth of embedment of the piles. However, this system could have likely extended the serviceability of the road for a few more years by preventing the tilting of the piles.

This case study illustrates how engineering judgments can sometimes go wrong when all the site conditions are not carefully investigated and analyzed. Review of aerial photographs would have been prudent from the inception of the investigation in 1983 but this was not done at the time. This review would have given insight into the instability problems along the creek and could have obviated the use of the pile wall or indicated that deeper piles would have been required. However, in hindsight, while this may have been the most sensible approach, very often there are social factors that influence engineering decisions. In this case, there was a need to have this roadway open to traffic at all times, and as such quick action was required.

The lesson learned from this failure is that the resolution of geotechnical problems requires the careful understanding of all the factors that influence the behaviour of a site and requires not only examining the location of failure but the surrounding terrain. This can be supported by reviewing aerial photographs and an understanding of the site geology. This failure resulted in the subsequent recognition that review of aerial photographs was an important aspect of landslide site evaluation irrespective of whether a site failure from observation is judged to be simple or complex.

The relationship between field measurements and soil behavior in TNEC deep excavation

Abdolreza Osouli, S.M. ASCE[1], Youssef M.A.Hashash[2], M. ASCE,

[1] Graduate researcher, Department of Civil and Environmental Engineering, University of Illinois at Urbana-Champaign, 205 North Mathews, Urbana, IL 61801; Email address: aosouli2@illinois.edu
[2] Associate Professor, Department of Civil and Environmental Engineering, University of Illinois at Urbana-Champaign, 205 North Mathews, Urbana, IL 61801: Tel.: 1-217-333-6986; Fax: 1-217-333-9464; Email address: hashash@illinois.edu

ABSTRACT

In urban deep excavations, instruments are placed to monitor deformations and to control construction and reduce the risk of excessive and potentially damaging deformations. Data collected from these instruments are also used in inverse analyses to develop improved soil models suitable for representing soil response during excavation. The second author has introduced a new inverse analysis approach, SelfSim, that utilizes measured excavation performance to extract the underlying soil behavior. A systematic study with synthetically generated set of instrument measurements using an idealized soil profile shows that in addition to the measurements of lateral wall deflections and surface settlement, inclinometers further away from the wall are equally important in extracting soil behavior.

In this paper a deep excavation case history is studied to implement the findings of the numerical study. The 18-story Taipei National Enterprise Center (TNEC) building with five basement levels was constructed in 1991. The measured lateral deformations of inclinometers in the wall and 22-m further away from the wall are used to extract the constitutive model through the inverse analyses approach. The extracted constitutive model is used in predicting the other inclinometer measurements located at different distances from the wall and surface settlements behind the wall. The predicted behavior of the excavation and its agreement with measurements at the site are discussed in detail.

INTRODUCTION

Due to influence of excavations on surrounding structures, monitoring programs are commonly set up at excavation sites to evaluate the design assumptions, determine causes of movements, adjust the construction procedure, and evaluate stability. Several empirical and numerical methods have been developed based on precedent and field observations to construct safe excavations (Peck 1969; O'Rourke 1981; Clough et al. 1989; Clough and O'Rourke 1990). Current modeling approaches are not always able to interpret and thus fully benefit from the rich soil stress–strain data implied by field observations which results in limited integration of numerical modeling with the observational approach. Hashash et al. (2006) introduced a robust and efficient approach to extract soil behavior using SelfSim framework by integrating field observations and numerical modeling. SelfSim is an inverse analysis framework that implements and extends the autoprogressive algorithm proposed by Ghaboussi et al. (1998). It extracts behavior of materials facilitated by the use of a continuously evolving neural network (NN) material model, rather than calibrating properties of conventional constitutive models. In SelfSim at a given excavation stage two complementary analyses are performed. First, the force boundary condition is applied to get stresses. Second, the measured field deformations (displacement boundary) are imposed on the model to get strains. The extracted stress-strain pairs are used to re-train the NN material model until the two analyses give similar results (Hashash et al. 2006). Song et al. (2007) in a systematic study showed the influence of various instrumentation layouts in extracting soil behavior from monitored excavations. The study demonstrated that having inclinometers further away from the wall can improve the extracted soil behavior by capturing the small strain non-linearity in soil stiffness.

This study applies the findings of Song et al. (2007), in which synthetic data was used, to the 20-m deep excavation of Taipei National Enterprise Center (TNEC), in Taipei. SelfSim is used to extract soil behavior from inclinometer measurements in the wall and 22 m away from the wall. The extracted model is used to predict the soil behavior elsewhere in the excavation.

LEARNING FROM INCLINOMETERS - NUMERICAL STUDY

A numerical study was conducted to simulate a typical braced excavation to study the inclinometer layout effect on extracted behavior. Inclinometers are placed at different locations within the excavation site (Fig. 1).

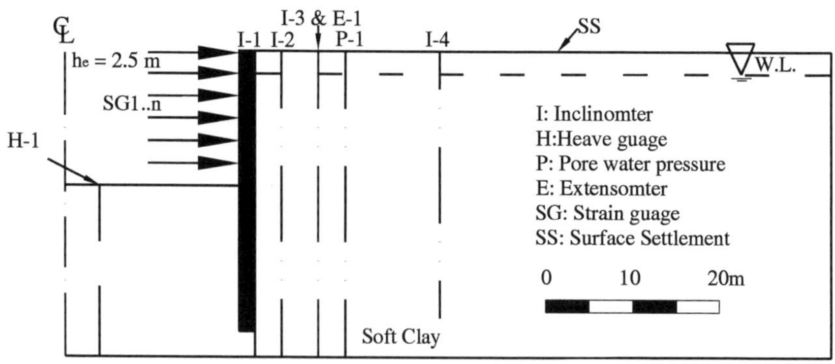

Fig. 1 Dimension of the excavation and instrument locations

Other instruments such as extensometer, piezometer, surface settlement points, strain gauges, and heave gauges are installed for reference comparison of extracted soil behavior. In this 15.0-m deep model excavation, a 0.9-m thick concrete wall is used, supported by struts. The assumed soil profile includes three sub layers of 50-m soft clay,

30-m stiff clay and 40-m very stiff clay. The ground water table is 2.5 m below the ground surface. The simulated measurements used as target are generated using a finite element model of the excavation with MIT-E3 soil constitutive model (Whittle 1987) representing the constitutive soil behavior. The inverse analysis is conducted using SelfSim, whereby the measured response from instruments is learned while the underlying soil behavior is extracted.

SelfSim learning was conducted using lateral wall deflection measurements only and the extracted soil model is then used in an FE simulation to compute the excavation behavior. This analysis showed that the model captures wall deflections very well but does not provide a very good match for surface settlements (not used in SelfSim learning) (Song et al. 2007).

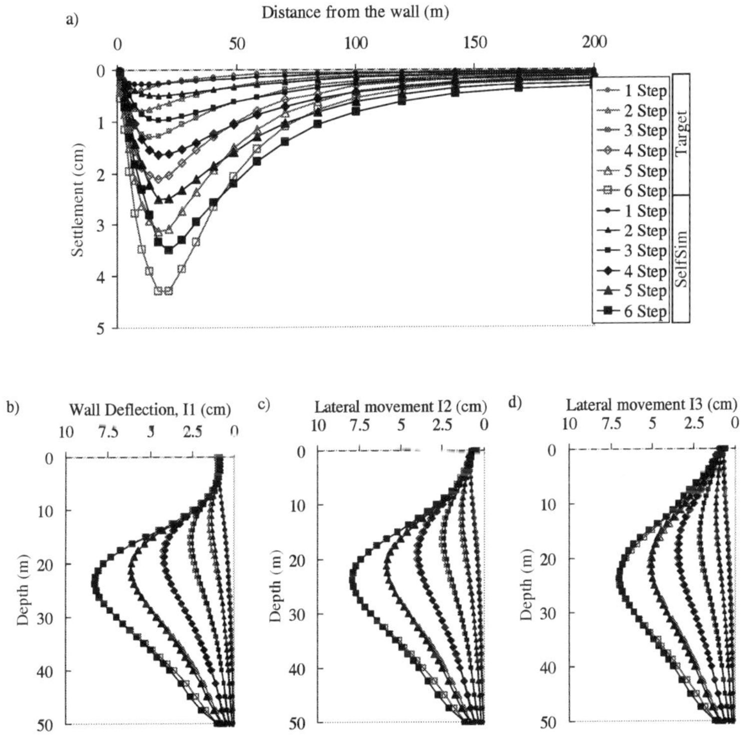

Fig. 2 Learning with wall deformations and lateral movement at I3; (a) Surface settlements, (b) Wall deflection (I1), (c) lateral movement at I2, (d) lateral movement at I3

Therefore, SelfSim learning is conducted using lateral wall deflection measurements and lateral deformations at I3. The extracted model is used in an FE simulation to compute the excavation behavior. Fig. 2 shows that not only the model captures the wall deformations and inclinometer measurements at I2, but also the surface settlement prediction is also improved. So it appears that inclinometers at some distance from the wall provide redundant measurements that can be used to obtain the surface settlements especially when settlement points are damaged and are no longer accessible.

LEARNING FROM INCLINOMETERS – CASE STUDY

The Taipei National Enterprise Center (TNEC) is an 18-story building with five basement levels (Ou et al. 1998). The site has slightly irregular shape and occupies an area of about 3500 m^2, as shown in Fig. 3.

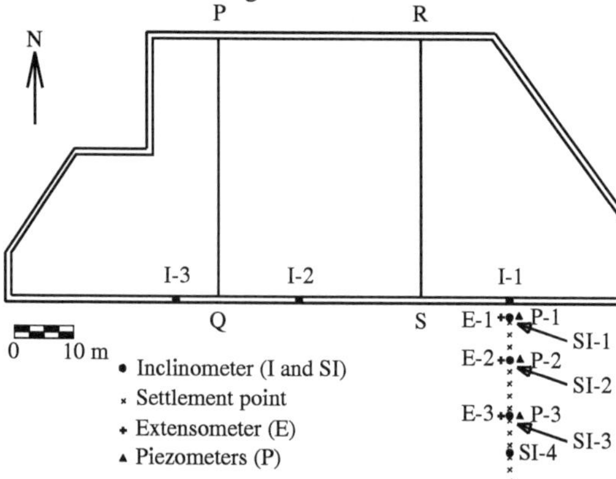

Fig. 3 TNEC plan view and Instrument locations, modified after Ou et al. (1998)

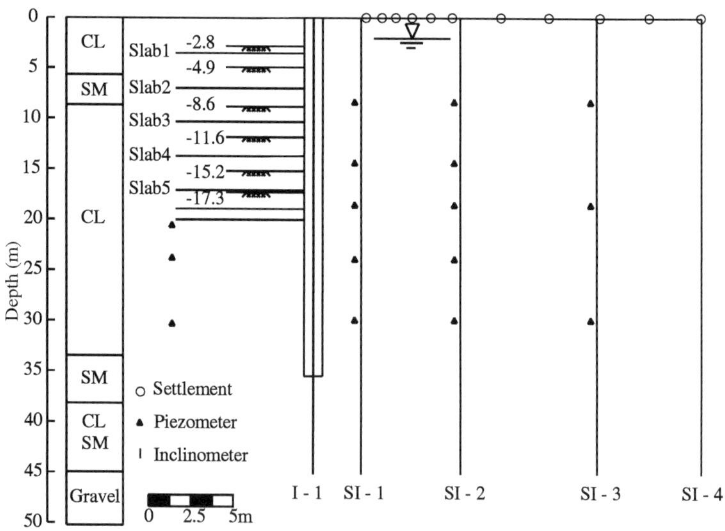

Fig. 4 Excavation section view

The 19.7-m deep top-down excavation with concrete floor slabs as support of the wall is extensively monitored with earth pressure cells on the wall, rebar stress meters on the reinforcement cage, inclinometers, heave gauges and settlement gauges (Fig. 3). A cross section of the excavation wall is shown in Fig. 4. The excavation site consists of six layers of alternating silty clay and silty sand deposits overlying a thick gravel formation. The third layer is 25-m-thick silty clay (CL), and it is mainly this layer that affects the excavation behavior in this case. The hydraulic conductivity (k) from one-dimensional consolidation tests is around $4x10^{-6}$cm/s (Ou et al. 1998). A gravel formation is located 45 m below the ground surface. Prior to excavation, the ground water table was 2 m below the ground surface. A 90-cm-thick and 35-m-deep diaphragm wall was used as the earth-retaining structure.

The TNEC site was excavated down to 19.7 m in seven main stages. However in this study the excavation site was modeled with five excavation stages down to a depth of 15.2 m depth due to slight shift of measured wall deformations after step five of the excavation that made the use of these measurements in inverse analysis difficult.

INVERSE ANALYSIS THROUGH SELFSIM LEARNING

A set of SelfSim learning analyses was conducted to extract the underlying soil behavior for TNEC excavation site. Prior to any learning, the NN material models are trained to reproduce linear elastic behavior. Computed deformations using elastic behavior training for all soil layers are shown in Fig. 5. The computed deformations are less than actual deformations and differ significantly from the measured values.

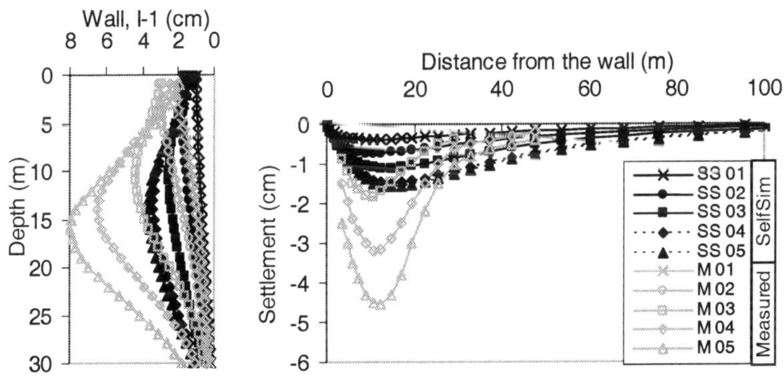

Fig. 5 Computed wall deformation (I-1) and surface settlement after prelearning with wall deformations only up to step 3 of excavation

SelfSim learning is then conducted using measured wall deformations up to the third step of the excavation. Computed and predicted deformations of the ground surface, wall and inclinometers SI-1, SI-2, SI-3 and SI-4 after five passes of SelfSim learning are shown in Fig. 6 and are an improvement over results reported in Fig. 5. While, the deformations of the wall, inclinometer SI-1 and SI-2 are reasonably matched with the measurements up to step 3 of the excavation, the surface settlement and deformations of inclinometers further away from excavation site (i.e. SI-3, and SI-4) differ significantly from the measured values. Since wall deformations up to step 3 were the only information used in SelfSim learning, the computed deformations for wall and inclinometers close to the wall are matched reasonably well. Other measurements are not well captured.

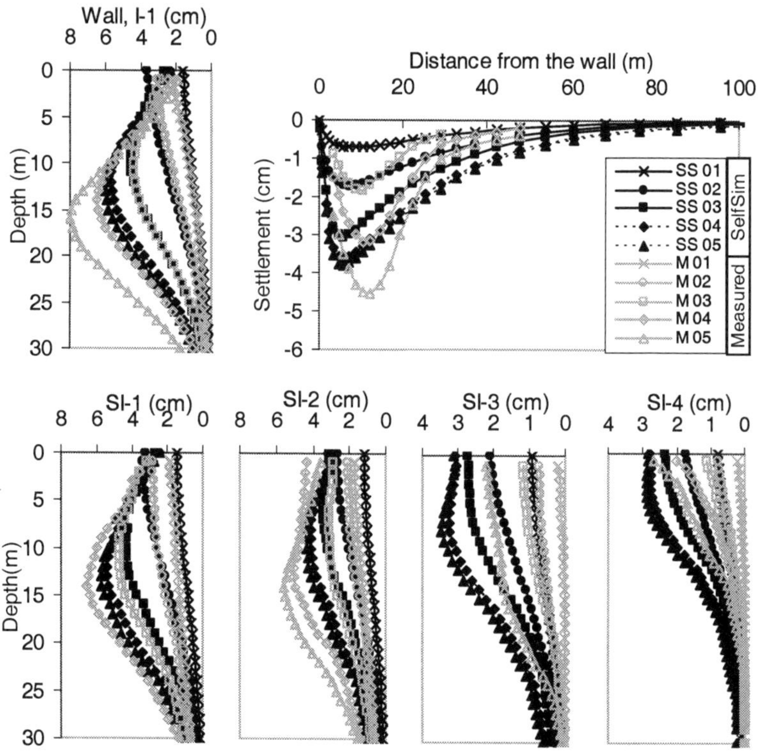

Fig. 6 Computed wall deformation (I-1), surface settlement, and inclinometers behind the wall (SI1..4), after 5 passes of SelfSim learning with wall deformations only up to step 3 of excavation

SelfSim learning is then continued with wall deformations and inclinometer deformations SI-4, 22 m away from the excavation wall using the database of SelfSim learning with wall deformations only. All 5 steps of measurement values are used. Fig. 7 shows the computed and predicted deformations of the excavation after six additional learning passes. It appears that the inclinometer further away from the excavation has important information about small strain nonlinearity of soil behavior since the computed settlements has improved significantly compared to Fig. 6. Note that surface settlement measurements were not used directly in the learning process. The predicted deformations of inclinometers SI-1, SI-2 behind the wall are reasonably matched with the actual measurements. The deformations at SI-3 are overpredicted. We are still trying to better understand the reason behind this over prediction.

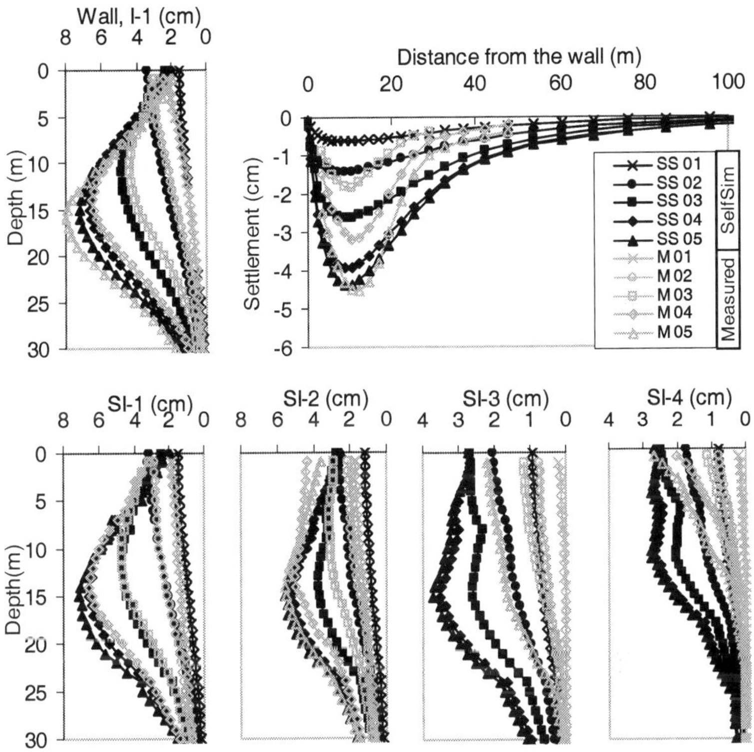

Fig. 7 Computed wall deformation (I-1), surface settlement, and inclinometers behind the wall (SI1..4), after 6 passes of SelfSim learning with wall deformations and inclinometer at 22 m from the wall (SI-4) up to step 5 of excavation using the database of SelfSim learning with wall deformations only

CONCLUSIONS

This paper presented the application of inclinometer measurements in extracting the soil behavior for TNEC excavation case study. This study shows that wall deformations provide essential information for learning of overall excavation behavior. An inclinometer placed within or in close proximity to the wall is important. The wall measurements are not enough to capture the small strain non linearity of soil behavior.

Additional inclinometers placed farther back from the wall provide supplementary information that can be used to extract the global excavation behavior. The SelfSim inverse analysis approach was able to extract sufficient information about the underlying soil behavior from the two sets of inclinometer measurements to provide a reasonable prediction of settlements. Installed inclinometers at some distance from the excavation wall provide redundancy to surface settlement measurement. In the event that settlement points are lost during construction, the inclinometer measurements can, via the SelfSim inverse analysis approach, provide reliable estimates of surface deformations.

ACKNOWLEDGEMENT

This material is based upon work supported by the National Science Foundation under Grant No. CMS 02-19123 under program director Dr. R. Fragaszy. The authors would like to thank Professor Chang-Yu Ou for facilitating access to the TNEC data. Any opinions, findings, and conclusions or recommendations expressed in this material are those of the writers and do not necessarily reflect the views of the National Science Foundation.

REFERENCE

Clough, G. W. and O'Rourke, T. D. (1990). "Construction induced movements of insitu walls". Design and Performance of Earth Retaining Structures. New York, NY, ASCE: 439-470.

Clough, G. W., Smith, E. M. and Sweeney, B. P. (1989). "Movement control of excavation support systems by iterative design". Foundation Engineering Proceedings Congress. Evanston, IL, ASCE: 869-884.

Ghaboussi, J., Pecknold, D. A., Zhang, M. F. and Haj-Ali, R. M. (1998). "Autoprogressive training of neural network constitutive models." *International Journal for Numerical Methods in Engineering* 42(1): 105-126.

Hashash, Y. M. A., Marulanda, C., Ghaboussi, J. and Jung, S. (2006). "Novel approach to integration of numerical modeling and field observations for deep excavations." *Journal of Geotechnical and Geoenvironmental Engineering* 132(8): 1019 - 1031.

O'Rourke, T. D. (1981). "Ground Movements Caused By Braced Excavations." *Journal of the Geotechnical Engineering Division* 107(GT9): 1159-1178.

Ou, C.-Y., Liao, J.-T. and Lin, H.-D. (1998). "Performance of diaphragm wall constructed using top-down method." *Journal of Geotechnical and geoenvironmental engineering* 124(9): 798-808.

Peck, R. B. (1969). "Advantages and limitations of the observational method in applied soil mechanics." *Geotechnique* 19(1): 171-187.

Song, H., Osouli, A. and Hashash, Y. (2007). "Soil behavior and excavation instrumentation layout". *7th International symposium on field measurements in geomechanics FMGM 2007.*, Boston, MA.

Whittle, A. J. (1987). "A Constitutive Model for Overconsolidated Clays with Application to the Cyclic Loading of Friction Piles". Civil Engineering. Cambridge, MA, MIT: 641.

Instrumentation and Monitoring of an MSE/Soil Nail Hybrid Retaining Wall

Timothy A. Wood[1], Priyantha W. Jayawickrama, M.ASCE, Ph.D.[2], and William D. Lawson, M. ASCE, P.E., Ph.D.[3]

[1] Research Associate, Center for Multidisciplinary Research in Transportation, Texas Tech University, Lubbock, TX 79409; timothy.a.wood@ttu.edu
[2] Associate Professor, Dept of Civil and Environmental Engineering, Texas Tech University, Lubbock, TX 79409; priyantha.jayawickrama@ttu.edu
[3] Assistant Professor, Dept of Civil and Environmental Engineering, Texas Tech University, Lubbock, TX 79409; william.d.lawson@ttu.edu

ABSTRACT: MSE/soil nail hybrid earth retaining walls provide a more economical design for applications in cut/fill situations than the traditionally used full height MSE and drilled shaft retaining walls. MSE/soil nail hybrid earth retaining walls use a soil nailed wall in the cut section and an MSE wall in the fill section. In spite of the significant cost savings they offer, hybrid walls have not seen widespread use primarily because of lack of understanding on wall design and performance. This paper describes an instrumentation and monitoring effort that was undertaken with the objective of improving our understanding of hybrid wall design and performance. In this project, two separate panels of a hybrid wall constructed in San Antonio, Texas were selected for instrumentation and monitoring. The first wall panel consisted of a 4.0m soil nail wall and a 5.4m MSE wall while the second wall panel consisted of 5.0m soil nail wall and a 4.4m MSE Wall. The instrumentation scheme for the wall included vibrating wire strain gages, vertical inclinometers, horizontal inclinometers, and tiltmeters. The data collected from these two instrumented wall sections provide valuable insight to the mechanisms controlling the performance of MSE/Soil Nail Hybrid Wall systems.

INTRODUCTION

Highway construction projects often require retaining walls to be built into "side hill" situations where the bottom portion of the wall is placed below existing ground, and the top portion placed above. Traditional design solutions used in side hill (or cut/fill) conditions have involved full height MSE walls and drilled shaft retaining walls. When full height MSE walls are used, significant amounts of earth material must be removed near the bottom portion of the wall to allow placement of the reinforced backfill. The excavated slopes must be supported with temporary shoring while the MSE wall is being constructed. The large amount of earthwork and the use of temporary shoring increase the overall cost of construction of this type of wall significantly. The other design alternative that has been used in side hill situations involves the use of drilled shafts to support lateral soil loads. The drilled shaft walls resist lateral earth pressure by cantilever action. The depth of embedment required generally varies from one to two times the wall height. Therefore this type of wall is quite expensive to construct.

An innovative and more economical design alternative that has been used more recently for side hill walls involves the use of a soil nailed wall in the cut section, and an MSE wall for the fill section (See Figure 1). Although such MSE/Soil Nail Hybrid wall systems have been found to be very cost effective, only a handful of walls of this type have been built until now. The limited use of hybrid walls is largely due to lack of clear guidance on the appropriate procedures to be used in the design of these structures. In other words, MSE/Soil Nail Hybrid Walls are still considered "experimental" by the industry. This paper describes a hybrid wall instrumentation and monitoring effort that was undertaken by Texas Tech University researchers under sponsorship of the Texas Department of Transportation (TxDOT). The above wall instrumentation work was completed as a part of a multi-year research project that was initiated with the objectives of investigating the design and performance of MSE/Soil Nail Hybrid walls with respect to wall deformation, internal stability and global stability.

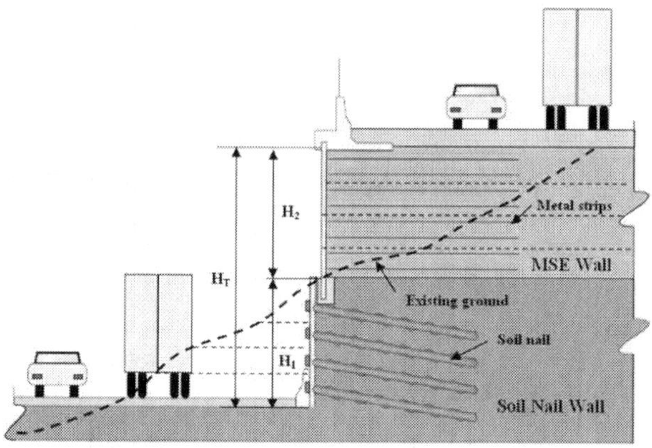

Figure 1 - Schematic of a MSE/Soil nail hybrid retaining wall

HYBRID WALL AT IH 410- INGRAM ROAD OVERPASS

The use of MSE/Soil Nail hybrid walls for cut/fill applications has been quite recent and, until now, very few walls of this type have been built in Texas. Review of literature revealed only one other detailed case study of this type of wall (Turner and Jensen, 2005). In the January of 2008, work began in connection with the construction of an MSE/Soil Nail Hybrid Wall at the IH-410 overpass over Ingram Road in San Antonio, Texas. This wall construction project offered a rare opportunity to instrument a hybrid wall during construction and monitoring of its performance. The wall is approximately 670m in length and is being built into the approach embankment of the west bound IH-410 overpass bridge at Ingram Road. The heights of soil nailed and MSE portions vary along the length of the wall. Two separate sections of the wall were selected for the purpose of instrumentation and monitoring. The first, Wall Section A, was located at Station 703+80. The height of the soil nail wall at this location is 4.0m and the height of the MSE wall is 5.4m MSE/soil nail hybrid earth MSE/SN ratio of 1.35. The second wall section, Wall Section B, is located Station 705+40. It has a 5.0m soil nailed wall and a 4.4m MSE wall yielding an MSE/SN ratio of 0.88. The two wall sections are shown in Figures 2(a) and (b).

DEVELOPMENT AND IMPLEMENTATION OF THE INSTRUMENTATION PLAN

The data of greatest interest in this instrumentation project included tensile loads in soil nails and deformations of the wall. The development of a suitable instrumentation plan to collect the necessary data required careful evaluation of several important factors. These include: (a) the type of data desired, (b) ruggedness of the instrumentation and its suitability for use in the construction environment, and (c) the need for redundancy so that the impact from possible instrumentation failure and consequent data loss could be minimized. The following sections provide a detailed description of the selection of the instrumentation, data collected, and significant findings on wall behavior.

Measurement of Tensile Forces in Soil Nails

Customarily, tensile forces in the soil nails are determined by using strain gages to measure tensile stresses in the bar and then calculating the nail forces by assuming a cracked grout section. Previous instrumentation projects have used two different types of strain gages: (a) electrical resistance type gages (i.e. foil gages), and (b) vibrating wire gages (VWGs). The electrical resistance foil gages have the advantage of providing instantaneous and continuous strain data at a low cost. However, prior experience with this type of gage shows that their ability to survive rough handling and the harsh environment that exists at a typical construction site is questionable. In particular, the authors' experience in previous instrumentation projects has shown that electrical resistance gages could yield greater than 60% rate of failure (Johnson, 2005). Based on this experience, the decision was made to use vibrating wire gages (VWG) in this project. Though VWGs are more expensive (nearly 10 times the price of foil gages) and only provide non-continuous or discrete data, they are far more durable than foil gages. To determine the stress in the soil nails, two types of vibrating wire strain gages were used. The first type of VWG is the spot welded gage. These gages were welded to steel tendons of the soil nails. In an effort to minimize effects from bending of the bars, spot welded VWGs were applied to both top and bottom of the bar at each gage location as shown in Figure 3(a).

The second type of VWG used for this project is the embedment type. These gages record the change in strain the concrete grout around the soil nail. The plates at either end of the gage lock into place in the grout, then the gage behaves the same as the spot welded gages. See Figure 3(b).

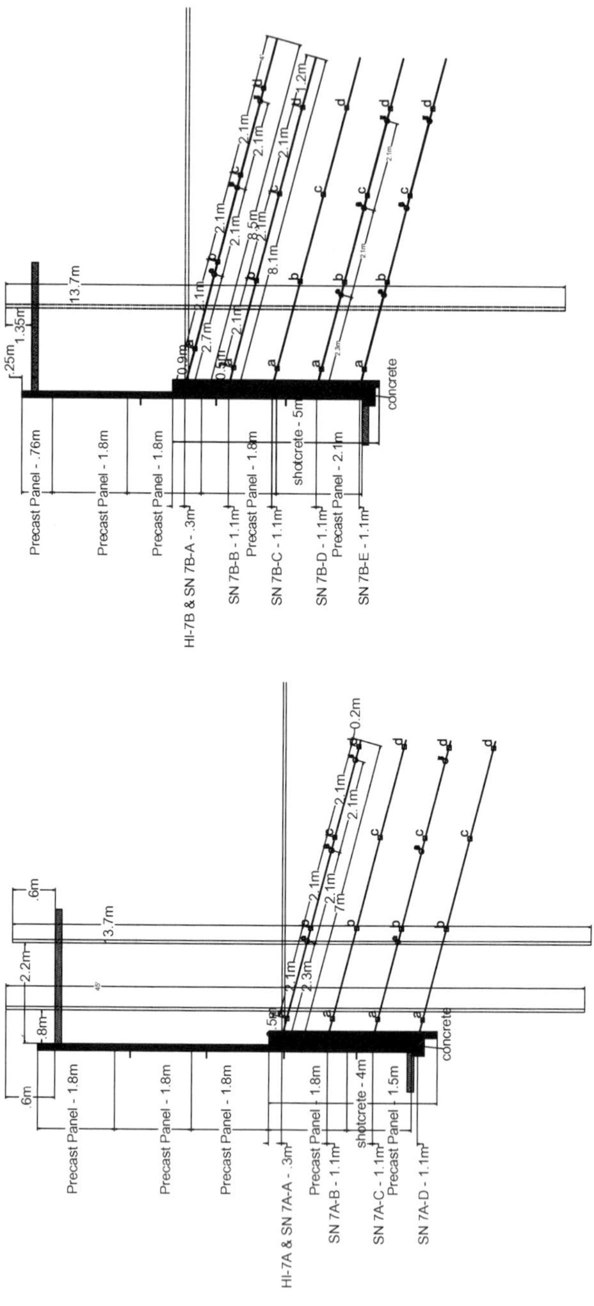

Figure 2. Hybrid wall sections selected for instrumentation and monitoring; (a) Wall Section A (b) Wall Section B

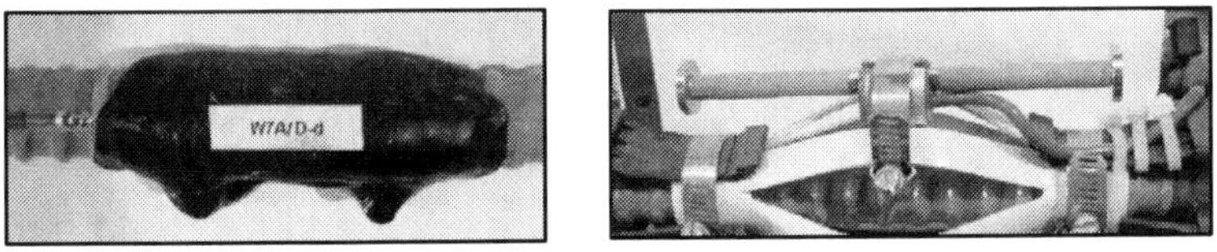

(a) (b)

Figure 3. Vibrating wire gages: (a) Spot welded VWGs (b) Embedment VWGs

The soil nails were prepared for placement before they were transported to the job site. Steel frames were welded from 75mm diameter pipe and a T-section to support the data acquisition boxes. Remarkably, all of the vibrating wire gages survived the placement and construction process, resulting in complete and accurate data throughout the construction of the wall.

The data acquisition system for collecting the data from the strain gages was built around the Campbell Scientific Inc. CR1000 data logger, the Geokon VWDSP, and 5 Geokon multiplexers. This system was programmed using Campbell Scientific's CRBasic programming environment. The program allowed for hot-swapable connections of VWG and tiltmeters. The data acquisition system was encased in two steel electrical boxes that were placed on steel frames. Initially each box was placed at the top of the wall above the soil nails. Once the soil nail portion of the wall was completed, the data acquisition system was moved to the base of the hybrid wall. The wires were run beneath the wall footing.

Measurement of Wall Deformation

Inclinometers were used as the primary method for measuring movement of the wall facing and the soil mass behind the wall. Two vertical inclinometers, placed at distances of 0.8m and 2.2m behind the wall facing were used to monitor lateral movements at Wall Section A. Wall Section B, however, could only accommodate one vertical inclinometer because of the presence of a horizontal drainage pipe at this location. This inclinometer was placed at a distance of 2.2m behind the wall. In addition, each wall section was provided with one horizontal inclinometer to monitor vertical movements of the retrained soil mass. These inclinometers were placed 0.3m below the soil nail/MSE interface. They extended 12.2m into the soil mass. All vertical inclinometers extended to approximately 4.5m below the elevation of the wall base.

While the inclinometer measurements served as the primary means of monitoring wall movement, measurements of lateral wall movements were also made with two other methods, surveying and vibrating wire tiltmeters, for verification purposes. Surveying involved the use of electronic distance measuring (EDM) techniques performed with a total station instrument. For this purpose, reflective tape was affixed to selected points on the wall facing and the distances to these points were measured from benchmarks that were established offsite. The surveying provided quick, repeatable measurements that were accurate within ±3mm. The vibrating wire tiltmeters were permanently mounted to the top of the shotcrete facing of the soil nail portion of the wall. They measured the angle of tilt at the location of the instrument. The data was recorded automatically using the vibrating wire data acquisition system.

SUMMARY OF DATA

The researchers were successful in implementing most aspects of the instrumentation program as originally planned. However, two major problems arose during implementation. The first problem occurred when the inclinometer probes and surveying equipment were stolen from the site during soil nail wall construction. This was remedied by purchasing new equipment to replace those stolen. Nevertheless, it resulted in an interruption in wall deformation data collection. The second problem occurred when the contractor did not follow the instructions and placed the precast concrete facing panels over the soil nailed wall without first coring them to allow access to the horizontal inclinometer casings. As a result of this mistake, no horizontal inclinometer data could be collected during MSE wall construction.

Both VWGs and inclinometers provided data that were consistent with expected trends. VWGs provided data on the development of stresses on soil nail tendons that were placed at different depths. They showed variation of stresses along the length of each tendon and at different stages of wall construction. The inclinometers provided data on movements that occurred within the soil mass at different stages of wall construction. Due to limitations in available space, only a few selected data plots are presented in this paper. The complete set of data can be found in the final research report prepared by the researchers for submission to the sponsor, TxDOT (Jayawickrama et al., 2008).

Tensile Stresses and Tensile Forces in Soil Nail Tendons

Figure 4(a) shows the typical tensile stress distribution that was recorded along the length of all four nail tendons in Wall Section A. Data in Fig. 4(a) represent one particular stage of wall construction, i.e. completion of both soil nail and MSE wall construction. Data show that nails near the top of the wall carry tensile stresses along the entire length while nails at the bottom tend to carry much higher stresses closer to the wall facing but hardly any stresses away from the wall. This trend is consistent with that observed in Wall Section B and with those reported in previous soil nail case studies. This trend is also in agreement with the general expectation that the line of maximum tendon load would coincide with the line of maximum stresses within the soil mass (or potential failure surface). Figure 4(b), on the other hand, shows tensile load within one selected tendon and how it changes from one construction stage to another. The data in Fig. 4(b) pertains to Nail A in Wall Section A. The significant impact that the MSE wall has on the nail tensile forces can be readily noticed in this data.

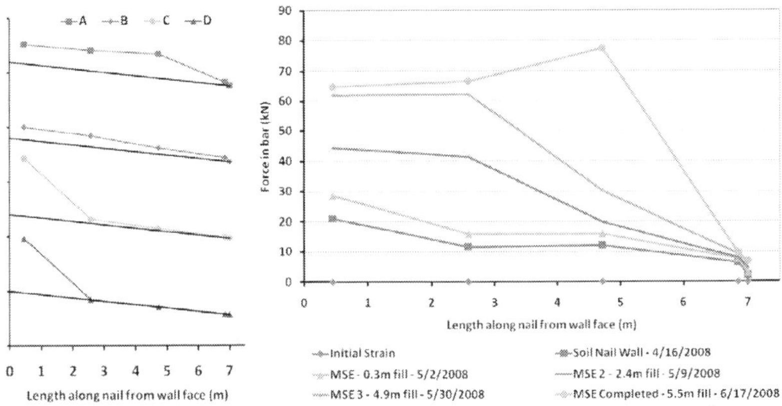

(a) (b)

FIG. 4 (a) Tensile stress distribution in soil nail tendons A through D in Wall Section A after placement of 5.5m of MSE fill, (b) Development of tensile forces in soil nail A, in Wall Section A at various stages of construction

Inclinometer Data

Figures 5(a) and 5(b) summarize the data collected from the inclinometers placed at 0.8m and 2.2m distances from the wall facing at Wall Section A. The resemblance between the two sets of data confirms that the two inclinometers are in agreement with each other and are showing similar deformation patterns within the soil mass. However, it can be noted that the deformations closer to the wall facing are significantly larger than those away from the wall facing. Another observation is that the MSE wall has had significant impact on the magnitude of the deformations within the nailed soil mass. The movements measured in this hybrid wall after the construction of the MSE wall are much larger than those measured in other soil nailed walls that carry little or no surcharge. A more detailed discussion of the data collected in this project can be found in Jayawickrama et al. (2008).

CONCLUSIONS

The MSE/Soil nail wall instrumentation and monitoring program described above has yielded data that will undoubtedly lead to better understanding of the behavior of this type of retaining walls. These data will be of great value in subsequent phases of the research that involve review of the design approach that is currently used for hybrid walls. The data collected at the two wall sections are in good agreement with each other and are consistent with expected trends. The data collected at various stages of construction clearly demonstrate the significant impact that the construction of the MSE wall on top of the soil nailed wall has on the tensile loads developed within the nail tendons as well as on the wall deformations.

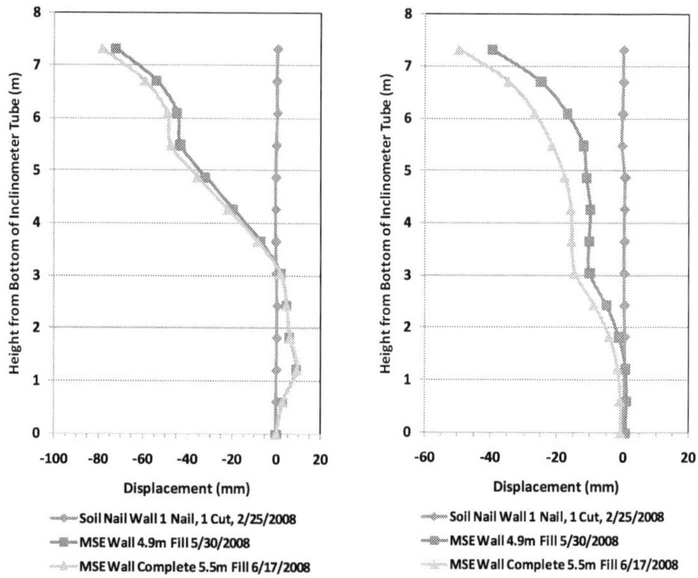

FIG. 5 (a) Inclinometer measurements made on Wall Section A: (a) Inclinometer placed 0.8m away from wall, (b) Inclinometer placed 2.2m away from wall

ACKNOWLEDGEMENTS

The authors wish to express their appreciation to the research sponsor, the cooperative research program between the Federal Highway Administration (FHWA) and the Texas Department of Transportation (TxDOT).

REFERENCES

Jayawickrama, P.W., Abdulrahman, A., Wood, T. A., Wei, Y. and Johnson, J. (2008), "Development of a Design Procedure for MSE/Soil Nail Hybrid Retaining Walls," Final Research Report (Draft), Texas Department of Transportation, October 2008.

Johnson, James (2005), "Field Monitoring of a Hybrid Retaining Wall," Master's Report, Texas Tech University, December 2005.

Turner, John P. and Wayne G. Jensen (2005), "Landslide Stabilization Using Soil Nail and Mechanically Stabilized Earth Walls: Case Study," *ASCE Journal of Geotechnical and Geoenvironmental Engineering*, Feb. 2005, pp.141-150.

Permanent Excavation Support in Urban Areas Using Cutter Soil MixingTechnology: Elliott Avenue Case History, Seattle, Washington

Roberto A. Lopez[1], M. ASCE, P.E., Andy Majewski[2], and Terry Harvey[3]

[1]Manager, Ground Improvement Division, Malcolm Drilling Company, Inc., 3524 Breakwater Avenue, Suite 108, Hayward, California 94545; rlopez@malcolmdrilling.com
[2]Project Engineer, Seattle District, Malcolm Drilling Company, Inc., 8701 South 192nd Street, Kent, Washington 98031; amajewski@malcolmdrilling.com
[3]Project Manager, Ground Improvement Division, Malcolm Drilling Company, Inc., 3524 Breakwater Avenue, Suite 108, Hayward, California 94545; tharvey@malcolmdrilling.com

ABSTRACT

In one of the first applications of Cutter Soil Mixing (CSM) technology in the United States, a 12 to 17.4 m (40 to 57 ft) permanent deep soil mixed wall was constructed to enable the excavation for an office building near downtown Seattle in difficult soil conditions with a high water table.

Unlike conventional slurry walls and diaphragm walls that utilize concrete, soil mixing relies on mixing the soils in situ with a cement and bentonite slurry to create a soil-cement wall. Cutter Soil Mixing technology utilizes two sets of vertically mounted cutting wheels rotating about a horizontal axis to produce rectangular panels of treated soil. By overlapping the soil mix panels, a continuous rectangular wall is constructed, as opposed to circular columns created with conventional single-axis or multiple axes deep soil mixing systems.

The Elliott Avenue project located just north of downtown Seattle at the base of Queen Anne Hill required permanent excavation support to depths of up to 17.4 m (57 ft) in mixed soil conditions consisting of loose sands and gravels, underlain by stiff plastic clays and very dense glacial till. Because the project is located on the water front, the groundwater table at the site was found at a depth of 0.6 m (2 ft) from construction grade. It was thus necessary to support the saturated loose sands and gravels with a robust earth retaining system. Cutter Soil Mixing (CSM) was selected as the method of choice for several reasons, namely: price and schedule relative to a concrete secant pile wall; the ability of CSM to construct a permanent, high quality cutoff wall in the dense gravels and stiff plastic clays; the capacity of the cutter mixing tool to key into the glacial till; and, the ability of CSM to produce a soil-cement material with a minimum strength of 1.38 MPa (200 psi) and a maximum permeability of 5×10^{-6} cm/sec.

BACKGROUND AND GEOTECHNICAL CONDITIONS

The area where the new Elliot Avenue building will be constructed is adjacent to Elliot Bay (part of Puget Sound) and was at one time the old Elliot Bay shoreline. Over the years the site has been filled so that the current shoreline is now located 90 to 120 m (300 to 400 ft) to the west of the project site (Kleinfelder 2007). As such, parts of an old buried sea wall, rip rap, timber piles and other obstructions were anticipated, and actually encountered, during construction of this project.

Soil conditions consisted of 3 to 4.5 m (10 to 15 ft) of loose sandy fill with gravel, wood debris and concrete obstructions. Underlying the fill were the original beach deposits, 3 to 6 m (10 to 20 ft) in thickness, consisting of very loose to medium dense silt, silty sand, clean sands and gravels. Below the beach deposits lies a stiff to hard, glacially overconsolidated clay layer, with SPT blowcounts ranging from 20 to 30 blows per foot and moderate to high plasticity, with 25 to 40 Plasticity Indices. Underlying the clays lies a very dense glacial till layer of silty sands with gravel with SPT values consistently higher than 50. The water table was found at a depth of 0.6 to 1.2 m (2 to 4 ft) below ground surface. To compound the high water table and difficult drilling conditions, the groundwater at the site was found to be contaminated with hydrocarbons and chlorinated solvents.

DESIGN

It was important for the shoring system not only to guarantee the stability of the excavation, but to keep the excavation dry and limit pumping of contaminated groundwater. Not only would this have led to expensive handling of the pollutants, but excessive dewatering would have likely induced settlement to the adjacent streets and the active Burlington Northern railroad line immediately west of the site.

The original design called for a concrete secant pile wall. Malcolm Drilling Company had already successfully installed such a wall further south along Elliott Avenue. This design had utilized circular shafts drilled in an overlapping pattern, with wide flange beams placed every other shaft and tieback anchors installed for lateral loads.

Malcolm Drilling Company proposed a soil cement wall constructed with the Cutter Soil Mixing technique as a means to reduce cost and shorten schedule, while at the same time complying with the technical requirements of the project and keeping the overall permeability of the system low. This technique was executed using a CSM tool able to cut a 2.4 m (7.9 ft) by 0.6 m (2 ft) rectangular panel. The overlap between primary and secondary panels was 0.15 m (0.5 ft), resulting in an effective cutting

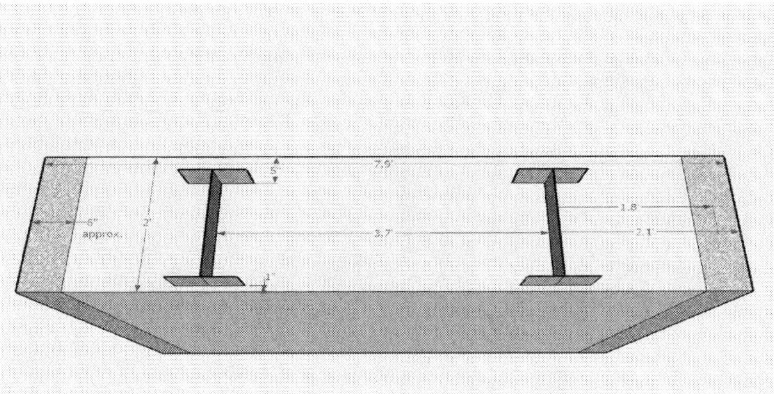

FIG. 1. Standard wall panel detail.

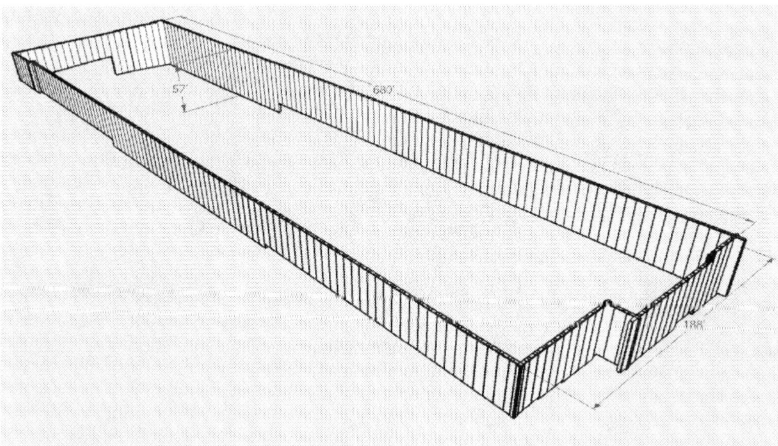

FIG. 2. Overall excavation rendering.

length of 2.25 m (7.4 ft). This in turn allowed for placing two wide flange beams in each panel, at 1.12m (3.7 ft) on center.

Depth of the panels was controlled by the need to key into the underlying clays and glacial till to minimize seepage into the excavation, and by the need to cantilever portions of the wall. A minimum clay embedment depth of 2.3 m (7.5 ft) was selected to ensure low permeability of the wall. A single row of tiebacks was used on the northern, eastern and southern cuts for lateral loads, with design loads of 89 to 578 kN (20 to 130 kips), and total anchor lengths of 17 to 33 m (55 to 110 ft). All tiebacks were placed above the water table to further reduce the risk of groundwater flow into the excavation. No tiebacks were used on the western wall due to the presence of the railroad tracks. See Figures 1 and 2 for wall panel details and a rendering of the excavation.

INSTALLATION PROCEDURE

Cutter Soil Mixing, CSM, is a new soil mixing technology developed in 2003 (Fiorotto et al 2005) and first used in North America in 2005 (Wilson et al 2008). Cutter Soil Mixing was developed out of diaphragm wall technology and utilizes two sets of counter rotating vertical cutter wheels, as opposed to conventional soil mixing technology that employs one or multiple axes tools that rotate horizontally. The wheels cut the surrounding soil, while at the same time blending the injected cement-bentonite slurry with the in situ soil, producing a high quality soil-cement material. Figure 3 graphically depicts the CSM construction procedure.

There are several advantages to using a CSM system to construct a retaining wall over conventional soil mixing. For one thing, the mixing tool cuts vertical rectangular panels, which are more efficient for retaining wall geometry than tangent or secant columns, thus optimizing both mixing energy and binder. Secondly, the CSM system allows complete instrumentation inside the cutter gearbox support frame to read and control in real time the x, y and z coordinates of the cutting head. This inclinometer system, coupled with the advantage of a steerable tool, provides assurance of complete overlap between panels. Finally, the wheels are equipped with cutting teeth capable of drilling and mixing even in stiff ground and keying into bedrock.

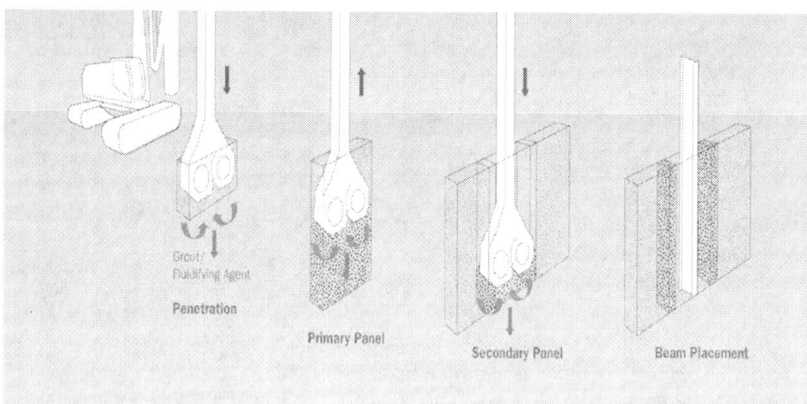

FIG. 3. Schematic of the Cutter Soil Mixing construction procedure.

QUALITY CONTROL / QUALITY ASSURANCE

The quality control program began with preproduction laboratory testing to identify a suitable grout slurry system. Test panels were then constructed and wet grab samples were taken for both permeability and unconfined compressive strength testing. Coring was performed on selected panels with mixed results due to the consistent presence of gravels. Excavating of test panels was conducted, but only a small portion of these were visually inspected due to the high water table. Once production commenced, quality control was provided by real time measurements during the mixing process. These measurements were recorded and provided to the Owner's representatives on a daily basis. Figure 4 shows a quality control log of a typical CSM panel constructed for this project.

Quality assurance was provided by the Owner's representatives. This consisted of wet grab samples taken from every other panel and tested for unconfined compressive strength. Approximately 25% of the samples were also tested for permeability using the ASTM falling head test. Figure 5 is the graphed result of the unconfined compressive strengths on grab samples. The average curves show the effect of early transport to the lab on the strength of soil mix samples.

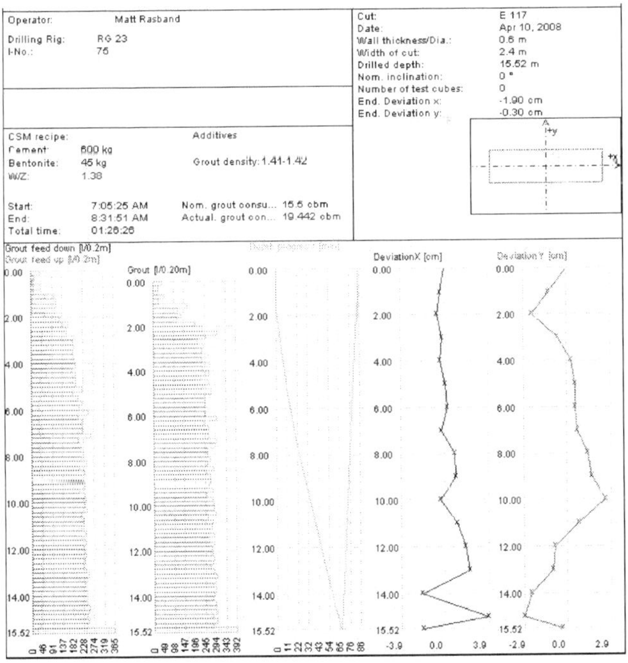

FIG. 4. Daily production report for the Elliott Avenue project.

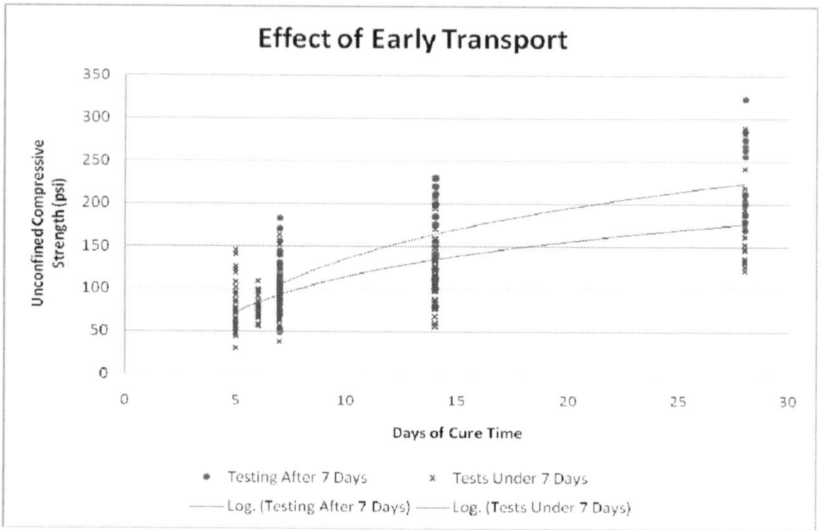

FIG. 5. Effect of early transport of samples on unconfined compressive strength

FIG. 6. CSM rig in operation.

FIG. 7. Excavated soil mixed wall along Elliott Avenue.

The lower line (red line) represents the average unconfined compressive strength of samples transported to the lab within seven days after being taken. The higher line (blue line) is the average unconfined compressive strength of samples that cured in a dedicated field trailer for at least seven days prior to transporting them to the materials testing lab. As evidenced by the results, proper handling and transporting of samples clearly affects the strength. The authors recommend that samples be kept in a cool field trailer for a minimum of seven days prior to transporting them to the lab. If possible, for large jobs, samples should be tested in the field to avoid transporting them altogether.

Twenty eight permeability (falling head) tests were conducted, with results ranging from 1×10^{-6} to 8×10^{-8} cm/sec, well within the permeability criteria of 1×10^{-5} cm/sec. Measured flow into the excavation was on the order of 19 to 38 liters per minute (5 to 10 gpm), also satisfying the allotted amount of 64 liters per minute (17 gpm).

Figure 6 shows the Cutter Soil Mixing tool as it is being extracted after installation of a panel along the east wall. Figure 7 shows the completed excavation and exposed soil mixed wall along Elliott Avenue. Note the dry, open excavation allowed by this procedure.

CONCLUSIONS

Cutter Soil Mixing technology was successfully used to construct a permanent excavation support system in difficult soil conditions with a high water table, achieving soil cement unconfined compressive strengths in excess of 1.38 MPa (200 psi) and permeabilities below 5×10^{-6} cm/sec. This technology was successful in limiting the amount of dewatering in a site contaminated with hydrocarbons and chlorinated solvents and was able to support an urban excavation using 12 to 17.4 m (40 to 57 ft) high walls with only one row of tiebacks, placed intentionally above the water table to further limit ground water infiltration. The method proved to be fast and cost effective when compared to other earth retention systems and should be considered as a viable alternative in future deep excavations.

ACKNOWLEDGEMENTS

The authors are grateful to the following persons and organizations intimately involved with this project: Martin Selig Real Estate, Lease Crutcher Lewis, David

Cotton and Robert Plum of Kleinfelder, Brian Robinson of KPFF, and John Kvinsland, Project Manager of this job for Malcolm Drilling Company.

REFERENCES

Fiorotto, R., Stotzer, E., Schopf, M., and Brunner, W. (2005). "Cutter Soil Mixing (C.S.M.) An Innovation in Soil Mixing for Creating Cut-Off and Retaining Walls." *Proceedings International Conference on Deep Mixing, Recent Advances and Best Practices*, Stockholm, Sweden.

Kleinfelder, Inc. (2007). Preliminary Geotechnical Report 635 Elliott Project, Seattle, Washington. March 12, 2007

Wilson, B., Scholte, J., and Atkinson, M. (2008). "Vancouver Island Conference Centre, Cutter Soil Mixer Transforms Site." Deep Foundations Institute Magazine, Summer 2008, pp 8 – 11.

Tension Testing of Grouted Strand Anchorages

Matthew J. Niermann[1], P.E., M. ASCE, Thomas D. Richards, Jr.[2], P.E., M. ASCE

[1] Sr. Research Engineer, Schnabel Foundation Company, 45240 Business Court, Suite 250, Sterling, Virginia 20166; matt@schnabel.com
[2] Chief Engineer, Nicholson Construction Company, 12 McClane Street, Cuddy, Pennsylvania 15031; trichards@nicholsonconstruction.com

ABSTRACT: Cement grout encasement is the most common means of providing corrosion protection for strand anchorages for earth retention applications in the United States. The grout surrounds the strand and wedges and completely fills the wedge pockets. The effect of the grout on the performance of the anchorage is unknown and the ability to develop significant additional capacity above the lock-off load has not been verified by testing.

A series of tests were conducted to determine if neat cement grouted strand anchorages can reliably develop load carrying capacity above the load at which they are locked off. The tests considered a range of lock-off loads, grout strengths, and surface condition of the wedges (clean or rusty). Baseline testing of ungrouted anchorages was also performed. Test results indicated that grout with compressive strength less than 38 MPa (5500 psi) did not adversely affect the ability of anchorages to develop load carrying capacity above the lock-off load, while very high strength grout with compressive strength greater than 81 MPa (12,000 psi) adversely affected performance. Rusty anchorages also performed poorly.

INTRODUCTION

Gr. 270 prestressing steel strand is a common tendon material for construction of rock and soil anchors. The strand is connected to structures using an anchor head that consists of a wedge plate and wedges. Most wedges in use today are 3-part wedges that are held together by a rubber or steel ring. The wedges are smooth on the outside and have serrations (teeth) on the inside. Wedges are designed to slide into a tapered pocket in the wedge plate. As they slide inward, the wedge teeth bite into the strand and develop mechanical interlock that provides load capacity (see Figs. 1 and 3).

PTI (2004) requires that anchor heads develop at least 95% of the specified minimum ultimate tensile strength (SMUTS) of the prestressing strand. In addition, PTI (2004) requires that wedges be locked off (seated) at a minimum load of 50% SMUTS. This ensures the wedge teeth bite deeply enough into the strand to prevent strand slip through the wedges, if future service loads exceed the lock off load.

PTI (2004) also requires that anchor heads for permanent and/or aggressive conditions be protected against corrosion. In the United States, the most common method of corrosion protection at the anchor head is grout encasement. The fluid grout fills the wedge pockets and any voids around the wedges, including the space below the wedge plate.

It has been postulated that hardened grout encapsulating the wedges and wedge pockets may adversely affect the ability of the anchor head to carry service loads higher than the lock off load. The concern is that the grout may limit the movement of the wedge into the wedge pocket and/or prevent the teeth from biting deeper into the strand. Either mechanism could limit the ability of the wedge to develop higher loads.

This paper presents the results of full scale tension tests performed on grouted strand anchorages. The purpose of the testing was to:

1. Determine if grout encasement affects the ability of anchor heads to develop load capacity greater than the lock-off load.
2. Determine if changes to the PTI (2004) recommendations for grout encasement are warranted.

WEDGE MECHANICS

The mechanics of wedges is described by Chacos (1993). Free body diagrams of the wedge and strand are given in Fig. 1. The normal force (R_S) on the strand is a function of wedge angle and friction coefficient between the wedge and wedge plate. The wedge angle is determined by the manufacturer. The friction coefficient between the wedge and wedge plate varies with surface conditions. The radial force (R_A) generated between the wedge and the wedge plate decreases exponentially with increases in the friction coefficient (see Fig. 2).

Wedge teeth are surface hardened and are designed to bite into strand to develop load capacity. The higher the radial force, the deeper the teeth will bite. This has been demonstrated by full scale testing of strands locked off at various percentages of SMUTS. (Unpublished data, Lang Tendons 2001)

Once a strand has been locked off at a certain load, it has not been established whether the wedge must bite deeper into the strand in order to develop additional load carrying capacity. If the wedge must bite deeper, than anything preventing that may interfere with developing additional load carrying capacity above the lock-off load.

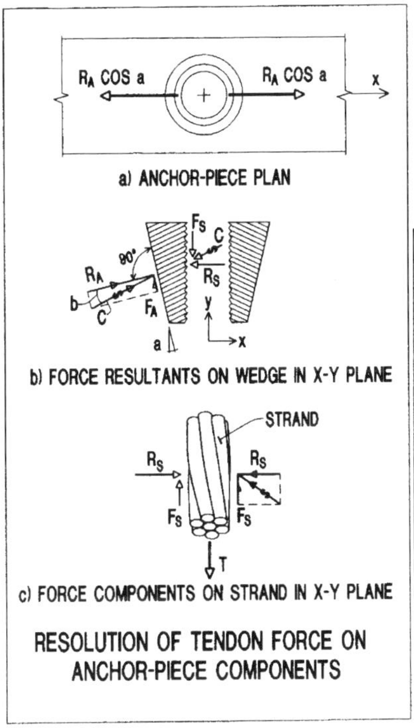

FIG. 1. Resolution of Forces. (Chacos 1993, with permission from Post Tensioning Institute)

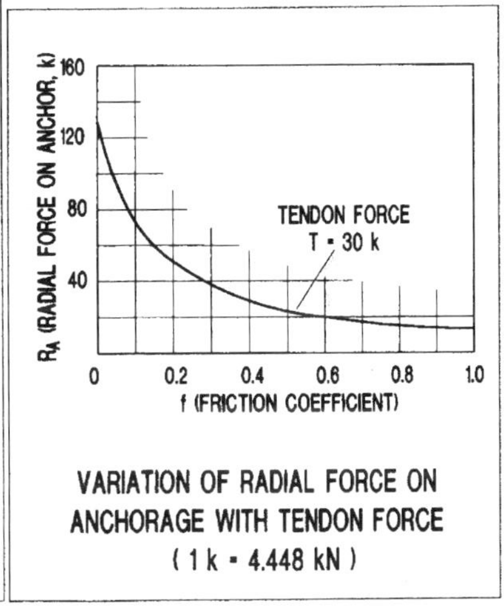

FIG. 2. Radial Force vs. Friction. (Chacos 1993, with permission from Post Tensioning Institute)

FULL SCALE TEST PROGRAM

The purpose of this test program was to evaluate the effect of neat cement grout on the load carrying capacity of grout encased anchor heads. Strand tendons were tensioned and locked off at various loads in a test frame. The anchor heads on one end of the tendons were grouted. Test loads were applied in increasing increments from the other end while the grouted end was monitored for movement.

Test Equipment

The test frame was 4.7 m (15 ft) long and was equipped with bearing plates on both ends. Both ends of the frame were accessible. The bearing plates had 86 mm (3.4 in) diameter center holes designed to receive 4-strand wedge plates. A steel pipe trumpet on the back of each bearing plate and a steel cap on the front of each bearing plate enabled the anchor head to be fully encased with grout. The grout completely filled the space in the wedge pockets between the wedge segments and the space between the wedge teeth and the strand.

A jack chair with integral wedge stop plate was used during stressing. The jack chair sat over the wedge plate and provided bearing for the test jack. The wedge stop plate was adjusted to provide the correct lock-off load for each set of strands. The wedge stop plate positions the wedges on the strand, but does not force the wedges into the pockets of the wedge plate, a practice known as power seating. In previous phases of testing (Bogdan 2005), power seating did not have a significant impact. Movements of the strand, wedges, and wedge plate were measured with Starrett dial indicators. These dial indicators had a resolution of 0.03 mm (0.001 in).

Test Procedure

The test procedure was as follows:

1. Load 4 pieces of strand into each position of the test frame. Install the wedge plate and 4 sets of wedges at each end per position. Tap the wedges into the wedge plate pockets while minimizing the slack in each strand.
2. Set position of wedge stop plate inside jack chair based on desired lock-off load.
3. At the anchor head location to be grouted ("grouted end"), set jack chair with integral wedge stop plate, hydraulic test jack, stressing wedge plate, and 4 sets of reusable stressing wedges.
4. Stress all 4 strand simultaneously to 70% SMUTS and reduce to desired lock-off load.
5. Perform individual lift-off tests on each strand to determine actual lock-off load for each strand. Monitor rate of pressure gain to determine when wedges have lifted out of pockets in wedge plate.
6. Cut strand about 50 mm (2 in) from wedge plate, install cap, and fill with neat cement grout.
7. After grout has cured, remove cap and chip away grout to expose anchor head.
8. Set dial indicators to measure movement of strand, wedge, and wedge plate.
9. Move to opposite end ("stressing end"). Set jack chair, jack, stressing wedge plate, and one set of reusable stressing wedges.
10. Apply load to one strand at a time. Begin at lock-off load and increase load in increments of 5% SMUTS. Record movements at grouted end at each load interval. Continue loading until strand fails in tension or slips out of the grouted anchor head.

Test Variables

The following variables were considered:

- Anchor head supplier
- Anchor head surface condition
- Grout strength
- Lock-off load

In addition, testing was performed on clean, ungrouted anchor heads from both suppliers in order to obtain baseline performance data.

Anchor heads were supplied by two companies. Both anchor head types consisted of a 4-strand wedge plate and 4 sets of 3-part wedges. Both wedge plates were ductile iron castings. The first supplier's wedge plate pockets had no burrs and were in the as-cast condition. The second supplier's wedge plate pockets had been deburred, resulting in 20-40% of the pocket surface area having a machined surface.

The anchor heads were tested in both clean and rusty conditions. Clean anchor heads were left in as-delivered condition from the suppliers. Clean anchor heads had a light coating of oil on both the wedges and wedge plates (see Fig. 3). Rusty anchor heads were prepared by degreasing and then spraying the wedges and wedge plates with salt water. The anchor heads were placed outside for several days until the entire wedge plate and approximately 50% of the wedge surface area were covered with rust (see Fig. 4).

FIG. 3. Clean Anchor Head from Supplier #1 in "As-Supplied" Condition.

FIG. 4. Rusty Wedges from Supplier #1.

Grout strength was varied by mixing neat cement grouts of varying water/cement ratios and allowing the grout to cure for different lengths of time. Type I/II Portland cement was used for all grout. FX-32 water reducing admixture from Fox Industries, Baltimore, Maryland, was used to aid mixing and enhance the flow of grout into voids and interstices of the wedges and strand. The grout was placed through a hole in the anchor head cap and completely encased the wedge plate and gaps in the wedges. Grout cubes were taken per ASTM C 109 to measure compressive strength. The planned lock-off loads ranged from 35% to 50% SMUTS. The actual lock-off loads ranged from 32% to 52% SMUTS.

TEST RESULTS

Overall Performance

Table 1 presents the test results.

TABLE 1. Test Matrix and Results.

Supplier	Surface Condition	Grout f'$_c$ (MPa) (psi)	Quantity Of Strands	Lock-Off Loads (% SMUTS)	Performance
#1	Clean	No grout	12	32 to 38	No slip
#2	Clean	No grout	4	40 to 43	No slip
#1	Clean	36 (5,200)	16	35 to 50	No slip
#2	Clean	38 (5,500)	16	34 to 52	No slip
#1	Clean	81 (12,000)	8	33 to 50	Slip[a]
#2	Clean	81 (12,000)	8	34 to 51	No slip
#1	Rusty	53 (7,700)	15	27 to 54	Slip[b]

NOTES:

(a) – Eight individual strands were tested. One strand locked off at 33% SMUTS completely slipped out of the anchor head at a load just over 85% SMUTS (see Fig. 6). One strand locked off at 47% SMUTS had an outside wire and the king (center) wire slip about 6 mm (0.25 in) at a load of 95% SMUTS.

(b) – Fifteen individual strands were tested. All 15 strands slipped through the anchor head.

Clean anchor heads with no grout exhibited no slip of the strands through wedges. All strands were tensioned to failure at loads ranging from 90% to 100% SMUTS. Failure occurred when one or two wires broke underneath the reusable stressing wedges (not the grouted end). See Fig. 5 for typical view of anchor head.

Clean anchor heads with 36 to 38 MPa (5,200 to 5,500 psi) grout also exhibited no slip of the strands through the wedges. Strands were tensioned to maximum loads ranging from 90% to 107% SMUTS. Failure occurred when one or two wires broke underneath the reusable stressing wedges.

Clean anchor heads with 81 MPa (12,000 psi) grout exhibited mixed results. One strand slipped completely through the wedges (see Fig. 6). Another strand had 2 of the 7 wires slip about 6 mm (0.25 in) relative to the other wires. Except for the strand that slipped completely through, all strand were tensioned to failure at loads ranging from 92% to 104% SMUTS. Failure occurred when one or two wires broke underneath the reusable stressing wedges.

Rusty anchor heads with 53 MPa (7,700 psi) grout exhibited very poor performance. Every strand slipped completely through the wedges. Eight of the 15 strand slipped immediately after reapplying the lift-off load at the stressing end during the incremental stressing process. Performance improved with higher lock-off loads, but no anchor heads developed more than 80% SMUTS. Only 15 strand were tested, rather than 16, because one strand could not be locked off due to rust.

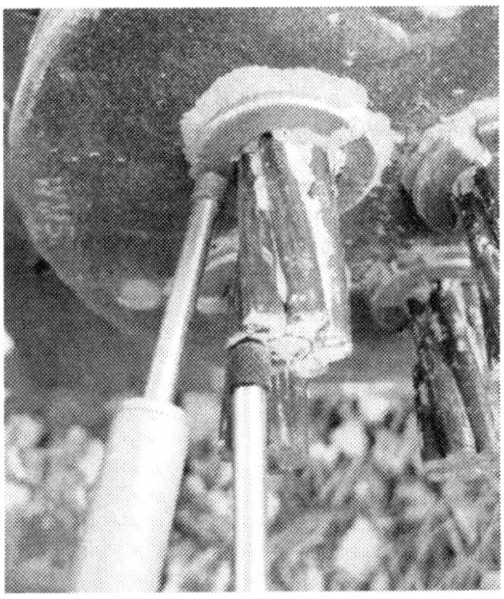

FIG. 5. Typical Grouted Anchorage with Dial Indicators in Position.

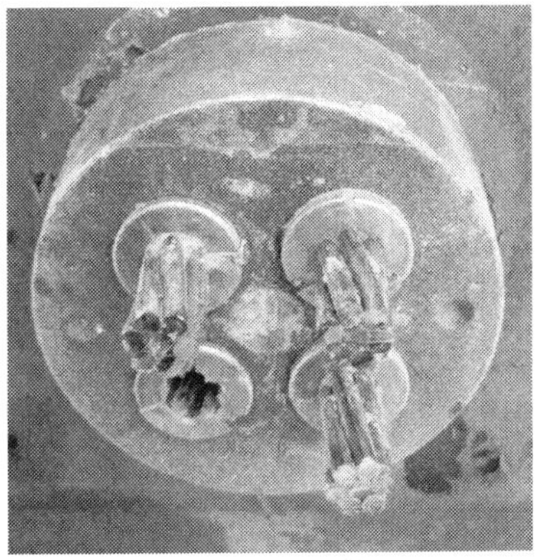

FIG. 6. Slipped Strand [Supplier #1, Clean, 33% SMUTS Lock-Off Load, f'_c = 81 MPa (12,000 psi)].

Wedge Movements

Testing showed that wedges are pulled deeper into the wedge plate pockets when the strand is subject to increasing loads above lock-off load. Although ungrouted anchor heads display the most wedge travel, wedges that have been surrounded with neat cement grout still travel into the pocket. Wedge travel decreases with increasing grout strength (see Fig. 7).

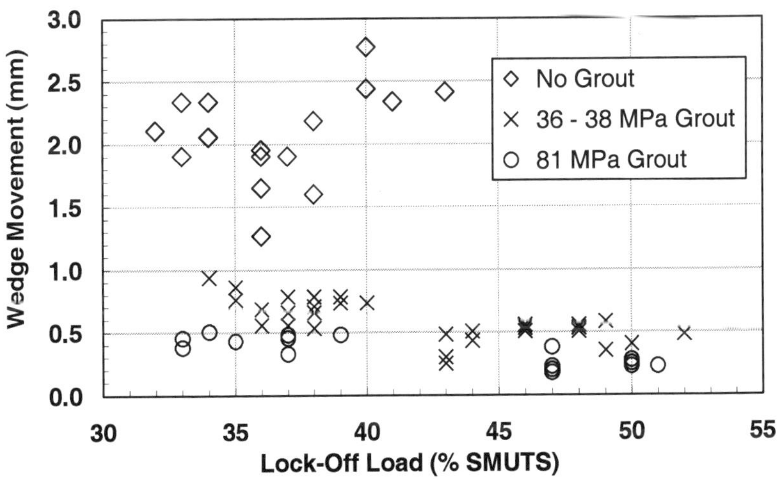

FIG. 7. Total Wedge Movement vs. Lock-Off Load at 85% SMUTS Stressing Load for Clean Anchor Heads

Strand Movements

Relative movement between strand and wedge was determined by subtracting wedge movement from strand movement. For all clean anchor heads (grouted and ungrouted), the total relative strand movement at 85% SMUTS did not exceed 0.40 mm (0.016 in). This movement represents about one-third of the 1.2 mm (0.045 in) tooth spacing of the wedges. Relative strand movement in clean ungrouted anchor heads was about 4 times greater than in clean grouted anchor heads.

CONCLUSIONS AND RECOMMENDATIONS

- Normal strength neat cement grout [f'_c less than 38 MPa (5500 psi)] restricts wedge travel but does not affect the ability of anchor heads to develop at least 95% SMUTS when locked off at or above 35% SMUTS.
- Very high strength grout [f'_c greater than 81 MPa (12,000 psi)] restricts wedge travel more than normal strength grout and adversely affects the ability of anchor heads to develop 95% SMUTS, especially when lock-off loads below 50% SMUTS are used.
- Rusty wedges are unacceptable and should not be used, regardless of grouting practices.
- Clean ungrouted anchor heads locked off at 35% SMUTS will develop at least 95% SMUTS.
- The PTI (2004) recommendations are appropriate. Normal strength neat cement grout may be used for filling anchorages without affecting the load carrying capacity of the anchor head.

ACKNOWLEDGEMENTS

This research was supported by the Association of Drilled Shaft Contractors and member companies DYWIDAG-Systems International USA Inc., Williams Form Engineering Corp., and Con-Tech Systems Ltd. Their support is gratefully acknowledged.

REFERENCES

Bogdan, Lucian I. (2005) "Tension Testing of the Grouted Anchorage Assembly for Multistrand Anchors" *Proceedings of the 30th Annual Conference on Deep Foundations*, Chicago, IL: 281-289.

Chacos, Gregory P. (1993) "Wedge forces on Post-Tensioning Strand Anchors," *PTI Technical Notes* (Issue 2) Post Tensioning Institute, Phoenix, AZ.

PTI (2004) "*Recommendations for Prestressed Rock and Soil Anchors, Fourth Edition*," Post Tensioning Institute, Phoenix, AZ.

3D Numerical Analyses of Geosynthetic Encased Stone Columns

Majid Khabbazian, Stud.M.ASCE

Graduate Student and GSI Fellow, Dept. of Civil and Environmental Engineering, 301 DuPont Hall, University of Delaware, Newark, DE 19716. E-mail: majid@udel.edu

Victor N. Kaliakin, M.ASCE

Associate Professor, Dept. of Civil and Environmental Engineering, 301 DuPont Hall, University of Delaware, Newark, DE 19716. E-mail: kaliakin@udel.edu

Christopher L. Meehan, A.M.ASCE

Assistant Professor, Dept. of Civil and Environmental Engineering, 301 DuPont Hall, University of Delaware, Newark, DE 19716. E-mail: cmeehan@udel.edu

ABSTRACT

In very soft soils, the use of stone columns can be problematic due to the lack of adequate lateral confining pressure. In these situations, to provide the required lateral confining pressure and to increase their bearing capacity, stone columns can be encased by a suitable geosynthetic. Using a high-strength geosynthetic for confinement not only increases the strength of a stone column, but also prevents lateral squeezing of the column due to movement in the very soft surrounding soil. This paper describes 3D finite element analyses carried out to simulate the behavior of a single geosynthetic-encased stone column in a soft clay soil using the computer program ABAQUS. The influence of geosynthetic stiffness, column diameter, and the stiffness and friction angle of the column material were studied in the numerical analyses.

INTRODUCTION

Stone-columns (also known as granular columns or granular piles) have been extensively used to improve the bearing capacity of soft soils and to reduce the settlements of flexible structures such as storage tanks or earthen embankments. They are also commonly used for application as deep foundation elements for structures that can safely tolerate the amount of displacement necessary to mobilize the strength of the column. Granular columns under compressive loads experience different failure modes such as bulging (Hughes et al. 1975), general shear failure (Madhav and Vitkar 1978), and sliding (Aboshi et al. 1979). However, in soft clays the most common failure mode for stone columns is bulging (Madhav and Miura 1994).

Bulging occurs in very soft soils due to the lack of required lateral confining pressure. In these situations, to provide the required lateral confining pressure and to increase the bearing capacity, stone columns are encased by a suitable geosynthetic. Using a high-strength geosynthetic for confinement not only increases the strength of a stone column, but also prevents lateral squeezing of the column due to movement in the very soft surrounding soil. Sharma et al. (2004) conducted experimental tests to investigate the effect of geogrid reinforcement on bulging and load-carrying capacity of a single stone column in soft clay. Murugesan and Rajagopal (2006 and 2007) performed model tests and numerical analyses to study the behavior of a single geosynthetic-encased stone column with a limited zone of soil influence (a tributary area approach to column group behavior). In the numerical analyses, Murugesan and Rajagopal (2006) performed axisymmetric analyses and assumed continuum elements for the geosynthetic without considering the behavior of the interface between different materials (this paper addresses this phenomenon by using interface elements in the numerical model). Lee et al. (2007) investigated the failure mechanism and load carrying capacity of individual geogrid encased stone columns by model tests. Alexiew et al. (2005) described the design principles, technologies, and procedures for geotextile encased stone columns and emphasized the importance of the tensile modulus of the geotextile that is used for column confinement.

This paper describes 3D finite element analyses that were carried out to simulate the behavior of a single geosynthetic-encased stone column in a soft clay soil using the computer program ABAQUS (Hibbitt et al. 2007). The influence of geosynthetic stiffness, column diameter, and the friction angle and stiffness of the column material are studied in the numerical analyses.

NUMERICAL ANALYSES

Finite element analyses were performed using the program ABAQUS (Hibbitt et al. 2007). Fig. 1 shows a typical finite-element mesh used in the analyses. In all of the numerical analyses that were performed, the thickness of the clay soil and the length of the stone column are assumed to be 5 m. It is also assumed that the soil and pile are underlain by a rigid layer. The lateral extent of the soft soil around the stone column was selected such that the effects of the vertical boundary conditions on the calculated results were minimal. As shown in Fig. 1, when the radius of the stone column is 0.5 m the overall radius of the cylinder is selected to be 2.5 m. At the bottom boundary of the finite-element mesh, the displacements are set to zero in the z direction. The displacements in the x and y directions are set to zero on the circumferential boundary of the soft soil zone.

The finite-element mesh is developed using 6-node linear triangular prism elements for both the stone column and soft soil. The stone column is modeled using a linear elastic-perfectly plastic model with Mohr–Coulomb failure criterion. The Mohr–Coulomb model is defined by five parameters: friction angle (ϕ), effective cohesion (c'), dilatancy angle (Ψ), effective Young's modulus (E), and Poisson's ratio (ν). The parameters used in the numerical analyses are summarized in Table 1. Mohr-Coulomb parameters used in the numerical analyses are similar to the typical values used by other researchers (e.g. Guetif et al. 2007, Ambily and Gandhi 2007).

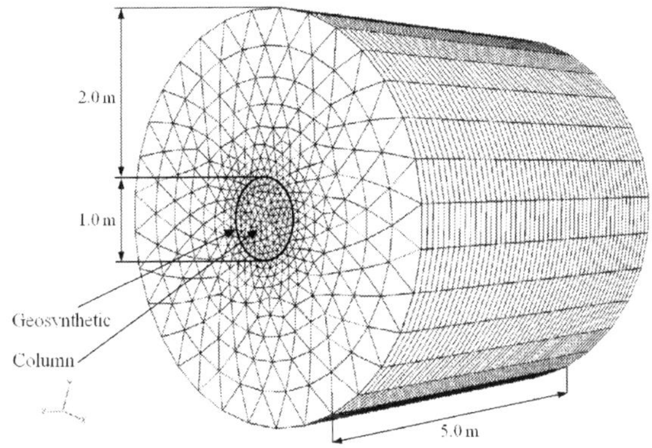

FIG. 1. Typical finite-element mesh used in the analyses

The soft soil was modeled as a modified Cam Clay material. Five material parameters were used in the model, namely the slope of the swelling line (κ), the slope of the virgin consolidation line (λ), the void ratio at unit pressure (e), slope of the critical state line (*M*), and Poisson's ratio (ν). The modified Cam Clay parameters used correspond to those obtained for experimental data on soft Bangkok clay (Balasubramian and Chaudhry 1978). These parameters are tabulated in Table 1.

The geosynthetic was modeled using 4-node quadrilateral, reduced integration membrane elements. The geosynthetic was assumed to be an isotropically linear elastic material, with an assumed Poisson's ratio of 0.3 (Liu et al. 2007). Alexiew (2005) documented that design values of tensile modulus (J) between 2000-4000 kN/m were required for the geosynthetic used to encase stone columns on a number of different projects. Consequently, a tensile modulus of 3000 kN/m was used in the numerical analyses. The elasticity modulus (*E*) of the geosynthetic was derived from the relationship $J = Et$, where t is the thickness of geosynthetic, which was assumed to be 5 mm for all of the numerical analyses that were performed.

Interface elements, with two sets of parameters, were used to model interaction behavior between the geosynthetic and the stone column, and between the geosynthetic and the surrounding soft soil. A Mohr-Coulomb failure criterion with zero cohesion was used for the interface elements. The coefficient of sliding friction between the geosynthetic and the stone column and between the geosynthetic and the soft soil was assumed to be 0.4 and 0.1, respectively.

Table 1. Material parameters

Part	Model	φ deg.	c kPa	Ψ deg.	E Mpa	ν	κ	λ	M	e
Stone column	Mohr-Coulomb	40	10	0	60	0.30				
Soft soil	Modified Cam Clay					0.20	0.09	0.51	1.00	2.00
Geosynthetic	Linear Elastic				600	0.30				

NUMERICAL RESULTS

In the numerical analyses, in order to determine the stress-displacement behavior on top of the geosynthetic encased stone column, soil nodal points corresponding to the top of the column were subjected to a series of vertical downward displacements. During these downward displacements, the resultant stress on top of the column was recorded, allowing the stress-displacement curve to be drawn accordingly.

Fig. 2a shows the stress-displacement response for a single stone column having the parameters shown in Table 1. From Fig 2a, it can be seen that as the settlement increases, the resultant stress on top of the column increases accordingly. For example, at a settlement of 25 mm (a common serviceability criteria), the mobilized vertical stress on top of the column is equal to 155 kPa.

The lateral bulging of the stone column at settlements of 25, 50, and 100 mm is shown in Fig. 2b. It is observed that maximum bulging occurs at a depth approximately equal to one diameter of the stone column and minimum bulging occurs beyond a depth of 3 m. This is similar to the findings of Lee et al. (2007), who performed model tests and observed that bulging failure occurs at a depth of 1 to 2.5 times the diameter of the stone column.

Fig. 2c shows the resultant hoop tension force along the length of the stone column at settlements of 25, 50, and 100 mm. The largest values of hoop tension force are observed to occur over the first 1 m of depth (one diameter of the stone column), which is consistent with the zone of maximum column bulging. Beyond this zone, the hoop tension force decreases up to a depth equal to 3 m (three times the diameter of the stone column), after which point the variations in hoop force are relatively insignificant. In general, this is consistent with Murugesan and Rajagopal (2006), who reported that a considerable amount of hoop tension force occurs from the ground surface up to a depth equal to twice the diameter of the stone column.

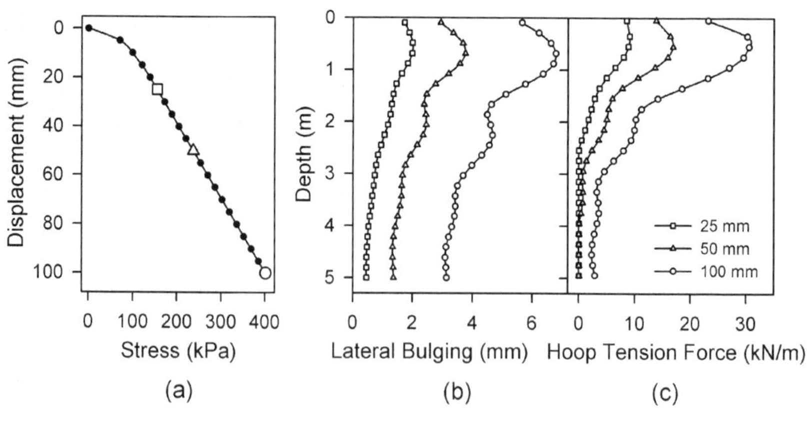

FIG. 2. Stone column response: (a) displacement vs. stress, (b) depth vs. lateral bulging, and (c) depth vs. resultant hoop tension force

PARAMETRIC STUDY

In order to investigate the influence of a number of the input parameters on the behavior of the geosynthetic encased stone column, a series of parametric analyses were performed. In these parametric analyses, only one parameter is changed and all of the other parameters were held constant at the base values listed in Table 1.

Influence of stiffness of geosynthetic

To investigate the influence of the stiffness of the encasement on the performance of the stone column, the encasement stiffness was varied over a wide range of values from 300 to 10000 kN/m. Fig. 3 shows the stress-displacement curve for different values of encasement stiffness. As shown in Fig. 3, the performance of the stone column improves significantly with an increase in stiffness of the encasement. The mobilized stresses in the stone column at a settlement of 25 mm are shown in Fig. 4 for different stiffnesses of the geosynthetic. Figs. 3 and 4 show that the load-carrying capacity of stone columns strongly depends on the value of encasement stiffness. For example, at settlement of 25 mm, the mobilized stress on top of the stone columns with encasement stiffness of 300 and 10000 kN/m is 61 and 284 kPa, respectively.

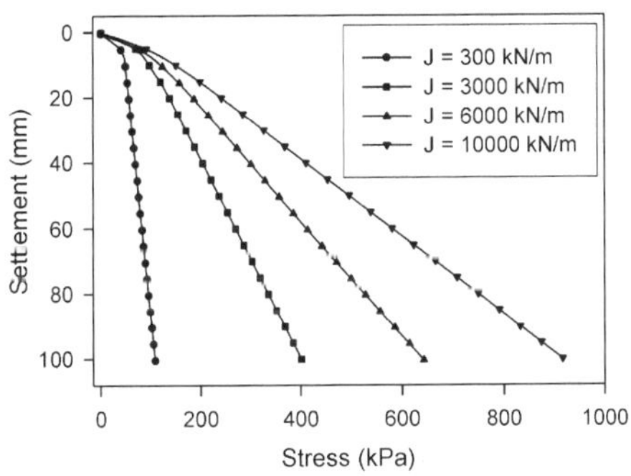

FIG. 3. Stress-displacement response of stone column with varying encasement stiffness

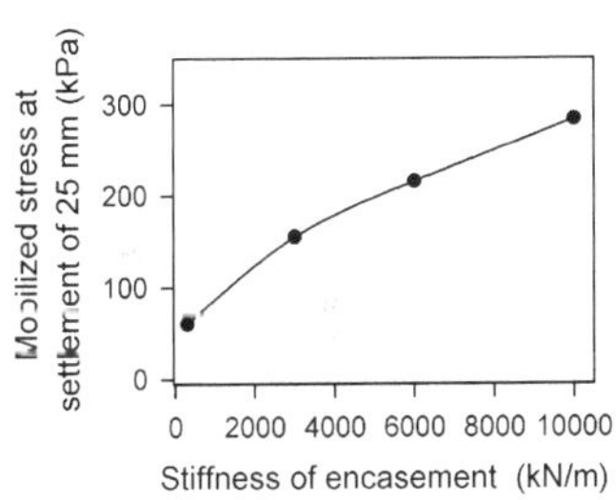

FIG. 4. Response of stone column with different values of encasement stiffness

Effect of friction angle and stiffness of stone column material

Three analyses were performed with different values of stone column friction angle (35, 40, and 45 degrees). The resultant stress-displacement curves are presented in Fig. 5. The performance of the stone column improves by increasing the friction angle of the stone column material, as expected. The effect of the stone column friction angle is, however, substantially less than the influence of encasement stiffness. For example, by increasing the friction angle of stone column from 35 to 45 degrees, the mobilized stress on top of the stone column at 25 mm settlement changes from 131 to 183 kPa.

Stress-displacement curves for three different stone column elasticity moduli are compared in Fig. 6. As evident from this figure, when the elasticity modulus of the stone column increases from 20 to 60 MPa, the resultant stress on top of the stone column increases. However, when the elasticity modulus is increased from 60 to 100 MPa, the change in the performance of the stone column is negligible.

As a result, to improve the performance of the stone column, it appears that it is more efficient to choose encasement with a higher stiffness than to increase the friction angle and stiffness of the stone column materials. It is possible that, if a high-strength stone column is required, improving only the quality of the stone column materials will not be sufficient to achieve the necessary pile capacity. However, more analyses are needed to confirm that this is the case.

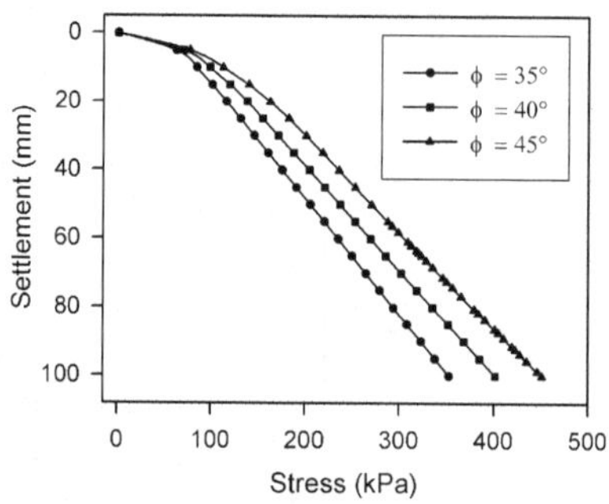

FIG. 5. Stress-displacement response of stone column with different values of stone column friction angle

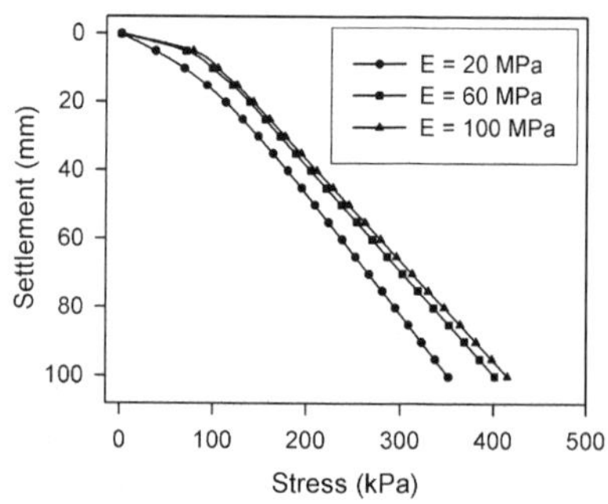

FIG. 6. Stress-displacement response of stone column with different values of stone column elasticity modulus

Influence of stone column diameter

Murugesan and Rajagopal (2007) studied the load capacity of encased stone columns through laboratory model tests and concluded that the stresses developed in stone columns decrease with an increase in the column diameter. For this study, this phenomenon was explored by varying the column diameter as shown in Fig. 7 (0.5, 0.8, 1.0, 1.2, and 1.5 m). For each of the columns that were analyzed, the radial extent of soft soil around the stone column was varied to minimize the effect of the boundary condition on the column behavior. As the diameter of the stone column increases, the load-carrying capacity of the column decreases at a decreasing rate (Fig. 7). For example, by changing the diameter of a stone column from 0.5 to 0.8 m, the mobilized stress decreases as much as 21%; however, this decrease is about 8% for a change in stone column diameter from 1.2 to 1.5 m. Based on the results shown in Fig. 7, using small diameter encased stone columns is a more efficient use of construction materials than large diameter stone columns. However, there is a trade-off in additional cost and time required for installation of multiple smaller columns. This trade-off is often partially countered or otherwise offset by difficulties that are commonly encountered when excavating larger diameter holes. Another potential disadvantage of smaller diameter columns is that they can be more prone to a buckling failure mechanism if the lateral resistance provided by the soil is low and the slenderness ratio of the column is large. These additional factors need to be accounted for by the design engineer when optimizing the design for encased stone columns for a specific project.

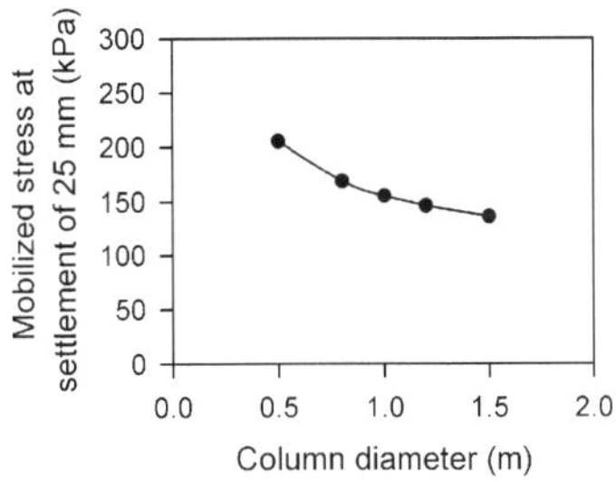

FIG. 7. Response of stone column with different values of column diameter

CONCLUSION

Three-dimensional numerical analyses were performed to study the effects of geosynthetic stiffness, column diameter, and the stiffness and friction angle of the column materials on the load-carrying capacity of encased stone columns. The following conclusions were developed as a result of the numerical analyses that were performed:

1. Maximum bulging and hoop tension occur at a depth equal to about one diameter of the stone column and minimum values are observed after a depth of three column diameters.
2. The stiffness of the encasement has a considerable effect on the load-carrying capacity of encased stone columns.
3. Improving the parameters of stone column materials (friction angle and stiffness) increases the load-carrying capacity of a given stone column; however, it is more efficient to select encasement with higher stiffness rather than to improve the stone column material.
4. Increasing the diameter of an encased stone column decreases the load-carrying capacity of the column. When the diameter of the modeled stone column was changed from 0.5 m to 1.5 m, the resultant stress on top the column at a settlement of 25 mm decreased as much as 65%.

ACKNOWLEDGMENT

The first author gratefully acknowledges the Geosynthetic Institute (GSI) for providing partial financial support for his graduate education.

REFERENCES

Aboshi, H., Ichimoto, E., Enoki, M. and Harada, K. (1979) "The composer-a method to improve characteristics of soft clay by inclusion of large diameter sand columns."

Proceedings of the International Conference on Soil Reinforcement: Reinforced Earth and other techniques, Paris, Vol. 1, p 211-216.

Ambily, A.P. and Gandhi, S.R. (2007) "Behavior of stone columns based on experimental and FEM analysis." *Journal of Geotechnical and Geoenvironmental Engineering*, ASCE, 133(4), p 405-415.

Alexiew D., Brokemper D., and Lothspeich, S. (2005) "Geotextile Encased Columns (GEC): load capacity, geotextile selection and pre-design graphs." *Proceeding of the Geo-Frontiers 2005 Congress*, Austin, Texas, p 1-14.

Balasubramian, A. S., and Chaudhry, A.R. (1978) "Deformation and Strength Characteristics of Soft Bangkok Clay." *Journal of the Geotechnical Engineering Division*, ASCE, 104(9), p 1153-1167.

Lee D., Yoo, C., and Park, S. (2007) "Model tests for analysis of load carrying capacity of geogrid encased stone column." *Proceeding of the 17th International Offshore and Polar Engineering Conference*, Lisbon, Portugal, p 1632-1635.

Guetif Z., Bouassida M., and Debats, J.M. (2007) "Improved soft clay characteristics due to stone column installation." *Computers and Geotechnics*, 34(2), p 104-111.

Hibbitt, Karlsson and Sorensen Inc. (2007). ABAQUS user's manual, Version 6.7, Pawtucket, R.I.

Hughes, J.M.O., Withers, N.J., and Greenwood, D.A. (1975) "A field trial of the reinforcing effect of a stone column in soil." *Geotechnique*, 25(1), p 31-44.

Liu, H. L., Charles W. W. Ng, and Fei K. (2007) "Performance of a Geogrid-Reinforced and Pile-Supported Highway Embankment over Soft Clay: Case Study" *Journal of Geotechnical and Geoenvironmental Engineering*, 133(12), p 1483-1493.

Madhav, M.R., and Vitkar, P.P. (1978) "Strip footing on weak clay stabilized with granular trench." *Canadian Geotechnical Journal*, 15(4), p 605-609.

Madhav, M.R., and Miura, N. (1994) "Soil improvement Panel report on stone columns." *Proceedings of the 13th International Conference on Soil Mechanics and Foundation Engineering*, New Delhi, India, Vol. 5, p 163-164.

Murugesan, S., and Rajagopal, K. (2006) "Geosynthetic-encased stone columns: Numerical evaluation." *Geotextiles and Geomembranes*, 24(6), p 349-358.

Murugesan, S., and Rajagopal, K. (2007) "Model tests on geosynthetic-encased stone columns." *Geosynthetics International*, 14(6), p 346-354.

Sharma, R.S., Kumar, B.R.P., and Nagendra, G. (2004) "Compressive load response of granular piles reinforced with geogrids." *Canadian Geotechnical Journal*, 41(1), p 187-192.

The Application of D-M Method in Embankments on Deep Soft Clay

YE Guan-Bao[1], Chang Hong-Tao[2], XU Chao[3] , Liao Xing-Yue

[1]Professor, Key Laboratory of Geotechnical Engineering, Department of Geotechnical Engineering, Tongji University, Shanghai, 200092, P. R. China; yeguanbao@vip.citiz.net

[2]Ph.D. Candidate, Key Laboratory of Geotechnical Engineering, Department of Geotechnical Engineering, Tongji University, Shanghai, 200092, P. R. China; cht1978@gmail.com

[3]Professor, Key Laboratory of Geotechnical Engineering, Department of Geotechnical Engineering, Tongji University, Shanghai, 200092, P. R. China; c_axu@mail.tongji.edu.cn

[4]Ph.D. Candidate, Key Laboratory of Geotechnical Engineering, Department of Geotechnical Engineering, Tongji University, Shanghai, 200092, P. R. China; liaoxingyue@gmail.com

ABSTRACT: This paper presents a new method to improve embankment on deep soft clay in China. The method, briefly named D-M method, combines the prefabricated vertical drains (PVD) and the cement deep mixing method (DMM) together to consolidate deep soft soil foundation. The short deep mixing columns improve the shallow soft soil and long PVDs accelerate the dissipation of excess pore pressure in the deeper soft soil layers. To evaluate the efficiency of the D-M method, a numerical study was conducted using the design parameters obtained from the Huaian-Yancheng Highway project by the 2-dimensional finite element software Plaxis. Results showed that the new method succeeds in consolidating deep soft soils below the deep mixing columns and improving the stability of the highway embankment.

INTRODUCTION

Quite often, e.g. while constructing highway, there arises a need for the construction of embankments on deep soft clay. These kinds of clay are often highly compressible and with lower strength and permeability. The unfavorable geological condition brings a lot of challenges to the construction of embankment. In the past decades, several techniques are available for enhancing the stability of embankments. The prefabricated vertical drains (PVD) and the cement deep mixing method (DMM) are the two most commonly used techniques. PVD can increase the rate of consolidation and settlement under the overburden pressure. It is also less expensive compared to the other methods. However, PVD method needs longer time for a project, especially on deep soft clay. DMM improves the stability of embankment and decreases the total settlement. Unfortunately, this method can only improve the shallow soft soil (≤20m), and the higher cost is another major convern of using

DMM. Obviously each of the two methods alone has some shortcommings for embankment construction on deep soft clay. This study proposed a new method, in which PVD and DMM were used in combination to consolidate the deep soft soil foundation in China (briefly called D-M method) (XU and YE 2006).

For easy to understand the mechanism of D-M method, the paper provides a numerical model of a test embankment project in Eastern China.

CONCEPTUAL MODEL OF D-M METHOD

The D-M method conceptually is the combination of DMM and PVD. There lies the issue of the layout of the deep mixing columns and the PVDs in a plane view. We can locate the short columns and the PVDs following different patterns on the basis of project need. In section, the short columns can only improve the shallow soft soil and the PVD can penetrate through the whole soft soil when the thickness of soft soil is not too large. When PVD can not penetrate through the soft clay layer for the limitation of machine, the length of PVD should be calculated by the thickness of the compressible layer.

The foundation profile of D-M can be divided into three layers from top to bottom (refer to Fig. 2). The top part, where short columns and PVDs working together, is the composite layer. The middle part with PVD alone is the consolidation layer. And the bottom part under the end of PVD is the untreated layer.

The D-M composite foundation, which is made up of the short columns and soil between columns, can increase the bearing capacity and composite modulus of the foundation. PVD penetrates through the composite layer and extends to the deeper soft soil layer so as to shorten the drainage path for consolidation layer and untreated layer. The shorter drainage path makes the consolidation of deep soil faster. During the filling course of embankment, the excellent stability of composite layer makes it is possible to speed up the rate of filling in theory. Furthermore, the composite layer and consolidation layer can complete most of the consolidation settlement during the preloading period of embankment construction, and the requirement of the post-construction settlement can be meet. Along with the consolidation process of soft soil, the bearing capacity of foundation is improved gradually and the foundation becomes stronger.

FINITE ELEMENT MODELING

What should be pointed out here is that the embankment is a kind of strip uniform load and extends very far in longitudinal direction. So the finite element model of embankment can be simplified as two-dimensional plane finite element through the equivalent substitution (Hird et al. 1992; Hird et al. 1995). The analyses presented here were performed using the finite element program Plaxis.

The deep mixing columns can be simplified as sheet piles in longitudinal arrangement with lower elastic modulus. The thickness of sheet piles is the same as the diameter of the DMM columns with the same length, that is, the deep mixing columns and a part of soil are replaced by sheet piles. The simplified method can be calculated by the area replacement ratio or the column spacing. In this paper, the equivalent replacement is based on the area replacement ratio. It is assumed that there is no change for the pre and post axial stiffness of material. The equivalent elastic modulus of sheet piles, E_e, can be estimated based on the following equation (LI and HONG 2004, CHEN and YANG 2005).

$$E_e = \frac{2E_p s + E_s(\sqrt{3}dl - 2s)}{\sqrt{3}dl} \quad (1)$$

Where E_p is the elastic modulus of the deep mixing column, E_s is the weighed average elastic modulus of each soil layer in the range of the pile length, s, d, and l are the cross-sectional area, the diameter and the spacing of the deep mixing columns, respectively.

In the D-M method, PVD acts as the drainage path for the whole foundation. Chai equivalent the effect of PVD to a higher mass hydraulic conductivity of the subsoil in the vertical direction on a macroscale(Chai et al. 2001). Based on this concept, a parameter called equivalent vertical hydraulic, k_{ve}, is put forward and a simple method for modeling PVD improved subsoil is proposed. This method simplifies the three-dimensional drainage consolidation as two-dimensional plane solution and the analysis of PVD improved subsoil is the same as that of the unimproved case. In the proposed method, the parameters are simple and have the definite physical meaning and the maximum calculated error can be satisfied with the limit of project design. The paper follows the simple method to model PVD improved subsoil. The equivalent vertical permeability k_{ve} can be expressed as

$$k_{ve} = \left(1 + \frac{2.5l^2}{\mu D_e^2}\frac{k_h}{k_v}\right)k_v \quad (2)$$

$$\mu = \ln\frac{n}{s} + \frac{k_h}{k_s}\ln(s) - \frac{3}{4} + \pi\frac{2l^2 k_h}{3q_w} \quad (3)$$

Where l =drainage length, D_e =diameter of unit cell, k_h =horizontal hydraulic conductivities of the nature soil, k_v =hydraulic conductivity in the vertical direction, k_s = horizontal hydraulic conductivities of the smear soil, $n = D_e/d_w$ (d_w =diameter of drain), $s = d_s/d_w$ (d_s =diameter of smear zone), q_w =discharge capacity of PVD.

Test Embankment on Soft Clay at HY Highway, China

The highway is located in Eastern China and most of the embankment is on top of deep soft clay. To get reliable data and experience to guide the design and construction, a full-scale test embankment improved by D-M method with a total length of 100m is constructed and investigated.

The modeled embankment was 5.4m in height, 14m in crest width and 22m in bottom width. The embankment construction history was simplified from eighteen to five steps (Fig.1).

The modeled subsoil was 26m below the ground surface, and 60m horizontally from the embankment centerline. The displacement boundary conditions were used, that is, horizontal displacement was fixed on both ends of the model (u_x=0), and both horizontal and vertical displacement were fixed at the bottom bondary (u_x=u_y=0).

The adopted drainage boundary conditions were ground surface and bottom line (sandy silt). The left and right boundaries were impermeable. The 15-node unit was used in the mesh and the properties of the subsoil are listed in Table 1.The groundwater level is about 1m below the ground surface.

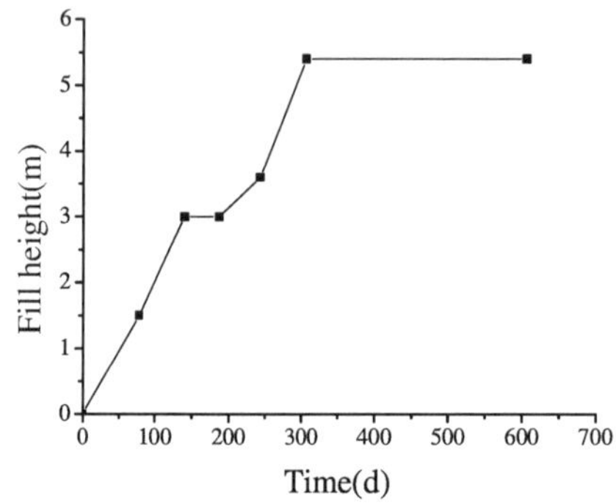

FIG. 1 Course of Embankment Construction

Table 1 Model Properties of Materials

Layer	*h* (m)	γ (kN/m²)	E (MPa)	ν	c (kPa)	Φ (°)	k_h (m/d)	k_v (m/d)	k_{ve} (m/d)
C	2.0	17.6	1.8	0.42	13.0	10.0	6.82E-04	5.04E-04	8.38E-04
MSC□	4.8	16.5	1.2	0.44	7.4	8.3	1.09E-04	8.38E-05	3.92E-04
SS□	2.7	19.3	4.3	0.33	2.9	28.8	1.00E-02	7.70E-03	1.28E-02
MSC□	5.0	16.8	1.5	0.43	8.2	8.9	1.94E-04	1.02E-04	6.82E-04
SC□	3.9	19.5	3.8	0.36	32.8	13.7	7.96E-04	6.12E-04	1.21E-03
SS□	---	19.8	6.5	0.32	3.3	29.6	1.30E-02	1.00E-02	-----
SC	0.4	16.8	35.0	0.30	0	35.0	-----	-----	-----
Fill	5.0	20.0	30.0	0.31	20	24.0	-----	-----	-----
DMM		21.0	105.0	0.28	-----	-----	-----	-----	-----

Note: C=clay; MSC□=mucky silty clay of the second layer; SS□=silty sand of the third layer; MSC□= mucky silty clay of the fourth layer; SC□=silty clay of the fifth layer; SS□= silty sand of the sixth layer; SC=sand cushion; h=thickness of clay layer; γ=bulk density; E= elastic modulus; ν=Poisson ratio; c=cohesion; Φ=internal friction angle; k_h, k_v =hydraulic conductivity in horizontal and vertical direction from laboratory test; k_{ve}=equivalent vertical hydraulic conductivity.

In the model of D-M method, the length of the columns was 8m and the spacing was 0.5m. The length of PVD was 17m and the dimension of the cross-section was 100mm×4mm. The discharge capacity of PVD was 100m³/a.

To analyze and compare the reinforcement effect of the D-M method, the paper presents two other methods to improve the foundation. One is the application of PVD alone, in which the length of PVD was 17m. The other is DMM used alone, in which the length of columns is 17m, too.

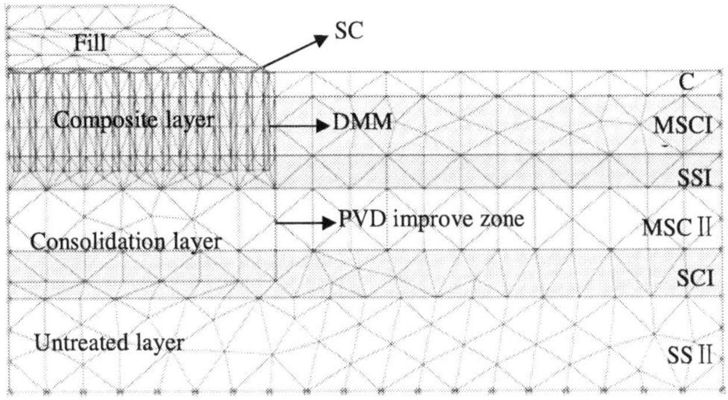

FIG. 2 D-M Finite-Element Mesh in Analysis

RESULTS AND COMPARISONS

In this section, the results obtained from the finite element analyses are compared with the results obtained from field trials (FS) together. Fig.3 shows the comparison of the total settlement for a point below the center line of embankment, for which is the representative point of largest settlement, between the finite element analyses and field trial. It can be seen from Fig.3 that there is a close agreement, especially in the beginning phase, between the observed and the computed total settlement of the D-M method. The void ratio and the permeability coefficient are invariable in Mohr-Coulomb model, but in fact, they will become smaller with the course of consolidation. As a result, the computed total settlement is slightly larger than the field trial result. The total settlement of the D-M method is obviously less than PVD, which demonstrats that the effect of short columns is considerable. The settlement of the D-M method is much larger than that of the DMM, which is demonstrated that the most of settlement of D-M method is attributed to the consolidation layer and the untreated layer.

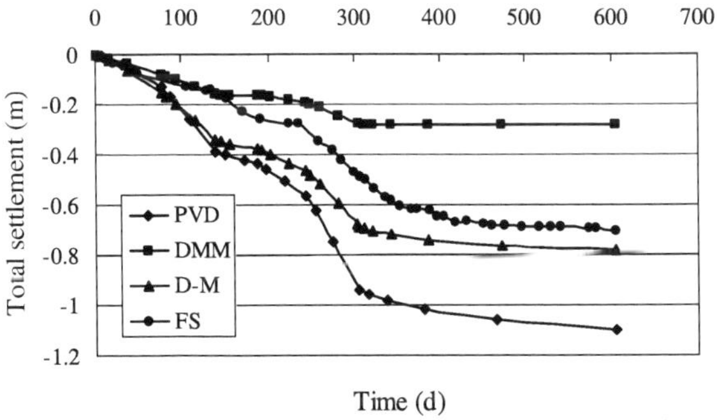

FIG. 3 Total Settlement Curves of Foundation

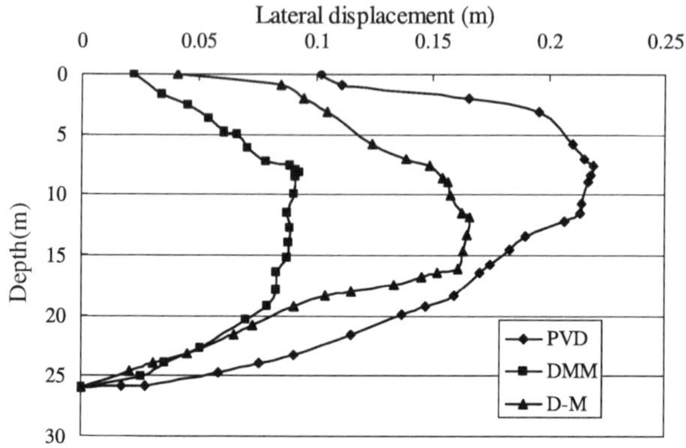

FIG. 4 Horizontal Displacement Profile

Fig.4 shows the horizontal displacement of the profile (located in the clay layer close to the toe of the embankment). It can be seen from Fig.4 that the horizontal displacement of DMM is smallest and the lateral displacement of PVD is largest. The largest horizontal displacement for PVD is 0.219m, which occurred at the depth of 7.6m. However, the largest horizontal displacement depth for D-M and DMM is 11.9m and 8.13m respectively. Certainly the largest horizontal displacement of DMM , which is only 0.093m, is much smaller than the other two methods. This result demonstrates the effect of columns in the restriction of lateral displacement for clay foundation. At the same time, the stability of the foundation is improved because the depth of largest horizontal displacement is deeper.

Fig.5 and Fig.6 show the variation of excess pore pressure during embankment construction at the depth of 3m and 13m respectively. For the simplified stages of loading (Fig.1), there were two peaks in all of curves. The first peak was caused by the first two stages of loading in the first 140 days. Then there was a break of 47 days in construction for bad weather. The last two stages of loading were responsible for the second peak. In Fig.5, it can be seen that the excess pore pressure of PVD is much larger than the other two. The reason for the phenomenon is that the columns support most of the vertical load and the soil between columns only surports a minor part of the load. The length of columns in DMM is 17m, which is twice as long as the columns in D-M and penetrated through the soft clay layer. So the load taken by the columns in DMM is much larger than in D-M and the excess pore pressure in DMM is smaller than in D-M. Fig.6 shows the variation of excess pore pressure at the depth of 13m, in which the curves are very close and similar for D-M and PVD. Due to the existence of long columns, the excess pore pressure of DMM is still smaller than the other two. Both of the graphs show the rate of excess pore pressure dissipation for DMM is the smallest due to the longer drainage path. In D-M, the short columns enlarge the stiffness of shallow ground and PVD shorten the path of drainage. As a result, the excess pore pressure dissipated faster in the clay foundation improved by D-M than the other two methods for the existence of vertical drainage gallery and short columns.

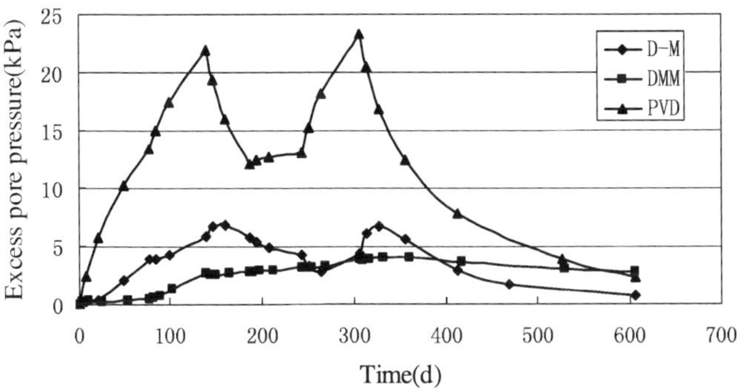

FIG. 5. Excess Pore Pressure Curves at 3m Depth

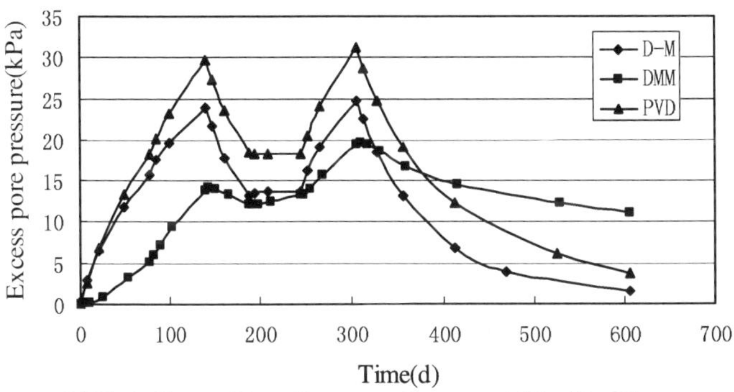

FIG. 6. Excess Pore Pressure Curves at 13m Depth

DISCUSSIONS

As mentioned previously, the D-M method can improve the stability of foundation, and meanwhile accelerate the dissipation of excess pore pressure. The method utilizes the inherent pre-loading term effectively in Eastern China and decrease the post-construction settlement. Compared with DMM, D-M method can cut down the expense for the shorter columns. The use of PVD can accelerate the excess pore pressure dissipation in deeper soft soil layer. Compared with PVD, D-M method improves the stability of foundation and increase the stiffness of the shallow ground for the existence of short columns during embankment construction. The load of embankment was transferred by the short columns to the deeper soft soil layers and the pore water flows through PVD to ground surface. Thus the high excess pore pressure dissipates and the degree of consolidation of deeper soft clay becomes larger.

CONCLUSIONS

A new method to improve the deep soft clay in China is proposed in this paper. The method combines the DMM and PVD method and overcomes the disadvantage of the two methods. The results of field trail and numerical modeling show that the method is an effective and economic method for deep soft clay.

In the D-M method, the short columns improve the stability and increase the stiffness of the shallow foundation and transfer the vertical load to the deeper clay layer. PVD provides the drainage gallery in vertical direction and shortens the drainage path. Thus the construction can be speed up and meanwhile decrease the post-construction settlement.

There exists an inherent pre-loading period during the highway construction in Eastern China. The D-M method can make effective use of the period to make excess pore pressure dissipate and shorten the construction period. In summary, the D-M method is a new effective method to improve the embankment on deep soft clay and can be a choice for similar soils in the future.

ACKNOWLEDGMENTS

This study was performed with financial support from Shanghai Leading Academic Discipline Project (Grant NO. B308.), the National Natural Science Fund (Grant NO. 50678130) and National high-tech research and development projects （863） (Grant NO. 2007AA11Z117).

REFERENCES

XU Chao, YE Guan-bao. (2006). "Research on mechanism of combined improvement of soft soils based on field monitoring." *Chinese Journal of Geotechnical Engineering*. 28(7): 918-921.

Hird, C. C., Pyrah, I. C., and Russell, D. (1992). “Finite element moelling of vertical drains beneath embankments on soft ground.” *Geotechnique,* London, 42(3):499–511.

Hird, C. C., Pyrah, I. C., Russell, D., and Cinicioglu, F. (1995). “Moelling the effect of vertical drains in two-dimensional finite element analyses of embankments on soft ground.” *Can. Geotech. J.,* Ottawa, 32: 795–807.

LI Tao, HONG Bao-ning. (2004). "Finite Element Analysis of Bearing Capacity of DJM Composite Foundation Considering the Influence of Transition between Piles and Soil." *Journal of Disaster Prevention and Mitigation Engineering*. 24(3):289-292.

CHEN Fu-quan, YANG Min. (2005). "Numerical analysis of piles influenced by lateral soil movement due to surcharge loads." *Chinese Journal of Geotechnical Engineering*. 27(11): 1286-1290

Jin-Chun Chai, Shui-Long Shen, Norihiko Miura, and Dennes T. Bergado. (2001). "Simple Method Of Modeling PVD improved Subsoil." *Journal of Geotechnical and Geoenvironmental Engineering*. N. 2001: 965-972.

Effect of Sand Columns on the Load Response of Soft Clays

Tarek Maakaroun[1], Shadi S. Najjar[2], and Salah Sadek[3]

[1]Graduate Student, Department of Civil and Environmental Engineering, American University of Beirut, PO.Box 11-0236 Riad El-Solh 1107-2020; PH (961) 70974251; email: ttm03@aub.edu.lb
[2]Assistant Professor, Department of Civil and Environmental Engineering, American University of Beirut, PO.Box 11-0236 Riad El-Solh 1107-2020; PH (961) 3108853; email: shadi.najjar@aub.edu.lb
[3]Professor, Department of Civil and Environmental Engineering, American University of Beirut, PO.Box 11-0236 Riad El-Solh 1107-2020; PH (961) 3850785; email: salah@aub.edu.lb

ABSTRACT

Stone or sand columns in soft clays have been used to increase the load carrying capacity, reduce the settlement, and decrease the generation of excess pore water pressure during loading. The main objective of this paper is to investigate the effect of sand columns on the undrained shear strength of soft clayey material. For this purpose, consolidated undrained triaxial tests were performed on normally consolidated Kaolin specimens that were reinforced with sand columns and tested at confining pressures of 100, 150, and 200 kPa. The parameters that were varied are the diameter of the sand columns, the height of the column, and the confining pressure. Results indicate substantial improvements in undrained shear strength reaching 75% for a maximum area ratio (ratio of area of sand column to area of specimen) of 17.8%. The data collected supports the hypothesis of a "critical column length" that is greater than about six column diameters, beyond which the increase in undrained shear strength due to the presence of the sand columns becomes negligible. Finally, the data did not show any clear indication of the effect of the confining pressure on the improvement in the undrained shear strength of the "reinforced" clay specimen.

INTRODUCTION

Due to the wide-spread development of urban cities around the world, ground modification techniques are becoming essential in improving the mechanical properties of soils to carry typical structural loads at sites considered marginal. In particular, the insertion of stone or sand columns into soft clay has been shown to have a positive effect on the load carrying capacity of the clay, resulting in a composite soil mass that has greater shear strength and improved stiffness compared to the unreinforced clay. In addition, stone columns generally act as vertical drains in the clay thus accelerating the dissipation of excess pore water pressures that are generated during loading. As such, stone columns currently stand as one of the most viable and practical techniques for improving the mechanical properties of soft clays.

Several researchers have conducted laboratory experiments to understand the behavior of sand columns in weak cohesive soils. Hughes and Withers (1974), Sivakumar et al. (2004), and Ambily and Ghandi (2007) conducted experiments in which single sand columns were installed into soft clays and tested by direct loading of the sand columns. Other researchers (Miur-Wood et al. 2000; Mc. Kelvey et al. 2004; Sivakumar et al. 2004; Black et al. 2006; Ambily and Ghandi 2007; and Black et al. 2007) conducted tests in which both single sand columns and column groups were loaded together with the surrounding clay. In a number of these studies, sand columns of different lengths were used to investigate the effect of the column penetration on the improvement in the load-carrying capacity of the specimens. The majority of the studies mentioned above were conducted in large one dimensional loading cells, while only a few were performed under full triaxial conditions. Results of the experimental investigations indicate that the mode of failure is characterized by lateral bulging of the sand column, particularly in the top 4 to 5 diameters along the height of the column. Specimens with short partially-penetrating columns generally resulted in no significant improvement in the load carrying capacity of the specimen. In addition, as the reinforcement area ratio increased, both the stiffness and the shear strength of the specimen increased.

Despite the experimental effort that is summarized above, there is currently a need for rational and reliable design methods for designing foundations that are supported by clay that is reinforced with stone or sand columns. The main objective of this paper is to investigate the effect of sand columns on the undrained shear strength of soft clayey material. For this purpose, consolidated undrained triaxial tests (CU) were performed on back-pressure saturated normally consolidated Kaolin specimens that were consolidated and tested at confining pressures of 100 kPa, 150 kPa, and 200 kPa. The parameters that were varied in the study are the area ratio of reinforcement, the penetration depth of the sand column, and the confining pressure. The analysis of the results is restricted to improvements in the undrained shear strength of the soft clay as a result of the sand column inclusion for different confining pressures.

TEST PROGRAM

A total of 17 consolidated undrained triaxial tests were performed on one-dimensionally consolidated Kaolin specimens having a diameter of 7.1 cm and a length of 14.2 cm. Single sand columns having diameters of 2cm and 3cm with height penetration ratios, H_c/H_s (height of sand column to height of specimen) of 0.5, 0.75 and 1 were installed in pre-drilled holes in the center of the clay specimens. The area replacement ratio, A_c/A_s, defined as the area of the sand column to the area of the clay specimen was equal to 7.93% and 17.8% for clay samples that were reinforced with 2cm and 3cm diameter columns respectively. Reinforced clay specimens were back-pressure saturated using a back pressure of 310 kPa and isotropically consolidated under effective confining pressures of 100, 150, or 200 kPa. In all the test specimens, the measured "B" value was greater than 0.96 indicating an adequate degree of saturation. Samples were then sheared undrained with a strain rate of 1% per hour. All tests were terminated at a maximum axial strain of about 10%. The program of testing is summarized in Table 1.

SAMPLE PREPARATION AND TESTING

Kaolin slurry was prepared by mixing Kaolin clay powder with water at a water content of 100% (1.8 times the liquid limit of 55.7%). From the resulting homogeneous mix, slurry was transferred into each of four custom-fabricated 1-dimensional consolidometers (Fig. 1). Dead weights were then added in stages to consolidate the samples to a maximum vertical effective stress of 100 kPa. During consolidation, water was allowed to drain from the top and bottom of the consolidometers through 2 porous stones. Primary consolidation required a period of about 7 days to be completed. The consolidated specimens were then removed from the PVC pipes by splitting the two halves of the pipes (Fig. 2a). The height of the specimen after consolidation was about 18 cm. As a result, the sample had to be trimmed to a final height of 14.2 cm (twice the diameter of 7.1 cm). The water content after consolidation was found to be relatively uniform (about 53%) throughout the depth of the sample. The variation of bulk density for different specimens had an average of 16.13 kN/m^3 and a standard deviation of 0.16 kN/m^3, while the average void ratio was equal to 1.34 with a standard deviation of 0.03. The relatively small variations observed in the water content, density, and void ratio of the clay specimens after consolidation indicate that friction in the 1-D consolidometers was minimal and that the specimen preparation procedure was generally repeatable.

Single sand columns were installed in clay specimens using the procedure described below. First, the trimmed clay specimen was encased with a PVC split mold and a prefabricated hand auguring machine was used to drill a hole in the center of the specimen (Fig. 2b). Frozen sand columns of different diameters and lengths were installed into the predrilled holes. The columns were constructed from Ottawa sand at a density of about 17.3 kN/m^3 and a water content of about 20%. These sand columns were prepared (prior to freezing) by pouring 3 layers of dry Ottawa sand in circular geosynthetic fabrics which were inserted in a glass tube having the same inner diameter as the sand column and vibrated using an electric vibrator. After vibration was complete, water was added to the sand column through the geosynthetic encasement to achieve a water content of about 20%. The sand column was then allowed to freeze for 24 hours after which the geosynthetic fabric was cut and detached from the sand column. The frozen sand column was then inserted into the predrilled hole and allowed to thaw (Fig. 2c). As a final step, reinforced clay specimens were transferred to the triaxial cell and back-saturated using a back pressure of 310 kPa followed by isotropic consolidation in a "TruePath" automated triaxial system. Specimens were then sheared undrained at a constant strain rate of 1 % per hour while measuring excess pore water pressures at the bottom of the specimen and axial deformations at the top.

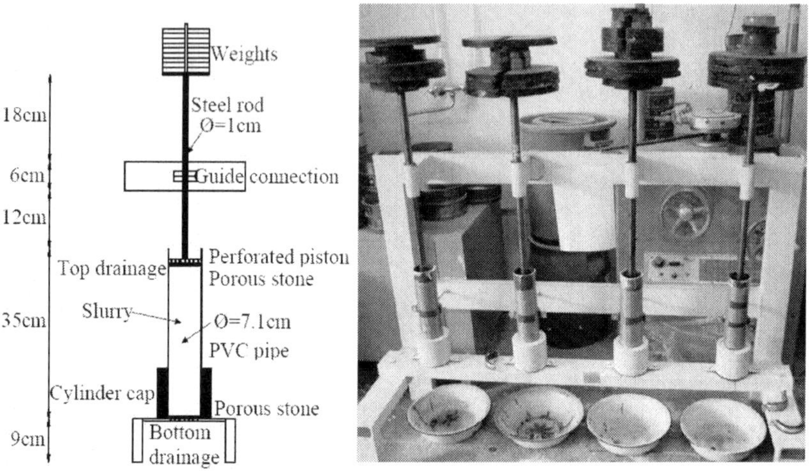

FIG. 1. Custom fabricated 1-dimensional consolidometers

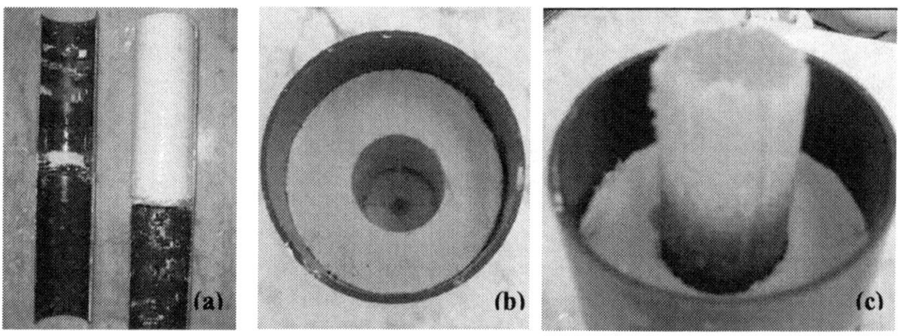

FIG. 2. Photographs of (a) clay specimen after 1-D consolidation, (b) predrilled 3-cm diameter hole, and (c) Insertion of frozen sand column in clay

TEST RESULTS

The test results are presented in Table 1 and Fig. 3 and include deviatoric stress and pore water pressure versus axial strain plots for control specimens and specimens that were reinforced with sand columns of different area ratios and penetration ratios. Results also include an analysis of the effect of relevant parameters on the improvement in the undrained shear strength of the clay. It should be noted that in all the discussion presented below, it was assumed that the sand column and surrounding clay act as a single element with homogeneous distributions of stresses and strains.

Stress-Strain behavior

Curves showing the variation in the deviatoric stress and the pore water pressure with axial strain for a confining pressure of 150 kPa are shown on Fig. 3. These results are representative of tests conducted at other confining pressures. In all the cases, pore water pressures increased as the deviatoric stresses reached their maximum values at axial strains of about 5 to 6%. For samples reinforced with sand columns, results on Fig. 3 indicate a consistent increase in the deviatoric stress and a consistent decrease in the pore water pressure at failure as the penetration depth and area ratio of the sand columns increased. These results are in line with the experimental test results reported in Black et al. (2007) and Sivakumar et al. (2004). The decrease in excess pore water pressures during shearing of reinforced clay can be attributed to the possible dilatational tendency of the sand columns, which is expected to increase as the area replacement ratio and the penetration depths increase.

Effect of Area Ratio and Height Penetration Ratio on Undrained Shear Strength

For fully penetrating columns, results in Table 1 indicate that for area replacement ratios of 7.9% and 17.8%, the increase in undrained shear strength ranged from 13 to 19.5% and from 67.5 to 75%, respectively. These increases in the undrained shear strength are substantial especially that some previous studies reported that the area ratio has to be as high as 20 to 40% before there are significant improvements in strength (Sivakumar et al. 2004). For columns with a penetration ratio of 0.75, the corresponding increases in strength ranged from 8.6% to 14.5% and from 20.1% to 34.9% for area ratios of 7.9% and 17.8%, respectively.

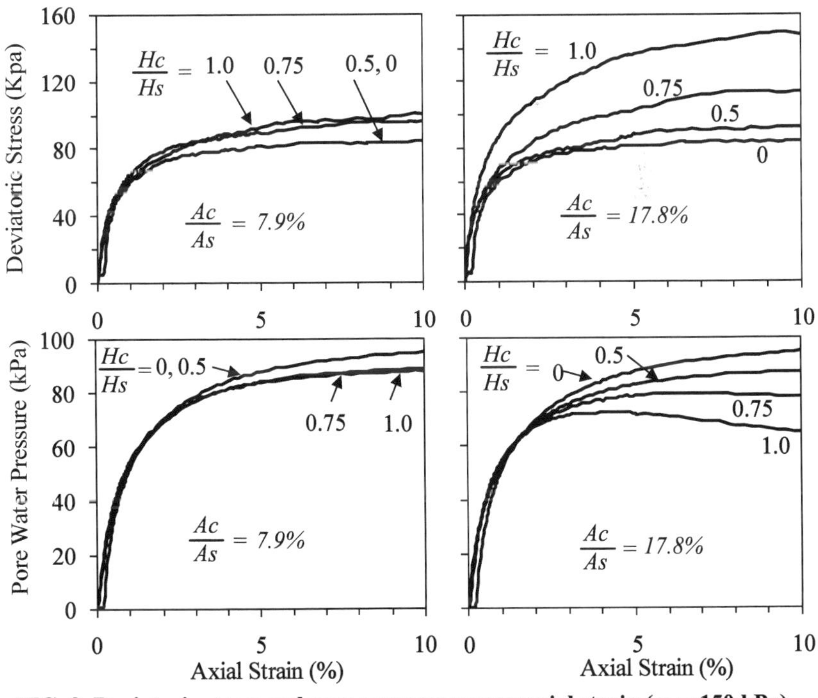

FIG. 3. Deviatoric stress and pore pressure versus axial strain (σ_3 = 150 kPa)

TABLE 1. Test Program and Results

No. of tests	Confining pressure σ_3 (kPa)	Diameter of sand column (cm)	Area ratio A_c/A_s (%)	Height of sand column (cm)	Height penetration Ratio H_c/H_s	Undrained shear strength (kPa)	Increase in undrained strength (%)	Reduction in excess pore water pressure (%)
1		0	0	0	0	32.4	-	-
2		2	7.93	10.65	0.75	35.2	8.6	6.5
3	100			14.2	1	36.6	13.0	16.5
4		3	17.8	10.65	0.75	38.9	20.1	20.2
5				14.2	1	56.7	75.0	30.3
6		0	0	0	0	42.1	-	-
7				7.1	0.5	42.8	1.7	1.2
8		2	7.93	10.65	0.75	48.2	14.5	3.7
9	150			14.2	1	50.3	19.5	7.7
10		3		7.1	0.5	46.0	9.3	8.3
11			17.85	10.65	0.75	56.8	34.9	17.9
12				14.2	1	73.9	75.5	31.4
13		0	0	0	0	55.1	-	-
14		2	7.93	10.65	0.75	60.1	9.1	8.1
15	200			14.2	1	65.8	19.4	14.4
16		3	17.85	10.65	0.75	71.2	29.2	17.6
17				14.2	1	92.3	67.5	31.7

An analysis of the results in Table 1 indicate that the relative increase in undrained strength when comparing partially penetrating columns (H_c/H_s=0.75) to fully penetrating columns was much more pronounced in the case where the area ratio was 17.8% (increase from 20.1% to 75% for 100 kPa confining pressure) compared to the case where the area ratio was equal to 7.8% (increase from 8.6% to 13.0% for 100 kPa confining pressure). This disproportionate increase in strength indicates that the improvement in undrained shear strength may not only be a function of the height penetration ratio, but also of the ratio of the column height to the column diameter. This dependence has been studied by other researchers who proposed the idea of the "critical column length" beyond which the column will not have any positive effect on improvements in capacity. In these studies, it is suggested that a column length greater than five diameters may no longer participate in increasing the load carrying capacity of soft cohesive clays. This is possibly due to the fact that bulging of the sand columns will generally be concentrated in the upper portions of the column.

To investigate the possible dependency of the increase in undrained strength on the ratio of the column height to column diameter, the percent increase in undrained strength was plotted on Fig. 4 against the ratio of the column height to diameter for all the tests conducted in this study and for all confining pressures. Average trend lines or curves were plotted through the data for the two area ratios used. In addition, data points from tests conducted by Sivakumar et al. (2004) and which show improvements in undrained shear strength due to the inclusion of 3.2-cm diameter sand columns in Kaolin clay are also plotted on the same figure for comparison.

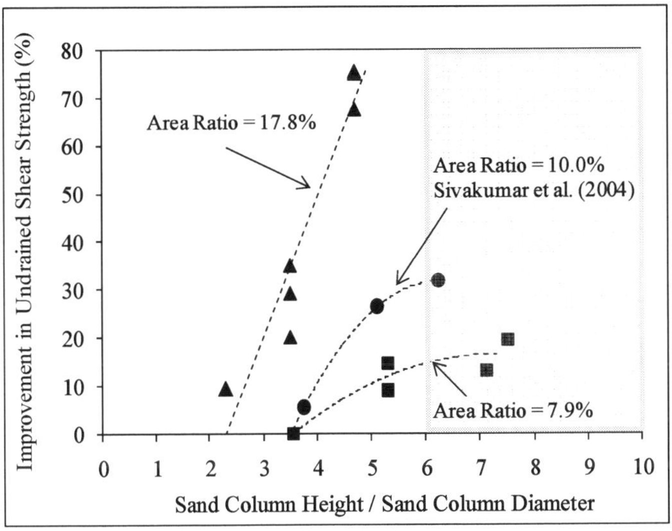

FIG. 4. Effect of ratio of column height to diameter on undrained shear strength

The data on Fig. 4 supports the hypothesis of a "critical column length" beyond which the increase in undrained shear strength becomes relatively negligible. This "critical column length" is greater than six column diameters as indicated by the grey area in Fig. 4. The existence of this critical column length provides a plausible explanation for the large relative increase in strength that was observed when the height of the sand column was increased from 10.65 cm (H_c/H_s=0.75) to 14.2 cm (H_c/H_s=1) for specimens that were reinforced with an area ratio of 17.8%. In these tests, the maximum ratio of the column height to column diameter was still less than 4.7 for the fully penetrating column, a ratio which is still smaller than the critical column ratio of about 6. In comparison, the ratio of the column height to column diameter was equal to 7.1 and 5.3 respectively for the fully penetrating and partially penetrating columns (Hc/Hs=0.75) for specimens reinforced with an area ratio of 7.9%. The improvement in undrained shear strength in these specimens leveled out at a maximum improvement of about 17%, but kept increasing to a maximum value of about 72% for the specimens that were reinforced with an area ratio of 17.8%.

Effect of Confining Pressure

Results presented in Table 1 do not show a clear trend with regards to the effect of confining pressure on the level of improvement in undrained shear strength. For fully penetrating columns, the average improvement was equal to 17.3% ± 3.74% and 72.7% ± 4.48% for area ratios of 7.9% and 17.8%, respectively. The variations about the mean values can be attributed to the effect of the confining pressure in addition to other factors which include testing errors and variability in soil properties between specimens. However, these variations can be considered to be relatively small given the general objectives of the research that is being conducted. More testing and analyses will be required if the effect of the confining pressure is to be quantified and included in future models that are aimed at predicting the behavior of soft clays that are reinforced with sand columns and subjected to undrained loading.

CONCLUSIONS

Based on results from 17 consolidated undrained triaxial tests that were conducted in this study, the following conclusions can be drawn regarding the effect of sand columns on the undrained load response of soft clay:

1. For fully penetrating columns and for area replacement ratios of 7.9% and 17.8%, the increase in undrained shear strength ranged from 13 to 19.5% and from 67.5 to 75%, respectively. These increases are substantial given the relatively small area ratios used.
2. The data support the hypothesis of a "critical column length" which is greater than about six column diameters, beyond which the increase in undrained shear strength due to the presence of the sand columns becomes negligible.
3. The data did not show any clear indication of the effect of the confining pressure on the improvement in undrained shear strength. The variations about the average improvement values are relatively small and can be attributed to the effect of the confining pressure in addition to other factors which include testing errors and variability in soil properties between specimens.
4. It is recommended that field tests should be conducted in the future to confirm the conclusions presented in this paper.

REFERENCES

Ambily, A.P., Gandhi, S.R., (2007). "Behavior of stone columns based on experimental and FEM analysis." *Journal of Geotechnical and Geoenvironmntal Engineering* 133 (4): 405-415.

Black, J., Sivakumar, V., Madhav, M.R., and McCabe, B. (2006). "An improved experimental set-up to study the performance of granular columns." *Geotechnical Testing Journal* 29 (3): 1-7.

Black, J.V., Sivakumar, V., and McKinley, J.D. (2007). "Performance of clay samples reinforced with vertical granular columns." *Canadian Geotechnical Journal* 44: 89-95.

Hughes, J.M.O., and Withers, N.J. (1974). "Reinforcing of soft cohesive soils with stone columns." *Ground Engineering* 7 (3): 42-49.

McKelvey, D., Sivakumar, V., Bell, A., and Graham J. (2004). "Modeling vibrated stone columns in soft clay." *Journal of Geotechnical Engineering* 157 (3): 137-149.

Muir Wood, D., Hu, W., and Nash, D.F.T. (2000). "Group effects in stone column foundations: Model tests." *Geotechnique* 50 (6): 689-698.

Sivakumar, V., McKelvey, D., Graham, J., and Hughus, D. (2004). "Triaxial tests on model sand columns in clay." *Canadian Geotechnical Journal* 41: 299-312.

Load-Displacement Compatibility Analysis of a Low-Height Column-Supported Embankment

Michael P. McGuire, P.E.[1], George M. Filz, P.E.[2], and Marcio S.S. Almeida[3]

[1]PhD candidate, Dept. of Civil and Environmental Engineering, Virginia Polytechnic Institute and State Univ., Blacksburg, VA 24060-0105. Email: mcguirem@vt.edu
[2]Professor, Dept. of Civil and Environmental Engineering, Virginia Polytechnic Institute and State Univ., Blacksburg, VA 24060-0105. Email: filz@vt.edu
[3]Professor, Graduate School of Engineering, Federal University of Rio de Janeiro, Rio de Janeiro, Brazil. Email: almeida@coc.ufrj.br

ABSTRACT

This paper presents instrumentation results and analyses of a low-height, geosynthetic-reinforced, column-supported embankment constructed in Brazil. The approach involves back analyzing relevant properties of the embankment fill and geogrid reinforcement and applying the Load Displacement Compatibility (LDC) approach to calculate geogrid deflections in a unit cell of a square pile cap array. The results show that the LDC approach is consistent with the observed response of the embankment.

INTRODUCTION

Geosynthetic-reinforced, column-supported embankments (GRCSEs) are constructed over soft ground to accelerate construction, improve embankment stability, control total and differential settlements, and protect adjacent facilities from the settlement that would otherwise be induced by the new embankment load. The geosynthetic reinforcement generally consists of one or more layers of geogrid or high-strength woven fabric that are placed at the base of the embankment near the elevation of the top of the columns or pile caps. In addition to resisting lateral spreading, the purpose of the reinforcement is to aid in the transfer of embankment loads to the columns. One of several important design considerations for GRCSEs is limiting the strain and vertical deflection of the geogrid under service conditions. For instances when the embankment height is low relative to the clear spacing between adjacent columns, excessive levels of strain developing within the geosynthetic reinforcement can lead to surface expression of differential settlements between the columns and the soft soil at the foundation level.

This paper examines a full-scale GRCSE constructed in the Barra da Tijuca District of Rio de Janeiro, Brazil. This embankment was designed such that the clear spacing between columns exceeded the supported embankment height. A portion of this low-height embankment was instrumented to allow for measurement of the vertical deflection and strain of the reinforcement. Analyses and comparisons with deflection and strain measurements are presented.

CASE STUDY

A low-height GRCSE was constructed in 2004 on a pre-existing fill overlying an approximately 10 m thick layer of very soft clay in the Barra da Tijuca District of Rio de Janeiro. The roughly 2 m thick existing fill was placed in 1996 as part of the original plan to construct the embankment without columns by surcharging and accelerating consolidation with vertical drains. Lawsuits delayed construction, and eventually a decision was made to construct the embankment as a GRCSE. The second phase of embankment construction increased the embankment height, H, by roughly 1 to 1.3 meters. A portion of the GRCSE was instrumented and used as an experimental site, with three types of pile cap arrangements, as shown in Figure 1. For all three arrangements, square precast concrete piles with a width of 0.18 m were used to support the pile caps. At Locations A and B, the piles were driven in a square array with a 2.5 m center-to-center spacing, s, and square pile caps with a side length, a, of 0.8 m were used. At Locations C and D, 8.3 m long and 0.8 m wide cap beams were used with center-to-center spacings of 2.5 and 3.5 m, respectively. For all pile cap arrangements, the clear spacing, s - a, exceeds the embankment height of 1 to 1.3 m. Additional details regarding the experimental site can be found in Spotti (2006), Almeida et al. (2007), and Almeida et al. (2008).

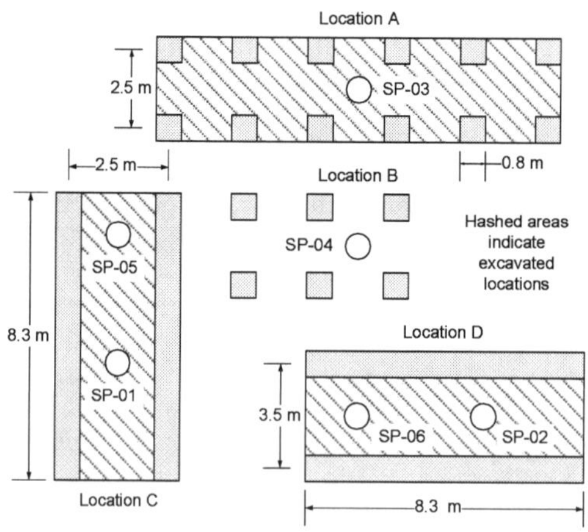

Figure 1: Layout of experimental site in the Barra da Tijuca District

A single layer of Huesker FORTRAC 200/200-30 polyester biaxial geogrid was placed over the pile caps with an underlying cushioning layer of non-woven geotextile to reduce abrasion. Field and laboratory testing was performed on the embankment material and underlying soft clay, and the details can be found in Spotti (2006). The embankment material used for both the 1996 and 2004 phases of construction consisted of a clayey sand compacted to at least 95% of the standard Proctor maximum and having a compacted unit weight, γ, of about 18 kN/m^3. The soft clay has a unit weight of about 12.5 kN/m^3, an undrained strength of 10 to 20 kPa, and a plasticity index of 120 to 250. The clay compressibility was assessed using conventional laboratory consolidation tests, which showed that the clay is normally consolidated with a representative compression ratio of 0.4 to 0.5. A stiff residual soil stratum underlies the clay layer.

To investigate the influence of subgrade support on load transfer mechanisms within the GRCSE, 1 m deep excavations were made prior to placement of the geogrid and embankment fill at Locations A, C, and D. An attempt to keep the geogrid taut was made by loading the edges with fill, although photos taken at the site indicate that some portions of the reinforcement did sag prior to placement of the embankment fill. Compaction of the lower lifts of fill material over the unsupported geogrid was performed with care using light equipment.

Instrumentation consisted of settlement plates (SP) installed at the level of the reinforcement at the locations shown in Figure 1 and strain gauges installed at Location A. Additional details regarding instrumentation can be found in Spotti (2006) and Almeida et al. (2007). Readings were taken over a period of 188 days after embankment construction. In all cases, the time history of the data indicates that the instrumentation readings had reached constant values during the monitoring period, which suggests that the soft clay had, for all practical purposes, completed primary consolidation.

ANALYSES

The settlement plate readings taken at Locations C and D were used to develop a single representative set of property values for the embankment material and geogrid reinforcement. These values were then used with the LDC approach (Filz and Smith 2006a) to produce calculated values of vertical geogrid deflection, which were compared to the settlement plate measurements.

Back Analysis of Embankment and Reinforcement Parameters

The method of back analysis was used to obtain representative properties of the embankment material and the geogrid reinforcement. The parameters varied in the back analyses were the shear strength properties (ϕ and c) of the embankment fill, the lateral earth pressure coefficient, K, used in the soil arching analysis (Terzaghi 1943, Russell et al. 2003), and the tensile stiffness, J, of the geogrid reinforcement. The back analyses were done in two dimensions because the cap beams are much longer than the clear span at Locations C and D. The two-dimensional approximation and the lack of subgrade support at both locations allows for direct application of

Terzaghi's (1943) expressions for estimating the vertical soil pressure acting on a downward yielding surface between two fixed supports. Using the current set of notation, the expression for calculating the net vertical pressure acting on the reinforcement, p, is

$$p = \frac{(s\text{-}a)\gamma\text{-}2c}{2K\tan\phi}\left(1 - e^{-2HK\tan\phi/(s-a)}\right) \tag{1}$$

The 2D parabolic approach provided by Carlsson (1987) in Rogbeck et al. (1998) was used for the load-deflection response of the geogrid. The tension, T, developed within the reinforcement can be determined using

$$T = \frac{p(s-a)}{2}\sqrt{1+\frac{1}{6\varepsilon}} \tag{2}$$

Stress-strain compatibility can be incorporated by substituting $\varepsilon = T/J$ into Eq. 2 yielding

$$24T^3 - 6\ddot{K}_g^2 - \ddot{K}_g^2 J = 0 \quad \text{where } \ddot{K}_g = p(s-a) \tag{3}$$

Eq (3) can be solved for T, which gives the strain using J. The maximum vertical deflection, d, for the assumed parabolic deformed shape of the reinforcement can be obtained from the reinforcement strain using $d \approx (s\text{-}a)\,(3\varepsilon/8)^{0.5}$.

Table 1 lists the final settlement plate readings at Locations C and D along with the corresponding embankment height at the settlement plate locations.

Table 1: Settlement plate readings 188 days following construction

Location	Settlement Plate	Embankment Height (m)	Measured Settlement (m)
C	SP-01	1.10	0.32
C	SP-05	1.14	0.22
D	SP-06	1.25	0.40
D	SP-02	1.28	0.37

The preceding expressions were used to calculate the geogrid deflections at the locations provided in Table 1. Initially, the shear strength parameters of the embankment fill, the earth pressure coefficient (K), and the geogrid stiffness were selected based on the available laboratory testing and the geogrid manufacturers specifications. These values were then incrementally adjusted independently such that good agreement between the predicted and measured values was achieved. It was not possible to choose one set of parameter values that simultaneously yielded good agreement at all four locations. It is believed that initial slack in the geogrid reinforcement at the location of SP-01 lead to a measured deflection that incorporates the release of the slack under vertical loading in addition to the deflection produced by the developed strain. Justification for this lies in the fact that the measured deflection at SP-01 exceeds the calculated deflection, even for the case of no stress reduction, i.e., $p = \gamma H$. Parameter values that provided good agreement between the predicted and measured geogrid deflections at SP-05, SP-06, and SP-02 are given in the first row of Table 2.

Table 2: Relevant properties of geogrid and embankment

Fill material	ϕ	c (kPa)	K	J (kN/m)
Clayey sand	37	11	0.8	1600
Equivalent c = 0 soil	68	0	0.8	1600

Because the Adapted Terzaghi method used by the LDC approach for estimating the load-displacment behavior above the reinforcement was developed for a cohesionless embankment material, it was necessary to find equivalent c = 0 shear strength parameters for the fill. Examination of Eqs. (1) through (3) shows that all combinations of ϕ, c, and K that produce the same of value of p are entirely equivalent with respect to calculation of ε and d. Figure 2 shows values of ϕ and c that all produce the same value of p for a value of K = 0.8, and that a logical tradeoff between ϕ and c exists. Thus, even though the ϕ value for the equivalent c = 0 soil is unreasonably high, the equivalent c = 0 soil produces the same strain and deflection in the reinforcement.

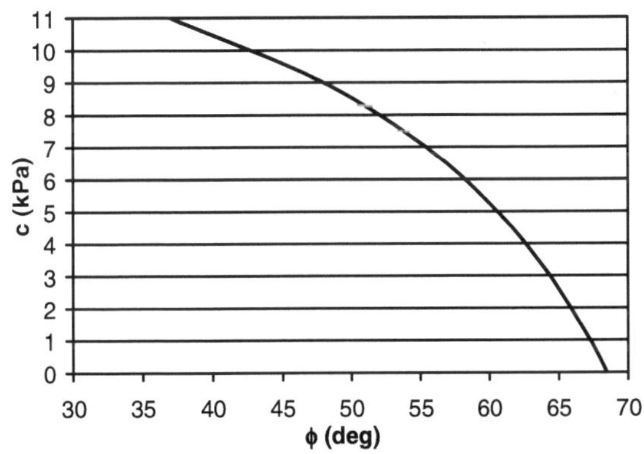

Figure 2: Equivalent shear strength parameters for the clayey sand fill

Table 3 shows the comparison between the measured and calcuated geogrid deflections at locations SP-05, SP-06, and SP-02 using the equivalent c = 0 shear strength values, and it can be seen that the agreement is good. On this basis, the values listed in Table 2 were used to calculate the performance at Locations A and B, as described in the next section.

Table 3: Comparison of predicted and measured deflection values

Settlement Plate	Measured Deflection (m)	Predicted Deflection (m)
SP-05	0.22	0.23
SP-06	0.40	0.39
SP-02	0.37	0.39

Analysis of Unit Cell of Square Pile Cap Array

The back analyzed parameter values for the embankment fill and geogrid were applied to Locations A and B using the LDC approach as implemented in an Excel spreadsheet named GeogridBridge 1.1 to predict the vertical deflection of the geogrid.

Load Displacement Compatibility (LDC) Approach

The LDC method is based on load and displacement compatibility among the embankment, columns, geosynthetic reinforcement, and foundation soil. Differential settlement between the columns and the foundation soil is determined as the difference between the vertical elastic deformation of the column and the settlement of the foundation soil, which includes linear compression of existing granular soil, if present, and nonlinear compression of clay. Friction between the settling foundation soils and the columns is accounted for by considering the interface friction angle and the lateral pressure produced by column installation as well as Poisson effects from the embankment load determined according to a concentric cylinder model (Davis and Poulos 1974).

The load-deflection response of geosynthetic reinforcement is obtained from axisymmetric numerical analyses of an annular membrane. The load-deflection relationship for the column or pile cap penetrating up, relatively, into the embankment is assumed to be linear up to a maximum load condition. The linear part is approximated using the linear solution for displacement of a rigid circular loaded area on a semi-infinite mass (Davis and Poulos 1974). The maximum load condition on the column area above the reinforcement level is established by setting a lower limit on the value of total vertical pressure acting down on the reinforcement using the Adapted Terzaghi Method with an earth pressure coefficient, K, equal to 0.8 for this particular case, which falls between the values of 1.0 used by Russell and Pierpoint (1997) and 0.5 used by Russell et al. (2003).

The simultaneous solution among the load-displacement compatibility relations for the embankment fill, geosynthetic reinforcement, columns, and foundation soils is performed using an iterative technique implemented in a Microsoft Excel® spreadsheet entitled GeogridBridge 1.1. Additional details regarding GeogridBridge 1.1 the LDC approach can be found in Filz and Smith (2006a,b).

Analysis of a 3D Unit Cell with Subgrade Support

Application of the LDC approach to Location B using the back-calculated property values from Locations C and D yield the results shown in Table 4. The good agreement between measured and calculated magnitudes of geogrid deflection in Table 4 indicates that the parameter values from the 2D back analyses without subgrade support are also applicable to a 3D unit cell that includes the influence of subgrade support.

Table 4: Comparison of predicted and measured deflection values

Location	Settlement Plate	Measured Deflection (m)	Deflection Predicted from LDC Analysis (m)
B	SP-04	0.10	0.11

Analysis of a 3D Unit Cell without Subgrade Support

Table 5 presents the result of the LDC analysis at Location A. The condition of no subgrade support was simulated in the GeogridBridge spreadsheet by modeling the foundation soil as a granular elastic material and incrementally softening the system by reducing the soil modulus to a value approaching zero. Instrumentation at this unit cell included strain gauges in the geogrid. Using the relationship between vertical deflection and strain that is intrinsic to the parabolic deflection assumption, the vertical deflection of the geogrid can be determined both directly from the settlement plates and indirectly using the measured reinforcement strain. The fact that the deflection measured by the settlement plate is significantly larger than the calculated value suggests that the reinforcement was not initially taut at this location. This is further supported by the fact that the deflection determined from the strain gauge readings is significantly lower than the calculated value. Vertical equilibrium requires that the vertical component of the tension developed within the reinforcement counteract the vertical pressure produced by the overlying embankment. If the geogrid is allowed to sag prior to fill placement, the vertical component of the developed tension is larger, which means that less tension, and therefore strain, is necessary to satisfy equilibrium than for reinforcement without initial sag.

Table 5: Comparison of predicted and measured deflection values

Location	Settlement Plate	Deflection Measured from Settlement Plate (m)	Deflection Derived from Strain Measurements (m)	Deflection Predicted by LDC (m)
A	SP-03	0.36	0.12	0.19

CONCLUSIONS

From the analysis of the experimental site in the Barra da Tijuca District, it was found that reasonably accurate calculations of geogrid deflection for low-height embankments can be made. The data set is limited, but a single set of geogrid and embankment fill properties, along with laboratory determined values of compressibility of the soft clay, yielded calculated deflections in good agreement with the measured deflections at four settlement plate locations with different pile cap configurations and subgrade support conditions. The analyses also demonstrate the significant contribution of subgrade support in limiting geogrid deflection, even when the subgrade is very soft. For conditions at Locations A and B, the calibrated calculations show that subgrade support reduces deflections by about 50%. The conclusion that subgrade support effects are considerable even for low height embankments is consistent with the findings of Rogbeck et al. (1998).

Results from the experimental site also show the significance of maintaining tautness in the reinforcement during embankment construction. If slack or wrinkling is initially present, the differential settlement between the foundation soil and the columns will not induce strain, and therefore tension, in the reinforcement until the slack is removed. This means that the reinforcement layer is providing no benefit to the load-transfer system until strain begins to develop.

The LDC approach as implemented in the GeogridBridge 1.1 spreadsheet was found to be capable of calculating accurate geogrid deflection for a condition with subgrade support and simulate a condition of no subgrade support.

REFERENCES

Almeida, M.S.S., Marques, M.E.S., Almeida, M.C.F., and Mendonca, M.B. (2008). "Performance on two low piled embankments with geogrids at Rio de Janeiro" *Proceedings, First Pan American Geosynthetics Conference and Exhibition,* Cancun, Mexico, pp. 1285 - 1295.

Almeida, M.S.S., Ehrlich, M., Spotti, A.P., and Marques, M.E.S. (2007). " Embankment supported on piles with biaxial geogrids" *Journal of Geotechnical Engineering*, Institution of Civil Engineers, ICE, UK, Vol. 160, No. 4, pp. 185-192.

Filz, G.M., and Smith, M.E. (2006a). "Design of bridging layers in geosynthetic-reinforced, column-supported embankments." *Contract Report* VTRC 06-CR12, Virginia Transportation Research Council, Charlottesville.

Filz, G.M., and Smith, M.E. (2006b). "Net vertical loads on geosynthetic reinforcement in column-supported embankments." *Proceedings, Geo-Denver: New Peaks in Geotechnics*, Soil Improvement, *Geotechnical Special Publication 172* (CD-ROM), ASCE, Reston, VA

Poulos, H.G., and Davis, E.H. (1970). *Elastic Solutions for Soil and Rock Mechanics*, John Wiley & Sons, Inc., New York, NY.

Rogbeck, Y., Gustavsson, S., Sodergren, I., and Lindquist, D. (1998). "Reinforced piled embankments in Sweden - design aspects." *Proceedings, Sixth International Conference on Geosynthetics*, Vol. 2, Rowe, R.K. (ed.), Atlanta, Georgia, pp. 755-762.

Russell, D., Naughton, P.J. and G. Kempton, G. (2003). "A new design procedure for piled embankments." *Proceedings of the 56th Canadian Geotechnical Conference and 2003 NAGS Conference*, Vol. 1, Winnipeg, MB, pp. 858-865.

Spotti, A. P. (2006). Monitoring results of a piled embankment with geogrids (in Portuguese). ScD Thesis, COPPE/UFRJ, Rio de Janeiro, Brazil

Terzaghi, K. (1943). *Theoretical Soil Mechanics*, J. Wiley & Sons, Inc., New York, NY.

Analysis of Deep Jet Grouting Field Trial in Clay

Chu E. Ho[1], Sc.D., Member, ASCE

[1] Associate, Arup, 155 Avenue of the Americas, New York, NY 10013, chu.ho@arup.com

ABSTRACT: This paper describes the investigation and back-analysis of a full scale jet grouting field trial using the triple fluid technique, to study the performance of jet grout column formation in soft marine clays. Cored samples obtained from the grouted soil mass indicated that although jetting in the marine clay was satisfactory, the lower levels encountered stiff desiccated clay, resulting in a smaller column size being formed. The cutting performance was evaluated using an analytical model for jetting with a compressed air shroud. The additional jet penetration distance achievable in using a compressed air shroud was estimated to be approximately 160%. The importance of obtaining accurate soil strength profiles for the soils to be grouted, including stress history, is emphasized.

INTRODUCTION

A deep jet grouting field trial using triple fluid technique was conducted to study the feasibility of forming interlocking jet grout columns in marine clays. The depth of production jet grouting was 11m to 20m below existing ground surface. The jetting trial was executed at the actual depth of production grouting so that the soil properties and jetting conditions at depth are accurately simulated in the trial. The trial columns penetrated marine clays of soft to medium consistency in the upper sections but encountered a stiff desiccated clay stratum in the bottom sections. The potential lack of jet grout column interlock within the desiccated clay was investigated by coring through the grouted soil mass for inspection and testing of the grouted product. This paper presents the interpretation of the trial results and discusses the evaluation of jet cutting performance using an analytical framework developed for jetting with a compressed air shroud.

JET GROUTING FIELD TRIAL

The jet grouting trial involved the execution of a grout block consisting of eighteen 1.8m diameter columns (T1 to T18) in a triangular grid spacing of 1.55m (Figure 1). The columns were installed in the sequence indicated by their increasing numbers. Two columns were executed within a day, one in the morning and one in the afternoon. Each column took 2 to 3 hours to complete. Cutting was achieved using a single water jet with a compressed air shroud, while grouting was executed using two grout jets located diametrically across at the lower level. The diameter of the water nozzle was 2.2mm and that for the two grout nozzles were 4.5mm. The diameter of the air nozzle surrounding the water nozzle was 10mm. The upper and lower jets were located 750mm apart. The diameter of the monitor was 89mm. Full details of the trial are described in Ho (1995). The jetting parameters adopted for the trial are summarized in Table 1.

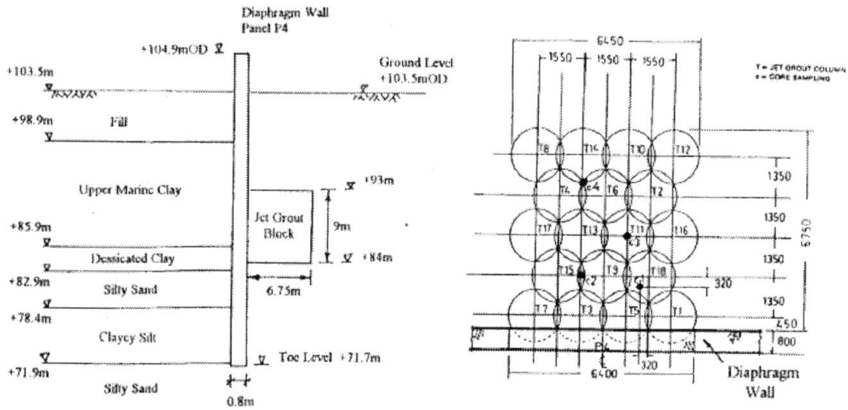

FIG. 1. Design layout of jet grouting trial

TABLE 1. Parameters for jet grouting trial

Jet Grout Column	Water Pressure (bar)	Air Pressure (bar)	Grout Pressure (bar)	Withdrawal Rate (mins/m)	Rotation speed (rpm)
T1	450	10	100	8	9
T2	450	10	100	8	9
T3	380 to 400	6 to 7	80	9	9
T4	450	10	100	9	9
T5	380 to 400	6 to 7	60 to 80	9	9
T6	450	10	80 to 100	9	9
T7	400	6.5	80 to 100	9	9
T8	430	6.5	80	9	9
T9	460	6.5	80	9	9
T10	450	6.5	70	8	9
T11	450	6.5	65	9	9
T12	450	6.5	60	8	9
T13	450	6.46	80	9	9
T14	450	6.5	80	9	9
T15	450	6.5	50	9	9
T16	450	6.5	60	9	9
T17	450	6.5	60 to 70	8	9
T18	450	6.5	60 to 70	8	9

Each grouted column was 9m long, executed through an upper layer of soft marine clay and terminating about 2m into the stiff desiccated clay stratum. Cores were obtained through the grouted soil mass at four locations, indicated as C1 to C4 in Figure 1. Core C1 was located 452mm from the theoretical centerline of column T18, C2 was located at the double intersection between columns T9 and T15, C3 was located at the theoretical centerline of column T11 and C4 was located at the triple intersection between columns T4, T6 and T14. Table 2 summarizes the condition of the recovered cores.

TABLE 2. Condition of recovered core samples

Coring Point	Jet Grout Column	Core No.	Core Run (m)	Total Core Recovery (%)	Length of Grout (m)	Length of Stiff Clay (m)	Remarks
C1	T18	1	1.5	95.3	1.42		
		2	1.5	100	1.5		Poor core recovery for Core No.4
		3	1.5	66.7	1.0		
		4	1.5	3	0.045		
		5	1.5	30		0.45	
		6	1.5	100		1.5	
C2	T9/T15	1	1.5	93.3	1.4		
		2	1.5	63	0.95		
		3	1.5	90	1.35		
		4	1.5	83.3	1.25		
		5	1.5	82	0.43	0.8	
		6	1.5	76.4		1.15	
C3	T11	1	1.5	96	1.44		
		2	1.5	0			No core recovery for Core No.2 to 6
		3	1.5	0			
		4	1.5	0			
		5	1.5	0			
		6	1.5	0			
C4	T4/T6/T14	1	1.5	66.7	1.0		
		2	1.5	83.3	1.25		Core material highly fractured, very weak
		3	1.5	100	1.5		
		4	1.5	46.7	0.7		
		5	1.5	93	0.28	1.12	
		6	1.5	67	0.075	0.93	

Although C3 was located in the middle of column T11, the core recovery was extremely poor, except for the first 1.5m core run. The quality of the grout in the first core run was however extremely good, with an unconfined compressive strength of 3.8 MPa being indicated by laboratory testing. It would have been expected that the grout located at the center of a column will be of the best quality, with the quality of grout progressively deteriorating towards the edges of the column due to poorer mixing as the jetting energy becomes depleted with increasing distance from the nozzle. It was suspected that the loss of sample in this case was caused by poor drilling technique, and not due to poor grout quality.

It was observed from C1, C2 and C4 that consistently no grout was present in the cores at levels below elevation +86mOD to +87mOD, where stiff desiccated clay was encountered. Continuous intact cores of desiccated clay were obtained for the last 1.95m in C1. It can be inferred from C1 that the reach of the jet in the desiccated clay was limited to less than 452mm from the theoretical centerline of the jetting monitor. The grout cores in C2 located at a double intersection indicated reasonably good core recovery within the marine clay. In contrast, the grout cores obtained from C4 located at a triple intersection were weak and highly fractured. Nevertheless, the presence of grout in the triple intersection of the columns suggests that the reach of the jet in the marine clay was of the order of 900mm from the centerline of the jetting monitor. The monitor used in the jetting trials was 89mm in diameter, therefore it can be deduced that the jet penetration distance from the nozzle was less than 407.5mm in the desiccated clay and at least 855.5mm in the marine clay.

ANALYTICAL MODEL

Recent research has shown that the cutting distance of a traversing jet in cohesive soil can be reasonably predicted assuming bearing failure at the tip of the jet (Ho 2005). The fundamental relationship for bearing failure at the tip of the jet is given by:

$$\frac{l_j}{d_n} = 6.25\sqrt{\frac{(P_i - P_o)}{q_{bu}}} \tag{1}$$

where l_j = maximum jet cutting distance from the nozzle, d_n = nozzle diameter, P_i = pressure at the nozzle inlet, P_o = pressure at the nozzle outlet and q_{bu} = ultimate soil bearing resistance. Laboratory experiments have indicated that soil bearing resistance to jetting can be represented by $q_{bu} = N_c s_u$, where $N_c = 2.4$ is a bearing capacity term corresponding to soil failure conditions at the jet tip and s_u = undrained shear strength (Ho 2005). The diameter of the jet grout column is given by $D = 2l_j + d_m$, where d_m = diameter of the jetting monitor. The prediction of jet penetration distance using Equation 1 has been shown to be excellent for clays, silty clays and cemented silts, but is conservative for silty sands (Ho 2007).

Equation 1 has been developed primarily for single fluid jet grouting. For triple fluid jet grouting, the cutting water jet is surrounded by a high velocity compressed air shroud and the mechanics of interaction between the cutting jet and air shroud is highly complex. A rigorous treatment of the physics involved is beyond the scope of this paper. An approximate method is presented herewith as a tool for interpreting the jet grouting trial data using the work of Forstall and Shapiro (1950). Laboratory experiments have suggested that the basic potential core length of the central jet (x_o) is increased by a factor of 3λ when the velocity of the ambient medium (u_o) is uniformly increased relative to the exit velocity of the central jet (v_o), where $\lambda = u_o/v_o$. Hence the resultant potential core length (x_o') is given by $x_o' = x_o + \Delta x_o$, where, $\Delta x_o = M\, x_o$ and $M = 3\lambda$. The corresponding centerline velocity (v_{xc}) decays according to the following relationship beyond x_o'

$$\frac{(v_{xc} - u_o)}{(v_o - u_o)} = \frac{x_o'}{x} \tag{2}$$

In extending the above formulations to the problem of triple fluid jet grouting, it is assumed that the air jet diffuses at a much faster rate than the central water jet due to its greater density difference relative to the surrounding slurry such that at large distances from the nozzle, u_o becomes small and Equation 2 reduces to

$$\frac{v_{xc}}{v_o} = \frac{x_o'}{x} \tag{3}$$

Correspondingly, the actual increase in potential core length (Δx_o) will be smaller with M being much less than the value of 3λ. The average stagnation pressure at the tip of the jet is given by

$$p_{xa} = \tfrac{1}{2}\rho_j {v_{xc}}^2 \tag{4}$$

At the limit of jet penetration (at $x = l_j$), the average stagnation pressure becomes equal to the ultimate bearing resistance of the soil ($q_{bu} = 2.4s_u$). It can be shown that under these conditions,

$$\frac{l_j}{d_n} = 6.25(1+M)\sqrt{\frac{(P_i - P_o)}{q_{bu}}} \tag{5}$$

ANALYSIS OF JET CUTTING PERFORMANCE

The marine clay is highly plastic with a sensitivity of about 2 to 6, while the desiccated clay is a silty clay of intermediate to high plasticity. Sampling of the desiccated clay proved to be difficult due to its stiff to hard consistency, hence reliable laboratory test results in this material are limited. Figure 2 summarizes the properties of the marine clay and desiccated clay obtained from laboratory testing. The properties of the marine clay are reasonably uniform over the range of depths investigated. The average bulk density and plasticity index are 15.36 kN/m^3 and 54.1% respectively. Based on the limited data available, bulk density and plasticity index of the desiccated clay can be taken to be 18.65 kN/m^3 and 38% respectively.

In-situ vane shear tests carried out in the marine clay showed that the field vane shear strength increased with depth in accordance with the relationship $s_{uv} = 1.75$ z (Figure 3), where s_{uv} is the field vane shear strength and z is the depth from the ground surface. The linearly increasing shear strength profile suggests that the marine clay is largely in a normally consolidated state. Based on Bjerrum (1972), a correction factor ($\mu = 0.78$) was applied to s_{uv} to obtain the undrained shear strength (s_u) of the marine clay. The desiccated clay is an overconsolidated soil, resulting from the drying out of the lower stratum of marine clay in the geological past. Figure 3 shows that SPT blow counts in the desiccated clay fall in the range of 2 to 30, with a very wide scatter. An average value of about 11 blows/0.3m was obtained for this material with a standard deviation of about 6. This implies that the character of the desiccated clay is highly variable. The undrained shear strength was estimated using the correlation $s_u = 4.5N$ proposed by Stroud (1974), for a plasticity index of 38% for the desiccated clay.

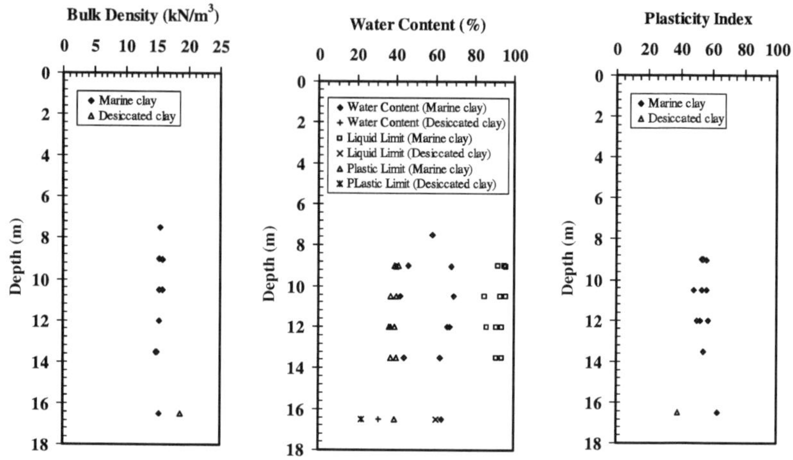

FIG. 2. Properties of marine clay and desiccated clay.

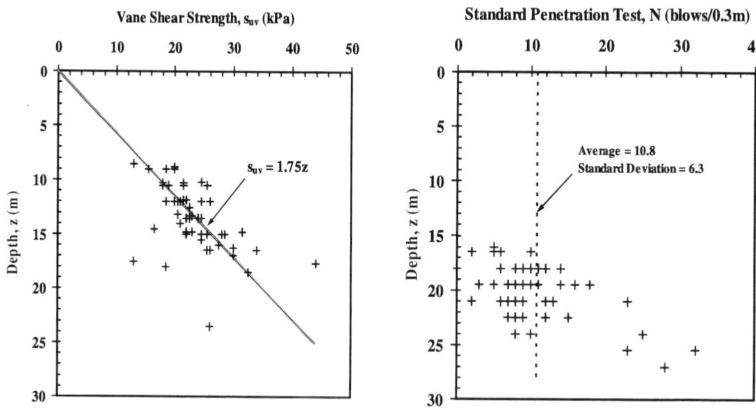

(a) Marine clay (b) Desiccated clay

FIG. 3. Shear strength of marine clay and desiccated clay

Equation 5 was used to predict the cutting distance of the jet. Based on measurements by Croce and Flora (2000), a reduction factor ($\alpha = 0.8$) was applied to the pressure at the pump, P_{pump} to obtain $P_i = \alpha P_{pump}$. To allow for the additional pressure build up at the nozzle outlet so as to maintain upward flow of the slurry, a multiplying factor ($\beta > 0$) was applied to the hydrostatic pressure head P_{static} such that $P_o = (1 + \beta)P_{static}$, where $P_{static} = \rho_{slurry}gz$, and ρ_{slurry} = slurry density in borehole, g = acceleration due to gravity and z = depth of the nozzle. Ho (2008) indicated that jet penetration distance is insensitive to the value of β, due to the fact that the value of P_i is always several orders of magnitude higher than P_o. In the present analysis, $\beta = 1$ was adopted.

Figure 4 compares the measured column dimensions with those predicted using Equation 5. Predictions were based on field vane shear tests for marine clay and SPT blow counts for desiccated clay. An alternative computation was made using the expression $s_u = 0.22\sigma_v'(OCR)^{0.8}$ where σ_v' is the effective overburden stress and OCR is the overconsolidation ratio of the soil (Ladd 1991). Figure 4 shows that the predicted column dimensions for the normally consolidated marine clay (OCR = 1) decreased with depth as the shear strength increased. The minimum diameters computed from both approaches match the dimension of 1800mm inferred from the field trial for a value of M = 1.6, implying that the additional jet penetration distance achieved was about 160% when an air shroud was used, which is very significant. The same value of M was used for jetting in the desiccated clay. It was found that the predicted diameter for the desiccated clay based on the averaged data from field tests (N = 11 blows/0.3m) was much larger than the 904mm inferred from the trial.

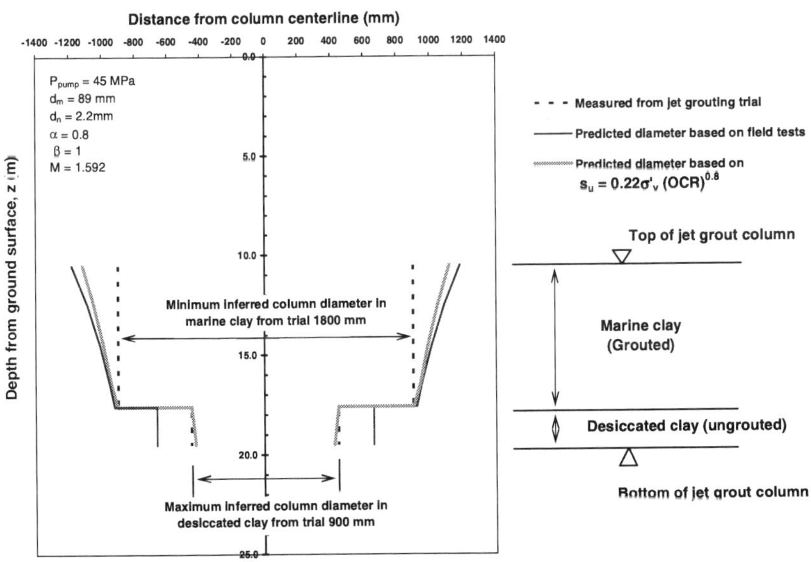

FIG. 4. Predicted and measured column dimensions

In order to match the maximum inferred column diameter of 904mm, a back-figured value of N = 26 blows/0.3m was found to be more appropriate for the desiccated clay encountered. A similar back-analysis using $s_u = 0.22\sigma_v'(OCR)^{0.8}$ yielded an OCR of 6 for the desiccated clay. Research testing using the flat dilatometer reported by Chang (1986) at three other sites indicated that OCR in the desiccated clay can vary between 2 to 7 depending on stress history. The above analyses demonstrate that local variations in soil characteristics within individual soil types can have a significant impact on the column size that can be achieved. Adopting site-averaged geotechnical data may result in misleading conclusions and specific soil investigation programs must be implemented to obtain an accurate description of the soil to be grouted.

CONCLUSIONS

A full scale jet grouting field trial using triple fluid technique was conducted at the actual depth of production grouting to study the performance of jet grout column formation in soft marine clays. Cored samples obtained from the grouted soil mass indicated that the deeper core runs encountered stiff desiccated clay and no grout was formed. An analytical model was used to evaluate jetting performance with a compressed air shroud. The enhancement of the jet penetration distance due to the compressed air was estimated to be about 160%. The importance of obtaining accurate shear strength profiles of the soil to be grouted, including an appreciation of stress history of the soil, was demonstrated for realistic interpretation of column size.

REFERENCES

Bjerrum, L. (1972). "Embankments on soft ground." *Proc. Conf. on Performance of Earth and Earth-Supported Structures*, (2), ASCE: 1-54.

Chang, M. F. (1986). "The flat dilatometer test and its application to two Singapore clays." *Proc. 4th Int. Geot. Sem. on Field Instrumentation and In-situ Measurements*, Singapore, NTI: 85-101.

Croce, P., and Flora, A. (2000). "Analysis of single-fluid jet grouting." *Geotechnique* 50 (6), Thomas Telford, pp. 739-748.

Forstall, W., and Shapiro A. H. (1950). "Momentum and mass transfer in coaxial gas jets". *Journal of Applied Mechanics*, APM-13, ASME: 399-408.

Ho, C. E. (1995). An instrumented jet grouting trial in soft marine clay. *Verification of Geotechnical Grouting,* GSP No.57, ASCE: 101-115.

Ho, C. E. (2005). *Turbulent fluid jet excavation in cohesive soil with particular application to jet grouting*, Doctoral Thesis, Massachusetts Institute of Technology, Cambridge, Massachusetts, 456 p.

Ho, C. E. (2006). "Prediction of jet grout column diameter in cohesive soil." *Proc. 10th Int. Conf. on Piling and Deep Foundations*, Amsterdam, DFI: 465-474.

Ho, C. E. (2007). "Fluid-soil interaction model for jet grouting." *Grouting for Ground Improvement: Innovative Concepts and Applications.* GSP No.168, ASCE .

Ho, C. E. (2008). "Investigation of parameter variability on jet grout column formation". *2nd BGA Int. Conf. on Foundations, Dundee*: 1357-1368.

Ladd, C. C. (1991). "Stability evaluation during staged construction". *J. Geot. Engg.*, 117 (4), ASCE: 540-615.

Stroud, M. (1974). "The standard penetration test in insensitive clays and soft rocks." *European Symp. on Penetration Testing*, Swedish Geot. Soc., Stockholm.

Assessing the Effectiveness of Compaction Grouting Using Seismic Methods

By: Khamis Y. Haramy[1], Justin T. Henwood[1], and Tom Szynakiewicz[2]

[1] Geotechnical Engineer, Central Federal Lands Highway Division – Federal Highway Administration, 12300 West Dakota Ave., Lakewood, Colorado 80228

[2] Project Manager, Hayward Baker, 11575 Wadsworth Blvd, Broomfield, CO 80020

ABSTRACT: Grouting has been successfully used for decades as a ground improvement technique to stabilize and retrofit historic structures. Over the years, new technology, equipment, and chemicals have improved the feasibility of grouting over a wide range of applications. However, near real-time, in situ control methods to assess the effectiveness of ground improvement following subsurface grouting are lacking. Currently, the most commonly used quality assurance method is to obtain Standard or Cone Penetration Test (SPT or CPT) measurements before and after ground treatment at pre-defined locations, typically between column/impact/injection locations to observe changes in relative density of in-place materials. All these methods are based on point measurements at specific locations and do not represent the overall treated ground mass.

A volumetric, non-destructive, surface seismic evaluation method, that uses a state-of-the-art algorithm, was recently developed to collect and analyze seismic data describing the subsurface conditions. This relatively rapid and inexpensive seismic method was utilized in Rocky Mountain National Park, Colorado, during rehabilitation of a historic placed-rock armored slope structure located below roadway grade along the outboard lane. Compaction grouting was selected to stabilize roadway settlement caused by piping of a failed culvert within the fill while preserving the historic and aesthetic character of the structure. The seismic method proved effective in producing accurate pre- and post-grouting volumetric difference images that depicts the effect of grouting, including the resultant grout columns and ground densification characteristics.

INTRODUCTION

Ground improvement techniques have become increasingly accepted by engineers as a means of mitigating soils with low bearing capacity or structures with high settlement potential. Specifically, ground improvement functions are to: increase materials shear or frictional strength, fill voids, improve soil density, create a water cut-off, and control settlement. These functions are generally accomplished by compaction, consolidation, reinforcement, or grouting.

Compaction grouting is a subsurface treatment technique that involves the injection of low-mobility grout under a designed pressure and flow-rate which fills voids and densifies the soil mass, in-place. The grouting is typically achieved by sequencing the grouting from primary to secondary borings located along the treated area at various locations. For compaction grouting to be effective, the in-situ stresses within the treated soils must be sufficient to allow the grout to displace the soil in the horizontal plane and the flow rate should be controlled so that the pore pressure is adequately dissipated. The grout forms a "bulb" at the injection point ultimately producing a grout column, thus displacing and densifying the surrounding soil.

An injection grid is designed with a specific spacing pattern based on the geotechnical subsurface investigation results. The grout is pumped in primary and secondary holes with certain criteria, including pressure, grout volume, ground movement, etc. to meet the project needs. The primary holes are to improve compaction and density of the soil within the treatment zone, while the secondary holes are for densifying any remaining soft zones, improving the soil compressive strength between columns, and verifying improvement.

Near real-time, volumetric, in-situ control methods to assess the effectiveness of ground improvement following grouting are lacking. Currently the most commonly used quality control methods include procedural inspection, testing to ensure proper mix design/injection rates, and verification of ground improvement using conventional SPT and CPT methods. Additionally, detailed data recording and analysis of important grouting parameters have been utilized on sensitive projects.

Mitigation of historic structures is costly, especially if the structure must be reconstructed. Compaction grouting is a cost effective method to improve the internal stability of a structure, without adversely impacting the structural or aesthetic aspects of the structure. This paper discusses a case history involving rehabilitation of a placed rock armored slope structure using compaction grouting and an advanced, non-destructive evaluation (NDE) seismic method for quality control assessment of grouting effectiveness.

CASE HISTORY

To rehabilitate a distressed historic embankment fill armored with dry-laid rock along Trail Ridge Road, a compaction grouting program was completed. The site was located within Rocky Mountain National Park, near Estes Park, Colorado.
The rock slope armor is 20-meters-high, 40-meters long, and is comprised mainly of large (> 0.5 meter) angular rock placed at a slope ratio of approximately 1.25H:1V with large voids between and behind many of the rocks near the top of the slope. Figure 1 is a photo detailing the site. This site is comprised of a pavement section, roadway embankment, stone guard wall, dry-laid rock slope armor, and a 600-mm culvert near the center of the site. The asphalt pavement is approximately 500-mm thick, near the middle of the outboard lane, and cracking appears to define a crescent shape extending to near the centerline. Settlement within this region was measured at over 100 mm/year. This

Figure 1. The Armored-Rock Slope Structure at Rocky Mountain National Park.

settlement is indicative of a fill failure caused by piping of a failed culvert, with the greatest amount of settlement near the guard wall. Significant periodic groundwater level fluctuations would likely be required to cause this degree of settlement of the embankment fill.

An internal culvert video inspection showed a break in the culvert at about 9 m from the inlet which, given the slope and skew of the culvert, would be approximately beneath the edge of the roadway near the stone guard wall. Based on this, groundwater flows through the subsurface material are believed to have caused piping within the fill and resultant settlement and slumping of the pavement. Gradual settlement at this site is a maintenance nuisance and, if not properly remediated, continued settlement and slope movement could adversely impact the historic armored slope and would possibly require road closure. Therefore, mitigation measures to preserve the historic character of the slope and to provide an increased margin of safety against slope failure were essential.

Planned mitigation measures included:

- **Drainage Improvements:** Drainage improvements were required at the site, including reducing surface water infiltration at the inboard ditch and repairing/replacing the culvert.
- **Embankment Stabilization:** Drainage improvements alone were not sufficient for mitigating the observed distress. Compaction grouting was recommended to stabilize the embankment materials, allowing the historic rock armor to be left in place.
- **Armor Stabilization**: Elimination of embankment movement leading to elimination of further damage to the rock armor. However, to improve the longevity of the rock armor, filling the voids behind the rocks and assuring greater element interlocking was also recommended. This was accomplished by partial or complete grouting of the face. Grout was designed and applied so as not to be significantly visible on the face.

A construction project was undertaken to stabilize the structure. The culvert was lined to prevent further piping and to keep the water away from the fill, the embankment rock armor materials were stabilized by injecting grout in the voids, and a comprehensive compaction grouting program was completed to densify the fill. The compaction grouting layout included injection holes along four longitudinal lines, spaced at 2.0 meters across the roadway width. Within each longitudinal line, grout injection holes were spaced at 4.0 meters, and staggered 2.0 meters from the nearest point in adjacent longitudinal lines, as shown in Figure 2. Injection holes were

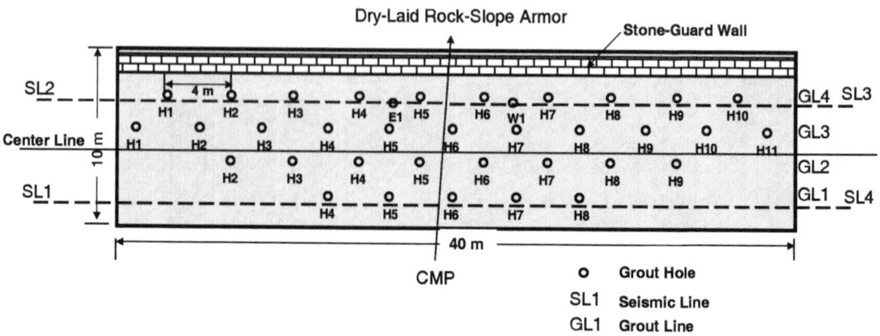

Figure 2. Compaction Grouting and Seismic Lines Layout.

terminated at a depth in which bedrock was encountered, generally between 2.0 and 9.0 meters below the paved surface. If competent bedrock was encountered above 1.0 meter, no grouting was required. Two injection holes (E1, W1) were added during construction near the culvert in areas where significant volumes of grout were injected in adjacent holes.

Grout was injected from the bottom of each injection hole under consistent pressure and continued as the casing was drawn from the hole in 300 mm increments. Figure 3 shows the compaction grouting conducted at the site. When one of the following criteria occurred, the grout pipe was raised to the next increment:

Figure 3. Compaction Grouting at the Site.

- An established grout pressure of 1,400 kPa at the gauge on top of the drill casing had been achieved.
- The horizontal displacement of the stone guard wall exceeded 20 mm/m of exposed face height.
- The horizontal displacement of the armored slope exceeded 40 mm/m of exposed face height.
- The injected grout volume exceeded 0.25 m^3/300 mm interval of grouted zone.
- The upward heave movement was in excess of 100 mm, as indicated on the ground surface.

The total measured injection volumes of grout for each of the holes is presented in Figure 4. The figure shows that most of the injected grout volume in lines 1 and 2

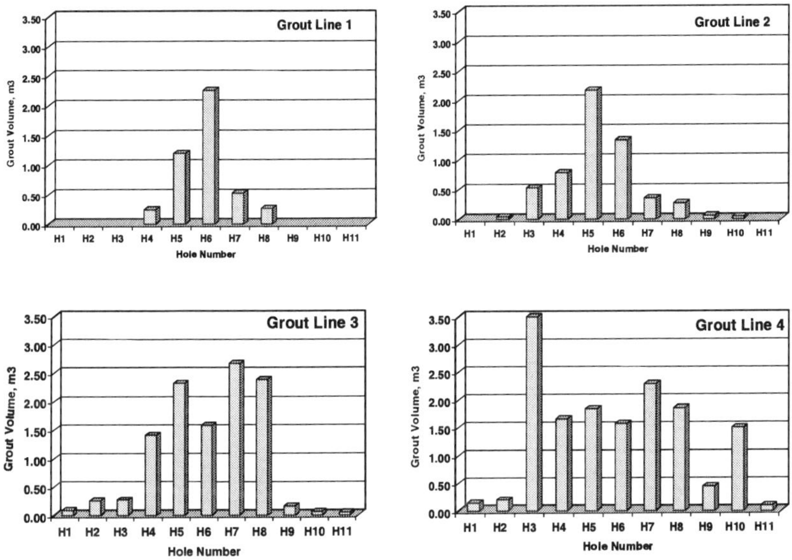

Figure 4. Measured Injection Grout Volume in each Hole

was near the location of the existing damaged culvert, between 1.2 and 2.5 m^3 per hole (with the exception of H3 of L4). In general, Lines 3 and 4 shows a high volume of injected grout, exceeding 1.2 m^3 per hole, extended a distance of approximately 8.0 meters on either side of the culvert. Although not represented in the figure, injection holes E1 and W1 received about 2.3 m^3 of injected grout each. This confirms the presence of piping and voids along the length of the culvert and specifically near the outer edge of the pavement beneath Line 4. The maximum volume of grout was injected near where the culvert break was found. It is believed that a sufficient amount of grout was pumped to fill all voids and to stabilize the roadway.

GROUTING EFFECTIVENESS DELINEATED USING A VOLUMETRIC, HIGH DEFINITION IMAGING TECHNIQUE

A newly developed, non-destructive volumetric evaluation method, known as High Definition Imaging (HDI), was utilized to collect and analyze pre- and post-grouting seismic data during rehabilitation of the rock-armor slope. HDI is a relatively inexpensive quality-control method for reconstructing subsurface properties using seismic energy propagating from surface-located sources through soil/rock materials to surface-mounted receivers. This method allows data collected using multiple conventional 2D or 3D seismic refraction surveys to be processed into a volumetric 3D assessment of subsurface conditions.

The first step in ground imaging is to collect seismic data. Geophones are deployed on the surface, and a source, such as a sledge hammer, is used to generate a surface seismic wave. The travel time of the seismic wave from each source to each receiver is determined for tomographic inversion. Although HDI can process data collected from seismograph systems using the SEG-2 file format, specialized data collection hardware that can be used in true 3D data collection have been developed and integrated into HDI to simplify the data collection process and significantly improve data quality. This system uses advanced digital-seismic sensors with six-dimensional accelerometers for collecting accurate data for enhanced imaging.

Superior tomographic analysis software was also developed allowing engineers to acquire data and produce accurate images of the subsurface in near real-time. In many cases, the seismic data can be processed and output data can be generated while in the field. This stand-alone system records 3D seismic data on source impact with automatic source location. This lightweight, compact system can be rapidly deployed on a variety of surfaces and is easily operated by one person. The software automatically generates hundreds of animated slides, with the 3D velocity contour volume of the imaged area and cross sections along several planes available in the X, Y, and Z directions. Images are produced in near real-time providing immediate confidence in seismic survey results, and allowing adjustments of receiver placement to "zero-in" on regions of interest.

The next step after data collection is to determine the travel time of a seismic wave from each source to each receiver. This can often be a highly skilled and labor intensive process, requiring signal filtering and manual travel time picking. HDI incorporates time, frequency, and velocity filters, and uses advanced algorithms for pattern matching analysis of the signal spectrogram and envelope to automatically determine consistent travel time picks. HDI uses a multi-dimensional B-Spline Interpolation Network that accounts for waveform shape, source/receiver distance, time, adjacent picks, and filter distortion when automatically determining first arrivals based on user-provided picking examples.

Once accurate travel time picks have been determined, tomographic inversion is performed. Waves are propagated from each source and each receiver to determine regions in the velocity model having the most significant effect on the travel time. This technique quickly converges to the solution of minimum travel time error with minimum velocity variation, providing consistent results.

The final step in ground imaging is to display the results in a compact form usable by engineers. HDI generates side-by-side perspective and parallel projections with cross-sections in the x, y, and z directions and 3D velocity contours for visualization and measurement of results.

At the investigated site, the seismic survey included the ground volume bounded by the inboard roadway shoulder (retained fill) and armored slope structure. Four seismic lines were established within the survey area, as previously shown in Figure

2: two lines (SL2 and SL3) along the edge of the roadway approximately 500 mm from the guard wall and two (SL1 and SL4) along the inside edge of the inboard lane approximately 500 mm from the roadway edge. Each seismic line contained 24, 4.5 Hz, uniaxial plate-mounted geophones spaced on 1.5-meter centers. Each surveyed line contained a total of 21 source locations spaced randomly along the roadway, generated by a 9-kg sledge hammer. The surveyed source/receiver array resulted in several thousand raypaths intersecting the structure and retained fill, an adequate coverage within the limits of the investigation.

Each source/receiver pair was mapped to the corresponding seismic signal collected with a Geometrics StrataView seismograph and stored in SEG-2 format. The arrival times were determined using the HDI automated picking algorithm. Source/receiver arrival times were then used to construct the 3D seismic velocity tomograms of the armored slope and retained fill. The 3D velocity model was initialized to a homogeneous average velocity for accurate, unbiased reconstruction of the medium.

The 3D velocity tomogram was then sliced and contoured in both perspective and parallel views for visualization of the velocity structure before and after grouting. Velocity difference plots allow volumetric changes to be easily viewed, indicating where in the raypath-defined survey volume changes in ground velocity occurred. Due to the limitations of the length of this paper, only velocity difference plots at random locations are presented. Figure 5 illustrates vertical cross-sections of the 2D velocity difference tomogram, pre-and post-grouting, at various locations and orientations (x,y,z) within the structure. The figure shows the level of ground improvement accomplished through grouting techniques employed at the site.

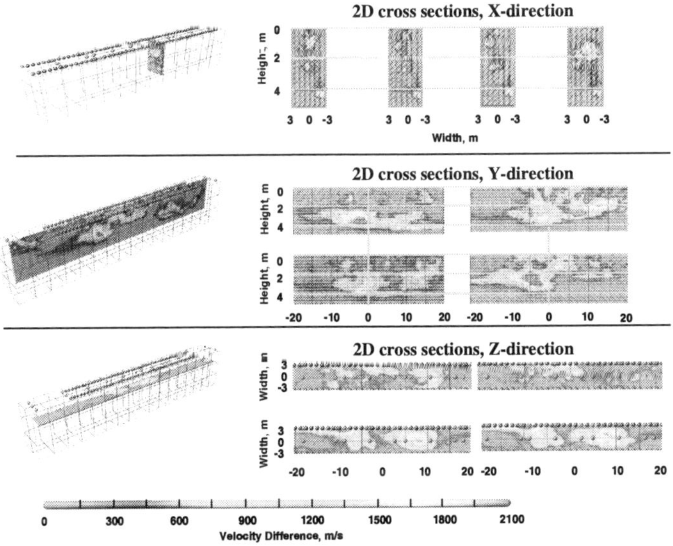

Figure 5. Vertical Cross-sections of the 2D Velocity Difference Tomograms in the X, Y, and Z Directions

The HDI software package and hardware components were developed by Summit Peak Technologies (www.summitpeak.net) for true 3D imaging. The Central Federal Lands Division of the Federal Highway Administration has tested and evaluated beta versions of the software and hardware packages. Release of final software and hardware packages are expected at a later date.

SUMMARY AND CONCLUSIONS

A 3D velocity-difference contour of the subsurface separating zones of relative velocity increase is presented in Figure 6. HDI's ability to produce 3D volumetric images of the improved area before and after grouting in conjunction with the ability to produce 2D cross sectional images at any location vastly enhanced the quality assurance and quality control of compaction grouting at this site. The seismic data was in agreement with the point measurements of grout injection volumes.

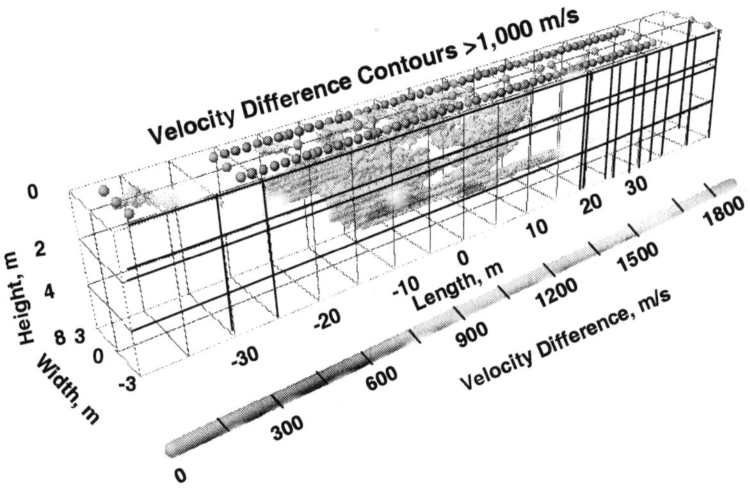

Figure 6. Three-Dimensional Volumetric Image of the Velocity Differences Contour Greater than 1,000 m/s

The HDI method can be an effective tool for use with grouting methods to enhance the quality assurance methods and to monitor volumetric grout migration within the improved soil mass in near real-time.

Grouting of Deep Foundations at the Thames River Bridge

Walter E. Kaeck[1] M. ASCE, P.E., Frederick C. Rhyner[1] M. ASCE, P.E., Hugh Lacy[1] M. ASCE, P.E. and Michael Quasarano[1] A.M. ASCE, E.I.T.

[1]Mueser Rutledge Consulting Engineers, 225 W. 34th Street, New York, NY; wkaeck @mrce.com; frhyner@mrce.com; hlacy@mrce.com; mquasarano@mrce.com

ABSTRACT: This paper describes an unusually deep grouting program that was undertaken to arrest the unexpected settlement that occurred during construction of an Amtrak bridge over the Thames River in Connecticut. During construction, a caisson started to settle and tilt when new, adjacent piles were installed. A remedial grouting program was performed to arrest the settlement. The target grout zone was the permeable and erratically-dense sand and gravel stratum beneath the caisson, at a depth of about 40 to 52 m (130 to 170 ft) below the river surface. Grout holes were placed around the perimeter of the caissons as well as through each caisson. A computerized data acquisition system was used to provide real time monitoring of grout injection parameters and rapid evaluation of the effectiveness of grouting as work progressed. Over one million gallons of microfine cement grout were pumped during the course of 11 months work. Cores of the grouted soils were obtained to confirm grout coverage and properties of the grouted soil. Eventually, the grout stabilized the pier and halted the creep movement. This paper describes the grouting program, grout mixtures, effects of various admixtures, and quality control measures.

INTRODUCTION

The moveable bridge carrying the Amtrak railroad over the Thames River in Connecticut was constructed in 1918. It is a five span bridge, approximately 423 m (1,387 ft) long and carries two tracks, the center span being a bascule. The clever design for the new bridge replaced the bascule with a lift span using a strategy to widen the piers and construct the lift-span towers while the tracks remained in full service. The 1918 bridge was constructed with timber crib caissons filled with tremie concrete. The Pier 2 caisson supported the pivot works and was 12 m (40 ft) wide by 30 m (99 ft) long. Pier 3 supported the rest pier and was 9 m (30 ft) wide by 30 m (99 ft) long. The caissons extended 40 m (130 ft) below water, through a stratum of soft to medium stiff organic silt, and were founded at the surface of a stratum of coarse to fine sand and gravel with trace silt, cobbles and boulders. Recent borings showed that the sand and gravel was about 9 to 12 m (30 to 40 ft) thick and was erratic in density, from loose to very dense. Underlying this stratum was a 3 m (10 ft) thick layer of boulders and cobbles, in turn underlain by gneissic schist bedrock.

The design for the new lift span called for widening Piers 2 and 3 to the south by 7.3 m (24 ft) using four, 90 cm (36 in.) diameter open-end pipe piles penetrating to bedrock to support the reinforced concrete extension. Since the piles were only 6 m (20 ft) from the existing caisson, they were drilled into place to minimize disturbance to the surrounding soil. After drilling to the top of rock, the piles were driven home using a 400,000 J (300,000 ft pound) drop hammer. The piles were instrumented with PDA equipment to verify an ultimate pile capacity of 8,900 kN (1000 ton). During drilling, Pier 2 began to settle and tilt when the tips of the piles advanced below the bearing level of the caisson. At the completion of drilling, the top of the caisson had tipped approximately 12 mm (0.5 in.) to the south. During the two days of PDA testing on piles at Pier 2, it moved an additional 25 mm (1 in.) to the south and continued creeping thereafter, but at a slower rate. Construction immediately stopped as the tilting of the structure impeded the opening and closing of the bascule span.

Amtrak became concerned as to how much the subsequent construction activities would cause the pier to further settle and tilt and not only impact the operation of the existing bridge, but also jeopardize the new lift span. The writers were engaged at this time to help understand and solve the problem. Amtrak authorized three new tasks: first, an instrumentation system to better understand the pier movements; second, a prediction of settlement, tilting and pile reactions; and finally, a remedial grouting program to arrest the movement. The first two tasks are described in a separate paper. This paper describes the grouting.

GROUTING PROGRAM

Grouting focused on the southern half of the caisson where settlement was occurring and bearing pressures in the sands were greatest, about 670 kPa (7 tsf). Grouting was divided into five work areas, Areas A through E (Fig. 1).

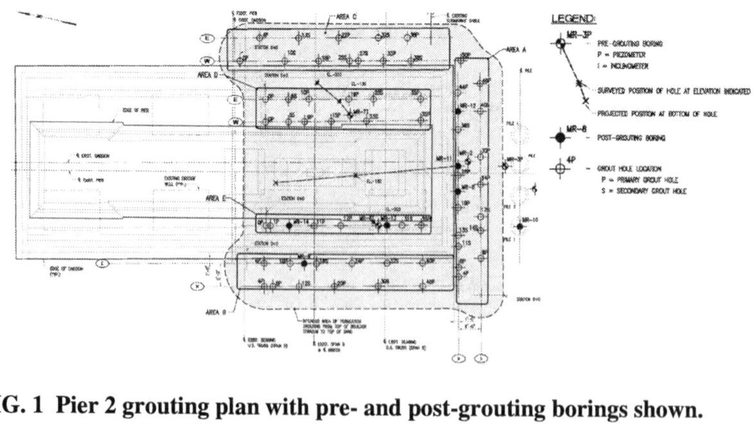

FIG. 1 Pier 2 grouting plan with pre- and post-grouting borings shown.

Areas A, B and C each consisted of two lines of grout holes along the south, west and east sides of the caisson respectively. These perimeter grout holes were intended to create a grout curtain that would contain the grout when grouting beneath the caisson. Grout holes in Areas D and E penetrated the caisson to facilitate grouting in the soils beneath the caisson. Grout holes were generally spaced eight feet on center except in Areas D and E where grout holes were primarily located in the dredge wells to avoid drilling through steel reinforcement when penetrating the caisson.

The target grout zone was the permeable and erratically-dense sand and gravel stratum that supported the caissons. The stratum extended from a depth of about 40 to 52 m (130 to 170 ft) below the river surface. Compaction grouting was selected as it would focus improvement to the weaker soil zones and horizons most affected by previous construction operations. Compaction grouting involves the controlled injection of a very stiff, mortar-like grout at high pressure into discrete soil zones. This low mobility grout remains in a homogeneous, expanding mass, which displaces and thus compacts and strengthens the adjacent soil. However, the plan included the flexibility to change to permeation grouting if compaction grouting proved ineffective. Permeation grouting involves the filling of the pore spaces of the soil with a high mobility grout for the purpose of strengthening the soil formation.

Drilling and grouting in open water around the piers required special considerations. The Contractor constructed work platforms around the perimeter of each pier to facilitate access for drilling and grouting. The grout plant was mobilized on a floatable barge that was anchored off the end of the pier. All weather enclosures (Fig. 2) were erected on the barge to protect materials and facilitate grouting in inclement weather. Grouting was performed six days per week and 12 hours per day.

FIG 2. Grout plant in all weather enclosure.

FIG. 3. Drill rig on Area C work platform.

GROUT HOLE DRILLING

Grout holes were drilled using two diesel–hydraulic track mounted rotary drill rigs equipped with top mounted hydraulic percussion hammers (Fig. 2). Rotary percussive duplex methods were used to limit soil removal and consequent potential of additional pier movement during drilling. This method consisted of an inner 63 mm (2.5 in.) diameter drill string with button bit and outer 88 mm (3.5 in.) diameter casing that were simultaneously rotated and driven into the ground. The outer casing stabilizes the hole during advance while the inner drill string and bit removes the soil cuttings from the hole. Soil cuttings were removed from the hole using internal flush whereby drilling wash water is circulated down the inner drill string and returns carrying the soil cuttings in the annular space between the drill string and outer casing. Similar methods were used in drilling through the caisson. Various drill bits including tri-cone roller bits, button bits and ballistic bits were substituted to improve penetration through the timber framing, steel reinforcing and steel cutting shoe where encountered in the caisson.

Initial grout holes at Pier 2 penetrated to the top of rock. When inclinometer data revealed that movements in the soils beneath the caisson were limited to the sand and gravel stratum, drilling of subsequent grout holes was stopped at the top of the boulder stratum at about 52 m (170 ft).

COMPACTION GROUTING EFFORT AND RESULTS

Grouting was initiated at Pier 2 using compaction grouting methods. Compaction grouting was performed using the outer casing for grout delivery to the bottom of the hole. Casing was withdrawn using the drill rig mast supplemented by the drills reverse percussion feature and hydraulic jacks reacting off the fixed work platforms where necessary. Grouting proceeded in two foot increments (stages) from the bottom of the hole to the bottom of the caisson. The grout injection rate was typically controlled within a range of 0.03 to 0.045 m^3/min (1 to 1.5 ft^3/min). In each increment, grout was pumped until essential refusal was observed as indicated by a sustained grout pressure of 5500 to 7000 kPa (800 to 1,000 psi) at the grout head.

The grout mix consisted of Portland cement, aggregate and sufficient water to produce a slump of between 25 and 50 mm (1 and 2 in.). The aggregate consisted of concrete sand mixed with ground limestone to provide a source of fines. Cement was added at roughly 20% by weight of the aggregate.

Compaction grouting was performed in eight grout holes between December 7, 2006 and January 9, 2007. Grout takes in individual grout holes ranged from 0.5 to 2.8 m^3 (18 to 100 ft^3) with almost all of the grout take occurring immediately below the bottom of the caisson. However, inclinometer data demonstrated that ground movement was occurring within the sand layer at significant depth below the caisson. Grout escaping into the overlying organic silt rather than compacting the sand and gravel was also suspected and was attributed to the wide range in grout takes. As a result, grouting methods were changed to permeation grouting in accordance with grouting plan provisions.

PERMEATION GROUTING PROCESS

Microfine cement was used in permeation grouting to maximize grout penetration into the variable sand and gravel deposits and limit the number of grout holes since drilling through the caisson and underlying soils was difficult and time consuming. The majority of the grouting was performed using Nittetsu Super Fine cement. U.S. Grout's Type V Standard microfine cement was also used.

Nittetsu Super Fine cement is a blast furnace slag based cement with a maximum particle size of 10 microns and Blaine Fineness greater than 9,000 cm^2/gram. The Blaine Fineness is a measure of the specific surface area of cement particles per unit volume of cement. Trial testing in the laboratory and field with grouts of different water:cement (w:c) ratios demonstrated that a grout mix with a w:c ratio of 4:1 by weight and Mighty 150 dispersant added at 1% by weight of cement provided the best combination of grout penetration, strength and material economy. Dispersants (super-plasticizers) are added to microfine cements to enhance particle dispersion during mixing and thereby reduce grout viscosity and bleed when fluid and improve grout strength and durability when solidified in the ground. However, testing revealed that the grout was unstable with an observed bleed of more than 50% at two hours. A stable grout was necessary to reduce cement filtration during pressure injection and avoid subsequent segregation of the cement in the soil pores following injection. This was particularly important given the relatively wide spacing between grout holes. The testing also demonstrated that the slag based grout took about 24 hours to set with bleeding continuing during this period. In response to these observations, a diutan gum blend was added to the mix at 0.75% by weight of water to improve grout stability. Diutan gum is a high molecular weight biopolymer that acts a thixotropic agent and significantly enhances a grouts resistance to pressure filtration. The diutan gum blend consisted of diutan gum with ground limestone added to improve additive dispersion during mixing. The gum addition was able to reduce bleed to about 10% after 24 hours.

Grout was mixed on the barge using a high shear mixer with double tub arrangement to provide a continuous grout supply. The mixed grout was transferred to an agitator tank from where it was distributed to up to three pumps for simultaneous grouting of several holes. Moyno pumps were used in initial grouting but quickly replaced by higher capacity plunger style pumps with 75 mm (3 in). diameter rams when high grout takes and low backpressures were observed in the sands and gravels. Grout was tested daily for unit weight, Marsh funnel viscosity and bleed to confirm consistent preparation and mixing of the grout.

Permeation grouting was performed using sleeve port pipes (tube-á-manchettes) consisting of 38 mm (1.5 in.) diameter PVC pipes with 6 mm (0.25 in.) holes (ports) spaced 0.75 m (2.5 ft) on center. Sleeve port pipes were implemented in grouting because they provide more uniform grout treatment with depth. Grout is targeted via individual ports to each soil horizon unlike other methods that can allow grout to travel to only the most permeable soil zones. The use of sleeve port pipes also facilitated access for multiple re-grouting attempts where necessary.

Initial grouting with microfine cement grout confirmed the viability of permeation grouting with grout flow in the sands and gravels occurring under little or no backpressure. Comparable trials with ordinary Portland cement as a cheaper alternative grout showed much higher pressures and limited grout flow due to the larger particle size being filtered and clogging the soil pores around the grout hole. The high initial grout takes led to extended pumping times and grout stages. In response, multiple grout pumps were mobilized to allow concurrent grouting of several grout holes. Grout volumes in perimeter holes were also limited to 3800 liters (1000 gallons) per 0.75 m (2.5 ft) grout stage to increase the rate of grout coverage around the perimeter of the pier and stop or at least reduce the rate of on-going pier movements as quickly as possible. This volume equates to an eight foot grout radius and theoretical overlap of adjacent primary grout holes assuming a soil porosity of 25%. The sleeve port pipes facilitated additional grouting at any stage, if necessary.

Grout flow was typically increased in increments with the maximum flow at any stage adjusted as necessary to avoid hydro-fracturing of the ground. Maximum flow rates of 38 to 45 L/min (10 to 12 gpm) were observed when grouting in the sands and gravels. Grouting in primary holes at each stage continued until either refusal occurred or a prescribed volume of grout was placed. Grouting in secondary holes was typically continued to refusal. Refusal criteria were defined to limit excessively long grouting stages where limited grout take was occurring, reduce wear on equipment and still achieve verification of the effectiveness of the grouting program.

PERMEATION GROUTING – MEASUREMENT OF RESULTS

A computerized data acquisition system consisting of an electromagnetic flow meter, pressure transducer and data logger was used to provide real time monitoring and control of grout injection parameters (flow rate, pressure and volume). The real time displays allowed rapid evaluation of grouting effectiveness as the work progressed and avoided unnecessary grouting if hydraulic fracturing of the ground were to occur.

Grouting generally proceeded from the perimeter of the caisson toward the interior. The outer grout lines in Areas A, B and C were grouted prior to grouting the inner grout lines. This was followed by grouting beneath the caisson in Areas D and E. Grouting within each area proceeded in an alternating sequence with primary grout holes grouted first followed by grouting of secondary grout holes spaced intermediate between the primary grout holes. The alternating pattern was used to verify the effectiveness of grouting. Lower grout takes and higher pressures in secondary holes confirmed the effectiveness of grouting. If not observed, additional grouting was performed to ensure overlap between adjacent grout holes.

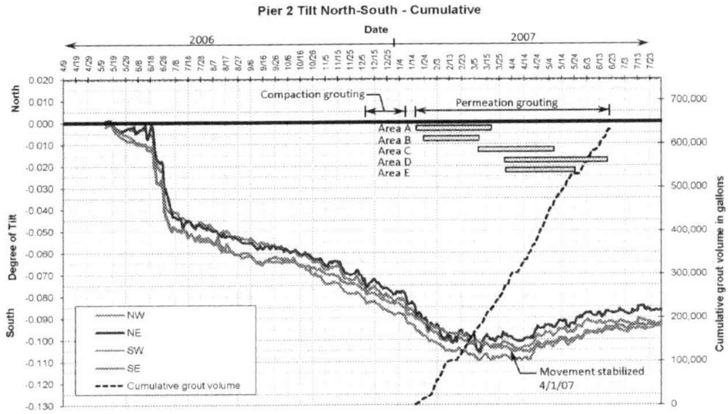

FIG. 4. Pier movement from tilt meters versus grouting progress.

Monitoring data (Fig. 4) from tiltmeters and inclinometers was used to judge necessary grout quantities and verify the effectiveness of grouting. The monitoring data showed a marked decrease in the rate of movement of Pier 2 near the end of March 2007 indicating that initial stabilization of the pier was achieved. At that time, about one million liters (270,000 gal) of grout was placed to strengthen the sands around the perimeter of the pier (Areas A, B and C). An additional 1.3 million liters (350,000 gal) of grout was subsequently injected beneath the pier between April and June 2007 to achieve more uniform bearing and long term stability of the pier.

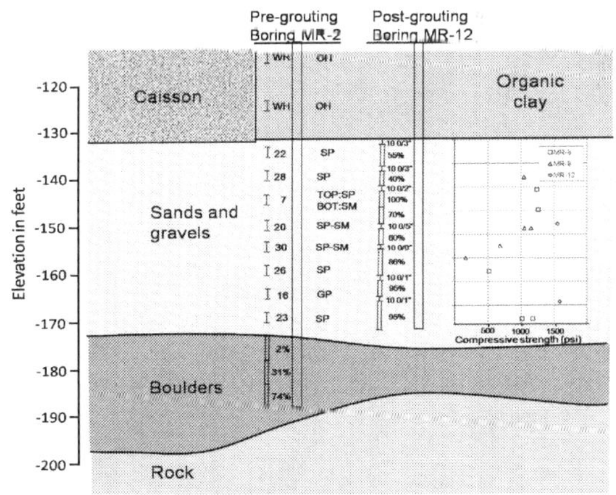

FIG. 5. Comparison of pre-grouting and post-grouting boring results

Borings were also made with conventional Standard Penetration Testing alternated with HQ wireline coring to recover samples of the grouted soil (Fig. 5). The boring results demonstrated effective permeation of the microfine cement grout into the variable sands and gravels with cores of the grouted soil (Fig. 6) exhibiting compressive strengths of 3,500 to 10,000 kPa (500 to 1500 psi).

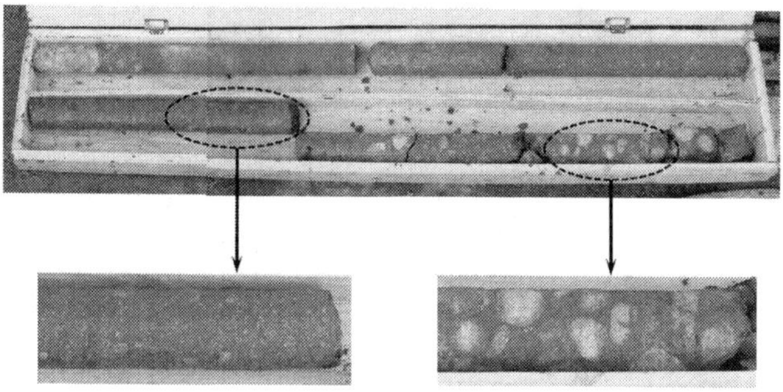

FIG. 6. Cores of grouted soil.

Permeation grouting was also performed at Pier 3 to mitigate potential differential movements between Piers 2 and 3 due to the additional loads applied by new bridge construction. Final grouting volumes in the sands and gravel beneath Pier 3 were about 70% of the volume injected at Pier 2 which is roughly proportional to the respective ground volumes beneath the pier footprints.

CONCLUSIONS

Grouting successfully halted the creep movements in the sands beneath Pier 2, but only after more than half the total volume of grout had been placed and after the grout achieved strength. The slag based microfine cement has an extended cure period, more than one month.

Microfine cement was necessary to permeate the sands – ordinary Portland cement did not work, probably because the cement particle sizes led to clogging of pore spaces. A high water to cement ratio (4:1) was used to promote permeation and increase the effective radius for each grout hole. The compressive strengths of grouted sands ranged from 3,500 to 10,000 kPa (500 to 1,500 psi).

Real time monitoring was essential. It enabled more accurate measurement of grout pressures and takes, and more precise targeting of grout zones. It led to faster and better decisions to control the grout program.

REFERENCES

Richter J., Finch P. and Martin M., "Replacement of Amtrak's Thames River Movable Bridge", Proceedings, AREMA 2008 Annual Conference, Sept. 2008.

Kolsch E., Kaeck W. and Rhyner F. "Thames River Piers Stabilization", Proceedings, AREMA 2008 Annual Conference, Sept. 2008.

Instant Solidification of Soft Ground Horizontally Using Jet-grouting

S.-L. Shen[1], C.-Y. Luo[2], Y. Bai[3], Y.-H. Kim[4], and S.-J. Peng[5]

[1]Professor, Department of Civil Engineering, Shanghai Jiao Tong University, 800 Dongchuan Road, Shanghai, 200240, China; PH (86) 21-54742254; e-mail: slshen@sjtu.edu.cn
[2]Lecturer, ditto; e-mail: cyluo@sjtu.edu.cn
[3]Chief Engineer, Shanghai Urban Construction Group Inc, 500 Fushan Road, Shanghai, 200122, China; PH (0086) 021-63017388; e-mail: baiy@sucg.com.cn
[4]Korea Foundation Technology Construction (KFT) Co. Ltd., 1505-29, ShinHan Deview B/D 4F, Kwanyang-dong, Anyang-City, Korea; PH(82)02-786-9200; email: kftenc@gmail.com.
[5]Shanghai Second Municipal Engineering Co. Ltd., 1565 Jiaotong Road, Shanghai 200065, China

ABSTRACT: Jet-grouting is usually used in underground construction in the Shanghai soft deposits. In some cases, jet-grouting is conducted horizontally. However, immediately after the mixing, the soil-cement admixture is in a fluid state, which will flow out from the hole. In order to solve this problem, Twin-Jet Technology was invented to solidify soft soil instantly. Two types of binders with cement and accelerator such as sodium silicate are jetted under high pressure of about 10 to 40 MPa and mixed with in-situ soil to create the instant solidification. A field test was conducted in the soft clayey deposit in Shanghai. Field and laboratory results showed that the admixture can be gelled within 10 to 20 seconds. The diameter of the jet-grouted column reached 1.2 m and the strength after 14 days reached 1.2 MPa.

INTRODUCTION

Jet-grouting is a soil treatment technique to stabilize soft ground by mixing cement slurry with in-situ soil. The cement slurry is jetted out under high pressure from a nozzle on a rotating rod and injected into the ground with high velocity and high pressure. The high speed fluid erodes the in-situ soil and mixes the eroded soil with cement slurry to form a soil-cement column. The shear strength of the soil-cement column can reach several MPa (Maswoswe, 2003; Ho et al., 2005).

Jet-grouting technology was initially invented by Nakanishi in 1968, and was known as chemical churning pile (CCP) method with single rod for single fluid injected (Brill et al., 2003; Burke, 2004). Then, a number of advances in jet-grouting construction method and equipment were made, including two-way or double rod method, triple rod method (TSG), and multi-rod method (SSS-MAN). In order to create larger diameters of columns, SuperJet (RJP) grouting technique was proposed in 1988, in which the diameter of the solidified column can reach 5m.

The first erosion of the soil in most jet-grouting methods is performed by high pressure water, which causes the following problems: i) admixture of cement and soil is in fluid state due to the large amount of water injection. Thus it requires several hours to days to gel and get strength for the admixture. ii) When applied in sand layer under the ground water, it is easy to cause quicksand in which the sand and cement may be separated before hardening. iii) When jet-grouting is conducted horizontally, the fluid admixture may flow out from the grouting hole and thus increase the difficulty of horizontal construction. These disadvantages restrict the application of horizontal jet-grouting in sands.

On the basis of triple-rod method, this paper proposed a new jet-grouting technique for instant solidification of soft ground using two types of binders, which is called as Twin-Jet Technology (Kim, 2008; Shen et al., 2008). The technique is based on the common understanding that the admixture of cement and its accelerator can be gelled in several to tens of seconds. Two types of binders (cement and its accelerator) are jetted under high pressure and mixed at the moment when they are ejected from nozzle. Then the admixture is mixed with the in-situ soil to achieve the instant gelling of the soft ground.

FUNDAMENTAL CONCEPTS OF TWIN-JET GROUTING

It has been known that accelerating additives, such as sodium silicate, can speed up the gel of cement. The admixture of cement and accelerator can be gelled in several to tens of seconds while the gelling time of unmixed cement is up to several hours. The gelling time of the admixture will vary with the mixing ratio of accelerator to cement. Therefore, choosing the appropriate mixing ratio plays a decisive role in the improvement effect of jet grouting when two types of binders are utilized. Figure 1 illustrates the relation between the mixing ratios of sodium silicate

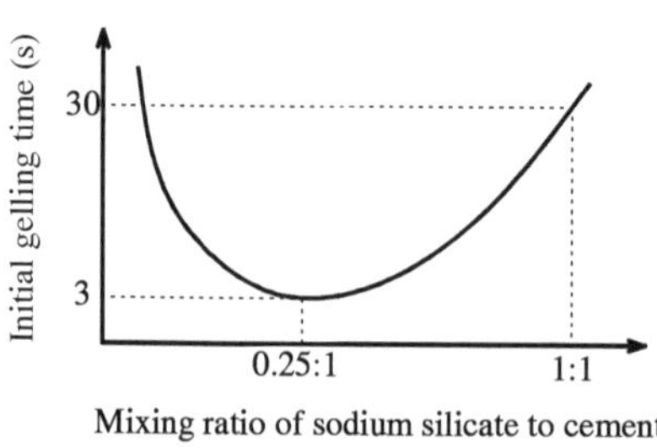

FIG. 1 Mixing ratio of sodium silicate over cement vs. gel time

to cement and the initial gelling time of cement-silicate admixture. The water-cement ratio of cement slurry is generally 1:1 (mass ratio). The concentration of sodium silicate is from 10 to 30.4 baume degree. According to Figure 1, the initial gelling time decreased rapidly following the increasing content of sodium silicate while total amount of sodium silicate was less. The initial gelling time was shortest (about 3~5 seconds) when the mixing ratio of sodium silicate to cement is 0.25:1. Then the initial gelling time went up to about 30 seconds as the mixing ratio reached 1:1.

EQUIPMENTS FOR TWIN-JET GROUTING

Figure 2 schematically illustrates the composition of equipment for horizontal twin-jet grouting, including the drilling and jetting system, the grouting system and the liquid transporting system. The drilling and jetting system consist of a horizontal drill rig, a triple rod, a three-channel swivel, and an injection monitor with multiple nozzles (as shown in Figure 2(b)). The grouting system includes a high pressure pump for injecting highly pressurized grout, a low pressure pump for feeding accelerator fluid, an air compressor for generating pressurized air and other facilities employed for these pumps. The grouting facilities are connected to the triple rod through the three-channel swivel.

Figure 3 shows the triple rod, which is used for drilling and transporting the pressurized liquid and air flow to the monitor. The rod is composed of three coaxial steel pipes, including the inner pipe (first pipe), the second pipe and the outer pipe (third pipe). The inner pipe is connected to the high pressure pump for transporting high pressure grout or water. The second pipe is connected to the low pressure pump for pumping accelerator fluid. The air compressor is connected to the outer pipe, through which the pressurized air can be transported to the monitor.

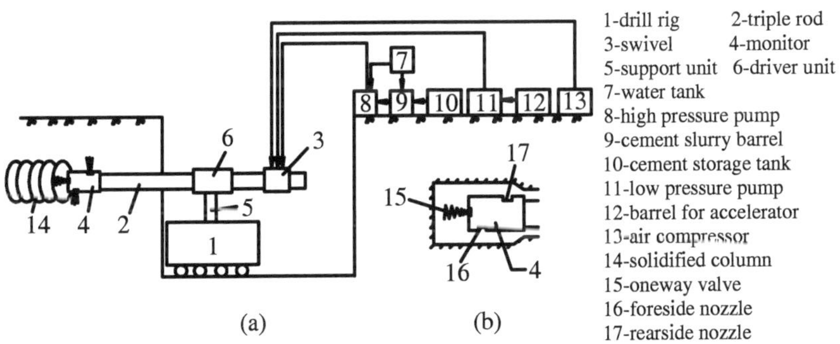

FIG. 2 Composition of the equipment for horizontal twin-jet grouting

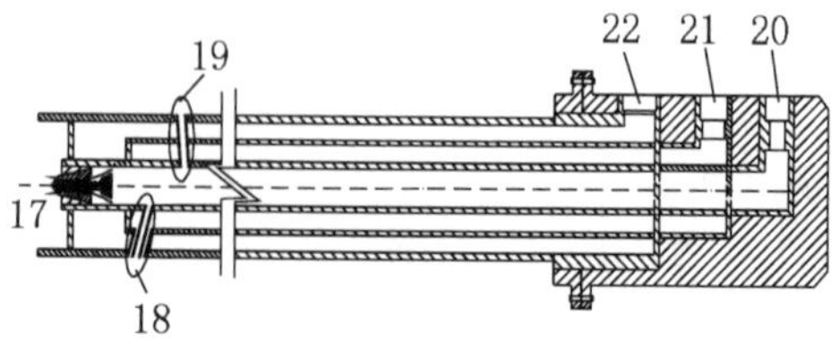

17-oneway valve 18-foreside nozzle 19-rearward nozzle
20-inner pipe 21-second pipe 22-outer pipe

FIG. 3 Illustration of triple drilling-jetting rod

Figure 3 also illustrates the configuration of the monitor, which is mounted at the top end of the triple rod. The monitor is equipped with an oneway valve at the top and two injection nozzles on opposite side. The oneway valve is connected to the inner pipe and can automatically control the liquid flow through the pressure. The foreside nozzle has dual outlets which are connected to the inner pipe and the second pipe, respectively, for injecting high pressure grout and accelerator fluid simultaneously. The rearside nozzle also has dual outlets connected to the inner pipe and the outer pipe, respectively, to jet the pressurized liquid and air at the same time. The first erosion of in-situ soil is performed by high pressure grout shrouded by pressurized air erupted from the composite nozzle. Then high pressure grout and low pressure accelerator fluid are coaxially jetted out from the foreside nozzle to erode the ground for the second time and mix with the eroded soil simultaneously.

Figure 4 illustrates the detailed configuration of the foreside nozzle and the rearside nozzle. The foreside nozzle reduces gradually to several millimeters while the inner outlet recesses into the outer outlet. Within the recession distance of the inner outlet, the outer outlet is converted into funnel-shaped, which can prevent the flowback of high pressure grout from the outlet for accelerator fluid. The rearside nozzle also becomes narrow gradually to several millimeters near the nozzle exit. The slurry was jetted out from the inner outlet and is surrounded by the pressurized air erupted from the outer outlet, which can improve the eroding ability of grout and therefore enlarge the diameter of the solidified column.

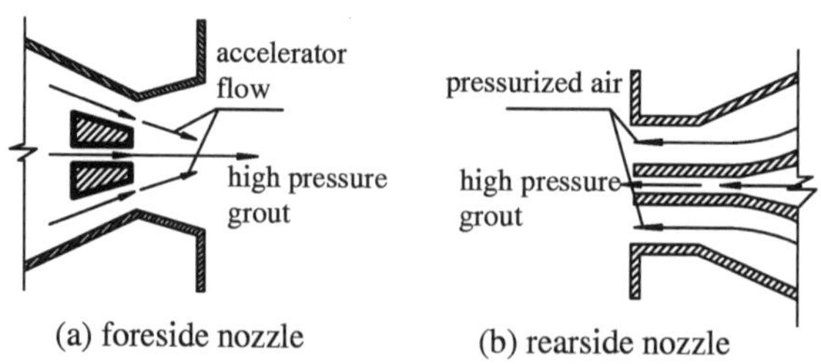

(a) foreside nozzle (b) rearside nozzle

FIG. 4 Illustrations of multiple nozzles

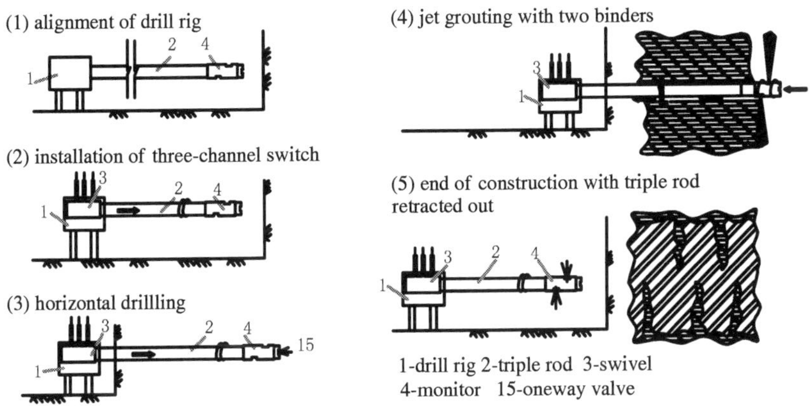

FIG. 5 A typical working process of horizontal twin-jet grouting

CONSTRUCTION PROCEDURE

Prior to installation of the soil-cement columns, jet-grouting parameters should be predetermined. Among them, the amount of cement and accelerator applied per unit soil are the key factors for instant solidification of the ground. To decide the optimal proportion of soil-cement-silicate, laboratory tests will be performed with the soil samples extracted from the site, according to the requirements of initial gelling time and strength of the admixture. Then detailed operation parameters should also be determined, on the basis of the target column diameter, soil properties, stress state and strength of in-situ soil, content of binders. Figure 5 demonstrates the working process of twin-jet grouting in application to horizontal ground improvement, including the following steps:

Step 1: assembling of equipment

The twin-jet grouting equipments are first installed and emplaced at the predesigned position, including the drilling and jetting facilities, the grouting facilities and the liquid transporting pipe system. The pressure range and displacement of the pressurization equipments should meet the design requirements. The pipe system must be well sealed, while the hoses and nozzles for jetting binders and air are kept smooth. A test running should be carried out to check the condition of twin-jet grouting machines before they are put into construction.

Step 2: alignment of drill rig

The horizontal drill rig must be placed evenly and firmly. The drill rod should be adjusted to the predetermined orientation. The position of drill holes must be set out strictly according to the design scheme with the errors limited within the allowable range.

Step 3: drilling

The drill rod will be rotated into ground at the predesigned position of the target column, and then moved forward at the designed velocity until the target length is reached. In this process, the drill rod is slowly rotated forward while at the same time water is injected through the middle channel to form high speed fluid flow. The high speed water is jetted out from the nozzle at the tip of the monitor and erodes the in-situ soil for the first time. The grouting pressure is generally 2 to 5MPa, which is determined according to the soil properties. When the torque induced on the drill rod is large, the grouting pressure needs to increase to loosen the in-situ soil.

Step 4: twin-jet grouting

Drilling and grouting is stopped when the drill rod reaches the designed length. The oneway valve at the top nozzle of the monitor is automatically closed by pressure controlled sensor. Then the drill rod is rotated and retracted backward from the drill hole slowly while opening the control valves to transport the binders in the following order: the air valve, the valve for accelerator, the grout valve. The grout shrouded by pressurized air is first jetted out from the nozzle at the rear of the monitor to erode the ground. Then the admixture of cement and accelerator is injected into soil from the foreside nozzle of the monitor. The high pressure grout shrouded by the accelerator fluid will erode the in-situ soil for the second time. The construction of rotary twin-jet grouting is based on design parameters including grouting pressure, retracting rate, rotation rate.

Step 5: end of construction for one column

As the drill rod is retracted to the predesigned position, the valves are closed according to the following order: the valve for accelerator, the grout valve, the air valve. Then pumps and other facilities need to be shut down to finish the construction of the column. After the construction of one column, the drill rod is moved to the entry position of next column and then work steps, from the second step to the fifth step, are repeated to perform the construction of the next column. This process keeps repeating until the target improvement area is entirely treated.

CASE HISTORY

The shield arrival shaft in Baili Station of Shanghai Metro Line 11 is located to the northwest of Shanghai. Horizontal jet-grouting at the entrance site of shield machine was conducted in this project to avoid ground subsidence. Field investigation showed that the soil condition in the construction site was very weak, consisting mostly of fill soil and very soft clay. The fill soil in upper layer was of loose structure with high water content and was in a fluid plastic to soft plastic state. The very soft clay was distributed from 7.0 m to 15.0 m below surface, in which the water content was within 45~50% and the unconfined compression strength was 40 kPa. Horizontal jet-grouting was carried out in this layer at depth from 10 m to 14 m below the ground surface. The designed length of jet-grouting columns was 15 m and the target diameter was 800 mm. The target strength of the solidified column was 1.0 MPa. Two binders used in this project were ordinary Portland cement and sodium silicate, which are harmless to the natural environment. The concentration of the sodium silicate was 30 to 40 baume degree.

Laboratory tests of mixing ratio with in-situ soft soil were performed. Table 1 lists the physical parameters for the soil-cement admixture to reach the target strength. Jetting parameters used in this project are as follows: grouting pressure P_m = 12 MPa; inner diameter of nozzles d_0 = 2.5 mm; cycle number N = 2; moving rate of nozzles V_{tr} = 500 mm/min; rotation rate of triple rod R_s = 12 rpm; grout flow rate Q = 88 *l*/min. The jetting pressure for accelerator fluid was 2 to 3 MPa and flow rate was set as 20 *l*/min. The air pressure was 0.7 MPa and flow rate was 6 to 7 m^3/min.

Field and laboratory tests were carried out to verify the quality of jet-grouted columns. Core samples of the jet-grouted columns were extracted and the unconfined compressive strength was determined by UC tests in laboratory to detect the column quality. The diameter of the jet-grouted columns was determined by coring and cone penetration tests. The results indicated that the initial gelling time of the admixture was 10 to 20 seconds and the jet-grouting columns were well constructed and uniformly mixed. The results of coring and compressive strength tests showed the strength of the jet-grouting columns after 14 days curing reached 1.2 MPa. The column diameter was confirmed as at least 1.2 meter. The quality and diameter of jet-grouting columns completely met the design requirements.

TABLE 1. Physical parameters for the admixture of twin-jet grouting

Water-cement ratio	Cement content per treated volumn (kg/m^3)	Special gravity of slurry	Initial gelling time (s)	Content of sodium silicate per treated volumn (*l*/m^3)
1:1	243.7	1.57	10-20	81

CONCLUSIONS

Twin-jet grouting technology mixing two types of binders (cement and its accelerator) with the in-situ soil simultaneously can achieve the instant solidification of soil-cement admixture. The effect of this technology in soft ground improvement was verified by its application in the soft clayey deposit in Shanghai. The advantages of twin-jet grouting technology are as following:

(1) Admixture of soil-cement can be gelled within 10 to 20 seconds, which makes

it possible to apply in sandy soil layers under groundwater.

(2) High pressure liquid jetted out from the two nozzles can erode the ground twice to creat larger diameter columns so that less jet-grouting columns were required.

(3) Work efficiency is evidently increase by about 3 times, and thus the cost of construction is decreased.

(4) The application in a horizontal work in Shanghai indicated that the diameter of solidified column reached to 1.2 m while the strength of the column after 14 days reached to 1.2 MPa, which completely meet the design requirements.

(5) Twin-jet grouting technique is simply operated with low cost and high efficiency, which will extend its application in China.

ACKNOWLEDGEMENTS

The research work described herein is supported by Shanghai Leading Academic Discipline Project (Project Number: B208) and the Science and Technology Commission of Shanghai Municipality (08201201600). This work was also supported by the Research Foundation of Shanghai Urban Construction Group Inc. Shanghai Second Municipal Engineering Co. Ltd., China carried out the field construction of horizontal jet-grouting in Shanghai Metro line 11. These financial supports are gratefully acknowledged.

REFERENCES

Brill, G.T., Burke, G.K., and Ringen, A.R. (2003). "A ten-year perspective of jet grouting: advancements in applications and technology." *Grouting 2003*, GSP No. 120, ASCE, Reston/VA: 218-234.

Burke, G.K. (2004). "Jet grouting systems: advantages and disadvantages." *GeoSupport 2004*, GSP No. 124, ASCE, Reston/VA: 75-87.

Ho, C.E., Lim, C.H., and Tan, C.G. (2005). "Jet Grouting Applications for Large-Scale Basement Construction in Soft Clay." *Geo-Frontiers 2005, Innovations in Grouting and Soil Improvement 136* (CD-ROM), ASCE, Reston, VA.

Kim, Y.-H. (2008). "Technological Manual of Twin-Jet." *Korea Foundation Technology* (KFT) E&C, Seoul, Korea. (in Korean)

Maswoswe, J.G. (2003). "QA/QC for jet grouting in deep Boston blue clay central artery/tunnel project." *Grouting 2003*, GSP No. 120, ASCE, Reston/VA: 236-247.

Shen, S.L., Kim, Y.H., Hong, Z.S., and Bai, Y. (2008). "Equipments for rapid solidification of soft ground through jet-grouting using two typers of binder." China Patent, Patent number: 200810033981.4. (in Chinese)

Jet Grouting to Increase Lateral Resistance of Pile Group in Soft Clay

Kyle M. Rollins[1], Matthew E. Adsero[2], and Dan A. Brown[3]

[1]Prof. Civil & Env. Engrg. Dept., Brigham Young Univ. Provo, UT rollinsk@byu.edu
[2]Construction Engr., Exxon-Mobil Development Co., Houston, TX matthew.adsero@exxonmobil.com
[3]Prof. Civil Engrg. Dept., Auburn Univ., Auburn, AL BROWND2@auburn.edu

ABSTRACT

Lateral load tests were performed on a full-scale pile cap in clay before and after construction of eight 1.5 m diameter jet grout columns to a depth of 3 m around the pile group. Jet grouting with a cement content of about 400 kg/m^3 (20% by weight) increased the average compressive strength of a soft, plastic clay from 40 to 60 kPa to an average of 4500 kPa. The lateral resistance was increased by 2200 kN or 177% and the initial stiffness was increased by 400%. About 65% of the increased resistance could be accounted for by passive pressure and side/base shear on the jet grout mass; however, the remaining 35% increase must be due to the interaction between the piles and the strengthened soil. Jet grouting provides a method to significantly increase the lateral resistance of pile group foundations at costs much lower than typical structural approaches.

INTRODUCTION

The lateral resistance of pile groups in soft clay is often critical to the seismic design of bridges and high-rise structures. Typically, when analyses indicate that the lateral resistance of a foundation is inadequate, additional piles, drilled shafts or micro-piles are added to increase the lateral resistance. Furthermore, an expanded pile cap or connecting beam is often provided to structurally connect the new piles to the existing pile group. While this approach produces the required lateral resistance, it is also relatively expensive and time consuming.

An alternative approach is to use soil improvement techniques to increase the strength and stiffness of the surrounding soil and thereby increase the lateral resistance of the pile group. The improved zone could be relatively shallow because the lateral resistance of piles is typically transferred within 5 to 10 pile diameters. This relatively simple approach has the potential of being more cost-effective and reducing construction time, but almost no tests are available to guide engineers in evaluating the actual effectiveness of this approach. In addition, numerical models to evaluate this approach have not been validated. To provide basic test data, full-scale lateral pile group load tests were performed on a nine pile group before and after treatment with jet grouting.

SOIL CONDITIONS

A generalized soil boring log at the test site is provided in Fig. 1. The depth is referenced to the top of the excavation which was 0.76 m above the base of the pile cap as shown in the figure. The soil profile consists predominantly of cohesive soils; however, some thin sand layers are located throughout the profile. The cohesive soils near the ground surface typically classify as CL or CH materials with plasticity indices of about 20 as shown in Fig. 1. In contrast, the soil layer from a depth of 4.5 to 7.5 m consists of interbedded silt (ML) and sand (SM) layers. The water table is at a depth of 0.60 m.

The undrained shear strength is also plotted as a function of depth in Fig. 1. Undrained shear strength was measured using a miniature vane shear test or Torvane test on undisturbed samples immediately after they were obtained in the field. In addition, unconfined compression tests were performed on most of the undisturbed samples. The undrained shear strength was also computed from the cone tip resistance using the correlation equation

$$s_u = (q_c - \sigma)/N_k \tag{1}$$

where q_c is the cone tip resistance, σ is the total vertical stress, and N_k is a coefficient which was taken to be 15 for this study. The undrained shear strength obtained from Eq. (1) is also plotted versus depth in Fig. 1 and the agreement with the strengths obtained from the Torvane and unconfined compression tests is reasonably good. Nevertheless, there is much greater variability. The drained strength in the interbedded sand layers is not plotted. The CPT data, as well as the Torvane and unconfined compression tests, indicate that the undrained shear strength decreases rapidly from the ground surface to a depth of about 2 m but then increases with depth. This is typical of a soil profile with a surface crust that has been overconsolidated by desiccation as is the case in this situation.

PILE GROUP CHARACTERISTICS

The pile group consisted of nine test piles which were driven in a 3 x 3 orientation with a nominal center to center spacing of 0.9 m. The tests piles were 324 mm OD pipe piles with a 9 mm wall thickness and they were driven closed-ended with a hydraulic hammer to a depth of approximately 13.4 m below the excavated ground surface. The steel conformed to ASTM A252 Grade 2 specifications and had a yield strength of 400 MPa based on the 0.2% offset criteria. The moment of inertia of the pile itself was 11,613 cm^4; however, angle irons were welded on opposite sides of two to three test piles within each group which increased the moment of inertia to 14,235 cm^4.

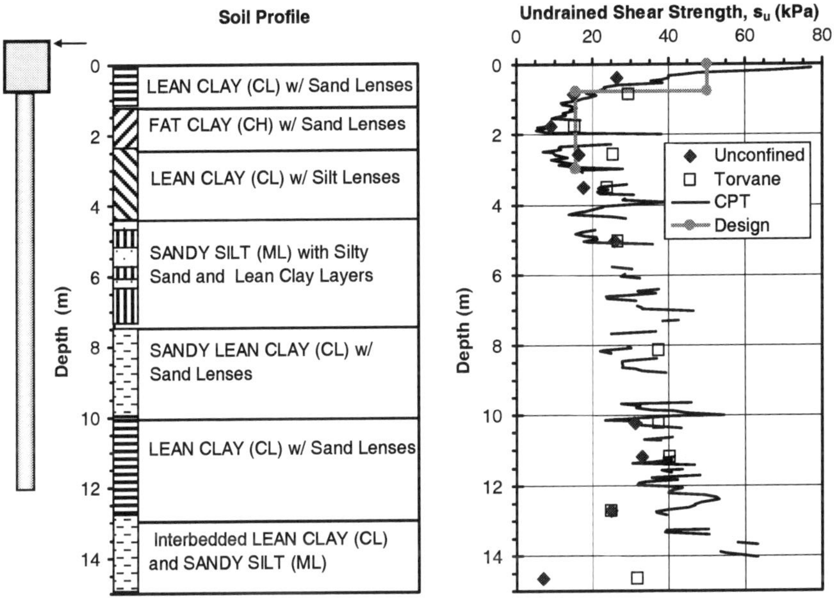

FIG. 1. Soil profile and undrained shear strength profile for the test site.

A steel reinforcing cage was installed at the top of each test pile to connect the test piles to the pile cap. The test piles typically extended about 0.6 m above the base of the pile cap and the reinforcing cage extended 0.7 m above the base of the cap and 2.7 m below the base. The steel pipe pile was filled with concrete which had an average unconfined compressive strength of 34.5 MPa.

A pile cap was constructed by excavating 0.76 m into the virgin clay. The concrete was poured directly against vertical soil faces on the front and back sides of each pile cap. This construction procedure made it possible to evaluate passive force against the front and back faces of the pile caps. In contrast, plywood forms were used along the sides of each cap and were braced laterally against the adjacent soil faces. This construction procedure created a gap between the cap sidewall and the soil so that side friction would be eliminated. Steel reinforcing mats were placed in the top and bottom of each cap. A corbel 0.55 m tall and 1.22 m wide was constructed on top of each cap to allow the actuator to apply load above the ground surface without affecting the soil around the pile cap.

TESTING PROCEDURE AND INSTRUMENTATION

Lateral pile group load tests were conducted using 2700 kN hydraulic actuators to apply load to the pile groups. Similar pile groups provided reactions for the applied load. The reaction groups were located 10 m away from the test pile group to minimize interaction effects. The lateral load tests were carried out with a displacement control approach with target pile cap displacement increments of 3, 6,

13, 19, 25, and 38 mm. During this process the actuator extended or contracted at a rate of about 40 mm/min. In addition, at each increment 10 cycles with a peak pile cap amplitude of ±1.25 mm were applied with a frequency of approximately 1 Hz to evaluate dynamic response of the pile cap. After this small displacement cycling at each increment, the pile group was pulled back to the initial starting point prior to loading to the next higher displacement increment.

4.1 Pile Group Testing Sequence

Plan and profile drawings showing the layout of the pile group for Tests 1 and 2 are provided in Fig. 2. Tests 1 and 2 were performed to provide a baseline of the lateral load behavior of the pile caps in virgin soil conditions prior to any treatment. Test 1 was conducted by pulling cap 1 to the left using the actuator while the untreated native soil was in place to the top of the pile cap. At the completion of test 1, the pile cap was pulled back to zero deflection, but after the actuator load was released some residual deflection remained.

Prior to Test 2, the soil immediately adjacent to the opposite face of the pile cap was excavated by hand to create a 0.3-m wide gap between the pile cap face and the adjacent soil as shown in Fig. 2. This excavation eliminated passive force against the pile cap for the subsequent test. After excavation was complete, which required less than an hour to accomplish, Test 2 was carried out by pushing the pile cap to the right using the actuator. The testing was performed using the same procedure described previously. Test 2 was designed to provide an indication of the passive force provided by the unsaturated clay soil against the pile cap.

Prior to Test 3 eight 1.5 m diameter jet grout columns were constructed around the pile group as shown in Fig. 2(a). The jet grout columns extended to a depth of 3.0 m below the base of the pile cap and created a soilcrete block around the pile group with dimensions of approximately 3.2 m x 4.57 m in plan. Although this pile group had previously been tested, the jet grouting process likely created a relatively virgin soilcrete mass without any gaps around the piles or cap. The actuators then pushed the pile cap to the left. Comparisons between this test and Test 1 in virgin clay make it possible to determine the improvement provided by the jet grout treatment.

JET GROUTING PROCEDURE

Plan and profile views of the jet grout columns around pile cap 2 are shown in Fig. 2. A total of eight 1.5-m diameter jet grout columns were installed beneath and around the pile cap. Four of the columns were installed at the periphery of the pile cap while an additional four were installed through the cap itself as shown in Fig. 2. During construction of the pile cap and corbel, four 0.15-m diameter PVC pipes were placed in the pile cap between the rebar to provide access for the jet grout drill rods. For retrofit projects these access holes would have to be drilled through the pile cap. The target diameter of the jet grout columns was 1.5 m. The jet grout columns were spaced at approximately 0.9 m center-to-center left to right and at 1.5 m center-to-center from top to bottom. This likely produced a 0.6 m overlap of the columns in the direction of loading with no expected column overlap perpendicular to the direction

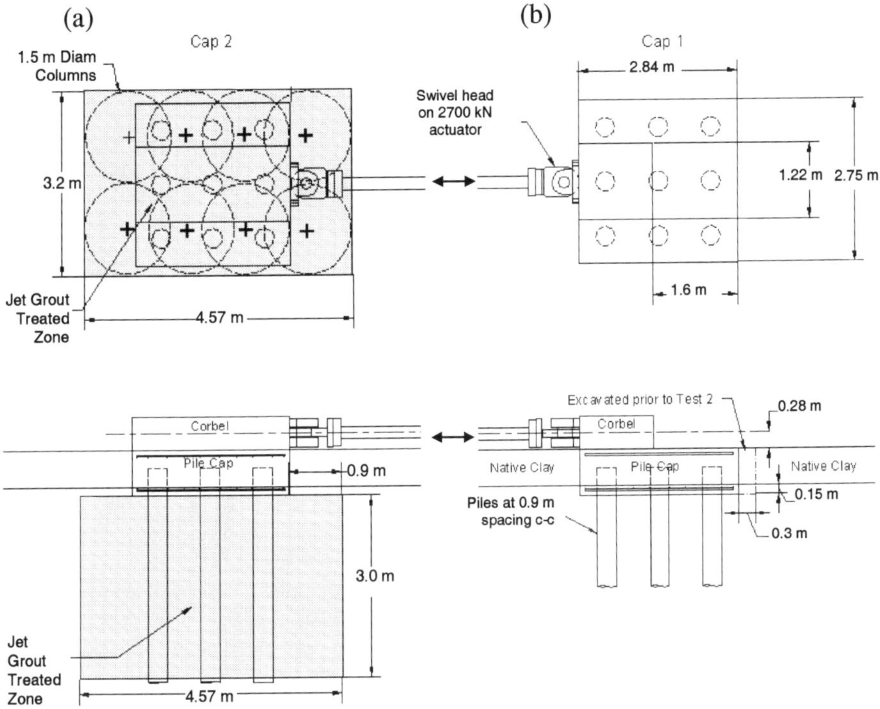

FIG. 2. Plan and profile drawings of the pile group (a) after jet grout treatment and (b) in virgin clay before jet grout treatment.

of loading. As can be seen in Fig. 2, nearly the entire volume of soil beneath the pile cap was treated to a depth of 3 m below the bottom of the pile cap. In addition, the grout treatment extended about 0.9 m beyond the front and back ends of the cap and somewhat beyond the cap on the top and bottom sides.

A double fluid jet grouting technique was employed to form the grout columns and each of the columns was constructed with identical installation parameters. The jet grout drill head was initially advanced to the base of the treatment zone, 3 m below the pile cap, using water jets and a drilling bit located at the bottom of the drill rod. Subsequently, the drill head was rotated and pulled upwards at a constant rate, while cement slurry was injected at a specified pressure and flow rate from the inner orifice of the drill nozzle. Concurrently, compressed air was injected from the outer orifice of the drill nozzle to form a protective shroud around the slurry jet and improve the erosive capacity of the cement slurry jet. The grout slurry mix had a specific gravity of 1.52, which is equivalent to a 1:1 water to cement ratio by weight. Throughout the jet grouting process, the flow rates, pressures, pull rate and drill rod rotation rate and specific gravity were controlled by a computerized system which also monitored and recorded these parameters. These parameters are summarized in Table 1. Based on the column diameter, flow rates, pull rates and rotation rates, the cement content for the jet grout columns would be expected to be about 400 kg/m^3. It can be seen that the pull rate is greater than the rotation speed. Thus, one rotation of the high pressure nozzle occurred for each 30 mm lift.

TABLE 1. Summary of jet grout treatment parameters.

Column Length	3 m
Estimated Column Diameter	1.5 m
Grout Specific Gravity	1.52
Grout Pressure	41.37 Mpa
Grout Flow Rate	340 Liters/min
Rotation Speed	7 rpm
Pull Rate	20 cm/min

The unconfined compressive strength of the soilcrete produced by the jet grouting process was evaluated using hand-mixed wet grab samples as well as core samples. Although there was significant scatter to the data, which is typical for soilcrete columns installed using jet grouting, there is a trend of increasing strength with curing time. Prior to treatment, the mean compressive strength of the untreated clay was only 40 to 60 kPa. Two weeks after jet grouting, the mean compressive strength of the wet grab samples had increased to about 3000 kPa; after four weeks the strength had increased to about 4500 kPa. These strength gains are typical for jet grouting applications (Burke, 2004). The average strength from two cored samples was about 3170 kPa, which is about 30% lower than the strength obtained from the wet grab samples. The strength from the core samples is likely more representative of in-situ conditions and is attributable to the poorer mixing produced by the jet grouting process relative to the hand mixing employed with the wet grab samples.

LOAD TESTING RESULTS AND ANALYSIS

Fig. 3 presents plots of the load-displacement curves for pile cap 1 in virgin clay before excavation (Test 1) and after excavation (Test 2) of the soil immediately adjacent to the front face of the pile cap. A comparison between the two curves indicates that the difference, attributable to passive resistance on the pile cap, is approximately 220 kN. The full passive force develops after a displacement of about 20 mm or 2.5% of the cap height.

Based on the measured passive force (P_p) the average undrained shear strength (s_u) of the upper 0.76 m of the soil profile was back-calculated using the equation

$$P_p = 0.5\gamma z^2 B + 2 s_u z B \tag{1}$$

based on Rankine theory for undrained conditions where γ = total unit weight of the clay = 18.37 kN/m^3, z = depth of the pile cap = 0.76 m, B = width of the pile cap = 2.74 m. Based on this back-analysis, the undrained shear strength in the upper 0.76 m of the soil was found to be about 50 kPa. This shear strength is higher than that measured by the unconfined compression testing, but within the range predicted by the correlation with the CPT cone tip resistance as shown in Fig. 1.

Fig. 3 also provides a comparison of the load-displacement curves for cap 1 during test 1 (virgin clay) and cap 2 for test 3 after the jet grouting improvement. With the jet grout improvement, the pile cap resisted 3475 kN compared to the 1253 kN resisted by the pile cap in the virgin clay at a displacement of 38 mm. This represents an increase of about 2200 kN or 177% in the lateral resistance provided by the pile group. It is also important to evaluate the increased stiffness due to the jet grout. Prior to treatment, the secant stiffness of the load-displacement curve at a displacement of 2.5 mm was 140 kN/mm while after jet grout treatment the stiffness increased to 700 kN/mm. This represents an increase in stiffness of about 400%.

The lateral resistance of the soilcrete block produced by jet grout treatment was computed by adding the passive force on the back of the block to the shear forces on the sides and base of the block. This calculation was made using the back-calculated undrained shear strength of 50 kPa in the upper 0.76 m of the profile and an average undrained shear strength of 15.5 kPa in the zone from 0.76 m to 3 m based on the soil strength testing as shown in Fig. 1. This approach can account for 1430 kN or about 65% of the measured increase in lateral resistance. The additional 35% of resistance must, therefore, be a result of more complex soil-pile interaction between the piles and the jet-grout strengthened soil. This interaction is currently being evaluated using finite element modeling techniques.

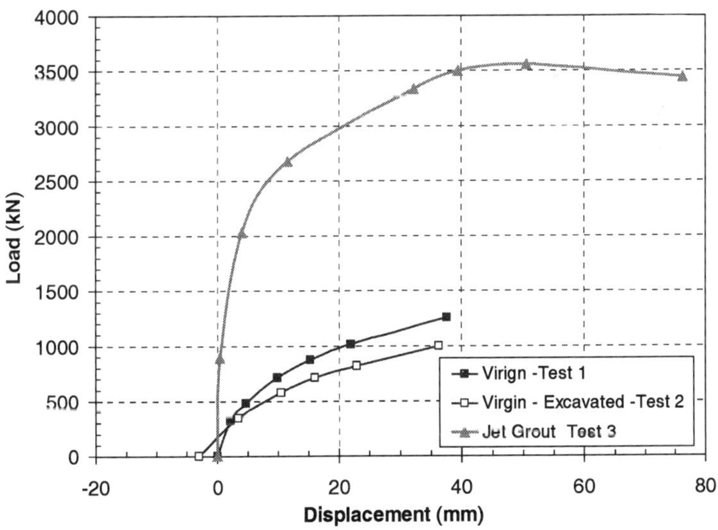

FIG. 3. Lateral load-displacement curves for pile groups in virgin clay before and after excavation of the soil adjacent to the pile cap along with curve for clay treated with jet grout columns.

COST CONSIDERATIONS

A complete cost assessment is beyond the scope of this paper. Nevertheless, some rough assessments are possible. Based on the lateral load test in untreated soil (Test 2), the piles in the group carried an average load of about 113 kN at a displacement of 40 mm. Therefore, an additional 20 piles would be necessary to produce the 2200 kN of increased resistance provided by the jet grout treatment. In addition, a larger pile cap would be required for the pile group. Based on typical unit costs, the jet grouting alternative would be significantly less expensive than the piling alternative neglecting mobilization costs (Adsero, 2008). Even considering mobilization costs, which are typically higher for jet grouting than pile driving, the total cost would still have been lower for the jet grouting alternative. Of course, mobilization costs become less important for large projects.

CONCLUSIONS

1. Jet grouting with a cement content of approximately 400 kg/m^3 (20% by weight) was able to increase the compressive strength of a soft, plastic clay from a value between 40 to 60 kPa to an average of 4500 kPa. This result is consistent with previous experience.
2. Construction of eight 1.5 m diameter jet grout columns around the nine pile group increased the lateral pile group resistance to 3475 kN relative to the 1253 kN resistance for the pile group in untreated virgin clay. This represents an increase in lateral resistance of 177%.
3. Jet grouting treatment of the pile group also increased stiffness from 140 kN/mm to 700 kN/mm, an increase of 400%.
4. About 65% of the increase in lateral resistance can be accounted for by passive force and shear resistance on the treated soilcrete block around the pile group, while an additional 35% must be a result of increased soil-pile interaction.
5. Jet grouting provides an opportunity to significantly increase the lateral resistance of existing pile group foundations at a cost that is economically viable with alternatives such as deep foundations.

ACKNOWLEGEMENTS

Funding for this study was provided by the US Transportation Research Board though NCHRP Project 24-30. Andrew Lemer was the project manager. This support is gratefully acknowledged; however, the conclusions represent those of the authors and do not necessarily represent those of the sponsors.

REFERENCES

Burke, G. (2004). "Jet Grouting Systems: Advantages and Disadvantages". *ASCE Geotechnical Special Publication No. 12*, pp. 875-886.

Adsero, M.E. (2008). Effect of Jet Grouting on the Lateral Resistance of Soil Surrounding Driven-Pile Foundations. *MS Thesis*, Civ. & Env. Engrg. Dept., Brigham Young Univ., Provo, UT, 245 p.

Cold-Weather Concreting Technology for Ground Modification

John L. Daniels[1], M.ASCE, P.E., Rajaram Janardhanam[2], James R. Starnes[3], Nicholas DeBlasis[4] and Koyett Miles[5]

[1]Associate Professor, Department of Civil and Environmental Engineering and Global Institute for Energy and Environmental Systems, The University of North Carolina at Charlotte, 9201 University City Boulevard, Charlotte, NC 28223, Tel. 704-687-3248, Fax. 704-687-6953, Email: jodaniel@uncc.edu [currently: Program Director and AAAS Science and Technology Policy Fellow, Directorate for Engineering, Division of Engineering Education and Centers, U.S. National Science Foundation, Arlington, VA 22230, 703-292-5360, jdaniels@nsf.gov]
[2]Professor, Department of Civil and Environmental Engineering, The University of North Carolina at Charlotte, 9201 University City Boulevard, Charlotte, NC 28223, Tel. 704-687-3248, Fax. 704-687-6953, Email: rajaram@uncc.edu
[3]Graduate Student, Department of Civil and Environmental Engineering, The University of North Carolina at Charlotte, 9201 University City Boulevard, Charlotte, NC 28223, Tel. 704-687-3248, Fax. 704-687-6953, Email: njdeblas@uncc.edu
[4]Undergraduate Student, Department of Civil and Environmental Engineering, The University of North Carolina at Charlotte, 9201 University City Boulevard, Charlotte, NC 28223, Tel. 704-687-3248, Fax. 704-687-6953, Email: jrstarne@uncc.edu
[5]Undergraduate Student, Department of Physics, Benedict College, 1600 Harden Street, Columbia, SC 29204, Tel. (803) 253-5000, Fax. (803) 253-5167, Email: MilesK97@benedict-tigers.org

ABSTRACT: This paper describes the application of cold-weather concreting technology to soil stabilization with cement. For concrete, the typical approach involves the use of chemical additives to depress the freezing point, accelerate the curing process and initiate other changes that allow cementitious reactions to proceed at low temperatures. Likewise, cementitious reactions are also critical to the performance of cement soil stabilization. To that end, a set of experiments was conducted to evaluate the influence of three commercially-available chemical additives on the unconfined compressive strength of a local soil stabilized with 6% cement by weight. In particular, the chemical additives were tested at three dosage levels, with strength measurements taken at 1, 7 and 28 day curing periods, and two different curing temperatures. The control (unmodified) mixture resulted in an average 7-day strength of 295 kPa (42.8 psi) when cured at -2°C (28.4 °F). For the same mixture at the same curing temperature, the average 7-day strength increased to 1318 kPa (191 psi), 1050 kPa (152 psi) and 1332 kPa (193 psi) for the Gilco, Polarset and Daraccel brand chemical additives, respectively. As tested, the incremental cost associated with using these additives varied from 8-41%, making them a viable alternative to extensive undercutting and/or geosynthetics.

INTRODUCTION

When the subgrade soils are deemed unsuitable in their present condition for a given design purpose, the two general options are to undercut/remove or stabilize/strengthen. It is often cost effective to leave the soils in place and focus on stabilization prior to construction of the overlying structure. The term "soil stabilization" may refer to lowering the plasticity/swell potential of clays, reducing compressibility or providing dust suppression. There are additives which accomplish all of these tasks and increase strength, however, the focus of this paper is primarily on the use of cement to increase strength. Cement has long been used to improve soils, with perhaps the first documented case being in 1935 where a roadbed for a state highway near Johnsonville, South Carolina was improved by blending the local soil with cement (PCA 2005). However, the use of cement to improve soils is generally restricted during cold-weather, as low-temperatures tend to prevent sufficient strength development (Daniels and Janardhanam 2007).

Similar concerns have prompted considerable research in the field of cold-weather concreting. For example, if concrete freezes prior to achieving much strength gain, it can irrecoverably lose more than half its total potential strength (Korhonen 2002). For these and other reasons, the American Concrete Institute requires concrete to be mixed and maintained above 5°C (41°F) unless specific precautionary measures are employed (ACI 1988).

Korhonen (2002) investigated a variety of chemical additives for use in allowing concrete to cure under cold conditions. These additives were found to both lower the freezing point and accelerate the rate of cement hydration. Increasing the rate of cement hydration manifests itself in terms of increased strength as a function of time. Poor early strength is achieved for unmodified specimens cured under cold conditions, however by 28 days, most of the modified mixtures met or exceeded the unconfined compressive strength of the control specimen, a trend which continued through 56 days.

The extent to which commercially available additives allow fresh concrete to cure under cold and subfreezing conditions was investigated under laboratory and field conditions (Korhonen et al. 2004). At the outset of their research project, one goal was to develop a mixture that allows concrete to cure down to an internal temperature of -5 °C. Based on historical weather data, the impact of achieving this goal would be to extend the construction season dramatically. In North Carolina, for example, this would essentially mean year-round construction. In essence, a minimum of 3-4 months more construction would be allowed throughout the U.S. if temperatures may be allowed to drop to -5°C instead of typical +5°C limitations. More specifically, the project objective was "to develop combinations of commercially available, currently approved, and commonly used admixtures that could be used to produce cold weather concrete that behaves like regular concrete at the time of placement and appearing 'seamless' to the laborer at the end of the concrete chute."

With the above constraints, there were a variety of chemical additives which were deemed appropriate. Details on the metrics by which these requirements are evaluated (e.g., what constitutes adverse long-term performance) may be found in the source document (Korhonen et al. 2004). Their data clearly indicates the ability of the additives to allow cementitious reactions to occur at temperatures of +5°C and -4°C with no subsequent strength loss when tested (or further cured) at higher temperatures. In general, the additives performed considerably better than the control. In the long term (56 days), the additive-modified concrete yielded strengths that were 30-80% better than the control.

As a follow up to the previously noted study, Korhonen and Semen (2005) evaluated the use of a chemical additive in constructing an unreinforced section of a parking apron at the Grand Forks Air Force Base in North Dakota – during subfreezing conditions. After 24 hours, the compressive strength was found to be 15.9 MPa (2300 psi). The apron was deemed ready for traffic after two days where the temperature ranged from 0 to -5°C.

As the foregoing would suggest, research focused on the placement of concrete under cold and sub-freezing conditions has matured considerably. It would appear that the approach of using chemical additives to depress the freezing point, accelerate the curing process and initiate other changes that allow cementitious reactions to proceed at low temperatures should also work for soil-cement stabilization. To that end, the objective of this paper is to present selected results of an experimental evaluation of the unconfined compressive strength of soil-cement mixtures as a function of curing time, temperature, chemical additive and dosage level.

MATERIALS AND METHODS

Three chemical additives were selected and procured from the W.R. Grace Company, namely Polarset, Darracel and Gilco Accelerator. All three of these chemicals are in aqueous form. The key ingredients include a mixture of calcium nitrate and calcium nitrite; calcium chloride and triethanolamine; and calcium nitrate for the Polarset, Daraccel and Gilco products, respectively. Note that this research is not bound by the same corrosivity constraints as reported above for concrete additives (where the concern is for exposure to reinforcing elements). In the lab, these chemical additives were mixed with the molding moisture to obtain the desired level of compaction. Table 1 shows the dosage rate used for each chemical, a value based on the manufacturer's recommended use rate for concrete applications. For field application, it is envisioned that the additives may be blended into the soil with rotary mixers (e.g. a RS-425) and water trucks in conjunction with typical field moisture conditioning. To evaluate their effectiveness in soil stabilization, these additives were tested with a cement stabilized soil, namely a mixture of 6% Type I Portland cement and a local soil. The amount to Portland cement used was based on NCDOT practice and experience which shows this to be a sufficient, if not necessarily optimized, amount for commonly encountered soils. The local soil was classified according to AASHTO as A-2-4 and USCS as SP, as shown in Table 2. As a preliminary metric of performance, unconfined compression (UC) strength testing (ASTM D2166) was conducted on mixtures, in replicate, with and without the chemical additives. All unconfined compression samples were made from Harvard Miniature (HM) Device compaction molds. The HM device has a spring loaded tamper, which was applied 10 times for each of three layers. Details on the HM device may be found in Bowles (1992). The resulting specimen dimension was 3.3 cm in diameter and 7.2 cm in length. The optimum moisture content (ω_{opt}) and maximum dry unit weight (γ_{dmax}) for the soil-cement mixture was found to be 13% and 18.4 kN/m^3, respectively. Compaction curves were not developed for samples prepared with the chemical additive, so it is possible that these values change with chemical addition. Generally, materials exhibit higher UC strength when prepared near optimum conditions, so the most likely effect, if any, is that the UC strength (at optimum) of chemically modified samples may be even higher than reported herein. The influence of chemical addition on ω_{opt} and γ_{dmax} will be explored in future work. This work was exploratory and simply focused on establishing general feasibility.

Specimens were tested in replicate (twice) after curing for 1, 7 and 28 days, without soaking. Controlled temperature chambers (Cincinnati Sub-Zero, Model Z16) were used where the curing temperature was set at +2°C or -2°C, as appropriate. Other temperature conditions are reported in Daniels and Janardhanam (2007). In all cases, samples were placed within sealed ziplock bags, which were labeled according to the temperature and extent of chemical modification. Even when sealed, however, moisture escapes the specimen and condenses on the interior of the bag. Naturally, it is the moisture in the specimen that matters, not the total amount retained in the bag and so in this regard desiccation did occur and may have influenced the results, albeit uniformly.

For specimens cured under freezing conditions, a delay of 12 hours was imposed between removing the samples from the chamber and strength testing. This was done to allow frozen pore fluid to melt, so that strength measurements reflect the extent of cementitious reactions, not frozen soil.

Table 1. Dosage rate used for each chemical additive

Dose Range	Gilco	Polarset	Daraccel
Level 1	0.00697 mL / g cement	0.0044 mL / g cement	0.0065 mL / g cement
Level 2	0.01498 mL / g cement	0.01303 mL / g cement	0.01628 mL / g cement
Level 3	0.01954 mL / g cement	0.01954 mL / g cement	0.02605 mL / g cement

Table 2. Classification details for local soil

Property	Value
Liquid Limit (%)	23.8
Plastic Limit (%)	17.0
D_{60} (mm)	2.95
D_{30} (mm)	0.65
D_{10} (mm)	0.24
AASHTO Classification	A-2-4, Silty or Clayey Gravel and Sand
USCS Classification	SP, Poorly graded sands

RESULTS AND DISCUSSION

The results (average of replicate tests) of the unconfined compression strength tests are given in Figures 1-6. Figures 1-2, 3-4 and 5-6 provide the results for the Gilco, Polarset and Daraccel, respectively, at curing temperatures of 2 °C and -2 °C.

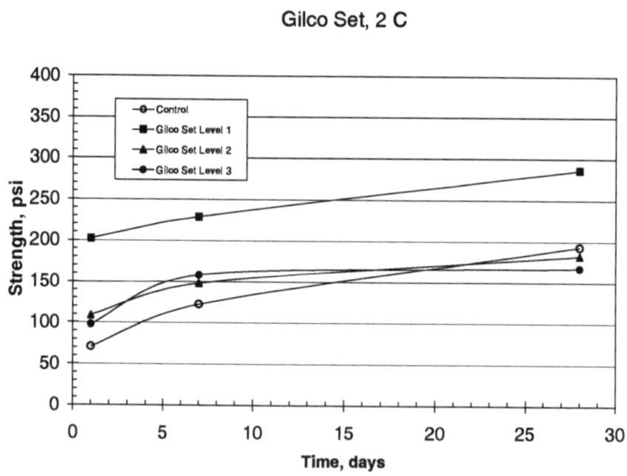

FIG. 1. Unconfined compressive strength for soil-cement with and without Gilco additive, +2°C cure

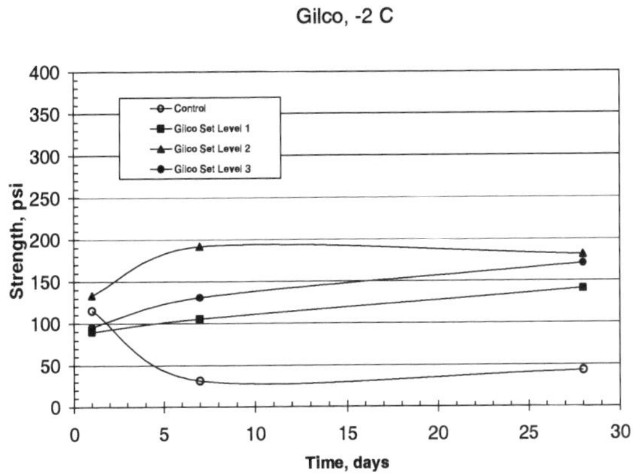

FIG. 2. Unconfined compressive strength for soil-cement with and without Gilco additive, -2°C cure

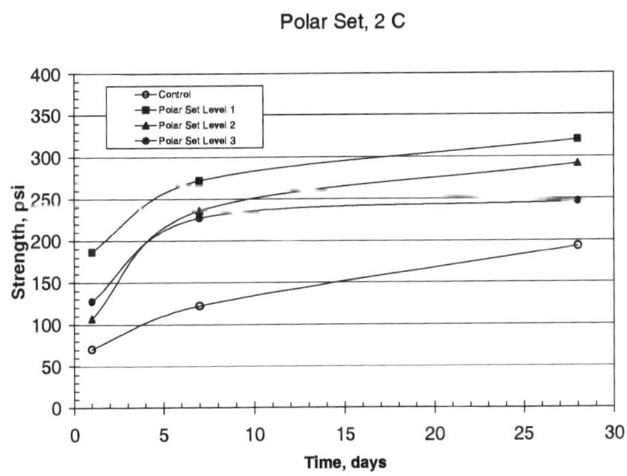

FIG. 3. Unconfined compressive strength for soil-cement with and without Polarset additive, +2°C cure

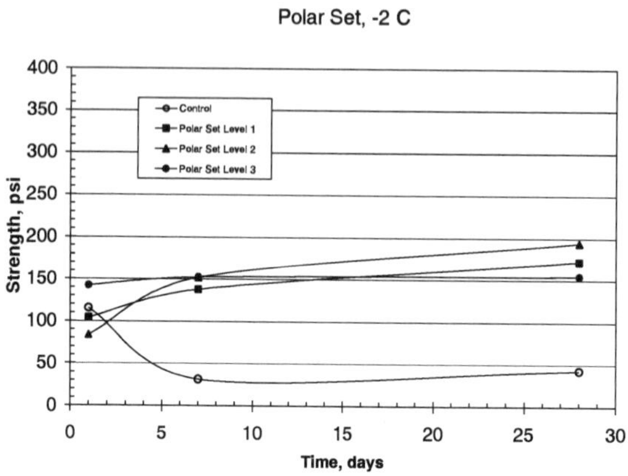

FIG. 4. Unconfined compressive strength for soil-cement with and without Polarset additive, -2°C cure

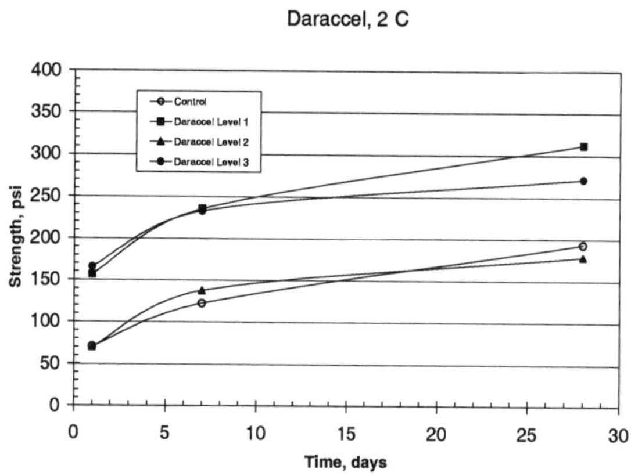

FIG. 5. Unconfined compressive strength for soil-cement with and without Daraccel additive, +2°C cure

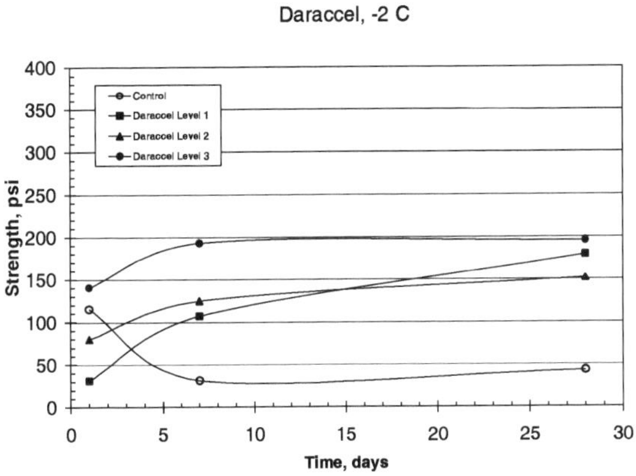

FIG. 6. Unconfined compressive strength for soil-cement with and without Daraccel additive, -2°C cure

Unmodified samples did not increase with time when cured at -2 °C, a result attributed to the freezing process interfering with reaction progress. The results for specimens cured at +2 °C are more intuitive, as all strengths increase with curing time. In any case, the general influence of additives at any temperature is to increase strength markedly, although the results are a function of the dosage applied. Interestingly, in the case of Daraccel, the strength improved dramatically at the lowest dosage level for the 2°C cure, while increasing the dosage level resulted in performance not much different than the control. Similarly, other additive and temperature combinations revealed that more is not always better in terms of strength development. Such sensitivity would be a concern in the field, particularly with the very low levels of amendment.

While achieving strength at low temperature is the stated goal, it is critical that the associated costs be reasonable. Benchmarking of costs may be made between two boundary conditions. The lower benchmark is the cost of cement stabilization, i.e., it would be ideal for a proposed additive to have enhanced performance without any additional cost. The upper benchmark is based on the cost associated with current limitations on cold weather construction. In other words, even if additives cost significantly more than the use of cement alone, they may well be cost effective when compared to the cost of delayed construction. Based on typical project data, an appropriate acceptable incremental cost for a proposed additive is taken as 50%. In other words, the proposed solution should not exceed conventional cement stabilization by 50%. This increment derives in part from consideration of the next alternative to cement stabilization, which is often to undercut the material and replace with stone and fabric. Review of these costs for two different states in the southeast indicates the cost of the undercut/replace option to be higher by greater than a factor of two. Keeping costs within 50%, therefore, is likely to be cost-effective. By way of an example, consider the cost of conventional cement stabilization for a soil with a dry density of 15.7 kN/m^3 (1600 kg/m^3). If the cost for Type I Portland cement is taken as \$0.10/kg, and the soil is modified with 6% (by dry weight of soil), then the cost per unit volume (material costs only) is given as:

$$1600\frac{kg-soil}{m^3}\cdot 0.06\frac{kg-cement}{kg-soil}\cdot\frac{\$0.10}{kg}=\$9.60/m^3$$

When considering costs, an advantage of using the proposed additives for cement is that no special handling procedures are required and engineers, contractors and owners are already familiar with cement stabilization. The additives are simply blended with the molding moisture content, generally as a function of the amount of cement added. Polarset (the most expensive additive tested) costs ~\$1.85/liter and the recommended dosage is 240 – 1800 ml per 45 kg of cement. The highest concentration of Polarset tested herein is 887 mL per 45 kg of cement. At this level, the incremental cost of treating a unit volume of soil with 6% cement (as previously calculated) may be given by:

$$1600\frac{kg-soil}{m^3}\cdot 0.06\frac{kg-cement}{kg-soil}\cdot\frac{887mL-Polarset}{45kg-cement}\cdot\frac{1l}{1000mL}\cdot\frac{\$1.85}{l}=\$3.50/m^3$$

As such, using Polarset, at the level indicated, to amend the cement stabilized soil costs \$9.60 + \$3.50 = \$13.10/m^3. This represents a 36% increase over the cost of cement stabilization alone. Similarly, using the least expensive additives (Daraccel and Gilco, \$1.05/liter) at the lowest concentration tested herein (Daraccel at 295 mL / 45 kg of cement) yields total costs and percent incremental increases of \$10.26/m^3 and 6%, respectively. So, the range of additives, as tested, increase the cost of conventional stabilization from approximately 6 to 36%.

CONCLUSIONS

The authors have evaluated the literature on the use of cold weather concrete additives and suggest that this technology is applicable to soil stabilization. Concrete and cement stabilization are virtually identical in terms of the key mechanism of strength gain. Results of the work presented herein indicate that chemical additives may be used to achieve higher strengths under low temperatures, as compared to unmodified samples. This work establishes that the approach is indeed viable, however work remains to translate this information into a usable framework.

ACKNOWLEDGEMENTS

This work was sponsored, although not necessarily endorsed, by the North Carolina DOT.

REFERENCES

ACI (1988) Cold weather concreting, *Report ACI 306R-88*, Detroit, MI, ACI.

Bowles, J.E. (1992) Engineering Properties of Soils and their Measurement, McGraw-Hill, NY

Daniels, J.L. and Janardhanam, R. (2007). "Cold-weather subgrade stabilization" *Soil Improvement* (GSP 172) ASCE, Reston/VA: 1-10.

Korhonen, C.J. (2002) "Off the shelf anti-freeze admixtures" *Technical Report TR-02-7,* U.S. Army CRREL, Hanover, NH.

Korhonen, C.J. (2004) "Extending the season for concrete construction and repair. Phase I: Establishing the technology" *Technical Report TR-04-2*, U.S. Army CRREL.

Korhonen, C.J. and Semen, P. (2005) "Placing antifreeze concrete at the Grand Forks Air Force Base" *Technical Report TR-05-9.* U.S. Army CRREL, Hanover, NH

PCA (2005) "Soil-cement technology for pavements: Different products for different applications" Portland Cement Association, Skokie, IL, 4 p.

Geotechnical Properties of Solidified Sludge by Mixing Cement and Calcium-Bentonite

Cheng Lin[1], Wei Zhu[2], and Jie Han[1]

[1]Civil, Environmental, and Architectural Engineering (CEAE) Department, the University of Kansas, 2150 Learned Hall, 1530 W, 15th Street, Lawrence, KS 66045, USA
[2]College of Environmental Science and Engineering, Hohai University, Nanjing 210098, China

ABSTRACT: China generates up to 4 million ton dried sewage sludge every year. How to reasonably handle this huge amount of hazardous waste has become a concern. However, sewage sludge is hard to handle due to its poor geotechnical properties, such as high water content, low compressibility, and low resistance to environmental change. The use of solidification/stabilization (S/S) into sewage sludge can improve not only its geotechnical properties but also its environmental properties (e.g. stabilization of the pollutants in the sludge). Therefore, S/S can make the waste re-usable as a construction material or pre-treat it prior to landfill disposal. In this study, ordinary Portland cement (OPC) and calcium-bentonite were used as the S/S for local sewage sludge from the Nanjing Wastewater Treatment Plant in China. Geotechnical properties of the solidified sludge were studied in terms of unconfined compressive strength (UCS) and durability after the wet and dry cycles. The test results show that in a short curing time, cement alone-S/S sludge had a low compressive strength but a significant improvement was observed when bentonite was added to replace some amount of cement. In a long curing period, cement played a major role in increasing the compressive strength as compared to bentonite. In addition, initial water content of sewage sludge influenced the UCS, which should be taken into account before S/S. The test results of the dry and wet cycles indicate that bentonite was beneficial to improve the durability of solidified sludge when the ratio of bentonite was less than that of cement. Considering both UCS (>100kPa) and dry/wet cycle (<30% relative mass loss) requirements, the ratios of sewage sludge, bentonite, and cement should be approximately 5:1:2.

INTRODUCTION

Sewage sludge is a by-product generated during treatment of waste water, usually composing of a high water content and a large amount of hazardous substances. China produces the dried sewage sludge up to 4 million tons per year and continues increasing the production amount (Dai et al. 2006). As a result of urbanization, industrial development, and increasing awareness of environmental protection, there is a growing concern about what to do with the ever increasing amount of sewage sludge in China. However, sewage sludge is usually hard to handle since it often has poor mechanical properties, such as poor self-support and low resistance to external environmental stresses (e.g., stresses induced during wet and dry cycles). In addition, sewage sludge has different complicated hazardous components, which easily lead to second pollution to the ambient environments. Landfilling of sewage sludge is a common method; however, poor geotechnical properties of sewage sludge often cause potential threat to landfill stability. For example, low strength of sewage sludge can result in the collapse of landfill. The change of ambient environment can also reduce the strength and break the integrity of landfill body, which eventually reduces the stability of landfill. As a consequence, it is necessary to improve geotechnical properties of sewage sludge before landfilling it. Solidification/stabilization (S/S) is considered as one of the cost-effective and easily operated methods for addressing problems of sewage sludge. This method can be used not only to improve the geotechnical properties of sewage sludge but also to control the environmental pollution from sewage sludge. Furthermore, the solidified products can also be utilized as construction materials (Gurjar 2001; Lim et al. 2002; Zhu 2007).

S/S is a process employing binders and additives to reduce the hazardous nature of waste by converting the waste and its constituents into a relatively stable form (Lo 1996). "Solidification" refers to the conversion of sludge from a semi-solid state into a solid state (i.e., forming a monolith solid). "Stabilization" means the product is insoluble, which in turn reduces the solubility, mobility, and toxicity of the waste (LaGrega et al. 1994). Portland cement is the most commonly used reagent in the S/S of hazardous waste since it is economical and has more performance data for references (Adaska 1991). However, when cement alone is used to solidify sewage sludge, the results are always not satisfactory unless a great amount of cement is added. Therefore, additives are used to replace some amount of cement during the process of S/S of sewage sludge in order to enhance and optimize cement S/S mixtures. Bentonite has been considered as a favorable additive since its high adsorption ability of adverse hazardous substances can enhance cement S/S effect. Montgomery (1988 and 1991) and Lo (2000) showed that organic bentonite as an additive contributed to cement hydration by reducing the interference of hazardous materials. Jun (1997) indicated that sodium bentonite could also be a good choice to improve cement hydration during S/S of hazardous wastes.

In this study, a relatively inexpensive bentonite (i.e., calcium bentonite) was used as an additive to replace some amount of cement during the S/S process of sewage sludge. The geotechnical properties of the solidified sewage sludge, in terms of its unconfined compressive strength and durability after dry and wet cycles, were studied to evaluate the feasibility of this S/S technology.

MATERIALS AND METHODS

Materials

The materials used in this study were ordinary Portland cement (OPC), calcium-bentonite, and dewatered sewage sludge. OPC was Grade 32.5 according to the Chinese standard (GB 175 1999) and was obtained from a local cement plant. Calcium-bentonite was obtained from the Nanjing Tangshan Bentonite Plant, which is a fine grinded powder with grain size < 75μm and has a strong adsorption ability of pollutants and water with cation exchange capacity (CEC) of 79.2 mmol/100g. Calcium-bentonite was chosen because of its low cost and large distribution in China in comparison with sodium-bentonite and organic-bentonite. The cost of calcium-bentonite is approximately half of OPC used in the experiment. The sewage sludge, sampled from the Jiang-Xin Zhou Wastewater Treatment Plant in Nanjing City, was dewatered using the centrifuge method. The basic characteristics of the sludge are shown in Table 1.

Table 1. Basic Characteristics of Raw Sludge

pH[1]	*Water content (%)*	*Density (g/cm)*	*Organic matter (%)*	*Gravity Ratio*	*Plastic Index (%)*
7.07	357	1.05	42.8	1.61	256.1

[1] measured at a sample: distill water ratio of 1:10.

Unconfined Compressive Strength

The solidification/stabilization experiment was carried out by mixing various amounts of calcium-bentonite with sewage sludge and then adding Portland cement to solidify the mixtures. After mixing the materials uniformly, the solidified samples were put into molds (with a diameter of 3.9 cm and a height of 8cm) at a temperature of 20° C and relative humidity of 100%. These samples were unmolded after 24 hours and stored in a curing room. After the curing time, the specimens were tested for UCS.

To assess the effectiveness of bentonite as an additive, bentonite was used to substitute for half amount of OPC in each of three groups, which have initial water content of the raw sludge at 314%. These samples were cured for 7 days. The mix design of these samples is shown in Table 2.

Table 2. Mix Design of Sludge, Bentonite, and Cement (First Experiment)

Mix	*A01*	*A02*	*B01*	*B02*	*C01*	*C02*
Sludge, bentonite and cement	5:0:1	5:0.5:0.5	5:0:2	5:1:1	5:0:4	5:2:2

Another mix design is shown in Table 3, in which the bentonite ratio was increased from 0 to 3 when the cement ratio was kept 1, or the cement ratio was raised from 0.5 to 3 when the bentointe ratio was set to 1. The influences of the initial water content of the raw sludge and the curing time on the UCSs were studied by using two water contents at 314 and 357% and curing times at 7, 28, 60, and 120 days.

Table 3. Mix Design of Sludge, Bentonite, and Cement (Second Experiment)

Mix	*Sewage sludge*	*Bentonite*	*Cement*
B0C1	5	0	1
B0.5C1	5	0.5	1
B1C1	5	1	1
B2C1	5	2	1
B3C1	5	3	1
B1C0.5	5	1	0.5
B1C2	5	1	2
B1C3	5	1	3

Durability under Dry and Wet Cycles

Durability tests were performed to evaluate resistance of a solidified/stabilized product to degradation due to external environmental stresses. Since the temperatures in the south areas of China are mostly above the freezing point during the whole year, the impact of freeze and thaw was not included in this study. Therefore, dry and wet cycling was conducted to assess the environmental stresses on the solidified sludge. The solidified sludge underwent repeated cycling during the testing. Each specimen was subjected to heating at 60±2°C for 12 cycles based on the ASTM standard (ASTM D: 4843-88 2004). According to the standard, the criterion of 30% weight loss was used as a rejection ground for each durability test.

RESULTS AND DISCUSSION

Unconfined Compressive Strength (UCS)

Effect of bentonite and cement content

The effect of bentonite on the strength of the solidified sludge was evaluated by replacing half amount of cement. The test results on this effect are shown in Fig. 1. It is shown that cement alone-solidified specimens had exceptionally low UCSs at 7 days, such as A01, B01, and C01. However, when half cement was substituted by bentonite in three groups, the compressive strength remarkably increased by approximately 2 to 5 times, such as A02, B02, and C02. The test results show that adding bentonite can effectively increase the compressive strength and reduce the consumption of cement, thus reducing the cost of treatment for S/S of the sewage sludge.

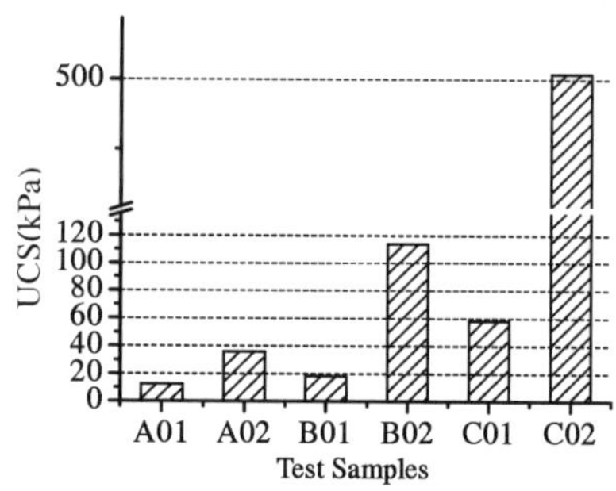

Fig. 1 Effect of bentonite on UCS by substituting half cement

Influence of initial water content of sewage sludge

The UCSs of solidified specimens at two initial water contents were presented in Fig. 2. At the same ratio, the specimens at a higher initial water content (i.e., 357%) gained much less compressive strengths than those at the initial water content of 314%. In addition, the influence of the initial water content on the compressive strength can be minimized by increasing the cement or bentonite content in the solidified sludge. This effect can be examined by comparing the compressive strengths of specimens with two initial water contents at various mixing ratios in Fig. 2.

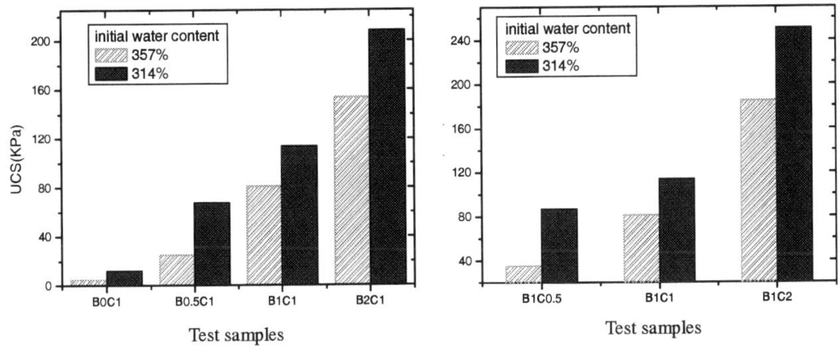

(a) (b)

Fig. 2 Influence of initial water content on UCS

Influence of curing time

Figure 3 shows the results of the compressive strength with the curing time at various mixing ratios of sludge, cement, and bentonite. It is shown that the compressive strength of the solidified sludge increased with the curing time. The increase of bentonite had little effect on the increase of the compressive strength versus time as shown in Fig. 3(a). However, cement significantly increased the compressive strength as shown in Fig. 3(b). To better understand the variation of the compressive strength with time, the percent of increase in the compressive strength from 28 to 120 days was calculated and is shown in Table 4. Table 4 shows that after 28 days, the solidified sludge still had not fully develop its strength. The maximum increase for Mix B1C3 reached 200% from 28 to 120 days. A reasonable benefit of bentonite in the strength development in a longer time was also observed, for example, B0.5C1. These results show that there was still a noticeable increase in UCS of the S/S sludge with time and cement had more influence on the development of the strength.

Table 4. Percentage of increase in UCS between 28 and 120 days

Mix	*B0C1*	*B0.5C1*	*B1C1*	*B2C1*	*B3C1*	*B1C0.5*	*B1C2*	*B1C3*
% increase in UCS	107.8	135.3	62.2	34.6	12.4	37.4	29.1	226.2

Durability

Durability of the specimens was evaluated using dry and wet cycle tests. The mass loss was accumulated after each dry and wet cycle. The cumulative mass loss in Table 5 was calculated based on 12 dry and wet cycles.

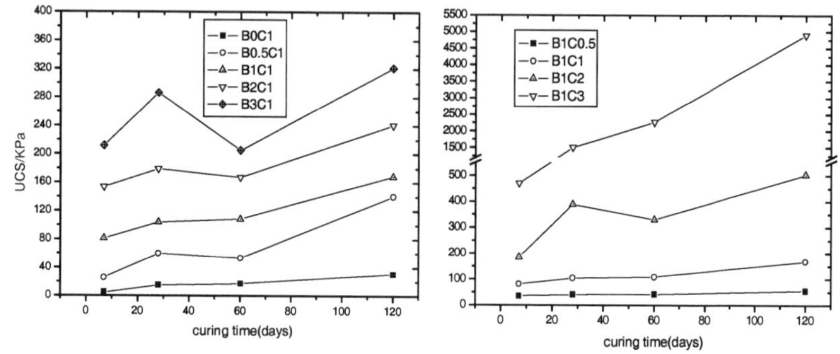

(a) (b)

Fig. 3 Effect of curing time on UCS

Quality of solidified sludge

Table 5 shows that among 8 specimens, B0.5C1, B1C1, B1C2, and B1C3 passed the requirement based on the ASTM standard (ASTM D: 4843-88 2004). Other specimens failed due to high mass losses or breakage during the cycling. The cement-alone treated specimen (B0C1) exceeded its mass loss over 30% without breakage. When bentonite was added, the mass loss decreased below 30%, for example, B0.5C1 and B1C1. However, a further increase of bentonite led to the breakage of the samples, for example, B2C1 and B3C1. The above results indicate that bentonite might have a positive or negative effect on the stability of the solidified sludge under dry and wet cycling. The ratio of bentonite to cement in this study was less than 1:1. At a certain amount of bentonite, an increase of cement significantly improved the durability of the solidified sludge as demonstrated for specimens, B1C0.5, B1C2, and B1C3. This result implies that cement could contribute to the integrity of the solidified sludge even in a harsh environmental condition.

Table 5. Average cumulative mass loss

Mix	*B0C1*	*B0.5C1*	*B1C1*	*B2C1*	*B3C1*	*B1C0.5*	*B1C2*	*B1C3*
% mass loss	32.9	23.8	19.3	Break	Break	Break	3.0	2.7

Volume of solidified sludge

Dry and wet cycles could change the volume of the solidified sludge. This change would be detrimental to the properties of the solidified sludge. In Fig. 4, the specimen with a letter "T" was a tested sample after dry and wet cycles while that with a letter "C" was a control sample without drying process. Figure 4(a) shows a significant decrease in the volume after dry and wet cycles. However, when cement and bentonite were increased to a certain ratio, the volume change was minimized as shown in Fig. 4(b).

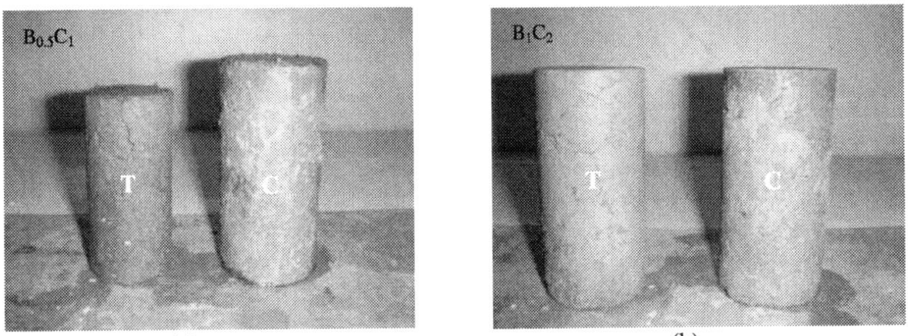

(a) (b)

Fig. 4 Volume Change of specimens after wet and dry cycles

To evaluate the volume change of samples, percent of volume loss is defined as the volume reduction after dry and wet cycles divided by the original volume in percentage. The percents of volume loss for B0C1, B0.5C1, B1C1, B1C2, and B1C3 were 40, 37, 34, 3, and 1%, respectively. These results show that both cement and bentonite were beneficial to the reduction of the volume change after dry and wet cycles. However, cement played a more important role in maintaining the volume of the specimen than bentonite. When the cement content reached a certain level, such as B1C2, the volume change was prevented.

It was determined in this study that the optimum mix design of sludge, bentonite, and cement was approximately 5:1:2 by weight to minimize the mass loss and the volume change. In addition, the sample with 5:1:2 can meet the requirement for the UCS of 100kPa as shown in Fig. 3.

CONCLUSIONS

Geotechnical properties of solidified sewage sludge by mixing cement and calcium-bentonite were investigated in this study in terms of its UCS and durability after dry and wet cycles. The following conclusions can be made:

(1) Bentonite increased the UCS of the solidified sludge in a short time and reduced the amount of cement, thus reducing the cost of treatment.

(2) An increase of the initial water content of the sewage sludge reduced the compressive strength of the solidified sludge. This negative effect could be compensated by adding

more cement and bentonite.

(3) The UCS of the solidified sludge varied greatly with the cuing time and continued increase after 28 days, especially at a higher cement content.

(4) Cement could significantly improve the durability of the solidified sludge. Bentonite could also improve the durability of solidified sludge at a lower dosage than cement; however, it is detrimental to the durability at a higher dosage than cement.

(5) The test results of UCS and durability show that the optimum mix of sludge, bentonite, and cement was 5:1:2 by weight.

REFERENCES

ASTM D 4843-88(2004). *Standard Test Method for Determining the Wetting and Drying Test of Solid Wastes.*

Adaska, W. S., Tresouthick, S. W., and West, P. B. (1991). "Solidification and stabilization of wastes using Portland cement." *Portland Cement Association*, Skokie, Illinois.

Chinese Standard GB 175 (1999). *Silicate Cement and Ordinary Portland Cement.*

Dai, J.-Y., Chen, L., Zhao, J.-F., and Ma, N. (2006). "Characteristics of sewage sludge and distribution of heavy metal in plants with amendment of sewage sludge." *Journal of Environmental Sciences*, 18(6), 1094-1100.

Gurjar, B. R. (2001). *Sludge Treatment and Disposal*, A.A. Balkema, the Netherlands.

Jun, K. S., Shin, H. S., and Paik, B. C. (1997). "Microstructural analysis of OPC/silica fume/Na-bentonite interactions in cement based Solidification of organic-contaminated hazardous waste." *Journal of Environmental Science and Health*, Part A, A32(4), 913-928.

LaGrega, M. D., Buckingham, P. L., and Evans, J. C. (1994). *Hazardous Wastes Management*, McGraw-Hill, NJ.

Lim, S., Jeon, W., Lee, J., Lee, K., and Kim, N. (2002). "Engineering properties of water/wastewater-treatment sludge modified by hydrated lime, fly ash and loess." *Water Research*, 36(17), 4177-4184.

Lo, I. M. C. (1996). "Solidification/stabilization of phenolic waste using organic-clay complex." *Journal of Environmental Engineering*, 122(9), 850-855.

Lo, I. M. C., Tang, C.I., Li, X.D., and Poon, C.S. (2000). "Leaching and Microstructural Analysis of Cement-Based Solidified Wastes." *Environmental Science and Technology*, 34(23), 5038 - 5042.

Montgomery, D. M., Sollars, C. J., and Sheriff, T. S. (1988). "Organophilic clays for the successful stabilisation/solidification of problematic industrial wastes." *Environmental Technology Letters*, 9, 1403-1412.

Montgomery, D. M., Sollars, C. J., and Perry, R. (1991). "Optimization of cement-based stabilization/solidification of organic-containing industrial wastes using organophilic clays." *Waste Management and Research*, 9(1), 21-34.

Zhu, W., Lin, C., Li, L., and Ohki, T. (2007). "Solidification/Stabilization (S/S) of Sludge Using Calcium-Bentonite as Additive." *Environmental Science*, 28(5), 1020-1025. (in Chinese)

Compressive Strength of Soil Improved with Cement.

Saroglou I. Haralambos, Ph.D

Department of Geotechnical Engineering, School of Civil Engineering, National Technical University of Athens, 7-9 Iroon Polytechniou, 15780, Zografou, Athens, Greece. saroglou@central.ntua.gr

ABSTRACT

The soil type can affect significantly the effect of cement stabilisation. In order to investigate this, five types of soils were mixed with cement at various cement proportions in order to evaluate the large variation of compressive strength values. The compressive strength of the cement treated soils was determined for a curing period of the cement equal to 7 and 28 days. It was obvious that the soil type is a controlling factor on the rate of increase of compressive strength with increasing cement content. Other factors affecting strength are the curing time of cement and the optimum water content.

1. INTRODUCTION

The use of cement as an improving agent in various applications in ground engineering is quite common. The improvement is achieved through a mixing procedure with the soil (and other additives). Soil - cement mixing has been increasingly used in the past in a number of applications. Primarily, it has been used as an improvement material for the stabilisation of sub-bases in road foundations (Marsellos et al., 1986). Additionally, in some developing countries, it has been extensively used for the production of compacted soil-cement blocks, which are used for the construction of masonry buildings (Reddy et al., 2007). Cement (and lime) is also used for the stabilisation and solidification for the remediation of contaminated soils. Soil improvement by the use of cement can also be achieved by means of other techniques, e.g. deep soil mixing, jet grouting and construction of lime or cement-stabilised soil columns.

The parameters which influence significantly the strength of a cement-stabilised soil are the sand and fines content, the mineralogy, the particle packing - density, the liquid limit, the water content, the pH, the amount of added cement and curing time (Anagnostopoulos & Chatziangelou, 2008). In the present study, different soil types were mixed with cement at various cement proportions and compacted using the standard Proctor test at the maximum dry density. The compressive strength of the cement treated soils was determined for a curing period of the cement equal to 7 and 28 days. It was obvious that the soil type is a controlling factor on the rate of increase of compressive strength with increasing cement content and curing time. Similar studies on sandy soils (Consoli et al., 1998), have proved that cement addition to soil increases stiffness, brittleness and peak strength.

2. LABORATORY TESTING

Treated Soil Types

The soil samples used in the present study are classified, according to the Unified Soil Classification System as: a) two non-plastic silty sands (SM); b) two non-plastic, poorly graded silty gravels (GP-GM); c) a plastic clayey gravel (GC); d) a clay and a silt of low plasticity (CL, ML). The grain size distribution of each soil type as well as their plasticity index (PI) and specific gravity of solids (G_s) are given in Table 1.

Table 1. Grain size distribution for tested soils

Soil Type	Coarse Gravel (%)	Fine Gravel (%)	Sand (%)	Fines (%)	Plasticity Index (PI)	Specific gravity (kN/m^3)
CL	6.5	3.6	18.4	71.5	22	2.75
ML	0.0	0.7	69.7	29.6	N.P.	2.73
SM (1)	22.0	18.8	41	18.2	N.P.	2.66
SM (2)	0.3	2.9	81.4	15.7	N.P.	2.71
GP-GM (1)	39.5	26.0	28.4	6.1	N.P.	2.74
GP-GM (2)	19.0	39.6	42.1	8.3	N.P.	2.73
GC	21.0	30.3	34.1	14.6	15	2.72

Testing Procedure

Cemented specimens were prepared with cement contents of 3%, 5% and 7% by weight of dry soil and cured for 7 and 28 eight days. The samples were prepared by mixing soil and water, with the addition of Portland cement (CEMII/ B-M (P-L) 32.5 N).

Samples with cement content of 5% were initially compacted in three equal layers. The soil-cement preparation procedures conform to ASTM D558-04, Method B. According to this method the soil that passes a ¾ inch (19.0 mm) sieve, corresponding to fine gravel or smaller, is mixed with cement. Most soil-cement construction specifications cover soil gradation limit maximum size to 3 in. (coarse gravel) or less. The values of maximum dry density and optimum moisture content for the soil-cement mixtures are listed in Table 2.

At the second stage, samples with different cement contents were compacted into a 101.6 mm diameter and 116.4 mm high cylindrical mold and prepared at the optimum moisture content (ASTM D1632-00), as determined from the compaction tests. During the preparation of the stabilized soil specimens, cement content between 3% and 7% by weight of dry soil was used.

The specimens were then extruded, trimmed and subjected to unconfined compression tests. The compressive strength of the soil-cement mixtures was determined according to ASTM D1633-00 Method A. This test method is used only on soils with 30% or less retained on the ¾ inch (19.0 mm) sieve. The tests were performed under a constant stress rate.

3. RESULTS

Compaction Characteristics

The moisture-density (compaction) curves for the cement treated soils were determined using the method described in ASTM D558-04, Method B. The cement content was equal to 5%. Figure 1 shows the variation of the maximum dry density with the moisture content for the tested samples, while their values are summarised in Table 2. The maximum dry density decreases with the optimum water content, as it can be seen by the best-fit line for all tested samples.

Silty gravel (GP-GM) and clayey gravel (GC) have higher dry density values, while clays (CL) and silts (ML) have lower values. The difference of the maximum dry density of the two sands stems from the higher percentage of gravel of the sample SM(1), thus its compaction characteristics are closer to the gravel soils mixtures. Additionally, the sand sample SM(1) has higher density than SM(2).

The effect of the cement content on the compaction curves (moisture – density relationships) was not investigated in the present study. Aiban et al. (1998) concluded that there is no or little difference in the dry density between the untreated soil and cement-soil (marl) mixtures, by conducting tests with cement content between 3% and 7%.

Table 2. Compaction characteristics of soil-cement mixtures

Soil Type	Optimum Moisture Content, OMC (%)	Maximum dry density, ρ_d (Mg/m^3)
CL	17.6	1.71
ML	13.1	1.90
SM (1)	8.4	2.09
SM (2)	14.6	1.76
GP-GM (1)	6.7	2.28
GP-GM (2)	5.2	2.29
GC	7.6	2.12

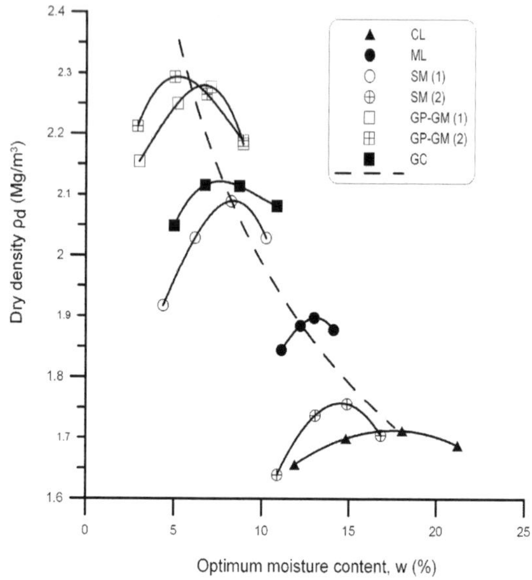

FIG 1. Compaction curves for soil-cement mixes, 5% cement

Strength Improvement

Considering the results from tests on all soil types (as presented in Table 3), it is evident that the compressive strength of the cement treated soil increases with cement content in the soil-cement mixture. The data in Figure 2a clearly indicate that there is an increase, almost described by a power curve, in the compressive strength with the cement content for curing time of 7 days. The increase is more pronounced for samples that have higher percentage of gravel (GP-GM and GC samples). On the contrary, fine soil-cement mixtures (CL or ML) show a slight increase in compressive strength with increasing cement content. Figure 2b shows the compressive strength of the soil-cement mixtures after a curing period of 28 days.

Based on the strength tests executed in the sandy soil–cement mixtures, with increasing cement/soil ratios, it was found that the compressive strength increases more persistently with cement content for soils having a lower percentage of sand and that the slope of the increase in strength is higher as the dry density of the mixture increases or the OMC decreases. The strength increase for the sand sample SM(1) is given by equation $q_u = 1.045\ C^{0.589}$ (r^2=0.99) and $q_u = 1.267\ C^{0.590}$ (r^2=0.99) for curing period of 7 and 28 days respectively.

Based on the strength tests executed in the gravely soil–cement mixtures, it is obvious that the increase in compressive strength of the clayey gravel sample (GC) is much lower than this of the silty gravel sample (GP-GM). Since the density of these samples is similar, this difference is due to the fact that sand and gravel are more efficiently cemented than silt and clay particles.

Table 3. Unconfined compressive strength of soil-cement mixtures

Soil Type	Unconfined Compressive strength, q_u (MPa) Curing time 7 days (28 days)		
Cement content	3%	5%	7%
CL*	0,69 (0,96)	1,05 (1,39)	2,08 (2,67)
ML	1,04 (1,68)	1,57 (2,51)	2,35 (3,54)
SM (1)	2,00 (2,42)	2,69 (3,29)	3,29 (3,99)
SM (2)	0,52 (0,70)	0,78 (1,22)	1,04 (1,46)
GP-GM (1)	0,52 (1,05)	3,91 (4,57)	4,85 (6,68)
GP-GM (2)	0,17 (1,39)	0,78 (1,90)	6,69 (4,51)
GC	1,39 (2,34)	2,61 (2,60)	3,12 (3,06)

* The soil-cement mixture was prepared at 4%, 6% and 8%.

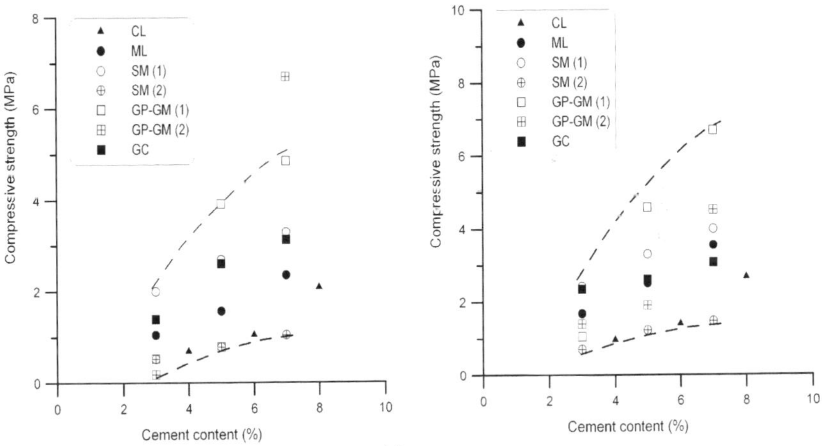

a) b)

FIG 2. Plot of compressive strength with cement content for treated soils, a) curing time 7 days, b) curing time 28 days.

The data in Figure 2a and 2b indicate that there is a power increase in the compressive strength of the treated soils with cement content, both for curing time of 7 and 28 days. Aiban et al. (1998) performed tests on a stabilised marl road base and concluded that beyond a cement content of 5% the increase in unconfined compressive strength is only marginal. This limit was not determined in the present study.

The change in compressive strength of the soil-cement mixtures is dependant upon the grain soil distribution. The strength decreases with the percentage of fines for all cement contents, but this decrease is more evident for cement contents equal to 5% and 7%. This behaviour is independent from the curing time of the cement. Figure 3 presents the decrease of compressive strength in relation to the fines percentage of the tested samples. The compressive strength of the mixtures increases with the percentage of gravel.

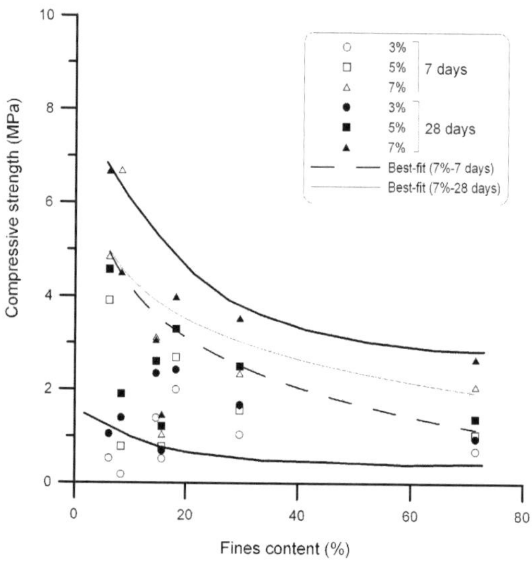

FIG 3. Influence of fines content (F) on compressive strength for different cement contents and curing periods.

The compressive strength of the soil-cement mixtures decreases with optimum water content and this is more evident for those with cement contents equal to 5% and 7%. The decrease is described by a power-fit line and given by equation $q_u = 41.2\ OMC^{-1.173}$ (r^2=0.77) for 7% cement and curing period of 7 days, while for a curing period of 28 days it is given by equation $q_u = 18.02\ OMC^{-0.739}$ (r^2=0.50).

Stress – strain curves

The tangent modulus values of the soil-cement mixtures generally increase with cement content. The modulus of elasticity of sand- and gravel-cement mixtures (samples SM1 and GC) was higher than that of clay- and silt-cement mixtures (samples CL, ML). Gravel-cement mixtures present the highest values. The stress – strain curves for the different curing periods are similar, apart from the difference in peak stress.

Figure 4 shows the stress – strain curves of the tested soil-cement mixtures for cement content equal to 3% and curing time equal to 28 days. It is clear that the strain at peak stress is larger for clay (CL), silt (ML) and silty sand (SM2). The silty gravel-cement mixtures (GP-GM) showed inconsistent stress – strain plots occasionally.

The difference in strength of the sand-cement mixtures is due to the different grain distribution curves of the two sands and the difference in their density. Sample SM2 has lower density than SM1 due to the poor packing of particles while it has lower gravel content.

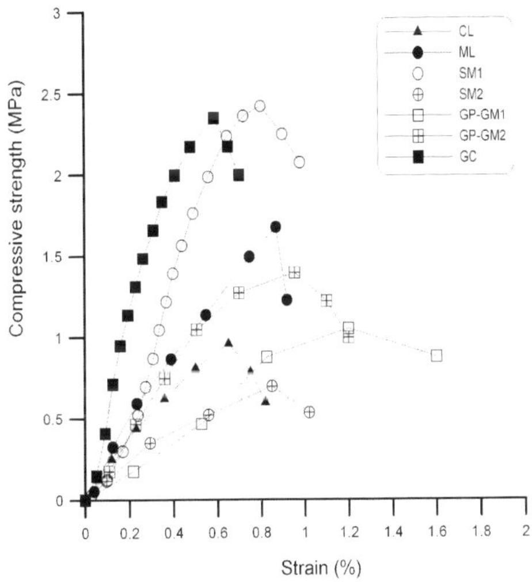

FIG 4. Stress – strain curves for tested soils at cement content 3%.

4. CONCLUSIONS

The test results indicated that the soil type influences significantly the increase of unconfined compressive strength due to cement addition. The addition of cement to all soil types increased the compressive strength and stiffness of the soil-cement mixture.

The increase of compressive strength with cement was more pronounced for samples that have higher percentage of gravel (GP-GM and GC samples) and less for more clayey or silty samples (CL, ML). The efficiency of bonding soil particles with cement is higher in coarser soils since the grain contact area that needs to be cemented is more limited than in finer soils. Moreover, in clayey soils, the water/cement ratio is higher and the cement is interacting with clay particles, thus the effect of cement on strength increase is influenced by a number of parameters.

The compressive strength of soil-cement mixtures almost doubled between a curing period of 7 and 28 days. The change in compressive strength of the soil-cement mixtures is dependant upon the grain size distribution of the soil samples. Moreover, uniformity of the soil sample plays a significant role. The strength decreases with the percentage of fines for all cement contents, but this decrease was more evident for cement contents equal to 5% and 7%. Additionally, the strength of the soil-cement mixtures decreases with optimum water content. Concerning the stiffness of the treated soils, the modulus of elasticity of sand- and gravel-cement mixtures (samples SM1, GC) was higher than that of clay- and silt-cement mixtures (samples CL, ML).

5. ACKNOWLEDGEMENTS

The author would like to thank soil-testing laboratory "General Consulting Ltd." and especially Mr. Kamouzis E. for the valuable assistance in the execution of the tests.

6. REFERENCES

Aiban, S.A., Wahhab, H.A, Al-Amoudi, O.S.B., Ahmed, H.R. (1998). "Performance of a stabilized marl base: a case study."*Construction and Building Materials*, 12: 329-340.

Anagnostopoulos, A. C., Chatziangelou, M. (2008). "Compressive Strength of Cement Stabilized Soils. A New Statistical Model." *Electronic Journal of Geotechnical Engineering*, 13: 1-10.

ASTM D558-04. (2004) "Standard Test Methods for Moisture-Density (Unit Weight) Relations of Soil-Cement Mixtures."

ASTM D1633-00. (2000) "Standard Test Methods for Compressive Strength of Molded Soil-Cement Cylinders."

Consoli, N., Prietto, P., Ulbrich, L. (1998). "Influence of fiber and cement addition on behaviour of sandy soil." *Journal of Geotechnical and Geoenvironmental Engineering*, 124 (12): 1211-1214.

Marsellos, N., Christoulas, S., Kolias, S. (1986) "Use of fly-ash in roadworks." *Bull. of Public Works Research Center*, 3-4: 113 – 129.

Venkatarama Reddy, B.V., Richardson, L., Nanjunda Rao, K.S. (2007). "Optimum soil grading for the soil-cement blocks." *Journal of Materials in Civil Engineering*, ASCE, 19 (2): 139-148.

Strength Difference between Clam-Shell and Long-Reach Excavator Constructed Cement-Bentonite Self-Hardening Slurry Walls

Paul J. Axtell[1], P.E., Timothy D. Stark[2], Ph.D., P.E., and John C. Dillon[3], P.E.

[1]Dan Brown and Associates, paxtell@danbrownandassociates.com
[2]University of Illinois, tstark@uiuc.edu
[3]US Army Corps of Engineers, Kansas City District, john.c.dillon@usace.army.mil

ABSTRACT: Cement-bentonite (c-b) self-hardening slurry walls are being constructed as a seismic retrofit to stabilize the downstream slope of Tuttle Creek Dam in Manhattan, Kansas. A full-scale test program was conducted to evaluate various slurry mixes along with two construction techniques capable of creating the required c-b transverse shear walls. The two methods tested were a crane-operated mechanical clam-shell excavator (CS) and a long-reach track-hoe (LR). The results of unconfined compression tests on hardened c-b samples constructed by CS and LR techniques were measured and the comparison revealed different unconfined compressive strengths (UCS). This paper presents UCS for both construction techniques and sample type, i.e., cored and wet-grab samples. The cause(s) of the differences in UCS and the impact on the remedial measures are discussed.

INTRODUCTION

Tuttle Creek Dam, located on the Big Blue River in the Kansas River Basin, is part of a system that provides a comprehensive plan for flood control and other functions in the Missouri River Basin. The dam was designed and constructed by the US Army Corps of Engineers, Kansas City District. It is located about 10 km north of the city of Manhattan in eastern Kansas. The embankment is 2,300 m long and about 43 m high. The crest width is 15.2 m and the base width varies from about 430 to 490 m. The top of the dam is at elevation 353.3 m while the original ground surface varies in elevation from about 310 to 313 m across the valley. Tuttle Creek Dam is a rolled earthfill dam; details of the fill zones and construction of the dam can be found in Lane and Fehrman (1960).

The main seismic source zones are the Nemaha Ridge uplift zone and the Humboldt Fault zone. The maximum credible earthquake (MCE) is a magnitude 6.6 event at 20 km with a return period of about 3000 years. The peak horizontal ground acceleration, PHGA, of the MCE is 0.30g mean and 0.56g mean plus one standard deviation. The threshold liquefaction event is a magnitude 5.7 with a return period of about 1700 years. The Kansas City District found that rehabilitation of the liquefiable foundation sands is required to prevent an uncontrolled release of the reservoir during or after the design ground motion.

As part of the required seismic rehabilitation, transverse shear walls are being constructed through the embankment and foundation soils near the downstream toe of the dam. The walls are oriented perpendicular to the crest of the dam to resist deformations induced by slope instability following the design seismic event. Large deformations at the downstream toe are not acceptable because of the presence of a fragile pressure relief well system. This relief well system provides vital underseepage pressure relief during operation of the reservoir.

The test program was considered both a test and production treatment covering approximately 10 percent of the required stabilization of the downstream stabilization project. Replacement walls were added as required between walls after initial construction to assure the designed strength was achieved. The construction contractor (Treviicos South) worked closely with the Corps to identify the most effective technologies, materials, and configurations to be used in the test program and subsequent design and full production at part of the Construction Manager at Risk contract.

CONSTRUCTION DETAILS

Thirty-eight test walls were constructed in this test program. Twenty-two walls were constructed using the CS while 16 were constructed with the LR. Four of the CS walls included a higher slag cement component than used in the LR walls, and therefore are not included in this comparison, leaving only 18 CS walls to include in the dataset. Tremie-placed plastic concrete was attempted on two walls, but was abandoned after trench stability problems were observed during construction. Four self-hardening slurry mixes were trialed with each construction method. The four cement-to-water ratios used were 0.3, 0.4, 0.45, and 0.5 with 5 percent bentonite, as shown in Table 1, along with the number of walls built with each method utilizing a respective mix.

The walls were excavated and simultaneously backfilled with self-hardening c-b slurry supplied by an on-site batch plant. A Liebherr HS855DH crane was used for the CS walls while a Koehring 1466 excavator was used for the LR walls. The CS walls were predominantly constructed by excavating three 4 m long primary bites/excavations, followed by two 1 m wide secondary bites/excavations. Necessarily, the LR walls were built in a single phase. Walls constructed with the CS were generally completed in about 1.5 days, while walls constructed with the LR took slightly less than a day. Neither method included a 24-hour operation.

The water used in the c-b mix was obtained from a well screened in the foundation sand just beyond the downstream toe of the dam. The cement used was MaxCem® which was obtained from Lafarge, and was composed of 50 percent Portland cement and 50 percent ground granulated blast furnace slag cement. The bentonite used was Wyo-Ben Hydrogel®. A dispersing agent admixture Lamsperse-HS, available through Lamberti, was also used in the mix. The bentonite was mixed with water in a high speed mixer and allowed to hydrate under continuous circulation for 8 to 12 hours. The cement was then mixed with the bentonite slurry in a high speed mixer. All mixes include a 5 percent bentonite component.

The walls are 13.7 m long, and 21 m deep with a 3 m clear space between them. The CS walls are 122 cm wide while the LR walls are 91 cm wide. The extra CS width was required to provide extra mass as a 91 cm wide clamshell was not large enough to efficiently excavate to the specified depths due to the relatively high density of the fluid slurry. The clear space between the walls is required to allow continued through seepage in the pervious drain downstream of the core and underseepage in the coarse-grained foundation soils. Restricting either of these seepage paths would result in unacceptable build-up of pore pressures and possibly slope instability.

Table 1. Various construction methods and corresponding mixes.

Number of CS Walls	Number of LR Walls	Cement-to-Water Ratio	Lamsperse-HS (%)	Theoretical Unit Weight (kN/m^3)
4	4	0.3	0.3 to 0.6	11.8
7	4	0.4	0.4 to 0.8	12.3
4	4	0.45	0.45 to 0.9	12.5
3	4	0.5	0.5 to 1.0	12.7

SOIL PROFILE

A working platform was constructed on the downstream slope of the dam to facilitate construction of the walls. The platform was constructed by: 1) removing the existing fill to expose the pervious drain fill; 2) importing and placing sand (SP); and 3) placing approximately 60 cm of road sub-base for a working surface. The only portion of the embankment the walls are in contact with is the pervious drain downstream of the core, which lies above the natural cohesive blanket (ML and CL). The pervious drain fill is composed of dense dredged SP soil, and is approximately 4.6 m thick.

The soils in the alluvial foundation of the dam consist of 2.4 to 8.2 m of silt and clay underlain by sand, silty sand, and gravely sand to a depth of 12.2 to 24.4 m. The silt and clay form a natural cohesive soil blanket over the more permeable sands. This natural cohesive blanket is an important component of the seepage control system for the dam, as are the pressure relief wells at the downstream toe. The sand deposits vary in thickness from about 7.6 to 18.3 m and can be separated into two distinct zones. The upper zone consists of a 4.6 to 6.1 m thick loose fine to medium sand (SM, SP, and SW) and the lower zone consists of a 7.6 to 9.1 m thick dense coarse to gravelly sand that increases in grain-size with depth (SP, SW, GP, GW). Due to the alluvial nature of the foundation deposit, multiple lenses of cohesive soil exist within the coarse-grained layers. The upper sand zone was determined to be potentially liquefiable during the design ground motion. The bedrock consists of alternating layers of shale and limestone (Permian age); however, the walls do not penetrate bedrock.

SAMPLING AND TESTING

Sampling and testing was performed at the on-site batch plant, on the fluid c-b slurry in the excavations via wet-grab samples, and on the hardened c-b slurry via core drilling. Wet-grab sampling was conducted on each wall at various depths. Wet-grab samples were cast in 7.6-cm by 15.2-cm cylinders and stored in a 100 percent humidity curing room until UCS testing was performed. Eventually, all samples (plant, wet-grab, and core) were stored submerged in water after it became apparent the samples were degrading with time in a 100 percent humidity curing room. This degradation was not apparent for samples cured for 28 days, but was visually noticeable for the 60 day samples and older.

Coring was conducted on the majority of the walls. Initially, coring was conducted with the Geobore system (double-barrel wireline) producing 10 cm diameter samples. Due to a significant strength discrepancy between the wet-grab and core sample strength, the coring operation was altered to a split-barrel coring device in an effort to reduce sample disturbance and micro-fracturing.

Based on an independent laboratory investigation of the proposed mixes, and verified by full-scale field measurements, relatively minor strength increases can be expected beyond 90 days for these materials. The ratio of slag cement to Portland cement, the type of Portland cement, and the amount of bentonite included in the mix likely control the rate of strength gain.

The changes in sample storage and coring operation requires two independent data sets be developed, one for each set of techniques (the changes in storage and coring were instituted at the same time). Accordingly, the data reported herein was sampled and stored in the following manner:

1. 28-day UCS tests (Figures 1 through 4)
    a. Coring with the Geobore system conducted between 14- to 28-days after panel construction.
    b. Wet-grab samples cured in 100 percent humidity room for 28 days.
2. Long-term UCS tests (Figure 5)
    a. Coring with split-barrel device a minimum of 120 days after panel construction.
    b. Strength testing performed immediately after coring or within two weeks after coring (submerged storage in the interim).

INTERPRETATION OF TEST RESULTS

The UCS results indicated that the strength varies with respect to several variables in the construction and testing process, resulting in a complex dataset. The influencing factors, among others, appear to be: 1) depth, 2) construction method, 3) sample type, 4) cement/water ratio, and potentially 5) slag content.

The measured strength of both wet-grab and core samples with depth for the CS and LR construction methods are shown in Figures 1 through 4 for the four c/w ratios of 0.5, 0.45, 0.4, and 0.3, respectively. From these figures, it appears the core sample strength varies more with depth than the wet-grab sample strengths. The variation with depth of the core strengths is likely a result of various material types being excavated in addition to the duration which the soil particles remain in suspension within the fluid slurry. The wet-grab strengths do not appear to be as depth-dependent, suggesting there is an effect of continued soil particle settlement within the fluid slurry, or perhaps the confining stress during curing (higher in the fluid walls for the core samples) alters the strength.

Another complicating factor is the construction method which appears to influence the measured strength of the core samples, but not the wet-grab samples. While much scatter exists in the data, it seems evident that the core strength for the LR samples is greater than that of the CS core samples. The CS and LR construction methods likely leave different amounts of suspended soil in the fluid slurry, the results of which are probably more likely to affect core strength than wet-grab strength.

The differences between wet-grab and core strength is more pronounced at higher c/w ratios. The results shown in Figure 4 for the mix with c/w ratio of 0.3 indicate very similar strengths between the two sample types, as opposed to the 0.5 mix shown in Figure 1. This suggests the density of the fluid slurry has an effect not only on the settlement of suspended soil particles, but also cement particles.

Further evaluation of density and sand content measurements on wet-grab samples, as well as density measurement of core samples could likely yield insight into the peculiar differences in core and wet-grab strength at various c/w ratios. While careful coring was conducted, as well as using multiple coring operations, the potential of micro-fracture of samples during coring must also be considered. The samples tested herein reached peak strength around one percent axial strain, indicating a relatively brittle material, as expected. The post-peak or ultimate strength measured on core samples via isotropic-consolidated, undrained triaxial tests with pore-pressure measurements was taken into consideration during the design of these walls for the seismic retrofit because some cracking of the walls is expected during the design ground motions.

To quantify the observed strength difference between LR and CS core samples, the average long-term strengths at various c/w ratios are shown in Figure 5. The definition of "long-term" refers to samples that were cored at least 90 days after construction of the walls. Necessarily, the average includes samples at various depths. As noted earlier, the strength of cored samples does vary with depth. However, the average trends suggest the strength of LR constructed walls exceeds that of CS walls by 20 to 52 percent at various c/w ratios.

Although a good data set was not developed for higher slag contents, it should be noted a 75 percent slag mix (as opposed to the 50 percent slag mix) was used on four CS walls (two with a cement-to-water ratio of 0.3 and two with a cement-to-water ratio of 0.4). Their average long-term UCS were 1.65 and 2.46 MPa for the 0.3 and 0.4 mixes, respectively; a significant increase than those values shown in Figure 5 for 50 percent slag mixtures.

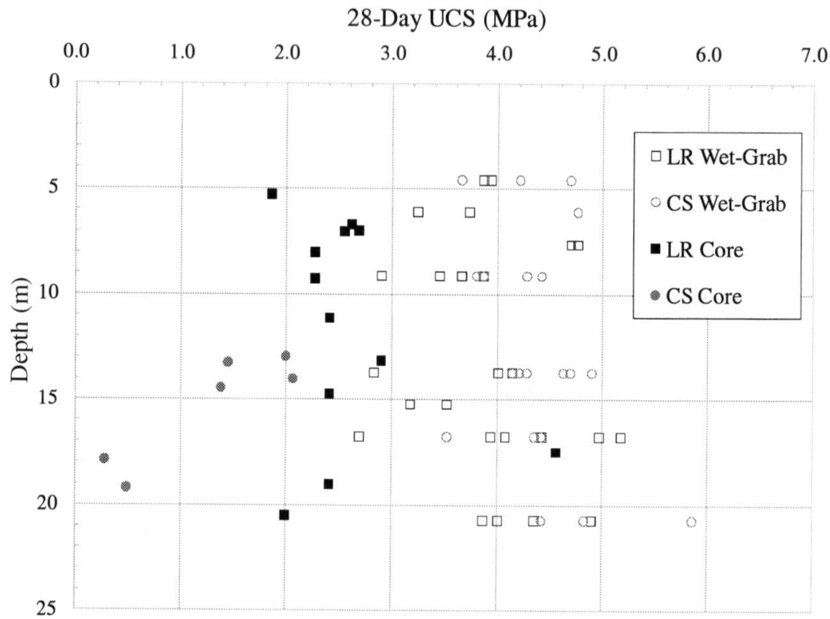

Figure 1. UCS for cement-to-water ratio of 0.5.

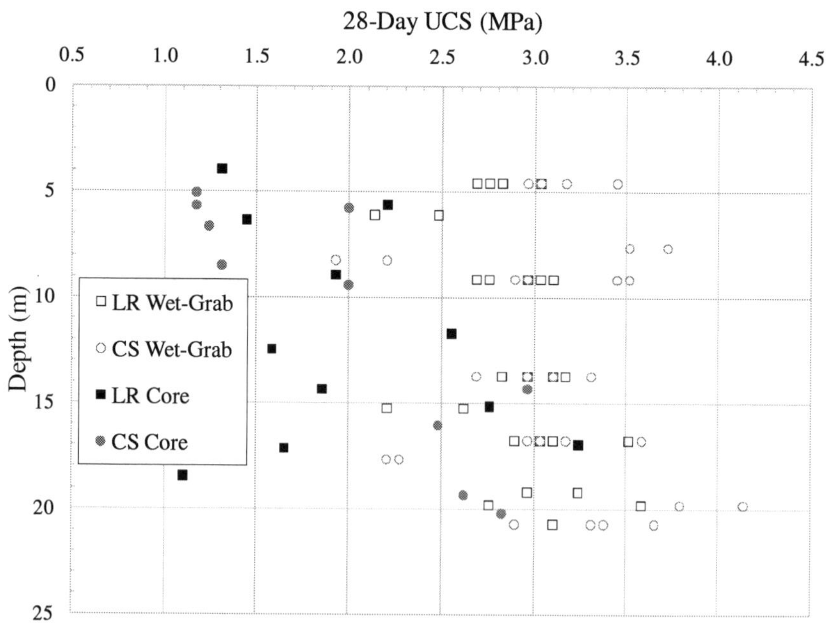

Figure 2. UCS for cement-to-water ratio of 0.45.

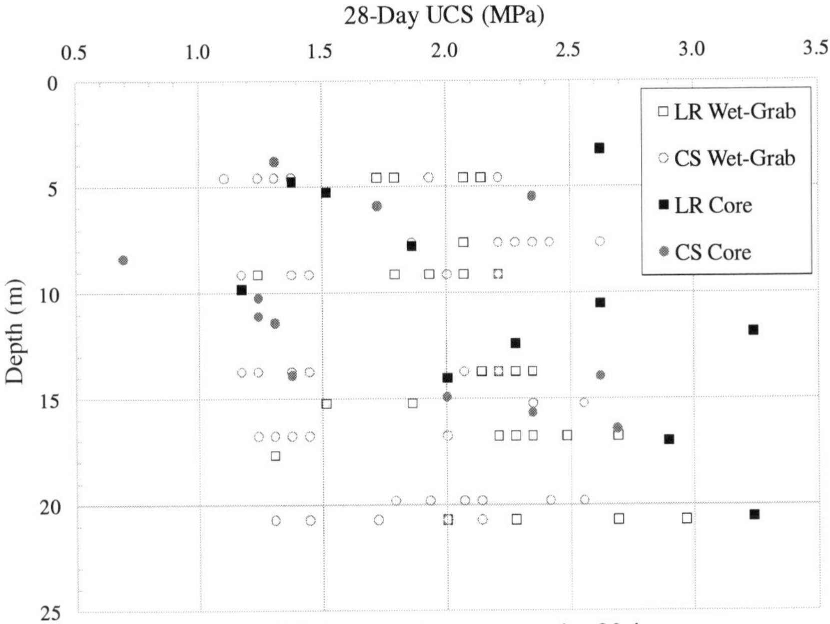

Figure 3. UCS for cement-to-water ratio of 0.4.

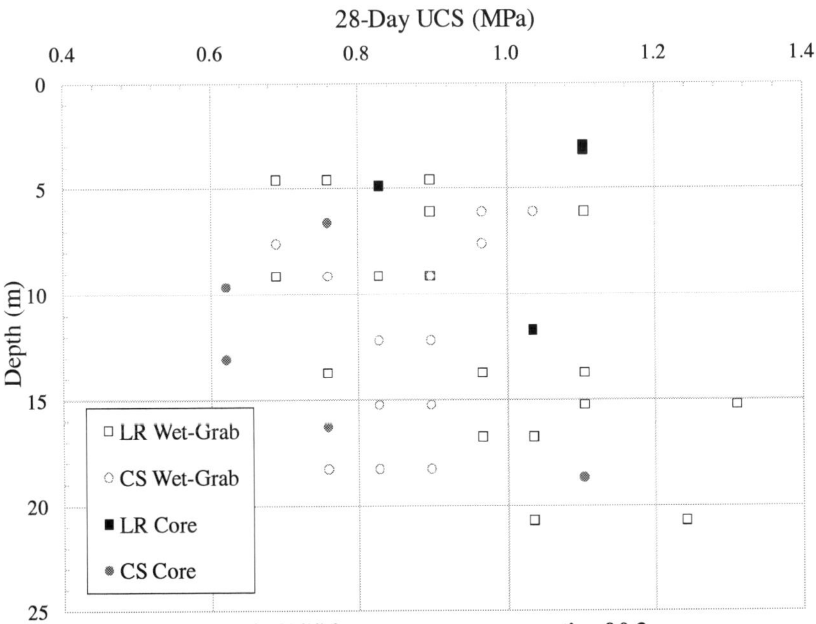

Figure 4. UCS for cement-to-water ratio of 0.3.

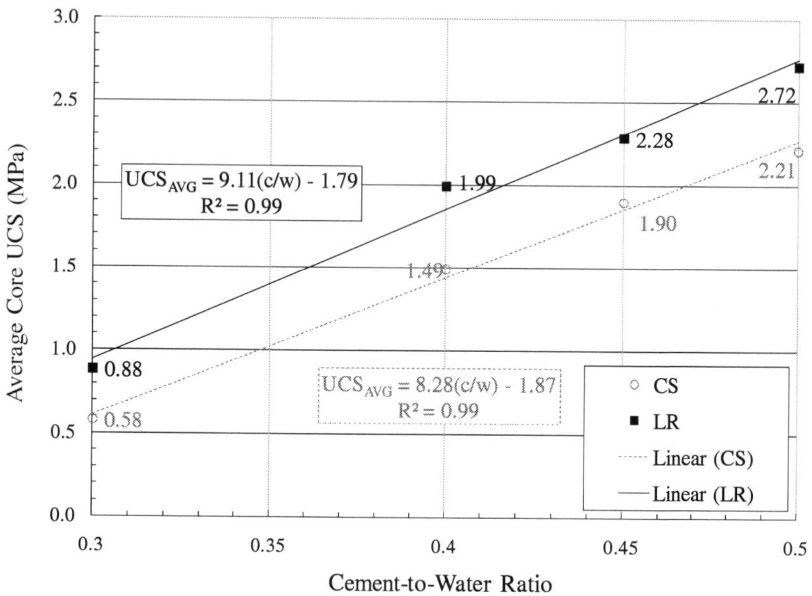

Figure 5. Average long-term UCS of core samples.

CONCLUSIONS

Cement-bentonite (c-b) self-hardening slurry walls are being constructed as a seismic retrofit of the downstream slope of Tuttle Creek Dam. A full-scale test was conducted to evaluate LR and CS construction techniques. The results of UCS tests on hardened c-b samples constructed by CS and LR techniques revealed different UCS for these construction methods. On average, walls constructed with the LR method appear to yield stronger core samples than those constructed with the CS method. The sample type, i.e., cored versus wet-grab samples, yield significantly different strengths at high c/w ratios, and essentially no difference at low c/w ratios.

ACKNOWLEDGMENTS

The contents of this paper are the authors' and do not necessarily reflect those of the represented entities. The authors acknowledge the support provided by the U.S. Army Corps of Engineers – Kansas City District, Joe Topi, Francke Walberg, Bill Empson, and the expertise of the contractor, Treviicos South.

REFERENCES

Lane, K.S., and Fehrman, R.G. (1960). "Tuttle Creek Dam of Rolled Shale and Dredged Sand." *J. Soil Mechanics and Foundation Division*, ASCE, 86(SM6), . 11-34.

Behavior of Disconnected Pile Foundation System

Jung-In Choi[1], BEngr, Ki-Hoon Min[2], BEngr, Sung-Ho Kim[3], M.S., P.E., Oh Sung Kwon[4], Ph.D., Myoung Mo Kim[5], Ph.D., M.ASCE

[1]Ph.D Student, Department of Civil Engineering, Seoul National University, San 56-1, Shinlim-Dong, Kwanak-Ku, Seoul, 151-742, Korea, next2u00@snu.ac.kr
[2]Deputy General Manager, POSCO Engineering & Construction Co.,Ltd., Dae Rung B/D. 5F, 826-20 Yeoksam-Dong, Gangnam-Gu, Seoul, 135-769, Korea, khmin@poscoenc.com
[3]President, JIN YOUNG Engineering & Consultants Co., LTD., 2nd, Shinwoo B/D. 1239-12, Gaepo-Dong, Gangnam-Gu, Seoul, 135-960, Korea, shkim@jinyoungenc.com
[4]Ph.D., Technology Research Institute, Daelim Industrial Co., Ltd., 146-12, Susong-Dong, Jongno-Ku, Seoul, 110-732, Korea, kwonos1@daelim.co.kr
[5]Prof., Department of Civil Engineering, Seoul National University, San 56-1, Shinlim-Dong, Kwanak-Ku, Seoul, 151-742, Korea, geotech@snu.ac.kr (corresponding author)

ABSTRACT: In the design of a foundation, the settlement of the foundation may exceed allowable design criteria even with a competent bearing stratum. In such a case, a pile foundation system may be adopted using piles as a settlement reducing component. In this paper, Disconnected Pile Foundation (DPF) system, which installs disconnected piles underneath the footing and uses the piles as ground reinforcements, is studied as a cost effective design method against the classical pile foundation system. To this end, large size loading tests were carried out on weathered ground, changing area replacement ratio and length of piles. The results indicated that the settlement of the reinforced ground was reduced by 46~85% from that of the unreinforced original ground. The correlating formula between the area replacement ratio and the load bearing ratio of piles was derived from the test results and numerical analysis. From the correlation, a design method determining the size and the quantity of the disconnected piles to enhance the bearing capacity of original ground to the desired value was proposed based on settlement criteria.

INTRODUCTION

A pile foundation is used to supplement a shallow foundation either by increasing the bearing capacity of the foundation or reducing the settlement of it under design load.

The pile foundation however would result in high stress concentration on the pile head and localized high bending moment on a footing. On the other hand, Disconnected Pile Foundation (DPF) system is a new concept solution, which uses a disconnected pile as a reinforcing element, not as a structural element, in increasing ground stiffness (Wong et al. 2000).

By disconnecting the piles from the footing, the stress level on a pile head can be reduced by spreading the working load on the footing to the surrounding ground, and this, in turn, would lead to the reduction of the required number of piles to meet the design load and thus save cost considerably.

Studies of the DPF system have been performed by many researchers and it has been confirmed that DPF system can reduce settlement and increase bearing capacity, effectively (e.g., Wong et al. (2000), Cao et al. (2004), Abdrabbo et al. (2004), Lee et al. (2007)). Wong et al. (2000) reported that the disconnection of piles from the footing does not significantly affect the piles effectiveness in reducing footing settlements and bending moments. But the design method applying this system to practice has yet to be developed. In this study, the behavior of the DPF system was analyzed by carrying out large plate load tests on weathered soil ground of SPT(N), 15~20, nine times by changing the area replacement ratio and the pile length. The results of the field load tests were simulated by numerical analysis and the design method of the DPF system was proposed on the basis of the parametric studies.

DESCRIPTION OF LOAD TEST

The ground of the test location was of weathered soil with SPT(N) value of 15~20 which is classified as SP by the Unified Soil Classification. The specific gravity of the soil is 2.66, in place, and the dry unit weight, 18 kN/m^3. The groundwater table was located at 18m depth from the ground surface.

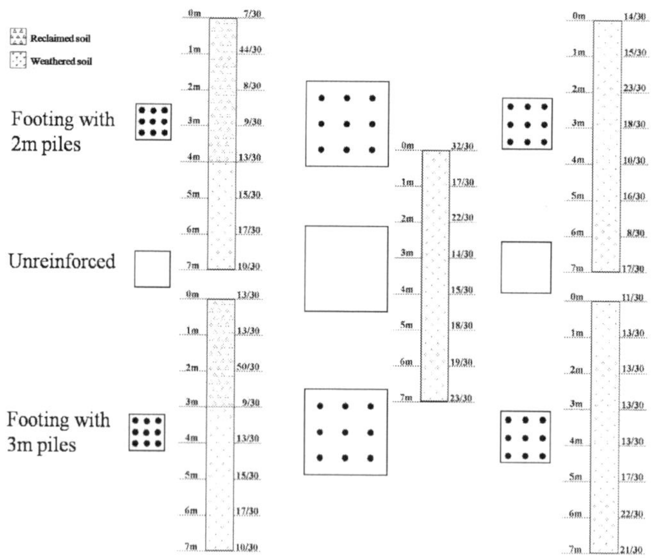

FIG. 1. Plan view of test site and soil profile

Large plate load tests were carried out nine times for three different sizes of footings of 0.65m, 0.9m, 1.5m in width and 0.6m in thickness made of steel-reinforced concrete.

The load tests were performed in three ways for the each size of the footing. The three cases were as follows: one unreinforced case, one reinforced case with piles of 2m length, and one case with piles of 3m length, respectively. The pile diameter was 100mm. The piles were made of unreinforced concrete. A total of nine cast-in-place piles were installed under the footing regardless of the footing size. The area replacement ratio was 16.7%, 8.7% and 3.1% for 0.65m, 0.9m and 1.5m square footing, respectively.

A compacted gravel layer of 10cm thickness was installed between the piles and the footing. Figure 1 shows the plan view of the load test.

The allowable bearing capacity of the unreinforced ground was estimated as $30t/m^2$ from an empirical equation and the N values. The test pressures were applied up to 300% of the allowable pressure for unreinforced ground and 450% for reinforced ground following ASTM D1143-82.

TEST RESULTS AND ANALYSIS

The load test results are presented in Figure 2. In the Figure 2, it is seen that the footings of 0.65m - and 1.5m - width with 3m piles are stiffer and bear greater loads than those with 2m piles. But in the case of the 0.9m size footing, the trend is reversed due to the local variability of the soil profiles

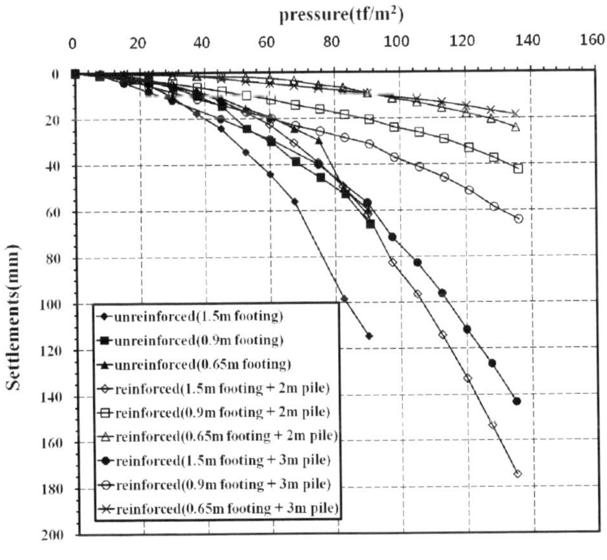

FIG. 2. Pressure-settlement curves

The ratios of settlement reduction by the disconnected piles are shown in Table 1 under the footing pressure of 90tf/m^2, whose magnitude is three times the assumed allowable bearing capacity of the original ground. In the table, it is shown that the settlement of the reinforced ground was reduced by 46~85% from the settlement of the original ground. Thus, it is evident that the disconnected piles are effective in reducing settlements of the footings.

Table 1. Settlement Reduction Ratios for DPF system

Footing size (m)	Pile length (m)	Settlement at 90tf/m^2 loading stage (mm)	Settlment reduction ratio[1)](%)
0.65	-	59.77	-
	2	8.97	85
	3	9.21	85
0.9	-	61	-
	2	20.5	66
	3	31.81	48
1.5	-	114.5	-
	2	61.72	46
	3	56.58	51

[1)] 100 - (settlement of reinforced case / settlement of unreinforced case) x 100

NUMERICAL ANALYSIS

For the case of field load tests, it was almost impossible to execute a sufficient number of tests for a parametric study due to the high cost and long testing time. In this study, the numerical analysis was conducted to develop a design method of the DPF system. The results of actual field tests were compared with those of the numerical analysis to confirm the validity of the numerical method. After the confirmation, a parametric study was conducted for simplified cases of DPF with a single disconnected pile. From the results of the parametric study the relationship between the pile area replacement ratio and the load bearing ratio of the pile was proposed.

Numerical Analysis Method

The analytical program FLAC3D (Ver. 3.10) was utilized to simulate the field load tests. The width of the numerical model was five times the footing width and the depth was two times the pile length. Input data used for the numerical analysis are shown in Table 2. The numerical analyses were conducted for three different soil conditions and three different area replacement ratios. Table 3 shows the numeric test programs. The elastic modulus of the soil was changed according to the soil conditions.

Table 2. Input Data for Numerical Analysis

Type of Material	Depth (m)	Model	Elastic modulus [1](Pa)	Friction Angle(°)	Cohesion (t/m^2)	Unit Weight (t/m^3)
Upper Weathered soil	0~0.5	M-C[2]	1.4e7	38	3	1.8
Middle Weathered soil	0.5~3	M-C	1.8e7	38	3	1.8
Lower Weathered soil	3~6	M-C	2.1e7	38	3	1.8
Compacted Layer	-	M-C	2.5e7	40	0	2.0
Concrete (Pile & Footing)	-	Elastic	2e10	-	-	2.5

[1]Elastic modulus = 7.5+0.8N(MPa) – Anagnosopoulos (1990)
[2]Mohr-Coulomb

Table 3. Numerical Test Programs

Soil Condition	N value[1]	Pile Diameter (mm)	Area Replacement Ratio, a_r[2] (%)
Dense	>30	100	3
		200	12
		300	28
Medium	10~30	100	3
		200	12
		300	28
Loose	<10	100	3
		200	12
		300	28

[1]The ranges of N value according to the soil conditions were suggested by Peck(1974)
[2]Area replacement ratio = area of the piles / area of the footing

Numerical Analysis Results

Figure 3 shows the measured settlements together with the estimated settlements by the numerical method for 1.5m square footing cases, which were either unreinforced or reinforced by a total of 9 concrete piles of 3m length. It is seen from the figure that the load settlement curves of the field measurement and numerical analysis show good agreement.

Figure 4 shows the area replacement ratio versus the fraction of load taken by piles for three different soil conditions. It can be seen that the load carried by piles decreases as the soil condition improves. Table 4 shows the relations between the area replacement ratio (a_r) and the fraction of load taken by piles (R) for three different soil conditions, which were obtained from the regression analysis.

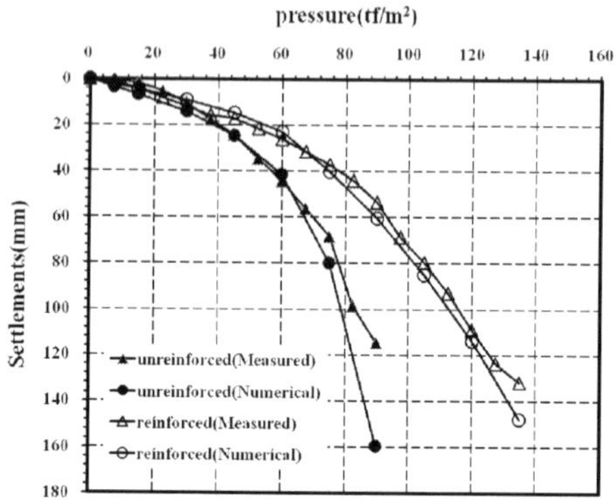

FIG. 3. Comparison between measured and analysis results for unreinforced and reinforced cases of 1.5m square footing

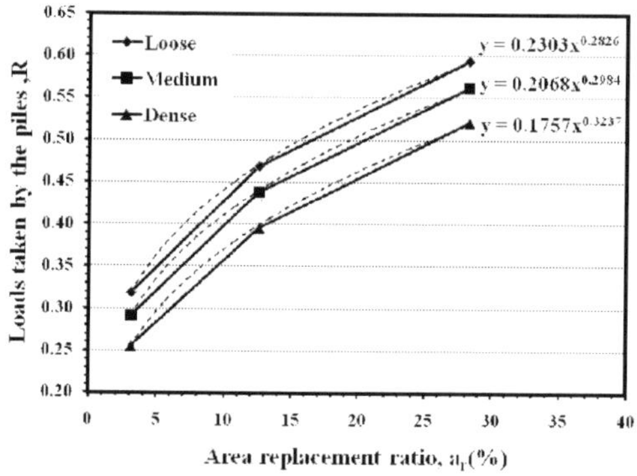

FIG. 4. Area replacement ratio versus fraction of loads taken by pile (all soil)

Table 4. Empirical Relations between Area Replacement Ratio and Fraction of Load Taken by Piles for Three Different Kinds of Soil Conditions

Soil Condition	N value	Empirical relations between [1)]R and a_r
Dense	>30	$a_r = Exp(3.086 \ln R + 5.36)$
Medium	10~30	$a_r = Exp(3.356 \ln R + 5.29)$
Loose	<10	$a_r = Exp(3.534 \ln R + 5.19)$

[1)]R = Fraction of loads taken by pile = Load carried by the piles (P_{pile}) / total applied l oad ($P_{soil}+P_{pile}$)

DETERMINATION OF AREA REPLACEMENT RATIO FOR DESIGN OF DPF SYSTEM

To apply the DPF system to practical cases, the area replacement ratio satisfying the design bearing capacity must be determined. To achieve this, the estimation method of the area replacement ratio that will yield the target allowable bearing capacity is suggested based on the settlement criteria.

The estimation procedures are as follows:

1. Determine the fraction of load(R) to be taken by the disconnected piles using Eq. 1 which is suggested by POSCO E&C (2008).

$$R = \frac{q_{target} - q_{soil}}{q_{target}} \quad (1)$$

where q_{target} is design bearing capacity and q_{soil} is allowable bearing capacity of original ground, respectively.

2. Determine the allowable bearing capacity of the original ground (q_{soil}) using Eq. 2 suggested by Bowles (1977) according to the tolerable settlement of a superstructure.

$$q_{net(all)}(kN/m^2) = 11.98(N_1)_{60}\left(\frac{3.28B+1}{3.28B}\right)^2 F_d\left(\frac{S_e}{25}\right) \quad (B > 1.22m) \quad (2)$$

where $(N_1)_{60}$ = corrected standard penetration number, B = foundation width, in m, F_d = depth factor = $1+0.33(D_f/B) \leq 1.33$, S_e = tolerable settlement, in mm

3. Estimate the area replacement ratio that will yield the target allowable bearing capacity using the equations shown in Table 4.

EXAMPLE FOR DESIGN OF DPF SYSTEM

Table 5 show examples of the required area replacement ratio that will yield various target allowable bearing capacities for various soil conditions. The allowable settlement of the superstructure was assumed to be one inch.

Table 5. Area Replacement Ratio Giving the Target Allowable Bearing Capacity for Each Soil Condition

Soil Condition	N value	Allowable bearing capacity of original soil (tf/m^2)	Target allowable bearing capacity q_{target} (tf/m^2)			
			20	30	40	50
			Area replacement ratio to obtain q_{target}			
Dense	32	38	-	-	0.02 %	3 %
Medium	20	24	-	1 %	10 %	22 %
Loose	8	10	16 %	43 %	65 %	82 %

CONCLUSIONS

From the field tests and the numerical analysis performed in this study, conclusions are as the followings.

1. The field test results showed that the settlement of the reinforced ground was reduced by 46~85% from that of the original ground. Therefore, it is verified that the DPF system is an effective method of ground reinforcement.

2. The relationship between the area replacement ratio and the fraction of loads taken by a pile was proposed from the results of parametric study using a numerical method.

3. A method of estimating the area replacement ratio that will yield the target allowable bearing capacity was suggested for three different states of soil conditions.

ACKNOWLEDGMENTS

The authors wish to acknowledge the financial support by POSCO Engineering and Construction Co., Ltd. and SNU SIR BK21 research Program funded by Ministry of Education & Human Resources Development.

REFERENCES

Abdrabbo, F. M., Abouseeda, H. M., and Gaaver, K. E. (2004) "Performance of raft with pile-settlement reducers." *International Conference: Future Vision and Challenges for Urban Development*: 1-18.

Anagnostopoulos, A. G. (1990) "The compressibility of cohesionless soils." *Geotechnik*, No.3

Bowles, J.E. (1977) *Foundation Analysis and Design*, 2nd ed., McGraw-Hill, New York.

Cao, X. D., Wong, I. H., and Chang, .M. F. (2004) "Behavior of model rafts resting on pile-reinforced sand." *Journal of Geotechnical and Geoenvironmental Engrg.* 130(2): 129-138.

Lee, Y. S., Hong, S. H. (2007) "Settlement-reducing effects of disconnected piled raft foundation system." *Proceeding of the 13th Asian Regional Conference on Soil Mechanics and Geotechnical Engineering*: 273-288.

$FLAC^{3D}$. (2006) "User's manual." Minneapolis, MN: Itasca Consulting Group Inc

Peck, R. D., Hanson, W. E., and Thornburn, T. H. (1974) *Foundation Engineering*, John Wiley and Sons, Inc., New York.

POSCO E&C. (2008) "Development of design method of disconnected pile foundation system – Final report."

Wong, I. H., Chang, M. F., and Cao, X. D. (2000) "Raft foundations with disconnected settlement reducing piles.", *Design application of raft foundations and ground slabs*, Chap. 17, Thomas Telford, London: 469-486.

Innovative Soil Reinforcement Method to Control Static and Seismic Settlements

Mike Majchrzak[1] GE, MASCE, Tom Farrell[2] GE, MASCE, Brian Metcalfe[3], PE MASCE

[1]Geotechnical Principal Engineer, Kleinfelder , Inc., 1701 Koll Center Parkway, Pleasanton, CA 94566 USA, mmajchrzak@kleinfelder.com. [2]President, Farrell Design-Build Companies, Inc., 3025 Venture Road, Placerville, CA 95667 USA, tom@farrellinc.com. . [3]Lead Engineer, Geopier Foundation Company, Inc., 150 Fairview Road, Ste 335, Mooresville, NC 28117 USA, bmetcalfe@geopier.com.

ABSTRACT: With increasing levels of seismic awareness and design-level earthquake accelerations, Geotechnical Engineers are challenged to find cost effective solutions to address both static and seismically induced settlement in high seismic areas. In Salinas, California a relatively heavy, one and two-story structure was recently constructed in an area of relatively deep, intermittent, layers of liquefiable soil resulting in the potential for excessive total and differential settlements.

A traditional foundation system of driven piles with structural slab similar to the new surrounding buildings was initially considered. Although this option addressed static settlement control and the potential for excessive seismically induced differential settlements, the solution was not considered economically viable and posed a risk to adjacent historic, unreinforced masonry structures. Alternatively, an innovative soil reinforcement method consisting of a stiffened geo-grid reinforced engineered fill over reinforced soil utilizing an evenly spaced Rammed Aggregate Pier® system was used to support the building. The reinforced soil mat was designed to control static settlement from the building loads and reduce seismically induced differential settlements to within allowable project tolerances.

INTRODUCTION

Engineer's increased awareness of potential liquefaction and seismic settlement related damages to buildings and infrastructure are driving project costs to all new highs. Tough decisions are required regarding risk-based performance of structures after a major earthquake. Geotechnical Engineers have a limited database of proven and cost effective mitigation measures for damages caused by liquefaction and seismic settlements. Such mitigation measures include:

- Driven piles thru loose liquefiable soil bearing in deeper soil and rock,
- Mass excavation and replacement with engineered fill,

- Densification by dynamic compaction or vibration,
- Reinforcement by high modulus aggregate piers or cement columns,
- Stiffened grade-beam foundation systems with structural slabs,
- Drainage by wick drains or aggregate piers, and
- Rammed Aggregate Pier system support of a geo-grid reinforced fill.

A geo-grid reinforced fill supported over a Rammed Aggregate Pier system was used to support a structure in Salinas California which is the subject of this paper.

PROJECT DESCRIPTION

The structure is a theater building for the Maya Cinema 14-Plex in Downtown Salinas, California. Directly adjacent buildings consist of one and two story historic, unreinforced masonry construction on shallow footings, and a newly constructed adjacent parking garage founded on driven piles. The theater building has a footprint area of about 5,110 square-meters, with masonry wall loads of 8.8 kN/m and column loads of 445 to 1,232 kN. The design earthquake has a moment magnitude of 7.3 with a corresponding peak ground acceleration of 0.5g. The site is located within 22 km of the San Andreas fault, Monterey Bay-Tularcitos, and Rinconada faults.

Utilizing the design earthquake and bored soil conditions, the total earthquake-induced settlement as a result of liquefaction was estimated to be about 50 to 100 mm. Static and dry sand seismic settlements added to the concern of building damage.

LIQUEFACTION POTENTIAL, STATIC AND SEISMIC SETTLEMENT

The primary geotechnical concerns for this site were the presence of near surface undocumented sand fill and potentially liquefiable sand. The undocumented fills were found in the upper 5.2 to 6 m below the ground surface. The potential liquefiable soil consisted of loose sand below the groundwater table below about 8.2 m below the ground surface. Liquefaction potential was based on a design earthquake with moment magnitude of 7.3, with corresponding peak ground acceleration (PGA) of the site of 0.50g. The depth to which liquefaction is estimated to occur varies between about 7.5 to 15 m, and is discontinuous, varying in thickness, depth and lateral extent. Estimates of the resulting settlement should all of the various layers liquefy was performed based on the recommendations by Youd (1996 and 1998). Estimates from both the CPTs and borings performed at the site indicated an estimated accumulated settlement of 50 to 100 mm. Differential settlements were based on the difference in computed total settlement. Based on Martin and Lew (1999), differential settlements may be taken as half of the total settlements between adjacent supports.

In addition to potential settlement due to liquefaction below the groundwater level, settlement was also anticipated due to seismic densification of the dry sand fill above the groundwater level, as well as static settlement of the fill due to the planned buildings loads. It was estimated that the dynamic densification would range from about 6 to 12 mm, and the settlement due to the building loads could be about 50 mm.

Potential total settlements could be on the order of about 100 to 165 mm.

DEEP FOUNDATION ALTERNATIVE

The initial approach to control settlements was to support the structure on a deep foundation that extended below the depth of the potential liquefiable soil. Typically, precast, prestressed concrete piles would be used with lengths of about 33.5 m with a 1,468 kN allowable load. Because of potential differential settlement between the pile supported building columns (limited settlement) and the subgrade supported building slab (potential for significant settlement), a structural slab would be required. Concerns were also expressed about the damaging impacts of pile driving on the adjacent historic, unreinforced masonry structures, as well as the additional foundation costs and the impact of the driven piles schedule on the aggressive project schedule.

ALTERNATIVE SOIL IMPROVEMENT / SOIL MAT SYSTEM

Due to high cost of piling, scheduling, and adjacent buildings, an innovative soil reinforcement method beneath the site was considered and subsequently used for foundation support, see FIG. 1. This reinforced soil mat system would need to address the undocumented fill in the upper portions of the site, reduce installation-caused settlement on the adjacent historic properties, and provide a mechanism to reduce differential settlement caused by potential liquefiable sands at the site. To address the effects of these potential damaging differential settlements, it was recommended that the Rammed Aggregate Pier (RAP) ground improvement system overlain by a geo-grid reinforced soil cap be installed to support a conventional shallow foundation system for the building. It was the intent of the RAPs and reinforced soil cap to both strengthen the fill soils beneath the floor slab and building columns, as well as to create a reinforced and stiffened soil mat over the potentially liquefiable soil layers to reduce damaging differential settlements.

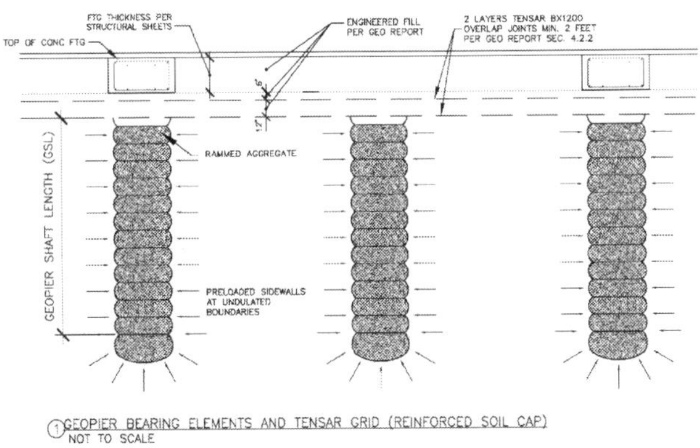

FIG. 1 Cross Section of the Installed Soil Mat System

REINFORCED SOIL MAT DESIGN

In designing a soil mat, the thickness and the method of the soil improvement is critical. The intent of the soil mat foundation system was not to eliminate total settlement, but to reduce differential static and seismic settlement to approximately 20 mm over 15 m. For this project, the design of the soil mat consisted of the following components:

- Highly densified Rammed Aggregate Piers installed to depths of 5.5 m below the slab-on-grade to provide a stiffened and densified soil mass within the loose fill to reduce dynamic densification. RAPs were evenly spaced at 3.3 m on center and concentrated under walls and columns.
- Composite strengthening and densification of a substantial thickness of soil above the potential liquefiable soil layers with RAPs to increase the arching capability of the soil mat system.
- Installation of a geo-grid reinforced fill above the RAPs near the surface and just below the footings and below the slab-on-grade to reduce differential settlement between the RAPs.
- Increasing the tensile strength of the soil mat system by having a two layer Tensar BX1200 geo-grid reinforced fill near the surface of the site.

Given the above criteria, and using a Finite Differential Method (FDM) computer analysis to model the soil mat's effectiveness, a soil mat foundation system was installed as follows:

- Shallow foundations consisting of isolated and continuous footings embedded a minimum of 0.6 m below the bottom of the building slab,
- A 0.45 m thick geo-grid reinforced soil mat below the footings containing geo-grids at the bottom and in the middle of the compacted sandy soil,
- Installation of 0.75 m diameter RAPs on 3.3 m spacing throughout the entire building envelope. The RAPs extended a minimum of 4.6 m below the geo-grid reinforced fill. RAP ramming installation also increases the density of sandy soil below the RAP drill depth to an estimated depth of 0.9 m. As a result, the depth of RAP improvement extended about 5.5 m below the bottom of the geo-grid reinforced fill portion of the soil mat.

The depth and spacing of the RAPs were based on using the increased stiffness of the improved area due to the RAPs themselves, the increase in densification between the RAPs that would increase strength and strain characteristics with resulting increase in arching of the soil between the RAPs, the increase in density of the mass to reduce dynamic settlement above the groundwater level, and the decrease in settlement due to building loads. An iterative analysis using FDM was performed. In addition, reduction of the amount of differential settlement would be further reduced with the geo-grid reinforced mat. This was modeled based on the methodology by Han (1999), and Han and Gabr (2002). The geo-grid analysis resulted in an estimated reduction in by about half. In calculating the spacing and depth of the RAPs, consideration was given to soil arching as it relates to densification of the sand, the shear strength of the sand, and the shear strength between the RAPs and the sand. In the analysis, consideration was given to providing higher densification of the sands within about 1.5 diameters from the center of the RAPs. This distance is based on experience with stone columns and other ground improvement methods in sands, which are susceptible to ramming/vibration induced compaction. The increased densification allowed for shorter distances for arching of the soil. Typical values for soil arching are about three times the spacing. Given that the estimated distance (approximately of 1.8 m) of the unimproved portion of the soils between the RAPs, the vertical distance would need to be about 5.5 m.

RAMMED AGGREGATE PIER® SYSTEM

Rammed Aggregate Pier systems are a ground improvement technique that consists of replacing and/or displacing loose/weak soil with highly densified, thin lifts of compacted aggregate. RAP construction is well defined in the literature (Majchrzak, 2004, Farrell, et al 2008). RAPs provide for a stiffened, composite soil mass where higher concentrations of stress are attracted to the stiff RAP elements compared to the soft matrix soil. The stress to the RAP is a function of the relative stiffness between the RAP element and matrix soil and area replacement as shown in Eq. 1.

$$q_g = \frac{qR_s}{R_a(R_s - 1) + 1} \qquad [1]$$

Where q is the applied maximum footing bearing pressure, q_g is the stress to the top of the RAP, R_s is the stress concentration factor, and Ra is the area replacement of the RAPs. The stress concentration factor, R_s, is equivalent to the ratio of the RAP stiffness to the matrix soil stiffness at a given bearing stress. The matrix soil stiffness may be estimated through traditional geotechnical analysis. The RAP stiffness is confirmed by modulus load testing on an individual RAP (Fox and Cowell, 1998).

During ramming, the beveled hammer forces crushed rock laterally into the sidewalls of the drilled shaft. This ramming action increases the lateral stress and strengthens the matrix soil, thus providing additional stiffening and increased normal stress perpendicular to the perimeter shearing surface. In seismic areas, the stiffer RAP elements attract and absorb high shear stresses thereby reducing the cyclic shear stresses to the matrix soil (Green, et al 2008, Girsang, et al, Baez and Martin 1993).

The intent of the RAP element installation for this project was to control structural static load settlement, reduce seismic settlement of the loose sand fill, and to increase the composite soil mass stiffness to reduce surface manifestation of deep settlements below 7.6 m. The engineered fill and structural loads were supported by arching to the top of the RAP elements. Control of static settlement was determined using Eq. 2:

$$s_{uz} = \frac{q_g}{k_g} \qquad [2]$$

where q_g is the stress applied to the RAP and k_g is the stiffness modulus of the RAP (Fox and Cowell 1998). The stiffness of the RAP and the settlement within the RAP reinforced zone was verified by full scale modulus testing of an individual RAP using load testing procedures in general conformance with ASTM D-1143. The load test was performed on a concrete cap embedded a depth of 0.6 m, consistent with the average bottom footing elevation, and located directly over the RAP. The results of the modulus test, shown in FIG. 2, confirmed the stiff RAP element with 5 mm of deflection at the design test stress of 880 kPa.

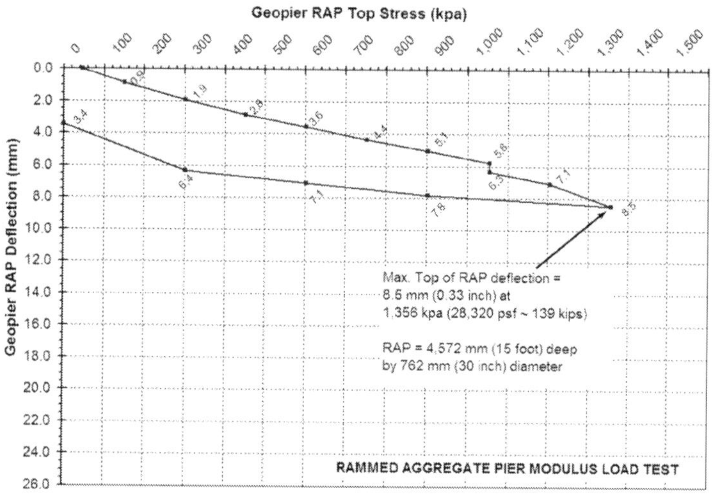

FIG. 2. Rammed Aggregate Pier Modulus Load Test

PRE AND POST CPT TESTING

Cone Penetrometer Testing (CPT) was performed before and after installation of the RAPs. The generalized soil profile and results of CPT testing are shown in FIG. 3. Prior to RAP installation, the Standard Penetration Test sampler had equivalent blow counts of less than 20 blows for 0.3 m to depths greater than 7.5 m. After installation of the RAPs, additional CPTs were performed to monitor the increase in density between the RAPs. The additional CPTs were located at 1 foot intervals between two adjacent RAPs spaced at 3.3m on center. The CPTs showed a significant increase of 150 to 200 percent of the original equivalent SPT N_{60} blow counts. The CPT results also showed significant increase in soil density between the RAPs.

CONCLUSION

The project site is underlain by loose sand fill above the groundwater level, and potential liquefiable sands below the groundwater level. The settlement from these soil layers could have a significant impact to the theater building. A typical

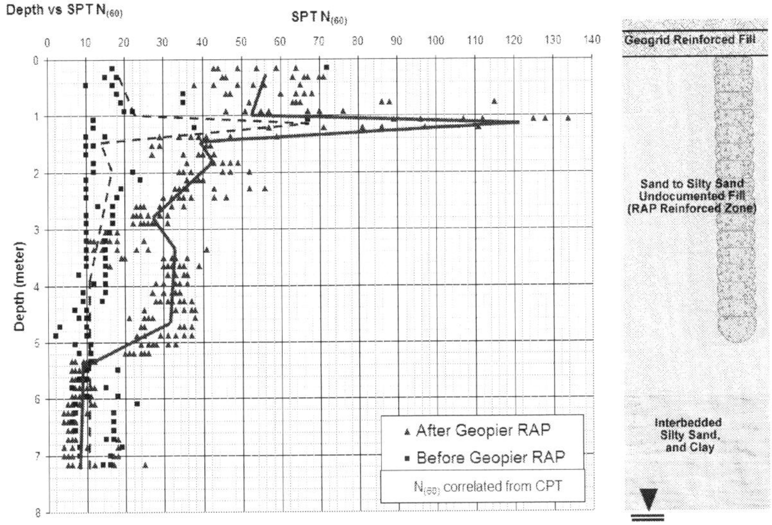

FIG 3. Pre and Post CPT Testing at Maya Cinemas in Salinas, CA

foundation system of driven piles was considered to support the building and slab. The deep foundation system was found to be costly, could not meet the aggressive project schedule, and could have a damaging impact on adjacent historic structures. An alternative system of a shallow foundation supported on a reinforced soil mat was considered and used. The soil mat consisted of two components that included a 0.45 m thick geo-grid reinforced fill supported on an evenly spaced Rammed Aggregate Pier system throughout the building footprint. The 0.75 m RAPs and resulting improvement extends to a minimum of 5.5 m below the geo-grid reinforced fill at an even spacing of 3.3 m on center. The design of the spacing and depth of the RAPs was based on the composite stiffness of the system and a conservative estimate of the lateral extent of densification resulting from installation of the RAPs.

The results of the modulus load test on a Rammed Aggregate Pier and the pre- and post- CPT testing of the soil demonstrated a significant improvement in the stiffened crust and higher than anticipated increase in soil densification between the RAPs. The improvements resulted in replacement of portions of the upper sands with stiffer and less compressible elements using the RAPs, increased density of the soils between the RAPs that increased its arching capability, stiff geo-grid reinforced fill above the RAPs, and finally lower settlement and higher shear strengths of the soil mat system. Specifically, the settlement due to the building loads would be reduced from about 50 mm to 5 mm (based on the improved stress/strain of the soil), the dynamic settlement above the groundwater level reduced to near zero, and the propagation of the liquefaction induced differential settlement estimated to be about 25 to 50 mm would be reduce to approximately 20 mm over a horizontal distance of approximately 15 m at the ground surface (based on further FDM evaluations; the RAPs would reduce the differential settlement by about 20 mm, and the geo-grid soil mat would reduce that by about 10 mm). It was concluded that this system would be capable of controlling static settlements and reducing the effects from dynamic densification in the sand fill above the groundwater as well as reduce the overall differential settlement associated with liquefaction occurring at depths below the reinforced soil mat system.

ACKNOWLEDEMENTS

The authors appreciate the support of Geopier Foundation Company, Kleinfelder, Farrell Design Build and the design team in developing and implementing this innovative soil mat system.

REFERENCES

Baez J.I., and Martin, G.R. (1993). "Advances in the Design of Vibro Systems for the Improvement of Liquefaction Resistance". *Proc. 7th Annual Symp. of Ground Improvement, 1-16.*

Fox, N.S. and Cowell, M.J. (1998). *Geopier Foundation and Soil Reinforcement Manual.* Geopier Foundation Company, Inc., Scottsdale, AZ.

Farrell, T.M., Fitzpatrick, B., Kenney, W.M., (2008). "Uplift Testing of Rammed Aggregate Pier Systems". *Proc. Geotechnical Earthquake Engineering and Soil Dynamics IV.* Sacramento, CA.

Green, R.A., Olgun, C.G., and Wissmann, K.J. (2008). "Shear Stress Redistribution as a Mechanism to Mitigate the Risk of Liquefaction". *Proc. Geotechnical Earthquake Engineering and Soil Dynamics IV.* Sacramento, CA.

Girsang, C.H., Gutierrez, M.S., and Wissmann, K.J. (2008). "Modeling of the Seismic Response of the Aggregate Pier Foundation System". *Proc. Geo-Support 2004* Orlando, FL.

Han, J. (1999), "Design and Construction of Embankments on Geosynthetic Reinforced Platforms Supported by Piles", Proc. 1999 ASCE-PaDOT Geotechnical Seminar, Central Pennsylvania Section, ASCE – Pennsylvania

Han, J. and Gabr, M. A. (2002), "Numerical Analysis of Geosynthetic-Reinforced and Pile-Supported Earth Platforms Over Soft Soil," Jour. Geotech. and Geoenv. Engr., ASCE, Vol.128, No. 1, pp. 44-53.

Majchrzak, M., Lew, M., Sorensen, K., and Farrell, T. (2004). "Settlement of Shallow Foundations Constructed Over Reinforced Soil: Design Estimates vs. Measurements." *Fifth Intl. Conf. on Case Histories in Geotechnical Engineering*

Martin, G.R., and Lew, M. 1999, "Recommended Procedures for Implementation of DMG Special Technical Publication 117, Guidelines for Analyzing and Mitigating Liquefaction in California,"Southern California Earthquake Center, University of Southern California, March, 63 p.

Youd, T.L. et al, "Liquefaction Resistance of Soils: Summary Report from the 1996 NCEER and 1998 NCEER/NSF.

Numerical Analysis of Consolidation of Soft Ground Improved by the DJM-PVD Combined Method

Lei Chen[1], Songyu Liu[2], Jie Han[3], and Dingwen Zhang[2]

[1]Institute of Geotechnical Engineering, Southeast University, Nanjing 210096, China; currently visiting scholar at the University of Kansas
[2]Institute of Geotechnical Engineering, Southeast University, Nanjing 210096, China;
[3]Department of Civil, Environmental, and Architectural Engineering, the University of Kansas, Lawrence, KS 66045, USA

ABSTRACT: Dry Jet Mixed (DJM) columns and Prefabricated Vertical Drains (PVDs) have been combined to improve soft ground for the Huai-yan Highway in China. This method could make the mixing of DJM columns easier and more homogeneous since the drainage of water through pre-installed PVDs reduces the excess pore pressure induced by the installation of DJM columns. In addition, the strength of the improved ground can be further quickly increased when the embankment load is applied at the existence of the PVDs. A two-dimensional plane-strain numerical model was adopted to analyze the consolidation of the DJM-PVD improved ground under the embankment load. The DJM columns and the PVDs in the treated ground were simply converted to DJM walls and PVD walls to obtain an equivalent plane-strain model. For comparison purposes, the conventional PVD improved ground, the DJM improved ground, and the unimproved ground were also analyzed. The numerical results showed that the DJM-PVD improved ground consolidated the fastest among these four types of ground. The stress concentration on the columns allows excess pore-water pressure in the surrounding clay to dissipate more quickly than that in the PVD improved ground.

INTRODUCTION

Dry Jet Mixing (DJM) is a widely used soil improvement technique in the world due to its merits, e.g. easy and rapid construction and relatively small vibration. For road and railway embankments, this technology is expected to reduce the settlement and increase the stability of soft ground. In recent years, DJM has been creatively combined with other soil improvement methods. One good example is the combined DJM-PVD method, i.e., Dry Jet Mixing and Prefabricated Vertical Drains (Liu et al., 2005; Liu et al., 2008). This method can make the mixing of DJM columns easier and more homogeneous since the pre-installed PVDs allow quick dissipation of the excess pore pressure induced by the installation of DJM columns. In addition, the strength development of the improved ground can be further accelerated due to the existence of the PVDs under the embankment load. Considering the strength gain in both DJM columns and the surrounding soils, the spacing of DJM columns can be properly increased to lower the cost.

The DJM-PVD method has been successfully used in the soft ground improvement for the Huan-Yan Highway in China. Consolidation property is considered as an important guidance in design and practice for soft ground, so this study was focused on the numerical analysis of the consolidation of the combined DJM-PVD treated soft ground.

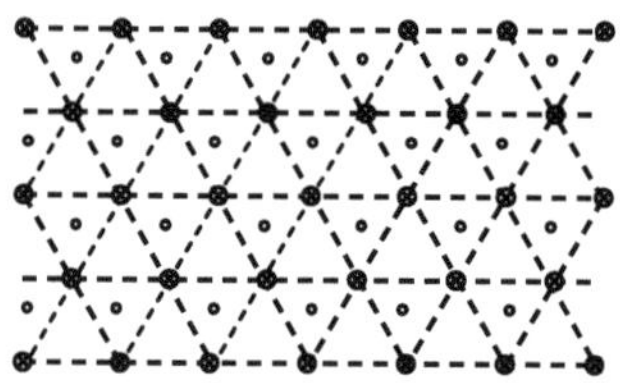

FIG. 1. Schematic plot of the combined DJM-PVD method

TWO-DIMENSIONAL MODEL OF DJM-PVD COMPOSITE FOUNDATION

Equivalent Plane-Strain Conditions

For the purpose of two-dimensional numerical modeling of the DJM-PVD composite foundation, the axisymmetric unit cells of DJM columns and PVDs need to be converted to an equivalent plane-strain model. In this conversion, the spacing of DJM columns and PVDs is not changed.

For PVDs, the conversion method based on the equivalent plane-strain permeability approach proposed by Chen et al. (2004) was adopted. This method was developed based on Barron's consolidation theory and the matching technique of axisymmetric and plane-strain consolidation analytical solutions under the following assumptions: (a) the average degree of consolidation from the axisymmetric model equals to that from the plane-strain model at the same depth, (b) an equal-strain condition, (c) following Carrilo's consolidation theory (d) considering the influence of the stress level on consolidation behavior, (e) the smear effect included in the equivalent permeability in the plane-strain model without any smear zone, and (f) the vertical permeability of the soil equal to the horizontal permeability. The conversion equations can be expressed as follows:

$$\frac{k_{pw}}{k_{aw}} = \frac{2(1+\nu)}{3} \cdot \frac{n-1}{n^2 - s^2} \quad (1)$$

$$\frac{k_{px}}{k_{ax}} = \frac{4(1+\nu)}{9} \cdot \frac{(n-1)^2}{n^2 \cdot \mu} \quad (2)$$

$$\mu = \frac{n^2}{n^2 - s^2} \ln\frac{n}{s} - \frac{3n^2 - s^2}{4n^2} + \frac{k_{ax}}{k_s} \frac{n^2 - s^2}{n^2} \ln s \quad (3)$$

where n is the diameter ratio of the unit cell to the well, s is the diameter ratio of the smear zone to the well, ν is Possion's ratio, $k_{pw}, k_{px}, k_{aw}, k_{ax}$ are the permeability of PVDs and the horizontal permeability of the soil in the plane-strain and axisymmetric models, respectively, and k_s is the permeability of the smear zone in the axisymmetric model.

Deng (2005) proposed a simplified method to obtain an equivalent stiffness of DJM walls in the plane-strain model from DJM columns using the following formula:

$$E_p^{cal} = E_{sz}(1 - \frac{\pi \cdot D}{4 \cdot d}) + E_{pz} \cdot \frac{\pi \cdot D}{4 \cdot d} \quad (4)$$

where E_p^{cal} is the equivalent modulus of the DJM wall, E_s and E_p are the moduli of the soil and the DJM columns, D is the diameter of the columns, and d is the spacing of columns.

Assuming the columns are impermeable, the equivalent vertical permeability of DJM walls, k_p^{cal} can be expressed as

$$k_{px}^{cal} = (1 - \frac{\pi \cdot D}{4 \cdot d})k_{sx}, \quad k_{pz}^{cal} = (1 - \frac{\pi \cdot D}{4 \cdot d})k_{sz} \quad (5)$$

where k_{sx} and k_{sz} are the horizontal and vertical permeability of the soil.

Models and Parameters

To verify the numerical model for the DJM-PVD composite foundation, a case study was selected. Tables 1 to 5 list all the parameters used in this study for the test site at the Huai-yan Highway in China. In the model, the modified CamClay model was used to simulate the soil properties, DJM and PVD were treated as linearly elastic materials, and embankment fill and cushion were modeled using the Duncan-Chang model. The interface between the columns and the soil was modeled using the Desai interface element.

The foundation comprised of three layers with the dimensions shown in FIG. 2. The ground water level was at -1 m deep. The bottom of the underlying layer was impermeable. FIG. 3 shows the loading stages for the embankment construction.

Table 1. Material properties of the soils

Soil	OCR	λ	κ	μ	Γ	P_c' (kPa)	M	$k_x = k_y$ (m/d)	
								Before conversion	After conversion
Crust		0.08	0.02	0.35	2.2	50	1.117	5.2x10^{-3}	0.78x10^{-3}

Soft soil	1.1	0.2	0.04	0.4	2.7	-	1.007	3.8×10^{-4}	0.57×10^{-4}
Underlying layer	1.2	0.024	0.006	0.35	1.7	-	1.177	3.8×10^{-4}	

Table 2. Material properties of the DJM columns and the DJM walls

	E(MPa)	μ	$k_x = k_y$ (m/d)
Before conversion	150	0.25	1.52×10^{-8}
After conversion	29	0.25	3.12×10^{-4}

Table 3. Material properties of the embankment (Deng, 2005; Wu, 2005)

Name	γ (kN/m³)	K	n	R_f	C (kPa)	φ(°)	μ	kx, ky (m/d)
Fill	19	150	0.4	0.8	30	28	0.30	1
Cushion	16	160	0.8	0.6	0	34	0.28	

Table 4. Material properties of the interface (Deng, 2005; Wu, 2005)

K_N (kPa/m)	K_s (kPa/m)
1.6×10^8	2×10^4

Table 5. Material properties of the PVDs

E(MPa)	μ	$k_x = k_y$ (m/d)	
		Before conversion	After conversion
2	0.32	4.3	0.11

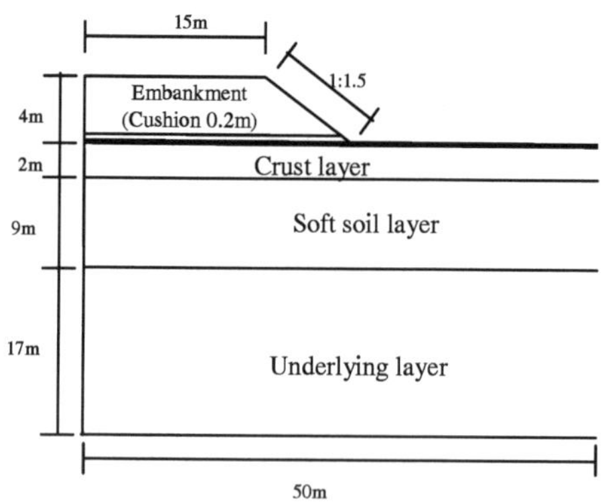

FIG. 2. Dimensions of the model.

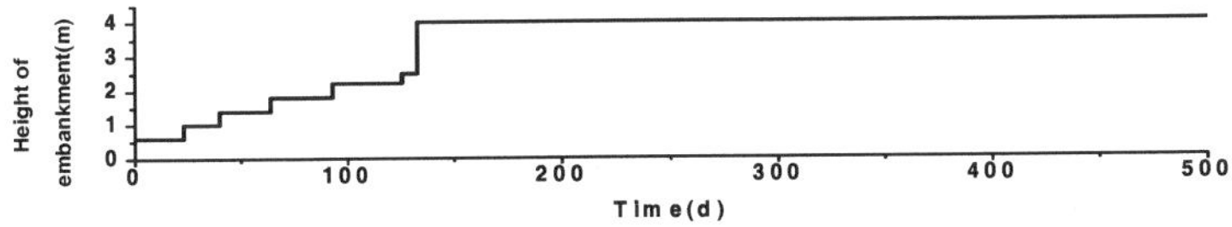

FIG. 3. Loading stages

RESULT AND ANALYSES

Excess Pore Pressure and Degree of Consolidation

FIGURES 4 and 5 show the comparisons between the measured and calculated excess pore pressures at different depths and the degrees of consolidation of the composite foundation. The numerical results matched well with the measured data. FIGURE 4 shows that the measured excess pore pressure increased rapidly for the first and last loading stages but decreased gradually during these stages. This phenomenon can be explained that at the beginning of the loading, the soil carried most of the load. After that, more and more load transferred to the DJM columns through the cushion and the stress concentration ratio of the column to the soil increased. The last stage loading was relatively large and could not be distributed much to the DJM columns immediately, therefore, there was an obvious increase of the excess pore pressure in the soil.

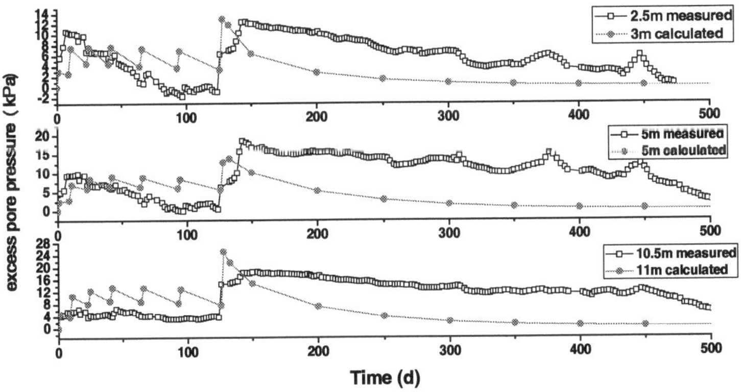

FIG. 4 Excess pore pressure dissipation

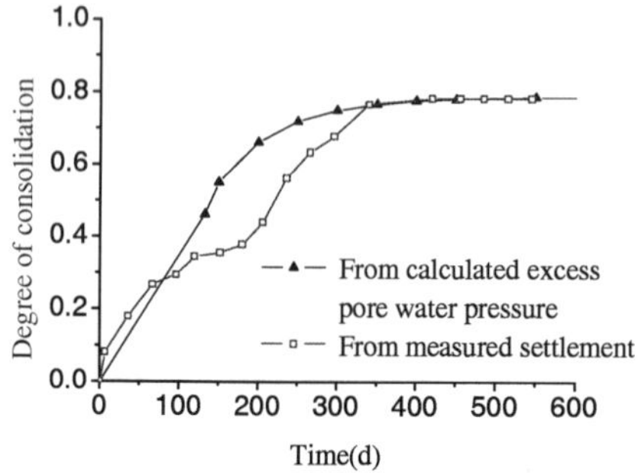

FIG. 5 Degree of consolidation of the DJM-PVD foundation

Comparison of Different Improvement Methods

Using the same material properties, equivalent plane-strain conversion method, and numerical modeling approach, the DJM treated, PVD treated, and untreated foundation can be modeled as easily as the DJM-PVD treated ground. Considering the difference in the installation quality (Wu, 2005), the modulus of the DJM columns in the DJM treated ground used in the model before conversion was 100MPa, which was lower than that in the DJM-PVD treated ground. The results of the above analyses were compared.

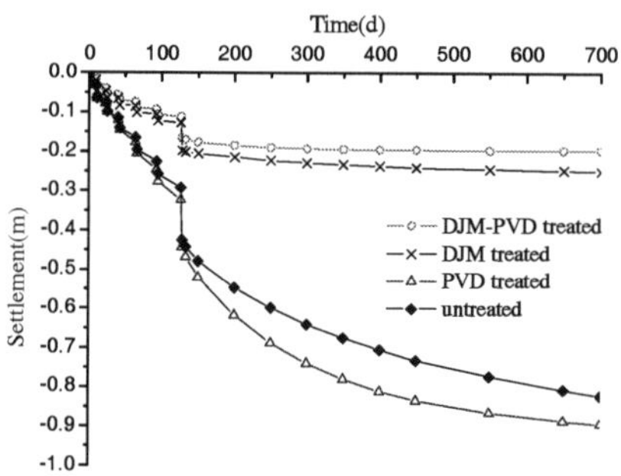

FIG. 6 Settlement-time curves of different foundations

FIGURE 6 shows the settlement-time curves of the embankment under different conditions. It is shown that DJM columns significantly reduced the settlement of the composite foundation to 1/4 of that without DJM. FIGURE 6 also shows that PVDs accelerated the consolidation of the soft foundation. The settlement of the PVD treated ground developed faster but tended to become closer to that of the untreated ground at the end of consolidation.

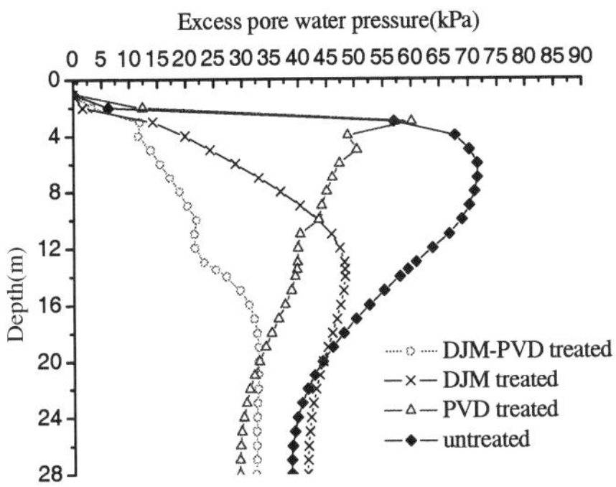

FIG. 7 Excess pore pressure with depth after the construction of the embankment (133 days)

FIGURE 7 shows that DJM columns had more effect on the excess pore pressure in the upper layer by carrying more load and reducing the load on the surrounding soil. Different from DJM, PVDs played a more important role in the underlying layer by reducing the drainage path of the water. Therefore, it is obvious that the combined DJM and PVD foundation could accelerate the dissipation of the excess pore pressure and the consolidation of the ground as shown in FIGS. 8 and 9.

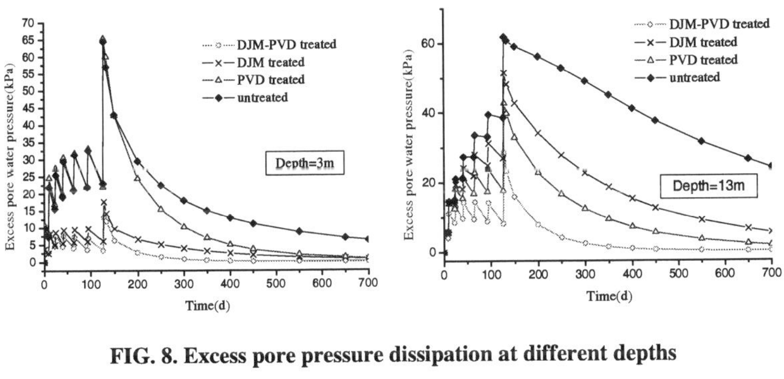

FIG. 8. Excess pore pressure dissipation at different depths

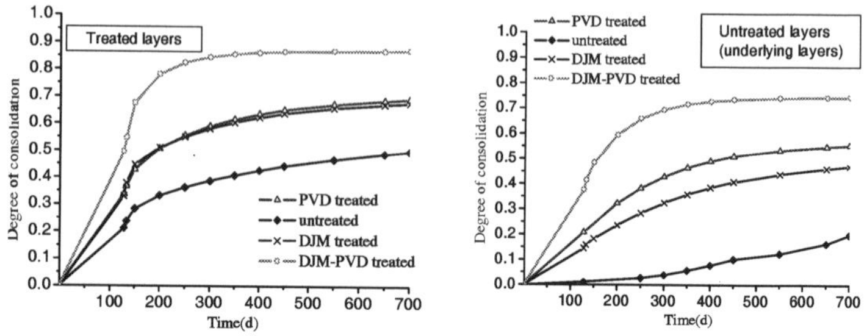

FIG. 9. Degree of consolidation of different layers

CONCLUSIONS

This study shows that the equivalent parameter approach in the two-dimensional plane-strain model reasonably simulated the field behavior of the combined DJM-PVD foundation. The numerical results demonstrated that DJM-PVD could accelerate the consolidation of the treated ground. The benefits of this ground improvement technique are: (1) PVD could reduce the drainage path for the groundwater and (2) the higher stiffness of the DJM columns enhanced the transfer of the embankment load from the soil to the columns.

REFERENCES

Chen, L.H., Chen, Z.Y., and Li, G.X. (2004). "Algorithm of equivalent plane strains for analyzing sand-drain foundations by the finite element method." *Journal of Civil Engineering,* 37(6): 82-86, in Chinese.

Deng,Y.F. (2005). *Interaction Principle between Deep Mixing Columns and Surrounding Soils and Determination Method of Column Spacing*. Ph.D. Dissertation, Southeast University, Nanjing, China, in Chinese.

Liu, S.Y., Han, J., Zhang, D.W., and Hong, Z.S. (2008). "A New DJM-PVD combined method for soft ground improvement." *Geosynthetics International Journal*, 15(1), 43-54.

Liu, S. Y., Wu, Y. K., Hong, Z. S., and Du, G. Y. (2005). "DJM-PVD combined method innovation and practice in soft ground improvement." *Proceedings of Int. Conf. on Deep Mixing Best Practice and Recent Advances* (CD-ROM), Stockholm, Sweden: 477-486.

Wu, Y.K. (2005). *Mechanisms and Applications Treated Soft Soil by Drain and Dry Jet Mixing*. Ph.D. Dissertation, Southeast University, Nanjing, China, in Chinese.

Taxiway Embankment Construction on Soft Ground Using In-situ Instrumentation and Staged Construction

By: Xavier C. Barrett[1], PE, and Thomas R. Wells[2], EI, M-ASCE, and Richard C. Wells[3], PE; F-ASCE

[1]Geotechnical Group Manager, Trigon | Kleinfelder, Greensboro, NC, xbarrett@kleinfelder.com; [2]Staff Professional, Trigon Kleinfelder, Greensboro, NC, twells@trigoneng.com , [3]Vice President, Corporate Development, Trigon Kleinfelder, Greensboro, NC, rwells@kleinfelder.com

ABSTRACT

The Piedmont Triad International Airport (PTIA), Greensboro, NC is undergoing an expansion as a result of the construction of a regional hub for Federal Express. Part of the construction includes an embankment over an alluvial floodplain in a wetlands area. A hybrid ground modification technique was used to stabilize the soft ground, consisting of geogrids, wick drains and rock toe berms. An instrumentation program was implemented to evaluate the changing properties of the alluvial soils, lateral displacements of the embankment foundation soils, settlement of the embankment, determine the fill placement rates or appropriate waiting periods, and to prevent slope failures from occurring. This paper highlights key features of the design and instrumentation system. It describes the project conditions and considerations, the instrumentation system that was installed, the results obtained during construction, and practical issues that were addressed during construction.

BACKGROUND INFORMATION

Part of the construction of the associated Taxiway E includes the construction of the embankment over a wetlands area in an alluvial floodplain. The final grades in this area require the construction of an earth embankment up to 18 meters (60 feet) in height over the existing soft ground. The footprint of the taxiway alignment is restricted by a Corps of Engineers agreement and other water quality standards, including wildlife habitat requirements imposed by state agencies. The wetland water quality and flow could not be impacted by the crossing, requiring that the potential for slope failures be minimized.

SUBSURFACE CONDITIONS

The subsurface exploration indicated the ground profile to consist of an upper zone of very soft to soft alluvial soil underlain by a residual soil profile, with groundwater within 0.6 meter (2 feet) of the ground surface. The alluvial soils within the wetland area extend to a depth of about 8 meters (27 feet) below the existing ground surface. In general, the alluvial soils consist of either sandy silts or silty sands with varying amounts of mica and clay. Standard Penetration Resistance values obtained in the alluvial soils ranged from Weight of Rod (WOR) to 10 blows per foot (bpf), with the majority of values less than 5 bpf. Undrained shear strengths in the fine grained soils measured between 0.96 kPa (20 psf) to 48.9 kPa (1000 psf), with the majority of values less than 24.5 kPa (500 psf). A photo of the site after clearing, but prior to construction, is provided in Figure 1 below:

FIG 1. View of Wetlands Area From the Northwest

PREDICTED PERFORMANCE

Settlement predictions for the alluvial soils below the embankment were based on the results of the soil test borings and dilatometer testing. The analyses indicated settlement magnitudes ranging from 0.8 meters (2.5 feet) to 1.2 meters (4 feet) within the central portion of the embankment area. Settlements of 15 cm (6 inches) or less were predicted at the toe of the rock berms.

Slope stability analyses were performed with the embankment at various fill heights, without improvement of the alluvial soils. The analyses indicated calculated factors of safety of less than 1.0 once the fill reached a height of about 10 feet.

STABILIZATION DESIGN

Various slope configurations, excavation, and ground improvement scenarios were investigated, since failures were predicted due to the soft alluvial soils beneath the embankment. A brief description of the initially considered scenarios within the wetlands area is provided below:

- Complete removal and replacement of all soft alluvial materials.
- Partial removal and replacement of the soft alluvial materials.
- The use of stone columns under the toe of the slopes.
- Sequencing the fill placement and construction over a long period of time.
- Installation of wick drains and stabilization berms.

The first four scenarios did not fit within the desired construction budget or schedule, as dictated by the construction schedule for the Federal Express facility. The chosen option included the use of vertical wick drains to improve drainage and decrease the consolidation time to allow shear strength improvements within the alluvial soils. Temporary rip rap berms were used beyond the toe of the final slopes in the wetlands to obtain the needed resisting forces for embankment stability purposes. The berm materials outside the slope toe were removed in later stages and used for protection of the embankment faces. The timeframe required for this scenario was estimated to be 2 years, as the construction contract included a minimum waiting period of 30 days between 1.5 meter (5 feet) fills. Of the 2 year contract schedule, 1 year consisted of waiting periods. A sketch of the embankment cross section for this scenario is provided in Figure 2 below:

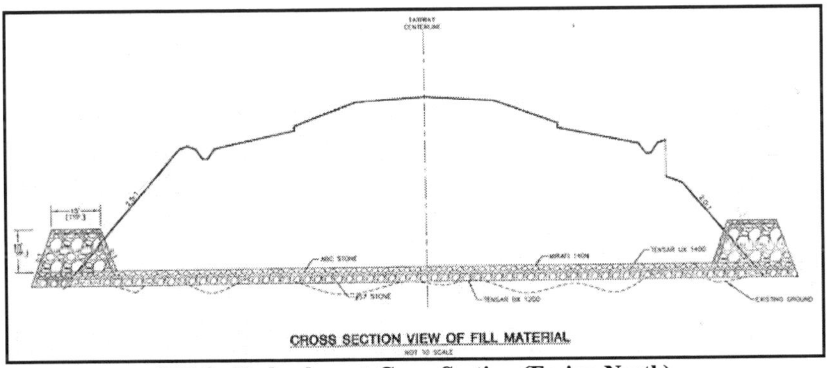

FIG 2. Embankment Cross Section (Facing North)

INSTRUMENTATION PROGRAM CONCEPT AND INSTALLATION

An instrumentation program was considered to be a necessity, to monitor the settlement magnitude, pore pressure buildup and dissipation in the alluvial materials, and lateral and vertical movement of the soils beyond the embankment toe with respect to filling rates. The instrumentation program consisted of piezometers, slope indicator pipes, settlement conduits, and vertical benchmarks. The required contractual waiting periods between fill placement stages could only be reduced based on the results of the instrumentation monitoring program.

Piezometers were installed at four locations at each station; one below the rip rap berm, two at the toe of the slope at varying depths, and one under the centerline of the embankment. The piezometer tips were pushed into the fine grained soils in order to monitor the soil possessing the potential for the greatest increases in pore pressure. For the purposes of this project, the pore pressure B values ($B = \Delta\mu/\Delta\sigma_1$) were used to assess the increase in excess pore pressure compared to the increase in overburden pressure as filling progressed.

Slope indicator pipes were installed approximately 9 meters (30 feet) from the toe of the rip rap berm. The pipes were installed to a depth of approximately 1.5 meters (5 feet) into the underlying residual soil for fixity.

A total of eight (8) horizontal settlement monitoring conduits were installed approximately 20 cm (8 inches) below the surface of the working platform. Four conduits were installed on each side of the taxiway centerline. Due to the presence of the haul road, the lengths of the conduits varied, and did not generally extend to the centerline.

Benchmarks were installed approximately 6 meters (20 feet) from the toe of the slope. The benchmarks were monitored to determine the magnitude of vertical ground movements (heave) beyond the toe of the embankment.

CONSTRUCTION

The ground improvements within the floodplain area of Taxiway E included clearing the area of above-ground vegetation and selective grubbing; the rootmat was allowed to remain due to the very low shear strength of the underlying alluvium. Prior to the beginning of construction, a temporary haul road was constructed through a portion of the left side of the embankment footprint in the alluvial floodplain. The haul road was not removed prior to instrumentation installation. The presence of the haul road restricted the location of some of the instrumentation components.

Over the cleared ground, a stabilized working platform was constructed. This working platform design section consisted of a combination of biaxial geogrids, geotextile filter fabric, and 0.6 meter (2 feet) of clean, uniformly graded crushed stone. After the working platform was constructed, wick drains were installed to facilitate drainage of the alluvial soils. The wick drains were installed utilizing a five feet triangular grid pattern through the alluvial soils, into the underlying residual material. Once the wick drain installation was completed, a layer of uniaxial geogrid was placed followed by 30 cm (1 foot) of well graded crushed stone, to provide a stable working platform for filling operations. A photograph of the construction of the working platform is provided in Figure 3 below:

FIG 3. Construction of Working Platform

Once filling operations began, approximately 0.6 meters (2 feet) of new fill were placed each 3 working days. Filling was suspended during the winter months due to compaction difficulties resulting from high soil moisture contents. At other times, filling was suspended due to rainfall. Typically, up to 4 days were lost per significant rainfall event, due to the soil drying effort required to obtain compaction.

INSTRUMENTATION RESULTS

The instrumentation was monitored beginning August 6, 2006. At the time this paper was prepared (August 2008), less than 60 cm (2 feet) of embankment fill remained to be placed. The results were used to control the rate of filling based on actual measurements instead of relying on estimated waiting periods, thus accelerating the construction schedule. For this project, the end result was the elimination of the waiting period altogether, saving 1 year from the contract schedule. The only delays experienced were weather-induced.

The piezometers indicated that the pore pressure B values ($B = \Delta\mu/\Delta\sigma_1$) increased to no more than 0.5, as compared to a value of 1.0, considered to be the critical value for the soil types at the project site. Measurements obtained during weather delays indicated that the excess pore pressure increases generally dissipated within a time period of about 10 days after the cessation of filling operations, with the majority of the dissipation occurring within 3 days. When the rate of pore pressure dissipation is compared with the contractual waiting period and filling rate, it becomes apparent that the filling rate lags behind the pore pressure changes. A typical pore pressure plot is included as Figure 4 below.

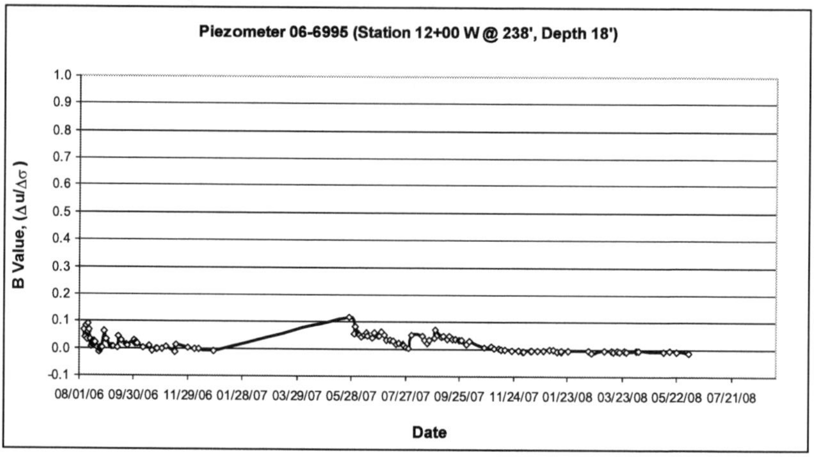

FIG 4. Typical Pore Pressure Dissipation Plot

The slope indicator measurements indicated lateral movements of less than 1.5 cm (0.6 inch) during the timeframe reported in this paper.

The horizontal settlement monitoring conduit measurements indicated settlement magnitudes in general agreement with the predicted values, based on the height of fill placed. Measured settlements under the central portion of the embankment ranged from 0.33 meters (1.1 feet) to 0.81 meters (2.7 feet). Measured settlements at the toe of the rock berm ranged from 8 cm (3 inches) to 24 cm (9.5 inches). The ratio of measured settlements to predicted settlements ranges from approximately 0.38 to 0.67. This is attributed to the heterogeneity of the subsurface profile, the conservatism of the preliminary analyses, and likely, the presence of the haul road that was installed some time before the instrumentation. A typical settlement plot is included as Figure 5 below.

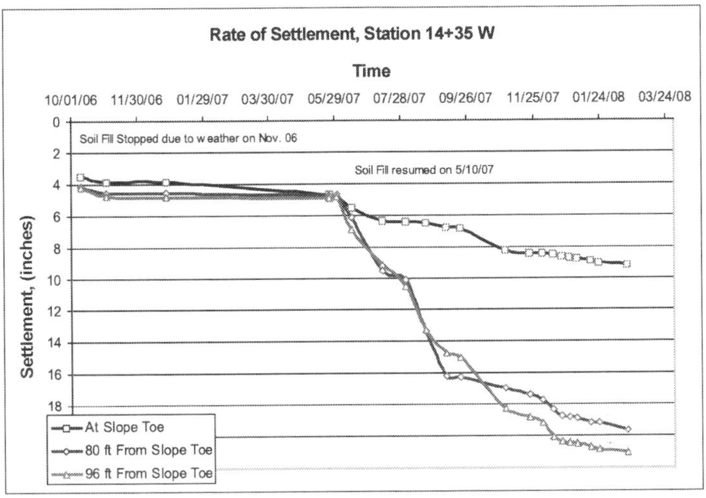

FIG 5. Typical Settlement-Time Plot

Measurements on the heave points did not indicate the occurrence of significant vertical movement. Vertical movements of less than 1.5 cm (0.6 inch) were measured for the timeframe reported in this paper.

PRACTICAL ISSUES

During the design phase, the issues most important to the project team were cost, the construction schedule, and the risk of slope failures into the wetlands area. By implementing an instrumentation program, the least cost alternative could be used and the construction schedule potentially shortened. By using the instrumentation results to control the rate of fill placement, the risk of slope failures into the wetlands was reduced.

An unexpected benefit of the instrumentation program was that while monitoring the construction of the working platform, it was noted that the surface of the first phase of the working platform exhibited better stability than expected. As a consequence, the depth of the upper layer of well graded crushed stone was decreased by 30 cm (1 foot), saving the owner approximately 20,000 tons of crushed stone and 10 days off the construction schedule.

ACKNOWLEDGEMENTS

Special thanks go to Mr. Ted Johnson of the Piedmont Triad International Airport, Mr. Kevin Baker of Baker Associates, and Mr. Carl Ellington, P.E. of Talbert & Bright, who have been involved through the design and construction phases of this project. In addition, thanks to Dr. J. Brian Anderson, P.E., Professor of Civil

Engineering at the University of North Carolina at Charlotte, and Dr. Manuel Gutiérrez, P.E., for their Plaxis finite element analyses. We would also like to thank Mr. Fran Furfaro of Blythe Construction Inc., for all his help in installing the instrumentation in the field. Additional thanks to Lisa Stafford for her word processing and formatting help.

REFERENCES

Fell, R., Hunter, G. 2003. Prediction of Impending Failure of Embankments on Soft Ground. *Canadian Geotechnical Journal*, Vol. 40: 209-220.

Han, et al. 2004. Evaluation of Deep-Seated Slope Stability of Embankments over Deep Mixed Foundations. *Proceedings of GeoSupport Conference: Innovation and Cooperation in the Geo-Industry*, Jan. 29-31, 2004, Orlando, Florida

Hayes, J.A. 1990. "The Marchetti Dilatometer and Compressibility", Southern Ontario Section of the Canadian Geotechnical Society. Seminar on "In Situ Testing and Monitoring", September 1990.

Kamal, Abdul Aziz A., Lane, Pauleen A., and Heshmati, Ali A.R. Parametric Study of Reinforced and Unreinforced Embankment on Soft Soil. *Proceedings of 13th ACME Conference*, March 21-22, 2005, University of Manchester.

Ladd, C.C. 1991. Stability Evaluation during Staged Construction. *Journal of Geotechnical Engineering* Vol. 117 (No. 4): 540-615.

INSTRUMENTATION AND MONITORING OF MSE WALLS SUPPORTED ON THE RAMMED AGGREGATE PIER® SYSTEM

Terry Micnhimer, PE[1], Jorge R. Parra, PhD, PE[2], and Tommy Williamson[3]

[1]Senior Geotechnical Engineer, Tolunay Wong Engineers, 10710 S. Sam Houston Parkway West, Suite 100, Houston, TX 77031 tmicnhimer@tweinc.com
[2]Lead Engineer, Geopier Foundation Company, Inc., 150 Fairview Road, Suite 335, Mooresville, NC 28117, jparra@geopier.com
[3]President, Geopier Foundation Company-Houston, 17602 Sierra Creek Lane, Humble, TX 77346 tlw90@aol.com

ABSTRACT: As part of the upgrade plans for the U.S. Highway 90 and Highway 6 Interchange, the Texas Department of Transportation (TXDOT) proposed the construction of a series of bridges and ramps. The bridge abutments and access ramps required the construction of Mechanically Stabilized Earth (MSE) retaining walls with heights of up to 8.2-m (27-feet). The weak clay foundation soils along the wall alignment presented global instability, settlement, and bearing capacity challenges. As an alternative to massive over-excavation and replacement or preloading with a surcharging, the design team and the Texas Department of Transportation selected a *Rammed Aggregate Pier®* solution. The Rammed Aggregate Pier system (RAPs) increases the factors of safety for bearing capacity and global stability as a result of the high angle of internal friction achieved during ramming and reduces the magnitude and time of settlement by increasing the overall stiffness of the foundation soils and providing a drainage pathway for dissipation of excess pore water pressure. The performance of the Rammed Aggregate Pier-supported walls was monitored using a suite of geotechnical instrumentation consisting of vertical and horizontal inclinometers, vibrating wire piezometers, and Sondex vertical settlement gauges located near the critical sections. The instrumentation was monitored for approximately one year and indicated acceptable performance.

This paper discusses the results of the monitoring program of an MSE wall supported by the Rammed Aggregate Pier system. This work is of particular significance because it is the first MSE wall support application performed in Texas and the first instrumented wall project supported on Rammed Aggregate Piers funded by the Federal Highway Administration (FHWA).

INTRODUCTION

As the Texas Department of Transportation continues with its $40 million dollar, 3-year plan of infrastructure development and improvement, construction at the US-90A and State Highway 6 Interchange in Sugarland, Texas, resulted in the need for support of new MSE walls for ramps and bridge abutments.

The foundation soils along the wall alignments were predominantly soft to medium stiff clay in the upper 9-m (30 feet) representing a challenge for achieving the factors of safety for global stability and bearing capacity required by the design team.

The Rammed Aggregate Pier reinforcing elements improve the bearing capacity of the foundation soils by providing significant increases in the composite shear resistance (foundation soils + RAP) because of their high angle of internal friction developed during the ramming process. Additionally, the piers act as vertical drains reducing the potential for slope instability caused by excess pore water pressure in the slope.

Monitoring of the MSE walls at two ramps and two bridge abutments was scheduled to occur for a period of 24 months following RAP installations. Based on the positive performance and quick stabilization of the walls, and small amount of vertical and horizontal wall movement and quick dissipation of excess pore pressures measured with the different instrumentation, the monitoring period was reduced from 24 months to less than 12 months.

PROJECT DESCRIPTION AND SUBSURFACE CONDITIONS

Table 1 summarizes the scope of work for areas requiring ground improvement. The MSE wall selected for this project was a Tricona Retained Wall System consisting of segmental precast panels (4.2 square meters) connected to steel reinforcement. The design of the segmental wall followed AASHTO codes.

Table 1. MSE Wall Scope of Work using Rammed Aggregate Piers

Location	Wall Length m (ft)	Maximum wall height (H_{wall}) m (ft)
North Abutment	79 (260)	7.3 (24)
South Abutment	69 (227)	8.2 (27)
North Ramp	108 (353)	7.3 (24)
South Ramp	108 (353)	6.7 (22)

The design requirements for the MSE wall related to ground improvement are summarized in Table 2.

Table 2. Wall Design Requirements

Design Criterion	Requirement
Global Stability	FS = 2.0 (drained and undrained conditions)
Bearing Capacity	FS = 1.3 (drained and undrained conditions)
Post-Construction Settlement	Less than 5-cm occurring within 6-months following end of construction

The geotechnical investigation was performed in 2001 by HVJ Associates, Houston, TX. The soil conditions consists of soft to medium stiff clay extending to depths of approximately 9-m to 12-m underlain by loose sandy silt to silty sand to depths of 15.2-m to 18.3-m, underlain by medium dense sand to silty sand to the maximum explored depth. Groundwater (GWT) was encountered at a depth of approximately 3-m. Figure 1a shows natural moisture content (w%) and effective stress shear strength parameter values with depth along with compressibility parameter values. Figure 1b shows undrained shear strength (S_u) with depth along with classifications.

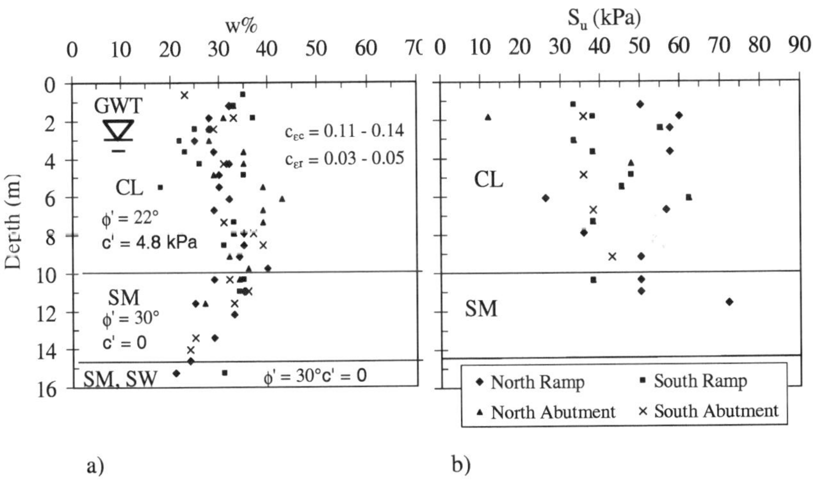

Fig. 1. Soil characteristics a) moisture content, compressibility and strength parameter values vs depth, b) Undrained shear strength vs depth.

RAP DESIGN APPROACH FOR MSE WALL SUPPORT

Rammed Aggregate Pier construction consists of a three step procedure as described in the literature by Fox and Cowell (Fox, N.S. and Cowell, M.J. (1998)) The design approach to address global stability and bearing capacity under MSE walls consists of increasing the composite shear strength parameter values within the Rammed Aggregate Pier-reinforced zones. The composite shear strength parameter values are estimated as a weighted average (FHWA 1999, Barksdale and Bachus 1983, Mitchell, 1981, Fitzpatrick and Wissmann, 2002). The reinforced zones with composite shear strength parameter values were modeled as rectangular areas directly beneath the MSE wall and backfill using the computer software GSlope version 4 by Mitre Software Corporation. The approach is based on two-dimensional limit equilibrium analyses using Bishop's method to determine factors of safety.

The design approach to estimate settlement within the RAP-reinforced zone (upper zone) and the area beneath the upper zone (lower zone) is based on a spring analogy and a two-layer approach as described by Lawton et. al. (1994), Lawton and Fox (1994). A modulus test is performed early during construction following general procedures for pile static compression tests (ASTM, 1994) to verify the assumed stiffness modulus value for the RAP element. The time for consolidation settlement is estimated taking into account radial drainage to the RAP elements and using Barron's approach (1948). Settlement of the soil below RAP-reinforced zone is estimated using conventional geotechnical approaches and includes an influence factor of unity.

The RAP design engineered for this project to satisfy the design requirements indicated previously in Table 2 consisted of closely-spaced RAPs along the wall face and within the reinforced zone to address bearing capacity, global stability and settlement control. The RAPs were installed to a maximum depth (H_s) of 7.6 m in most areas to reinforce through the soft clay and intersect any possible deep seated slip surface. Bearing capacity and global stability issues typically govern and dictate the spacing along the wall face and reinforced zone. These two design requirements normally do not govern the spacing required further back behind the main reinforced zone. For that reason, reinforcing elements at a wider spacing can be afforded for settlement control and consolidation time acceleration purposes only. Figure 2 summarizes the typical pier spacing and shaft lengths.

The project included a total of 1411 drilled RAPs for the four walls. PCI, the licensed installer, performed the installation with two working crews in approximately eight months. RAPs were installed during Phase I from July, 2005 through January, 2006. During Phase II, RAPs were installed from October, 2006 through January, 2007. Fill placement at Phase I started on January, 2006 through February, 2006.

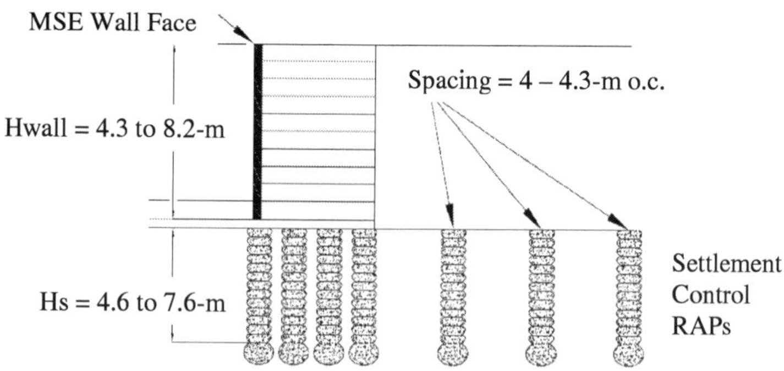

Fig. 2. Rammed Aggregate Pier design layout

The stiffness of the RAP elements was measured in the field to verify the initial spring stiffness assumed for design. A modulus test was performed at every wall location as shown in Figure 3. The design was based on the assumption of a RAP stiffness modulus value of 27 MN/m^3 (100 pci) at the design pressure of 862 kPa (18000 psf). The stiffness modulus values measured during the performance of the modulus tests exceeded the initially assumed value by at least 50 percent. The tests consistently indicated that the point of maximum curvature occurred at the design stress or at slightly higher stress.

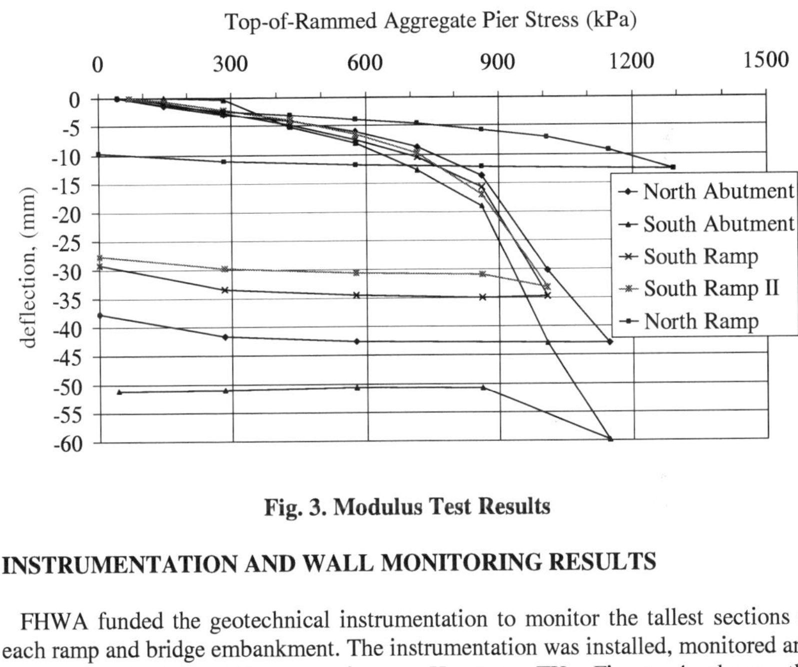

Fig. 3. Modulus Test Results

INSTRUMENTATION AND WALL MONITORING RESULTS

FHWA funded the geotechnical instrumentation to monitor the tallest sections at each ramp and bridge embankment. The instrumentation was installed, monitored and plotted by Tolunay Wong Engineers, Houston, TX. Figure 4 shows the instrumentation location for the south bridge abutment. The instrumentation consisted of horizontal inclinometers located perpendicular to the center line of the embankment extending over the subgrade and over the tops of some of the piers from one wall face to the other. Also, vertical inclinometers and Sondex settlement gauges were installed outside and adjacent to the wall face. Vibrating wire piezometers were installed at the middle of the embankment in grouted vertical nests. The vertical inclinometers and Sondex gages were both installed to a depth of 15.2 m (50 ft).

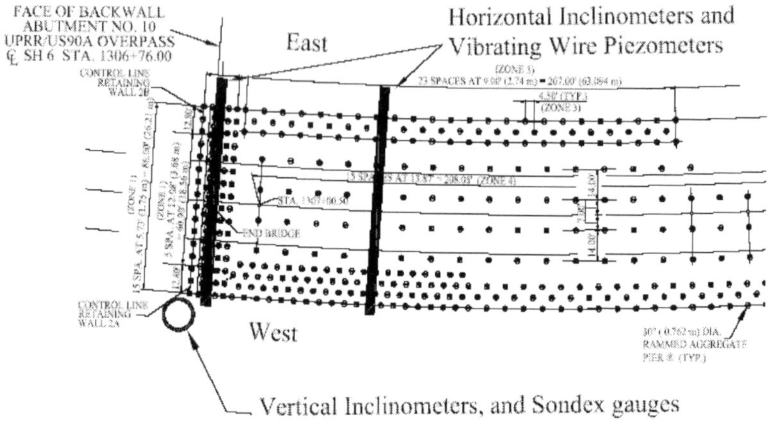

Fig. 4. Geotechnical Instrumentation Layout

The piezometer readings at the South Ramp are representative of the trend in excess pore pressures at all four structures and are shown in Figure 5. The piezometric readings registered minimal change in excess pore pressure during earth fill placement because of the radial drainage component to the closely spaced RAP elements. Pile driving operations in the bridge abutment early during construction explain the highest peak on the curve on December, 2005. Regional rise in groundwater elevation due to heavy rain activity was responsible for the gradual rise in piezometric readings observed after June, 2006.

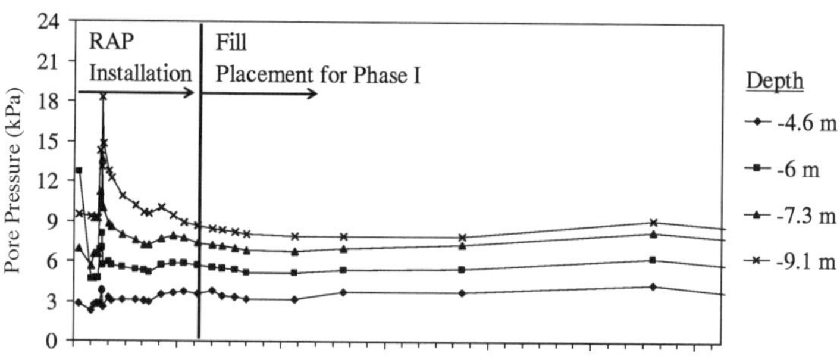

Fig. 5. Piezometric Readings at the South Bridge Abutment.

Figure 6 shows vertical a deformation profile across the bottom of the south ramp wall. Overall vertical settlement for all four structures was on the order of 5-cm (2-

inches) or less, mostly occurring during construction. Post-construction settlement was less than 1.5-cm (0.5-inch). Disturbance at ground surface from construction equipment passing near the wall face produced the apparent excessive deformation shown on the left hand side of the curve.

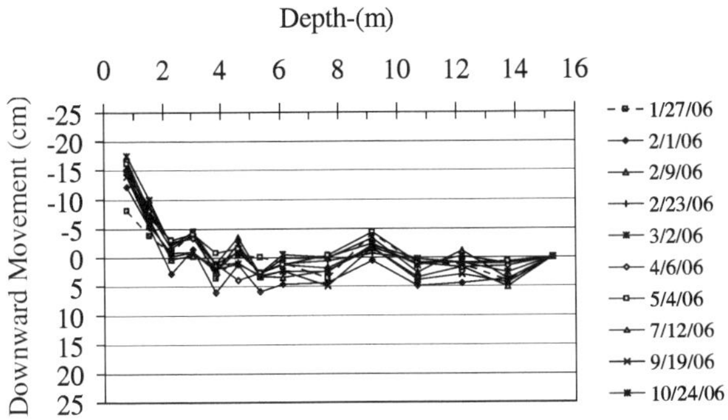

Fig. 6. Sondex settlement reading for the South Ramp.

Figures 7a and 7b show the maximum and minimum horizontal soil displacements measured near the wall face for all four structures, respectively, taken from February, 2006 through June, 2007.

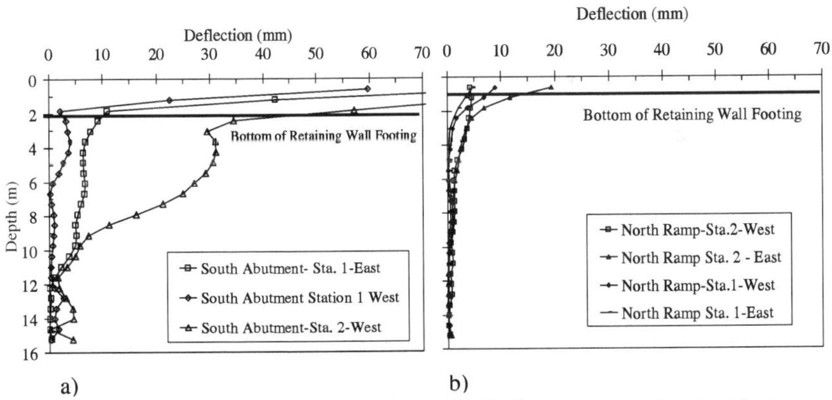

Fig. 7. Inclinometer Readings. a) Lateral displacement at South Abutment, b) Lateral displacement at North Ramp.

The lateral soil displacement for the majority of the walls was consistent with the readings at the North Ramp and generally on the order of 1.25-cm (0.5-in). The maximum lateral soil displacement was approximately 3.2-cm (1.25-in) and occurred in front of the South Abutment wall which had the maximum wall height and softer soil near ground surface. The apparent excessive deformation measured near bottom of wall footing corresponds to disturbance caused by construction equipment operating near the monitoring station.

SUMMARY AND CONCLUSIONS

Rammed Aggregate Pier soil reinforcing elements were selected to increase the factors of safety against bearing capacity, global stability and settlement control under MSE walls in compressible and unstable foundation soils. Modulus tests along with a monitoring program testified that the performance of the MSE reinforced by Rammed Aggregate Piers satisfied the design criteria established by the design team. The post-construction settlement following wall construction was significantly less than the original design. This is explained by the stiffening and improvement of the foundation soils and because of the practically immediate excess pore pressure dissipation afforded by radial drainage to the RAP elements allowing both shear strength increase, reduction in compressibility of the matrix soil and faster settlement as the wall was constructed.

ACKNOWLEDGMENTS

The authors acknowledge the efforts of the Rammed Aggregate Pier installer, PCI, of Reinbeck, Iowa, for their help, support, and coordination during the monitoring program. Mr. Stanley Yin with TXDOT is also acknowledged for his support and initiative.

REFERENCES

ASTM (1994). "Standard Test Method for Piles under Static Axial Compression Load", Annual Book of ASTM Standards, Volume 4.08, Soil and Rock, pp1-11

Barksdale, R.D., Bachus, R.C. (1983). "Design and Construction of Stone Columns, Volume I". Report No. 1 FHWA/RD 83/026, 210 pp

Barron, R.A., (1948). "Consolidation of fine-grained soils by drain wells". ASCE, Volume 113, pp 718 - 742.

Federal Highway Administration (1999). Ground Improvement Technical Summaries, Volume II. Demonstration Project 116. Publication No. FHWA-SA-98-086.

Fox, N.S. and Cowell, M.J. (1998). Geopier Foundation and Soil Reinforcement Manual. Geopier Foundation Company, Inc., Scottsdale, AZ.

Lawton, E. C., and Fox, N.S. (1994). "Settlement of Structures Supported on Marginal or Inadequate Soils Stiffened with Short Aggregate Piers". Proceedings, Vertical and Horizontal Deformations of Foundations and Embankments. College Station, TX.

Mitchell, J.K. (1981). "Soil Improvement: State of the Art". Session 12, Tenth International Conference on Soil Mechanics and Foundation Engineering, Stockholm, Sweden, June 15-19.

Wissmann, K., et.al. (2002). "Improving global stability and controlling settlement with Geopier soil reinforcing elements", Proceedings, 4th International Conference on Ground Improvement. Kuala Lumpur, Malaysia.

Effects of relative creep of geosynthetic-reinforcements on the responses of geosynthetic MSE walls

Huabei Liu[1], Member, ASCE, Xiangyu Wang[2], Erxiang Song[3]

1. Assistant Professor, Department of Civil Engineering, City College of New York/CUNY, 160 Convent Ave., New York, NY 10031. Email: hbliu@ce.ccny.cuny.edu
2. Ph.D. student, Department of Civil Engineering, Tsinghua University, Beijing 100084, China
3. Professor, Department of Civil Engineering, Tsinghua University, Beijing 100084, China

ABSTRACT

The creep of geosynthetics leads to the decrease of long-term strength and the increase of deformation. More importantly, the influences of creep of geosynthetics are also affected by the creep properties of soils. In this paper, the effects of creep of geosynthetic-reinforcements are investigated using Finite Element procedure, on the context of its relations to the creep of backfill soils. The nonlinearity and creep of both geosynthetics and backfills were considered in the procedure. A model MSE walls at a height of eight-meter was analyzed using the Finite Element program ABAQUS. Geosynthetic reinforcements with different rates of creep were analyzed to investigate the effects of relative creep on the responses of the model MSE wall. It is found from the analysis that the relative creep rate between geosynthetic reinforcement and backfill soil is very important to the responses of the model wall. Both the reinforcement loads and the stress states in the soils were influenced by the relative creep rates of the two materials.

INTRODUCTION

The Load Resistance Factor Design of MSE walls using geosynthetic reinforcements is complicated by the time-dependent properties of geosynthetics. Unlike metallic reinforcements, geosynthetic reinforcement experiences creep deformation under sustained load but its load tends to relax if its strain is restrained to develop. If clean granular backfill soil is used in geosynthetic MSE walls, the backfill soil exhibits negligible creep deformation, hence the straining of geosynthetic reinforcements would be restrained and the maximum reinforcement load in the wall would decrease. The horizontal displacement of MSE wall would also cease to develop during its early service life. Such behavior has been reported from field observation and full-scale testing (e.g., Allen et al. 2002) as well as numerical simulation (e.g., Liu and Ling 2007). However, if marginal backfill soil with cohesive fine content is used as backfill, the backfill soil would also exhibit considerable creep deformation under sustained loading. The response of MSE walls would then be dependent on the relative creep rates of the backfill soil and reinforcement. The development of horizontal displacement, the variation of reinforcement load and the change of stress states in the backfill are much complicated and deserve extensive investigation.

In this study, Finite Element procedure was used to preliminarily investigate the effects of relative creep rates of backfill soil and reinforcement on the response of MSE walls. A model MSE walls with clayey silt backfill was analyzed but geosynthetic reinforcements with different creep rates were used. The results of the different cases were compared and analyzed.

FINITE ELEMENT MODEL

An eight-meter high model MSE wall using geosynthetic reinforcements was analyzed in this study using the Finite Element program ABAQUS (Abaqus, inc., 2004). The setup of the model wall is shown in Fig. 1. Plane strain condition was assumed in the analysis. The model wall was assumed to have a vertical concrete block facing. The backfill soil was assumed to be a clayey silt with a unit weight of 20 kN/m^3. A friction angle of 30 and a cohesion of 16 kPa were assumed for the backfill soil, considering the fact that backfill soil in a MSE wall is generally in an unsaturated state that can lead to moderate cohesion in the soil. The reinforcement length L = 5.6 m (0.7H) and the reinforcement spacing S = 0.6 m, which was obtained using FHWA design procedure and assuming a rupture strength of 55 kN/m based on a real HDPE geogrid.

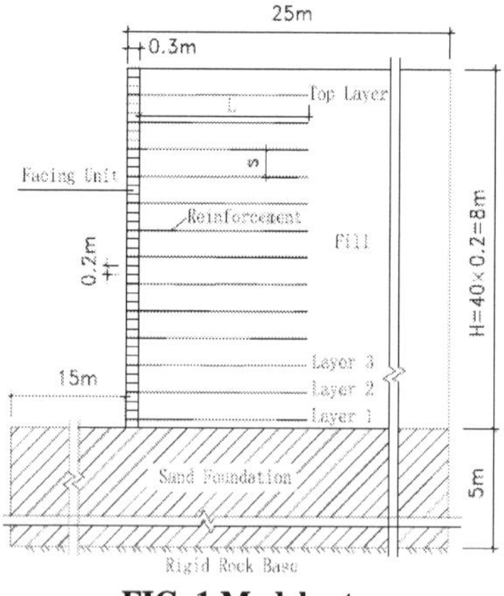

FIG. 1 Model setup

The backfill soil was modeled using a modified Drucker-Prager Creep model. In the original Drucker-Prager Creep model, which is a model inherent in ABAQUS, linear elastic behavior is assumed. In the present study, it was modified to consider the nonlinearity of soil based on the well-known hyperbola formula. The tangent shear modulus is defined as:

$$G = G_{\max} / (1 + R_f' G_{\max} |\gamma^n| / \tau_f)^2 \quad (1)$$

in which $G_{\max} = G_0 \cdot p_a \cdot (p / p_a)^{n_G}$, γ^n is the time-independent shear strain, R_f' is the failure ratio, τ_f is the yield shear stress of soil, and p is the effective confining stress. The bulk modulus of the soil is assumed to be only dependent on the confining stress:

$$K = K_0 \cdot p_a \cdot (p / p_a)^{n_K} \quad (3)$$

But its value must follow the restriction on Poisson's ratio. The yielding of soil is assumed to follow the Drucker-Prager Criterion, the yield surface of which can be expressed in p-q space as (Abaqus, Inc., 2004):

$$f = q - p \tan\beta - d = 0 \quad (3)$$

in which β and *d* are related to the friction angle ϕ and cohesion c, respectively. Non-assoicated flow rule can be considered with dilation angle ψ different from the friction angle ϕ. The creep strain rate follows the Singh-Mitchell creep model (Abaqus, Inc., 2004):

$$\dot{\bar{\varepsilon}}^c = A \cdot \exp(\xi \cdot \bar{\sigma}^c) \cdot (t_1 / t)^m \quad (4)$$

in which $\bar{\sigma}^c = (q - p\tan\beta)/(1 - 1/3 \cdot \tan\beta)$.

The modified Drucker-Prager Creep model can model the nonlinear, elasto-plastic and creep behavior of soils. Although it is obvious that it cannot describe the dependence of strength on stress path, since plane strain condition was assumed in this study and the strength parameters of soil at plane strain conditions can be used in the analysis, the results of Finite Element Analysis can still reproduce the behavior of MSE walls. In the present study, the backfill soil was modeled using $G_0 = 550$, $K_0 = 500$, $n_G = n_K = 0.5$, $R_f' = 0.9$, ϕ=30°, c = 16 kPa, $\psi = 0°$, $m = 0.85$, $A = 3.5 \times 10^{-6}\,\mathrm{s}^{-1}$, $\xi = 0.11\,\mathrm{kPa}^{-1}$, and $t_1 = 1\,\mathrm{s}$. With this set of parameters, the soil exhibits moderate creep rate, as can be seen in Fig. 2a, which shows the creep behavior of the backfill soil in triaxial creep test with σ_3= 50 kPa. It is understood that cohesion of backfill soil could change with time due to rain fall. However, during the majority of service life, cohesive backfill soils can maintain certain cohesion. In this study, it is assumed that that the cohesion of backfill soil does not change in the service life of GRS walls. The effects of cohesion change will be investigated in a later stage.

The geosynthetic-reinforcement was modeled using the elasto-plastic visco-plastic bounding surface model in Liu and Ling (2005, 2007), which was incorporated into ABAQUS by the authors. The model parameters for a real geogrid, HDPE Geogrid 1 in Liu and Ling (2007) were used to simulate the nonlinear and creep behavior of geosynthetic reinforcement in Case 5. In order to investigate the effects of reinforcement creep, for the other cases, different viscous parameter η's were used: Case 1: $\eta = 2.2\times10^{25}$ $(kN/m)^{1/n}$, Case 2: $\eta = 2.2\times10^{13}$ $(kN/m)^{1/n}$, Case 3: $\eta = 2.2\times10^{11}$ $(kN/m)^{1/n}$, Case 4: $\eta = 2.2\times10^{10}$ $(kN/m)^{1/n}$, and Case 6: $\eta = 2.2\times10^{8}$ $(kN/m)^{1/n}$. In case 1, η is very large, indicating that the reinforcement creep was negligible and the reinforcement can be considered to be non-viscous. The comparison of reinforcement creep is shown in Fig. 2b.

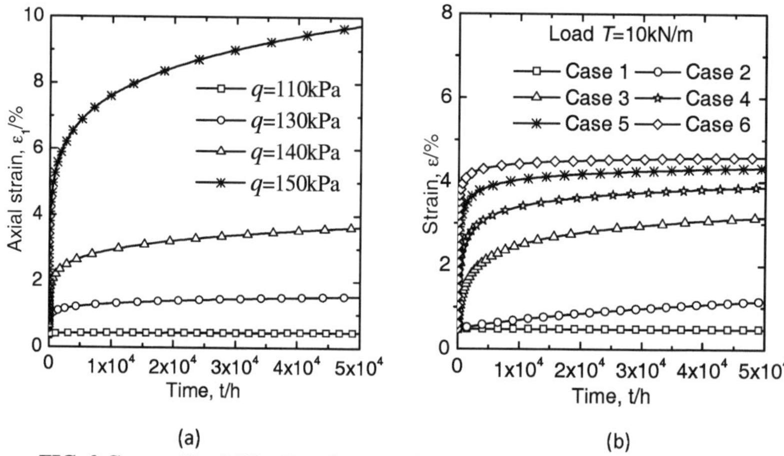

FIG. 2 Creep of backfill soil and geosynthetic reinforcements (a) backfill soil; (b) geosynthetic reinforcement

The foundation soil was assumed to be a dense sandy soil that followed nonlinear elastic behavior described by Eqs. (1) and (2), with K_0 = 1100 kPa, G_0 = 1250 kPa, The facing was assumed to be linear elastic with a unit weight of 24 kN/m^3. The interfaces between backfill and facing as well as those between facing blocks were simulated using thin-layer element following Mohr-Coulomb failure criterion. Thin-layer elements also existed between reinforcement layer and backfill soil, but their constitutive properties follow those of backfill soils. Hence the time-dependent interaction between reinforcement and backfill soil can still be considered. No degradation on the strength of soil-geosynthetic interface was considered in the present study considering the large aperture-size of most geogrids used in practice (e.g. Hatami and Bathurst 2006). The reinforcement was assumed to be pinned to the facing blocks.

Altogether six cases were analyzed in order to investigate the effects of reinforcement creep on the response of MSE walls.

RESULTS OF FINITE ELEMENT ANALYSIS

Horizontal Facing Displacement

Fig. 3a shows the horizontal facing displacements at the end of construction (EOC) and five years afterwards. It can be seen that with the same backfill soils, with an increase in the reinforcement creep rate, the horizontal displacement increased to some extent at EOC, but the increase was much more significant five years afterwards. It can also be seen that the horizontal displacement increased with time, even for the case with non-viscous reinforcement (Case 1) which is due to the creep deformation of backfill soils. However, with larger reinforcement creep, the horizontal displacement developed more significantly with time. This trend is better demonstrated in Fig. 3b, which shows the development of horizontal displacement with time at h = 3 m.

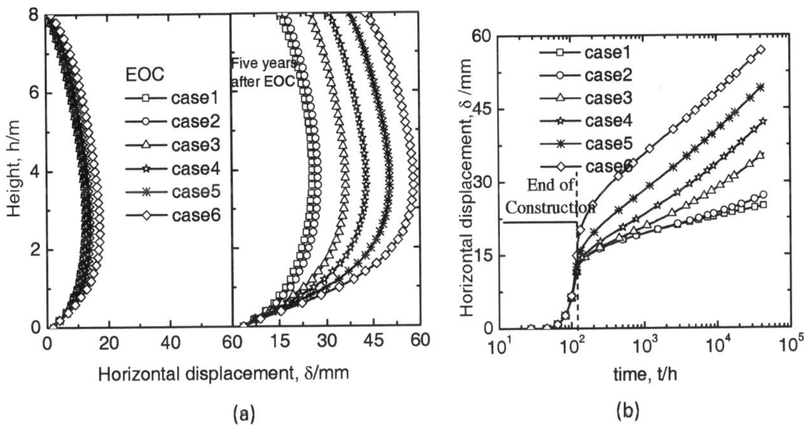

FIG. 3 Horizontal facing displacements: (a) comparison of displacements at EOC and five years afterwards (b) development of displacement at h = 3 m

Reinforcement Load

In contrast to the horizontal facing displacements, reinforcement load increased significantly if non-viscous reinforcement was used, as shown in Fig. 4. At the end of construction, the difference in the reinforcement loads can already be witnessed. With very large creep rate (Cases 5&6), even during construction, the reinforcement already experienced considerable load relaxation, which lead to smaller load in comparison to the cases with smaller reinforcement creep. After construction, for the cases with large reinforcement creep, the load relaxation continued to develop in the lower and mid portion of the wall and five years afterwards, and the loads in Cases 3, 4, 5 and 6 were approximately the same, but in the non-viscous case (Case 1), the loads were much larger. Figs. 4b&c show the development of reinforcement loads in Reinforcement Layers 2 and 13. Reinforcement Layer 2 was the one with the largest reinforcement load in the wall, while Layer 13 was the top layer. The speed of load relaxation decreased with a decrease in the reinforcement creep rate. From Fig. 4a, it can also be seen that with larger reinforcement creep rate, the distributions of reinforcement load along the wall height became uniform. The upper reinforcement layers were stretched due to increasing shear deformation of soil at the top portion and their loads increased slightly, as can be seen in Fig. 3. For the non-viscous case, on the other hand, the reinforcement loads in all layers continued to increase with time, owing to the stretching of reinforcement induced by the creep deformation of backfill soil.

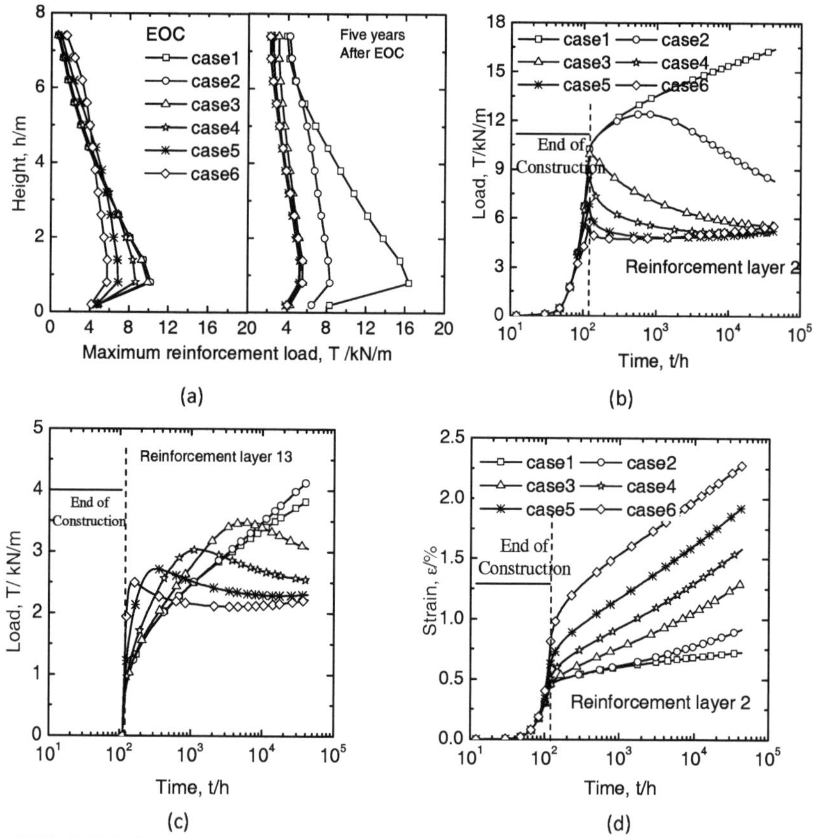

FIG. 4 Reinforcement loads: (a) comparison of reinforcement load at EOC and five years afterwards; (b) development of reinforcement load in Layer 2; (c) development of reinforcement load in layer 13; (d) development of reinforcement strain in Layer 2

Fig. 4d shows the development of maximum strain in Reinforcement Layer 2. As expected, Case 6 exhibited the largest strain increase, although its load relaxed the fastest. Due to the large creep reinforcement creep rate in Case 6, the reinforced soil mass continued to deform and the increase of soil deformation was larger than its creep deformation. Load was thus transferred from the reinforcement layer to the adjacent soil, leading to load relaxation in the reinforcement.

Stress State in Backfill Soil

The stress state in the backfill soil is discussed indirectly using the distribution of plastic shear strain in this paper. The plastic octahedral shear strain γ_{oct}^{p} is used to illustrate the development of soil yielding.

Fig. 5 compares the distribution of plastic shear strains in Case 1 and Case 6 at EOC and five years afterwards. Clearly, in Case 1, the yielding of backfill soil did not developed due to the restriction of reinforcement. On the other hand, in Case 6, not only that the plastic region was much larger than that of Case 1 at EOC, it continued to develop with time, due to the transfer of load from the reinforcement to the backfill soil.

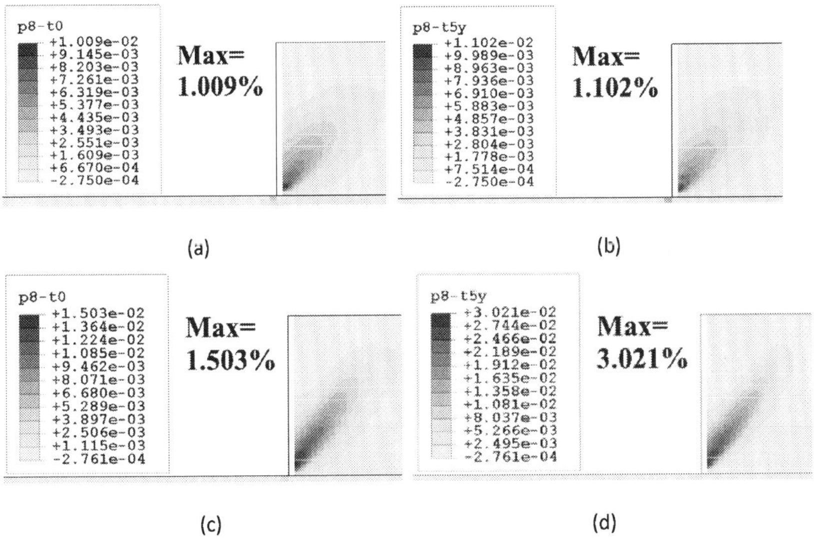

FIG. 5 Plastic regions (γ_{oct}^{p}) in the backfill soil: (a) Case 1, EOC; (b) Case 1, 5 years; (c) Case 6, EOC; (d) Case 6, 5 years

DISCUSSIONS AND CONCLUSIONS

From the results of Finite Element analysis, it can be seen that not only the deformation of GRS walls were significantly affected by the relative creep of reinforcements and backfill soil, the load and stress distribution in the system was also dependent on it. With the same backfill soil, larger reinforcement creep would lead to larger wall deformation, and the backfill soil would carry more load while the reinforcement load tended to relax with time. On the other hand, reinforcement with very small creep could restrict the creep deformation of backfill soil, hence leading to smaller wall deformation, but the reinforcement load could increase with time, depending on the creep deformation of backfill soil.

The results from this study indicates that in the Load Resistant Factor Design of geosynthetic-reinforced MSE walls, it is necessary to take into account the relative creep rate of reinforcement and backfill soil, especially if marginal backfill soil with cohesive fine contents is used. As the first step, while calculating long-term reinforcement load, it might be more rational to take into account the effects of relative creep, and increase or reduce reinforcement load based on the relative creep rates.

ACKNOWLEDGEMENT

This study was carried out by the support from National Natural Science Foundation of China awarded to the first author when he was working at Tsinghua University in China (Grant No. 50408002). The support is gratefully acknowledged.

REFERENCES

Abaqus, Inc (2004), *ABAQUS/Standard user's manuals*, Version 6.4, Providence, Rhode Island.

Allen, T. M., and Bathurst, R. J. (2002). "Observed long-term performance of geosynthetic walls, and implications for design." *Geosynthetics International*, 9(5–6), 567–606.

Hatami, K., and Bathurst, R.J. (2006). "A numerical model for reinforced soil segmental walls under surcharge loading." Journal of Geotechnical and Geoenvironmental Engineering, ASCE, 132(6), 673-684.

Liu, H., and Ling, H.I. (2005). "Constitutive modeling of the time-dependent monotonic and cyclic behavior of geosynthetics." Geosynthetics and Geosynthetic-Engineered Soil Structures, H. I. Ling et al. Ed., Columbia University, New York, NY, pp. 281-302.

Liu, H., and Ling, H. I. (2007). "A unified elastoplastic-viscoplastic bounding surface model of geosynthetics and its applications to GRS-RW analysis." *Journal of Engineering Mechanics, ASCE*, 133(7), 801-815.

Instrumentation of MSE Wall Containing Laterally Loaded Drilled Shafts

Matthew C. Pierson[1], Robert L. Parsons P.E[2], Jie Han P.E.[3], James J. Brennan P.E.[4], and Christina Vulova[5]

[1] Graduate Student. Department of Civil, Environmental, and Architectural Engineering, University of Kansas 1530 W 15th St., Room 2150 Lawrence, KS 66045; PH 785-393-2427; email: matthewp@ku.edu

[2] Associate Professor, Department of Civil, Environmental and Architectural Engineering, University of Kansas, 1530 W 15th St., Room 2150, Lawrence, KS 66045; PH (785) 864-2946; email: rparsons@ku.edu

[3] Associate Professor, Department of Civil, Environmental and Architectural Engineering, University of Kansas, 1530 W 15th St., Room 2150, Lawrence, KS 66045; PH (785) 864-3714; email: jiehan@ku.edu

[4] Assistant Geotechnical Engineer, Kansas Department of Transportation, 2300 Van Buren, Topeka, KS 66611-1195; PH (785) 296-3008; email:brennan@ksdot.org

[5] Application Technology Manager, Tensar International Corporation, 5883 Glenridge Dr., Suite 200, Atlanta GA 30328; PH (404) 250-1290; email:cvulova@tensarcorp.com

ABSTRACT

Construction, instrumentation, and lateral load testing of eight 0.9m (36in.) diameter columns solely supported by the geogrid-reinforced backfill behind a 6m (20 ft) tall MSE wall was conducted for the Kansas Department of Transportation (KDOT). This paper describes the instrumentation of the wall and the performance of that instrumentation. Instrumentation included 24 pressure cells, 16 inclinometer locations, 112 strain gauges, 20 telltales, 84 photo target locations on the wall facing, and load cells and LVDTs associated with lateral load and response. A detailed discussion of the photogrammetric process used to remotely measure wall deflections through the use of cameras and AutoCAD is presented. A discussion of the relative benefits of the different methods of instrumentation is included.

INTRODUCTION

This paper contains a description of the instrumentation of a 43 m (140 ft) long, 6 m (20 ft) tall mechanically stabilized earth (MSE) wall with eight shafts penetrating the reinforced backfill. The shafts were laterally loaded to failure in the direction of the wall facing. Instrumentation consisted of 24 pressure cells, 16 inclinometer locations, 112 strain gauges, 20 telltales, 84 photo target locations on the wall facing, and load cells and LVDTs associated with lateral load and response (Figure 1). A detailed description of the photogrammetric process, which was used to analyze wall facing and internal displacements during lateral load testing, is included. The costs and relative benefits of the types of instrumentation are also discussed.

PROJECT DESCRIPTION

Lateral load testing was conducted on eight 0.9 m (3.0 ft.) diameter concrete shafts, supported by and contained within the reinforced mass of an MSE wall (Figure 1). The wall was built using modular block units and uniaxial high density polyethylene punched-drawn geogrid. The completed wall was 6 m (20 ft) tall including 0.9 m (3.0 ft) of embedment. It was divided into five test sections 4.6 m (15 ft) in length and one group test section 13.7 m (45 ft) in length. Four of the shafts had depths equal to the full height of the wall and were 0.9, 1.8, 2.7, and 3.7 m behind the back of the wall facing. Measured from the center of the shafts to the back of the facing blocks, one shaft was 1.8 m from the wall facing and extended through 75% of the full wall height (4.6 m). Three shafts spaced 4.6 m (15 ft) apart and 1.8 m (6 ft) from the facing were tested as a group. The test sections were separated by slip joints, where geogrid and blocks were cut to isolate movement within the section being tested.

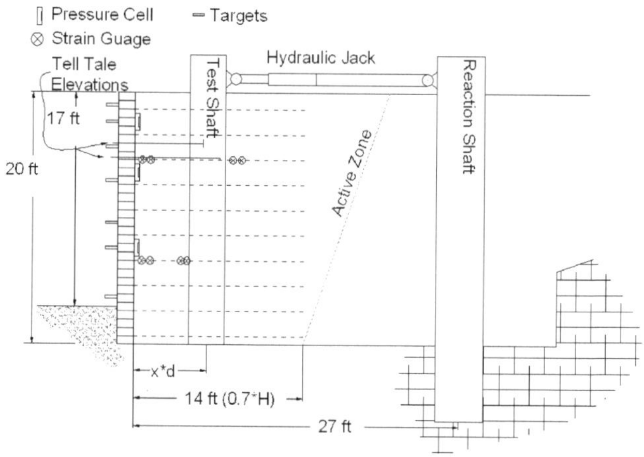

FIG. 1 MSE wall test sections.

INSTRUMENTATION

The instrumentation of this project was selected to monitor the following: load and displacement of the shafts; wall displacements both inline with loading and off center of loading; pressures behind the wall facing; strains within the geogrid; and movements within the wall mass. All of the data collected are being used to calibrate three dimensional computer models. Load cells, hydraulic pressure transducers, inclinometers and Linear Variable Displacement Transducers (LVDTs) were used to measure the response of the shafts. Total pressure cells (TPC) mounted directly behind the facing in front of the shafts, strain gauges attached to reinforcement, telltales, photogrammetry, and excavation and inspection were used to measure the response of the MSE wall system. All strain gauges and pressure cells measurements in the area of the test were automatically recorded every twenty seconds to a Campbell Scientific CR 1000 data logger.

Shaft Instrumentation

Instrumentation of the shafts consisted of the data logger and instrumentation provided by Applied Foundation Testing. It was used to monitor hydraulic pressure, load cells, and LVDTs. The pressure was monitored directly from the hydraulic manifold. On all but the group test, load cells were attached in line with the hydraulic jacks (Figure 2). For the group, each of two hydraulic jacks had a load cell placed directly inline. Also on the group a single load cell was placed between the center shaft and the loading beam. During all but the group test, each test shaft and each reaction shaft was fitted with two LVDTs at different elevations to produce an initial slope. During the group test each reaction shaft and each of the two hydraulic jacks was fitted with an LVDT. Two LVDTs were also fitted to the loading beam to describe its movements. All LVDTs that were not fitted to a hydraulic jack were supported by a reference beam. An inclinometer casing was included within each test shaft for determination of displacement and bending of the shaft.

FIG. 2 Typical test setup for single shafts

Strain Guages

112 strain gauges were attached to the geogrid in 5 test sections on 4 different layers of grid. Pairs of strain gauges were located on the top and on the bottom of the geogrid at up to six different distances from the wall facing. For protection the wires were run through small flexible tubing toward each slip joint. From the slip joint the wires were encased in PVC pipe and run from the slip joint to the data logger.

Inclinometers

During each test an inclinometer was lowered into the test shaft every other load step. Inclinometer measurements provided information on the horizontal movement and rotation of the shaft that could be used to determine internal fill movement and provide guidance for the development of p-y curves. Readings were also taken after the final loading and the unloading steps on the test and reaction shafts. For the group test two inclinometer casings were also placed right behind the face of the wall and spaced evenly between the three test shaft centerlines. Inclination readings were taken during every other interval at these locations as well to describe the wall movement during loading and unloading.

Telltales

Two types of telltales were used on this project. The first type of telltales were constructed using 6.4 mm (1/4 in.) diameter all-thread passing through 12.7 mm (1/2 in.) electrical conduit. The all-thread was then attached to geogrid by drilling a hole in the transverse bar of the geogrid, and fasteners. The second type of telltale was constructed out of 12.7mm electrical conduit passing through 19.1 mm (3/4 in.) conduit. The 12.7 mm conduit had a plate welded to the end to fix the end in the fill. The telltales were monitored with the same technique as the wall targets.

Total Pressure Cells

Total pressure cells (TPC) were located on the back of the wall, directly in front of each shaft at 2.3, 4.2, and 5.4m (7.7, 13.7, and 17.7 ft) elevation. The TPC wires were run through the slip joints directly to the data logger. All but three cells were recessed in a concrete backer block that provided a solid surface for the back of each cell. For comparison purposes, three cells did not have a backer block to evaluate the potential benefit provided by the backer block. Each cell had a sand bag in front to protect the surface from damage due to the aggregate backfill (Figure 3). The bag was made of non-woven geotextile, folded in half and stapled. The goal was to have 25mm (1 in.) of sand in front of the pressure cell to distribute the load to the pressure cell plate more evenly. This was achieved by placing the empty bag in front the cell and placing crushed stone next to the bag. The bags were filled with a small amount of sand and tamped to reach the 25mm (1 in.) thickness. This process was repeated until each block was covered with crushed stone.

Photogrammetry

Wall facing and telltales were monitored with photogrammetry. Eighty four facing blocks had a target attached. Columns of targets were attached to the wall facing directly in front of each shaft. A row of targets was also installed along the entire wall at an elevation of 5.4 m (17.7 ft). A fixed digital camera was used to capture images of the wall facing, including the targets, as it deflected. These images were then sequentially imported into AutoCAD to analyze the deflection of the wall. The accuracy provided by the camera averaged 1 mm (0.04 in).

FIG. 3 Left: Covered pressure cell Center: Cell and protective sand bag.

Targets were made of 2.54 cm (1 in.) white PVC and attached to the wall facing with concrete anchors and 2.54 cm (1 in.) PVC end caps. Each target has a black center that was 15.2 cm (6 in.) long, with white on either end to help distinguish the scale (Figure 4). The target scales also automatically compensated for a camera position that was not perfectly perpendicular to target movement because the apparent length of the scale in the image was reduced proportionally.

A digital single lens reflex (SLR) camera with a 10 megapixel resolution was used with an 18-70mm lens. The lens and resolution both play a role in the accuracy of the measurements.

Prior to testing, the precision achievable with the camera and lens was evaluated by taking images of gradational checkerboard targets from fixed distances. It was determined that the camera and lens combination at a distance of 4.6 m (15 ft) and zoomed such that 6 m of elevation was visible was able to distinguish movements as small as 1 mm.

An external shutter switch or trigger was used to avoid handling the camera. The focus and the zoom were fixed during the test to avoid changing the scale of the images. Maintaining focus on targets at different distances simultaneously was addressed using aperture settings in the camera that enabled the distance in focus to be increased at the cost of decreased brightness.

Analysis of the images was done using AutoCAD. The first image of each test representing the initial position of the wall was first rastered into a drawing. A fixed reference point that was easily identifiable in each subsequent image was then chosen. Choosing a large reference in the center of the image was determined to be the most useful due to possible changes in zoom or focus that could skew subsequent images. Each target's scale was outlined in the first image to establish the initial position of the target (rectangular box in Figure 4). This box was also used to determine the scale of each target. The scale was then input into a spreadsheet for later use. The next image was then rastered into CAD and placed behind the previously drawn lines. The fixed reference in the image was then checked and corrected for location as well as scale. Each target was then checked for movement. If movement occurred then one edge of the target was marked (vertical lines for each measurement in Figure 4). The distances from the initial box to these lines using the previously determined scale were calculated and recorded.

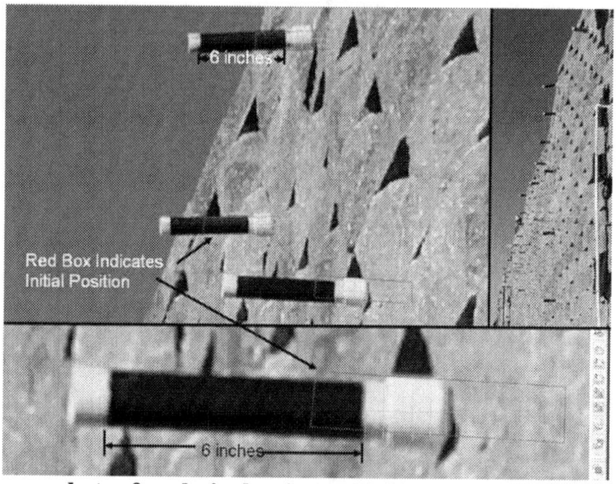

FIG. 4 Screen shots of analysis showing red initial position and green current position.

The advantages of photogrammetry were that it provided a relatively inexpensive and uncomplicated field measurement of multiple targets and telltales simultaneously. The most significant disadvantage is that accuracy decreased as distance from the camera increased. Another disadvantage was that movement could not be directly measured parallel to the camera. There was also a significant amount of post-processing time to obtain wall facing displacements. Maintaining the camera in a fixed position, to avoid constant centering of the fixed reference during post-processing, helped reduce processing time.

RELATIVE BENEFITS OF MEASUREMENTS

The relative benefits of each method of instrumentation varied based on the objective. The monitoring of the shaft was a priority and for this reason redundancy was designed into the shaft instrumentation. Some primary and secondary redundant systems included: the load cell and the hydraulic pressure transducer for load, multiple LVDTs and inclinometers for shaft deflection, and targets and inclinometers for wall deflection. In each case the more accurate measurement system was mentioned first, but the secondary system could have been used if necessary. One unfortunate consequence of using the inclinometers was the additional time required to take measurements, averaging about 7 minutes per shaft, per reading. All of these systems worked well and provided useful information.

Pressure cells provided valuable information to be used in design and computer modeling. No alternative or redundant system was available so it is difficult to assess the quality and value of this system. It was determined through comparison of pressure cells with and without a backer block that a backer block was not necessary to obtain more accurate readings. The pressure cells were effective in detecting an increase in stress during wall construction and loading.

The strain gauges that were attached to the geogrid also provided very useful information, including daily changes in strain with temperature fluctuations. Installation of the instrumented sections of geogrid was somewhat difficult due to the number of gauges and the length of wire required. The wires had a tendency to tangle and passing as many as fifty wires through one 76 mm diameter PVC pipe proved difficult. A below average (20%) strain gauge failure rate was experienced on this project.

The use of telltales proved to be highly effective when combined with the photogrammetric method as shown in Figure 5. Construction of the equipment was simple and inexpensive. The photogrammetric process was highly preferred for monitoring wall facing movements and internal mass movements.

A summary of observations regarding the various types of instrumentation is presented in Table 1. Additional results produced from these methods are presented in Parsons, et al (2009).

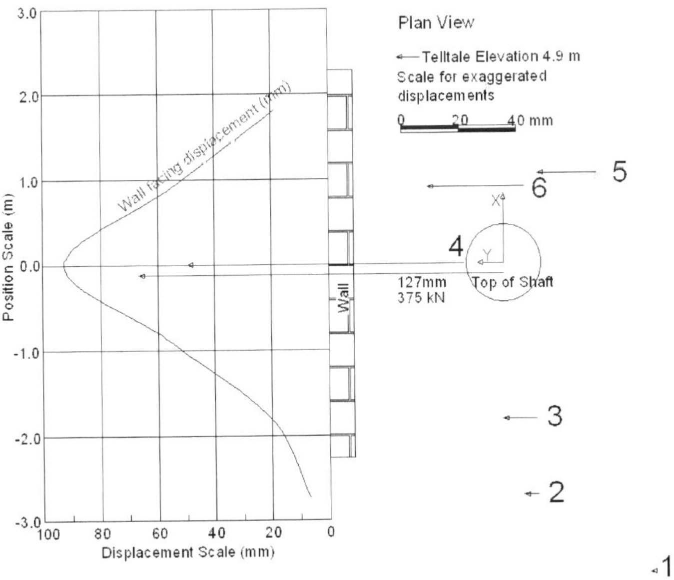

FIG. 5 Wall facing and telltale displacements at 127 mm of top of shaft movement

TABLE 1 Summary of Instrumentation

Item	Cost $/Item	Installation Effort	Data Recording	Post Processing Effort
Target[1]	1	Low	Instant	High (45min/test)
Inclinometer Casing[2]	160/shaft	Low	7min/shaft	Moderate
Pressure Cells[3]	620	Moderate	Instant	Low
Strain gauges[3]	15	Moderate	Instant	Low
Telltales[1]	20	Low	Instant	High (45min/test)
LVDT's[3]	400	High*	Instant	Low

** Due to required mounting structure*
[1]Camera and Tripod Required
[2]Data Logger Required
[3]Data Acquisition System Required

CONCLUSIONS

Appropriate instrumentation is necessary to maximize the value of physical testing. Often more then one method is available to acquire the required data. Specific conclusions reached as a part of the research include:

- Photogrammetry has proven to be very efficient and effective at measuring wall and telltale displacements.
- Displacements as measured from wall targets, telltales, and LVDTs were consistent.
- The use of a backer block behind pressure cells for support is not required to detect pressure accurately.
- Strain gauges on geogrid can be useful, but challenging to install.
- Inclinometers were useful for monitoring shaft behavior and for redundancy, but caused delays during testing.

Additional information about construction, testing, and results related to this MSE wall project can be found in Pierson (2009).

Acknowledgements

The authors wish to acknowledge the financial and logistical contributions of the Kansas Department of Transportation and Tensar International. The people of the KDOT Bureau of Materials and Research and District One Maintenance were particularly helpful in completing this project. The authors also wish to acknowledge the assistance of Dan Brown and Associates in designing the shafts and testing protocol. The dedication of all those involved is greatly appreciated.

REFERENCES

Pierson, M. C., Parsons, R. L., Han, Jie, Brown D.A (2009) "Capacity of Laterally Loaded Shafts Constructed Behind the Face of a mechanically Stabilized Earth Block Wall." Prepared for the Kansas Department of Transportation 90 Pages.

Parsons, Robert L., Pierson M.C., Han, Jie, Brennan J.J., Brown D.A. (2009). "Lateral Load Capacity of Cast-in-Place Shafts behind an MSE Wall." *Submitted to Foundation Engineering Congress and Equipment Expo* (IFCEE'09), March 2009, Orland FL.

Reinforced Soil Slope (RSS) Failure and Reconstruction
US 70 Design-Build Project, Ruidoso to Riverside, New Mexico

Elizabeth M. Smith[1], PE, GE, M. ASCE and Brendan R. Fisher[2], PE, LEG, M. ASCE

[1]Principal, National Transportation Manager, Terracon Consultants, Inc., 5307 Industrial Oaks Blvd., Austin, TX 78735; emsmith@terracon.com
[2]Principal, Fisher & Strickler Rock Engineering, LLC, 38579 SE River St., Suite 9, Snoqualmie, Washington, 98056; bfisher@fsrengineering.com

ABSTRACT: A major geotechnical slope failure occurred at Sta. 1663 on the US 70 Design-Build project in New Mexico a year after construction. The 1663 reinforced soil slope was 7.6 m (25 ft.) high and constructed at 0.5H:1V. A slope stability-type failure occurred, with movement associated only with severe rain storms. This case history study will describe the project site conditions, design methods, failure monitoring and analysis, and reconstruction of the RSS. The authors were involved in all aspects of design, construction, forensic evaluations, reconstruction, and claims resolution. The failure is notable because following reconstruction, the contractor was successful in a dispute claim based upon differing site conditions. This paper describes the range of project evolution from design, failure monitoring, forensic evalutions, and reconstruction, to the dispute resolution process and findings.

INTRODUCTION

The US 70 project, located in the Hondo Valley between the towns of Ruidoso Downs and Riverside, was the first design-build highway project in the State of New Mexico. The US 70 alignment winds along the foothills of mountains to the north and the rivers Rio Ruidoso and Rio Hondo to the south. A key requirement of the project was to preserve the rural character of the Hondo Valley and minimize environmental impacts to the rivers. The highway design, as a result, required extensive cut slope excavation on the north side of US 70, and retaining walls and reinforced soil slope (RSS) construction on the south side, to keep the new highway footprint from significantly affecting sensitive environmental resources.

RSS DESIGN, SITE CONDITIONS, AND CONSTRUCTION

The planned RSS at Sta. 1663 ranged in height from 2.4 to 6 m (8 to 20 ft.) at a 1:1 (horizontal:vertical) slope inclination above the Rio Hondo. A down slope of varying heights and inclinations was located at the toe of the proposed RSS slope and ended at the bank of the Rio Hondo. Photo 1 shows an aerial view of the site, with the approximate RSS location noted.

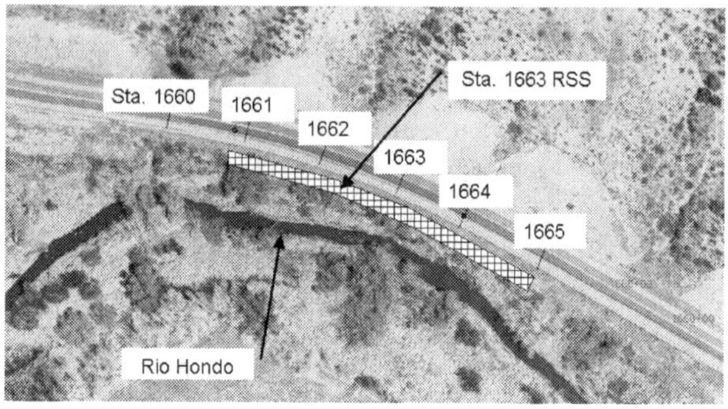

Photo 1: Aerial View of RSS Location

The geotechnical investigation for the site consisted of drilling two soil borings adjacent to the deepest parts of the roadway fill along this slope. The subsurface investigation in 2002 indicated the geology in the area was terrace deposits and alluvium overlying Yeso formation mudstone; groundwater was not encountered in the borings drilled at this site. No signs of slope instability were noted during the site reconnaissance.

For the critical design slope, the geotechnical engineers evaluated stability and settlement at Sta. 1663, where the tallest part of the 1:1 RSS was planned above an existing 2.7 m (9 ft.) high 1:1 slope. The design included a toe key of excavated and recompacted soil below the new RSS. Following completion of the design, environmental restrictions adjacent to the Rio Hondo reduced the distance to the toe of the RSS slope. The project designers steepened the RSS to fit within the allowable area at a ½:1 inclination; the geotechnical engineers re-evaluated the steeper slope and calculated a safety factor of 1.3, which met the project design requirements and the RSS design criteria.

The contractor, Sierra Blanca Constructors (SBC), constructed the reinforced slope at Sta. 1663 in the summer of 2003. Because the toe of the RSS was very close to the edge of the river, SBC also constructed a riprap-filled trench at the toe of the RSS to protect it from river scour. The final configuration of the RSS had a front slope that ranged from ¼:1 to ½:1, which was steeper than the planned ½:1 face.

SLOPE DISTRESS

In January 2004, guardrail posts were installed along the top of slope through the asphalt shoulder; shortly thereafter, cracks were observed in the guardrail pad radiating from the posts. SBC performed survey monitoring of the slope, and no movements were observed between February and early August 2004.

Heavy rains in July and August, 2004 resulted in severe pavement cracking and rotational RSS movement. The project team excavated a test trench through the roadway to evaluate the conditions at the cracked area, and noted vertical cracks observed at the surface were about 5 cm (2 in.) deep and were located directly behind the top layer of geogrid.

FIELD INVESTIGATION

Additional field exploration activities began in August 2004. The team drilled seven additional borings to depths of 4.1 to 14.7 m (13.5 to 48.2 ft.) below the roadway. Boring, piezometer, and inclinometer locations are shown on Figure 3. The borings were used to establish a detailed stratigraphic log of the subsurface conditions at the RSS and to collect soil samples for laboratory testing. Piezometers were installed at selected drilling locations to allow groundwater measurement.

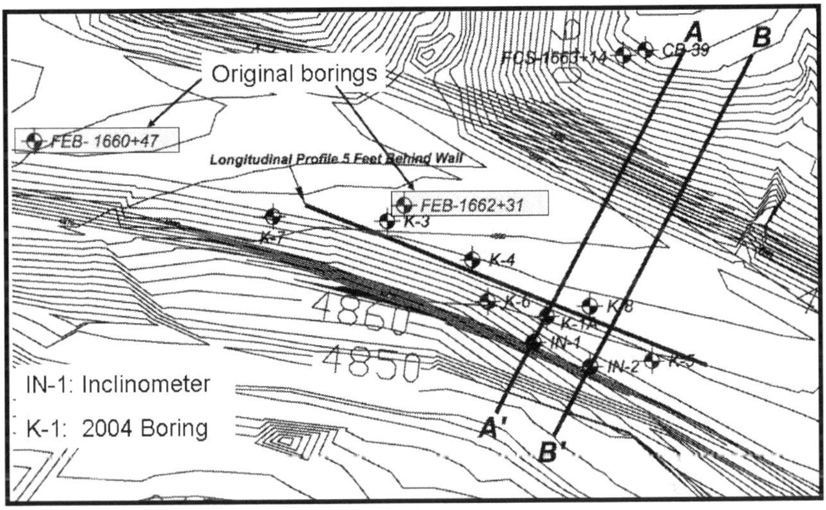

FIG. 3: Boring Location Plan

SLOPE MONITORING

In addition to the surveying points installed during January 2004, the project team constructed two inclinometers in August 2004. The inclinometer casing was set to depths below the RSS, with the principal grooves aligned in the south direction (toward the river). Figures 1 and 2 show the readings collected at Inclinometer 2 (Sta. 1663+65), which indicated the most movement.

The inclinometer readings, coupled with rain gauge readings in the area, indicated that major slope movements occurred following significant rainfall. These rainfall events correspond to points 1 through 4 on Figures 1 and 2. Between storms, the RSS essentially didn't move. The piezometer readings were consistently below the river water elevation, even after significant storms. The RSS movement appears to be rotational, with the RSS behaving as designed – as a reinforced block. About one month after installing and reading the inclinometers, the geotechnical engineer recommended removing the top five feet of soil behind the RSS to unload the slope and mitigate further movement. Survey data indicated movements slowed considerably after the top 1.5 m (5 ft.) of the slope was removed (the inclinometers were destroyed during this removal). The slope moved significantly to the south towards the river between late September and early October. The maximum cumulative displacement measured at the top of the inclinometers was approximately 8.9 cm (3.5 in.) at IN-2, and a clear failure surface was apparent just beneath the bottom geogrid of the RSS.

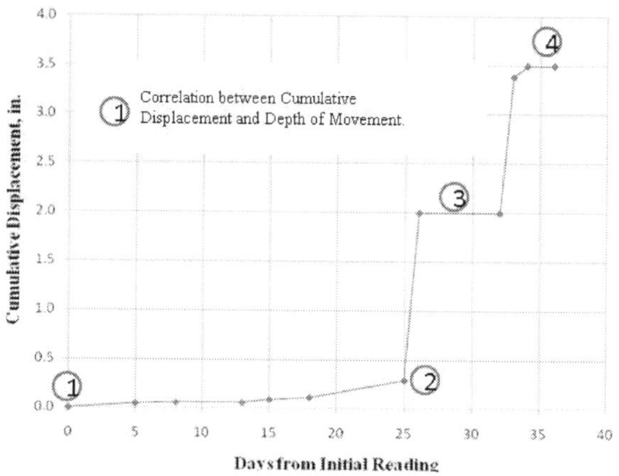

FIG. 1: Cumulative Displacement at IN-2

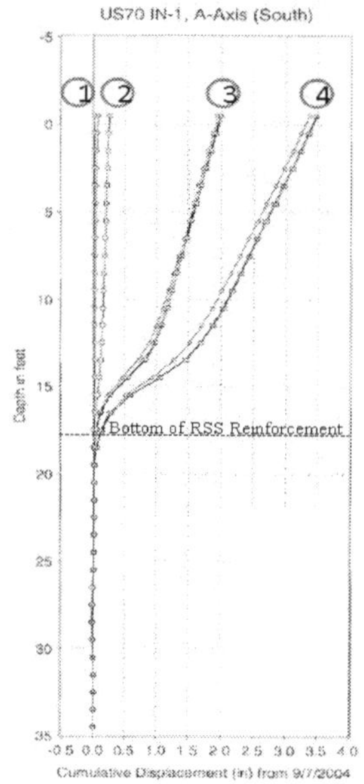

FIG. 2: Displacement vs. Depth at IN-2

BACK ANALYSIS

The additional subsurface explorations confirmed the interpretation made in 2002, although materials encountered below the RSS appeared to be either old fill or colluvium (CL and SM in Figure 4). After advancing the borings on site, establishing the geologic stratigraphy and groundwater conditions, the authors completed computer simulated slope stability (sensitivity) analysis to gain insight regarding the RSS failure mechanism(s) and remedial strategies.

The first part of the analysis involved estimating the reason for RSS movement. Readings from the inclinometers suggested movement just below the slope within a silty sand (SM) layer. These movements corresponded to periods of heavy precipitation and therefore, we hypothesized that water pressures behind the RSS

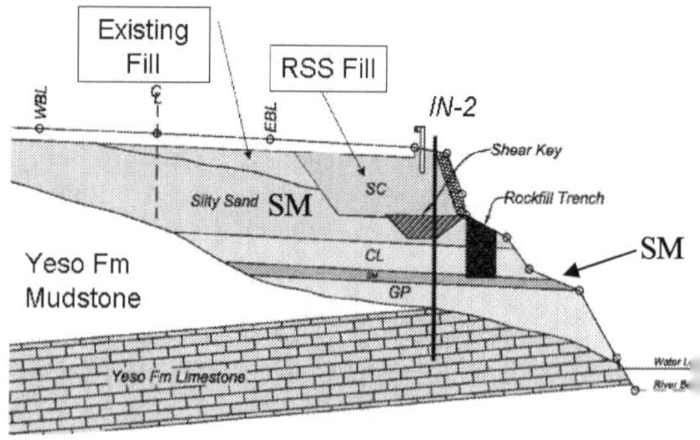

FIG. 4: Section B-B'

and/or uplift water pressures beneath the slope were prompting the slope to move towards Rio Hondo. We considered circular failures with shearing behind or through the RSS reinforcement zone and through the soil beneath the RSS. We also evaluated block-type failure scenarios. In each of the analyses, we allowed the computer program to consider only those surfaces that would pass through the zone of shearing as recorded in the inclinometers. We established that when water pressures were significantly increased and translational failures were modeled, the predicted safety factors approached one. When circular-type failures were modeled, the safety factors were greater than one, which suggested translational sliding of the RSS was more likely than a deep-seated circular failure. Our analysis concluded the RSS distress was the result of peak flows and saturation during heavy and prolonged rainfall that developed excess water pressure at the rear of the RSS.

SOURCES OF WATER INFILTRATION

Based on the sensitivity analysis, the slope as constructed should have been stable unless excess pore water pressures developed behind the RSS backfill. We reviewed the site topographic plans and as-built information and identified the following possible sources of water:

- Rapid river water rise during storms
- Surface runoff infiltration
- Subsurface infiltration from the north side of the roadway, whether from fractures in the bedrock or infiltration from the drainage ditch
- Possible subsurface features from unknown drainage features that were hidden by past construction
- Previous culverts in the area were to have been removed during previous construction, although as-built plans indicated at least one culvert may not have been removed (a note to remove the culvert was crossed-out on the as-built plans)

REPAIR AND RECONSTRUCTION

Our engineering evaluation provided us with several potential remediation strategies. After discussing various scenarios with the project team and considering the constructability of the possible remediation strategies, we reached the following conclusions:

- The inclinometer readings showed the correlation of slope movement to significant rainfall events.
- Movement of the RSS was likely because of an increase of water pressure behind the RSS.
- The origin of the water pressure was not the result of water level rise in the Rio Hondo to flood level.

Based on our engineering evaluations and conclusions, and a minimum factor of safety of 1.5, we developed reconstruction recommendations as shown graphically on Figure 5.

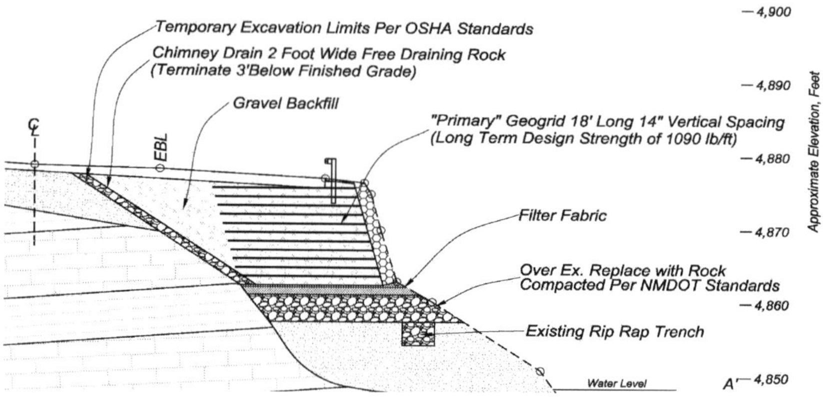

FIG 5: RSS Reconstruction Recommendations

Figure 6 shows the profile of the reconstructed RSS showing the ultimate bedrock surface and the fill section beneath the RSS. Figure 7 presents a typical as-built cross-section of the repair, with the original recommendations shaded back. During excavation, we observed about 0.6 to 0.9 m (2 to 3 ft.) of old fill overlying a layer of roots and 0.3 to 0.6 m (1 to 2 ft.) of silty fine sand immediately above bedrock. Both

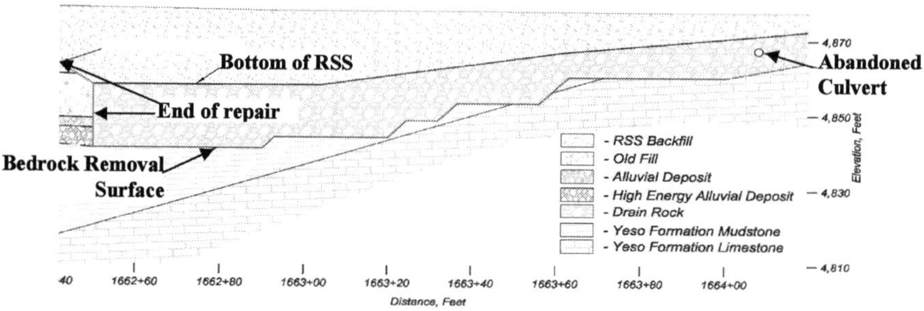

FIG 6: As-Build Longitudinal Profile

layers were dark gray to dark brown and contained significant quantities of organics. The old fill was recognizable as fill because of the presence of small wood fragments and other man-made debris. The underlying silty sand appeared to be alluvium because of its relatively homogenous clean sandy texture. It appears that this segment of the original road embankment involved placing fill directly over alluvium and river deposits on the north side of the Hondo River (to construct the south side of the embankment), which resulted in a slightly straightened riverbank in this area. This fill was apparently placed during the original road construction in the 1930's.

The contractor (SBC) excavated about 6 m (20 ft.) long segments to bedrock to create a keyway and to remove unsuitable material below the embankment. SBC lined the back and sides of the keyway excavation with filter fabric to reduce the potential for piping of fines from the native, finer-grained materials into the rock fill.

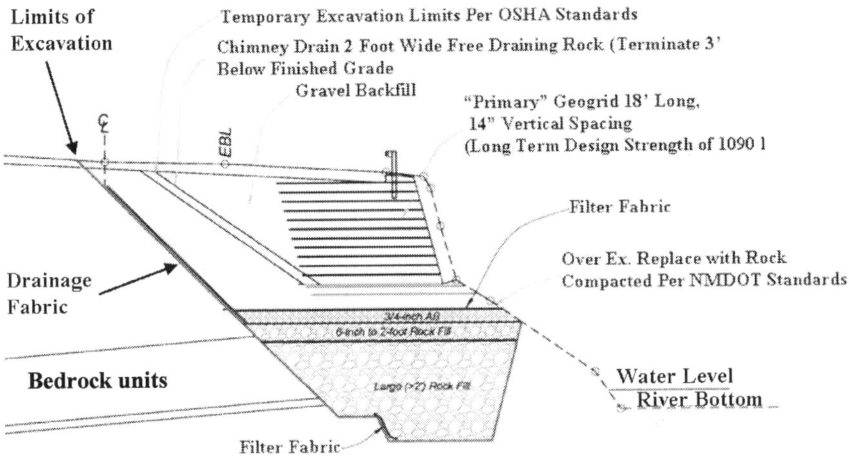

FIG 7: As Built Cross Section

Each keyway segment was backfilled with angular rockfill [typically greater than 0.6 m (2 ft.) size] to protect against undermining and erosion during peak river flows. Another 0.6 m (2 ft.) of smaller rockfill 15 cm to 0.6 m (6 in. to 2 ft.) size was placed over the large rockfill, followed by approximately a 0.3 m (1 ft.) thick layer of 1.9 cm (¾ in.) aggregate base (AB) compacted into the rock mass. The AB layer was smoothed to form a foundation for the new RSS, and SBC used filter fabric to separate the AB from the RSS fill.

On the back of the excavation, SBC placed a geocomposite drain fabric that extended vertically from just below the existing pavement down to the smaller rockfill layer. Once the drain fabric was in place, SBC reconstructed the RSS with a wire basket facing at a ½:1 stepped slope.

1664 CULVERT

During reconstruction of the RSS, an existing culvert was exposed at Sta. 1664+08 (see Photo 2). The culvert consisted of a 0.6 m (24 in.) pipe that extended from 7.9 m (26 ft.) north to 2.1 m (7 ft.) south of the US 70 centerline. The project plans from the 1930's show the location of the culvert. A small swale is visible at the top of the cut slope along the north side of the highway adjacent to the old culvert location. Seepage had been noted in the Rio Hondo riverbank at about this location when the slope distress was

Photo 2: Existing Culvert in Excavation

noted in August 2004. US 70 was widened in the late 1980's and the 1988 plans showed a culvert to be removed at about Sta. 1664. However, the construction note is crossed off of the as-built plan and no other information is shown regarding the disposition of the culvert. Based on the findings during reconstruction of the RSS, the culvert had not been removed and was likely a significant source of water pressure behind the RSS. SBC abandoned the culvert with grout during reconstruction of the RSS.

DISPUTE RESOLUTION

Our team concluded the cause of the RSS failure were lateral seepage pressure and unsuitable material beneath the old fill. The presence of excess groundwater pressures and unsuitable materials deep below the previous toe of slope were not anticipated at the time the project was designed. SBC therefore submitted a differing site conditions change order to recoup the cost of the RSS reconstruction. When the claim was rejected, SBC filed a dispute with the project's Dispute Resolution Board (DRB). A formal hearing was held to resolve the dispute that consisted of 30 minute presentations by each side (SBC and NMDOT), followed by questions from the Board. The parties were given 1 hour to prepare rebuttals, which were then presented to the Board. The Board conferred and reached their decision within about 2 hours after the final presentations.

The DRB found that the geotechnical investigation and design were appropriate and performed within the standard of care. They agreed that unsuitable materials should not have been present beneath the existing fill, and the presence of such unsuitable materials was the responsibility of the DOT. NMDOT contended during their presentation the 1664 culvert would have been crushed at the uphill side, and could not have been the cause of the water pressure behind the RSS. Because SBC did not fully expose the uphill end of the culvert, there was no proof that the culvert was indeed the culprit in causing the excess water pressures. The DRB concluded that control of groundwater was the responsibility of the contractor and denied that portion of the change order. The final result was SBC was reimbursed for about half the cost of repairing the RSS.

ACKNOWLEDGEMENTS

This work was performed while the authors were employed with Kleinfelder, Inc. Kleinfelder was a subconsultant to URS, the lead designer on the project under contract to SBC. We appreciate URS' and Kleinfelder's consent to present this paper. The graphics and photos presented were from the studies and reports prepared by Kleinfelder. Thanks to Sierra Blanca Constructors (a Joint Venture of Granite Construction Company, Joint Venture Sponsor; James Hamilton Construction Co., and Sundt Construction Inc.) for allowing us to share this information with the geotechnical community. We also thank Robert A. Meyers, PE, NMDOT State Materials Engineer and Gary Shubert, PE, District Engineer, NMDOT District 2.

Lessons Learned: Field Installation of Strain Gages on High-Strength Geotextile

Nicole A. Walsh[1], Stud.M.ASCE, Christopher L. Meehan[2], A.M.ASCE, and Dov Leshchinsky[3], M.ASCE

[1]Graduate Student, Dept. of Civil and Environmental Engineering, 301 DuPont Hall, Newark, DE 19716; e-mail: nikkihoy@udel.edu
[2]Assistant Professor, Dept. of Civil and Environmental Engineering, 301 DuPont Hall, Newark, DE 19716; e-mail: cmeehan@udel.edu
[3]Professor, Dept. of Civil and Environmental Engineering, 301 DuPont Hall, Newark, DE 19716; e-mail: dov@udel.edu

ABSTRACT

The Cherry Island Landfill (CIL), located in Wilmington, DE, is undergoing a major expansion to increase its service capacity. To increase landfill volume, a mechanically stabilized earth (MSE) wall is being constructed along three sides of the site. Very soft, compressible soils characterize the subsurface conditions, creating the potential for large settlements and widespread slope stability issues. To facilitate strength gain of the foundation soils, the proposed design includes prefabricated vertical drains (PVDs) and staged construction of the MSE wall. To reduce lateral spreading movements and improve the system's global slope stability, high-strength geotextile has been placed beneath the MSE wall. As a component of the project's instrumentation program, the geotextiles have been instrumented with strain gages to monitor the performance of the reinforcement. This paper discusses the methodology used to install and monitor the strain gages and presents lessons learned from gage performance to date.

INTRODUCTION

Located at the confluence of the Christina and Delaware Rivers, CIL became operational in 1985 and is currently operating near waste capacity. The Delaware Solid Waste Authority (DSWA), the CIL owner-operator, endeavors to vertically extend the landfill to increase its volume by constructing an MSE wall along three sides of the site. Upon completion, the MSE wall will be 18.3 m high, nearly 3.2 km long, and constructed with 1.5 million m^3 of fill (Espinoza et al. 2008).

CIL is founded on dredged material reclaimed from the adjacent Delaware River. Consequently, very soft, compressible soils characterize the subsurface conditions. With nearly indistinguishable boundaries, the compressible layers are comprised of dredge and older alluvial materials and are considered to be normally-consolidated to under-consolidated clayey silts. Their combined thickness varies between 18.3 m and 30.5 m, and undrained shear strengths range between 9.6 kN/m^2 and 47.9 kN/m^2 (Espinoza et al. 2008). Natural moisture content ranges from 58% to 74%, liquid limit ranges from 79% to 98%, and plasticity index ranges from 41% to 54%. These values integrate the results of field and laboratory testing associated with this project and other representative historic data from the site (Geosyntec 2004).

To accelerate consolidation and facilitate strength gain of the weak foundation soil, the project includes the installation of PVDs and staged construction. Settlement compensation berms were designed in anticipation of high settlements. To reduce movements associated with lateral squeeze, high-strength geotextile reinforcement was also incorporated into the design. An extensive instrumentation program includes inclinometers, vibrating wire piezometers, liquid settlement sensors, and settlement plates to monitor wall stability and to optimize construction stages (Espinoza et al. 2008). The attachment of strain gages to the high-strength geotextile completes the field instrumentation, and regular strain monitoring provides direct feedback of the geotextile's response to construction loading. Figure 1 illustrates a typical two-tier section of the proposed MSE wall and the associated instrumentation.

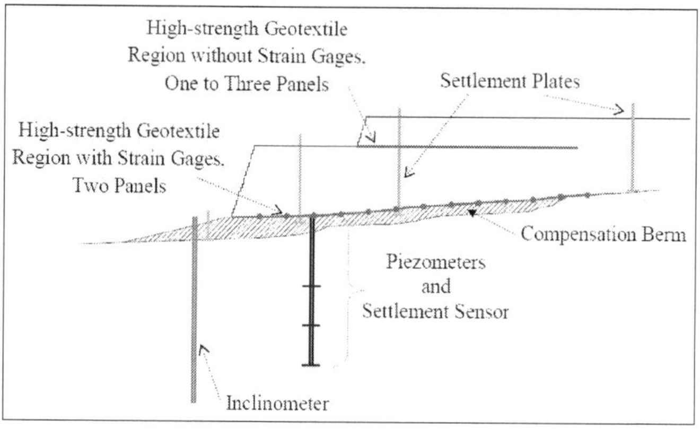

FIG. 1: Typical two-tier section of proposed MSE wall

HIGH-STRENGTH GEOTEXTILE

Following installation of PVDs and compensation berms, two layers of high-strength polyester geotextile were placed at the base of the MSE wall, separated by 45.7 cm of granular fill. Depending on the height of the wall and the associated tier level, an additional one to three layers of high-strength geotextile are proposed. Tencate Mirafi manufactured the geotextile, Geolon® PET1170/50, and reports a minimum ultimate tensile strength in the warp direction of 1168 kN/m, determined through ASTM D4595.

Four locations were selected for stress monitoring during MSE wall construction, for a total of eight geotextile panels. Each instrumented panel, approximately 42.7 m in length and 4.9 m in width, is outfitted with 13 strain gages, spaced 3.0 m on center.

Strains gages EP-08-40CBY-120, manufactured by Vishay Micro-Measurements, with a nominal gage length of 10.2 cm were used.

Strain in High-strength Geotextile

Field monitoring of installed strain gages is regularly performed with a FLUKE 189 multimeter. In the field, gage resistance is directly measured, but it is desired to correlate these measurements to field strain in the geotextile. For the same model of strain gage, Leshchinsky and Fowler (1990) demonstrated that strain elongation (%) equals 0.4 times the difference in gage resistance (ohms), relative to a measured initial gage resistance (ohms).

An experiment modeled after the procedure described in Leshchinsky and Fowler (1990) verified this relationship by monitoring changes in gage resistance upon elongation with a linear variable differential transformer (LVDT) calibrator. Two trials each produced linear plots with slopes of 0.38% strain/ Ω, nearly the same 0.4 factor measured by Leshchinsky and Fowler (1990). For purposes of this research, a factor of 0.38 % strain/ Ω was used to convert measured field resistances to strains.

After translating gage resistance to strain, it is next desirable to express strain as force. A series of wide-width tensile tests modeled after ASTM D4595 were performed on instrumented geotextiles at the University of Delaware to verify the manufacturer reported ultimate tensile strength and to determine a conversion of strain to force. Similar to the approach presented in Leshchinsky and Fowler (1990), specimens of 150 mm width were clamped at each end to roller grips and wrapped three revolutions around each grip. Tensile load was applied with an 890-kN Tinius Olsen loading machine with a rate of machine grip displacement of 3.2 mm per minute. At 0.2 Ω increments in resistance, applied load and elapsed time were recorded. Note that ASTM D4595 recommends loading 200 mm-wide specimens at 10% strain per minute (ASTM 2006). The actual tests were conducted at a non-constant rate of strain varying between approximately 0.5% and 0.25% strain per minute. Although the strain rate applied at the jaws is constant, the strain rate at the gage location varies because of slippage and stress redistribution at the roller grips.

Five trials yielded an average modulus of 104.7 kN/m/% strain, consistent with design values, and an average tensile strength of 803 kN/m, lower than the manufacturer's reported strength of 1168 kN/m. The lower tensile strength value is attributed to loading at a rate two orders of magnitude slower than the ASTM standard used by the manufacturer. Supplementary data and plots concerning calibration procedures are presented in Walsh (2009).

Strain Gage Installation

Similar to procedures described by Schimelfenyg et al. (1990), Fowler and Leshchinsky (1990), and Volk et al. (1994), the installation of gages involved a two-part process that began at the University of Delaware civil engineering laboratory and ended at the CIL field site. Figure 2 documents the installation procedures. Dow Corning 3145 RTV Adhesive/Sealant provided watertight, flexible bonding of gages to the geotextile.

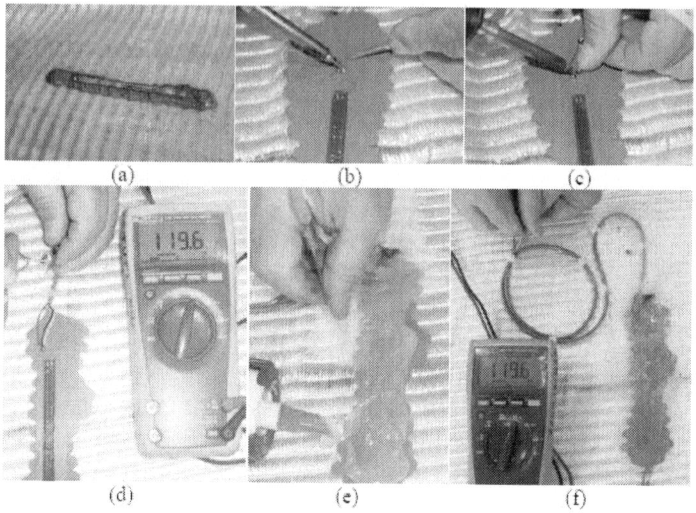

FIG. 2: Attachment of strain gages to high-strength geotextile: (a) gage attached with RTV; (b) gage soldered to terminal strip; (c) cable pigtail soldered to terminal strip; (d) continuity check following cable connection; (e) final RTV application to waterproof gage, and (f) stress relief loop and continuity check

In the field, instrumented panels were rolled out with the gages oriented perpendicular to the face of the MSE wall. To provide redundancy and independent monitoring, gages were wired individually. Individual cables were extended to accessible readout boxes where they were attached to rotary dials for data collection.

Cables were extended along the length of the reinforcement in two manners. As illustrated in Figure 3, the cables for Panels 3 and 7 were threaded through 5.1-cm PVC-pipe with elbow connections at each gage.

FIG. 3: Installation with PVC-pipe cable protection

Because the cabling could become damaged under the load of the MSE berm as settlement occurs, it was hypothesized that the conduit would provide protection to the cables, allowing them to move freely as settlement develops. Damage to the cables would prevent strain measurement but would not necessarily imply loss of the strain gage. Cables for the other six panels were laid in a serpentine fashion without PVC protection and with approximately 50% slack in anticipation of the high predicted settlements, similar to the field approach described by Volk et al. (1994).

Strain Gage Monitoring Observations

A total of 104 strain gages were attached to the high-strength geotextile. In addition to other details, Table 1 presents the number of gages operational at the time of field installation and the number of gages lost over time for each instrumented section.

Note that for Panel 1, 12 of the 13 original gages remained operational at the time of field installation. Although the gage remained functional, Panel 8 was damaged at a gage location during lay-down. Damage most likely occurred during either delivery to the site or lay-down of the panel. Non-instrumented geotextile is received from the manufacturer spooled onto steel pipes and rolled into place. Because of laboratory and equipment constraints, instrumented geotextile could not be respooled and was delivered to site folded onto pallets. Consequently, folded test panels required additional manpower and different equipment for final field lay-down, an approach which may have posed a higher, but necessary, risk to the installed strain gages.

Table No. 1: Detail Summary of Instrumented Geotextile Sections

<table>
<tr><th>Station</th><th colspan="2">24+75</th><th colspan="2">44+80</th><th colspan="2">53+70</th><th colspan="2">74+80</th></tr>
<tr><td>Panel #</td><td>1</td><td>2</td><td>3</td><td>4</td><td>5</td><td>6</td><td>7</td><td>8</td></tr>
<tr><td>Installation Date, mm/yy</td><td colspan="2">05/07</td><td colspan="2">01/08</td><td colspan="2">10/07</td><td colspan="2">08/07</td></tr>
<tr><td>PVC</td><td>No</td><td>No</td><td>Yes</td><td>No</td><td>No</td><td>No</td><td>Yes</td><td>No</td></tr>
<tr><td>Total Gages,[a] #</td><td>12</td><td>13</td><td>13</td><td>13</td><td>13</td><td>13</td><td>13</td><td>13</td></tr>
<tr><td>Gages Lost,[b, c] #</td><td>7</td><td>6</td><td>1</td><td>1</td><td>1</td><td>2</td><td>6</td><td>3</td></tr>
<tr><td>Range of Strain,[c] %</td><td colspan="2">0 – 1</td><td colspan="2">0 – 0.5</td><td colspan="2">0 – 0.5</td><td colspan="2">0 – 1</td></tr>
<tr><td>Anticipated Settlement, m</td><td colspan="2">2.4</td><td colspan="2">3.4</td><td colspan="2">3.4</td><td colspan="2">3.4</td></tr>
<tr><td>Total Settlement,[c] m</td><td colspan="2">1.5</td><td colspan="2">0.9</td><td colspan="2">1.2</td><td colspan="2">1.9</td></tr>
<tr><td>Differential Settlement,[c,d] m</td><td colspan="2">0.47</td><td colspan="2">0.58</td><td colspan="2">0.36</td><td colspan="2">0.26</td></tr>
<tr><td>Horizontal Movement,[c] m</td><td colspan="2">0.38</td><td colspan="2">0.21</td><td colspan="2">0.35</td><td colspan="2">0.63</td></tr>
<tr><td>Wall Height,[c] m</td><td colspan="2">10.0</td><td colspan="2">5.5</td><td colspan="2">9.6</td><td colspan="2">9.6</td></tr>
</table>

a. Following site delivery and prior to final field wiring
b. Gages lost after final field wiring; some data collected
c. Observed as of 07/2008
d. The difference between the minimum and maximum observed settlements in the instrumented geotextile sections referenced

During the field installation of Panels 3, 4, 7, and 8, the geotextiles overlapped into the Contractor's haul way. To prevent disruption of the hauling activity, the panels were fully extended then folded back accordion-style such that the impacted gages were accessible for final wiring. Upon redirect of the haul way days after installation, the Contractor extended the instrumented panels their full length. Of the four impacted gages, three gages immediately lost function. One cable was damaged by a bulldozer after foldout, but circumstances allowed for its immediate repair. It is not known how the other gages lost function.

In addition to the number of inoperable gages, several strain gages have presented unanticipated, apparently unreasonable behavior such as sudden increases in strain, high values of calculated strain, or fluctuating strain values. These observed behaviors may be in response to localized stress concentrations – rock or debris underlying gages – or water infiltration into the RTV encasement. The high strain readings are not supported by other instrumentation. Figure 4 presents calculated strain with distance from MSE wall for Panel 7, a bottom reinforcement layer installed with PVC pipe. Prior to failure, Gage A shows a sudden, high strain increase from March to April 2008, corresponding to a waiting period between fill stages. Note that the negligibly small negative strain values likely do not imply compressive forces acting on the geotextile.

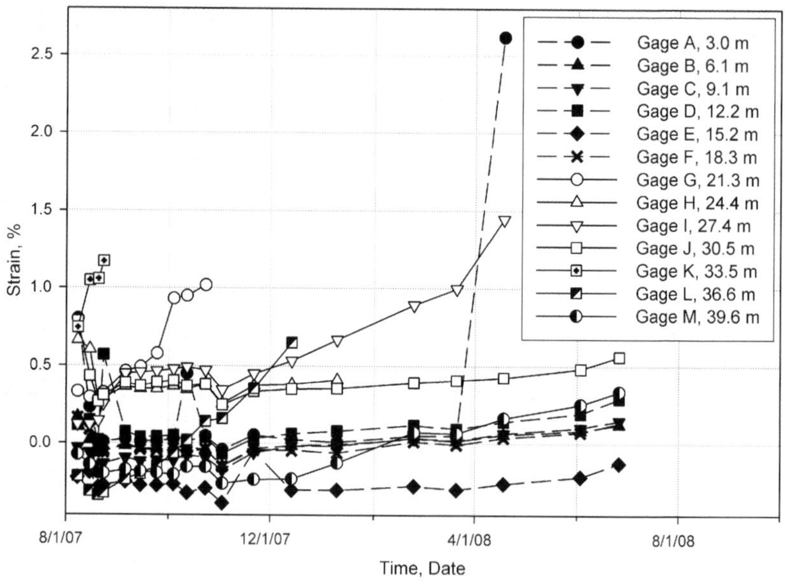

FIG 4: Panel 7 Strain: 0 Meters to 40 Meters from Wall

Recall that Panels 3 and 7 were installed with PVC-pipe to provide conduit protection to the cabling. An alternative hypothesis is that the PVC-pipe could break in response to differential settlement and damage strain gages and/or cables. Figure 5 presents calculated strain with distance from MSE wall for Panel 3, a bottom reinforcement layer also installed with PVC pipe. Once again, note that the negligibly small negative strain values likely do not imply compressive forces acting on the geotextile. Upon comparison of the data in Table 1 for Panels 3 and 7, Panel 3 has experienced less total settlement but greater differential settlement than Panel 7, more uniform strain behavior, and fewer gage losses. While this does not exclude damage induced by differential settlement, differential settlement and subsequent breaking of the PVC-pipe cannot be isolated as the primary mechanism for observed gage failure. Further review of Table 1 does not indicate that the PVC-pipe has provided additional protective value for the expense of time, labor, and material cost. Supplementary data and plots are presented in Walsh (2009).

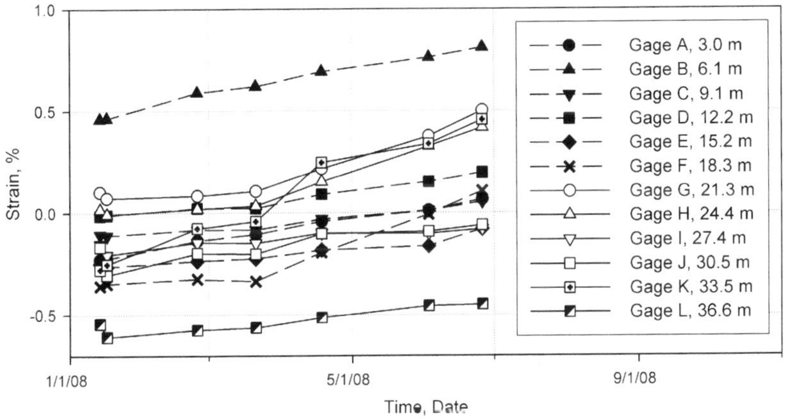

FIG 5: Panel 3 Strain: 0 Meters to 40 Meters from Wall

CONCLUSIONS

An MSE wall is being constructed over very soft, normally-consolidated to under-consolidated silt and clay. To facilitate strength gain and reduce soil movements associated with lateral squeeze, the design and ongoing staged construction includes PVDs and high-strength polyester geotextile reinforcement. An extensive instrumentation program has been implemented to monitor stability and optimize stages throughout construction. The following lessons have been learned:

- Remote attachment of gages reduces interruption of construction activity.
- Delivery and geotextile lay-down create occasions for gage damage. It is surmised that installation damage to instrumented panels can be reduced by respooling the geotextile onto pipe. This better replicates the delivery and lay-down methods of non-instrumented production rolls and eliminates the need for additional manpower and equipment.

- To prevent subsequent gage damage and failure, final field wiring of gages should be completed in one setting with the geotextile fully extended.
- At this stage in construction, data does not suggest that the PVC-pipe protection provides additional value for the expense of time, labor, and material cost.

The geotextile has been designed to withstand a maximum 6.6% strain. Gage resistance is monitored on a regular basis with a FLUKE 189 multimeter and converted to strain. To date, measured strain ranges between 0.00% and 1.00% with several gages exhibiting unlikely high strain. As construction proceeds, strain gage data will remain an important component of the project's instrumentation program as its reading directly indicates the state of stability of the MSE wall.

ACKNOWLEDGMENTS

Funding for this research project is provided by the Delaware Solid Waste Authority. Additional field data has been provided by GeoSyntec Consultants, Inc. and Sevenson Environmental Services, Inc. The authors would like to acknowledge the suggestions and assistance of Anne Germain, Lucas DeMelo, and Amy DeMell.

REFERENCES

ASTM (2006). *Geosynthetics*, Annual Books of ASTM Standards, Volume 4.13, American International, West Conshohocken, PA.

Espinoza, R. D., Lazarte, C. A., Germain, A. M., Houlihan, M. F. (2008). "Design Considerations for Expansion of an Existing Landfill Over Extremely Compressible Soils." *Geo-Strata*, 9(2), 38 – 43.

Fowler, J. and Leshchinsky, D. (1990). "Field installation of strain gaged geotextiles." *Proceedings of the First Materials Engineering Congress*, Denver, 360-369.

Geosyntec Consultants. (2004). "Volume 2: Geotechnical Characterization Report: Text, Tables, and Figures, Cherry Island Landfill Expansion Project, Wilmington, Delaware." Columbia, Maryland.

Leshchinsky, D. and Fowler, J. (1990). "Laboratory Measurement of Load-Elongation Relationship of High-Strength Geotextiles." *Geotextiles and Geomembranes*, 9(2), 145-164.

Schimelfenyg, P., Fowler, J., and Leshchinsky, D. (1990). "Fabric reinforced containment dike, New Bedford superfund site." *Fourth International Conference on Geotextiles and Geomembranes*, The Hague, 149-154.

Volk, J. C., Hunt, R. E., Leshchinsky, D., DiLoreto, R., and Collins, T. G. (1994). "Reinforced Embankment over Very Soft Soil: Grassy Sound Highway Embankment." *Fifth International Conference of Geotextiles, Geomembranes and Related Products*, Singapore, 41 – 48.

Walsh, N. (2009). "High-strength Geotextile: Strain Monitoring at Cherry Island Landfill." *Thesis to fulfill the requirement of the Master degree at the University of Delaware, Department of Civil and Environmental Engineering*, (in preparation).

A New Generation of Resilient Modulus Characterization of Unbound Materials

Zapata, Claudia[1], M. ASCE and Cary, Carlos[2]

[1]Assistant Professor, Department of Civil and Environmental Engineering, Arizona State University, P.O. Box 875306, Tempe, AZ 85287-5306; czapata@asu.edu
[2]Research Assistant, Department of Civil and Environmental Engineering, Arizona State University, P.O. Box 875306, Tempe, AZ 85287-5306; ccary@asu.edu

ABSTRACT

Pavement damage is heavily influenced by seasonal variations in the environmental regime and by externally applied loads. The state of the art methodology currently evaluated by pavement engineers focuses upon coupling the effects on pavement performance of moisture content/matric suction properties of unbound materials with the external stresses.

The literature review showed that several equations have been proposed to incorporate moisture variation effects on resilient modulus; however, the limited models available lack fundamental principles or have not been properly validated. A feasibility study on cohesionless granular base material is presented, whereas matric suction was controlled during drained and measured during undrained resilient modulus tests. It is apparent that certain modifications in the recommended stress state conditions of the NCHRP 1-28A protocol may be necessary when measuring matric suction due to the axis-translation needed during the test. Difficulties associated with the test include the current confinement fluid, which allows for air diffusion during the suction equilibration stage. Preliminary results show that the k_1-k_2-k_3 parameters used in the approved AASHTO Mechanistic Empirical Pavement Design Guide (ME-PDG) may be functions of the suction applied to the specimen during the test protocol.

INTRODUCTION

This paper has several key objectives: 1) summarize the state-of the art of advanced analytical models that include moisture content/matric suction influence on the resilient modulus; 2) describe the laboratory characterization of the resilient properties of unsaturated granular base materials; 3) outline difficulties associated with the testing methodology; and 4) present results of an initial feasibility study on a granular base material aimed at controlling/measuring matric suction during drained/undrained tests.

INCORPORATING MATRIC SUCTION IN THE PREDICTION OF M_R

It is well known that any increase in moisture content reduces the resilient modulus of the soil, and vice versa. The moisture content of the soil is intimately related to the matric suction, and its relationship is characterized by the soil-water characteristic curve (SWCC). As a result, any increase in matric suction results in a corresponding increase in the resilient modulus of unbound materials, and vice versa.

A literature review revealed at least three distinctive approaches when attempting to incorporate the effects of suction in the resilient behavior of the soil. The first one defines the moduli as a function of suction by translating the concept of effective stress for saturated soils to unsaturated soils as proposed by Bishop (1959) (Yang et al. 2005; Liang et al. 2008). The second approach is based on the concept of resilient modulus as a function of three stress variables, one being the suction (Fredlund and Rahardjo 1987). In this case, the matric soil suction is one of the components of the stress tensor. The third approach imposes a factor that takes into account the contribution of matric suction independently from the contribution of the externally applied stress (Parreira and Goncalves 2000; Witczak et. al 2000).

Fredlund et al. (1977) made an attempt to relate the modulus of a compacted subgrade to the stress state variables for unsaturated soils proposed by Fredlund and Morgenstern (1977). Two variables, that are functions of matric suction, define the linear relationship between the logarithm of resilient modulus and deviator stress: the slope of the line and the intercept. The modulus is then related to two stress variables: the deviator stress and suction. The authors suggest that the effect of confining pressure could either be ignored or a correction could be applied. The model proposed by Fredlund et al. (1977) is presented in Eq. 1. Even though the model was not fully validated with lab data, it follows a fundamental and sound approach.

$$log\ M_R = c_{ld} - m_{ld}\,(\sigma_1 - \sigma_3) \quad (1)$$

where, M_R is the resilient modulus, c_{ld} is the intercept, m_{ld} is the slope of the line on the semi-log plot, σ_1 is the major principal stress, and σ_3 is the minor principal stress.

Parreira and Goncalves (2000) presented a model describing the influence of suction on M_R of a lateritic subgrade in Brazil. In their study, statistical correlations between resilient modulus, suction and externally applied stress were analyzed. The best correlation found is presented in Eq. 2, where ψ_m is the matric suction:

$$M_R = 14.10\,\sigma_d^{0.782}\,\psi_m^{0.076} \quad (2)$$

Witczak et. al (2000) proposed a model that incorporates an environmental factor that represents the contribution of the change in moisture content and soil density to the resilient modulus evaluated at a particular externally applied stress. The model was based on an extensive literature review and was validated with ten soils of the State of Arizona. This model has been incorporated in the new AASHTO ME-PDG.

$$M_R = 10^{a+\frac{b-a}{1+EXP\left(\beta+k_S\cdot\left(S-S_{opt}\right)\right)}} \cdot k_1 \cdot p_a \cdot \left(\frac{\theta}{p_a}\right)^{k_2} \cdot \left(\frac{\tau_{oct}}{p_a}+1\right)^{k_3} \quad (3)$$

where, S is the degree of saturation, S_{opt} is the saturation at optimum conditions, θ is the bulk stress, τ_{oct} is the octahedral stress, p_a is the atmospheric pressure, and k_1, k_2, k_3, a, b, k_s are fitting parameters.

The model developed by Yang et al. (2005) explicitly incorporates the effect of both, the externally applied stress and the matric suction, on the prediction of resilient modulus, by using the effective stress approach. This model is based on the relationship proposed by Moossazadeh and Witczak (1981), where the Bishop's effective stress parameter represents the moisture condition of the soil. Eq. 4 presents the model proposed by Yang et al. (2005):

$$M_R = k_1\left(\sigma_d + \chi_m \psi_m\right)^{k_2} \quad (4)$$

where, χ_m is the Bishop's parameter, and k_1, k_2 are regression parameters.

Liang et al. (2008) proposed a similar model based on the effective stress approach. The model is a modified version of the expression proposed by Witczak and Uzan (1988), known as the Universal model. The model is presented in Eq. 5:

$$M_R = k_1 p_a \left[\frac{\left(\theta + \chi_m \psi_m\right)}{p_a}\right]^{k_2} \left[\frac{\tau_{oct}}{p_a}+1\right]^{k_3} \quad (5)$$

Finally, Khoury and Zaman (2004) showed that the resilient modulus of compacted clay exhibits a hysteretic behavior due to wetting and drying cycles. This hysteresis, similar to that observed in the SWCC relationship, confirms the strong correlation between resilient modulus and matric soil suction. All of these studies clearly show the relationship between resilient modulus and matric suction of the soil.

CHARACTERIZATION OF M_R FOR UNSATURATED MATERIALS

In order to characterize the resilient modulus of unsaturated materials, the matric suction needs to be controlled during drained tests and measured during undrained tests. Therefore, a system that simultaneously applies dynamic loading and controls the pore air and water pressures is needed.

A fully integrated and automatically controlled system capable of applying repeated cycles of a haversine-shaped load pulse was used for the M_R test during this study at Arizona State University. This electro-hydraulic system is capable of direct control/ measurement of pore water pressure at top/bottom of the specimen and of pore air pressure at the top of the specimen. Unlike a standard Triaxial system, which has only three ports to measure external confining pressure, and top and bottom pore water pressures; this system has two additional pressure ports that allows for pore air and inner pressure measurements. The bottom platen is equipped with a high air entry ceramic disk and can be modified to accommodate specimens up to 150 mm (6 in) in diameter. GCTS CATS v1.6 is the software available for the operation of the system. It is a user-friendly Windows 98, XP™ compatible software for automated test control. The NCHRP 1-28A protocol (Witczak 2004) was implemented in the software by the research team. The software handles the loading sequences for base/subbase materials, granular subgrades, and fine grained subgrades required in the protocol.

Characteristics of the Material. A typical granular base AB material used by the Maricopa County Department of Transportation (MCDOT) in Arizona was chosen for the testing program. Sieve analysis, specific gravity, compaction curves and SWCC tests were performed. The material was classified as A-1-a, non-plastic soil with an average of 7% passing US sieve # 200, 21% passing #40, 43% passing #4; and a maximum particle size of 25.4 mm. Compaction tests yielded a maximum dry density of 2.26 g/cm^3 and an optimum moisture content of 7%.

Soil Water Characteristic Curve Testing. A pressure membrane with an air entry value of 1,500 kPa was used for the determination of the SWCC. Five samples were compacted at optimum water content and maximum dry density. The samples were then saturated and placed in a saturated membrane. Air pressures of 16, 60, 140, 200 and 400 kPa were applied in order to obtain the drying path of the SWCC. The results were fitted with the Fredlund and Xing (Fredlund and Xing, 1994) model as shown in Fig. 1. Matric suction values corresponding to the optimum, saturated (wet of optimum) and optimum minus 2% (dry of optimum) moisture contents were interpolated. Matric suction values of 0, 4 and 235 kPa were obtained; which corresponded to degrees of saturation of 100%, 93% and 67%, respectively.

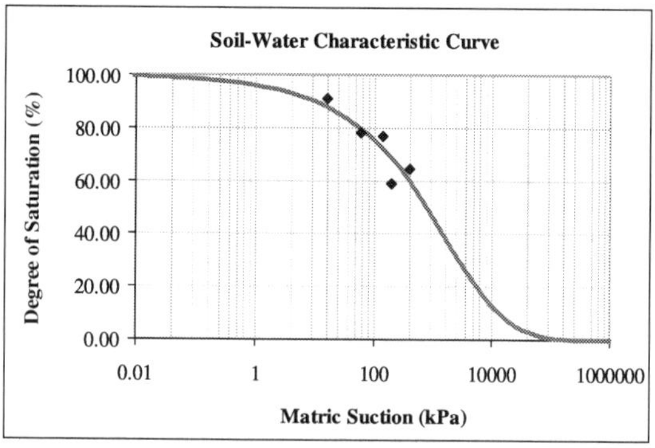

Figure 1. Soil Water Characteristic Curve for the AB Granular Base Material

Equilibration Stage and Resilient Modulus Testing. Clamp-mounted LVDTs were attached to the specimen at ¼ points, leaving a gage length of ½ of the specimen height as recommended by Andrei (2003). Subsequently, LVDTs were placed on the clamps in diametrical opposite positions. In order to comply as much as possible with the guidelines of the NCHRP 1-28A protocol for the determination of resilient modulus; the use of air confinement through an external line was installed. A conditioning confining pressure of 103.5 kPa was adopted for the equilibration phase, which corresponded to the initial confining pressure recommended in the Procedure Ia of the NCHRP 1-28A protocol. The specimen was then equilibrated to the suction needed to obtain the desired water content throughout the sample. The time of equilibration process can last from some days up to several weeks depending on the size of the specimen, the type of material to be tested, and the air entry value of the ceramic disk. Obviously, this factor will have to be evaluated in future studies.

The NCHRP 1-28A – Procedure Ia protocol was followed. Procedure Ia for base and subbase materials consists of one conditioning sequence of 1,000 repetitions and 30 loading sequences of 100 repetitions each, using a haversine-shaped 0.1-second load pulse followed by a 0.9-second rest period. Six specimens were tested under both drained and undrained conditions. Matric suctions of 0, 4, and 235 kPa were applied. The drained tests were performed under pressure or suction control, whereby the water volume of the specimen was allowed to vary and monitored during the test, while the air and pore water pressures were kept constant at the desired values. During the undrained tests, the pore air pressure was kept constant while the pore water pressure was monitored or measured.

PRELIMINARY M_R TEST RESULTS AND DIFFICULTIES ENCOUNTERED

Resilient modulus results obtained at a tpical state of stress in Arizona pavements for hot conditions (θ=303 kPa, τ_{oct}=110 kPa) along with the corresponding initial matric suction applied in the test, are presented in Table 1. The regression parameters were obtained by using the widely known modified Universal Model, proposed by Witczak and Uzan (1988). The Universal Model is expressed as follows:

$$M_R = k_1 \cdot p_a \cdot \left(\frac{\theta}{p_a}\right)^{k_2} \cdot \left(\frac{\tau_{oct}}{p_a}+1\right)^{k_3} \quad \text{......(6)}$$

Preliminary results corresponding to the 12 tests performed on AB material can be observed in Fig. 2. The data was fitted with the Witczak et. al model (Witczak et al. 2000). For the specimens tested at saturations lower than 80%, the M_R was found to be about 1.2 times greater than that at optimum condition. At saturated conditions, no significant difference was found. This result is attributed to the fact that the optimum degree of saturation of this particular sample was very close to 100%.

When comparing drained and undrained conditions, no difference in the M_R of the material was observed at lower degrees of saturation. This can be logically explained by the fact that pores in the material are not significantly filled with water and therefore, the pore water pressure buildup is not significant. Apparently, the state of stress of the sample is not affected by the drainage condition of the test if the suction is relatively high. At higher degrees of saturation, a difference in the resilience response in the order of 34,000 kPa was observed between the specimens tested under drained and undrained conditions. This result suggests that pore water pressure buildup further reduces the M_R of the material under undrained conditions.

Table 1. Resilient Modulus Test Results

Specimen		1	2	3	4	5	6
Drained	ψ_m (kPa)	13	0	0	4	235	235
	k_1	1886.1	2002.7	2010.9	1831.8	2144.5	2397.9
	k_2	1.108	0.776	0.827	0.680	0.726	0.581
	k_3	-1.063	-0.658	-0.705	-0.267	-0.275	-0.161
	S_e/S_y	0.235	0.527	0.345	0.250	0.241	0.246
	R^2	0.951	0.743	0.890	0.942	0.947	0.944
	M_R (kPa)	294,317	292,731	300,074	321,434	393,187	408,115
	M_R (psi)	42,687	42,457	43,522	46,620	57,027	59,192
Undrained	ψ_m (kPa)	0	0	4	4	235	235
	k_1	1285.7	1633.4	1391.1	1353.5	2252.7	2354.1
	k_2	0.721	0.825	0.772	0.826	0.655	0.584
	k_3	-0.252	-0.500	-0.315	-0.415	-0.256	-0.207
	S_e/S_y	0.170	0.138	0.167	0.151	0.215	0.212
	R^2	0.973	0.983	0.974	0.979	0.957	0.958
	M_R (kPa)	238,407	282,975	260,725	249,907	387,720	388,777
	M_R (psi)	34,578	41,042	37,815	36,246	56,234	56,387

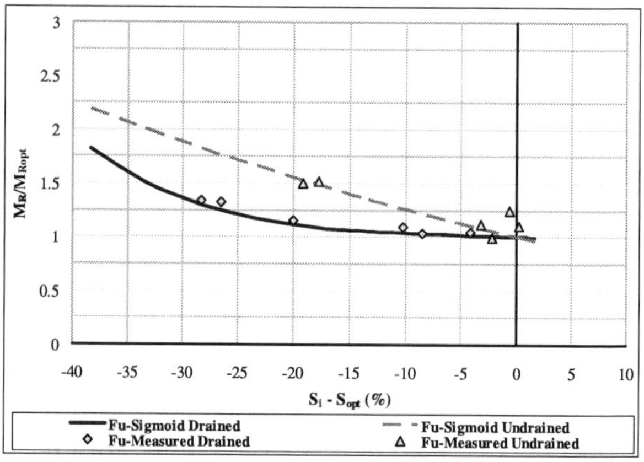

Figure 2. Resilient Modulus vs. Degree of Saturation

Fig. 3 shows the variation of matric suction during M_R testing under undrained conditions. A decrease in matric suction in the order of 40 kPa was observed in the specimens tested at low initial matric suction. Results from specimens tested at high initial matric suction show a drop of matric suction in the order of 110 kPa. These results clearly suggest a buildup of pore water pressure during dynamic loading.

Difficulties Encountered During M_R Testing with Suction Measurements. The equilibration period needed to reach the desired water content based on controlled matric suction is usually too long and a more practical method that allows for faster equilibration is required. Future studies will concentrate on evaluating the results by Perez (2006), who found that the equilibration can be achieved by bringing the specimen moisture content as close as possible to that needed to reach a particular suction value, provided a SWCC for the material is available.

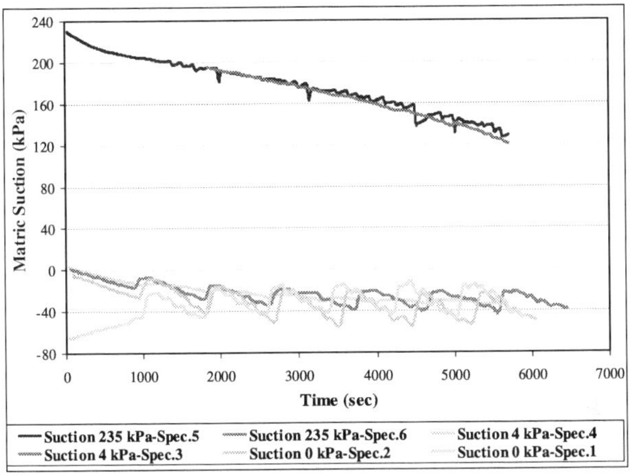

Figure 3 Variation of suction along the time for undrained M_R tests

The available 1-28A protocol calls for the use of air as the confining fluid for resilient modulus testing. It is widely recognized that air diffuses through latex membranes and generates instability in the suction during equilibration. Future studies will be aimed at matric soil suction control by either finding a more suitable material that impedes air diffusion or considering the use of water as confining fluid.

The axis translation technique is needed during the equilibration stage, especially when dealing with high suction levels. Due to the fact that resilient modulus protocols are based on a total stress approach, the contact pressure is specified as a percentage of the cell pressure. When applying the axis translation, the cell pressure needs to be increased and as a consequence, the contact stress specification is no longer valid. This problem could be easily avoided by considering the contact load as a percentage of the net confining pressure instead of the cell pressure.

CONCLUSIONS

Matric suction has a significant effect on the resilient modulus of the material and should be incorporated in the next generation of models. This initial feasibility study shows that different models have been proposed to incorporate the effects of matric suction in the M_R determination. The creation of a much larger material database needs to be developed before a more precise model can be recommended.

A modified Triaxial system is necessary to either measure/control matric suction during M_R testing. Testing difficulties are mainly associated with constraints of existing equipment and limitations of the available protocols. Preliminary results aiming at validating the influence of suction on the resilient modulus of unbound materials clearly show the effect of suction on M_R results.

ACKNOWLEDGMENT

This study was funded by MCDOT under the Project: *Pavement Design Research Towards the Implementation of the Mechanistic-Empirical Pavement Design Guide.*

REFERENCES

Andrei, D. (2003). Development of a predictive model for the resilient modulus of unbound materials. *Ph.D. Diss.*, Dep. Civil Eng., Arizona State Uni., Tempe, AZ.

Bishop, A. (1959). "The principle of effective stress" *Tek. Ukeblad,* 106(39):859-863.

Fredlund, D.G. and Morgenstern, N.R. (1977). "Stress state variables for unsaturated soils." *ASCE J. of the Geotechnical Eng. Division*, 103(5): 447-466.

Fredlund, D.G., and Rahardjo, H. (1987). "Soil mechanics principles for highway engineering in arid regions." *Transportation Research Record 1137, Soil Mechanics Considerations in Arid and Semiarid Areas*, TRB, NRC, Wa, DC: 1-11.

Fredlund, D.G., and Xing, A. (1994). "Equations for the soil-water characteristic curve." *Canadian Geotechnical J.*, 31(4): 521-532.

Fredlund, D.G., Began, A.T. and Wong, P.K. (1977). Relation between resilient modulus and stress conditions for cohesive subgrade soils." *Transportation Research Record 642*, Transportation Research Board, Washington, D.C.: 73-81.

Khoury, N.N., and Zaman, M. (2004). "Correlation among resilient modulus, moisture variation and soil suction for subgrade soils." *Transportation Research Record 1874*, Transportation Research Board, Washington, D.C.: 99-107.

Liang, R.Y., Rabab'ah, S. and Khasawneh, M. (2008). "Predicting moisture-dependent resilient modulus of cohesive soils using soil suction concept." *ASCE J. of Transportation Engineering*, 134(1): 34-40.

Moossazadeh, J. and Witczak, M.W. (1981). "Prediction of subgrade moduli for soil that exhibits nonlinear behavior." *Transportation Research Record 810,* Transportation Research Board, Washington, D.C.: 9-17.

Parreira, A.B. and Goncalves, R.F. (2000). "The influence of moisture content and soil suction on the resilient modulus of a lateritic subgrade soil." *GeoEng - An Int. Conf. on Geotechnical & Geological Engineering*, Melbourne.

Perez, N. (2006). "Development of a protocol for the assessment of unsaturated soil properties", *Ph.D. Dissertation*, Dep. of Civil Eng., Arizona State U., Tempe, AZ.

Witczak M.W., Andrei, D. and Houston W.N. (2000). "Resilient modulus as function of soil moisture." *Development of the 2002 Guide for the Development of New and Rehabilitated Pavement Structures, NCHRP 1-37 A, Tech. Report (Seasonal 1).*

Witczak, M.W. (2004). "Laboratory determination of resilient modulus for flexible pavement design." *Research Results Digest No. 285*, From *NCHRP 1-28A: Harmonized Test Methods for Laboratory Determination of Resilient Modulus for Flexible Pavement Design*, NCHRP.

Witczak, M.W. and Uzan, J. (1988). *The Universal Airport Design System, Report I of IV. Granular Material Characterization*. Dep. of Civil Eng., Univ. of Maryland.

Yang, R.R., Huang, W.H., and Tai, Y.T. (2005). "Variation of resilient modulus with soil suction for compacted subgrade soils." *Transportation Research Record 1913*, Transportation Research Board, Washington, D.C.: 99-106.

Field Evaluation of Fly Ash Stabilized Subgrade in US 12 Highway

Lin Li[1], M. ASCE, P.E., Onur Tastan[2], Craig H. Benson[3], F. ASCE, P.E., and Tuncer B. Edil[4], F. ASCE, P.E.

[1]Assistant Professor, Department of Civil and Environmental Engineering, Jackson State University, Jackson, MS; lin.li@jsums.edu
[2]Formerly graduate student, University of Wisconsin-Madison, Madison, WI; tastan@wisc.edu
[3]Professor and Department Chair, Department of Civil and Environmental Engineering, University of Washington, Seattle, WA; chbenson@u.washington.edu
[4]Professor, Department of Civil and Environmental Engineering, University of Wisconsin-Madison, Madison, WI, edil@engr.wisc.edu

ABSTRACT: This paper describes a case study where subgrade soils were stabilized with Class C fly ash to create a working platform during reconstruction of a 1.2-km section of rigid pavement in US 12 between Cambridge and Fort Atkinson, Wisconsin. The subgrade soils were blended with cementitious fly ash to increase its bearing resistance and stiffness. Resilient modulus (M_r), unconfined compression strength (q_u), soil stiffness gauge and dynamic cone penetrometer tests were conducted on the subgrade alone and fly-ash stabilized subgrade (FASS). Falling weight deflectometer (FWD) tests were conducted on the as-built pavement. After 14 d of curing, M_r of the field-mixed FASS ranged between 60 and 129 MPa and M_r of the laboratory-mixed FASS ranged between 115 and 167 MPa, whereas the M_r of subgrade was between 34 and 42 MPa. q_u of the laboratory-mixed FASS was four to nine times the q_u of the subgrade, while q_u of the field-mixed FASS was up to three times q_u of the subgrade. In situ stiffness and dynamic penetration index also illustrated that the addition of the fly ash and compaction increased the strength and stiffness appreciably. Moduli back-calculated from FWD tests showed that the modulus of the FASS varies seasonally.

INTRODUCTION

Fly ash stabilization of subgrade is a developing and promising technology (Edil et al., 2002, 2006; Bin-Shafique et al., 2004; Trzebiatowski et al., 2004, Wen et al. 2003; Li et al. 2008). For example, Edil et al. (2002) use fly ash to stabilized fine-grained subgrades in two short experimental sections in STH 60 at Wisconsin. Many states have active programs promoting the beneficial reuse of fly ash and other high volume industrial by products. However, the effectiveness of stabilizing subgrades with coal fly ash in full-scale applications is unavailable. The primary objective of this paper is to evaluate how fly ash increases the bearing strength and stiffness of the subgrade soils during construction and as the pavement is exposed to weather.

This paper describes a project where self-cementing Class C fly ash from a coal-fired electric power plant was used to stabilize subgrade during reconstruction of a 1.2-km section of US 12 highway between Cambridge and Fort Atkinson in Wisconsin (≈ 40 km east of Madison). The subgrade soil was graded using motor grader. Class C fly ash was spread uniformly on the surface using truck-mounted lay-down equipment similar to that described in Edil et al. (2002). The fly ash was mixed with the subgrade to a depth of 305 mm using the Wirtgen WR 2500S road reclaimer, with water being added during mixing using a water truck. This mixture, which contained 12% fly ash by dry weight, was compacted within 1-2 h by a tamping foot compactor followed by a vibratory steel drum compactor. The fly-ash stabilized subgrade (FASS) was cured for 7 d and then overlain with 150 mm of base layer with recycle asphalt and gravel material and 200 mm concrete pavement.

Prior to stabilization, samples of the subgrade were collected and tested to determine their index properties and how addition of fly ash would affect the M_r and q_u of the soil. Tests were also conducted on samples of the in situ stabilized soils to determine if similar improvements in properties were obtained during construction. DCP and SSG tests were used to measure the strength and stiffness near the surface of the stabilized subgrade. FWD tests were conducted to evaluate the overall improvement in stiffness achieved through stabilization.

MATERIALS

Disturbed samples of subgrade soil (≈ 20 kg each) were collected from a depth between 0-0.5 m at the fourteen stations during construction (Fig. 1). Tests were conducted on these samples to determine index properties, soil classification, water content and dry unit weight. Samples of the fly ash were also collected.

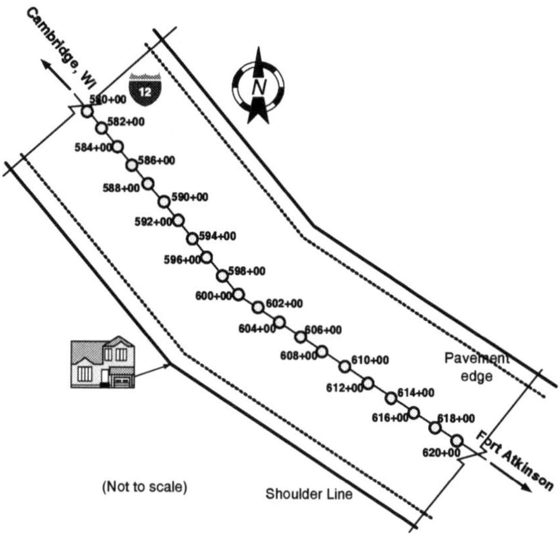

FIG. 1. Layout of stations at field site in US 12.

Subgrade

A summary of the properties of the subgrade is shown in Table 1. Particle size distribution curves for the subgrade are shown in Fig. 2. All fourteen soils are broadly graded. The subgrade consists of lean clay (CL), clayey sand (SC), and silty sand (SM) according to the Unified Soil Classification System. According to the AASHTO Soil

Classification System, most of subgrade soils at this site are A-7-6 with a group index (GI) larger than 10 and A-6 with GI larger than 2. Four of the coarse-grained subgrade soils are classified as A-2-6 (Stations 582+00, 590+00, 610+00, and 614+00) and have GI < 1.

TABLE 1. Physical properties and classifications of subgrade soils.

Station Number	LL[a]	PI[b]	GI[c]	Classification		w_N[f] (%)	γ_d[g] (kN/m^3)	SSG Stiffness (MN/m)	DCP Index (DPI) (mm/blow)
				USCS[d]	AASHTO[e]				
580+00	46	29	19	CL	A-7-6	-[h]	18.9	18.9	18.9
582+00	29	14	1	SM	A-2-6	4.8	21.3	21.3	21.3
586+00	40	27	6	SC	A-6	7.8	22.0	22.0	22.0
590+00	31	17	1	SM	A-2-6	14.7	18.2	18.2	18.2
594+00	65	49	29	CH	A-7-6	7.8	20.0	20.0	20.0
598+00	38	21	4	SC	A-6	10.8	20.1	20.1	20.1
602+00	42	23	10	CL	A-7-6	-	17.6	17.6	17.6
606+00	41	24	10	CL	A-7-6	12.2	18.6	18.6	18.6
610+00	27	15	1	SM	A-2-6	6.7	17.7	17.7	17.7
612+00	27	12	2	SC	A-6	14.3	17.6	17.6	17.6
614+00	26	11	0	SM	A-2-6	8.1	20.7	20.7	20.7
616+00	33	18	5	SC	A-6	6.0	20.0	20.0	20.0
618+00	36	22	6	SC	A-6	7.9	18.9	18.9	18.9
620+00	25	11	2	SC	A-6	7.5	21.3	21.3	21.3

[a]LL = Liquid Limit, [b]PI = Plasticity Index, [c]GI = Group Index, [d]USCS = Unified Soil Classification System, [e]AASHTO = Association of American State Highway and Transportation Officials, [f]w_N = In situ water content, [g]γ_d = Dry unit weight, [h] hyphen indicates test was not conducted.

Fly Ash

Fly ash from Columbia Power Station in Portage, Wisconsin was used for stabilization. Chemical analysis of the fly ash by X-ray fluorescence indicated that the calcium oxide (CaO) content is 23% and the $SiO_2 + Al_2O_3 + Fe_2O_3$ content is 56%. The loss on ignition is 0.7%. According to ASTM C 618, the Columbia fly ash is a Class C fly ash.

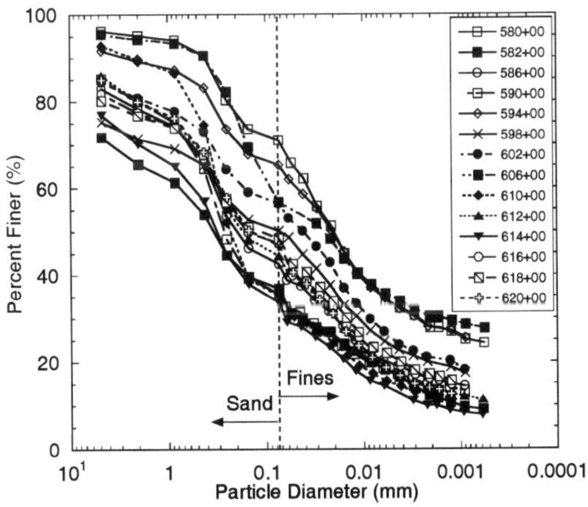

FIG. 2. Particle size distributions of subgrade.

Fly-Ash Stabilized Subgrade (FASS)

FASS was prepared by mixing Class C fly ash (12% by dry weight) with subgrade soils to a depth of 305 mm. Water content and unit weight of the compacted FASS were measured at each station using a nuclear density gage (ASTM D 2922) immediately after compaction was completed. Samples of FASS were also collected immediately after mixing for preparation of test specimens for M_r and q_u testing. These specimens were compacted to the measured density at each station, sealed in plastic wrap, and cured for 14 d at 100% humidity.

Specimens of FASS were also prepared in the laboratory using samples of the subgrade soils and 12% (dry weight) Columbia fly ash. The mixtures, referred as laboratory-mixed specimens, were prepared with optimum water content, and then cured using the same procedures employed for the field-mixed specimens. A similar set of specimens was prepared with subgrade soils only (no fly ash) using the same procedure, except for the curing phase.

METHODS

Laboratory M_r and q_u Tests

The M_r tests were performed in accordance with AASHTO T292 after 14 d of curing (FASS) or immediately after compaction (subgrade). The 14-d curing period is based on recommendations in Trzebiatowski et al. (2004), and is intended to reflect the condition when most of the hydration is complete (Edil et al., 2006). The loading sequence for cohesive soils was used for the FASS as recommended by Bin-Shafique et al. (2004) for soil-fly ash mixtures. Subgrade soils were tested using the loading sequence for cohesive soils. Two specimens of field-mixed FASS split horizontally after curing. These specimens were trimmed to an aspect ratio of 1 prior to testing. All other specimens had an aspect ratio of 2.

The q_u tests were measured on specimens of FASS after the resilient modulus tests were conducted. Only those specimens having an aspect ratio of 2 were tested. A strain rate of 0.21%/min was used for the q_u tests following the recommendations in ASTM D 5102 for compacted soil-lime mixtures. No standard method currently exists for q_u test of materials stabilized with fly ash.

Field Tests

Strength and stiffness of the FASS were measured with SSG, DCP, and FWD. Testing with the SSG and DCP were conducted directly on the FASS after 7 d of curing. The FWD testing was conducted after the concrete pavement.

The SSG tests were conducted in accordance with ASTM D 6758 using a Humboldt GeoGauge. DCP testing was conducted at each station in accordance with ASTM D 6951 using a DCP manufactured by Kessler Soils Engineering Products Inc. The dynamic penetration index (DPI) obtained from the DCP was computed as the mean penetration (mm per blow) over a depth of 150 mm. The FWD surveys were conducted using a Dynatest 8002E FWD following the method described in

ASTM D 4694. The FWD data were inverted using the program MODULUS v6.0 from the Texas Transportation Institute.

RESULTS

Resilient Modulus

Resilient moduli of subgrade, field-mixed FASS, and laboratory-mixed FASS are shown in Fig. 3a. These M_r correspond to a deviator stress of 21 kPa, which represents typical conditions within the base course of a pavement structure (Trzebiatowski et al. 2004). There is no systematic variation in M_r along the alignment, suggesting that the variability in M_r is more likely due to heterogeneity in material rather than systematic variations in site conditions or construction methods.

Comparison of the M_r for subgrade and FASS in Fig. 3a indicates that adding fly ash increased the M_r. For the subgrade, the M_r ranged between 34 and 42 MPa (mean = 38 MPa), where the field-mixed FASS had M_r 60~129 MPa (mean = 82 MPa) and the M_r of laboratory-mix FASS ranged 115~167 MPa (mean = 139 MPa).

In the laboratory-mixed FASS, there are only three specimens selected from Station 580, 586 and 616 to represent lean clay and clayed sand. Interaction of the two different soils with fly ash may vary depending on the mineralogical compositions (Edil et al. 2006). The cementitious reactions occurring in lean clay is slightly different from the reactivity of clayed sand. The range of measured M_r in the three laboratory-mixed FASS (115~167 kPa) showed that the variability of the M_r of the laboratory-mix FASS. The M_r of the field-mix FASS also was more variable than the M_r of the laboratory-mix FASS.

The M_r of the field-mix FASS is lower, on average, than the M_r of the laboratory-mixed FASS (82 MPa vs 139 MPa). This is consistent with the investigator's experience, which has shown that M_r of field mixtures of fly-ash stabilized materials typically lower than M_r of mixtures prepared in the laboratory (Li et al. 2008, Bin-Shafique et al. 2004). The effect of cementation of fly ash depends on the mineralogical compositions of soil, mixing water content, type and amount of fly ash, type of mixing, and curing period (Edil et al. 2006). The difference between the M_r of field-mixed FASS and laboratory-mixed FASS in this study should be related to the mixing condition and type of mixing. Bin-Shafique et al. (2004) attribute these differences in M_r to more thorough blending of soil and fly ash in the laboratory compared to the field, resulting in more uniform distribution of cements within the mixture.

Unconfined Compressive Strengths (q_u)

The q_u of the subgrade, field-mixed FASS, and laboratory-mixed FASS are shown in Fig. 3b. As observed for M_r, there is no systematic variation in q_u along the alignment. The q_u of the subgrade ranges from 92 to 138 kPa (mean = 112 kPa), the laboratory-mix FASS has q_u between 360 and 1120 kPa (mean = 663 kPa), and the q_u of field-mix FASS ranges 110~410 kPa (mean = 183 kPa). Thus, adding fly ash to the subgrade increased the q_u appreciably, although the q_u in the field was 72% lower, on average, than the q_u of the laboratory-mix FASS.

As observed for M_r, the range of measured q_u in the three laboratory-mixed FASS (360~1120 kPa) showed that the variability of the q_u of the laboratory-mix FASS. The q_u of the field-mix FASS also was more variable than the q_u of the laboratory-mix FASS. The relative low q_u of field-mixed FASS compared to q_u of laboratory-mixed FASS should be caused by a more thorough blending of soil and fly ash in the laboratory compared to the field, resulting in more uniform distribution of cements within the mixture.

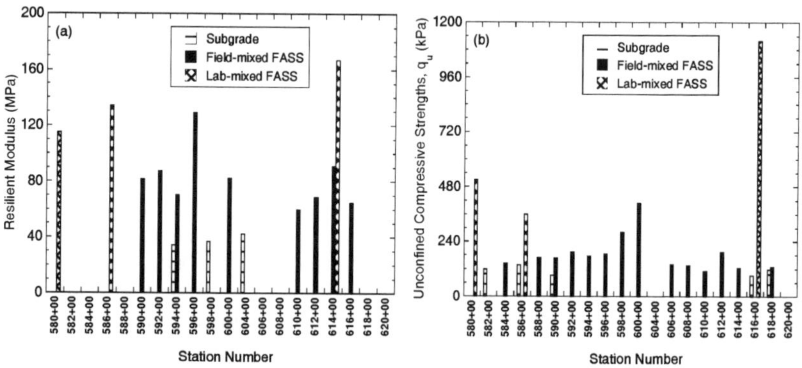

FIG. 3. Stiffness and strength subgrade and FASS (laboratory-mix and field-mix) after 14 d of curing: (a) M_r, and (b) q_u.

In situ Stiffness

In situ gained stiffness measured with the SSG and DPI measured with the DCP are shown in Fig. 4 for the subgrade and FASS after 7 d of curing. Addition of the fly ash and compaction increased the strength and stiffness appreciably, with the DPI decreasing from 33 to 15 mm/blow, on average, and the stiffness increasing from 8 to 20 MN/m, on average. The DPI and stiffness of the stabilized and compacted subgrade are also less variable than those of the subgrade.

FWD Deflections and Moduli

Maximum deflections from the FWD tests conducted with a 40-kN load are shown in Fig. 5a. Maximum deflection, which is measured at the center of the loading plate, is a gross indicator of pavement response to dynamic load. Tests were conducted in August 2004 (one month after construction), and the following spring (May 2005). Similar deflections were measured during both surveys, suggesting that the FASS had maintained its integrity even after exposure to freezing and thawing. Elastic moduli of the FASS obtained by inversion of the FWD data are shown in Fig. 5b.

The moduli drop between August 2004 and May 2005 (average mean 1324 MPa vs 725 kPa). These changes in modulus probably reflect changes in the water content of the FASS, which was highest in the spring and lower in the summer and fall.

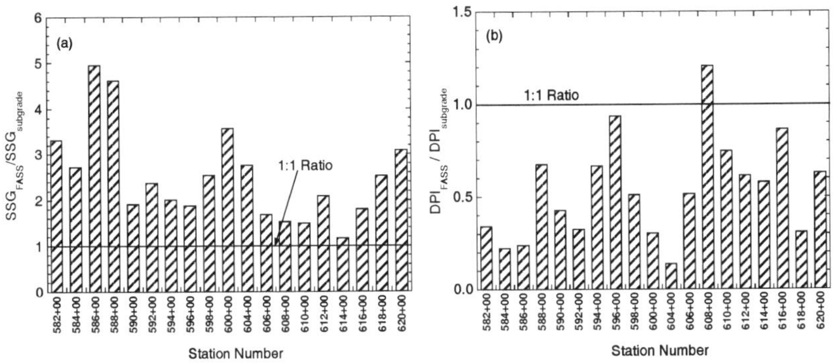

FIG. 4. In situ stiffness gain after fly ash stabilization after 7 d of curing (a) SSG, and (b) DPI.

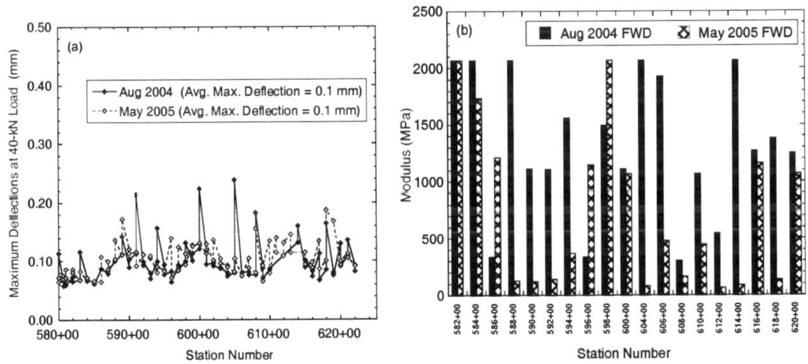

FIG. 5. Maximum FWD deflection for 40-kN load (a) and moduli of FASS obtained by inversion of FWD data (b).

SUMMARY AND CONCLUSIONS

The FASS had significantly higher M_r and q_u than the subgrade. In situ stiffness measured with the SSG and dynamic penetration index (DPI) measured with the DCP also illustrated that the addition of the fly ash and compaction increased the strength and stiffness appreciably. These finding suggest that fly ash stabilization of subgrade should be beneficial in terms of increasing pavement capacity and service life. However, the M_r and q_u of FASS mixed in the field were lower than those for FASS mixed in the laboratory (40% lower for M_r, 72% lower for q_u). Similar biases between mixtures prepared in the laboratory and field have been observed by others and need to be considered as one of factors when pavement design is based on data obtained by testing mixtures blended in the laboratory.

Moduli back-calculated from falling weight deflectometer tests conducted along the alignment at twice since construction showed that the modulus of the FASS varies seasonally, with the lowest moduli occurring in the spring when the pavement has thawed and has higher water content. Monitoring of the pavement will be conducted in the future to determine how the resistance to damage by freeze-thaw cycling changes.

ACKNOWLEDGEMENT

Financial support for this study was provided by the Wisconsin Department of Transportation (WisDOT). The conclusions and recommendations in this report are solely those of the authors and do not reflect the opinions or policies of WisDOT. Bulent Hatipoglu, Xiaodong Wang, and Jacob Sauer assisted with the project in the field and laboratory.

REFERENCES

Bin-Shafique, S., Edil, T.B., Benson, C.H. and Senol, A. (2004). "Incorporating a fly-ash stabilised layer into pavement design." *Geotechnical Engineering-ICE*, 157 (GE4): 239-249.

Edil, T.B., Acosta, H.A. and Benson, C.H. (2006). "Stabilizing soft fine-grained soils with fly ash." *Journal of Materials in Civil Engineering*, 18 (2): 283-294.

Edil, T.B., Benson, C.H., and Bin-Shafique, S. (2002). "Field evaluation of construction alternatives for roadways over soft subgrade." *Transportation Research Record*, 1786: 36-48.

Li, L., Benson, C. H., and Edil, T. B (2008) "Sustainable construction case history: fly ash stabilization of recycled asphalt pavement material." *Journal of Geotechnical and Geological Engineering*, 26: 177-187.

Trzebiatowski, B., Edil, T.B. and Benson, C.H. (2004). "Case study of subgrade stabilization using fly ash: State Highway 32, Port Washington, Wisconsin." *Beneficial Reuse of Waste Materials in Geotechnical and Transportation Applications*, 127: 123-136.

Wen, H.F., Tharaniyil, M.P. and Ramme, B. (2003). "Investigation of performance of asphalt pavement with fly-ash stabilized cold in-place recycled base course." *Transportation Research Record*, 1787: A27-A31.

Neuronet-Based Soil Chemical Stabilization Model

Yacoub Najjar[1], Chune Huang[2], Hakan Yasarer[3]

[1] Professor, Department of Civil Engineering, Kansas State University, Manhattan, KS, 66506, Phone: 785 532 5863 ; Fax: 785 532 7717; e-mail: ea4146@ksu.edu
[2] Geotechnical Engineer, Pillori Associates, P.A., Geotechnical Engineering, Secaucus, NJ, 07096 E-mail: chokIk8338@yahoo.com
[3] Graduate Student, Department of Civil Engineering, Kansas State University, Manhattan, KS, 66506. Phone: 785 317 4490; e-mail: hiy@ksu.edu

ABSTRACT: Adding chemical agent to stabilize problematic highway subgrade soil is a common engineering practice in the United States. Due to the fact that theoretical accomplishments in soil chemical stabilization lag far behind the engineering practice, laboratory testing, which is expensive and time-consuming, is almost always necessary to determine the effectiveness of the soil stabilizer in enhancing engineering properties of the soil. Over the years, large amount of valuable data from laboratory tests on stabilizing different soils with different chemical stabilizers was accumulated in the literature. Efforts to extract the relationships and associations from the existing test data in order to provide guidance for new soil chemical stabilization cases were carried out for many years, however, due to the technology (statistic regression) limitations, reliable models are still not available. In this paper, Neuronet (NN) approach to study soil chemical stabilization was introduced. NN model to predict the unconfined compression strength (UCS) of the stabilized soil was built based on the experimental data from stabilizing three representative Kansas embankment soils with five chemical stabilizers. The results showed that the trained NN model could precisely predict the UCS of stabilized soil. Furthermore, NN model enables us to study the significance of each input factor, thus providing a powerful tool for optimizing the mixture and construction design.

INTRODUCTION

Engineering properties of highway subgrade soils have dominant influence on the performance of the pavement. Weak or unstable subgrade layer will lead to premature failure of the pavement (www.pavement.com). Since natural soils are typically not adequate for supporting pavement during construction and service, soil chemical stabilization is often used to stabilize soils when mechanical methods of stabilization are inadequate and complete soil replacement is not cost effective.

Studies on the mechanisms of soil chemical stabilization were vigorously carried out in the past several decades. However, theoretical accomplishments from the previous studies are still not amenable to instruct new engineering practice. Therefore, laboratory testing is almost always necessary to determine the effectiveness of the soil stabilizer in reducing plasticity of the soil, increasing strength, determining the optimum mixture composition, and establishing density and curing time requirements. As a result, large amount of valuable data from laboratory tests on stabilizing different soils with different chemical stabilizers were accumulated in the literature (Examples: Arman and Saifan 1967; Wibawa and Rahardjo 2001).

Although engineering properties of soils on the earth vary with locations, soils bear local consistency and can be classified according to several classification systems, such as the AASHTO soil classification (Lindeburg, 2003). Therefore, if a comprehensive model, which is able to learn the relationships and associations from previous test data and to reliably predict the desired engineering properties for new soil chemical stabilization case, could be developed, it would become a powerful tool in facilitating the soil chemical stabilization in engineering practice.

Artificial neural network is an information-processing system based on mathematical models of human cognition and neural biology. ANN consists of many simple processing elements called neurons (or nodes) and connection links between the neurons. When the ANN processes information, the connection links are used to transfer signals between neurons. Each neuron evaluates its input signals to determine its output signal. The neuron output signal is then transmitted to all neurons, which are on the receiving side of the connection links originating in the transmitting neuron. Each connection link has an associated weight that multiplies the signal transmitted. Its physical massively parallel distributed structure and its ability to learn and generalize give ANN a distinguished computing power (Haykin 1999).

ANN MODEL DEVELOPMENT

Database

The experimental data used for the ANN development were obtained from stabilizing three representative Kansas embankment soils, designated as soil A, soil B and soil C (classified as A-7-6, A-6 and A-4, respectively according to AASHTO soil classification system) with five different chemical stabilizers, including Type I Portland cement, quick lime, class C fly ash, Cement Kiln Dust (CKD) and EMC SQUARED® Stabilizer (1000) (Banda, 2003). Soil A and B were stabilized by Portland cement, lime, fly ash, CKD at three different stabilizer contents and stabilized by EMC SQUARED® Stabilizer (1000) at three different moisture contents. Soil C was stabilized by Portland cement, fly ash, CKD at three different stabilizer contents and stabilized by EMC SQUARED® Stabilizer (1000) at two different moisture contents. Samples were compacted at 95% and 100% of standard Proctor density and then extracted by a hydraulic sample extruder. After extraction, the samples were placed in the plastic bags and then cured in a moist curing room at 25°C for 7, 14, 28 and 90 days (samples of stabilized Soil A were cured for additional curing periods of 2 days and 150 days). The EMC treated soil samples were first dried in the air before they were cured in the moist curing room. In order to differentiate this two curing condition, we denote the time the samples were dried in the air as dry curing period and the time the samples stay in the moist curing room as moist curing period. Two duplicates were tested for each condition. 792 valid data sets were collected from this experimental program. Further details, about this testing program, can be found in Banda (2003).

ANN Model Architecture

Based on the characteristics of the available experimental data, static backpropagation neural network was identified as the most suitable ANN type for developing an appropriate UCS prediction model. Static backpropagation neural network has been successfully used in the past (Najjar and McRyenold 2003; Najjar and Ali 1998; Huang et. al 2006) to solve problems that require the computation of a static function (i.e., a function whose output depends only upon the current inputs, and not on any previous inputs). The following are the experimentally-based fifteen potential input parameters:

1 Percentage passing no. 200 sieve for untreated soil (%)
2 Liquid limit of untreated soil (LI) (%)
3 Plastic limit of untreated soil (PL) (%)
4 Plastic Index of untreated soil (PI) (%)
5 Maxim dry density of untreated soil in standard proctor test (MDD) (kg/m^3)
6 Optimal moisture content of untreated soil in standard proctor test (%)
7 Percentage of cement in the mixture (by dry weight of untreated soil) (%)
8 Percentage of lime in the mixture (by dry weight of untreated soil) (%)
9 Percentage of fly ash in the mixture (by dry weight of untreated soil) (%)
10 Percentage of Cement Kiln Dust (CKD) in the mixture (by dry weight of untreated soil) (%)
11 Percentage of EMC SQUARD (1000) in the mixture (by volume of untreated soil)
12 Dry density of stabilized soil (MD) (kg/m^3)
13 Moisture content of stabilized soil (%)
14 Dry curing period (day)
15 Moisture curing period (day)

The only output parameter considered in this study is the unconfined compression strength (UCS) of the stabilized soil. For ANN mapping, one hidden layer was used between the input layer and output layer. The number of hidden nodes in the hidden layer was determined by adaptive training and on-line monitoring of accuracy measures on the testing datasets. This was done by varying the number of initial hidden nodes, in the hidden layer, until the network was able to best learn the patterns involved in the testing datasets. Fully connected internal structure, where every node in one layer connects to all the nodes in the next layer, was used. In order to determine the domain of input parameters for the optimal ANN model, effect of input parameters on the output and the performance evaluation criteria (statistical accuracy measures and graphical evaluation) are utilized to identify and distinguish the most important parameters that contribute to the best UCS prediction. In the initial step, all 15 potential input parameters were used to develop the desired UCS prediction model. Then, in second step, one parameter was eliminated to arrive at 14 inputs. It is worthy to mention that, based on the previous soil chemical stabilization knowledge, only the inputs that have high probability of having negative effect on the accuracy of ANN model prediction were eliminated from the input domain. If the statistical accuracy measures were improved by eliminating one input parameter, the effect of eliminating two input parameters at the same time was further investigated. This procedure was repeated until the statistical accuracy measures did not improve by eliminating more input parameters. The purpose of this procedure was to obtain the optimal ANN model with highest statistical accuracy and the least number of input parameters.

Model Training

In order to obtain the best UCS prediction model, the database used for training should embody all possible features and sub-features that the network is required to learn. For every trial ANN model structure, two stage training methodology was adopted in this study. In the first stage, the entire database (792 data sets) was divided into training, testing and validation sub-cases and then the least-error-structure is selected based on the statistical accuracy measures (i.e., Mean Absolute Relative Error (MARE), Maximum Absolute Relative Error (Max ARE), and Minimum Absolute Relative Error (Min ARE)) on the testing datasets. If the statistical accuracy measures, for the least-error-structure, for training, testing and validation data sets were found to be comparable, then the second stage of training is not necessary. Otherwise, this indicates that the developed net does not recognize some of the features in the database. Therefore, as a compromise, all data sets in the database are then used to re-train the least-error structure identified from stage one (i.e., the net that has already been identified, from stage one, via the complete training, testing and validation processes).

Model Selection

The best model is considered as the one with the highest statistical accuracy. Both statistical accuracy measures (MARE, max ARE and min ARE) and graphical accuracy measures (quantitatively reflected as points above/below the 45 degree line and standard deviation of error) criteria were adopted to select the desired optimal network models. It is noted that for each trial case, the statistical accuracy measures, mainly MARE were improved by training with all data at its optimal ANN structure. Therefore, the "training with all data" sub-case from each trial case was filtered out and summarized into Table 1 for further selection. Based on evaluation according to MARE, it was found that cases (3) and (6) are comparable, however, case (3) outperformed case 6 in all other statistical accuracy measures and graphical measures. Therefore, it was decided to select case (3) as the optimal ANN model for this study. This case is highlighted using bold font in Table 1. It is to be noted that this model was already validated on data sets that never used in training or testing tasks. Graphical comparison between actual and predicted UCS values using case (3) model is depicted in Figure 1. Excellent agreement between actual and predicted values is noted in this Figure. In order to demonstrate the convenience of using the developed ANN model, EXCEL application was created for the final neural network model listed in Table 1.

Table 1. Accuracy Measures for Developed ANN Models
(ANN model highlighted in bold is chosen as the final model)

trial case # / Statistical measures	(1)	(2)	**(3)**	(4)	(5)	(6)
# total datasets	792	792	**792**	792	792	792
# data above 45 deg line	434	390	**388**	419	443	402
# data below 45 deg line	358	402	**404**	373	349	390
MARE (%)	12.009	12.240	**9.979**	11.467	12.515	9.026
Max ARE (%)	284.020	81.642	**58.818**	142.140	372.684	133.511
Min ARE (%)	0.001	0.014	**0.020**	0.010	0.004	0.026
Standard deviation	18.251	14.291	**9.661**	14.641	22.354	10.231

(1) 15 inputs (all potential parameters included)
(2) 14 inputs (% passing no. 200 sieve not included)
(3) 14 inputs (liquid limit not included)
(4) 14 inputs (plastic index not included)
(5) 14 inputs (plastic limit not included)
(6) 13 inputs (liquid limit and % passing no. 200 sieve not included)

SIMULATIONS

It is convenient to use the developed ANN UCS prediction model to arrive at the optimal design of soil stabilization. As an example, the effect of curing time and stabilizer content on the UCS of cement-stabilized soil was investigated herein. The results are depicted in Figures 2 and 3.

Figures 2a, 2b, and 2c show that cement increases the strength of all three types of stabilized soils but the relationship between achieved strength and the moist curing time varies among them. For soil A, UCS increases with the curing time and reaches its maximum value at about 90 days and then decreases afterwards. For soil B, the pattern (increase-peak-decrease) is similar to stabilized soil A, but soil B reaches the highest strength at 60 days of curing time and the speed of UCS increase and decrease is much faster than soil A. UCS of soil C continues to increase beyond 90 days. Although the speed of strength increase for soil C lies between those of soil A and soil B, soil C will have higher maximum UCS than the other two soils because its strength increase could last a longer time. The strength-curing time relationship is consistent for different cement content for each stabilized soil.

Figures 3a, 3b, and 3c show that UCS of all three stabilized soils increases with the increase of cement content. In order to prepare the samples with optimal cement content, previous researcher's recommendations were utilized herein to optimize the mixture design (i.e., use of cement content of 5%, 7%, 9% for soil A and 7%, 9%, 11% for soils B and C). However, from Figure 3a, UCS of soil A increases until 9% cement and this trend continues afterward, which indicates that the optimal cement content for soil A lies beyond the recommended range. In other words, the recommended optimal cement content for soil A, as recommended from previous research, is actually not valid in this case. In contrast to stabilized soil A, the UCS curves increase gradually and level out at about 11% cement content, indicating that the optimal cement content lies around 11% cement for soils B and C. The previous research recommendations are somewhat applicable for soils B and C.

The above simulations illustrate that ANN-UCS prediction model is able to facilitate the proper determination of the optimal design of a soil chemical stabilization task. More importantly, the simulation can significantly reduce the number of laboratory tests needed to arrive at (or verify) the optimal soil-chemical stabilization design. For example, if it is desired to obtain the optimal stabilization parameters for a specific soil, then the ANN-based EXCEL simulator can be used to generate appropriate UCS vs. cement content responses at various curing times similar to those depicted in Figure 3b. The obtained responses will help identify the optimal stabilization parameters. Based on this information, limited and carefully planned tests can be performed to verify or fine tune the optimal stabilization parameters. Therefore, reducing the number of laboratory tests needed to properly arrive at the optimal parameters.

CONCLUDING REMARKS

In this study, a static NN of the backpropagation type was developed for predicting the UCS of chemically stabilized soil. Statistic measurements from ANN model predictions showed that the ANN predictions are quite accurate. This study also demonstrated the advantages of the NN approach in studying soil chemical stabilization. First, NN model was simply built up by training it on the available experimental data where no complex mathematical expressions were needed and no mechanics-based theories were adopted. Second, the trained NN model can simulate for new scenarios without the need for any additional experimental–based information. Third, developed NN model can reliably be used to predict UCS of stabilized soil as long as the input parameters fall within the applicable training ranges. Fourth, the developed NN model can easily be implemented into an EXCEL spreadsheet for convenient practical usage.

ACKNOWLEDGMENTS

Financial support received from the Kansas State University Transportation Center to conduct this research is highly appreciated.

REFERENCES

Arman, A. and Saifan, F.S. (1967). "The Effect of Delayed Compaction on Stabilized Soil-Cement". *Transportation Research Record 827*, Transportation Research Board, Washington, D.C, pp. 30-38.

Banda S.K. (2003). "Laboratory Evaluation of Chemical Stabilization of Kansas Embankment Soils", *Master's Thesis*, Kansas State University, Manhattan, Kansas.

Haykin S. (1999). "*Neural Networks: A Comprehensive Foundation*", Second Edition, Prentice Hall, Upper Saddle River, New Jersey 07458.

Huang, C., Najjar, Y. and Romanoschi, S. (2006) "Characterizing the Fatigue Life of Asphalt Concrete," *Intelligent Engineering Systems through Artificial Neural Networks*, Vol. 16, pp. 369-374.

Lindeburg, M.R. (2003). "*Civil Engineering Reference Manual for the PE Exam*". Professional Publications, Inc. Belmont, CA.

Najjar, Yacoub and McRyenold, Richard (2003) "Modeling the Joint Deterioration Behavior of PCC Kansas Pavements via Dynamic ANN Approach," *ASCE Geotechnical Special Publication # 123*, pp. 209-222.

Najjar, Yacoub and Ali, Hossam (1998) "CPT-Based Liquefaction Potential Assessment: A Neuronet Approach," *ASCE Geotechnical Special Publication No.75*, Dakoulas, P. et. al. (Editors), Vol. 1, pp. 542-553

Wibawa, B. and Rahardjo, H. (2001). "Stabilization of a compacted swelling clayey soil". Clay science of engineering, proceedings of the *International Symposium on Suction, Swelling, Permeability and Structure of Clays*, Is-Shizuoka, Shizuoka, Japan. pp. 11-13.

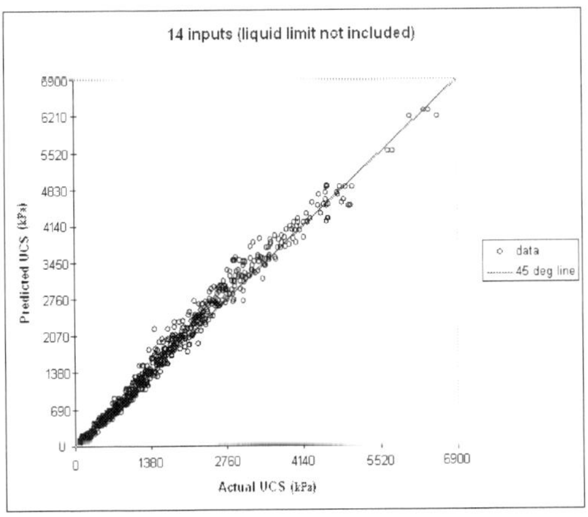

Figure 1. Predicted and Actual UCS for The Most Promising Network

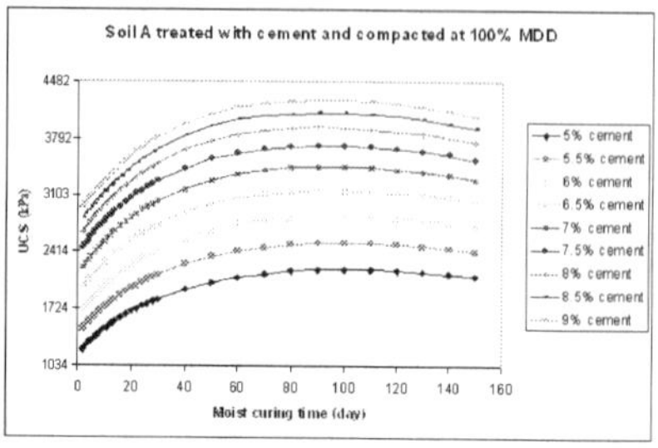

Figure 2a. UCS vs. Moist Curing Time for Soil A treated with Cement

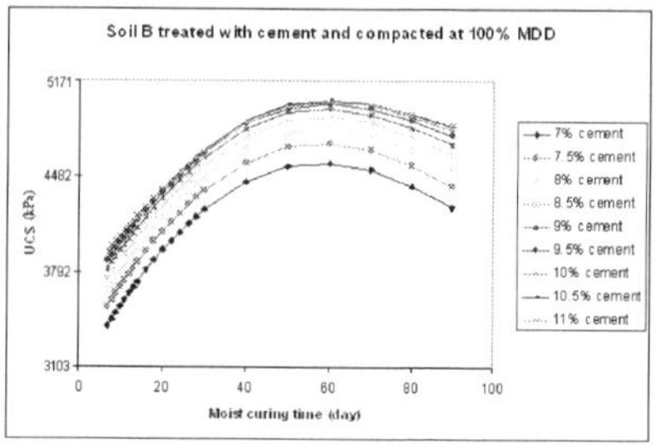

Figure 2b. UCS vs. Moist Curing Time for Soil B treated with Cement

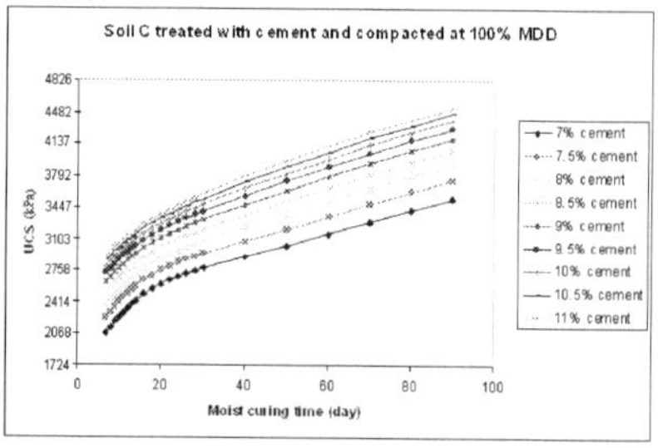

Figure 2c. UCS vs. Moist Curing Time for Soil C treated with Cement

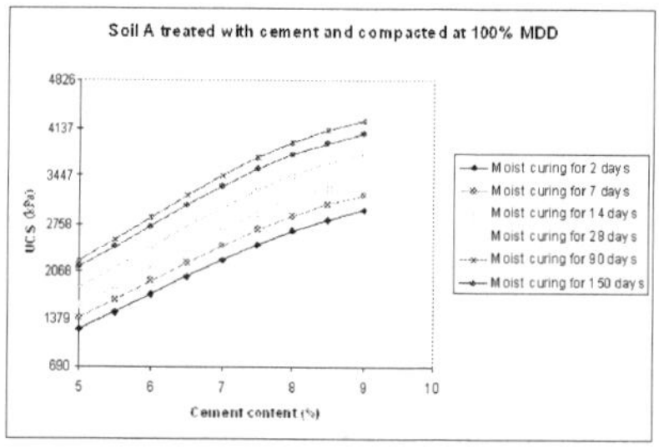

Figure 3a. UCS vs. Cement Content for Soil A treated with Cement

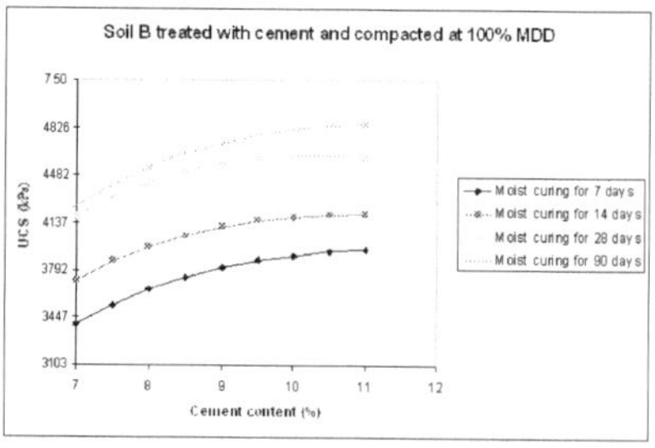

Figure 3b. UCS vs. Cement Content for Soil B treated with Cement

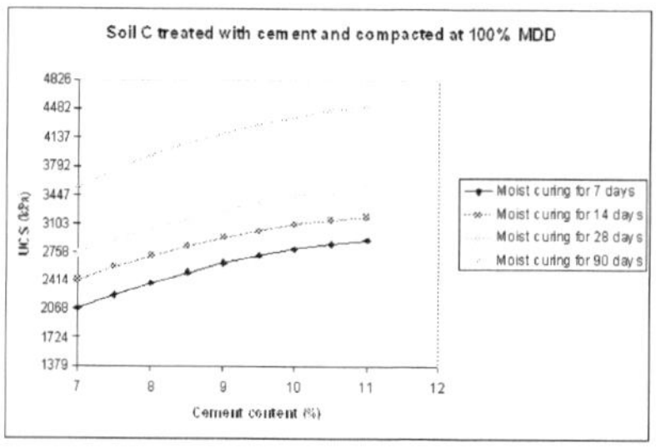

Figure 3c. UCS vs. Cement Content for Soil C treated with Cement

RESILIENT MODULUS PREDICTIVE MODEL FOR UNBOUND PAVEMENT MATERIALS

Dragos Andrei[1], Matthew W. Witczak[2], William N. Houston[3]

ABSTRACT

Data from 96 resilient modulus laboratory tests on four base materials and four subgrade soils was used to develop a predictive model capable of estimating changes in resilient modulus as a function of changes in the state of stress, moisture and density. Test specimens were compacted at optimum moisture/density conditions and then dried or soaked to various moisture conditions. Both standard and modified compactive efforts were used. Plastic, subgrade-type soils were especially affected by moisture. Non-plastic, base-type materials were more sensitive to changes in the state of stress.

M_R-MOISTURE MODEL

In contrast to the 1986 and 1993 AASHTO Design Guides, the Mechanistic-Empirical Pavement Design Guide (NCHRP 2004a) uses resilient modulus as input for all unbound pavement materials: base, subbase and subgrade. The accurate estimation of the resilient modulus is therefore crucial to the overall accuracy of the mechanistic-empirical predictive methodology. During development of the Guide, Witczak et al. (2000) conducted a synthesis of existing resilient modulus – moisture data available at the time in the technical literature and developed the following predictive model for implementation in the Guide:

$$M_R = 10^{a + \frac{b-a}{1+EXP\left(\beta + k_S\left(S - S_{opt}\right)\right)}} k_1 p_a \left(\frac{\theta}{p_a}\right)^{k_2} \left(\frac{\tau_{oct}}{p_a} + 1\right)^{k_3} \quad (1)$$

[1] Civil Engineering Department, California State Polytechnic University, Pomona, California 91768, Phone: 909.869.2487, Fax: 909.869.4342, Email: DAndrei@csupomona.edu

[2] Department of Civil and Environmental Engineering, Arizona State University, Tempe, Arizona 85287-5306, Phone: 480.965.2759, Fax: 480.965.0557, Email: Witczak@asu.edu

[3] Department of Civil and Environmental Engineering, Arizona State University, Tempe, Arizona 85287-5306, Email: Bill.Houston@asu.edu

Where:
M_R = resilient modulus at degree of saturation S (%);
a = regression parameter, minimum of $\log(M_R/M_{Ropt})$;
b = regression parameter, maximum of $\log(M_R/M_{Ropt})$;
M_{Ropt} = resilient modulus at maximum dry density and optimum moisture content;
β = calculated as a function of a and b by imposing the condition of a zero intercept ($M_R/M_{Ropt} = 1$ for $S = S_{opt}$):

$$\beta = \ln \frac{-b}{a} \tag{2}$$

k_S = regression parameter for the moisture and density effects;
$(S - S_{opt})$ = variation in degree of saturation expressed in decimal;
k_1, k_2, k_3 = regression parameters for the stress-dependent part of the model;
p_a = atmospheric pressure in units consistent with θ, τ_{oct};
θ, τ_{oct} = stress invariants.

Given the limited amount of data available from the literature, unbound materials were broadly divided into fine-grained and coarse-grained. Corresponding to the two categories, the values used in the Guide for the coefficients a, b and k_S are given in Table 1. Equation 1 has been successfully implemented in the Guide, however, a comprehensive laboratory test program was desired to validate the new model.

LABORATORY TEST PROGRAM

Four subgrade soils and four base materials were included in this study. The materials were selected and provided by the Arizona Department of Transportation (ADOT) and are presented in Table 2. The first part of the laboratory test program consisted of routine soil classification tests, CBR tests and moisture-density curves at two compactive efforts: standard and modified.

The second part of the testing program included compaction of the resilient modulus test specimens, moisture conditioning and testing in a triaxial apparatus according to the NCHRP 1-28A resilient modulus test protocol (NCHRP 2004b). For all subgrade soils, the vertical deformation was measured by LVDTs mounted on studs placed inside the test specimen during compaction. The technique is described in more detail by Andrei (2003). All subgrade soil specimens were cylindrical, 100 mm diameter by 200 mm height. Base material specimens were 150 mm diameter by 300 mm height, and vertical measurements were made using clamp supported, spring-loaded LVDT's.

In order to account for moisture changes similar to those occurring in the field, test specimens were initially compacted at optimum moisture and density conditions. Then, the specimens were either soaked in water or allowed to dry in open air. By periodically monitoring the weight of the specimens, the loss or gain of water was estimated. When the desired change in moisture was achieved, the specimens were sealed in a membrane and kept sealed for a minimum of 24 hours to achieve a uniform moisture distribution throughout the test specimen before testing.

In summary, for each of the 8 materials investigated, 12 test specimens were compacted at optimum moisture content and maximum dry density, 6 using standard compactive effort and 6 using modified compactive effort. Of each group of 6 specimens, 2 have been tested as compacted, 2 soaked and the remaining 2 dried prior to testing.

TEST RESULTS AND ANALYSIS

To graphically illustrate the effect of changes in moisture and density on modulus, a single state of stress at which all modulus values are calculated and plotted has to be chosen. This fixed state of stress should be the one of most interest, i.e. the state of stress expected to occur under traffic loading.

In his dissertation, Andrei (2003) estimated typical states of stress for 30 Arizona pavement sections selected from the LTPP database. As expected, the modulus of the asphalt concrete layer was found to strongly influence the state of stress both in bases and subgrade soils. To take into account the seasonal variation in the stiffness of the asphalt concrete layer, two typical E_{AC} values were considered: 345 MPa (50 ksi), corresponding to hot conditions and 6,895 MPa (1,000 ksi), corresponding to cold conditions. As a result, two typical states of stress, corresponding to the two conditions ("Hot" and "Cold") were used for base materials and another two for subgrade materials. These rough estimates of typical stresses for Arizona pavements are given in Table 3.

After each resilient modulus test, the raw test data was carefully analyzed, filtered and reduced to eliminate errors related to signal noise and sensor resolution (non-linearity error). A preliminary analysis was performed to generate k_1, k_2 and k_3 values for each individual test. With very few exceptions, the k_i parameters were found to be rational. Higher values of the k_2 parameter were found for bases compared to subgrades and higher values of the k_3 parameter were obtained for subgrades, at the same moisture content, when compared to bases, as expected.

Partial results for Phoenix Valley Subgrade are summarized in Table 4. Due to the size restraints for this paper, only results for this material will be discussed in more detail. Phoenix Valley Subgrade (PVSG) is a clayey sand, A-2 type material and can be considered representative of the group of 3, plastic, A-2 type subgrade soil, which includes the Flagstaff and Sun City subgrade soils (see Table 2).

The Phoenix Valley Subgrade showed great variations in modulus with moisture content. For specimens compacted with standard effort, at optimum moisture content the modulus was 48 MPa (7 ksi). When the specimens were allowed to dry, moisture content was reduced about 8 percent and the modulus increased to 1,379 MPa (200 ksi), more than 20 times higher than the value at optimum. When the specimens were allowed to gain water, in this case around 3%, the modulus dropped from 48 MPa (7 ksi) to 14 MPa (2 ksi). As a result of changes in stress, the modulus value ranged from 34 to 62 MPa (5 to 9 ksi) at optimum, 1,034 to 1,724 MPa (150 to 250 ksi) when dry and from 10 to 17 MPa (1.5 to 2.5 ksi) when soaked. The variation in modulus with degree of saturation in "Hot" conditions is illustrated in Figure 1. Full dots are used for specimens compacted with standard compactive energy and empty dots for specimens compacted with modified compactive energy.

Equation 1, the M_R-moisture model currently implemented in the Guide, was used to fit the resilient modulus test data obtained for each of the eight materials, however without success. A good fit could only be obtained when separating the data points by compactive effort, as shown in Figure 1. Two predicted curves, using two different sets of model parameters were obtained. As shown in Figure 1, at about 80% saturation (optimum), the measured resilient modulus corresponding to standard compaction is significantly lower than the value corresponding to modified compaction. Therefore Equation 1 would need to be modified to include a parameter related to compactive effort or density. The same observation was made for the other 7 materials included in the study. The difference between the "standard" and "modified" predicted curves was higher for the plastic, subgrade-type soils.

However, when plotting the same modulus values against gravimetric moisture content, all points seemed to fall on a unique curve, regardless of the compactive effort, as shown in Figure 2. This unique relationship can be modeled as a sigmoid curve of a form similar to Equation 1:

$$M_R = 10^{a+\frac{b-a}{1+EXP\left(\beta+k_w\left(w-w_{opt\_std}\right)\right)}} k_1 p_a \left(\frac{\theta}{p_a}\right)^{k_2} \left(\frac{\tau_{oct}}{p_a}+1\right)^{k_3} \tag{3}$$

Where:
w = gravimetric moisture content expressed in decimal
w_{opt_std} = gravimetric optimum moisture content corresponding to standard compaction energy in decimal;
k_w = regression parameter

The only difference between Equations 1 and 3 is that degree of saturation was replaced by gravimetric moisture content. Consequently, the regression parameter k_S was replaced by k_w. All other model parameters were explained under Equation 1.

Equation 3 was successfully used to model the resilient modulus behavior of the eight materials included in this study. Model parameters and goodness of fit statistics for each material are given in Table 5.

CONCLUSIONS AND RECOMMENDATIONS

Moisture was found to have an impact on the resilient modulus of unbound materials, especially on plastic subgrade-type soils (PVSG, FCSG and SCSG). For these materials, a 3 to 5% increase in moisture content from optimum conditions resulted in a 50 to 70% reduction in resilient modulus. Drying of the test specimens resulted in a significant increase in resilient modulus, in some cases ten-fold.

Compared to the plastic subgrade-type materials, non-plastic base-type materials (GMAB2, SRAB2, GLAB2, PRAB2 and YSSG) were less affected by moisture. Generally the modulus increased two or three times upon drying from optimum but no significant reduction in modulus was observed upon wetting.

The compactive effort was found to play a significant role in the variation of resilient modulus with degree of saturation. Therefore, a predictive model based on degree of saturation, like Equation 1, cannot be effectively used to fit the test data. However, by replacing degree of saturation with gravimetric moisture content, a unique $M_R - w$ relationship was observed, regardless of the compactive effort used. Based on this observation, a resilient modulus predictive model based on gravimetric moisture content and stress invariants is proposed as described by Equation 3.

It must be recognized that the study presented in this paper was based upon the M_R relationships for 8 different Arizona DOT materials. All of these materials were either an A-1-a or A-2 AASHTO classification group. It is, without question, a somewhat limited range of soils. As a consequence, one of the very major priorities for future research would be to greatly expand the testing program (and associated database) to as many different soil subgrade types as possible.

REFERENCES

Andrei D. (2003). "Development of a Predictive Model for the Resilient Modulus of Unbound Materials." Doctoral Dissertation, Arizona State University, May 2003.

National Cooperative Highway Research Program (2004a). "Guide for Mechanistic Empirical Design of New and Rehabilitated Pavement Structures." <http://www.trb.org/mepdg/guide.htm> (August 11, 2008)

National Cooperative Highway Research Program (2004b). "Laboratory Determination of Resilient Modulus for Flexible Pavement Design." Research Results Digest, January 2004 – Number 285, the National Academies, Transportation Research Board, Washington, DC

Witczak M.W., Andrei D. and Houston W.N. (2000). "Resilient Modulus as Function of Soil Moisture – Summary of Predictive Models." Development of the 2002 Guide for the Development of New and Rehabilitated Pavement Structures, NCHRP 1-37 A, Inter Team Technical Report (Seasonal 1), June 2000.

Table 1. Values of a, b and k_s for Coarse Grained and Fine Grained Materials

Parameter	Coarse Grained Materials	Fine Grained Materials	Comments
a	- 0.3123	-0.5934	Regression parameter
b	0.3	0.4	Conservatively assumed, corresponding to modulus ratios of 2 and 2.5
k_S	6.8157	6.1324	Regression Parameter

Table 2. ADOT Materials Investigated

Material ID	Source	Function	AASHTO Class.	Unified Soil Class.
PVSG	Phoenix Metro Area	Subgrade	A-2-4	SC
YSSG	Yuma Area	Subgrade	A-1-a	GP
FCSG	Flagstaff Area	Subgrade	A-2-6	SC
SCSG	Sun City	Subgrade	A-2-6	SC
GMAB2	Grey Mountain	Base	A-1-a	GW
SRAB2	Salt River	Base	A-1-a	SP
GLAB2	Globe	Base	A-1-a	SP-SM
PRAB2	Prescott	Base	A-1-a	SP-SM

Table 3. Typical States of Stress

AC Layer Temperature	E_{AC} in MPa (ksi)	Base θ in kPa (psi)	Base τ in kPa (psi)	Subgrade θ in kPa (psi)	Subgrade τ in kPa (psi)
Hot	345 (50)	303 (44)	110 (16)	90 (13)	34 (5)
Cold	6,895 (1,000)	90 (13)	48 (7)	34 (5)	7 (1)

Table 4. Regression Analysis Results for Phoenix Valley Subgrade (PVSG), Standard Compactive Effort

Material ID			Phoenix Valley Subgrade				Gs = 2.719	A-2-4/SC
Compactive Effort			STANDARD					
Optimum Moisture Content		%	11.3					
Maximum Dry Unit Weight		kg/m^3	1,970					
Moisture State			DRY		OPTIMUM		WET	
Test ID			RMJL01	RMJL06	RMAP15	RMMA04	RMMA17	RMMA18
Date			07/05/02	07/16/02	04/29/02	05/07/02	06/07/02	06/07/02
Moisture Content		%	3.7	3.2	10.7	10.6	13.6	14
Degree of Saturation		%	27.1	23.0	80.5	79.1	89.0	90.6
Dry Unit Weight		kg/m^3	1979.2	1973.9	1993.7	1989.9	1917.3	1911.4
Void Ratio		%	37.1	37.8	36.2	36.4	41.5	42.0
Number of data points			11	8	14	14	10	10
Stress Dependent M_R	*k1*	*	*17,245*	*18,903*	*807*	*937*	*142*	*144*
$M_R = k1 p_a(-/p_a)^{k2}(-_{oct}/p_a+1)^{k3}$	*k2*	*	*0.38*	*0.19*	*0.63*	*0.52*	*0.19*	*0.19*
	k3	*	*-1.09*	*-0.43*	*-0.83*	*-0.80*	*1.26*	*1.31*
	Se/Sy	*	0.64	0.45	0.49	0.60	0.61	0.64
	R^2	*	0.58	0.80	0.76	0.64	0.62	0.60
M_R @ Hot Conditions		kPa	1,214	1,651	59	71	20	21
M_R @ Cold Conditions		kPa	1,086	1,523	39	52	13	13
$W_i - W_{opt}$		%	-7.6	-8.1	-0.6	-0.7	2.3	2.7
$-_{di} - _{dmax}$		kg/m^3	9.2	3.9	23.7	19.9	-52.7	-58.6

* Dimensionless

Table 5. Model Parameters and Statistics for Equation 3

Material ID	AASHTO	USCS	a	b	k_w	β	k_1	k_2	k_3	$w_{opt\ std}$	Se/Sy	R^2
PVSG	A-2-4	SC	0.2	41.9	67.255	0.974	467	0.358	-0.686	11.3%	0.15	0.98
YSSG	A-1-a	GP	1.0	94.0	82.758	8.714	1,468	0.827	-0.864	11.0%	0.26	0.93
FCSG	A-2-6	SC	0.3	10.9	74.489	0.722	634	0.187	-0.855	19.0%	0.16	0.98
SCSG	A-2-6	SC	0.1	19.2	53.100	0.000	747	0.221	-0.104	11.3%	0.27	0.93
GMAB2	A-1-a	GW	0.0	2,096.4	2.559	-0.509	1,423	0.758	-0.288	6.7%	0.32	0.9
SRAB2	A-1-a	SP	0.6	2,096.4	22.401	2.666	1,170	0.919	-0.572	6.9%	0.16	0.97
GLAB2	A-1-a	SP-SM	0.7	2,096.4	35.787	2.981	1,032	0.830	-0.307	6.7%	0.18	0.97
PRAB2	A-1-a	SP-SM	1.0	2,096.5	144.223	8.711	1,092	0.784	-0.236	6.3%	0.21	0.95

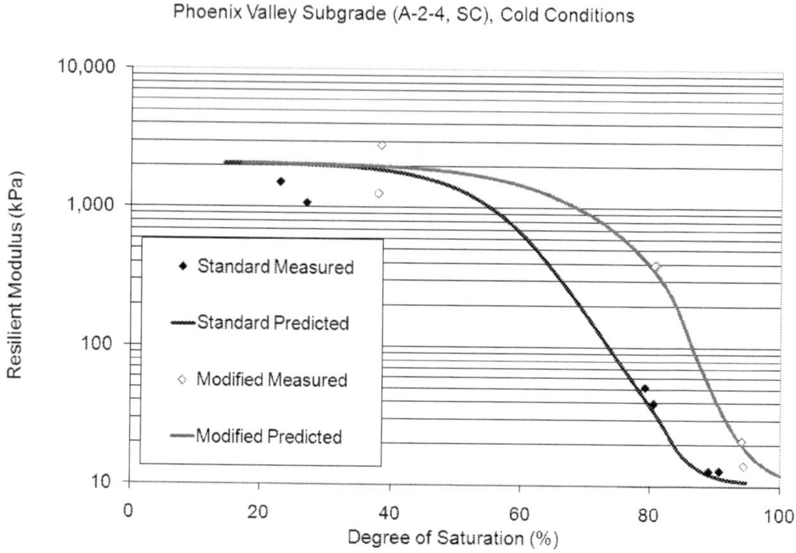

Figure 1. PVSG – M_R vs. degree of saturation, cold conditions

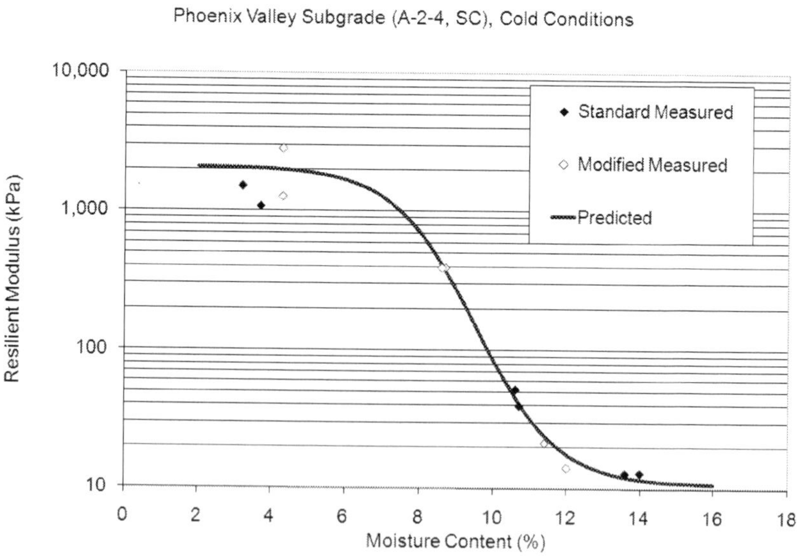

Figure 2. PVSG – M_R vs. gravimetric moisture content, cold conditions

Behavior of Geocell-Reinforced Granular Bases under Static and Repeated Loads

Sanat Kumar Pokharel[1], S.M. ASCE, Jie Han[1], P.E., M. ASCE, Dov Leshchinsky[2], Robert L. Parsons[1] P.E., M. ASCE, and Izhar Halahmi[3]

Department of Civil, Environmental, and Architectural Engineering, the University of Kansas, 1530 W. 15th street, Lawrence, KS, 66045-7609. Phone: 785-864-3016, Email: sanpok@ku.edu, jiehan@ku.edu , and rparsons@ku.edu

[2] Department of Civil and Environmental Engineering, the University of Delaware, Newark, DE19719. Phone: 302-831-2446, Email: dov@ce.udel.edu

[3] PRS Mediterranean Ltd., Europe-Israel Tower, 2 Weizmann St., Tel-Aviv 64239, Israel, Email: izharhal@012.net.il

ABSTRACT

Geosynthetics have been used for base reinforcement since 1970s. Numerous research has already been carried out for planar geosynthetic reinforcement but limited research has been conducted for three-dimensional geocell reinforcement. Literature review has also demonstrated a significant gap between the applications and theories of geocell reinforcement outlining the need for more research. This study was to investigate the behavior of reinforced bases using a single geocell under static and repeated loads on a loading plate. The experimental results show that the single geocell could increase the stiffness by approximately 50% and the maximum load by 100% as compared with those of the unreinforced base. The repeated test shows that the geocell-reinforced base had the percentage of elastic deformation increase with the number of cycles of the repeated load up to 95%.

INTRODUCTION

Geosynthetics have been used for subgrade improvement and base reinforcement since 1970s (Webster, 1979). Tingle and Jersey (2007) categorically pointed out the problems associated with maintaining low-volume unpaved roads with minimal funding and identified geosynthetic reinforcement as a possible means to deal with this condition. Use of geosynthetics as soil reinforcement has been reported to increase the overall stiffness and bearing capacity of the geosynthetic-soil composite. Geosynthetics are also found helpful in reducing settlement and rutting depth. For a given design condition, these improvements lead to a reduced amount of aggregate material and/or extending of the service life. Planar reinforcement, such as geogrid and geotextile, has been widely used in unpaved roads with established design methods (for example, Giroud and Noiray, 1981; Giroud and Han, 2004). Geogrid interacts with the granular base to mainly provide confinement in two dimensions and the geotextile mainly provides separation between base course and subgrade. Geocell, due to its three-dimensional configuration, is ideal for soil confinement. The idea of geocell was first introduced by the United States Army Corps of Engineers to provide lateral confinement to beach sand during 1970s (Webster, 1979). Different materials of geocell have been used in the field and/or research, such as metallic, polymeric, and paper cells. However, the commercial geocells currently available are composed of polymeric materials (most commonly high-density polyethylene (HDPE)). The polymeric cells are welded together to produce an interconnected honeycomb type of structure that provides lateral confinement, a tensioned membrane effect, and stress distribution to a wider area (Zhou and Wen, 2008). The majority of the research so far has focused on planar reinforcement for subgrade improvement and base reinforcement. Yuu et al. (2008) identified a significant gap between the applications and the theories for geocell reinforcement outlining the need for further research.

To understand the behavior of geocell-reinforced bases under static and repeated loads, plate load tests were carried out on reinforced granular bases using a single geocell and compared with an unreinforced base. This paper discusses the results of these plate load tests. A novel polymeric alloy type of geocell and poorly-graded Kansas River sand as the base material were used for these tests.

REINFORCEMENT MECHANISMS

For roadway applications, geotextiles have been mostly used for separation, drainage, and filtration and woven geotextiles are sometimes used for reinforcement as a tensioned membrane. Geogrids and geocells have been mostly used for reinforcement by providing confinement to base courses and subgrade. Lateral confinement, increased bearing capacity, and the tensioned membrane effect have been identified as the major geosynthetic reinforcement mechanisms (Giroud and Han, 2004). Geogrids provide lateral confinement through the interlocking between geogrid apertures and aggregates. Giroud and Han (2004) developed a design method for geogrid-reinforced unpaved roads considering the interlocking effect by the geogrid in-plane stiffness, the increased bearing capacity, the stress distribution, the relative stiffness of base to subgrade, the subgrade strength properties, and other traffic related factors.

In geocell-reinforced bases, three-dimensional geocells can effectively provide lateral confinement to infill materials. In addition, the friction between the infill material and the geocell wall and the action of the geocell-reinforced base as a mattress to restrain the subgrade soil from moving upward outside the loaded area provide the vertical confinement to the infill material and the subgrade. Therefore, the stiffness of geocell is important for the lateral and vertical confinement.

Mhaiskar and Mandal (1996) indicated that at a given height and a width to height ratio, the elastic modulus of the geocell was more important than the seam strength. Tests on single geocell-reinforced bases showed an increase in the resilient modulus: 16.5 to 17.9% for cohesive soils and 1.4 to 3.2% for granular soils (Mengelt et al., 2006).

Bathurst and Jarrett (1989) reported a stiffer geocell for a given mattress thickness and a designed rut depth could increase load bearing capacity by a factor of two. Shimizu and Inui (1990) found that the geocell could increase bearing capacity and the extent of increase correlated to the horizontal stiffness of the cell material. Inclusion of geocell in the granular bases could increase both the bearing capacity and the elastic modulus of the base by providing confinement to the infill material (Han et al., 2008). Pokharel et al. (2009) found that the behavior of geocell-reinforced sand depended on the initial shape and the elastic modulus of the geocell. Geocell reinforcement has also been reported to provide good improvement on resistance to repeated loads (Rea and Mitchell, 1978). Chang et al. (2008) found the dynamic modulus of subgrade reaction increased after 100 cycles of loading in a geocell-reinforced sandy soil.

PROPERTIES OF BASE AND GEOCELL MATERIALS USED IN THE TESTS

Kansas River sand was used as the granular base for the tests. Kansas River sand is poorly-graded and has a minimum void ratio of 0.354, a maximum void ratio of 0.583, and a specific gravity of 2.65 at 20°C. The mean grain size (d_{50}) of this sand was 2.6 mm.

The geocell used for the tests was made of novel polymeric alloy, which is characterized by flexibility at low temperatures similar to HDPE and elastic behavior similar to engineering thermoplastic. The geocells had tensile strengths 23.27 N/mm. The elastic modulus of the geocell at 2% strain was 620 MPa. The 2% strain was chosen because the measured strains in geosynthetics in the field were typically within this range. The geocell used in this study had two perforations of 100 mm^2 each on both pallets. The height and thickness of the geocell were 100mm and 1.1 mm, respectively. A single geocell was laid out in a near circular shape with a diameter of 205 mm. The selection of this shape was based on the earlier study by the authors (Pokharel et al., 2009). The stress-strain curve of this geocell is shown in FIG. 1.

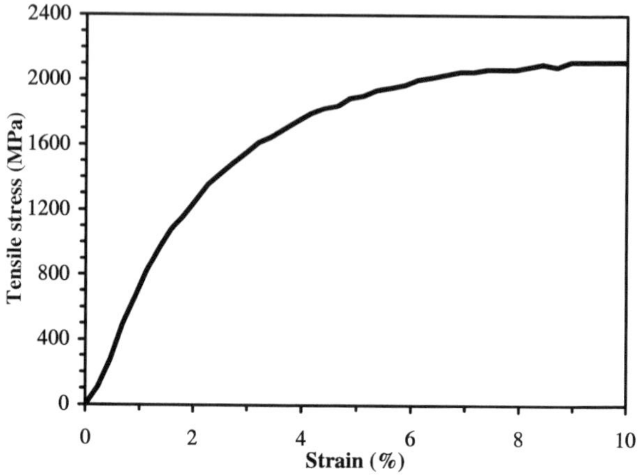

FIG. 1 Tensile strength of geocell

TEST PROCEDURE

Plate load tests were conducted in a loading apparatus designed and fabricated at Department of Civil, Environmental, and Architectural Engineering at the University of Kansas. The loading system had a 150 mm diameter air cylinder with a maximum air pressure of 2100 kPa and a 150 mm diameter loading plate. The test box was 605 mm by 605 mm square in plan. Tests were conducted to verify that single geocell within this box would not have any boundary effect. Single geocells with a height of 100mm were used for both static and repeated load tests. The geocell was placed in the center of the box at a near circular shape and infilled and embedded in the sand compacted to 70% relative density. Compaction was done in three layers, 50 mm each for the first two layers and 20mm for the top layer. For comparison purposes, unreinforced sand was also prepared in a similar way and tested under static loading until failure. For the geocell reinforced sand, both static and repeated loading tests were conducted. The static test was conducted by increasing the load until the geocell-reinforced sand failed. The repeated load test was only conducted on the reinforced sand at an applied pressure of 345 kPa (corresponding to approximately 70% of the pressure at failure under the static loading). Due to the low ultimate bearing capacity of the unreinforced sand, a repeated load test at the same pressure was impossible. The repeated load was applied at 1 cycle/minute for 150 cycles.

RESULTS AND DISCUSSIONS

Figure 2 presents the pressure-displacement curves of the unreinforced sand under static loading and the geocell-reinforced sand under static and repeated loading. It is shown that the unreinforced sand under static loading failed at 230 kPa while the single geocell-reinforced sand under static loading failed at 480 kPa. Therefore, the maximum load was increased by approximately 100% from the unreinforced to reinforced sand under static loading. The stiffness of the unreinforced and reinforced sands can be determined based on the slopes of the linear portions of the pressure-displacement curves. As shown in FIG. 2, the stiffness of the reinforced sand is approximately 1.5 times that of the unreinforced sand.

For the repeated loading, approximately 70% (345 kPa) of the failure load was applied on the geocell-reinforced sand. The first loading up to a load of 345 kPa was applied in the same manner as that in the static loading to verify the repeatability of the test results. The comparison clearly shows a reasonable repeatability.

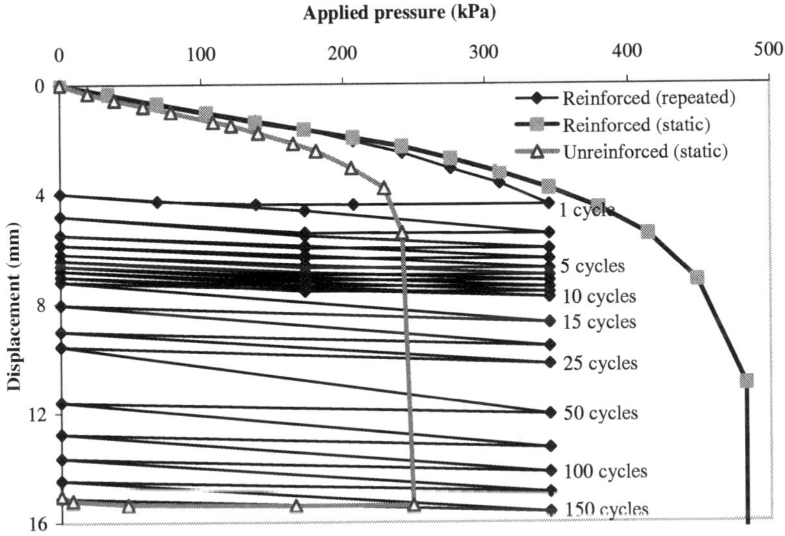

FIG. 2 Pressure-displacement curves under static and repeated loading

From the repeated loading test, the cumulative displacements at the applied pressures of 0 and 345kPa were measured and presented in FIG. 3. The cumulative displacement at 0kPa is the permanent deformation (also called plastic deformation). In the roadway applications, the permanent deformation is often reflected as a rut depth by a loaded wheel. The difference of the displacements at 0 and 345kPa is the elastic deformation, which is also plotted in FIG. 3. In other words, the total displacement is equal to the plastic deformation plus the elastic deformation. The percent of elastic deformation was calculated by dividing the elastic displacement to the total displacement induced by each load cycle. FIGURE 4 shows that the percent of elastic deformation increased with the number of the load cycle. It is shown that the initial loading had a low percent of elastic deformation, in other words, it had a high percent of plastic deformation. However, the percent of elastic deformation increased rapidly with the loading cycle and became relatively stable (80% or more) after 10 cycles. The elastic deformation after 150 cycles was 95.2% of the total deformation. The high percent of elastic deformation is beneficial to the service life of the road. FIGURE 4 also shows that it took approximately 150 cycles to reach a displacement (15mm), which corresponds to 10% the diameter of the loading plate. However, the same displacement was reached in one cycle for the unreinforced sand at 67% magnitude of the repeated load.

The above results and discussion have demonstrated the clear benefits of geocell reinforcement in terms of increased stiffness, increased bearing capacity, and reduced plastic deformation. It is expected that the geocell-reinforced sand can perform much better than the unreinforced base under traffic loading.

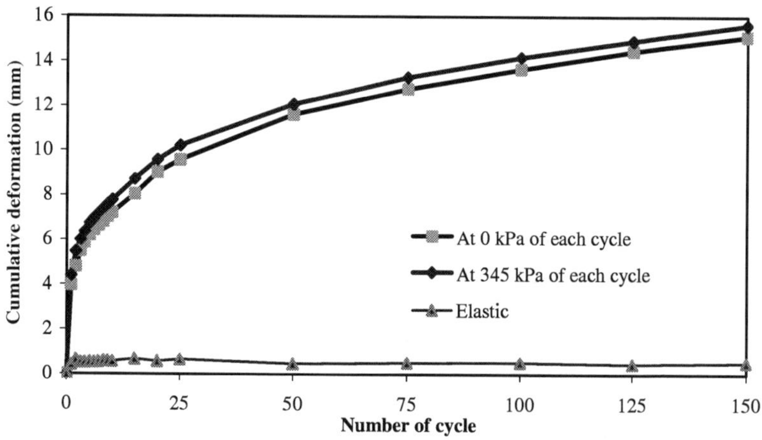

FIG. 3 Cumulative total and elastic deformation under repeated loading

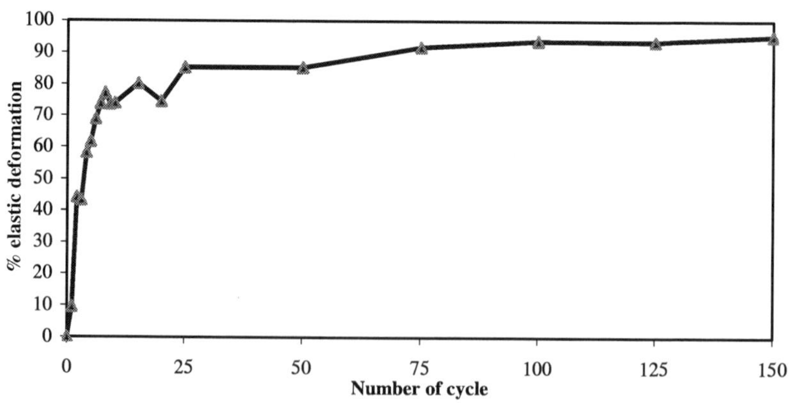

FIG. 4 Percent of elastic deformation under repeated loading

CONCLUSIONS

This paper presents the results of experimental work conducted to investigate the behavior of geocell-reinforced sand under static and repeated loading. Both static and repeated plate loading tests data were obtained. A single geocell was embedded in the base material to provide the confinement. Following conclusions can be drawn for this study:

1) Single geocell reinforcement improved the stiffness of the reinforced sand by a factor of 1.5 compared to the unreinforced sand.
2) Single geocell reinforcement increased the maximum load by two times from that of the unreinforced sand.
3) Single geocell reinforcement reduced plastic deformation and increased percent of elastic deformation under repeated loading. It only took 10 cycles to reach 80% or more of elastic deformation. The elastic deformation reached above 95% of the total deformation at the end of 150 loading cycles. This favorable behavior can be credited to the contribution of geocell.

ACKNOWLEDGMENTS

This research was funded jointly by the University of Kansas (KU), Transportation Research Institute from Grant #DT0S59-06-G-00047, provided by the US Department of Transportation – Research and Innovative Technology Administration and PRS Mediterranean, Inc. in Israel. Their support is greatly appreciated. The loading apparatus used in this research was designed and fabricated by Mr. Howard Jim Weaver, the lab manager in the Department of Civil, Environmental, and Architectural Engineering (CEAE) at KU. Undergraduate student, Mr. Milad Jowkar, in the CEAE Department at KU assisted in the lab test. The authors are thankful for their great help.

REFERENCES

Bathurst, R.J. and Jarrett, P.M. (1989). "Large-scale model tests of geocomposite mattresses over peat subgrades." *Transportation Research Record* 1188: 28-36.

Chang, D.T., Chang, C.H., Kou, C.H., and Chien, T.W. (2008). "Bearing capacity and resilient property studies for sandy soil with confinement of geocells. *Proceedings of Transportation Research Board 87th Annual Meeting (CD-Rom)*, January 13–17, 2008, Washington, D.C.

Giroud, J.P. and Han, J. (2004). "Design method for geogrid-reinforced unpaved roads. I. Development of design method." *ASCE Journal of Geotechnical and Geoenvironmental Engineering*, 130 (8): 775-786.

Giroud, J.P. and Noiray, L. (1981). "Geotextile-reinforced unpaved road design" *ASCE Journal of the Geotechnical Engineering Division*. 107(GT9): 1233-1254.

Han, J., Yang, X.M., Leshchinsky, D., and Parsons, R.L. (2008). "Behavior of geocell-reinforced sand under a vertical load." *Journal of Transportation Research Board*, in press.

Mengelt, M.J., Edil, T.B., and Benson, C.H. (2006). "Resilient modulus and plastic deformation of soil confined in a geocell." *Geosynthetic International.* 13(5): 195-205.

Mhaiskar, S.Y. and Mandal, J.N. (1996). "Investigation on soft clay subgrade strengthening using geocells." *Construction and Building Materials*, 10(4): 281-286.

Pokharel, S.K., Han, J., Leshchinsky, D., Parsons, R.L., and Halahmi, I. (2009). "Experimental evaluation of influence factors for single geocell-reinforced sand." Submitted for possible presentation and publication at the TRB 88th Annual Meeting, January 11 to 15, Washington, DC.

Rea, M. and Mitchell, J.K. (1978). "Sand reinforcement using paper grid cells." Regular meeting- Rocky Mountain Coal Mining Institute: 644-663.

Shimizu, M. and Inui, T. (1990). "Increase in the bearing capacity of ground with geotextile wall frame." *Geotextiles, Geomembranes and Related Products*, Den Hoedt (ed.), Balkema, Rotterdam: 254.

Tingle, J.S. and Jersey, S.R. (2007). "Empirical design methods for geosynthetic-reinforced low-volume roads." *Journal of the Transportation Research Board.* 1989(2): 91-101.

Webster, S. L. (1979). *Investigation of Beach Sand Trafficability Enhancement Using Sand-Grid Confinement and Membrane Reinforcement Concepts*. Report GL-79-20 (1). U.S. Army Engineer Waterways Experiment Station, Vicksburg, MS.

Yuu, J., Han, J., Rosen, A., Parsons, R.L., and Leshchinsky, D. (2008). "Technical review of geocell-reinforced base courses over weak subgrade." *Proceedings of GeoAmericas*, Cancun, Mexico, March 2 to 5, 2008: 1022-1030.

Zhou, H. and Wen, X. (2008). "Model studies on geogrid – or geocell-reinforced sand cushion on soft soil." *Geotextiles and Geomembranes*, 26(3): 231-238.

Laboratory Study on Recycled Building Waste Materials for Road Construction

Fen Ye[1], Xiao-Yang Jia[1], Tian-Tong Zhu[1], and Jie Han[2]

[1] Key Laboratory of Road and Traffic Engineering of the Ministry of Education, Tongji University, Shanghai, 201804, China, email: yefen@mail.tongji.edu.cn

[2] Department of Civil, Environmental, and Architectural Engineering (CEAE) Department, the University of Kansas, 2150 Learned Hall, 1530 W, 15th Street, Lawrence, KS 66045, email: jiehan@ku.edu

ABSTRACT: Saving of resources is a significant indicator to evaluate social and economic benefits of a highway project. The purposes of this study are to investigate the feasibility of using recycled building materials as base, subbase, and subgrade and provide guidelines for selection and performance evaluation of recycled materials in future projects

This study first considered the requirements for building waste materials which could be recycled. Laboratory tests were conducted to evaluate whether the waste materials fulfilled these requirements. Cement and High strength water stable Earth Consolidator (HEC) were employed as stabilizing/recycling agents. Orthogonal analysis was adopted to evaluate the effect of gradation, type, and quantity of stabilizing/recycling agents on unconfined compressive strength (UCS) and resilient modulus of recycled materials. The laboratory study shows that the use of the recycled waste materials as highway base, subbase, and subgrade is feasible and would fulfill its requirements.

INTRODUCTION

Urban development has increasingly generated waste materials, nearly 30 to 40% of which are from pavements and infrastructures, also called construction and demolition (C&D) wastes. Concrete rubble constitutes the largest proportion of the C&D waste. During the construction of brick-concrete structures, cast-in-place structures, and framework structures, it was reported that 500 to 600 ton waste materials are produced for every 10,000 m^2 construction process. In other words, the construction in China produces more than a hundred million ton building wastes annually. This number would be even higher if the wastes from building decoration and housing removal were included. Therefore, how to deal with these C&D wastes has become one of the top priorities for roadway engineers and researchers in many countries.

LITERATURE REVIEW

Building wastes are usually solid materials, which are by-products in the process of construction, maintenance, and demolition. Despite of different components, C&D wastes are often composed of soil, residue soil, broken concrete, mortar and brick, reinforced concrete stakes, metal, timber, building decoration waste, wrapper, etc.

In Australia, different types of recycled aggregates from C&D wastes are classified based on their properties, such as density, water absorption, percentage of fine content, content of organic materials, and others (Kotrayothar, 2007). Ordonez (2006) investigated the effect of cement and fiber on C&D waste-recycled materials using basic tests including gradation, specific gravity, Atterberg limits, and standard Proctor compaction tests, permeability, leaching, unconfined compression, and small-strain shear modulus tests. Kongshicong (2006) examined the possible use of recycled aggregates in structure concrete by conducting a comprehensive laboratory study. He employed a mix proportioning procedure to produce six series of concrete mixtures using different percentages of recycled coarse aggregates with and without the use of fly ash at four water-to-cement contents to gain a better understanding of their mechanical, microstructure, and durability properties. In addition, he studied the effects of recycled aggregate and steam curing on hardened properties of the recycled concrete. Movassaghi (2006) conducted a comparative study on the durability of three concrete mixtures containing new clean recycled concrete aggregate (NC-RCA), old de-icing salt contaminated, recycled concrete aggregate (OC-RCA), and natural aggregate as a control material. Movassaghi (2006) concluded that the durability and the strength of recycled concrete were dependent on the age of the recycled aggregate and recommended that the OC-RCA concrete should be used as a sustainable material in structural applications. Gottfredsen and Thogersen (1993) discussed the recycling of concrete in an aggressive environment and concluded that deleterious expansion and cracking could develop when recycled aggregate was used in concrete in the aggressive environment. They also found that adding fly ash to the recycled concrete delayed the expansion significantly when a reactive recycled aggregate was used. Wainwright et al. (1993) found that the properties of hardened concrete made with recycled coarse aggregates were inferior to those of a control mix (a mix made with all natural materials), and the strength of concrete made with recycled coarse sand was weaker than that of concrete made with natural sand.

RAW MATERIALS

Construction and Demolition Wastes

Construction and demolition (C&D) wastes include waste materials from pavements and houses. In this study, waste brick and concrete were used, which include

.

(1) There has been almost no gradation requirement for waste brick employed as a material for subgrade. Herein the waste brick is classified into four types: coarse brick (size from 19 to 37.5mm), medium brick (from 4.75 to 19.0mm), fine brick (from 2.3 to 4.75mm), and soil (less than 2.36mm).
(2) The waste concrete is considered as a material for subbase and can be divided into four types. It should fulfill the gradation requirements as listed in Table 1.

Table 1. Gradation of Recycled Concrete

Sieve Size (mm)	19 - 37.5	4.75 - 19	2.36 - 4.75	0 - 2.36
37.5	100%	100%	100%	100%
31.5	89.0	100	100	100
26.5	81.3	100	100	100
19	57.8	100	100	100
9.5	24.4	85.5	100	100
4.75	14.4	24.2	100	100
2.36	11.3	7.2	71.5	100
0.6	7.4	5.1	38.0	60
0.075	2.4	2.9	5.9	5.0

Stabilizing/Recycling Agents

Grade 32.5 cement and high strength water stable earth consolidator (HEC) can be employed as stabilizing/recycling agents. HEC, a kind of inorganic hydraulically cementing material, can be used to stabilize natural soil, special soil, soil and industrial waste residue. In the process of stabilization, the active component of HEC infiltrates into the phase interface of the basic units in the soil, which results in the activity of luminosilicate salt in the soil and then forms a polycrystalline structure. The HEC hydrate formation combined with basic units of the soil becomes a solidified unit. This stabilization would enhance the strength and moisture susceptibility of the material.

OBJECTIVE AND PROCEDURE

The aim of this study was to evaluate the feasibility of applying waste building materials (waste brick and concrete) for base, subbase, and subgrade, to identify key factors influencing the strength and stiffness of the recycled material in mix design, and to determine the optimum combination.

7-d unconfined compressive strength (UCS) and resilience modulus were adopted to evaluate the performance of the recycled concrete in this study. Orthogonal analysis method was employed to investigate the effect of gradation, the type of stabilizer, and the stabilizer content on the performance of the recycled brick and to determine the optimum combination.

RECYCLED BRICK

Orthogonal Experiment on 7-d UCS

Mixed level orthogonal analysis was used to investigate the recycled brick. The factors for analysis are: Factor A (Gradation), Factor B (stabilizer), and Factor C (stabilizer content). Factors A and C had three levels while Factor B had two levels. The three levels of Factor A are listed in Table 2. Factor B had two levels: cement and HEC. Factor C had three content levels: 4, 5, and 6%. The 7-d UCS of the sample for each combination was measured. Since Factor B had two levels, one level of Factor B should be repeated as the third level for Factor B according to the principle of quasi-level orthogonal analysis. In this study, the level of HEC was repeated once. The quasi-level orthogonal gauge head is listed in Table 3.

The test procedures followed the Chinese Standard "Test Methods for Materials Stabilized with Inorganic Binders for Highway Engineering (JTJ 057-94)". The optimum water content and the maximum dry density of each group were determined using the modified compaction test. Each specimen was prepared by compacting the material in the mold at the optimum water content, was cured under standard conditions for 6 days and was saturated on the seventh day. The UCS and resilience modulus then were determined and listed in Table 4.

Table 2. Factor A (Gradation) and Its Levels

No. of gradation	Coarse brick	Middle brick	Fine brick	Soil	Weight ratio of brick: soil
1	10%	10%	20%	60%	4：6
2	12.50%	12.50%	25%	50%	5：5
3	15%	15%	30%	40%	6：4

Table 3. Quasi-level Orthogonal Gauge Head

Factors / Levels	A (No. of gradation)	B (stabilizer)	C (stabilizer content)
1	1	HEC	4%
2	2	Cement	5%
3	3	HEC	6%

Table 4. Results of Orthogonal Tests

Factors / No. of mix	Gradation	Stabilizer	Stabilizer content (%)	Maximum dry density (g/cm3)	Optimum water content (%)	7d UCS (MPa)	7d resilient modulus (MPa)
1	1	HEC	4	1.614	10.50	1.2	274
2	1	Cement	5	1.638	10	1.5	244
3	1	HEC	6	1.642	10.50	2.6	405
4	2	HEC	5	1.618	13	2.9	435
5	2	Cement	6	1.603	10	2.7	409
6	2	HEC	4	1.614	11	1.9	331
7	3	HEC	6	1.626	12	3.2	289
8	3	Cement	4	1.591	15	2.3	299
9	3	HEC	5	1.604	14	2.7	304

According to the results of the orthogonal analysis, the order of the importance of the influence factors on the UCSs is the content of the stabilizer, the gradation of the recycled material, and the type of the stabilizer. The UCS of the sample reached its peak value when it was stabilized at 6% HEC using Gradation 3. Considering the economy of the mix, the specimens were prepared at 4% HEC using Gradation 3. These specimens had the maximum dry density, the optimum water content, the 7 d UCS, and the resilient modulus of 1.603g/cm^3, 13.5%, 1.86MPa, and 251MPa, respectively. The required 7d UCS of the subgrade was at least 1 MPa. Therefore, this mix fulfilled the requirement.

The required 7d UCS for a cushion was at least 1.8 MPa. All the mixes except Mixes 1 and 2 met this requirement as cushion materials. The UCS of Mix 7 was 3.2MPa, which met the requirement for bases of highways with Level 2 or below based on *the Chinese Technical Specifications for Construction of Highway Bases* (JTJ034—2000). This mix could also be used for highway bases subjected to less than 12 million ESALs. It is obvious that this mix can be used for subbase.

Table 4 shows that the mixes with Gradation 3 performed better than those with Gradations 1 and 2. All the mixes with the 7d UCSs higher than 2 MPa can fulfill the requirement for highway subbase.

Moisture Susceptibility

Since the waste brick had relatively low strength and density, and high water absorption ratio, its potential for moisture susceptibility should be checked. The test procedures for evaluating its moisture susceptibility are as follows. Two groups of specimens were prepared. The first group of specimens were cured under a standard condition for 7 days and then saturated for 7 days or one month. The second group of specimens were cured under a standard condition during all the curing period and saturated for the last day. The UCSs of these specimens under the above conditions are provided in Table 5.

Table 5. Test Results of Moisture Susceptibility

Time (d)	UCS after saturation, qu_a, (MPa)	UCS before saturation, qu_b, (MPa)	qu_a / qu_b (%)	Strength loss (%)
7	2.4	2.8	85.7	14.3
30	2.6	3.0	86.7	13.3

Table 5 shows that the strength loss after a 7 to 30 day-curing period ranged from 13.3 to 14.3%, which is acceptable for highway subbase. The experimental study also showed that the HEC–stabilized material had less moisture susceptibility.

RECYCLED CONCRETE

Recycled concrete was divided into four grades: coarse recycled aggregate (the size from 19 to 37.5mm), medium recycled aggregate (from 4.75 to 19.0mm), fine recycled aggregate (from 2.36mm to 4.75mm), and soil (less than 2.36mm). The proportion of each grade and the gradation ranges are provided in Table 6. The test results showed that, at 6.5% of No. 32.5 cement, the maximum dry density, the optimum water content, the 7d UCS, and the 7d resilient modulus of the mix were 1.914 g/cm^3, 13%, 4.2MPa, and 650MPa, respectively. This mix met the strength requirement for highway bases.

Table 6. Proportion and Gradation Range of Recycled Concrete (%)

Sieve Size (mm)	37.5	31.5	26.5	19	9.5	4.75	2.36	0.6	0.075
Upper limit	100	100	100	89	67	49	35	22	7
Lower limit	100	100	90	72	47	29	17	8	0
Composite gradation	100	96.2	93.5	85.3	68.4	43.5	32.2	19.1	3.5
Proportion	35.0				35.0		15.0	15.0	

COMBINATION OF RECYCLED BRICK AND CONCRETE

Since building wastes are often the mixture of waste brick and waste concrete, it is hard to separate them. Therefore, recycled materials were prepared by mixing of waste brick and waste concrete in proportions. The proportion of recycled materials used in this study was: coarse brick: medium brick: fine brick: medium recycled concrete: soil of brick = 1:1:2.5:1.5:4. The test results showed that at 6.0% No. 32.5 cement, the maximum dry density, the optimum water content, the 7d UCS, and the 7d resilient modulus of the mix were 1.686 g/cm^3, 14%, 3.1MPa, and 315MPa, respectively. This mix met the strength and stiffness requirements for highway subbase.

CONCLUSIONS

Construction and demolition wastes, which include waste brick and concrete, can be recycled for pavement applications by adding stabilizing/recycling agents. If the recycled aggregates are well graded, the recycled material could meet the strength requirements for bases and/or subbases. The following conclusions can be made from this study:

(1) The importance of influence factors on the UCS of the recycled material was identified on the order of the stabilizer content, the gradation, and the type of stabilizer.

(2) The UCS of the recycled material with the weight ratio of brick to soil of 6:4 at 4% HEC was 1.86MPa, which met the requirement for subgrade. The UCS of the specimen at 6% HEC was 3.2MPa, which met the requirement for the bases of highways of Level 2 or less and could withhold less than 12 million ESALs.

(3) When the gradation of the recycled concrete met the requirement for bases, the 7-d UCS of the recycled concrete at 6.5% cement was 4.2MPa, which met the strength requirement of highway bases.

(4) When the percent of the waste concrete over the total weight (brick plus concrete wastes) was 15%, the UCS of the recycled material at 6% cement met the strength requirement for highway subbases and bases of Level 2 or below.

REFERENCES

Gottfredsen, F.R. and Thogersen, F. (1993). "Recycling of concrete in aggressive environment." *Proceedings of the Third International RILEM Symposium on Demolition and Reuse of Concrete and Masonry, RILEM BULLETIN*, Odense, Denmark, 24-27 October, 309-317.

Kou, S.-C. (2006). Reusing Recycled Aggregates in Structural Concrete, Ph.D dissertation, Hong Kong Polytechnic University, Hong Kong, P.R. China.

Kotrayothar, D., Tam, V.W.Y., and Y.C., Loo (2007). "On the prevailing waste recycling methods: a Southeast Queensland study", *Fourth International Conference on Construction in the 21st Century (CITC-IV): Accelerating Innovation in Engineering, Management and Technology,* Gold Coast, Australia: 1-8.

Movassaghi, R. (2006). *Durability of Reinforced Concrete Incorporating Recycled Concrete as Aggregate (RCA)*, Ph.D dissertation, the University of Waterloo, Waterloo, Ontario, Canada.

Ordonez, C.A. (2006). *Characterization of Cemented and Fiber-reinforced RAP Aggregate Materials for Base/Subbase Applications*, MS thesis, the University of Texas, Arlington, Texas, USA.

Wainwright, P.J., Trevorrow, A., Yu, Y., and Wang, Y. (1993). "Modifying the performance of concrete made with coarse and fine recycled aggregates." *Proc. of the Third International RILEM Symposium on Demolition and Reuse of Concrete and Masonry, RILEM BULLETIN*, Odense, Denmark, 24-27 October, 319-330.

Influence of Support Conditions on Roller-Integrated Machine Drive Power Measurements for Granular Base

Pavana K.R. Vennapusa, S.M.ASCE, David J. White, A.M.ASCE, Heath Gieselman

Ph.D. Candidate, Wegner Professor of Civil Engineering and Director, Assistant Scientist, respectively, Earthworks Engineering Research Center (EERC), Department of Civil, Construction, and Environmental Engineering, Iowa State University of Science and Technology, 2711 South Loop Drive, Suite 4700, Ames, IA 50010, pavanv@iastate.edu, djwhite@iastate.edu, giese@iastate.edu

ABSTRACT

A controlled field study was conducted to evaluate the influence of underlying layer support conditions on roller-integrated machine drive power (MDP) measurements. Multiple layers of granular base material were placed over a section of wet/soft subgrade and a section of concrete base for comparison. The material was compacted in layers using a 12-ton vibratory padfoot roller. MDP and in-situ spot test measurements were obtained on each layer and the results were statistically analyzed. Results indicate that compaction layer MDP measurements are strongly dependent on stiffness and heterogeneity of the supporting layer. If relatively stable and homogenous support conditions exist, the effect is not statistically detectable. Although the compaction layer properties (as measured by in-situ test measurements) are relatively uniform, the MDP measurements tend to capture the variability of the underlying layers. Post-construction tests by carefully excavating the compaction layers showed significant improvement in stiffness of the granular base layers compared to the initial compaction layer measurements. The reason is attributed to possible densification of underlying layers during compaction of the layers above and an increase in lateral stresses during compaction.

1. INTRODUCTION

Efforts to implement roller-integrated compaction monitoring technologies (also referred to as intelligent compaction (IC) and continuous compaction control (CCC)) in earthwork quality control/quality acceptance (QC/QA) are increasing in the United States. Recent field studies (see White et al. 2007, Thompson and White 2008) demonstrated the use of machine drive power (MDP) as a reliable indicator of soil compaction for granular and cohesive soils. MDP is a power-based technology that monitors the mechanical performance of the roller during compaction which can be empirically related to soil strength and stiffness properties.

MDP measurements are generally assumed to be a response of soil conditions greater than the compaction layer, but conclusive data to support this assumption did not exist prior to this study. Insights on the influence of supporting layer properties on MDP compaction layer measurements is important as it affects the correlations developed to in-situ test measurements within the compaction layer. A test embankment was constructed using seven layers of granular base material placed over a section of wet/soft subgrade and a section of concrete base for comparison (Fig. 1). A 12-ton vibratory padfoot roller (Fig. 1) equipped with GPS to spatially reference the MDP measurements was used for compaction. Post-construction change in stiffness of the compacted granular layers was investigated through in situ testing.

FIG 1. (Left) CP563 padfoot vibratory roller, (right) wet/soft subgrade and concrete base at the bottom of the excavation

2. BACKGROUND

The use of MDP as a measure of soil compaction is a concept originated from study of vehicle-terrain interaction (Bekker 1969). The basic premise of determining soil compaction from changes in equipment response is that the efficiency of mechanical motion pertains not only to the mechanical system but also to the physical properties of the material being compacted. MDP is calculated as:

$$MDP = P_g - WV\left(\sin\theta + \frac{a}{g}\right) - (mV + b) \tag{1}$$

where P_g = gross power needed to move the machine (kJ/s), W = roller weight (kN), a = machine acceleration (m/s^2), g = acceleration of gravity (m/s^2), α = slope angle (roller pitch from a sensor), V = roller velocity (m/s), and m (kJ/m) and b (kJ/s) = machine internal loss coefficients specific to a particular machine (White et al. 2005). The second and third terms of Eq. (1) account for the machine power associated with sloping grade and internal machine loss, respectively. MDP is a relative value referencing the material properties of the calibration surface, which is generally a hard compacted surface (MDP = 0 kJ/s). Positive MDP values therefore indicate material that is less compact than the calibration surface, while negative MDP values would indicate material that is more compact than the calibration surface. The reference calibration surface for the machine used in this study was stiff compacted CL soil with average CBR = 25 and average E_{LWD-Z3} = 140 MPa.

3. TEST EMBANKMENT PREPARATION AND TESTING

A test embankment with plan dimensions of approximately 2.4 m x 24 m was prepared by excavating the existing glacial till subgrade material (classified as CL according to the USCS classification) to a depth of about 1.2 m below grade (see Fig. 2). Following excavation, about half of the excavation was scarified and moisture conditioned to a depth of about 0.3 m below the excavation to create a wet/soft subgrade layer. A 0.3 m thick concrete layer was continued in the remaining half of the excavation to create a stiff layer. Seven layers of granular base material (classified as SC-SM according to the USCS classification) with nominal loose lift thickness of about 0.30 to 0.35-m were then placed in the embankment. Compaction of each layer was performed with nine roller passes in vibratory mode with nominal frequency f = 30 Hz, amplitude a = 0.3 to 1.9 mm, and speed v = 3.2 km/h. The final compaction pass on each layer was performed at a = 0.3 mm.

Zorn 300-mm plate light weight deflectometer (LWD) and 1-m deep dynamic cone penetrometer (DCP) tests were performed after final pass on each layer, at six locations along the embankment. LWD tests were performed by excavating down to the bottom of the padfoot penetration, and following manufacturer recommendations to determine elastic modulus, E_{LWD-Z3} (Zorn 2003) (assuming Poisson's ratio ν = 0.4, shape factor f = 8/3). DCP tests were performed in accordance with ASTM D6951. California Bearing Ratio (CBR) was calculated from DCP index values from the upper 300-mm penetration. After completing tests on the final layer, the fill material was carefully excavated down to the surface of each underlying layer at two locations along the embankment (one location above the concrete base and the other location above the wet/soft subgrade). LWD tests were performed in the excavation on top of each layer to investigate post-construction change in stiffness of the granular layers.

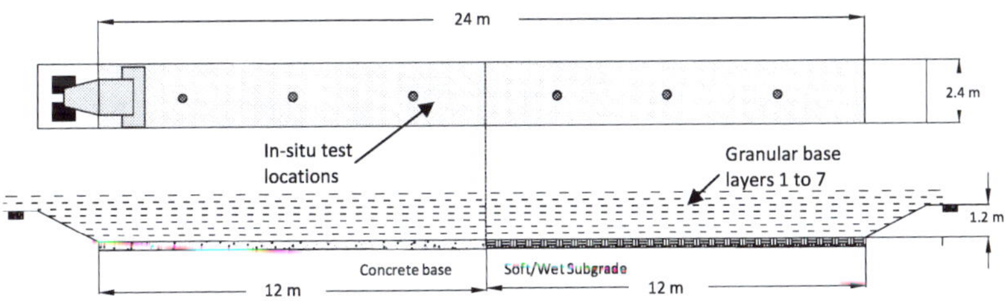

FIG 2. Plan and profile view of the test embankment

4. TEST RESULTS AND DISCUSSION

MDP values on layers 1 to 7 in comparison with E_{LWD-Z3} and CBR values are presented in Fig. 3. MDP values are shown as solid lines. E_{LWD-Z3} and CBR data are shown as discrete points. Both CBR and E_{LWD-Z3} measurement values tracked well with MDP measurement values on all layers. MDP data for layer 1 shows a clear distinction between the underlying support conditions with 2 to 10 kJ/s in the area underlain by concrete base and 20 to 38 kJ/s in the area underlain by wet/soft subgrade. MDP measurements on layers 2 appear to "bridge" the section with underlying wet/soft subgrade.

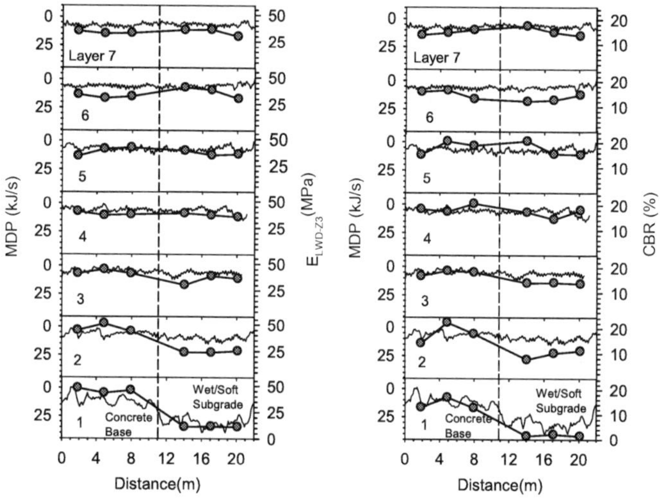

FIG 3. Comparison of MDP to E_{LWD-Z3} and CBR spot test measurements

Fig. 4 presents average MDP, E_{LWD-Z3}, and CBR values on each layer for the wet/soft subgrade and concrete base areas. The MDP plot shows changes from pass 1 to 9 separately for portions of the embankment underlain by concrete base and wet/soft subgrade. It is clear that for layers 1 and 2, stiff underlying support with concrete base helps facilitate compaction compared to soft subgrade support. The average MDP measurements over both concrete base and wet/soft subgrade converge and plateau at approximately 0.9 m to 1.0 (layer 4) above the bottom of the excavation. The ratio of average MDP in the area underlain by wet/soft subgrade to average MDP in the area underlain by concrete base at 0.3-m elevation (layer 1) was about 2.3. The ratio at elevations 0.6- (layer 2), 0.8- (layer 3), and 1.1-m (layer 4) decreased to approximately 1.6, 1.5, and 1.1, respectively.

Average E_{LWD-Z3} and CBR measurements taken on each layer over wet/soft subgrade and concrete base areas are also shown in Fig. 4. Similar to MDP, E_{LWD-Z3}, and CBR measurements over concrete base and wet/soft subgrade areas converge and plateau at approximately 0.9 and 1.0 m (layer 4) above the bottom of the excavation. The ratio of average E_{LWD-Z3} in the area underlain by concrete base to the average E_{LWD-Z3} in the area underlain by the wet/soft subgrade at about 0.3-m elevation (layer 1) was about 3.7. The ratio at elevations 0.6-, 0.8-, and 1.1-m decreased to approximately1.9, 1.2, and 1.0, respectively. Similarly, the ratio of average CBR in the area underlain by concrete base to the area underlain by wet/soft subgrade at 0.3-m elevation was about 8.7, and measurements at elevations 0.6-, 0.8-, and 1.2-m decreased the ratio to about 1.8, 1.4, and 1.1, respectively.

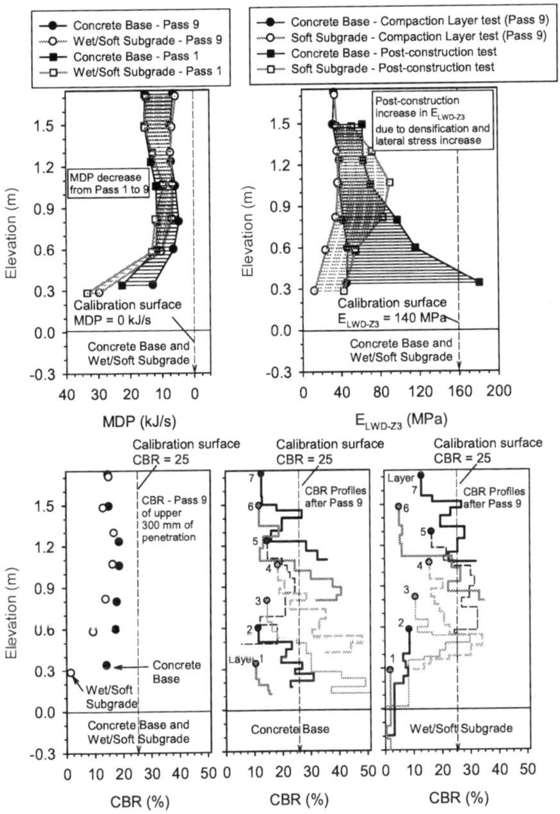

FIG 4. Average MDP, $E_{LWD\text{-}Z3}$, and CBR measurement values on all compaction layers

Statistical Analysis

The test results were analyzed statistically to characterize the influence of underlying support conditions on MDP, CBR and $E_{LWD\text{-}Z3}$ measurements. The analysis is performed by incorporating measurement values of the underlying layers into a linear regression model to predict measurement values at the surface. The significance of the underlying layer properties are selected based on statistical *p*- and *t*- values. The criteria for identifying the significance of a parameter included: p-value < 0.05 = significant, < 0.10 = possibly significant, > 0.10 = not significant, and t-value < -2 or > +2 = significant. The *p*-value indicates the significance of a parameter and the *t*-ratio value indicates the relative importance (i.e., higher the absolute value greater the significance).

Analysis of MDP measurements indicates that for layers 1, 2, 3, 4, and 5, MDP measurements on the immediate underlying layer are statistically significant. For layers 6 and 7, the effect of underlying layers is not statistically significant. Further, the statistical significance of the underlying layers seems to decrease as additional layers are placed in the embankment, which is evidenced by the decreasing *t*-value (Table 1). For example, the influence of layer 1 on layer 2 MDP is greater compared to the effect of layer 2 on layer 3 MDP, and so on. Interestingly, as the *t*-values decrease, the variability (COV) of MDP measurements along the embankment also decreases (Table 1). MDP values show high variability for layers 1 to 4 (COV = 38 to 51%), while they were less variable for layers 5 to 7 (COV = 7 to 30%).

Similar analysis for E_{LWD-Z3} and CBR measurements indicate that for layers 2 and 3, measurements on the immediate underlying layer are statistically significant. For layers 4, 5, 6, and 7, the effect of E_{LWD-Z3} and CBR of the underlying layers is not statistically significant. Similar to MDP measurements, as the *t*-value decrease, the variability (COV) of E_{LWD-Z3} measurements along the embankment also decrease (Table 13). E_{LWD-Z3} measurements on layers 1 and 2 showed relatively high variability (COV = 35 to 63%) along the embankment compared to layers 3 and above (COV = 7 to 13%). Similarly, CBR measurements on layers 1 and 2 showed relatively high variability (COV = 39 to 89%) along the embankment compared to layers 3 and above (COV = 10 to 15%).

TABLE 1. Statistical evaluation on the effect of underlying layer properties on compaction layer measurements

Surface layer (depth, m)	Mean (COV)			Effect of layer(s)	MDP		E_{LWD-Z3}		CBR	
	MDP	E_{LWD}	CBR		*t* value	*p* value	*t* value	*p* value	*t* value	*p* value
Layer 1 (0.29)	20.6 (51)	29.0 (63)	7.9 (89)	Concrete base/soft subgrade	—*	—*	—*	—*	—*	—*
Layer 2 (0.59)	8.1 (37)	35.4 (35)	13.9 (39)	Layer 1	11.1	<0.001	3.00	0.004	5.64	0.005
Layer 3 (0.80)	5.5 (38)	38.6 (13)	15.9 (15)	Layer 2	4.4	<0.001	6.61	0.003	4.72	0.009
				Layer 1	NS§		NS§	3.00	0.004	5.64
Layer 4 (1.06)	6.0 (43)	37.7 (5)	17.8 (11)	Layer 3	2.8	0.007	NS§		NS§	
				Layers 2, 1	NS§		NS§		NS§	
Layer 5 (1.26)	7.0 (30)	37.7 (9)	17.8 (15)	Layer 4	2.6	0.012	NS§		NS§	
				Layers 3, 2, 1	NS§		NS§		NS§	
Layer 6 (1.49)	7.0 (22)	34.1 (13)	14.2 (13)	Layers 5, 4, 3, 2,1	NS§		NS§		NS§	
Layer 7 (1.72)	5.9 (7)	33.8 (7)	15.1 (10)	Layers 6, 5, 4, 3, 2, 1	NS§		NS§		NS§	

*No measurements on the concrete base/soft subgrade, §NS – not statistically significant

The analysis results indicate that both roller and in-situ compaction measurements are influenced by unstable and heterogeneous supporting layer properties. If the underlying layer is relatively stable and homogenous (i.e., COV < 30% for MDP, and COV < 15% for E_{LWD-Z3} and CBR), the effect of underlying layer on the compaction layer measurements is not statistically detectable. MDP measurements showed the effect of underlying layer properties up to layer 5 (elevation 1.26-m), while E_{LWD-Z3} and CBR measurements show the effect only up to layer 3. This means that as although the compaction layer properties become relatively uniform, the MDP measurements tend to capture the variability of the underlying layers. Insights on the differences between MDP and in-situ test measurement influence is important for interpretation of relationships during field calibration.

Post-construction change in stiffness of granular base layers

Fig. 4 shows post-construction E_{LWD-Z3} measurements performed in the excavation. The results show the E_{LWD-Z3} values in the concrete base area increased by about 2 times for layer 6 to about 4 times for layer 1, while the values in the wet/soft subgrade area increased by about 1.5 times for layer 5 to 13 times for layer 1. This increase in E_{LWD-Z3} values is attributed to possible densification of underlying layers during compaction of the layers above, and also to the effect of increasing lateral stresses, which is locked in from compaction of overlying layers. The latter effect is well discussed in the literature (e.g. D'Appolonia et al. 1969). The CBR profiles presented in Fig. 4 show a significant increase in CBR as the layer above is placed and compacted (for e.g. see increase in layer 1 CBR from layer 2 CBR profile). This is an important aspect of compaction of granular materials that is not well documented in the literature.

5. SUMMARY AND CONCLUSIONS

Results from a controlled field study conducted to evaluate the influence of support conditions on MDP measurements on granular base material are presented in this paper. Some key findings and conclusions are as follows:

- Roller and in-situ compaction measurements are influenced by the stiffness and heterogeneity of the supporting layer conditions. If the underlying layer is relatively stable and homogenous (i.e., COV < 30% for MDP, and COV < 15% for E_{LWD-Z3} and CBR), the effect of underlying layer on the compaction layer measurements is not statistically detectable.

- Although the compaction layer properties (as measured by in-situ test measurements) are relatively uniform, the MDP measurements tend to capture the variability of the underlying layers. Differences between MDP and in-situ test measurement influence are important for interpretation of results during field calibration.

- Post-construction tests performed after careful excavation of compaction layers indicate significant (1.5 to 13 times) increases in stiffness of the granular layers. The reason is attributed to possible densification of underlying layers during compaction of the layers above and to the effect of increased lateral stresses due to compaction.

6. ACKNOWLEDGEMENTS

The authors would like to thank Caterpillar Inc. (CAT) for sponsoring this study. The findings, opinions, recommendations and conclusions expressed in this paper are those of the authors and do not necessarily reflect the views of the sponsor and administrators. The authors further acknowledge Tom Congdon, Allen Declek, Donald Hutchen, and Glen Feather at CAT, and Alex Buchanan, Andy Weber and Kevin McLaughlin at Iowa State University for their help with field and laboratory testing.

7. REFERENCES

ASTM D6951. (2003). "Standard test method for use of dynamic cone penetrometer in shallow pavement application", ASTM, West Conshohocken, PA.

Bekker, M. G. *Introduction to terrain-vehicle systems*. (1969). The University of Michigan Press: Ann Arbor, MI.

D'Appolonia, D.J., Whitman, R.V., and D'Appolonia, E. (1969). "Sand compaction with vibratory rollers," *J. Soil Mech. Found. Div*, ASCE, 95, No. SM1, 263–284.

Thompson, M., and White, D.J. (2008). "Estimating compaction of cohesive soils from machine drive power." *J. Geotechnical and Geoenvironmental Engrg.*, ASCE (in press).

White, D.J., Jaselskis, E., Schaefer, V., Cackler, T. (2005). "Real-time compaction monitoring in cohesive soils from machine response." *Transp. Res. Rec.*, 1936, Transportation Research Board, Washington D.C., 173–180.

White, D.J., Thompson, M., Vennapusa, P. (2007). *Field study of compaction monitoring systems: Self-propelled non-vibratory 825G and vibratory smooth drum CS-533E rollers*. Center for Transportation Research and Education, Iowa State University, Ames, IA.

Zorn, G. (2003). *Operating manual: ZFG2000*, Zorn Stendal, Germany.

Resilient Behavior of Cement-Fiber Treated Reclaimed Asphalt Pavement Aggregates

[1]Anand J. Puppala, [2]Sireesh Saride, [3]Ajay Potturi, [4]Laureano R. Hoyos

[1]Professor, Department of Civil Engineering, The University of Texas at Arlington, Texas 76019, anand@uta.edu
[2]Post Doctoral Fellow, Department of Civil Engineering, The University of Texas at Arlington, Texas 76019, sireesh@uta.edu
[3]Former Masters Student, Department of Civil Engineering, The University of Texas at Arlington, Texas 76019, pvakraju@gmail.com
[4]Associate Professor, Department of Civil Engineering, The University of Texas at Arlington, Texas 76019, lhoyos@uta.edu

ABSTRACT: The use of Reclaimed Asphalt Pavement (RAP) aggregates in road construction reduces both the amount of construction debris disposed of in landfills and the rate of natural resource depletion. This paper presents the results obtained from a series of repeated load triaxial tests conducted on both untreated and treated RAP aggregates. Test specimens were compacted at density corresponding to optimum moisture content with different dosages of ordinary Portland cement (Type I/II) with fibrillated polypropylene fibers. The resilient moduli of untreated aggregates varied between 180 and 340 MPa and that of cement-fiber treated aggregates varied between 200 and 580 MPa. Effects of cement and fiber dosages on the moduli enhancements of RAP aggregates are explained. Structural coefficients of these materials are also presented.

INTRODUCTION

A flexible pavement system consists of an asphalt surface layer, a base course, a subbase and the subgrade. The subbase layer is optionally employed. Whereas, the base course which is in between the surface layer and the subgrade plays a very prominent role in transferring the loads coming onto the surface layer to the natural soil subgrade. This ability to distribute load is primarily a function of thickness of base course, quality and the strength of the base course material.

Traditionally natural aggregates derived from a variety of source rocks have been used as a road base material. But the extraction of these natural aggregates resources is increasingly being constrained by urbanization, increased costs and environmental concerns. Thus, increased amounts of reclaimed materials are being used to supplement natural aggregates in road construction. The 1993 Federal Highway Agency (FHWA) and Environmental Protection Agency (EPA) report mentioned that approximately 73 million tons of asphalt pavement material was recycled each year, which amounts to about 80% of the asphalt removed from pavements each year. The use of reclaimed pavement materials in road construction could serve the purpose of reducing the amount of construction debris disposed of in landfills, reducing environmental disturbance and the rate of natural resource depletion.

Typically the asphalt pavement is either removed by milling of upper surfaces or full depth removal of the entire pavement section itself. Another form of RAP material can be produced by in situ recycling of the old pavements by pulverizing them first and then incorporating them into base courses with or without additives. These processes are referred as cold in-place and hot in-place recycling methods, with hot process requires the heating of the upper asphalt surface layers of typically 2 in. using a hot recycling machine. Cold in-place mixing can be carried out to different depths. The RAP materials produced from both the methods can be used in several civil engineering applications such as asphalt concrete aggregate, asphalt cement binder, granular base aggregate, stabilized base aggregate and embankment or fill material. The base material used in this research was a bonded base and subbase material produced by the process of blending crushed recycled construction/demolition waste/debris with cement additive.

Although recycled aggregates could be used as a base material in pavement construction applications, product variability (source dependent) and strength characteristics usually limit their application in road bases (Wilburn and Goonam, 1998). The high volume usage of RAP in base applications is, therefore, better ensured by meeting the minimum standards set by AASHTO for material performance. Most reclaimed asphalt pavement materials do not often meet the minimum requirements set by AASHTO. In such cases, stabilization with lime or cement allows the use of these low quality reclaimed asphalt pavement materials with the minimum required strength characteristics.

In this research, the main focus is on the aggregate application of RAP materials to support asphalt pavements (Flexible pavements). The main variable in the design of a flexible pavement structure is its thickness which is based on its structural support and the predicted level of traffic. The main criteria involved in the design of the pavement thickness are the magnitude of the imposed loads (Equivalent single axle load or ESAL is used to approximate the actual truck load with an 18 kip load using a fourth power formula) and the strength and resilient modulus of the subgrade material. Design also demands the structural number of the pavement which in turn calculated from the structural numbers of each sub base layers. To obtain the structural numbers the resilient properties of the respective layers are necessary.

In this study, the material used is a fiber-reinforced reclaimed pavement material tested at different dosage levels of Portland cement (Type I/II). The inclusion of fibers was to serve the purpose of enhancing the energy absorption capacity or toughness of the material and to retard the crack propagation process. Specimens with no fibers (control specimens) were tested at 0, 2, and 4% dosage levels of Portland cement and those with fibers (fiber-reinforced) were tested at 2 and 4% dosage levels of Portland cement. A comprehensive series of resilient modulus tests were conducted on the compacted specimens to evaluate the resilient moduli properties of cement-fiber treated reclaimed asphalt pavement materials using a repeated load triaxial test.

BACKGROUND

One of the earliest studies in the dynamic characterization of base and subbase materials was undertaken by Lotfi and Witczak (1985). Resilient moduli of five cement-treated base materials used were determined and evaluated. Specific values of layer coefficients based on the material moduli were evaluated for use in the design of flexible pavements. The M_R testing was performed at a range of stress levels varying from 40 to 320 psi. It should be noted, however, that the material used in this study was natural aggregate and not a recycled/reclaimed aggregate.

Janoo (1994) experimented with the use of reclaimed asphalt concrete as a base material. Test sections of different RAP materials were built by NHDOT near Concord off Interstate 89 for this study. Falling Weight Deflectometer (FWD) tests were conducted on the test sections and the deflections obtained were then used to back-calculate the layer modulus. The layer moduli were then used with the established relationships for determining the layer coefficients in the 1993 AASHTO guide for design of pavement structures. In another study conducted by Taha et al. (2002), RAP/Virgin aggregate mixtures were tested with different cement dosages. The modulus values were arrived at from the Unconfined Compression Strength (UCS) results using correlations between M_R and UCS. No resilient modulus tests were conducted in the laboratory environment. The layer coefficient charts included in the AASHTO guide for design of flexible pavements were used for obtaining the values.

Gnanendran and Woodburn (2003) conducted series of resilient modulus, CBR and UCS tests on cement, lime and fly ash stabilized reclaimed asphalt aggregates. These tests provided typical resilient moduli values of these aggregates, and enhancements in moduli values with respect to each of the chemical treatments.

Although reclaimed aggregate materials are widely used in various pavement applications, neither sufficient numbers of resilient modulus laboratory tests were conducted nor field FWD studies were attempted in determining the resilient modulus values. Also, it should be noted that each kind of the recycled material is distinct with respect to the others (basing on its source and other constituents involved) and hence comprehensive experimental characterizations using resilient modulus tests are analyzed for its qualification in pavement applications.

EXPERIMENTAL PROGRAM

The base material used in this research is a bonded base and subbase material produced by the process of blending crushed recycled construction waste. A series of basic and engineering tests have been conducted on this RAP base material. The basic tests included grain size distribution tests, specific gravity, and proctor compaction tests. Resilient Modulus tests using the cyclic triaxial setup were also conducted on the compacted RAP base specimens. All of the basic testing was done in accordance with the current TxDOT and ASTM standard testing procedures.

The specific gravity of the material was found to be 2.43. The sieve analysis indicates that about 99% of the material was retained on No. 200 sieve. Figure 1 presents the grain size distribution curve of the RAP base material. Compaction tests were conducted on the RAP base material to establish the optimum moisture content and dry unit weight relationships. TxDOT (Tex-113-E) procedure requires a compactive effort of 13.26 ft-lb/in^3. Based on this requirement, for a 4.54 kg (10 lb) weight of hammer and a drop height of 0.46 m (1.5 ft), it was determined to compact the specimen in three layers with 17 blows per layer for a specimen size of 10 cm in diameter and 11.6 cm in height. Figure 2 presents the dry unit weight and moisture content relationships of the RAP base material at 0, 2, and 4% cement treatments. The results of this compaction test were adopted in preparing the specimens for resilient modulus tests at the optimum moisture content and maximum dry unit weight.

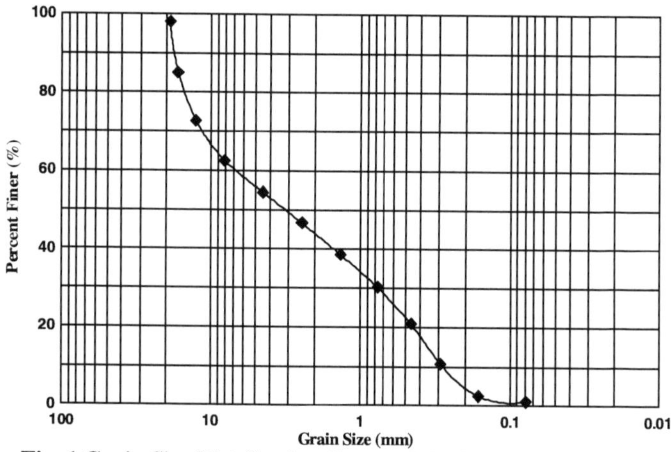

Fig. 1 Grain Size Distribution Curve of the RAP Base Material

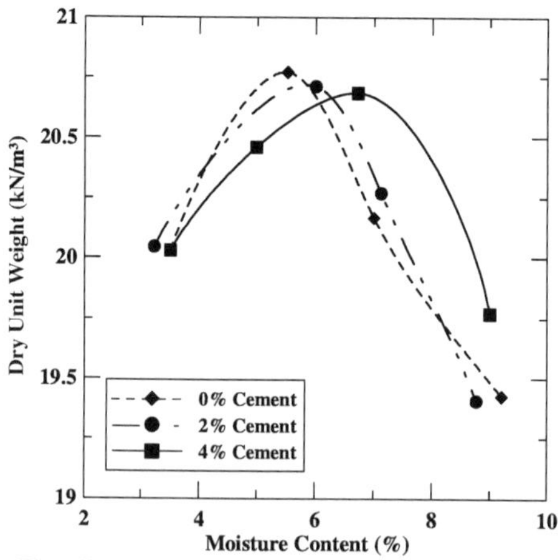

Fig. 2 Compaction Test Results for the Treated RAP Aggregate Material

Resilient modulus tests were performed in accordance with AASHTO Designation T 307-99. The stress levels used for testing the specimens are based upon the location of the specimen within the pavement structure as standardized by AASHTO for Base/Subbase materials. These tests were conducted on the RAP base specimens (both with and without fibers) compacted at the optimum moisture content, and at different dosage levels of the Portland cement. RAP base specimens without fibers are tested at 0%, 2% and 4% of the cement dosage levels, whereas the specimens with fibers are tested at 2%, and 4% of the cement dosage. The percent of fiber content used in this study was not varied and was kept constant at 0.15%. All of the specimens were consistently tested at a 7-day curing period.

RESULTS AND DISCUSSION

Figures 3 and 4 present the typical M_R results for the cement treated specimens both without and with fiber-reinforcement respectively. It can be observed that the resilient modulus increases appreciably with increasing percent of cement at every confinement. In addition, both confining and deviatoric stresses have major influence on the resilient behavior of cement-fiber treated RAP aggregates.

An increase in the deviatoric stress resulted in an increase in the resilient modulus of the material, which is attributed to stress hardening of the aggregate specimen at high deviatoric stresses. This deviatoric stress effect on moduli is more pronounced at low confining pressures than at high confining pressures. This is because, at high confining pressures, the granular cemented specimen is strong and hence does not respond to increased axial deviatoric stresses as it does at low confining pressures.

The resilient modulus of the cement aggregate materials increased with an increase in the applied confining pressure. Also, this increase in the resilient modulus value with respect to confining pressure is not noted in cement treated aggregate specimens, particularly those treated with higher cement contents. This is because the cemented aggregate material is strong and stiff at high cement contents and as a result they behave similar to a weak concrete material. Such materials will be unaffected by the applied confining pressures.

It can also be noted from Figure 3 that for a maximum confining pressure of 137.9 kPa, the addition of 2% cement increased the M_R by 32% when compared to the same resilient modulus of untreated specimens. For the same confinement, the addition of 4% cement increased the M_R by about 50% when compared to the same of untreated specimen. The percent moduli increase with respect to cement treatment of the present materials can be regarded as moderate at best. The maximum value of the resilient modulus achieved for a specimen treated with 4% cement is about 505 MPa. Thus cement appears to be an effective stabilizer for stabilizing the reclaimed asphalt pavement materials in achieving high strength and hence high resilient characteristics.

Cement treated samples with fiber reinforcements also showed the same trends, as those without fiber-reinforcement. With the addition of 0.15% fiber content to the cement treated RAP specimen, the resilient behavior is further improved. This improvement is slightly higher for cement treated specimens. The enhancements are attributed to the tensile fiber elements, which increased overall strength of the specimen mixtures by inducing more interlocking between RAP aggregate specimens.

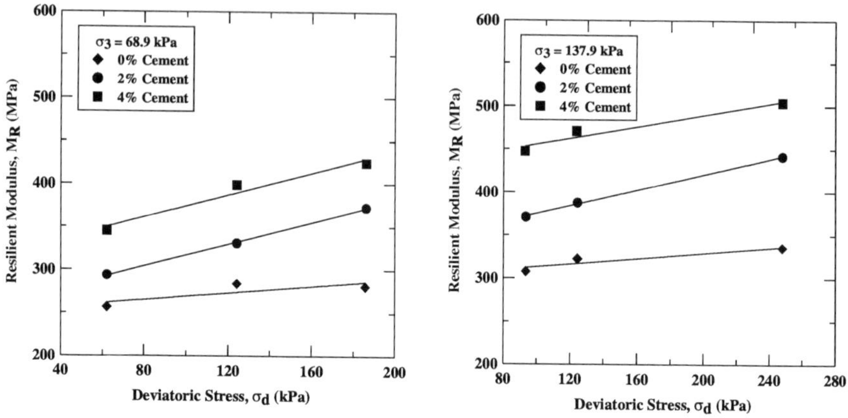

Fig. 3 Influence of Cement Treatment on Resilient Modulus Test Results

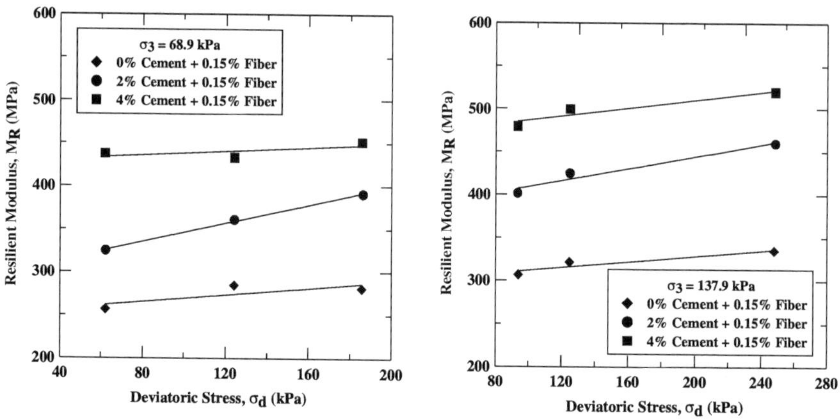

Fig. 4 Influence of Cement - Fiber (0.15%) Treatment on Resilient Modulus Test Results

Structural Coefficients

The resilient modulus tests showed that both confining and deviatoric stresses have shown a major influence on the resilient moduli values of the untreated, cement treated and cement-fiber treated reclaimed aggregate base materials. The combinations of deviatoric and confining stresses applied in the test sequence have been used to determine various average subgrade resilient moduli for different confining pressure conditions. If specific confining and deviatoric stresses of aggregate bases under pavement loading conditions are known, one can directly use the measured moduli from the same conditions. For the present research, average moduli, from the measured moduli at three different deviatoric pressures, is calculated for each confining pressure condition. The same step is repeated for each cement and cement–fiber dosage contents. In order to determine the structural coefficients, the following AASHTO (1993) equation 1 was used:

$$a_2 = 0.249 \times \log M_R - 0.977 \qquad (1)$$

Where a_2 is structural coefficient; M_R is the resilient modulus in psi. This equation was used along with the average resilient moduli to determine the a_2 values, and these results are summarized in Table 1.

The structural coefficient (a_2) values ranged from 0.17 to 0.24 and this range is in agreement with those materials reported by Janoo (1994) for New Hampshire DOT. As expected, the structural coefficient values increase with the confining pressure and they are higher for cement and cement-fiber treated aggregates. These values can directly be used in determining the base layer thickness in the design of flexible pavements. It should be noted that the structural coefficients determined by the above procedure still need to be verified with other methodologies including the field falling weight deflectometer measurements on both untreated and cement treated aggregate materials.

Table 1 Average Resilient Moduli and Structural Coefficients of Cement-Fiber Treated Aggregates

Notation	Confining Pressure (kPa)	Average Resilient Moduli (MPa)	Structural Coefficient, a_2
Untreated	20.7	199	0.13
	68.9	274	0.17
	134.7	321	0.19
2% Cement Treated	20.7	231	0.15
	68.9	332	0.19
	134.7	400	0.21
4% Cement Treated	20.7	247	0.16
	68.9	360	0.20
	134.7	430	0.22
2% Cement-Fiber Treated	20.7	280	0.17
	68.9	390	0.21
	134.7	475	0.23
4% Cement-Fiber Treated	20.7	315	0.18
	68.9	441	0.22
	134.7	500	0.24

SUMMARY AND CONCLUSION

This paper presents the results obtained from a series of repeated load triaxial tests conducted on both untreated and treated RAP aggregate specimens to evaluate its function as a pavement bases. A constant amount of fiber content (0.15%) was used to prepare the test specimens through out the testing program. Cement dosage was varied from 0 to 4% with 2% intervals.

The resilient modulus tests showed that both confining and deviatoric stresses have a certain influence on the resilient moduli values of the untreated, cement treated and cement-fiber treated reclaimed aggregate base materials. Cement –fiber treated samples showed the same trends. This enhancement is much higher for lower dosages of cement. This enhancement was attributed to the tensile fiber elements which increased overall strength of the specimen mixtures by inducing more interlocking between RAP aggregate specimens.

The structural coefficient, a_2, values of treated and untreated aggregates are calculated and these values ranged from 0.13 to 0.24. As expected, the structural coefficient values increase with an increase in the confining pressure and these values are higher for cement and cement – fiber treated aggregates. These values can directly be used in the design of flexible pavements.

ACKNOWLEDGMENTS

We acknowledge TXI, Inc for providing financial support to conduct this research. We also acknowledge Fred Brouillette, Dan Richwine and Cliff Ryan for their help with the research.

REFERENCES

American Association of State Highway Transportation Official. (1993). AASHTO Guide for Design of Pavement Structures, *American Association of State Highway and Transportation Officials*, Washington, D. C.

A Study of the Use of Recycled Paving Material-Report to Congress (1993), Federal Highway Agency and Environmental Protection Agency, FHWA-RD-93-147, EPA/600/R-93/095, p. 78.

Gnanendran, C. T. and Woodburn, L. J. (2003). Recycled Aggregate for Pavement Construction and the Influence of Stabilization. *Proceedings- Conference of the Australian Road Research Board*, v 21, 2003, p 1755-1768.

Janoo, V. C. (1994). Layer Coefficients for NHDOT Pavement Materials. Prepared for New Hampshire Department of Transportation and U.S. Department of Transportation, *Special Report 94-30*, September 1994.

Lotfi, H. and Witczak, M. W. (1985). Dynamic Characterization of Cement-Treated Base and Subbase Materials. *Transportation Research Record*, n 1031, p 41-48.

Taha, R., Al-Harthy, A., Al-Shamsi, K. and Al-Zubeidi, M. (2002). Cement Stabilization of Reclaimed Asphalt Pavement Aggregate for Road Bases and Subbases. *Journal of Materials in Civil Engineering*, v 14, n 3, May/June, 2002, pp 239-245.

Validations of Anisotropic Aggregate Base Behavior from Full-Scale Pavement Tests

Erol Tutumluer[1] M.ASCE and Jayhyun Kwon[2]

[1]Professor, Newmark Civil Engineering Laboratory, Department of Civil and Environmental Engineering, University of Illinois, 205 N. Mathews, Urbana, Illinois 61801; tutumlue@uiuc.edu
[2]Senior Pavement Engineer, Tensar International Corporation, 5883 Glenridge Dr. Suite 200, Atlanta, GA 30328; jayhyunkwon@tensarcorp.com

ABSTRACT: Low volume conventional pavement test sections with different base course thicknesses were recently constructed at the University of Illinois over a weak subgrade to evaluate pavement response and performance from accelerated loading. This paper presents the full-scale test study measured pavement responses used to validate the stress- and direction- (anisotropic) dependent unbound aggregate base behavior. Field validations involved comparing measured full-scale pavement test section responses with the predicted ones using both the isotropic and anisotropic aggregate base modulus characterizations. The nonlinear cross-anisotropic base characterization was needed to accurately predict several of the full-scale pavement section field responses and observations, such as the vertical deformation on top of subgrade, hot mix asphalt fatigue strains, and especially the large horizontal base course aggregate movements measured in the direction of traffic loading. The findings of this paper support the development of next generation of highway as well as airport pavement infrastructure by employing advanced technology and the detailed knowledge on the nonlinear, anisotropic aggregate base behavior gathered in the past 15 years of research activities.

INTRODUCTION

Low volume flexible pavements often built with thin hot mix asphalt (HMA) surfaces rely heavily on the superior performances of unbound aggregate base/subbase courses. Factors critical to the success of such pavement designs and construction alternatives include a willingness to invest in quality aggregates with superior properties along with paying significant attention to size and gradation. Significant advances made within the last 15 years in characterizing the actual stress- and direction- (anisotropic) dependent aggregate behavior under traffic loading have established a methodology to evaluate aggregate material property inputs in relation to the structural contributions of constructed granular layers affecting pavement performance, as summarized in a recent state-of-the-art review paper on anisotropic aggregate behavior by Tutumluer (2008). Accordingly, a proper consideration of both stress sensitivity and modulus anisotropy of unbound aggregate structural layers is essential for accurately predicting load distributions and their expected field performances. This would enable optimized use of available aggregate resources and bring convenience and confidence for building pavements with deep aggregate base/subbase courses in accordance with the successful practices of foreign countries in the area of long life flexible pavements built on aggregate structural layers and strong pavement foundations (Beatty et al., 2002).

From advanced triaxial tests conducted in the laboratory, unbound aggregate materials were clearly shown to be stiffer in the vertical compaction and repeated loading direction with the anisotropy of the resilient modulus (M_R) behavior adequately characterized from the developed stress-dependent modulus models (Tutumluer and Thompson, 1997; Tutumluer and Seyhan, 1999). Field validations of the anisotropic aggregate layer behavior have involved comparing measured full-scale pavement test section responses with the ones computed from Finite Element (FE) analyses of the pavement system using the anisotropic aggregate modulus characterizations (Tutumluer et al., 2003, Steven et al., 2007). Masad et al. (2006) compared AASHO road test pavement surface deflection measurements under wheel loads to FE predictions based on models that incorporated isotropic and anisotropic properties for the unbound base and subbase layers. The surface deflection predictions correlated best with the AASHO road test measurements when the horizontal moduli were about 30% of the vertical moduli in the unbound aggregate layers.

Low volume conventional flexible pavement test sections with different base course thicknesses were recently constructed at the University of Illinois over a weak subgrade having a California Bearing Ratio (CBR) of 4% and instrumented with sensors to evaluate pavement response and performance from accelerated loading. The objective of this paper is to compare the full-scale pavement study measured pavement responses with predicted ones for both the isotropic and anisotropic base course modeling scenarios in an effort to demonstrate the importance of considering properly stress- and direction- (anisotropic) dependent unbound aggregate base behavior.

FULL-SCALE PAVEMENT TESTS

An accelerated full-scale pavement test study was conducted at the University of Illinois Advanced Transportation Research and Engineering Laboratory (ATREL) from 2005 to 2007 (Al-Qadi et al., 2007). A total of nine instrumented full-scale flexible pavement test sections were designed and constructed to measure pavement responses, monitor pavement performances, and quantify the effectiveness of geogrid-reinforced flexible pavements. The sections consisted of reinforced as well as control sections. Laboratory and field tests were performed throughout the testing program for material characterization and quality control purposes. Test section variables examined in the full-scale testing study included two HMA (76 and 127 mm) and three granular base layer (203, 305 and 457 mm) design thicknesses and type and location of geogrid within the base course. The subgrade conditions were maintained similar for all pavement sections; the California Bearing Ratio (CBR) values were at less than 4%. No variations in the moisture content were measured throughout the response and trafficking tests.

The test sections were heavily instrumented with pressure cells, linear variable pressure transducers (LVDTs), and strain gauges to measure pavement response to vehicular loading. The instrumentation also included thermocouples and time domain reflectometry (TDR) probes to capture environmental changes and piezometers to measure pore-water pressure. For the response testing, a dual wheel load configuration with a tire pressure of 689 kPa applied individual wheel loads of 26.7, 35.6, 44.5, 53.4, and 62.3 kN on the pavement surface uni-directionally at 8 and 16 km/h using the mobile Accelerated Testing Loading ASsembly (ATLAS).

Because of proper drainage facilities constructed within the test sections, the subgrade moisture content was maintained almost constant at an average of 18.75% throughout testing. Recorded temperature profiles of pavement layers showed that the HMA surface layer was the most sensitive to temperature changes, as expected. To compare responses under same environmental conditions, all measured responses had to be adjusted to values that would result at a reference HMA temperature of 25°C adopted in this study.

The control sections had the deepest surface ruts formed from the beginning until the end of the traffic testing (Kwon et al., 2008). After both response and traffic tests were completed, pavement trenches were excavated for further examinations of the construction features and the test section performances. Rut profile data obtained from open trench measurements clearly indicated that all test sections had reached subgrade shear type failure. This failure was the most pronounced in the control sections and to a lesser degree in the geogrid reinforced sections.

MEASURED AND PREDICTED PAVEMENT RESPONSES

Response predictions made on the test sections designed with aggregate base thicknesses of 305 and 457 mm (unreinforced test sections B1 and D3, respectively, both having 76 mm thick HMA surfacing) are primarily used in this paper. Figure 1 shows the design thicknesses and the measured response types and locations of the B1 and D3 conventional flexible pavements. As-constructed layer thicknesses obtained from the post-forensic trench profiles had to be used in the nonlinear mechanistic model GT-PAVE FE analyses (Tutumluer, 1995) for predicting accurately the measured pavement responses. The as-constructed aggregate base thicknesses of the unreinforced control sections were generally thicker than those of the reinforced sections. This was partly due to the significant intermixing of the base aggregate and subgrade soil at the bottom of the base layer as clearly observed in B1 test section.

Another important post-trafficking/failure forensic investigation was conducted by pushing the Dynamic Cone Penetrometer (DCP) into the non-trafficked pavement foundation after coring the top HMA layer. The DCP test results indicated that the field stiffnesses of the control sections were not uniform throughout the depth because of the aggregate and subgrade fines intermixing at base-subgrade interface. Specifically, the top 229-mm subgrade was much weaker than the intended CBR 4%. Similar softer zones were also observed in the bottom 76-mm of the 381-mm thick as-constructed B1 base layer due to intermixing at the subgrade-base interface. Therefore, to accurately account for the variations in strength/stiffness throughout the layers in FE mechanistic response modeling, the B1 base and the subgrade layers of both B1 and D3 sections had to be divided into sublayers with the assignment of different model parameters and material property assignments based on the field observations.

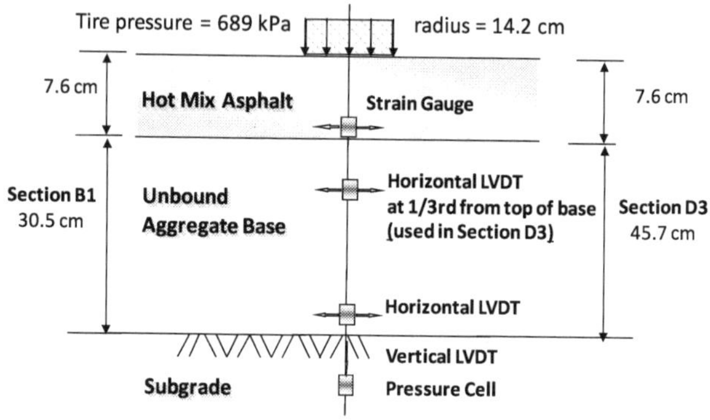

FIG. 1. Pavement test section geometries and measured response locations

Field validations involved comparing under the measured responses with the predicted ones using both the isotropic and anisotropic aggregate base modulus characterizations in the GT-PAVE FE analyses. An 800-element, 2533-node axisymmetric FE mesh was used to analyze the B1 and D3 conventional flexible pavement test sections as nonlinear elastic layered systems. The wheel load was approximated as a circular uniform pressure of 689 kPa with 142-mm radius of loading (see Figure 1). The nonlinear, stress dependent behavior of the silty subgrade and the unbound aggregate base moduli were properly considered while the HMA layer was modeled as linear elastic. Subgrade soil resilient moduli characterized by the bilinear model and the model parameters used in the bilinear model were assigned in accordance with a previous study by Thompson and Elliot (1985). The unbound aggregate layer was modeled as isotropic and cross-anisotropic to properly account for stress sensitivity and the significant differences between vertical and horizontal moduli and the in plane and out of plane Poisson's ratios (ν^h and ν^v).

The layer material properties used and the nonlinear model parameters assigned in modeling of the test sections and for the sub-layered base and subgrade are given in Table 1. Resilient modulus (M_R) model parameters Ki are given for the Uzan model (1985) in the granular base and for the bilinear approximation by Thompson and Elliot (1985) used in the subgrade. The cross-anisotropic modulus model parameters for the resilient moduli in vertical and horizontal directions, M_R^V and M_R^H, and shear modulus, G_R, were characterized from laboratory testing in accordance with the approach by Tutumluer and Thompson (1997).

Figures 2 and 3 show comparisons of the measured and predicted pavement responses for B1 and D3 test sections, respectively. The predictions were obtained only for the 689-kPa tire pressure case, which was also used in the trafficking of the test sections as described by Al-Qadi et al. (2007). Note that the measured LVDT responses corresponded to measured relative elastic displacements within the 150 to 200 mm lengths the LVDTs spanned. Accordingly, different vertical LVDT responses could be measured in different layers, such as those measured on the top of subgrade and the bottom part of the base layers, as indicated in Figure 2.

In test section B1, the anisotropic mechanistic model predictions are generally in better agreement with the measured responses than the isotropic predictions (see Figure 2). A number of pavement responses were predicted at the same time reasonably close to the measured values for all the different load levels studied. The predicted values of both subgrade vertical displacement and HMA strains compare reasonably well with the order of magnitudes of the measured responses except for the vertical subgrade pressure. The anisotropic model predictions were much better matching with the measured strain responses of the vertical LVDT on top of subgrade, which is an important critical pavement response controlling subgrade rutting. The vertical displacements predicted at bottom of base layer are still lower than the measured values. This indicates that the unbound aggregate used in the test section was actually much weaker under similar stress states. However, both the subgrade LVDT and the bottom base horizontal LVDT response predictions are in very good agreement.

In D3 test section with a thicker aggregate layer, the cross-anisotropic base characterization was again much better for predicting HMA strains and base course radial displacements, while the isotropic base characterization predicted somewhat closer subgrade vertical pressures to the measured ones at low load levels (see Figure 3). Note that the measured subgrade LVDT values in section D3 were not reliable to compare reasonably with the predicted responses. The anisotropic horizontal to vertical modulus ratios (M_R^h/M_R^v) predicted for the B1 section ranged from 0.11 to 0.12 throughout the base layer. This low ratio is in compliance with the constant K_4/K_1 ratio (see Table 1) estimated for nonlinear anisotropic material parameters and indicates a strong tendency of these somewhat low quality unbound aggregate layers to spread in the horizontal direction in accordance with the previous findings by Tutumluer and Thompson (1997). The good anisotropic model predictions obtained for the top and bottom base course radial deformations therefore suggest that the anisotropic model successfully captured the large movements under the applied loading.

SUMMARY

Full-scale pavement test sections, heavily instrumented with pavement sensors, were recently built for a geogrid base reinforcement study at the University of Illinois. Two of the unreinforced pavement test sections having 305 and 457 mm of unbound aggregate bases overlain by 76-mm thick hot mix asphalt (HMA) surfacing were analyzed with stress-sensitive, anisotropic (direction dependent) resilient modulus models in the base. Field validations of the anisotropic, nonlinear modeling approach were successfully accomplished by analyzing the test sections using GT-PAVE finite element program. Nonlinear cross-anisotropic model predictions matched better than isotropic analysis results. Better predictions achieved for subgrade deflections, HMA strains, and especially large base course radial deformations all at the same time clearly demonstrated the importance of both stress sensitivity and directional dependency (anisotropy) in the proper structural characterization of unbound aggregate layers.

Table 1 Material Properties and Model Parameters used in Modeling Responses of B1 and D3 Sections with Sublayering

Layer Type	Thickness (mm)	Initial Values					Vertical Modulus[1]			Horizontal Modulus[2]			Shear Modulus[3]		
		Vertical Modulus (MPa)	ν^v	Horizontal Modulus (MPa)	ν^h	Shear Modulus (MPa)	K_1 (kPa)	K_2	K_3	K_4 (kPa)	K_5	K_6	K_7 (kPa)	K_8	K_9
AC	102	7,653	0.35	-	-	-	-	-	-	-	-	-	-	-	-
Base															
B1 top	305	124	0.4	19	0.1	43	3,826	0.849	-0.196	370	1.318	-0.624	1,036	0.96	-0.302
B1 bottom	76	124	0.4	19	0.1	43	1,563	0.849	-0.196	148	1.318	-0.624	415	0.96	-0.302
D3	457	124	0.4	19	0.1	43	3,826	0.849	-0.196	370	1.318	-0.624	1,036	0.96	-0.302

Layer Type	Thickness (mm)	Initial Values		Bilinear Model Parameters[4] (Thompson and Elliott, 1985)					
		Resilient Modulus (MPa)	Poisson's Ratio ν	E_{RI} (kPa)	σ_{di} (kPa)	K_3	K_4	σ_{dll} (kPa)	σ_{dul} (kPa)
Subgrade (B1 and D3)									
top	229	10	0.45	3,447	43	75	1	7	145
middle	229	10	0.45	6,895	43	1,110	1	7	145
bottom	-	31	0.45	52,952	43	1,110	1	7	145

Uzan (1985) type granular material modulus models

[1] Vertical Modulus: $M_R^{\ v} = K_1(\theta / p_0)^{K2}(\sigma_d / p_0)^{K3}$

[2] Horizontal Modulus: $M_R^{\ h} = K_4(\theta / p_0)^{K5}(\sigma_d / p_0)^{K6}$

[3] Shear Modulus: $G_R = K_7(\theta / p_0)^{K8}(\sigma_d / p_0)^{K9}$

where θ: Bulk stress ($\sigma_1 + 2\sigma_3$) in triaxial conditions;
σ_d: Deviator stress ($\sigma_1 - \sigma_3$) in triaxial conditions;
p_0: Atmospheric pressure (~100 kPa);
K_i: Regression constants from triaxial test data.

[4] Bilinear subgrade model (Thompson and Elliott, 1985

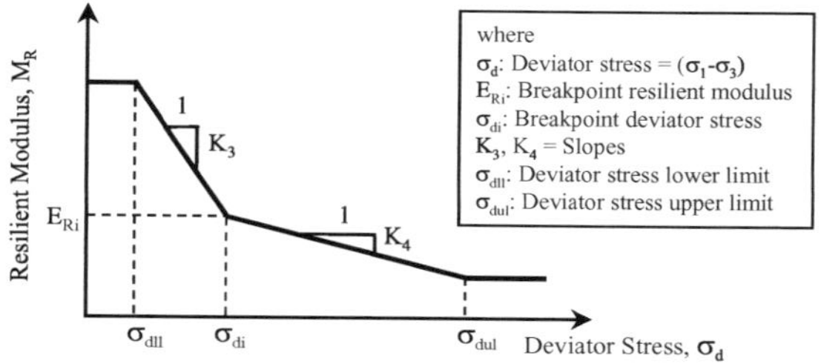

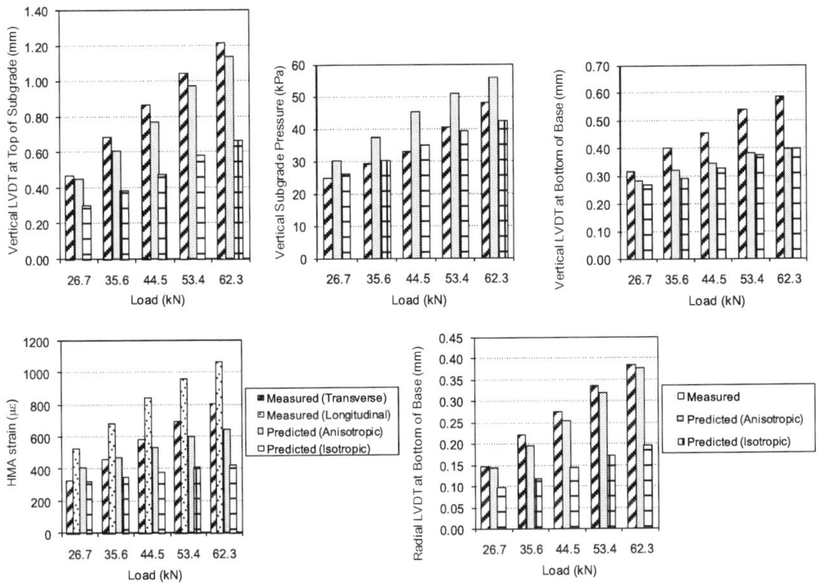

FIG. 2. Comparisons of measured and predicted B1 test section responses

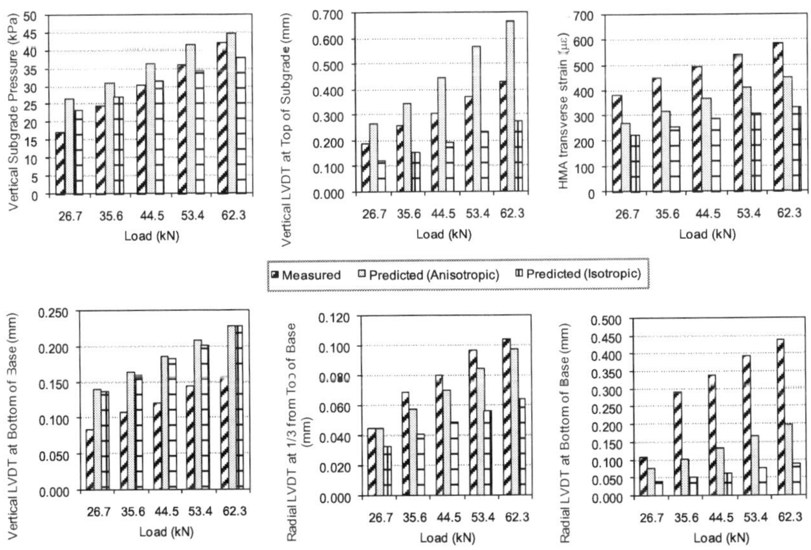

FIG. 3. Comparisons of measured and predicted D3 test section responses

ACKNOWLEDGMENTS

The full-scale pavement test study at the University of Illinois was sponsored by Tensar Earth Technologies, Inc. The authors would like to acknowledge significant research contributions of Drs. Imad Al-Qadi and Samer Dessouky. The assistance of J. Baek, P-J Yoo, E. Fini, J. Meister, M. Elseifi, B. Harkanwal, J. Anochie-Boateng, C. Montgomery, K. Jiang, and Z. Leng during pavement construction and instrumentation is also greatly appreciated. The content of this paper reflects the views of the authors, who are responsible for the facts and the accuracy of the data presented herein. This paper does not constitute a standard, specification, or regulation.

REFERENCES

Al-Qadi, I.L., Tutumluer, E., Dessouky, S., and Kwon, J. (2007). "Mechanistic response measurements of geogrid reinforced flexible pavements to vehicular loading." In CD-ROM Proceedings of the Geosynthetics 2007 Conference, Washington, D.C., January 16-19.

Beatty, T.L., Jackson, D.C., Dawood, D.A., Ford, R.A., Moulthrop, J.S., Taylor, G.D., Danchetz, F.L., Blanck, D.A., Fay, J.M., Keough, D.S., Rodriguez, L.M., Voth, M.D., and Webb, Z.L. (2002). "Pavement preservation technology in France, South Africa, and Australia." Publication No. FHWA-PL-03-001, Office of International Programs, FHWA/US DOT (HPIP), 400 7th St., SW, Washington, DC 20590.

Kwon, J., Tutumluer, E., Al-Qadi, I.L., and Dessouky, S. (2008). "Effectiveness of geogrid base reinforcement in low-volume flexible pavements," ASCE Geotechnical Special Publication No. 178, GeoCongress 2008: Geosustainability and Geohazard Mitigation, Edited by K.R. Reddy, M.V. Khire, A.N. Alshawabkeh.

Masad, S., Little, D.N., and Masad, E. (2006). "Analysis of flexible pavement response and performance using isotropic and anisotropic material properties." ASCE Journal of Transportation Engineering, Vol. 132, No. 4, 342-349.

Steven, B., Alabaster, D., and de Pont, J. (2007). "Elastic nonlinear finite element analysis of a flexible pavement subjected to varying falling weight deflectometer loads." Transportation Research Record: Journal of the TRB, No. 2016, Washington, D.C., 31-38.

Thompson, M.R., and Elliott, R.P. (1985). "ILLI-PAVE based response algorithms for design of conventional flexible pavements." Transportation Research Record 1043, TRB, National Research Council, Washington D.C., 50-57.

Tutumluer, E. (1995). Predicting behavior of flexible pavements with granular bases. Ph.D. Dissertation, School of Civil and Environmental Engineering, Georgia Institute of Technology, Atlanta, GA, September.

Tutumluer, E. and Thompson, M.R. (1997). "Anisotropic modeling of granular bases in flexible pavements." Transportation Research Record 1577, TRB, Washington D.C., 18-26.

Tutumluer, E. and Seyhan, U. (1999). "Laboratory determination of anisotropic aggregate resilient moduli using an innovative test device." Transportation Research Record, Journal of the Transportation Research Board, No. 1687, Washington, D.C., 13-21.

Tutumluer, E., Little, D.N., and Kim, S.H. (2003). "Validated model for predicting field performance of aggregate base courses." Transportation Research Record No. 1837, TRB, Washington D.C., 41-49.

Tutumluer, E. (2008). "State of the art: Anisotropic characterization of unbound aggregate layers in flexible pavements." To be published in the Proceedings GSP of the Symposium of Pavement Mechanics & Materials, ASCE EM08 Conference, Minneapolis, MN, May 18-21.

Uzan, J. (1985). "Characterization of granular material." Transportation Research Record No. 1022, TRB, Washington D.C., 52-59.

Drained Soil Modulus from Load Tests for Shallow Foundation Design

Sami O. Akbas[1], A.M.ASCE and Fred H. Kulhawy[2], Dist.M.ASCE

[1]Lecturer, Department of Civil Engineering, Gazi University, Celal Bayar Bulvari, Maltepe, Ankara, Turkey 06570; soakbas@gazi.edu.tr
[2]Professor, School of Civil and Environmental Engineering, Hollister Hall, Cornell University, Ithaca, NY 14853-3501; fhk1@cornell.edu

ABSTRACT: Elastic theory is used to back-calculate the drained secant soil modulus from loads and settlements measured in full-scale load tests on footings in axial compression. The moduli are evaluated at L_1 (the end of the initial linear region) and at a settlement of 1% of the foundation width. These drained soil moduli are normalized by standard penetration test results at each site, and the normalized values are correlated effectively with the soil stress history, stress state, and stress level. The resulting correlations will be useful for footing design.

INTRODUCTION

The settlement of shallow, spread-type foundations on cohesionless soils often is estimated by assuming that the soil behaves as an elastic material. However, the soil modulus is dependent on its initial density and stress history, and it is both nonlinear and stress-dependent. Therefore, estimating the "correct" value of soil modulus is a major source of uncertainty in settlement predictions. Soil moduli, which are back-figured from well-documented full-scale field load tests, can be very useful for understanding the principal features of this parameter for foundation design.

In this paper, an extensive database of 167 full-scale field load tests of footings on cohesionless soils under axial compression is used to obtain realistic values of the drained secant soil modulus (E_s), which then are normalized by the site standard penetration test (SPT) N values. These normalized values vary with the soil stress history, stress state, and stress level. Correlations are developed among these parameters, and pertinent equations are given for future design use.

DATABASE

The database used herein includes published case histories from 37 sites, with 167 axial compression footing field load tests conducted in cohesionless soil ranging from silt to gravel. Case histories were selected only if they have load-settlement curves that extend at least beyond the linear-elastic region. Also, to be representative of full-scale footing behavior, none were included with a footing width smaller than 0.25 m. Details that include the location of each test, a general description of the prevailing soil types encountered, and available geometrical and geotechnical properties are given elsewhere (Akbas and Kulhawy 2008). Key foundation parameters are summarized as follows: mean width (B) = 0.81 m, ranging from 0.25 to 4.0 m; mean length (L) = 0.85 m, ranging from 0.25 to 4.0 m; and mean depth (D) = 0.18 m, ranging from 0 to 2.5 m. For the soil, the mean effective stress friction angle in triaxial compression (ϕ'_{tc}) = 38.1°, ranging from 28.0 to 55.0°.

In-situ testing results are available for 123 of the 167 cases in the database. Only SPT data are available for 59 cases, and only cone penetration test (CPT) data are available for 3 cases. For these CPT cases, the q_c profiles were converted to equivalent SPT N profiles using the q_c/N relationship given by Kulhawy and Mayne (1990). For 36 cases, both SPT and CPT data are available. Where SPT results are available, N_{60} values were obtained by correcting the N values for field procedures to an average energy ratio of 60%. The energy efficiency of the SPT equipment was not specified, except for five cases, so correction factors for energy ratio (C_{ER}), borehole diameter (C_B), sampling method (C_S), and rod length (C_R) suggested by Skempton (1986) were used to estimate N_{60} values, considering the geographic location and the time frame of each reference.

SECANT SOIL MODULUS FROM THEORY OF ELASTICITY

The general equation for settlement (ρ) using the theory of elasticity is given by:

$$\rho = q\, B\, I / E_s \tag{1}$$

in which q = applied stress and I = settlement influence factor. Eq. 1 is modified as follows for homogeneous to Gibson soil profiles, finite layer thickness, foundation flexibility, and foundation embedment (Mayne and Poulos 1999):

$$\rho = q\, B_{eq}\, I_G\, I_F\, I_E\, (1 - \nu^2) / E_o \tag{2}$$

in which B_{eq} = diameter of circular foundation with same area; I_G, I_F, and I_E = correction factors for modulus variation with depth, foundation flexibility, and foundation embedment; ν = Poisson's ratio; and E_o = soil modulus directly beneath foundation base. The relationship between E_o and E_s is given by:

$$E_s = E_o + k_E\, z \tag{3}$$

in which k_E = rate of increase of modulus with depth (units of stress / length), and z = depth below ground surface. The values of E_o, and consequently E_s at a depth = B, were back-calculated for each footing in the database using Eqs. 2 and 3. The details are given in Akbas (2007).

Using Eqs. 2 and 3, only one value of E_s is obtained from a load test at a particular stress level. However, it is widely accepted that the soil stiffness or modulus is strain dependent, with larger values at smaller strains. Because of the highly nonlinear load-settlement behavior of footings in cohesionless soils, it is necessary to select E_s at an approximate strain level corresponding to that for the foundation during its design life. A practical issue that arises is how to estimate this strain level. Herein, the Vesic (1973) suggestion is used, in which strain levels are defined approximately by ρ/B.

By definition, Q_{L1} is the load corresponding to the settlement at the end of the initial linear region of the load-settlement curve (ρ_{L1}), or the largest load at which nominal elastic settlement occurs (Fig. 1). Therefore, it is appropriate to calculate E_s using Q_{L1} and ρ_{L1}. For typical footing widths in practice, a settlement of 1% B is between 10 and 30 mm, which is a common range for design settlements. Therefore, it also is useful to compute E_s at $\rho_{1\%B}$ and its corresponding load ($Q_{1\%B}$), which corresponds roughly to the design load at the upper limit of the serviceability limit state. These drained secant soil moduli are denoted E_{sL1} and $E_{s1\%B}$, respectively. Herein, both moduli are divided by the atmospheric pressure (p_a) to make them dimensionless.

CORRELATION OF SOIL MODULUS WITH IN-SITU TEST RESULTS

The back-calculated, dimensionless E_s values from the footing load tests were correlated with SPT results at each site that were averaged over the footing depth of influence (D_i). Where N increased or was constant with depth, D_i was computed from the Burland and Burbidge (1985) relationship. Where N showed a consistent decrease with depth, D_i was taken to be 2B. These N values were corrected to N_{60}, as described previously. They also were corrected for overburden stress to $(N_1)_{60}$, using the relationship suggested by Liao and Whitman (1986), with a maximum limit value of 2.

The basic results of the analyses are given in Table 1, which shows the soil group, number of tests, and the mean and coefficient of variation (COV), first for (E_{sL1}/p_a) normalized by N_{60} and $(N_1)_{60}$ and then $(E_{s1\%B}/p_a)$ normalized by N_{60} and $(N_1)_{60}$. For the entire database (n = 119, with soil groups A+B+C), the mean $(E_{sL1}/p_a)/N_{60} = 30.1$ (with COV = 90%) and the mean $(E_{s1\%B}/p_a)/N_{60} = 16.1$ (with COV = 78%). When normalized by $(N_1)_{60}$, the mean $(E_{sL1}/p_a)/(N_1)_{60} = 19.1$ (with COV = 86%) and the mean $(E_{s1\%B}/p_a)/(N_1)_{60} = 10.4$ (with COV = 69%).

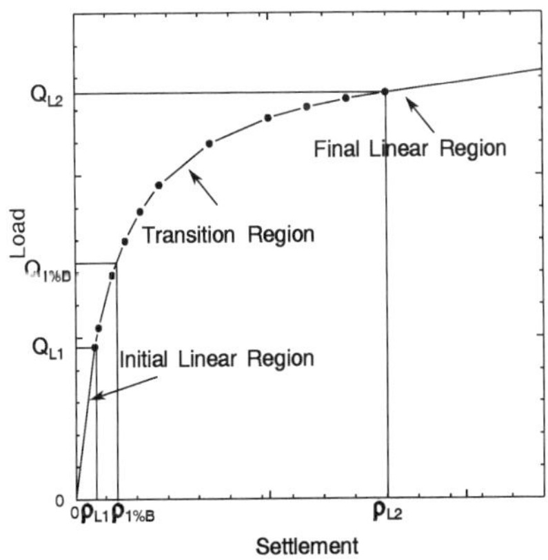

FIG. 1. Regions of a Typical Axial Load-Displacement Curve for a Footing

For the database, the mean ρ_{L1} = 1.91 mm, ranging from 0.15 to 9.60 mm, while the mean $\rho_{1\%B}$ = 8.1 mm, ranging from 2.5 to 40.0 mm. Since $\rho_{1\%B}$ is larger than ρ_{L1} (the elastic limit), it is on the nonlinear portion of the load-displacement curve and will have a lower E_s, in this case by almost one-half. When the moduli are normalized by $(N_1)_{60}$ instead of N_{60}, where $(N_1)_{60}$ is larger for shallow footings, the ratios are smaller by about one-third. However, the COVs are large and roughly comparable for all of these cases, decreasing slightly from E_{sL1} to $E_{s1\%B}$ and from N_{60} to $(N_1)_{60}$.

Table 1 also shows, in parentheses, the results when they are averaged per test site to eliminate potential bias from multiple tests at the same site. As can be seen, the means are about the same, but the COVs consistently are smaller.

Based on available data, Kulhawy and Mayne (1990) suggested that, on a first-order basis, values of $(E/p_a)/N_{60}$ could be estimated as follows: (a) 5 for sands with fines, (b) 10 for clean, normally consolidated sands, and (c) 15 for clean, overconsolidated sands. These values are lower than those in Table 1 for the entire database.

All attempts to date to correlate a modulus with N have shown considerable scatter, because N varies with many factors (e.g., Kulhawy and Mayne 1990) and E is also quite variable, as noted previously. For example, back-calculated values from drilled shaft load tests at Q_{L1} and ρ_{L1} show COV of 70-100% (Cushing and Kulhawy 2002). Also, for use in reliability-based design, Phoon and Kulhawy (1999) note that the COV for soil modulus is on the order of 30-90%. Values in Table 1 that are separated by soil group and averaged per site are at the low end of this range.

Effect of Stress History

When selecting the soil stiffness based on a correlation with penetration test results in cohesionless soils, it is necessary to consider the soil stress history. Research suggests (e.g., Jamiolkowski et al. 1988) that the large strains caused by the penetration of SPT, CPT, and DMT devices might erase the effects of stress and strain history in the soil to a large extent. However, the soil modulus is influenced strongly by the soil

Table 1. Back-Calculated Drained Secant Soil Modulus Normalized by SPT N (Averaged per Site)

Soil Group*	n	(E_{sL1}/p_a) / N_{60}		(E_{sL1}/p_a) / $(N_1)_{60}$		$(E_{s1\%B}/p_a)$ / N_{60}		$(E_{s1\%B}/p_a)$ / $(N_1)_{60}$	
		Mean	COV(%)	Mean	COV(%)	Mean	COV(%)	Mean	COV(%)
A+B+C	119	30.1	90	19.1	86	16.1	78	10.4	69
	(29)	(26.8)	(70)	(18.7)	(69)	(14.1)	(67)	(9.8)	(60)
A	40	12.9	44	8.3	45	7.9	28	5.2	35
	(14)	(13.8)	(42)	(9.2)	(38)	(7.4)	(21)	(5.5)	(29)
B	52	21.8	32	14.2	35	13.2	30	8.7	37
	(9)	(23.7)	(20)	(16.2)	(26)	(14.6)	(20)	(9.9)	(24)
C	27	48.5	41	39.9	41	24.3	27	19.7	35
	(6)	(58.1)	(24)	(38.9)	(19)	(28.7)	(23)	(19.3)	(21)

* - A: very recent, normally consolidated, cohesionless soils less than a few thousand years old and man-made cohesionless fills, excluding well-compacted structural fill

B: lightly overconsolidated, naturally aged cohesionless soils more than a few thousand years old

C: Heavily overconsolidated cohesionless soils and structural fills compacted under well-controlled conditions stress history, especially in the range of strains of practical interest. Therefore, even for a given unit weight, gradation, and stress or strain level, no unique correlation between penetration resistance and modulus can be expected for soil deposits of different stress histories.

In this study, soils were categorized into three groups by their stress history, as recommended by Berardi et al. (1991) and as shown in Table 1. Group A includes very recent, normally consolidated (NC), cohesionless soils less than a few thousand years old and man-made cohesionless fills, excluding well-compacted structural fill. Group B includes lightly overconsolidated (LOC), naturally aged cohesionless soils more than a few thousand years old. Group C includes heavily overconsolidated (HOC) cohesionless soils and structural fills compacted under well-controlled conditions. The groupings were based on the reported site geology or the soil preparation technique, together with the estimated site overconsolidation (OCR) values.

The mean and COV for the back-calculated E/N are given in Table 1 for groups A, B, and C. Compared with the entire database, these groupings result in substantial decreases in the COVs, from about one-half to two-thirds, to values between 27 and 45%, but otherwise the relative ratios noted above are roughly consistent. These groupings also show a very clear correlation with the soil stress history. The E/N increases from group A (NC) to group B (LOC) to group C (HOC), which should be expected. The increase ranges from factors of about 3 to 5, with a mean about 4.

For groups A through C, normalizing the secant modulus by $(N_1)_{60}$ instead of N_{60} generally increases the uncertainty slightly, as shown in Table 1. However, field and large-scale laboratory tests indicate that both stiffness and penetration resistance are influenced in a similar manner by the current level of the effective stress (Berardi et al. 1991). Considering a cohesionless deposit with constant relative density, it is expected that both N and E will increase with depth, since both are functions of the mean effective stress (Clayton et al. 1985). Considering these points, Stroud (1988) suggested using N values uncorrected for overburden effects when correlations are made between E and N. The modest adverse effect of normalization by $(N_1)_{60}$ shown in Table 1 supports this suggestion.

Effect of Stress State

The influence of the current effective overburden stress on the modulus was estimated using the following simple relationship suggested by Janbu (1963):

$$E_s = K_E\, p_a \{[\sigma'_{vo} + (\Delta\sigma'_v / 2)] / p_a\}^{0.5} \tag{4}$$

in which K_E = modulus number, σ'_{vo} = vertical effective overburden stress, and $\Delta\sigma'_v$ = change in vertical effective stress, computed using the Boussinesq solution up to Q_{L1} and $Q_{1\%B}$.

Following the format in Table 1, the back-calculated, dimensionless K_E/N values are given in Table 2 for the entire database and for groups A, B, and C. Basically, the same patterns of behavior are noted for K_E/N as for E/N. The means decrease going from L_1 to 1% B and from N_{60} to $(N_1)_{60}$, while the COVs are large. When separated by group, the COVs decrease dramatically, and the means increase from A to C.

When averaged per site, the means are about the same, but the COVs only decrease slightly. Based on these higher COVs, there appears to be a small advantage in using the E/N correlations.

Table 2. Back-Calculated Modulus Number Normalized by SPT N (Averaged per Site)

Soil	N	K_{EL1} / N_{60}		$K_{EL1} / (N_1)_{60}$		$K_{E1\%B} / N_{60}$		$K_{E1\%B} / (N_1)_{60}$	
Group		Mean	COV(%)	Mean	COV(%)	Mean	COV(%)	Mean	COV(%)
A+B+C	119	40.5	94	24.7	84	15.6	90	9.5	76
	(29)	(33.8)	(78)	(22.8)	(64)	(13.0)	(80)	(8.6)	(62)
A	40	21.8	68	12.4	61	8.4	35	5.3	29
	(14)	(21.2)	(59)	(15.8)	(65)	(7.5)	(36)	(5.4)	(31)
B	52	29.0	44	18.3	36	12.0	39	7.6	36
	(9)	(31.9)	(28)	(20.4)	(37)	(14.7)	(36)	(9.5)	(39)
C	27	82.5	53	48.7	46	21.8	35	13.1	29
	(6)	(66.1)	(54)	(42.6)	(35)	(23.3)	(31)	(12.2)	(27)

Effect of Stress Level

The load test data also were used to examine the variation of E/N with stress level. Instead of the original stress level definition used by Stroud (1988), which is the vertical effective stress divided by the soil compressive strength, the stress level was defined as the applied load (Q) divided by the interpreted failure load at the beginning of the final linear region of the load-settlement curve (Q_{L2}). This approach is useful because Q_{L2} is a consistent indicator of the footing capacity (Akbas and Kulhawy 2008), as well as the capacity of many types of drilled foundations, including augered cast-in-place (ACIP) piles, pressure injected footings (PIF), micropiles, and drilled shafts in cohesive and cohesionless soils, ranging from clays to coarse gravelly soils (e.g., Kulhawy 2004, Chen 2004).

Figure 2 shows the variation of $(E_s)/N_{60}$ with stress level (Q/Q_{L2}) for soil groups A (NC), B (LOC), and C (HOC). The E_s values shown were calculated directly from the reported loads and corresponding settlements along each test curve in the database. In all cases, E/N decreases with increasing stress level, and the amount of decrease increases with increasing overconsolidation. The r^2 values suggest that the regression equations can be used effectively to estimate the soil moduli for computing footing settlements using elastic theory, once the soil group is chosen carefully.

Note that, as a first approximation, the stress level can be considered as the reciprocal of the factor of safety. This approximation helps to establish the design stress level for modulus estimation. For reference, the average stress level at L_1 is about 0.14 (0.08, 0.21, and 0.12 for A, B, and C, respectively), while it is about 0.37 at 1% B (0.37, 0.42, and 0.31 for A, B, and C, respectively).

SUMMARY AND CONCLUSIONS

The drained secant soil modulus (E_s) was back-calculated, using elastic theory, from the results of full-scale, axial compression load tests on footings in cohesionless

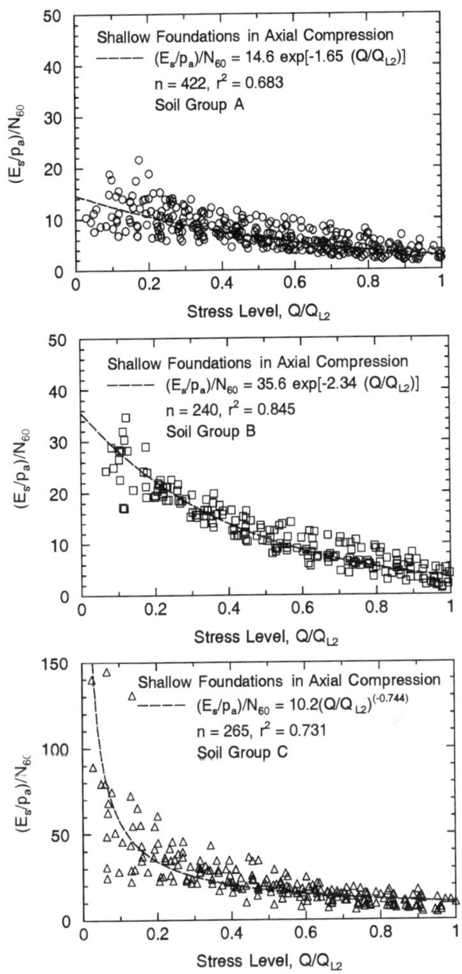

FIG. 2. Normalized E/N versus Stress Level

soils. The moduli were evaluated at L_1 (the end of the initial linear region) and at a settlement of 1% of the foundation width (B). These moduli then were normalized by the site SPT results. When the entire database is evaluated, the normalized moduli exhibit very large coefficients of variation (COV), which are consistent with values reported elsewhere. However, when the soils are grouped according to their approximate stress history, the COVs decreased significantly. These results suggest that overburden corrections are not necessary for correlations between E and N.

The influences of stress state and stress level on the normalized moduli also were examined. Correlations showed that the normalized moduli decrease as the stress level increases. When the soils are grouped by stress history, the correlations of the normalized moduli with the stress level are very good. The resulting correlation equations will be useful for estimating footing settlements on cohesionless soils.

REFERENCES

Akbas, S.O. (2007). "Deterministic and probabilistic assessment of settlements of shallow foundations in cohesionless soils", PhD thesis, Cornell Univ., Ithaca.

Akbas, S.O. and Kulhawy, F.H. (2008). "Axial compression of footings in cohesionless soils - part A: load-settlement behavior", (in review).

Berardi, R., Jamiolkowski, M., and Lancellotta, R. (1991). "Settlement of shallow foundations in sands: selection of stiffness on the basis of penetration resistance", *Geotech. Eng. Cong. 1991 (GSP 27)*, Ed. F.G. McLean, D.A. Campbell, and D.W. Harris, ASCE, New York, 185-200.

Burland, J.B. and Burbidge, M.C. (1985). "Settlement of foundations on sand and gravel", *Proc.*, Inst. of Civil Eng., 78(Pt 1), 1325-1381.

Chen, J.-R. (2004). "Axial behavior of drilled shafts in gravelly soils", PhD thesis, Cornell Univ., Ithaca.

Clayton, C.R.I., Hababa, M.B., and Simons, N.E. (1985). "Dynamic penetration resistance and the prediction of the compressibility of a fine-grained sand - a laboratory study", *Geotechnique*, 35(1), 19-31.

Cushing, A.G. and Kulhawy, F.H. (2002). "Drained elastic behavior of drilled shafts in cohesionless soils", *Deep Foundations 2002 (GSP 116)*, Ed. M.W. O'Neill & F.C. Townsend, ASCE, Reston, 22-36.

Jamiolkowski, M., Ghionna, V. N., Lancelotta, R., and Pasqualini, E. (1988). "New correlations of penetration tests for design practice", *Proc. 1st Intl. Symp. Penetration Testing (1)*, Orlando, 263-296.

Janbu, N. (1963). "Soil compressibility as determined by oedometer and triaxial tests", *Proc. European Conf. Soil Mech. Fndn. Eng. (1)*, Wiesbaden, 19-24.

Kulhawy, F.H. (2004). "On the axial behavior of drilled foundations", *GeoSupport 2004 (GSP 124)*, Ed. J.P. Turner and P.W. Mayne, ASCE, Reston, 34-51.

Kulhawy, F.H. and Mayne, P.W. (1990). "Manual on estimating soil properties for foundation design", *Report EL-6800*, EPRI, Palo Alto.

Liao, S.S. and Whitman, R.V. (1986). "Overburden correction factors for SPT in sand", *J. Geotech. Eng.*, ASCE, 112(3), 373-377.

Mayne, P.W. and Poulos, H.G. (1999). "Approximate displacement influence factors for elastic shallow foundations", *J. Geotech. Eng.*, ASCE, 125(6), 453-460.

Phoon, K.K. and Kulhawy, F.H. (1999). "Evaluation of geotechnical property variability", *Canadian Geotech. J.*, 36(4), 625-639.

Skempton, A.W. (1986). "Standard penetration test procedures and the effects in sands of overburden pressure, relative density, particle size, ageing and overconsolidation", *Geotechnique*, 36(3), 425-47.

Stroud, M.A. (1988). "Introduction to papers 1-9, part 2", *Proc. Penetration Testing in UK*, Thomas Telford, London, 29-49.

Vesic, A.S. (1973). "Analysis of ultimate loads of shallow foundations", *J. Soil Mech. & Fndns. Div.*, ASCE, 99(SM1), 45-73.

MICROSCALE MODELING OF THE SEISMIC RESPONSE OF SHALLOW FOUNDATIONS

U. El Shamy[1]

[1]Environmental and Civil Engineering Dept., Southern Methodist University, PO Box 750340, Dallas, TX 75275, E-mail: uelshamy@engr.smu.edu

ABSTRACT

This paper presents a microscale approach to analyze the seismic response of a spread footing system founded on a liquefiable granular soil deposit. The approach utilizes a three-dimensional transient fully-coupled hydromechanical model while taking into account the effects of soil-foundation interaction. The porous soil medium is modeled as a mixture of two interpenetrating phases, namely the fluid phase (water) and the particulate solid phase. The fluid is idealized as a continuum by using averaged Navier-Stokes equations that account for the presence of the solid particles. The Discrete Element Method (DEM) is employed to model the assemblage of these particles. The interphase momentum transfer is modeled using an established relationship that accounts for the dynamic change in porosity. The spread footing is idealized as a rigid block and its motion is described by the resultant forces and moments acting upon it. A computational simulation is conducted to investigate the response of spread-footings on a saturated granular deposit when subjected to a dynamic excitation. Results of the conducted simulation showed that the foundation sustained excessive settlement as the ground shaking progressed. The conducted simulation appears to capture essential dynamic response patterns typically observed in such systems.

INTRODUCTION

Liquefaction-induced settlement of shallow foundations is a major form of failure for buildings in saturated sand deposits. Examples of such failure were noticed during the 1964 Niigata, Japan earthquake. Despite of the importance of such foundation systems, there appears to be very little work done to investigate the response of such systems particularly on the computational front. This is perhaps due to the complexity of such systems as opposed to analyzing free-field deposits which has received a great deal of attention over the past two decades (e.g., Arulanandan and Scott, 1993). Most of the analytical framework available in the literature deals with the seismic response of shallow foundations on dry deposits or saturated soils (excited at low strain levels) within the context of the "equivalent linear approach" (see for example Mylonakis et al., 2006, for a comprehenssive compilation of work in this area). Seismic analysis of soil-foundation systems encompasses several issues such as soil nonlinearities and dynamic soil-foundation interaction as well as the development of pore-water pressure as shaking progresses and its impact on soil stiffness degradation, among other factors. Most of the available computational models rely on continuum-based approaches with constitutive relations that usually include parameters that are difficult to predict (Kramer, 1996).

The Discrete Element Method (DEM) emerged as a powerful tool to analyze granular materials (Cundall and Strack, 1979). Recent simulations of the seismic response of saturated granular deposits (free-field) have shown the seamless capabilities of DEM to capture essential characteristics that are commonly observed in physical modeling (Zeghal and El Shamy, 2004; El Shamy and Zeghal, 2005b). In this paper, a DEM-based approach is utilized to investigate the response of a saturated soil-foundation system to seismic excitations. As shown below, the approach inherently accounts for the modeling issues described above and thus provides a powerful tool to model such complex systems.

COMPUTATIONAL MODEL

Conceptually, in presented computational simulation, the mixture of solid particles and pore fluid is viewed as two interpenetrating media, namely the solid phase and the fluid phase. The fluid is idealized as a continuum by using a homogenized form of Navier-Stokes equations that account for the presence of the solid particles. These particles are modeled at a micro scale using DEM. Averaged fluid-particle interactions are quantified using a semi-empirical relationship. The shallow foundation is idealized as a rigid block by generating a group of particles glued to each other, and that subsequently interact with the surrounding deposit particles.

Fluid Phase

The averaged Navier-Stokes continuity and momentum equations for an incompressible inviscid pore fluid are given by (e.g., Jackson, 2000):

$$\frac{\partial n}{\partial t} + \nabla \cdot (n\bar{\mathbf{v}}_f) = 0 \tag{1}$$

$$\rho_f \left(\frac{\partial (n\bar{\mathbf{v}}_f)}{\partial t} + \nabla \cdot (n\bar{\mathbf{v}}_f\bar{\mathbf{v}}_f) \right) = -n\nabla \bar{p}_f \boldsymbol{\delta} - \bar{\mathbf{f}}_i + n\rho_f \mathbf{f}_g \tag{2}$$

where $n = n(\mathbf{x},t)$ is porosity (in which $\mathbf{x}$ and t are space and time coordinates), $\bar{\mathbf{v}}_f = \bar{\mathbf{v}}_f(\mathbf{x},t)$ is averaged fluid velocity vector, $\bar{p}_f = \bar{p}_f(\mathbf{x},t)$ is averaged fluid pressure, ∇ is gradient operator, ρ_f is fluid density, $\mathbf{f}_g$ is gravitational acceleration vector, and $\bar{\mathbf{f}}_i = \bar{\mathbf{f}}_i(\mathbf{x},t)$ is averaged fluid-particle interaction vector. The associated boundary conditions consist of fluid velocity and/or pressure constraints.

Averaged fluid-particle interactions may be quantified using a number of relationships. In this study, the semi empirical equation developed and calibrated by Ergun (1952) was employed.

Solid Phase

The discrete element method (Cundall and Strack, 1979) was used to idealize the assemblage of soil particles using distinct spheres. The motion of a particle p is dictated by the momentum equations:

$$m_p \dot{\mathbf{v}}_p = m_p \mathbf{f}_g + \sum_c \mathbf{f}_c + \mathbf{f}_d \tag{3}$$

$$I_p \dot{\boldsymbol{\omega}}_p = \sum_c \mathbf{r}_c \times \mathbf{f}_c \tag{4}$$

where $\mathbf{v}_p$ and $\boldsymbol{\omega}_p$ are translational and rotational velocity vectors (a superposed dot indicates time derivative), m_p is particle mass, I_p is particle moment of inertia, $\mathbf{f}_c$ refers to inter-particle force at contact c ($c = 1, 2, \cdots$), $\mathbf{r}_c$ is vector connecting the center of the particle to the location of the contact c, and $\mathbf{f}_d$ is drag force exerted by the fluid on the particle p which includes buoyancy and fluid-particle interaction terms. The inter-particle forces are dictated by contact laws which are direct functions of grain stiffness properties and relative movements at the contacts. The normal component was idealized using a Hertzian nonlinear spring stiffness which is connected in parallel to a viscous dashpot. The shear contact force was modeled using an elastic spring in series with a frictional slider. The shear and normal forces are related by a slip Coulomb model (Itasca, 2005). More details of the coupled fluid-particle model may be found in Zeghal and El Shamy (2004) and El Shamy and Zeghal (2005a).

COMPUTATIONAL SIMULATION

The proposed approach was used to investigate the response of a spread footing foundation system founded on liquefiable granular deposit using the commercial software PFC^{3D} (Itasca, 2005). The system is assumed to be composed of isolated square footings (2.4 m×2.4 m) equally spaced at 5.6 m in the lateral directions. The foundation depth is 1.2 m below ground surface and the soil profile at the site is composed of 1.2 m of dry granular soil underlain by a 9.6 m thick saturated granular material. The granular deposit is resting on a rigid bedrock.

Some assumptions were made in order to simulate such a boundary value problem based on micro-mechanical considerations. Periodic boundaries (El Shamy, 2004) were employed to simulate an intermediate spread footing of the system described above along with the portion of soil deposit "serving" this footing (Fig. 1). Furthermore, use was made of the high g-level concept commonly used in centrifuge testing to reduce the dimensions of the domain that needs to be filled with particles and to benefit from the shorter time scaling law within which the simulation has to be conducted. Finally, relatively large particle diameters were used to bring the number of particles in the system to a manageable size. In order to compensate for the employed high g-level and large particle size, viscous fluid was used in the simulation (Kutter, 1992).

The deposit shown in Fig. 1 was created as follows. Soil particles were generated at a larger space and then allowed to settle under gravity. Once settled, the high gravitational field (50 g) was introduced and the particles were allowed to reach equilibrium. Next, the particles occupying the space where the foundation would be installed were deleted and the foundation structure is formed as described in the next section. The system was then allowed to reach equilibrium under the static load from the foundation.

Modeling the Foundation System

The spread footing was modeled as a rigid block that is composed of clumped spherical particles. That is, regardless of the forces acting upon it, the block will not break apart. The contact forces between the clumped particles are not accounted for during the DEM calculation cycle. However, the interactions between the clumped particles and soil particles composing the deposit are considered. The spherical clumped particles were assembled to resemble near-realistic conditions between the foundation and the soil particles. The block was formed as a parallelpiped structure whose walls are made of particles

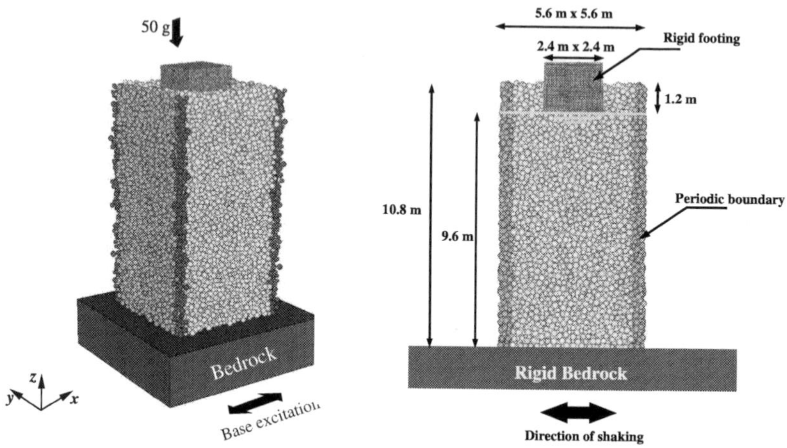

FIG. 1. Analyzed particulate deposit in the conducted simulation.

arranged next to each other. These particles were smaller than the particles composing the deposit to resemble the typical contact surface between a foundation and the underlying soil. The stress applied by the foundation on the underlying soil was about 100 kPa (a typical value for spread footing systems). Such stress was achieved by specifying the density of the clumped particles such that the weight of the block over its area produces the desired average stress.

Since the clumped particles in PFC^{3D} behave as a rigid body, translational and rotational motion equations are sufficient to describe its motion (Eqs. 5 and 6). The translational motion of the center of mass is described in terms of its position (x_i), velocity ($\dot{x}_i$) and acceleration ($\ddot{x}_i$). The rotational motion of the block is described in terms of its angular velocity and acceleration (Itasca, 2005).

$$F_i = M_{cl}(\ddot{x}_i - g_i) \tag{5}$$

$$M_i = \dot{H}_i \tag{6}$$

In the above equations, F_i is the resultant force (the sum of all externally applied forces acting on the clump), g_i is the gravitational acceleration vector, M_i is the resultant moment about the center of mass and $\dot{H}_i$ is the time rate-of-change of the angular momentum of the block.

The deposit and the foundation are then subjected to a dynamic base excitation that gradually increases to reach a maximum acceleration of 0.2 g. The fluid phase boundary conditions are such that the pressure is constant (atmospheric) at the water level and no flow across the impermeable base (the bedrock). Periodic boundaries along the four lateral sides of the domain were also enforced for the solution of averaged Navier-Stokes equations. Multiple solid and pore-fluid state variables were monitored during the course of the simulation. Table 1 summarizes the computational data for the solid and fluid

TABLE 1. Simulation data in model units.

Particles	
Diameter	3.8 mm to 5.8 mm
Shear modulus	2.9×10^9 N/m^2
Poison's ratio	0.15
Shear Stiffness	1×10^6 N/m
Critical damping ratio	0.10
Friction coefficient	0.5
Density	2650 kg/m^3
Number of particles	28,310
Foundation	
Width	48 mm
Length	48 mm
Height	25 mm
Number of clump particles (n_{cl})	2,208
Density of a clump particle (ρ_{cl})	1,169 kg/m^3
Diameter of a clump particle	2 mm
Fluid	
Density	1000 kg/m^3
Viscosity	4.75 Pa s
Boundaries	
Width	112 mm
Length	112 mm
Depth	130 mm
Computation parameters	
Time steps for DEM	0.5×10^{-6} s
Time steps for fluid	1×10^{-6} s
Number of fluid cells	$7\times7\times5$
Applied 'g' level	50

phases as well as other computational details. Results are presented in prototype units exclusively.

Saturated Soil Layer Response

Fig. 2 shows the computed accelerations for selected depth locations within the saturated portion of the deposit in the free-field and along the centerline of the footing. The acceleration time-histories suggest that the free-field deposit fully liquefied after about 6.5 s. On the other hand, the soil underneath the footing did not fully liquefy and only the top one third of the saturated layer experienced some de-amplification of acceleration as a result of pore-pressure buildup.

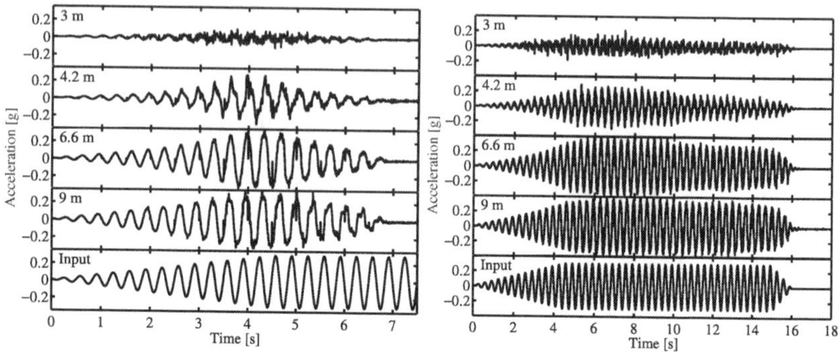

FIG. 2. Time histories for computed accelerations in the free-field (left) and underneath the foundation (right) at selected depth locations below the ground water level.

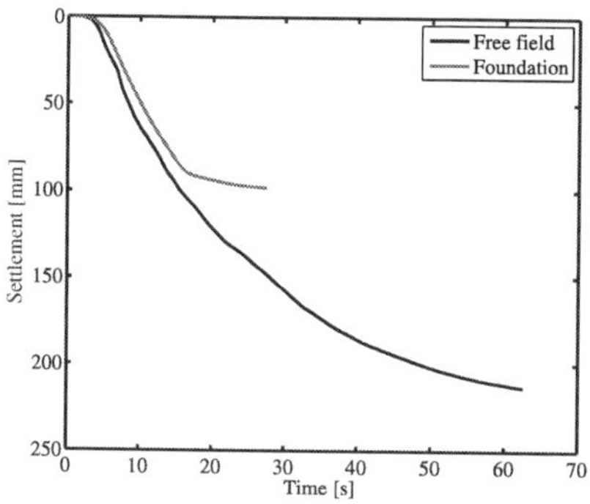

FIG. 3. Time histories of free-field and foundation settlement.

Foundation Response

The foundation experienced a settlement of about 100 mm compared to a final settlement of about 210 mm in the free-field (Fig. 3). Note that the free-field in this study is a deposit that does not include the foundation. On the other hand, the periodic deposit that simulates an interior footing in a foundation system produces more confining pressure from the footing itself and the surrounding footings. This relatively higher confining pressure in the vicinity of the footing makes the soil underneath it less sensitive to volumetric deformations than the free-field deposit and therefore resulted in less foundation settlement.

The induced forces and displacements of the footing were monitored during the course of the simulation. The moment at the center of the foundation base is plotted against footing rotation in Fig. 4a. As shaking progressed, the footing experienced accumulations of permanent rotation with a final amplitude of about 0.2 rad. It can also be seen that the

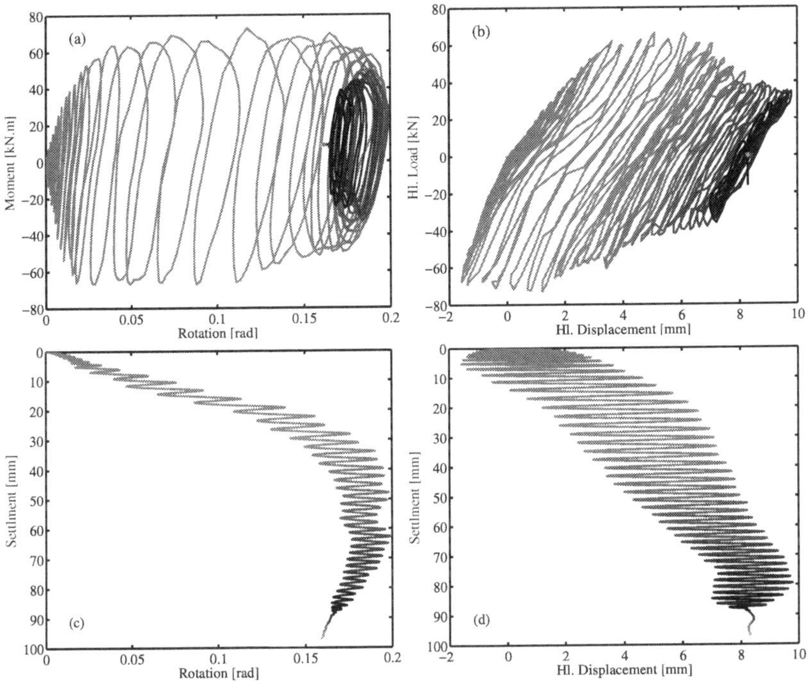

FIG. 4. Time histories for footing response: (a) moment-rotation relationship, (b) horizontal load-horizontal displacement relationship, (c) settlement-rotation relationship, and (d) settlement-horizontal displacement relationship.

rotational stiffness degraded as shaking continued. A permanent lateral displacement (sliding) of about 8 mm was accumulated by the end of shaking (Fig. 4b). The lateral load-sliding plot does not indicate a significant decrease in the lateral stiffness of the footing. The foundation settlement-rotation and settlement-sliding relationships do not suggest that any uplift of the footing took place during shaking (Figs. 4c and 4d).

CONCLUSIONS

A microscale computational model is presented to investigate the dynamic response of a coupled soil-foundation system. The computational framework accounts for soil nonlinearity as well as the possible buildup of pore pressure due to cyclic loading. In addition, the foundation is modeled as an embedded rigid block with frictional side walls and base. This block is allowed to undergo deformation in any of the associated six degrees of freedoms. All these features are presented in a fully-coupled transient formulation that appears to be a very promising tool to analyze soil-foundation systems subjected to seismic loads. Computational simulation of the seismic response of a foundation system on saturated sand indicates possible development of excessive settlement that is beyond typical allowable amplitudes. More results from the conducted simulation are currently being prepared and will be presented in a future publication.

REFERENCES

K. Arulanandan and R. Scott, eds. (1993). *Verification of numerical procedures for the analysis of soil liquefaction problems. Proceedings of the International Conference on the Verification of numerical procedures for the analysis of soil liquefaction problems.* A.A.Balkema, Davis, California.

Cundall, P. and Strack, O. (1979). "A discrete numerical model for granular assemblies." *Geotechnique*, 29(1), 47–65.

El Shamy, U. (2004). "A coupled continuum-discrete fluid-particle model for granular soil liquefaction," PhD Thesis, Rensselaer Polytechnic Institute, Troy, NY.

El Shamy, U. and Zeghal, M. (2005a). "A coupled continuum-discrete model for saturated granular soils." *Journal of Engineering Mechanics*, ASCE, 131(4), 413–426.

El Shamy, U. and Zeghal, M. (2005b). "A micro-mechanical study of the seismic response of saturated cemented deposits." *Journal of Earthquake Engineering*, 9. Special Issue 1 on Geotechnical Earthquake Engineering.

Ergun, S. (1952). "Fluid flow through packed columns." *Chemical Engineering Progress*, 43(2), 89–94.

Itasca (2005). *Particle Flow Code, PFC3D, release 3.1.* Itasca Consulting Group, Inc., Minneapolis, Minnesota.

Jackson, R. (2000). *The dynamics of fluidized particles.* Cambridge, U.K.; New York: Cambridge University Press.

Kramer, S. (1996). *Geotechnical earthquake engineering.* Prentice-Hall, Englewood Cliffs, NJ.

Kutter, B. (1992). "Dynamic centrifuge modeling of geotechnical structures." *Transportation research record*, (1336), 24–30.

Mylonakis, G., Nikolaou, S., and Gazetas, G. (2006). "Footings under seismic loading: Analysis and design issues with emphasis on bridge foundations." *Journal of Soil Dynamics and Earthquake Engineering*, 26, 824–853.

Zeghal, M. and El Shamy, U. (2004). "A continuum-discrete hydromechanical analysis of granular deposit liquefaction." *International Journal for Numerical and Analytical Methods in Geomechanics*, 28, 1361–1383.

DEM Study of a Shallow Foundation under Vertical Loading

Anil Bhandari[1] and Jie Han[1]

Civil, Environmental, and Architectural Engineering (CEAE) Department, the University of Kansas, Lawrence, Kansas. Email: banil@ku.edu and jiehan@ku.edu

ABSTRACT: Presence of firm soil near the ground surface favors the construction of a shallow foundation. The shallow foundation is also an attractive option when combined with ground improvement techniques (such as geosynthetic reinforcement) even in weaker soil. This study presents the bearing capacity estimation of a shallow foundation on a granular soil underlain by a rigid base using the Discrete Element Method (DEM), which is suitable for granular media. A physical model test of the shallow foundation on sand was conducted using a plate load test in the laboratory. The DEM model of the plate load test simulated the load-settlement behavior of the shallow foundation. The failure model of the foundation under this condition was investigated.

INTRODUCTION

Shallow foundations are commonly used for lightweight structures due to their economical advantages. However, the presence of weak bearing strata near the ground surface restricts designers from adopting the shallow foundation. Under this situation, over-excavation and replacement with or without geosynthetic reinforcement is one common solution. The bearing capacity of a shallow foundation is often evaluated using Terzaghi's bearing capacity formula or its subsequent modifications. Terzaghi's bearing capacity formula consist of three terms namely cohesion, overburden, and unit weight of the soil and size of the foundation. Each of these terms involves a different bearing capacity factor: N_c for the cohesion, N_q for the overburden, and N_γ for the unit weight of the soil.

Presence of a rigid base within the failure zone of a shallow foundation restricts the full development of a failure surface thus increasing the bearing capacity of the foundation. The rigid base can be a firm soil or bedrock or geosynthetic layer.

Wayne et al. (1998) pointed out that the shallow foundation fails above the geosynthetic reinforcement is one of the failure modes for the geosynthetic-reinforced foundation. The details of various methods to account for the increase in bearing capacity due to the existence of the rigid base can be found in Cerato and Lutenegger (2006 and 2007). The bearing capacity factors are often evaluated and back-calculated using laboratory and field tests. This study involved laboratory tests and the DEM simulation for evaluating N_γ. The DEM simulation approach permits a detailed understanding of the failure mechanism. This approach can also be extended to evaluate other bearing capacity factors.

LABORATORY TESTS

Poorly-graded sub-rounded river sand with a mean particle size (d_{50}) of 2.6 mm in diameter was used for all tests. The other properties of sand were: G = 2.65, C_c = 0.98, C_u = 2.73, γ_{min} =16.4 kN/m^3, and γ_{max} = 19.5 kN/m^3. The stress-strain behavior of the sand was obtained from the consolidated drained (CD) triaxial tests. For the triaxial tests, the samples were prepared at 72% relative density. The peak angle of internal friction of the sand was 37.3°, which was obtained from the triaxial tests.

A square wooden box, 0.67 m wide (a) and 0.12 m high (H), was used for plate load tests. The test sample was prepared by filling sand into the box with four equal layers, which were distinguished by marking at different depths. To control the density of the sample (relative density approximately 70%), the weight of sand in each layer was calculated and that amount of sand was compacted in the box by tamping until the base thickness achieved the requirement. A steel plate of 152 mm in diameter (B) was placed at the center of the sample and a vertical load was applied with an increment of 360 N (equivalent to 20 kPa). To permit the stable reading of settlement, each load increment was placed for five minutes before a settlement measurement was taken. The increment was continued until a failure of the foundation. The details of this test can be found in the paper by Pokharel et al. (2009).

NUMERICAL SIMULATION

Particle Flow Code (PFC2D) used in this study was developed by Itasca based on the discrete element method (Cundall and Strack, 1979). PFC2D uses rigid body and soft contact approaches. The soft contact approach allows deformation calculations at the contacts (Itasca, 2004). The calibration of the micro-mechanical parameters and the model preparation method are discussed below.

Calibration

The micromechanical properties of materials are usually calibrated using known responses of the materials in a macro level. Numerical biaxial tests and shear tests are commonly used to evaluate the microscopic properties of geo-materials (Chareyre and Villard, 2002; Zhang and Thornton, 2007). In this study, biaxial tests were used to calibrate the micro-parameters of the sand. The stiffness of the particle was calculated using Eq. (1) (Itasca, 2004; Potyondy and Cundall, 2004).

$$k_n = 2tE_c \tag{1}$$

where, k_n is the normal stiffness of the particles, t is the thickness of particles along the plane of paper, E_c is Young's modulus of the particle-particle contact.

The dimensions of the biaxial samples were 75 mm in width (B') and 150 mm in height (H'). Uniform sized cylindrical particles of 2.6 mm in diameter (d) were generated at a porosity of 0.16. The dimensions of the biaxial sample were similar to those used in the triaxial test. The ratio (H'/d) between the maximum dimensions of the sample (H') and the diameter of the particle was approximately 58. Evaluating the effects of the specimen size and the particle diameter in a direct shear test simulation, Jacobson et al. (2007) recommended the minimum H'/d ratio of 58 for a distinct shear band formation. Therefore, the dimensions used in the laboratory tests as well as numerical simulations were adequate to form the distinct shear band.

The biaxial test of the sand was simulated using the method outlined in PFC^{2D} manual. The parameters involved in the calibration are shown in Table 1. The laboratory tests and the numerical simulations were performed at three confining pressures: 69, 103, and 138 kPa. FIGURES 1 and 2 show the deviatoric stress vs. the axial strain and the volumetric strain vs. the axial strain. FIGURE 1 shows that the simulated peak deviatoric stress at the higher confining stress of 138 kPa was lower than the measured stress in the lab. This difference could be a consequence of assuming cylindrical shaped particles in the model to simulate the sub-rounded river sand. FIGURE 2 shows the dilatant behavior of the sand, in which the contraction of the sample was plotted as a negative volumetric strain and the dilation of the sample was plotted as a positive volumetric strain. Therefore, the samples prepared at a 72% relative density are dense.

Table 1. Micromechanical Properties for DEM Analysis.

Material	Parameters	
Sand	Young's modulus of the particle-particle contact, E(MPa^2)	60
	Ratio of normal to shear stiffness of particles, k_n/k_s	1.2
	Friction coefficient (μ)	0.55

From the DEM simulation and the laboratory tests, the deviatoric stress, q, vs. the mean effective stress, p' is plotted in Figure 3. The q-p' plot shows that the simulated q and p' are close to those measured in the laboratory tests. From this q-p' plot, the peak friction angle of the aggregate can be calculated as follows:

$$M = \frac{q}{p'} = \frac{6\sin\phi}{3-\sin\phi} = 1.5 \tag{2}$$

$$\phi = 37.3^o$$

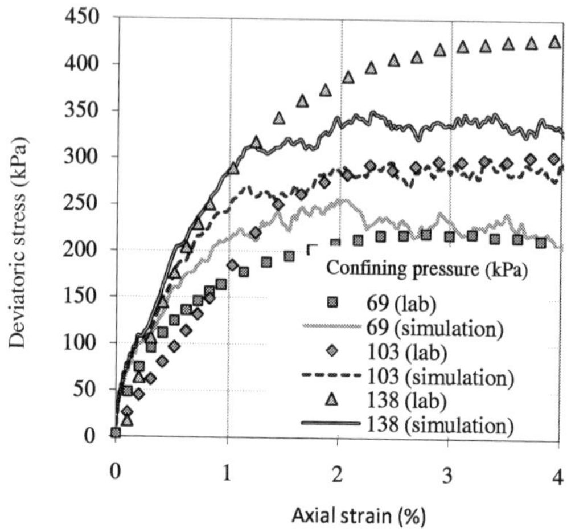

FIG. 1 Deviatoric Stress vs. Axial Strain

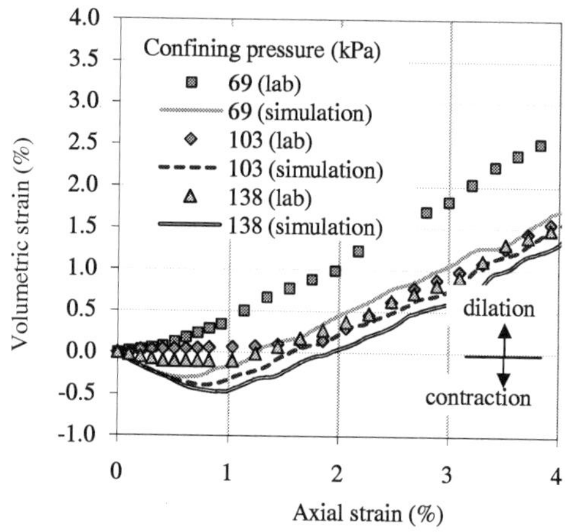

FIG. 2 Volumetric Strain vs. Axial Strain

Plate Load Tests

After the calibration of the micromechanical properties of the sand, a plate load test was simulated using PFC^{2D}. Similar dimensions of the box used in the lab were used in the simulation (FIG. 4). Radial expansion technique was used to generate the particle at a porosity of 0.16. In order to minimize the computation time, the size of

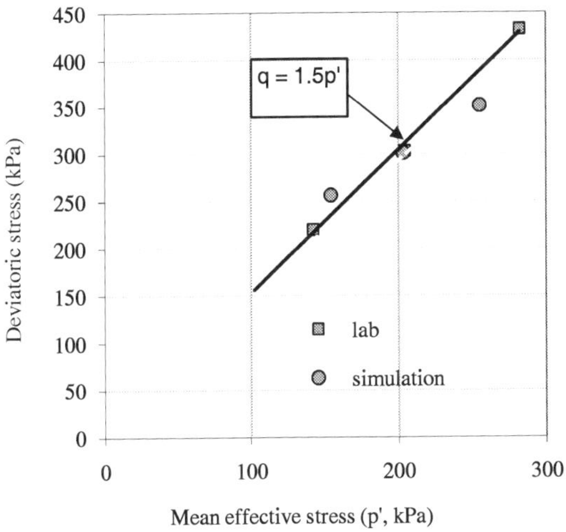

FIG. 3 Deviatoric Stress (q) vs. Mean Effective Stress (p')

the particles was increased twice as that used in the biaxial simulations. The peak angle of internal friction and the stiffness of the assembly remained unchanged by modifying the particle sizes and the dimensions of the biaxial sample at the same ratio. This up-scaling technique was also reported by Hainbuchner et al. (2003). The radial expansion technique was easy to implement and it provided a relatively homogenous assembly in the previous research of geosynthetic-reinforced base course under a wheel load (Bhandari and Han, 2008; Bhandari et al., 2008). The only demerit of this technique is the generation of larger initial stresses in the assembly which can be brought to the realistic initial stresses by moving side walls in the lateral direction using numerical servo mechanisms. Once a homogenous initial stress field was ensured, a plate was placed at the top center of the model and a load was applied with an increment of 360 N. At each load increment, the settlement was measured and the simulation was terminated at a total load of 5400 N (the maximum load in the lab).

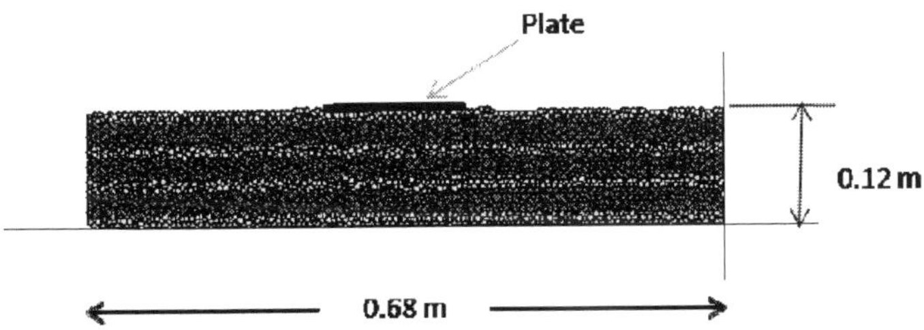

FIG. 4 DEM Model of a Plate Load Test

ANALYSIS OF RESULTS

Bearing Capacity

FIGURE 5 shows the vertical displacement curves of the foundations under the applied pressures from the lab and the DEM simulation. The calculated pressure-displacement curve of the foundation using the numerical simulation closely matched with that measured in the laboratory test. The ultimate bearing capacity of the dense sand obtained from the lab and the simulation were approximately 228 kPa.

The classical bearing capacity equation for a vertically loaded foundation resting on a horizontal bed of the cohesionless soil without any embedment can be expressed as:

$$q = \frac{Q}{BL} = 0.5\gamma B N_\gamma s_\gamma d_\gamma \tag{3}$$

where, γ is the unit weight of the soil; N_γ is the bearing capacity factor, s_γ is the shape factor, and d_γ is the depth factor.

To incorporate the effects of a rigid base on the bearing capacity of the foundation, Cerato and Lutenegger (2006) proposed the use of the same shape and depth factors as given by Tergazhi in Eq. (3) with modified N_γ^*. Using this approach, the modified N_γ^* for the tested foundation from the lab was 269, which is less than 340 obtained from the proposed design chart by Cerato and Lutenegger (2006) based on a friction angle of 37.3°.

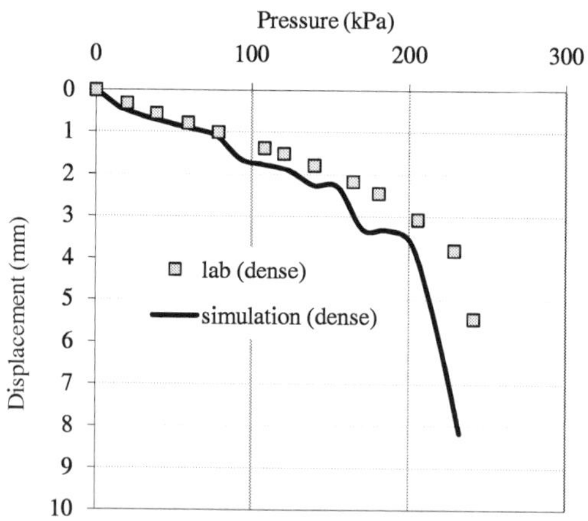

FIG. 5 Pressure vs. Displacement of the Shallow Foundations

Failure Mechanism

FIGURE 6 presents the displacement vectors for the shallow foundation with a limited depth, which shows a general shear failure mode. The displacement vectors of the particles were captured at the end of the plate load test simulation and the failure surface was added later. The failure surfaces were bounded by the bottom of the box. Since the friction between the particles and the walls was given the same value as that between the particles, the failure surfaces had to pass through the particles without touching the boundary walls. Furthermore, the trace of particle movement showed that the triangular zone immediately below the foundation did not cover the width of the foundation and was limited to a narrow zone below the foundation center. This small wedge under the footing was much smaller than that assumed by Terzaghi and others to derive the bearing capacity formulae. FIGURE 6 also shows the unsymmetrical movement of particles. Even though enough care was taken to ensure the homogeneity in the stress field and set zero displacement and velocity of the particles before the application of the plate load, the particles were randomly generated which contributed to the unsymmetrical movement of particles.

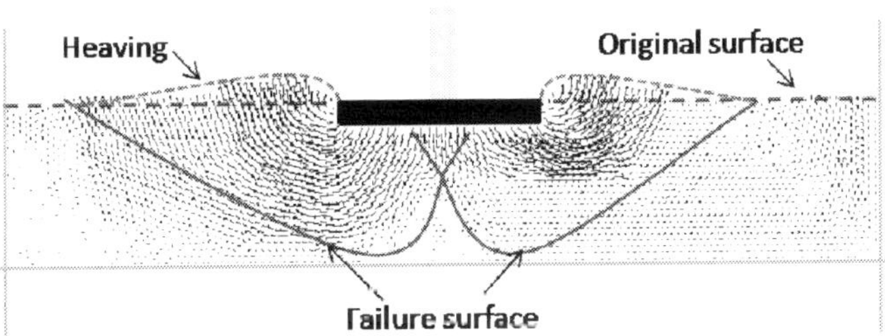

FIG. 6 General Shear Failure in Dense Sand (n=0.16) for H/B=0.8

CONCLUSIONS

This study presented the laboratory test and the DEM simulation of the shallow foundation underlain by a rigid base. The DEM simulation closely matched the laboratory test result. The particle movement in the DEM simulation clearly shows a general failure mode. The failure surfaces were bounded by the rigid base and the triangular wedge under the foundation was much smaller than that assumed by Terzaghi and others. The back-calculated bearing capacity factor, N_γ^*, is smaller than that obtained by Cerato and Lutenegger (2006).

ACKNOWLEDGEMENT

This research was funded by the University of Kansas, Transportation Research Institute from Grant # DT0S59-06-G-00047, provided by the US Department of

Transportation – Research and Innovative Technology Administration. This support is greatly appreciated.

REFERENCES

Bhandari, A., and Han, J. (2008). "Investigation of geotextile-soil interaction under a cyclic wheel load using the discrete element method." Submitted to *Geotextiles and Geomembranes.*

Bhandari, A., Han, J., and Parsons, R. L. (2008). "DEM analysis of geotextile-soil interaction under wheel loading." *Proceedings of the Research Symposium on the Characterization and Behavior of Interfaces (CBI)*, Atlanta.

Cerato, A. B., and Lutenegger, A. J. (2006). "Bearing Capacity of Square and Circular Footings on a Finite Layer of Granular Soil Underlain by a Rigid Base." *Journal of Geotechnical and Geoenvironmental Engineering, 132*(11), 1496-1501.

Cerato, A. B., and Lutenegger, A. J. (2007). "Scale effects of shallow foundation bearing capacity on granular material." *Journal of Geotechnical and Geoenvironmental Engineering, 133*(10), 1192-1202.

Chareyre, B., and Villard, P. (2002). "Discrete element modeling of curved geosynthetic anchorages with known macro-properties." *Proceedings of the Numerical Modeling in Micromechanics via Particle Methods*, Gelsenkirchen, Germany.

Cundall, P. A., and Strack, O. D. L. (1979). "A discrete numerical model for granular assemblies." *Geotechnique, 29*(1), 47-65.

Hainbuchner, E., Potthoff, S., Koneitzky, H., and Kamp, L. (2003). "Particle based modeling of shear box tests and stability problems for shallow foundations in sand." *Proceedings ofthe 1st International PFC Symposium on Numerical Modeling in Micromechanics via Particle Methods*, Gelsenkirchen, Germany.

Itasca. (2004). *Paticle Flow Code in Two Dimensions.* Minnesota: Itasca Consulting Group, Inc.

Jacobson, D. E., Valdes, J. R., and Evans, T. M. (2007). "A numerical view into direct shear specimen size effects." *Geotechnical Testing Journal, 30*(6), 1-5.

Pokharel, S.K., Han, J., Leshchinsky, D., Parsons, R.L., and Halahmi, I. (2009). "Experimental evaluation of influence factors for single geocell-reinforced sand." Submitted for possible presentation and publication at the TRB 88th Annual Meeting, January 11 to 15, Washington, DC.

Potyondy, D. O., and Cundall, P. A. (2004). "A bonded-particle model for rock." *International Journal of Rock Mechanics and Mining Sciences, 41*(8 SPECISS), 1329-1364.

Wayne, M. H., Han, J., and Akins, K. (1998). "The design of geosynthetic reinforced foundations." *Design and Construction of Retaining Systems, ASCE Geo-Institute Geotechnical Special Publication*, No. 76, edited by John J. Bowders et al., 1-18.

Zhang, L., and Thornton, C. (2007). "A numerical examination of the direct shear test." *Geotechnique, 57*(4), 343-354.

Analysis and Performance Monitoring of a Spread Footing Bridge Foundation

Joseph G. Bentler, P.E., M. ASCE[1], Derrick Dasenbrock, P.E., M. ASCE[2], and Megan J.L. Hoppe, E.I.T.[3]

[1]Geotechnical Engineer, American Engineering Testing, Inc., 550 Cleveland Ave N, Saint Paul, MN 55114, jbentler@amengtest.com
[2]Foundations Geomechanics Engineer, Minnesota Department of Transportation, 1400 Gervais Avenue, Maplewood, MN 55109, derrick.dasenbrock@dot.state.mn.us
[3]Staff Engineer, American Engineering Testing, Inc., 550 Cleveland Ave N, Saint Paul, MN 55114, mlee@amengtest.com

ABSTRACT

Detailed foundation analyses showed the feasibility of supporting the abutments of a two-span bridge on spread footings founded on overconsolidated glacial till. Embankments were used to preload both abutment areas prior to footing construction to reduce long-term settlements to tolerable amounts. The design-build team worked with the Minnesota Department of Transportation to develop a cooperative instrumentation program to monitor bridge performance during construction and provide data for comparison with the anticipated foundation response during the staged construction process. Instrumentation included optical survey targets, inclinometers, earth pressure cells, and tiltmeters. Monitoring data indicated overall good agreement between predicted abutment behavior and the measured response.

INTRODUCTION

A freeway bridge on the U.S. Trunk Highway (T.H.) 212 Design-Build Project in Minnesota was originally planned to be pile-supported. Foundation analysis revealed that compression of the underlying soils under the height of new abutment fill would cause significant downdrag forces at the abutments, resulting in a highly inefficient foundation design. A supplemental geotechnical investigation showed that spread footing foundations were feasible at the abutments.

However, the size of the abutments and approach embankments raised concerns over possible excessive differential settlement between spread-footing-supported abutments and the pile-supported center pier. Of similar concern was the potential for small foundation rotations to cause large movements at the bridge deck. Preload embankments were placed in both abutment areas prior to footing construction in an effort to reduce long-term settlements. Additional stress and settlement analyses were performed to model anticipated foundation response during the staged construction process. The design-build team, including Zumbro River Constructors (ZRC), HDR, Inc., and American Engineering Testing, Inc. (AET) worked with the Minnesota Department of Transportation (Mn/DOT) to develop an instrumentation program to provide data for comparison with the anticipated foundation response.

Bridge Description

The bridge, located about 40 km southwest of downtown Minneapolis, has two 47-m spans, one over Carver County Road (C.R.) 110 and one over Chaska Creek. The bridge is commonly referred to as the C.R. 110 Bridge. The abutments and center piers are cast-in-place, reinforced concrete, and the girders are 2.06-m deep prestressed concrete beams. There are two parallel cast-in-place decks, each carrying two lanes of the newly constructed T.H. 212. The abutment walls and footings are continuous across the width of the bridge. The walls are up to 11.6 m tall, and the footings are 7.3 m wide (Fig. 1). Approach embankment fills are up to 15 m.

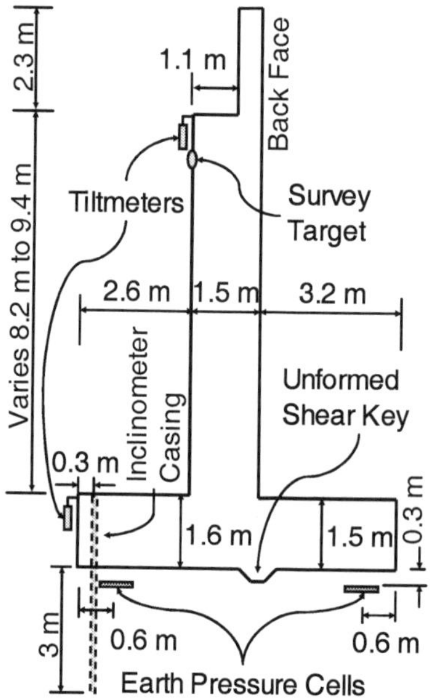

FIG. 1. Cross Section at abutment center with instrumentation.

Subsurface Investigation

Eight standard penetration test soil borings extending up to 24 m below the surface had been performed to estimate pile lengths for the bridge. To evaluate spread footing foundations, AET performed four seismic cone penetration test (CPT) soundings to depths of up to 21 m. A pore water pressure dissipation test was also performed. Soils at the abutments consisted of soft topsoil and alluvial clay to depths of about 2 to 4 m, over firm to hard sandy lean clay and clayey sand glacial till deposits interbedded with medium dense to very dense sands and gravels. The till and granular deposits became increasingly competent with depth, and measured shear wave velocities were in excess of 200 m/s. The cohesive till had an overconsolidation ratio (OCR) of at least 3 based on correlations with shear wave velocities (Mayne et al. 2001).

FOUNDATION ANALYSES

Elastic Settlement Analysis

Foundations settlements were initially calculated by elastic theory using the small strain shear modulus, G_0, determined for different soil layers from the measured shear wave velocities. Small strain Young's moduli were calculated assuming Poisson's ratio of 0.2 and then reduced using methodology given in Mayne et al. (2001). The computed elastic foundation settlement due to the design footing bearing pressures of 290 kPa was about 45 mm.

To reduce post-construction settlement to the required 25 mm, AET recommended preloading the abutment areas for an estimated duration of four months. Because the foundation soils were already significantly overconsolidated, the primary goal was not to increase the OCR, but rather to start the settlement expected under the final load condition, thereby reducing post-construction settlement of the bridge. Ultimate settlements of about 70 mm had been calculated for the preloading; therefore, a minimum of 45 mm of settlement was considered necessary prior to preload removal.

Preload and Stress Analysis

The design-build process gave ZRC the flexibility to place the foundation preload prior to completion of the bridge design. To maintain the project schedule, ZRC placed the preload embankments in December 2006. The original intent of the preload was to fill the approach embankments to full height at least 1.5 m in front of the planned abutment wall. However, the actual preload generally did not include full-height construction of the approach embankments behind the abutments, although it extended farther in front than required. In part because of the changed extents of the preload embankments, Mn/DOT requested additional analyses to show that bridge foundation performance would meet project requirements. Mn/DOT also had concerns that relatively small rotations of the foundations might cause large movements at the bridge deck due to the height of the abutments.

In February 2007, AET performed detailed foundation analyses for each abutment. In order to calculate both immediate settlements of the granular soils and time-dependent consolidation settlements, the geotechnical software FoSSA 2.0 (developed by ADAMA Engineering, Inc.) was employed. Stress computations utilized Boussinesq stress theory and considered simplified three-dimensional embankments; in addition, the software allowed the analysis of staged loading. Soils below depths of 24 m were assumed to be incompressible because of the hard tills and very dense sands and gravels encountered at that depth. Contact stresses below the abutment footings were assumed to be trapezoidal with the greatest pressure below the heel, as discussed in Bentler and Labuz (2006). Stresses were generally evaluated along a vertical profile located at the center of the abutment footings.

Below depths of 1.0 to 1.5 times the footing width (i.e. 7 to 11 m), the increase in vertical stress from the abutment foundation load was shown to be indistinguishable from the approach embankment loading. The stress increase at that depth was comparable to the expected stress distribution below a sloped embankment rather than a retaining wall; therefore, differential settlements were expected to be gradual. Analysis also showed that the calculated increase in vertical stress for the as-built preload embankments was slightly larger than that for the planned preloads (Fig. 2).

This was a direct result of the additional material placed at the front and sides of the embankments. It was also evident that the stresses from the preload would be greater than the final stresses from the abutment foundations and final embankments.

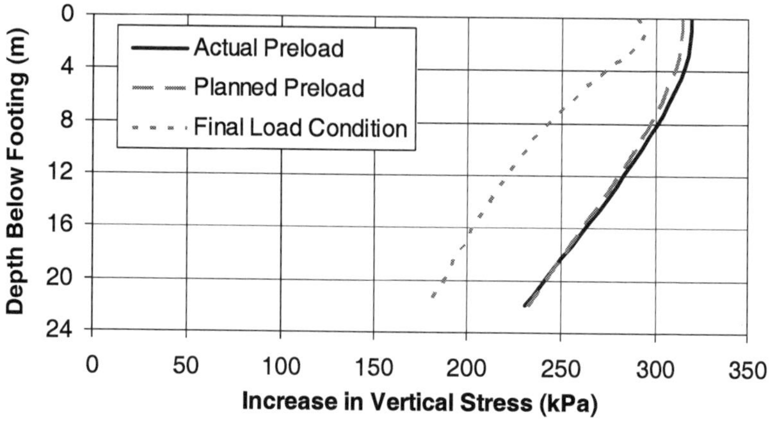

FIG. 2. Calculated increase in vertical stress below bridge abutment.

Time Rate of Settlement

To predict stresses and settlements at various stages during the construction process, AET incorporated laboratory consolidation test data from samples of the glacial till soils at a nearby spread-footing-supported bridge (T.H. 41 over T.H. 212). The glacial till soils encountered at the T.H. 41 site, which is about 2.4 km away from the C.R. 110 Bridge, had similar N-values, CPT tip resistances, and shear wave velocities. Coefficients of consolidation from the T.H. 41 laboratory tests agreed with the CPT pore water pressure dissipation test results at C.R. 110, which indicated values of about 0.001 to 0.002 cm^2/s. Settlement magnitudes for the C.R. 110 Bridge estimated using the nearby T.H. 41 Bridge's consolidation test data compared well with the elastic settlements estimated using only the seismic CPT data from C.R. 110.

Predicted Abutment Behavior

Estimated total ultimate settlements were about 45 mm, including settlement that would occur if the approach embankments and bridge were immediately constructed. However, due to the preload embankments and the incremental construction process, post-construction settlements were expected to be considerably less. After removal of the preload to begin footing construction, some rebound of the already overconsolidated soils was expected to occur. The rebound would theoretically continue until abutment construction reached the point at which the foundations were sufficiently loaded to induce settlement. As such, there was some risk that an excessive delay during abutment construction could have resulted in enough rebound that post-construction settlements would be problematic.

For the assumed distribution of contact stresses below the foundation, calculations indicated the differential settlement between the foundation toe and heel at the center of the abutment would be 20 mm. For the 7.3-m wide footing, this equates to 0.15 degrees rotation, which, for the overall 13-m tall abutment, would cause the top of the stem to move into the backfill 34 mm. However, if the stem were to move into the backfill, the horizontal earth pressure on the stem would increase, thereby increasing the overturning moment on the wall, resulting in increased stress below the footing toe and decreased stress below the heel, until equilibrium is reached. Hence, expected differential settlement between the footing toe and heel was less than 20 mm.

INSTRUMENTATION PLAN

The design-build team worked with Mn/DOT to develop an instrumentation program to monitor settlement, lateral movement, and rotation. These could be monitored with sufficient accuracy to assess abutment performance using inclinometers and a total station survey, so the design-build team was responsible for installation, monitoring, and data analysis of this quality control (QC) instrumentation. To provide independent quality assurance (QA) and more frequent readings, Mn/DOT installed earth pressure cells (EPCs) and tiltmeters, each with automated single-channel data loggers reading each sensor every 15 minutes.

After excavation for the footings, AET installed inclinometer casing to depths of about 3 m below each footing, at the approximate center and 0.3 m back from the front edge of the footing (Fig. 1). The casing was installed in a borehole drilled with hand augers, which limited their depth, but the casing was sufficiently deep such that the bottom would be stationary, allowing for measurement of footing rotation and translation. After the abutment walls were constructed, the contractor also mounted three reflective survey targets at the ends and center of each abutment wall, just below the girder seat (Fig. 1). The positions of the survey targets were monitored using a total station with an accuracy of about 3 mm.

Mn/DOT installed four 350 kPa vibrating wire EPCs, one below the toe and heel of each abutment footing at the abutment center (Fig. 1). Nests for the EPCs were excavated about 0.3 meters below the foundation elevation. The EPCs were bedded in and surrounded by hand-placed sand. Because the main goal of the EPCs was to make relative comparison of stresses below the footing heel and toe, a calibration of the EPCs in soil was not performed prior to installation. Rather, the EPC data was normalized based on assumed loadings from soil cover and the concrete footing.

Two vibrating wire tiltmeters were installed on each of the abutments, in the same cross section as the EPCs (Fig. 1). They were installed on the front face of the footing toe and on the front face of the abutment stem near the girder seats. The tiltmeters on the footings were enclosed in protective PVC cases. The intent of the tiltmeters was to assess differences in rotation between the footing and stem, and to provide an independent measurement of rotation for comparison with the inclinometer data.

MONITORING RESULTS

Settlements

Measured wall settlement was 18 to 28 mm from late fall 2007, after completion of abutment construction, through June 2008, when the bridge deck was cast. Figure 3 depicts the average abutment settlements, computed by averaging measured settlements from the three survey targets on each abutment. It is evident that settlements occurred quickly during backfilling. Based on the shape of the curves, post-construction settlements are expected to be minimal.

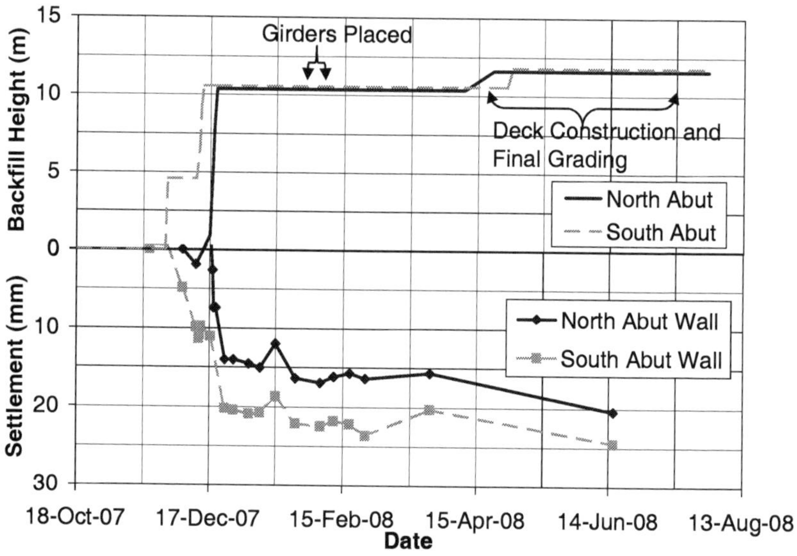

FIG. 3. Settlement of abutment walls after start of backfilling.

Footing Rotation

A maximum rotation of about 0.09 degrees into the backfill was measured by the inclinometers in mid-winter 2008; the tiltmeters independently recorded rotation of 0.08 degrees. This rotation would correspond to a movement of 20 mm at the bridge joints, 13 m above the base of the footings, if no elastic stem deflection occurred. However, tiltmeters near the top of each abutment showed less backward rotation of the stem, indicating at least some forward elastic deflection of the stem. It appears that construction on the bridge deck in spring 2008 caused relative rotation away from the backfill such that the measured footing rotation in July 2008 was 0.06 to 0.07 degrees into the backfill.

Horizontal Movement of Footing

Horizontal movement of the abutment footings was relatively minor until construction activities resumed in spring 2008 (Fig. 4). Since that time, an additional 2 to 3 mm of movement has occurred. Periodic inclinometer readings will continue to determine if the movement stops after construction is complete.

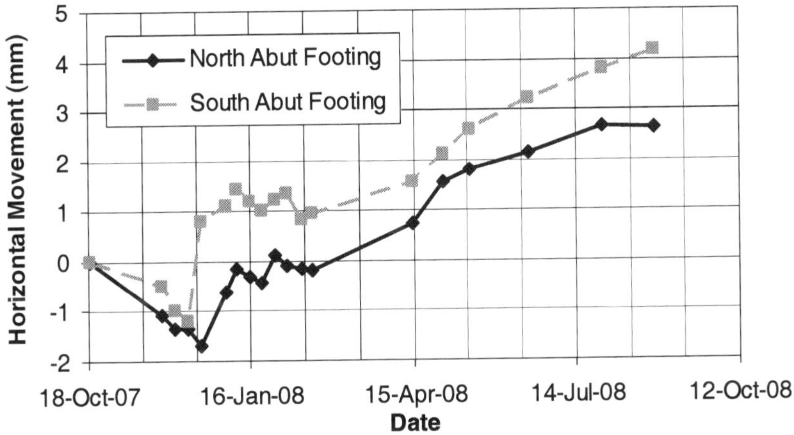

FIG. 4. Horizontal movement of abutment footings indicated by inclinometers through footings (positive movement is away from backfill).

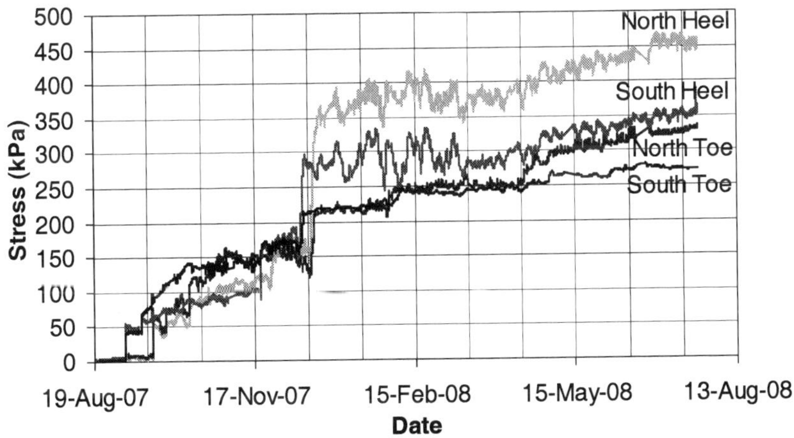

FIG. 5. Pressures measured by EPCs below the abutments.

Footing Contact Stress

The EPCs clearly showed stress increases associated with the construction process (Fig. 5). In addition, heel stresses were higher than toe stresses. The gross foundation design stresses were 290 kPa; the normalized EPC data suggest stresses in this range, although the error in EPC readings may be as much as 30 percent. The EPC data also indicated transient behavior not directly correlated to construction activities; this was likely due to temperature effects and moisture-related changes in backfill weight.

CONCLUSIONS

Preload embankments and spread footings provided an economical foundation system for the bridge. Detailed stress and settlement analyses were performed to predict foundation behavior. A combined QC/QA instrumentation program indicated reasonable agreement between predicted abutment performance and actual abutment movements, which were within required tolerances. Inclinometers and tiltmeters independently recorded about 0.1 degree of footing rotation into the embankment during backfilling. Tiltmeters also showed an overall slight backward rotation of the wall stem, resulting from the backward rotation of the footing and forward elastic deflection of the wall. The EPCs measured higher stresses under the foundation heel, corroborating the tiltmeter and inclinometer data, which indicated footing rotation into the backfill. This same phenomenon was observed on a previous instrumentation project for Mn/DOT (Bentler and Labuz 2006). Based on the rate of the measured settlements, future post-construction settlements are expected to be small.

The instrumentation not only provided information on the performance of the foundations but also justification for further use of spread footings for highway bridge support on Mn/DOT projects. The successful instrumentation program required considerable planning, as well as cooperation from the contractor. The design-build environment can be well suited to facilitate the close interaction required among designers, contractors, and agencies or owners.

ACKNOWLEDGEMENTS

The authors thank AET and Mn/DOT for direct support. The cooperation of ZRC and HDR is also gratefully acknowledged. The findings and conclusions presented in this paper are those of the authors and do not necessarily represent other parties.

REFERENCES

Bentler, J.G. and Labuz, J.F. (2006). "Performance of a cantilever retaining wall." *Journal of Geotechnical and Geoenvironmental Engineering*. 132(8): 1062-1070.

Mayne, P.W., Christopher, B.R., and DeJong, J. (2001). *Subsurface Investigations (Geotechnical Site Characterization)*. Rep. No. FHWA-NHI-01-031, Federal Highway Administration, Washington, D.C.

Effects of Footing Shape on Bearing Capacity of Rectangular Footings

Jing-Pei Li[1], Le-Yi Chen[2] and Fa-Yun Liang[3]

[1]Professor, Department of Geotechnical Engineering, Tongji University, Shanghai, 200092, P. R. China; lijp2773@mail.tongji.edu.cn
[2]Ph.D. Candidate, Department of Geotechnical Engineering, Tongji University, Shanghai, 200092, P. R. China; geotechbrian@gmail.com
[3]Associate Professor, Department of Geotechnical Engineering, Tongji University, Shanghai, 200092, P. R. China; fyliang@mail.tongji.edu.cn

ABSTRACT: Bearing capacity of rectangular footings was simulated with three-dimensional finite difference program. In the analysis, soil is modeled using Mohr-Coulomb model. Shape effects on bearing capacity of rectangular footing are evaluated without using the conventional superposition. To avoid error induced by superposition, cohesion and soil unit weight are normalized as one parameter. Numerical results show that bearing capacity of rectangular footing decreased with the increase of footing aspect ratio. The effects of footing shape are also different with the characteristics of soil. In general, effects of footing shape are more obvious for soils with high cohesion, while they are slight on soils with low cohesion.

INTRODUCTION

Ultimate bearing capacity may be estimated by various techniques, such as limit equilibrium (Terzaghi 1943; Meyerhof 1951), method of characteristics (Sokolovskii 1965; Bolton and Lau 1993), limit analysis (Shied 1954; Chen 1975; Michalowski 1997), numerical methods (Griffiths 1982; Frydman and Burd 1997). The above mentioned theories that are mainly based on plane strain conditions are only appropriate for footings with high aspect ratio (length to width), such as strip foundation in nature. However, many shallow foundations have a lower aspect ratio with a rectangular or square footprint. There is no exact solution available for the bearing capacity of a rectangular footing, even a simple case such as a surface footing resting on an isotropic or homogeneous deposit obeying an undrained failure criterion (Gourvenec, 2006).

For plane strain conditions, a slip plane develops only adjacent to the long side of the footing, while for three dimensional footing, slip plane usually developed around the footing, so the bearing capacity of a rectangular footing should be higher than a strip footing. Conventional bearing capacity theories for rectangular footings typically introduce some empirical modifications to the strip footing solutions to consider effects of footing shape. It can be expressed with Eq. (1).

$$p_u = cN_c s_c + qN_q s_q + \frac{1}{2}\gamma B N_\gamma s_\gamma \tag{1}$$

where the bearing capacity factors N_c, N_q and N_γ represent the effects of soil cohesion c, surcharge q, and soil unit weight γ, respectively; s_c, s_q and s_γ are shape factors corresponding to N_c, N_q and N_γ, respectively; and B is the width of the strip footing.

In order to obtain a rigorous solution of the above problem, the bearing capacity of the rectangular footing was simulated with a three-dimensional finite difference program in this paper. Computational results are presented in the form of dimensionless bearing capacity depending on a dimensionless parameter, soil weight parameter, and aspect ratio. Failure mechanism characteristics in limit states are also presented.

REVIEW ON BEARING CAPACITY OF RECTANGULAR FOOTING

Before the numerical analysis, a brief review is given to illustrate the available research as follows.

The first rigorous upper bound solution to a square and rectangular punch indentation of a perfectly plastic and purely cohesive material was presented by Shield and Drucker (1953); however, the solution is applicable only to clays in undrained conditions. Since three-dimensional (3D) mechanisms in frictional soils are very complex, there was no attempt in the following several decades until Michalowski (2001). Upper bound solutions of limit analysis require making an estimate of the shape and geometry of the most critical failure surface. In most cases, the selection of the failure mechanism strongly influences the quality of the solution. The deformation pattern is explicitly constrained in upper bound analyses, so the failure mechanisms can hardly be optimized. The solutions based on mechanisms proposed by Shield and Drucker (1953) and Michalowski (2001) are likely to overestimate the ultimate bearing capacity by a large margin. 3D rigorous upper- and lower-bound analyses of rectangular rough footings in clay and sand have been performed, optimizing by nonlinear programming (Salgado 2004; Lyamin 2007). Since the analyses of Salgado *et al.*(2004) and Lyamin *et al.* (2007) are more refined and the upper-bound and lower-bound results are quite close, the quality of the results are much better than those given by Shield and Drucker (1953) and Michalowski (2001).

There were also some attempts to estimate the bearing capacity of rectangular footings employing numerical techniques, such as FEM (Zhu and Michalowski 2005; Gourvenec et al. 2006) which did not explicitly constrain the pattern of deformation.

Different from upper-bound analyses, numerical techniques need not make hypotheses nor search for the position of the slip surface; therefore numerical techniques are easier and more efficient than the upper-bound method. The comparison between limit analysis and numerical analysis presented by Michalowski (2002) indicates that bearing capacity for square footings obtained by the latter method was significantly lower than that from the former approach.

Most of these methods for bearing capacity of rectangular footings present results in the form of Eq. (1). Terzaghi's assumption of superposition, although widely used, is still questionable since soil behavior in the plastic range is non-linear. The theoretical justification for using superposition has been investigated by Davis and Booker (1971). Michalowski's (1997) limit analysis results show a dependence of all coefficients on dimensionless parameters $c/\gamma B$ and $q/\gamma B$ in addition to φ. Once the soil weight is considered, the Prandtl failure mechanism is no longer capable of yielding exact results. Therefore, N_c and N_q in Eq. (1) are not strictly valid (Chen, 1975). Chen and Michalowski's work suggests that the use of superposition is not rigorous and leads to conservative estimates of the ultimate bearing capacity.

For a rough rectangular footing on surface of soil, dimensionless bearing capacity q_u/c depends on internal friction φ, dimensionless soil weight parameter $G=\gamma B/2c$ and shape parameter L/B (Chen, 1975). For smaller G, the material is equivalent to a soil with high cohesion; and for larger G, bearing capacity depends mainly on soil weight rather than cohesion. To avoid the error induced by superposition, the results were presented in form of dimensionless bearing capacity q_u/c, which depends on internal friction φ, dimensionless soil weight parameter G and shape parameter L/B rather than shape factors based on strip footing.

THREE-DIMENSIONAL ANALYSIS MODEL

Numerical simulations were carried out with $FLAC^{3D}$ (Itasca Consulting Group 1997). The program is very suitable for analyzing nonlinear behavior of materials and related instability and failure phenomena.

In numerical analyses, the bearing capacity was obtained by the swipe test, which was introduced by Tan (1990). In small-strain numerical analyses, once ultimate bearing capacity is reached, there is no further increase in bearing capacity with increased displacement, hence the plastic stiffness at failure is zero and there is no expansion of the failure locus.

Because of symmetry, only a quarter of the model is discretised (Fig. 1). The mesh extends $10B$ from edges of the footing and $7B$ beneath the footing. The vertical boundaries and the base are fixed in all three coordinate directions. The displacements of symmetry planes are fixed in the normal direction. All computations are performed for a rough soil-footing interface. The $FLAC^{3D}$ mesh is composed of eight-node brick elements, subdivided into two overlapping sets of constant-strain tetrahedral elements. Volumetric strain rates are averaged over each set of tetrahedrons to avoid the over constraint problems common in FEM plasticity formulations (Marti and Cundall 1982).

The elastic-perfectly plastic Mohr-Coulomb yield criterion was used in the numerical simulations. Young's modulus $E = 2 \times 10^5$ kPa and Poisson ratio $v = 0.3$

were assumed. (To ensure the small-strain, the Young's modulus E used in numerical simulation might be higher than that for real soils.) The dimensionless soil weight parameter varied from 0.1 to 100. Five three-dimensional models were created to investigate the shape effect on bearing capacity for $L/B = 1$ (square), 1.5, 2, 3 and 5.

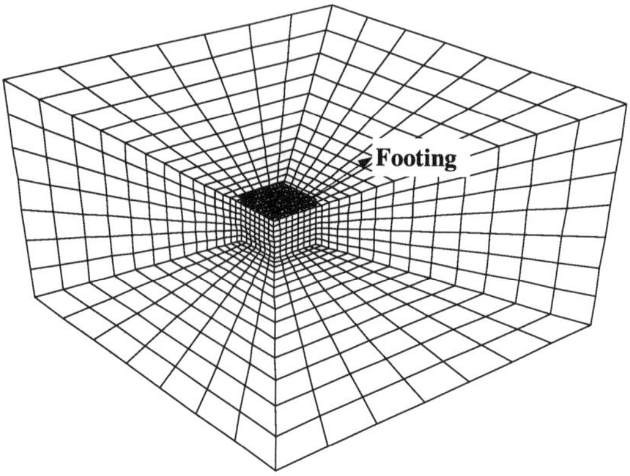

FIG. 1. The meshing of computational model (quarter symmetry)

RESULTS AND DISCUSSION

As soil unit weight is normalized as $G = \gamma B/2c$, the contributions of c and γ to bearing capacity were replaced by parameter G. The normalization avoided the error introduced by superposition. G also indicates the type of soil. A smaller G means the material has higher cohesion, while a bigger G means the material has lower cohesion.

Taking the example of $\varphi = 30°$, the normalized vertical bearing capacity was calculated for a range of footing aspect ratios from 1 to 5 and the normalized soil weight parameter from 0.1 to 100, and the results are shown in Fig. 2. In general, it was found that lower aspect ratio leads to higher bearing capacity for the same soil weight parameter G. It is also noticed that the extent of shape effects on bearing capacity vary significantly with G. The shape effect decreases with increasing G. For $G = 0.1$, the bearing capacity with $L/B = 1$ is 37.4% larger than that for $L/B = 5$; while for $G = 100$, the variation of bearing capacity for these different shapes was not more than 2%.

The results indicate that footing shape affects the bearing capacity of soil with high cohesion significantly and affects that of soil with low cohesion only slightly. The difference can be explained by the influence range of the two soils. Fig. 3 shows the influence range of footings in the two soils. Note that the influence range of the footing in the soil with $G = 0.1$ is much larger than that in the soil with $G = 100$. It is reasonable to conclude that the shape of the footing has a greater effect on bearing capacity where there is a large influence range (high cohesion and high G).

Conversely, where there is little influence range (low cohesion and high G), the shape of the footing has minimal effect. Put another way, the bearing capacity depends on the work of failure mechanism. For soil with high cohesion, the work around the footing takes a big proportion in bearing capacity. Therefore, shape of footing affects bearing capacity significantly. The range of failure mechanism in soil with low cohesion is not as large as highly cohesive soil, so the work of failure mechanisms is affected only slightly by footing shape. Generally speaking, a larger influence range leads to greater shape effect. The phenomena were also confirmed by Salgado *et al.* (2004), Zhu and Michalowski (2005) and Lyamin *et al.* (2007). Their works show that s_c is affected by shape significantly and s_γ reach to 1 for any shapes.

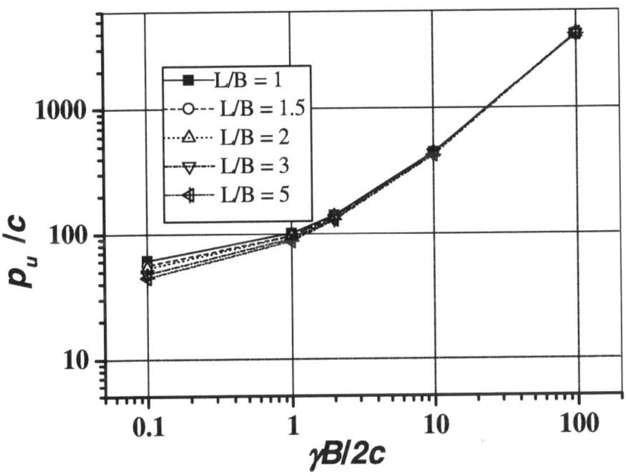

FIG. 2. Bearing capacity for rectangular footing of various shapes (φ = 30°)

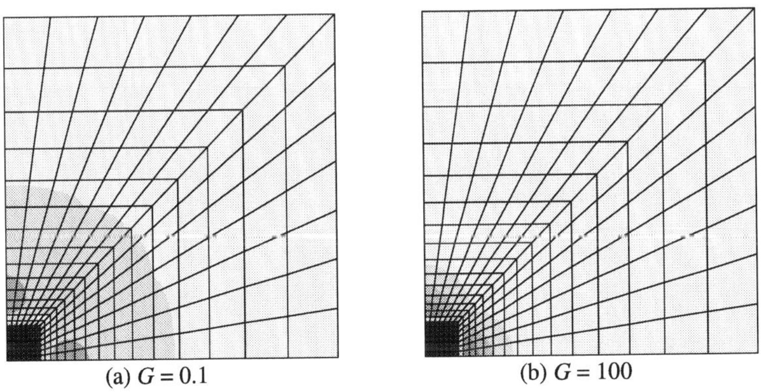

FIG. 3. Plan view of footing's influence range

CONCLUSIONS

In this paper, shape effects on bearing capacity of rectangular footing were investigated by three-dimensional numerical analysis. Soil weight, width of footing and soil cohesion are normalized as $G = \gamma B/2c$ to eliminate error induced by superposition. In general, bearing capacity of rectangular footings decreased with an increase in aspect ratio of the footing. The results show that the footing shape affects bearing capacity of soil with high cohesion obviously and has no significant effect on soil with low cohesion. In summary sensitivity of shape effect on bearing capacity can be expressed by influencing range.

ACKNOWLEDGMENTS

The work reported herein was supported by the Excellent Young Faculty Cultivation Foundation at Tongji University and Shanghai Education Committee of China (Grant No. TJU-07018). The authors wish to express their gratitude for the above financial support.

REFERENCES

Bolton M.D. and Lau C.K. 1993. "Vertical bearing capacity factors for circular and strip footings on Mohr-Coulomb soil." *Can. Geotech. J.*, Vol. 30(6): 1024-1033.

Chen, W. F. (1975). *Limit analysis and soil plasticity*. Elsevier, New York, N.Y.

Davis E.H. and Brooker J.R. (1971). The bearing capacity of strip footings from the standpoint of plasticity theory. *Proc. 1st Australian-New Zealand Conf. on Geomechanics*, Melbourne: 275-282

De Beer, E.E. (1970). "Experimental determination of the shape factors and the bearing capacity factors of sand." *Géotechnique*, Vol. 20: 387-411.

Frydman, S. and Burd, H.J. (1997). Numerical studies of bearing-capacity factor N_γ. *J. Geotech. Geoenviron. Eng.*, Vol. 123(1): 20-29.

Gourvenec, S., Randolph, M., and Kingsnorth O. (2006). Undrained bearing capacity of square and rectangular footings. *Int. J. Geomech.*, Vol. 6(3): 147-157.

Griffiths, D.V. (1982). Computation of bearing capacity factors using finite elements. *Géotechnique*, Vol. 32(3): 195-202.

Itasca Consulting Group. (2002). FLAC^{3D}, *Fast lagrangian analysis of continua in 3 dimensions, user's manual*, Minneapolis, Minnesota, USA.

Levin, E. (1955). "Indentation pressure of a smooth circular punch". *Q. Appl. Math.*, Vol. 13(2): 133–137.

Lyamin, A.V., Salgado, R., Sloan S. W., and Prezzi M. (2007). "Two- and three-dimensional bearing capacity of footing in sand." *Géotechnique*, Vol. 57(8):647-662.

Marti, J. and Cundall, P. (1982). "Mixed discretization procedure for accurate modelling of plastic collapse." *Int. J. Numer. Analy. Meth. Geomech.*, Vol. 6: 129-139.

Meyerhof, G.G. (1951). The ultimate bearing capacity of foundations. *Géotechnique*,

Vol. 2 (4), 301-332

Michalowski, R.L. (1997). "An estimate of the influence of soil weight on bearing capacity using limit analysis." *Soils Found.*, Vol. 37(4): 57-64.

Michalowski, R.L. (2001). "Upper-bound load estimates on square an rectangular footings." *Géotechnique*, Vol. 51(9): 787-798.

Michalowski, R. L. and Dawson E. M. (2002). "Three-dimensional analysis of limit loads on Mohr-Coulomb soil." *Foundations of Civil and Environmental Engineering*, Poznan University of Technology Press, Poland, Vol. 1:137-147.

Salgado, R., Lyamin, A.V., Sloan, S. W., and Yu, H. S. (2004). "Two- and three-dimensional bearing capacity of foundations in clay." *Géotechnique*, Vol. 54(5): 297-306.

Shield, R.T. (1954). "Plastic potential theory and Prandtl bearing capacity solution." *J. Appl. Mech.*, Vol. 21(2), 193-194.

Shield, R.T., and Drucker, D. C. (1953). "The application of limit analsis to punch-indentation problems." *J. Appl. Mech.*, Vol. 20: 453-460.

Sokolovskii, V.V. (1965). *Statics of Soil Media*. Pergamon Press.

Tan, F.S. (1990). "Centrifuge and theoretical modeling of conical footings on sand". PhD thesis, Cambridge University, UK.

Latest Highway Bridge Spread Footing Research Findings in Ohio

Teruhisa Masada[1], A.M. ASCE and Shad Sargand[2], M. ASCE

[1]Professor, Civil Engineering Department, Ohio University, Athens, OH 45701-2979; masada@bobcat.ent.ohiou.edu
[2]Russ Professor, Civil Engineering Department, Ohio University, Athens, OH 45701-2979; ssargand@bobcat.ent.ohiou.edu

ABSTRACT: Shallow foundations are more economical than deep foundations and are utilized to support many different types of structures in our society. However, one type of structure to which shallow foundations have not seen widespread applications in Ohio and elsewhere is highway bridges. This may be because most bridge engineers question shallow foundation's ability to maintain adequate load-carrying capacity in all environments. The research team at Ohio University continued its study of shallow foundations at highway bridge construction sites in Ohio. At one of these sites, A-2-4 and A-3a soil types were encountered in a borehole. The central pier footing was instrumented with sensors and monitored during construction. The field instrumentations consisted of multiple sensors and stations for recording contact soil pressure under the footing, settlement of the footing, and tilting of pier columns tied to the footing. A USGS quality bench mark was incorporated to establish a solid permanent bench mark at the site. The field performance data collected in the study provided further insights into how contact soil pressure, footing settlement, and columns/wall tilting were correlated with each other throughout various construction stages. The study also produced outcome on general reliability of a settlement prediction methods outlined in the AASHTO LRFD Design Specifications.

INTRODUCTION

Shallow foundations are typically at least 30% more economical than deep foundations and can be utilized to support many different civil engineering structures. However, shallow foundations have not been utilized often for supporting highway bridge structures. This is because bridge engineers generally believe that bridges supported on shallow foundations tend to develop settlement/scour problems and become high maintenance items in their infrastructure inventory. In order to dispel this common negative notion regarding shallow foundations among bridge engineers, their satisfactory field performance must be demonstrated through well planned and documented field case studies.

The authors have been conducting spread footing studies in Ohio for over fifteen years. Outcome of their previous research efforts on this subject was presented previously in journal publications (Sargand et al. 1999, 2003). The latest study, which was completed a few years ago for the Ohio Department of Transportation (ODOT) and the Federal Highway Administration (FHWA), had the following major objectives:

- Instrument and monitor field performance of spread footing foundations at two new highway bridge construction sites in Ohio; and
- Evaluate the reliability of the spread footing performance prediction methods outlined in the AASHTO LRFD Bridge Design Specifications.

LATEST FIELD CASE STUDY

Project/Site Descriptions

As part of the latest study, one spread footing foundation was instrumented with sensors and monitored throughout construction stages at a major highway bridge construction site in Ohio. At this site, a two-span bridge structure FRA-670-0380 was erected to allow crossing of High Street over Interstate Highway I-670 in Columbus, Franklin County, Ohio. Table 1 summarizes basic specifications of the bridge.

Subsurface exploration work was conducted at the site, using a truck-mounted rig with a safety hammer. A-2-4 and A-3a soil types were encountered in a borehole. Table 2 summarizes the soil boring log data. Neither groundwater table nor bedrock was encountered in the borehole.

Table 1. Basic Information for Highway Bridge FRA-670-0380

Bridge Spans	North Span: 102.9 ft (31.4 m) South Span: 100.2 ft (30.5 m)
Bridge Deck Width	78 ft (23.77 m)
Footing Width (B) & Length (L)	8.0 ft (2.44 m) & 40.0 ft (12.19 m)
Footing Thickness	3.0 ft (0.91 m)
Embedment Depth	5.0 ft (1.52 m)
Max. Design Bearing Pressure	2.0 tsf (192 kPa)

Table 2. Subsurface Exploration Data

Depth Below Bottom of Footing (ft)	Soil Descriptions	Ave. SPT-N (uncorrected)
0	Gray sandy silt w/ trace gravel (A-4a)	51
5.0 (1.5 m)	(same as above) (A-4a)	65
10.0 (3.0 m)	(same as above) (A-4a)	73
15.0 (4.6 m)	Gray gravel, some sand, and little silt (A-2-4)	63
20.0 (6.1 m)	(same as above) (A-2-4)	84
25.0 (7.6 m)	(same as above) (A-2-4)	70

Field Instrumentations Plan

The spread footing structure received multiple vibrating-wire pressure cells (for recording contact pressure), several survey monument points (for detecting vertical displacement or settlement), and one tilt-meter measurement station. The following sections describe each type of field instrumentations.

Instrumentations for Contact Pressure

Earth pressure cells (Model 4800E) were purchased from Geokon Inc. (New Lebannon, NH) and had a full scale of 100 psi (689 kPa) with resolution of 0.1 psi (689 Pa). Each pressure cell was attached to a small concrete block prior to the footing construction and placed sensitive face down against a sand bedding layer within the footing construction area. This arrangement was necessary to keep the pressure cell from being disturbed during concreting of the footing, to ensure that the pressure cells become integrated into the shallow foundation structure, and to minimize undesirable bridging effects on their pressure readings. Obviously, all the pressure cells are located over the bottom surface of the footing. A minimum of five (5) sensors was needed to gain insights into contact pressure distribution changes during construction.

Following any site visit, soil pressure was computed from the field readings coming from each cell using the following formula:

$$P = G(R_0 - R_i) + K(T_i - T_0) \qquad (1)$$

where P = soil pressure (psi); G = pressure calibration constant (psi/digit); R_0 ,R_i = initial, subsequent transducer frequency reading (digit); K = temperature calibration constant (psi/°C rise); and T_0, T_i = initial, subsequent transducer temperature (°C).

Instrumentations for Settlement

The footing received a total of five (5) settlement monitoring points installed over the top surface of the footing to detect both average and differential settlement movements. The locations of these points matched the locations of earth pressure cells underneath the footing for developing correlations between the soil pressure and settlement data. A 6-inch (152-mm) diameter PVC pipe coupler was embedded into concrete at each point, so that a solid PVC riser pipe could be installed to provide continued access to the point even after backfilling. During each visit, conventional optical surveying technique was employed to measure elevations of the points with respect to a USGS quality bench mark establish at the site.

Instrumentations for Pier Wall/Columns Rotation

A tilt measurement station was set up on each footing column. An accelerometer-based tilt-meter from Slope Indicator Inc. (Seattle, WA) was used at each tilt station to obtain tilt measurements taking place along the direction of the footing width. In any field visit, the wall/column tilting angle θ were computed using the field data through:

$$\theta = \sin^{-1}\left[\frac{(+R)-(-R)}{2}\right] \qquad (2)$$

where θ = angle of tilt (from true vertical; in degrees); +R = positive side accelerometer reading (digit); and –R = negative side accelerometer reading (digit).

Construction History

Table 3 summarizes the construction history data. It took a total of 151 days for the bridge to officially open to general traffic.

Table 3. Construction History Data

Stage No.	Stage Descriptions	Days Elapsed	Notes
0	Excavation of footing const. area	0	---
1	Concreting of footing	8	75 yd^3 (57 m^3) of concrete poured
2, 4	Concreting of (pier columns + cap)	21	32 yd^3 (24 m^3) of concrete poured
3	Backfilling over footing	33	Soil cover thickness 2.0 ft (0.6 m)
5	Barrier wall	57	10 yd^3 (7.5 m^3) of concrete poured
6	Girder & cross beams placement	84	400 kips (181,820 kg) of steel
7	Bridge deck construction	117	269 yd^3 (204 m^3) of concrete poured

Contact Pressure & Footing Settlement

A visit to the site was made to collect all sensor/instrument readings, typically one week from the time the footing underwent each major construction stage. Table 4 summarizes the average field measured contact pressure values at different stages of construction for the spread footing. The average contact pressure peaked at 2.85 tsf (273 kPa). Figure 1 presents three-dimensional views of the contact pressure distribution over the footing bottom face. This plot was made by locking the southwest corner of the footing at the coordinate origin and rotating the footing 90° increments to show the field measurements from various angles. The contact pressure distribution started out relatively uniformly during the early stages of construction but became progressively more uneven after Stage 4 (backfilling). The soil pressure mound developed a definite tilt from the north to the south. Stage 7 (bridge deck construction) induced the largest increase in the contact pressure, followed by Stage 6 (placement of beams).

Table 4. Summary of Field Measurements of Contact Pressure

Stage No.	Stage Description	Average Contact Pressure (tsf)	
		Cumulative	Increase
1	Footing construction	0.33 (31.6 kPa)	0.33 (31.6 kPa)
2	Pier columns construction	0.44 (42.1 kPa)	0.11 (10.5 kPa)
3	Backfilling over footing		
4	Pier cap construction	0.76 (72.8 kPa)	0.32 (30.6 kPa)
5	Barrier wall construction	0.96 (91.9 kPa)	0.20 (19.2 kPa)
6	Placement of girder & cross beams	1.45 (138.8 kPa)	0.49 (46.9 kPa)
7	Bridge deck construction	2.85 (272.9 kPa)	1.40 (134.1 kPa)
8	Bridge open to traffic	2.85 (272.9 kPa)	0.00

Table 5 summarizes the average footing settlement values detected at different stages of construction. The maximum average settlement was measured to be only 0.20 inches (5.1 mm). Figure 2 presents three-dimensional views of the footing settlement behavior. Examinations of this plot show that the footing tilted increasingly more downward toward the north side as the construction progressed. The maximum degree of tilt was close to 0.4%. Based on this table, Stage 7 (bridge deck construction) produced the largest increase in settlement, followed by Stage 2 (pier columns construction).

Table 5. Summary of Field Measurements of Footing Settlement

Stage No.	Stage Description	Average Contact Pressure (inches)	
		Cumulative	Increase
2	Pier columns construction	0.06 (1.5 mm)	0.06 (1.5 mm)
3	Backfilling over footing		
4	Pier cap construction	0.09 (2.3 mm)	0.03 (0.8 mm)
5	Barrier wall construction	0.05 (1.3 mm)	-0.04 (-1.0 mm)
6	Placement of girder & cross beams	0.11 (2.8 mm)	0.06 (1.5 mm)
7	Bridge deck construction	0.20 (5.1 mm)	0.09 (2.3 mm)

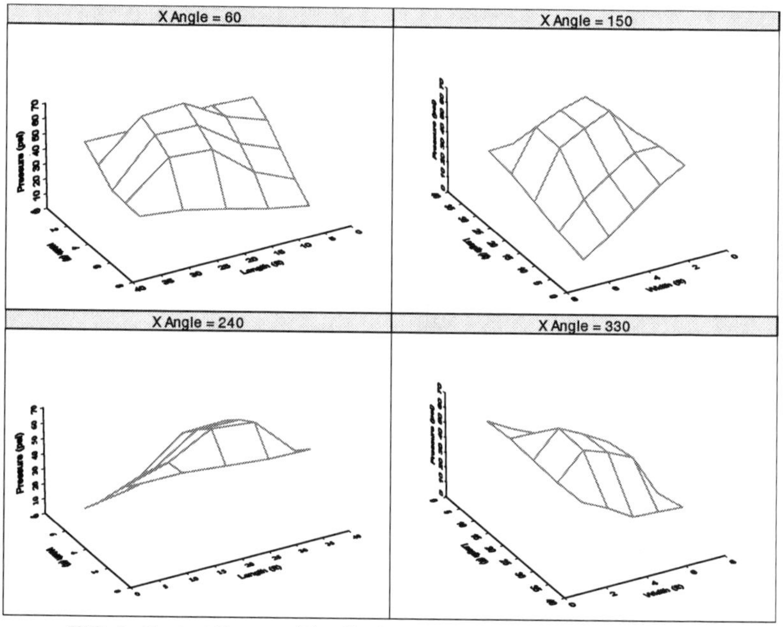

FIG. 1. Contact Pressure Distribution (Stage 8: Bridge Open to Live Loads)

Column Tilting

Two spans of unequal length met at the Central Pier footing at the FRA-670-0380 site. The north span was 102.9 ft (31.4 m), and the south span was 100.2 ft (30.5 m). This condition generated net overturning moment of 1,962 kip-ft (2,660 kN-m). By selecting Poisson's ratio of 0.3 and a soil modulus ranging from 340 to 588 tsf (33 to 56 MPa), the titling angle was estimated to vary between 0.16 and 0.28 degrees. This range complied well with the field measurements, which showed that both of the monitored columns inclined slightly (0.13° for east column, 0.24° for west column) toward the north.

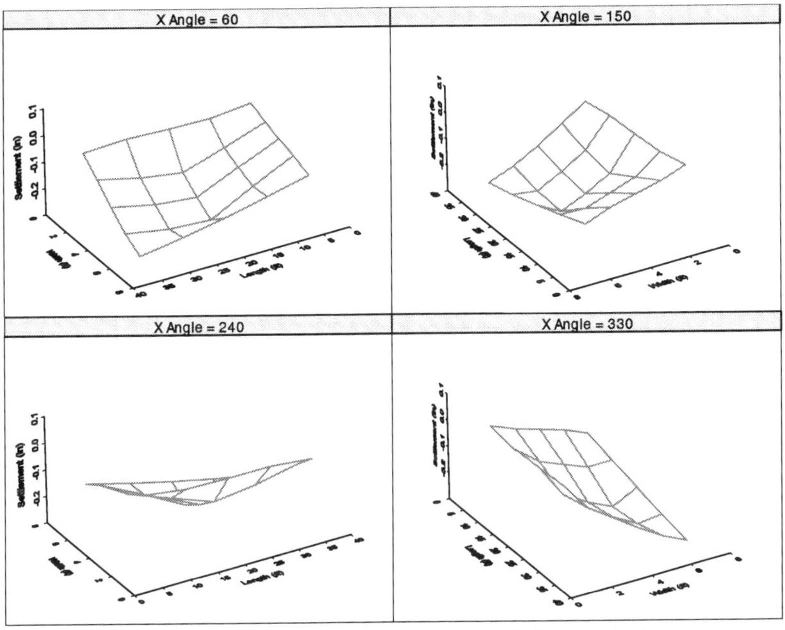

FIG. 2. Settlement (Stage 7: Bridge Deck Construction)

Correlations Between Field Performance Data

In general, the settlement data and column tilting data agreed with each other. Both the footing and columns tilted slightly toward the north, which indicates that the footing and columns acted as one rigid body. This tilting behavior also suggests higher contact pressure development on the north side. However, soil pressure measurements taken in the field were not compatible with the other performance measurements. This outcome may imply that settlement and column tilting data normally reflect global behavior of the footing structure while each soil pressure cell reading tends to be influenced not only by the global rigid body movement/rotation of the footing under the applied load but also by the stiffness of the bearing soil that exists underneath.

Reliability of Selected Footing Design/Analysis Methods

The American Association of State Highway & Transportation Officials (AASHTO) has been setting national and international standards for the design of bridges since 1931. In 1987, the AASHTO subcommittee reassessed U.S. bridge specifications and reviewed foreign design specifications and codes. A recommendation was made to adjust the conventional working stress design (WSD) so that the variability in the loadings and the material properties are reflected. With this new philosophy, the Load and Resistance Factor Design (LRFD) was implemented on the basis of recent developments in structural engineering and statistical methods. Section 10.6 of AASHTO LRFD Bridge Design Specifications document (2006) presented two methods for estimating the settlement performance of spread footings resting on cohesionless soils. The first method is based on the analysis of an elastic half space, and its key equation is given by Eq. 3:

$$S_e = \frac{q_o\sqrt{A}}{E_s\beta_z}\left(1-\upsilon_s^2\right) \tag{3}$$

where q_o = applied vertical stress; A = footprint area of footing (L·B); L = footing length; E_s = elastic modulus of soil; β_z = rigidity factor; and υ_s = Poisson's ratio of soil.

Guidelines on the rigidity factor, Young's modulus, and Poisson's ratio can be found in the works of the U.S. Department of the Navy (1982) and Bowles (1988).

The second method is based on the work of Hough (1959), which can provide immediate settlement of a footing on cohesionless soils. The Hough method calculates the settlement by:

$$S_e = \sum_{i=1}^{n}\Delta H_i = \sum_{i=1}^{n} H_c \frac{1}{C'}\log\left(\frac{\sigma'_o+\Delta\sigma_v}{\sigma'_o}\right) \tag{4}$$

where ΔH_i = elastic settlement of layer; n = number of soil layers within zone of influence; H_c = initial layer thickness; C′ = bearing capacity index; σ'_o = initial vertical effective stress; and $\Delta\sigma_v$ = increase in effective stress.

In the Hough method, the zone of influence is considered to extend to the depth equivalent to three times the footing width. When the soil in the influence zone is subdivided, each soil layer should be a maximum of 10.0 ft (3.0 m) thick. The correlation between the blow counts and the bearing capacity index C′ values is available and is based on the original results obtained by Hough (1959) and the modifications made by Cheney and Chassie (2000).

For each prediction method, a few variations in the value of the soil modulus (E_s or C′) chosen resulted in a range of settlement behavior. Table 6 shows the results. For the elastic method, the case with the maximum soil modulus value produced better outcome. For the Hough method, the case with the higher C′ value also produced better settlement predictions. Both methods shared a tendency to underpredict the actual settlement during the early stages and overpredict the field performance during the later stages. This implies that the soil modulus increased in the field under higher loads. The settlement prediction methods specified by the AASHTO LRFD design guidelines are reasonable, but further refinements may be needed.

SUMMARY AND CONCLUSIONS

The researchers at Ohio University continued their comprehensive study of spread footing foundations in Ohio. In their newest research project, funded by the Ohio Department of Transportation (ODOT) and the Federal Highway Administration (FHWA), one spread footing foundation existing at a major highway bridge construction site in Ohio was instrumented and monitored. Subsurface soil conditions at the site were favorable to the use of shallow foundations. This was evidenced by the fact that uncorrected SPT-N values are all well above 50 blows/ft. The spread footing field performance data were then compared with results provided by the geotechnical methods outlined in the recent AASHTO LRFD Bridge Design Specifications to assess general reliability of the LRFD design guidelines. The analysis of the footings at the site showed that they can be regarded as rigid structures. Also, the results of the current study indicated that the methods presented in the AASHTO LRFD Design Specifications (2006) were generally acceptable. According to a quick review of the most current version of the AASHTO LRFD Specifications (2008), descriptions of the elastic settlement prediction method and the Hough method have not been modified at all.

Table 6. Comparison Between Measured and Estimated Average Settlement Performance

Const. Stage No.	Stage Description	(Estimated/Measured) Settlement Ratio for:			
		Elastic Method w/E_s =:		Hough Method w/ C′ =:	
		$(E_s)_{min}$	$(E_s)_{max}$	164	236
2	Pier Columns Construction	0.82	0.33	0.46	0.32
3	Soil Backfilling	NA	NA	NA	NA
4	Pier Cap Construction	1.67	0.91	0.90	0.62
5	Barrier Wall Construction	5.69	2.15	2.09	1.45
6	Girder Beams Placement	4.62	1.75	1.53	1.06
7	Deck Construction	5.09	1.94	1.51	1.05
Average		3.58	1.42	1.30	0.90

[Note] $(E_s)_{min}$ = 250 tsf = 23.9 MPa, and $(E_s)_{max}$ = 650 tsf = 62.1 MPa. "NA" = Not Available.

REFERENCES

American Association of State Highway and Transportation Officials (2006, 2008), *AASHTO LRFD Design Specifications*, Section 10.6 – Spread Footing Foundations. Washington, D.C.

Bowles, J. E. (1988), *Foundation Analysis and Design*, 4th Edition, McGraw-Hill Book Co., New York, NY, 1004 pp.

Cheney, R. S. and Chassies, R. G. (2000), "Soils and Foundation Workshop Manual." National Highway Institute Publication NHI-00-045, Federal Highway Administration, Washington, D.C.

Hough, B. K. (1959), "Compressibility as the Basis for Soil Bearing Value." Journal of Soil Mechanics and Foundations Division, Vol. 85, No. SM4, American Society of Civil Engineers (ASCE), pp. 11-39.

Sargand, S. M., Masada, T., and Abdalla, B. (2003), "Evaluation of Cone Penetration Test-Based Settlement Prediction Methods for Shallow Foundations on Cohesionless Soils at Highway Bridge Construction Sites." Journal of Geotechnical and Geoenvironmental Engineering, Vol. 129, No. 10, American Society of Civil Engineers (ASCE), Reston, VA, pp. 900-908.

Sargand, S. M., Masada, T., and Engle, R. (1999), "Spread Footing Foundation for Highway Bridge Applications." Journal of Geotechnical and Geoenvironmental Engineering, Vol. 125, No. 5, American Society of Civil Engineers (ASCE), Reston, VA, pp. 373-382.

U.S. Department of the Navy (1982), "Soil Mechanics." Design Manual 7.1 – NAVFAC DM 7.1, Naval Facilities Engineering Command, Alexandria, VA, 348 pp.

DESIGN AND PERMITTING OF A FGD LANDFILL OVER AN EXISTING ASH RESERVOIR

Daniel G. Bodine[1], P.E., M.ASCE, Burak F. Tanyu[2], Ph.D.,A.M.ASCE, Paul J. Sabatini[1], Ph.D., P.E.,M.ASCE, Mohammad Ajlouni[3], Ph.D., P.E., and Dana E. Limes[3]

[1]Geosyntec Consultants, 1420 Kensington Road, Suite 103, Oak Brook, IL 60523;
[2]Geosyntec Consultants, 134 N LaSalle Street, Suite 300, Chicago, IL 60602; btanyu@geosyntec.com
[3]American Electric Power, 1 Riverside Plaza, Columbus, OH 43215

ABSTRACT

A large coal fired operating plant in the Ohio River Valley had a need to develop a disposal facility for residual waste from its newly installed Flue Gas Desulfurization (FGD) system. Siting studies were conducted resulting in the selection of a site located partially over an inactive fly ash reservoir (FAR 1) which was undergoing closure and partially over a remnant mine spoil disposal site. While the selected site would allow for environmental and operational advantages, significant engineering design and construction challenges had to be addressed.

This paper provides information on the challenges that relate to the design and permitting of this residual waste landfill. Specific issues addressed include: (1) boundary issues (i.e., existing mining highwall, valley stream, and 45-m high dam); (2) potential for differential settlements of the foundation subsoils and effects on leachate collection and piping systems; (3) design of a subsurface drainage system to collect and transmit consolidation waters; and (4) impacts to landfill drainage systems resulting from liquefaction of the underlying fly ash during the design seismic event.

INTRODUCTION

The newly installed FGD systems at this plant will produce residual waste and purge stream solids. The residual material that could not be utilized in off-site commercial operations would need to be placed in a Class III residual waste landfill (RWL). Based on the need to identify a location for the RWL, a siting study was performed with an objective that the site must be located as close to the coal fired operating plant as possible and still meet the Ohio Administrative Code (OAC) siting criteria for such facilities. Based on these objectives, a site location was chosen that is approximately 2.4 kilometers north of the operating plant facility, which is located near the town of Brilliant in Wells Township, Jefferson County, Ohio. The selected location is within a valley that is bound by a mining highwall on the southwest, a stream at the base of another highwall on the northeast, and a 45-m high earthen dam on the southeast (FAD 1). This valley is filled with 24 to 44 m of very loose, saturated fly ash (i.e., FAR 1) and an adjacent area with up to approximately 14-m thick mine spoil (Fig. 1).

This site location satisfied OAC siting criteria, however, it created several design and permitting challenges. The purpose of this paper is to present some of these challenges and the proposed solutions that were acceptable to Ohio Environmental Protection Agency (EPA).

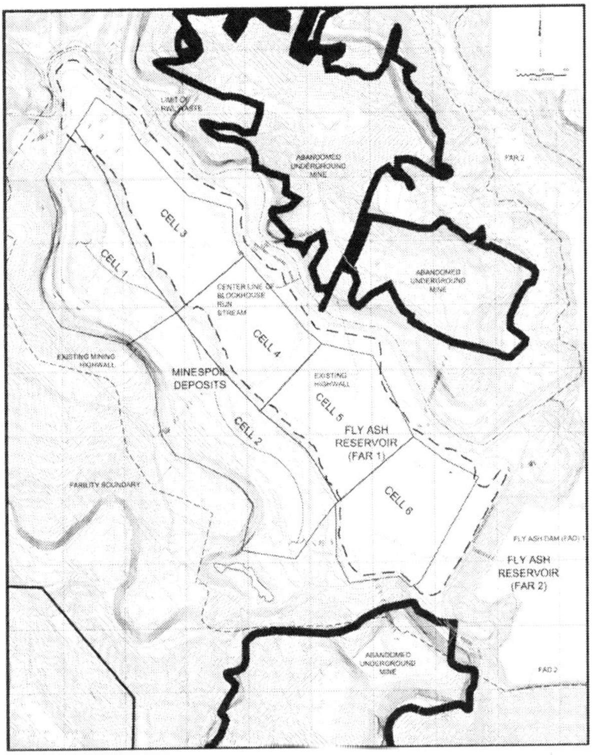

FIG. 1. Location of the Selected RWL Site.

LICENSING REQUIREMENTS

In the State of Ohio, a RWL must be permitted, designed and constructed in accordance with OAC 3745-30-05, OAC 3745-30-06 and OAC 3745-30-07. These rules present the permit application submittal requirements and design and construction requirements, including calculation requirements for the liner and cover systems, static and seismic stability analyses, liquefaction analyses, settlement analyses and leachate and surface water management systems. Relevant criteria related to the design considerations discussed herein include: (1) post-settlement grades of leachate collection systems should be greater than 0.5 percent; (2) site foundation soils should have a demonstrated factor of safety against liquefaction of at least 1.0; and (3) calculated factors of safety against hydrostatic uplift should be at least 1.2.

OVERVIEW OF THE DESIGN

To accommodate the anticipated yearly production rate, the footprint of the RWL was selected to be 51.5 hectares with a maximum permitted thickness (or height) of residual waste of approximately 60-m high with 3H:1V side slopes. The generalized section depicting the RWL components is shown in Fig. 2. The contouring fill layer ranged from 1.2 to 3.6-m thick and served as a foundation for the isolation layer and RWL liner system. The top of the contouring

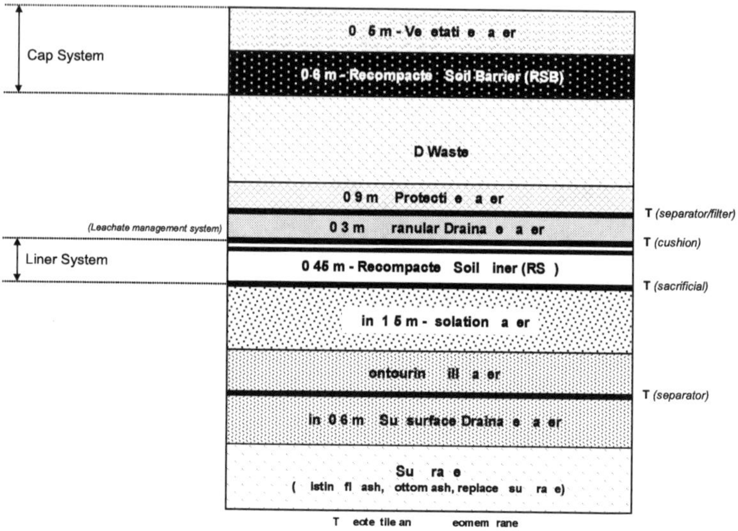

FIG. 2. Generalized section of the RWL.

fill layer was graded at 3% to develop a gravity-drained leachate drainage system for each cell.

The sequential development of the RWL was permitted with construction to occur in two primary phases: Phase 1 (Cells 1 and 2) would be developed along the excavated minespoil bench area, located southwest of FAR 1 and Phase 2 (Cells 3, 4, 5, and 6) developed over the FAR 1 existing ash (Fig. 1). During Phase 1, the existing minespoil deposits in Cells 1 and 2 will be excavated and replaced with a minespoil subgrade fill (comprising the same material) compacted in maximum 0.3-m thick loose lifts to criteria specified in the quality assurance/quality control (QA/QC) Plan. This criteria state that a method specification could be developed and approved after permit approval based on results of an onsite test fill. Also, as part of Phase 1 construction, the southwest highwall will be sloped and benched by blasting and face cutting such that a highwall liner system can be constructed and tied into the base liner system.

The highwall liner system comprises a 0.6-m thick (minimum) highwall drainage layer immediately adjacent to the highwall followed by a 1.5-m thick isolation layer, a 0.9-m thick recompacted soil liner (RSL), and an overlapping double sided geocomposite drainage layer. Constructing adjacent to the highwall increased site development costs but resulted in greater storage volumes for nearly all RWL cells and possible future vertical expansion.

Phase 2 will be developed over FAR 1 and will proceed from the northwest to southeast. The phasing schedule allows for a period of preloading with available soil fill over the FAR 1 ash to consolidate and strengthen the fly ash prior to the construction of Phase 2 cell liner and leachate management systems. Phase 2 site development will take place during the operation of Phase 1.

TECHNICAL AND LICENSING CHALLENGES ASSOCIATED WITH DESIGN

Boundary Issues

In addition to the highwall boundary issues discussed in the previous section, the limits of the

RWL would need to be away from underground mines and abandoned oil and gas wells. The location of the RWL within the valley did not leave sufficient room for leachate or surface water sedimentation basins. Since there was already an existing active fly ash reservoir (FAR 2), it was decided to utilize this reservoir for both RWL leachate and surface water treatment. This required a permit variance because the FAR 2 reservoir is not lined. This variance was approved, principally because the RWL leachate will be subject to pre-treatment prior to discharge to FAR 2.

Along the sloping portion of the northwest end of the RWL (Fig. 1), a zone of shallow cemented fly ash had been previously placed and capped. This zone was left in-place and required no preloading. A variance was not required and the northwest end of the RWL is to be constructed over top using the appropriate isolation layer and liner system.

The last boundary condition consisted of minimizing effects of the RWL on the existing FAD 1 (Fig. 1). An approximate 60-m setback from the crest on the dam was used to minimize potential effects of future waste loading on the current stability of FAD 1.

Settlement of Fly Ash within FAR 1

The placement of over 60 m of residual waste over the FAR 1 fly ash was anticipated to result in significant compression of the very loose and saturated existing fly ash (Fig. 3). Due to the geometry of the existing valley, it was anticipated that differential settlements of the fly ash would result in unacceptable tensile strains in the RSL and geomembrane components of the liner system and, more importantly, the potential for grade reversals on the pipes comprising the leachate collection system. Regulations required a post-settlement minimum grade of leachate collection pipes to be no less than 0.5 percent. Typically, this design requirement is addressed via conventional settlement computations using compression parameters developed for site specific soils.

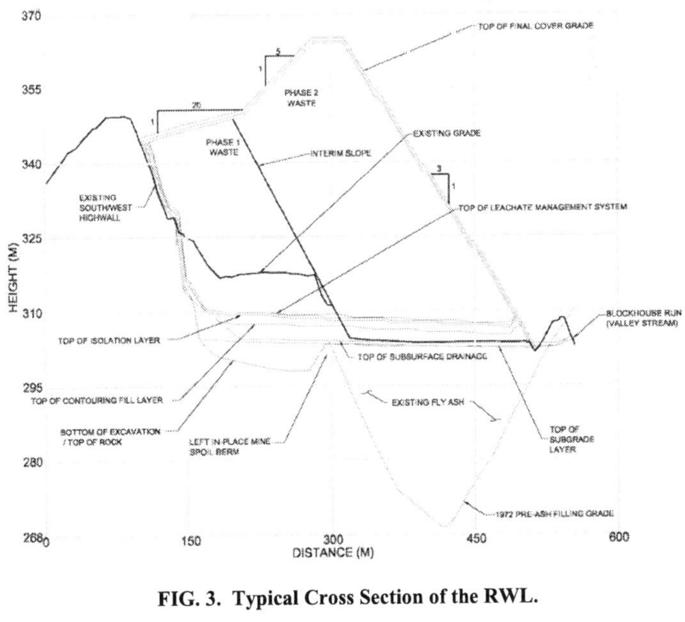

FIG. 3. Typical Cross Section of the RWL.

As previously introduced, to reduce settlements of the liner system, each landfill cell above FAR 1 is scheduled to be preloaded. Preliminary calculations indicated that a preload fill of approximately 9 m would be practical to construct and could result in a reduction of total settlement of the underlying fly ash of up to 0.6 m when subject to loads from the full height of the landfill. The preload would then be removed to construct the liner system prior to the placement of waste. The preload removal would pre-consolidate the in situ fly ash before landfill waste loads were applied thus reducing total and differential settlements.

Site-specific fly ash settlement parameters were developed from two laboratory testing programs. One program comprised a series of conventional one-dimensional incremental load (IL) consolidation tests on Shelby tube fly ash samples. Additional constant rate of strain (CRS) consolidation tests were also performed comprising consolidation tests performed in triaxial chambers wherein fly ash samples were mixed with distilled water and the resulting slurry was "poured" into the chamber to replicate original field placement. Testing loads up to 1,530 kPa were used to be consistent with anticipated loads from the full height of waste in the landfill. Compression parameters for the in situ fly ash from both test programs were nearly the same (Fig. 4).

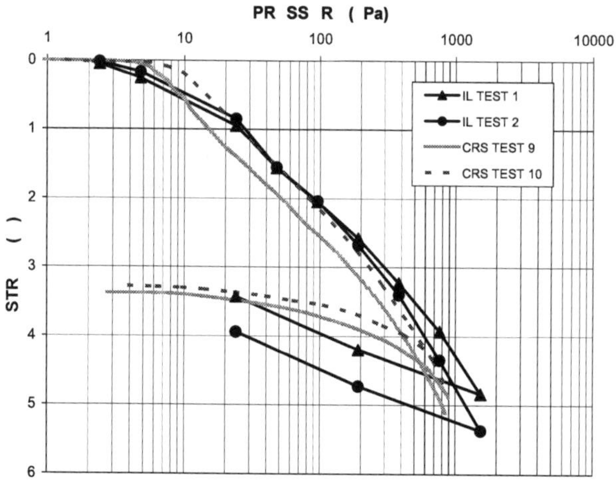

FIG. 4. Laboratory Consolidation Data for Fly Ash Samples.

Settlement computations were performed based on the parameters from the laboratory testing programs described above indicating computed maximum settlements due to final waste configuration loading on the order of 0.6 to 0.9 m. Due to the concern that the laboratory test results may not be sufficiently representative, a monitored preload fill (MPF) was built and instrumented. The MPF area was approximately 230 m by 150 m. Fill was placed to a height of 9 m and built with a combination of fly ash and minespoil material. A complete description of the MPF for this project is provided in Haydar et al., 2008.

The settlement and pore-water pressure response of the in-place fly ash was monitored using a suite of geotechnical instrumentation including surface settlement plates, a vertical extensometer, a settlement profiler, and pore-water pressure transducers. Settlement measurements were taken during the MPF construction and at the completion of each lift (Fig. 5). Monitoring results showed that measured settlements using settlement plates and the vertical extensometer were closely comparable and indicated a total vertical settlement of approximately 0.85 m for the total MPF height (corresponding to a total vertical stress of approximately 180 kPa). This measured settlement, coupled with previous design calculations based on laboratory testing data, confirmed that anticipated post-settlement leachate collection piping grades would be acceptable (i.e., remain greater than 0.5 percent) provided design grades for the leachate collection pipes ranged from 1.0 to 1.5 percent. Pore-water pressure dissipated quickly, approximately one week after completion of an incremental lift of fill.

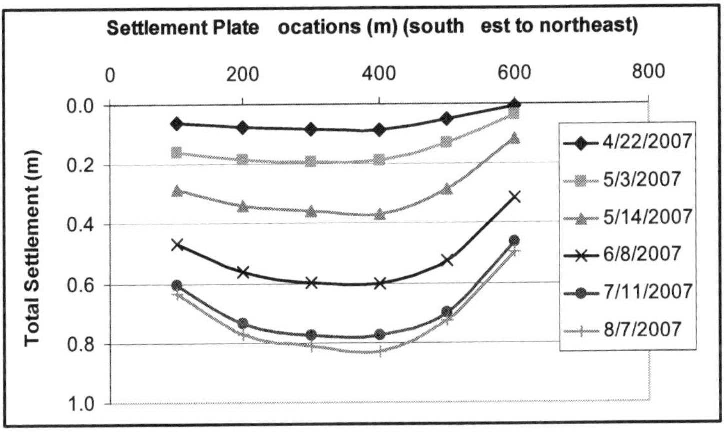

FIG. 5. Total Settlement Data Recorded from Settlement Plates.

Design of Subsurface Drainage System

As described in the previous section, portions of the landfill that will be constructed over FAR 1 will be preloaded. A subsurface drainage layer would be required to provide drainage for consolidation waters that may be generated within the saturated FAR 1 ash resulting from preloading and future waste loading. In particular, there was concern that unacceptable uplift pressures would develop just below the isolation layer (during preload removal) resulting in factors of safety less than 1.2, hence the need for a robust subsurface drainage system.

The subsurface drainage system consists of a 0.6-m thick bottom ash layer with internal gravity drainage collector pipes. For design, the maximum "thickness" of consolidation water was estimated and this water thickness was compared to the elevation of the bottom of the clayey soil isolation layer. A factor of safety against hydrostatic uplift of the isolation layer was computed.

The consolidation water thickness was determined based on the equation by Giroud and Houlihan (1995) as given below:

$$\frac{T_{max}}{L} = \frac{\sqrt{\tan^2\beta + q_i/k} - \tan\beta}{2\cos\beta} \qquad (1)$$

where: T_{max} = Maximum head due (or consolidation water thickness) above existing fly ash;

L = Half of the horizontal distance between two drainage collector pipes located within the subsurface drainage layer ;

β = Slope of the drainage layer;
q_i = Water impingement rate; and
k = Hydraulic conductivity of the subsurface drainage layer.

The hydraulic conductivity of the subsurface drainage layer was specified to be a minimum 1×10^{-3} cm/sec and the slope of the drainage layer was designed to be a minimum 0.5%. The drainage layer soil is expected to comprise site bottom ash with sand-sized particles. The drainage pipes were 150-mm diameter standard dimension ratio (SDR) 17, smooth, perforated, high-density polyethylene (HDPE) installed every 60 m.

The water impingement rate (q_i) was calculated based on Terzaghi's one-dimensional consolidation theory. An average rate of water impingement was computed as the ratio between the total settlement (during preloading) and the estimated time required to achieve this settlement. The maximum rate of settlement (and corresponding water impingement rate) was expected to occur during the relatively rapid placement of the preload. Data from laboratory testing and piezocone dissipation tests resulted in an assumed coefficient of vertical consolidation (C_v) of the existing FAR 1 fly ash of 4.5 m^2/day.

Calculations based on Eq. 1 indicated that water would rise approximately 1.2 m from its preconstruction elevation (which was generally at the elevation of the existing fly ash surface). Under this condition, the maximum computed water surface elevation within the subsurface drainage layer would not reach the bottom of the isolation layer. Therefore, hydrostatic uplift was not a concern and a factor of safety computation for hydrostatic uplift was deemed to not be necessary.

Liquefaction Potential of Fly Ash within FAR 1

Results of a comprehensive subsurface investigation indicated that zones of fly ash in FAR 1 are very loose, saturated, and cohesionless. As the fly ash reservoir will provide foundation support for the FGD landfill, the potential for liquefaction of the fly ash during the design earthquake had to be evaluated. Information on the liquefaction potential of fly ash is limited, especially for the relatively small design earthquake ground motions anticipated for this site, i.e., the design Peak Ground Acceleration (PGA) for this project was estimated as 0.08 g.

Robertson and Wride (1998) presented a method for evaluating liquefaction potential based on measured CPT parameters. The results of the evaluation indicated that fly ash that is between 0 and 9.5 m depth may be susceptible to liquefaction, with the exception of one CPT sounding that indicated a maximum total depth of potentially liquefaction susceptible fly ash of 17 m. While the Robertson and Wride method demonstrated that the fly ash is susceptible to liquefaction, it does not take into account whether there is opportunity for liquefaction to occur. That is, while the material may be susceptible to liquefaction, it would be overly conservative to assume that the material would liquefy without evaluating whether the design PGA was sufficiently large to cause liquefaction to occur.

A method proposed by Stark and Olson (2003) was used to evaluate the factor of safety against the triggering of liquefaction. This method was useful since it considers the effects of the earthquake ground motions and can consider the beneficial effects of ground consolidation and confinement due to the weight of the landfill. The approach is as follows:

- Conduct a slope stability back analysis of the prefailure slope geometry to estimate the static shear stress ($\tau_{driving}$) in the potentially liquefiable soil;
- Calculate the average seismic shear stress ($\tau_{ave, seismic}$) for each slice within the potentially liquefiable zone;

- Calculate the yield strength ratio ($s_{u(yield)}/\sigma_{vo}'$) of the potentially liquefiable soil; and
- Calculate the factor of safety against the triggering ($FS_{triggering}$) of liquefaction for each slice (within the zone of potentially liquefiable soil) as:

$$FS_{triggering} = (s_{u(yield)}) / (\tau_{driving} + \tau_{ave,\ seismic}) \quad (2)$$

If the slices along the critical failure surface have $FS_{triggering} \geq 1.0$, it is assumed that liquefaction is unlikely.

As part of the design, two interim waste slope conditions were considered critical: (i) 30-m high waste slope; and (ii) 15-m high waste slope. For these analyses, the liquefied fly ash was assumed to extend from approximately 1.8 m below the design groundwater elevation (i.e., the elevation of the temporal high phreatic surface) to a maximum assumed thickness of potentially liquefiable fly ash of approximately 21 m. For up to a 30-m high unbuttressed residual waste slope, $FS_{triggering}$ for all portions of the potential slip surface were greater than 1.0. Based on this, interim slopes were designed to be no higher than 30 m.

CONCLUSIONS

The permitting of a 60-m high RWL over an inactive fly ash reservoir and remnant minespoil disposal site presented significant challenges mostly associated with the relatively poor foundation provided by in place fly ash. From the outset of this project, the owner facilitated open discussion between the design team and the regulators to address specific concerns related to static and seismic stability, settlements, and effective drainage of leachate. The owner provided the necessary resources required to address site-specific technical challenges related to the lack of data on: (1) compressibility characteristics of in-place loose and saturated fly ash; (2) settlement performance from previous similar facilities constructed over fly ash; and (3) seismic performance of in-place fly ash subject to relatively modest design earthquake ground motions. This project demonstrated that perceived "fatal flaws" and complexities associated with adherence to specific regulatory criteria could be effectively addressed through sound engineering based on site-specific information and open communication with the regulatory agency. Data collected and evaluations performed as part of this project demonstrated that existing fly ash materials, while very loose and saturated, were significantly less compressible and more free draining than had previously been demonstrated or understood.

REFERENCES

Giroud, J.P. and Houlihan, M.F. (1995), "Design of Leachate Collection Layers", *Proceedings of the Fifth International Conference on the Environment*, Sardinia, Italy, pp. 613-640, October 1995.

Olson, S. M. and Stark, T. D. (2003), "Yield Strength Ratio and Liquefaction Analysis of Slopes and Embankments," *Journal of Geotechnical and Geoenvironmental Engineering*, ASCE, Vol. 129, No. 8, pp. 727-737.

Robertson, P.K and Wride, C.F. (1998), "Evaluating Cyclic Liquefaction Potential using the Cone Penetration Test," *Canadian Geotechnical Journal*, Vol. 35, pp. 442-459.

Haydar, M., Sabatini J., and Ajlouni, M. (2008), "Field-scale Instrumentation and Testing of a Monitored Preload Fill," *Proceedings of the 15th Great Lakes Geotechnical and Geoenvironmental Conference on Applications of Geotechnical Instrumentation for the Performance Evaluation of Constructed Facilities*, Carmel, Indiana, May.

Estimation of Stress History by Partial Piezocone Dissipation Tests

Beyong S. Lim[1], M. ASCE, PhD, PE, Mehmet T. Tumay[2], F. ASCE, PhD, PE, and Scott H. Slaughter[3], M. ASCE, PE

[1] Senior Project Engineer, URS corporation, 7389 Florida Boulevard, Suite 300, Baton Rouge, LA, 70806, USA; beyong_lim@urscorp.com
[2] Georgia Gulf Distinguished Professor Emeritus, Department of Civil and Environmental Engineering, Louisiana State University, 3304 CEBA, Baton Rouge, LA, 70803, USA; Visiting Professor, Department of Civil Engineering, Bogazici University, Bebek, Istanbul, Turkey; mtumay@eng.lsu.edu
[3] Principal Engineer, URS corporation, 7389 Florida Boulevard, Suite 300, Baton Rouge, LA, 70806, USA; Scott_Slaughter@urscorp.com

ABSTRACT: In a Piezocone Penetration Test (PCPT or CPTu), the probe can be stopped at a desired depth during penetration, and the excess pore pressure generated around the cone can be measured as dissipating u_t versus time, t. Using the rate of dissipation, it is possible to estimate the value of the soil deposit's coefficient of consolidation and thereby determine changes in hydraulic conductivity. Theoretically the dissipation is complete as u_t approaches in-situ equilibrium pore pressure, u_0. In fine-grained soils time required for complete dissipation or even 50% dissipation maybe very long, making the test impractical. The in-situ (equilibrium) pore pressure, u_0, can be estimated from a partial dissipation record requiring a shorter dissipation, by extrapolating the measured u_t data on an inverse time scale (i.e. u_t against 1/t). Based on this partial dissipation test, the stress history (or the status of the consolidation) of the soil deposit can be evaluated. However, no case study has been reported in open literature where the extrapolation method has been applied to estimate unknown stress history of soil deposits. In this paper, this methodology was applied to a dredge disposal area to obtain soil parameters related to stress history as well as soil deformation due to consolidation.

INTRODUCTION

As an *in situ* test, the piezocone penetration test (PCPT or CPTu) measures the soil's resistance and pore water pressure during intrusion into the ground to obtain subsurface soil characterization and the evaluation of design soil parameters. Lately, the device has been increasingly found to be versatile for evaluation of the consolidation and flow characteristics and undrained shear strength and other strength parameters for saturated fine-grained soils. To obtain the consolidation and flow characteristics, the dissipation tests by the piezocone are typically performed during the sounding. The test process requires stopping the piezocone and allowing the dissipation of the generated excess pore pressure around the cone. Using this rate of dissipation, the value of the soil deposit's coefficient of consolidation can be evaluated and thereby the change in hydraulic conductivity (i.e., permeability) is estimated. With these dissipation test results, the stress history (or the status of the consolidation) of the soil deposit can also be obtained. A dissipation test can be performed at any required depth by stopping the penetration and measuring the decay of generated pore water pressure with time.

Ideally, dissipation tests should continue until no further dissipation is observed or equilibrium pore water pressure is determined. However, it is very time consuming to dissipate completely and thus is not routinely performed. In practice, most dissipation tests have been typically incomplete and stopped before achieving equilibrium pore water pressure. With frequency, the dissipation tests are recommended to be continued to a degree of dissipation (U) of greater than 50 percent. Some researchers have estimated the *in situ* hydrostatic pore pressure; u_0, by using this partial dissipation record (Whittle et al., 2000). They have adopted an inverse time scale (1/t method) to extrapolate the partial dissipation data and estimated soil deposit's stress history or the degree of consolidation. In this paper, ten (10) dissipation tests in the field were analyzed to verify the effectiveness of the 1/t extrapolation method in estimating the status of consolidation. This dissipation data was obtained from tests at the site of a soil improvement project being undertaken in the south-west coastal area of South Korea.

CONCEPT OF INHERENT PORE PRESSURE IN DISSIPATION CURVE

Figure 1 shows a typical dissipation curve prior to normalization. The dynamic pore pressures measured by a piezocone consist of *in situ* hydrostatic pressure, u_0, and excess pore pressure generated by piezocone intrusion, Δu. The in situ hydrostatic pressure (u_0) can be divided into *original (or pure) in situ* hydrostatic pressure (u_w) and inherent pore pressure (u_f). The *original in situ* hydrostatic pressure (u_w) is usually obtained from unit weight of water (γ_{water}) x depth below groundwater elevation. The inherent pore pressure u_f may be defined as a residual excess pore pressure in the soil deposit due to ongoing consolidation (if underconsolidated). On the other hand, excess pore pressure generated by piezocone intrusion (Δu) can be computed from the difference between the measured total excess pore pressure (u_T) by the piezocone at any time t and *in situ* hydrostatic pore pressure (u_0). If the soil deposit is underconsolidated, an inherent pore pressure (u_f) can remain in the soil deposit even though Δu is fully depleted. The inherent pore pressure (u_f), originally imparted by changes in geologic overburden or because of recent construction loads, is separate from the excess pore pressure (Δu) generated by piezocone intrusion. Eventually, u_f will decay and converge to the *original in situ* hydrostatic pressure (u_w) with the associated consolidation process. As such, the remaining amount of u_f may be used as an indicator of the status of consolidation or the stress history of the soil deposit. Fully consolidated or overconsolidated soil deposits theoretically should have no remaining u_f.

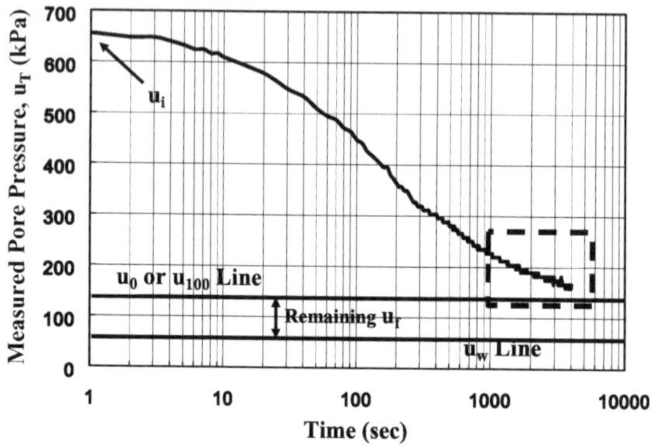

FIG. 1. Typical dissipation curve

In Figure 1, the last part of the dissipation curve, in the dotted rectangular box, approaches a straight-line relationship between dissipation pore pressure, u_T, and time. If we replot this segment of measured total excess pore pressure (u_T) versus an inverse time scale (1/t, or 1/time), the last part of the dissipation curve (i.e., the u_T versus 1/t data) typically appears as a straight line as shown Figure 2. By statistical linear extrapolation, the u_T intercept at time approaching infinity, u_{100}, (when geologic overburden or recent construction loads were activated) can be estimated.

The pore pressure intercept u_{100} can be greater than or equal to the *original in situ* hydrostatic pore pressure, u_w, depending on the status of consolidation of the soil deposit. If the soil is underconsolidated, this intercept will be greater than the *original in situ* hydrostatic pore pressure (u_w) and indicate *in situ* hydrostatic pressure (u_0), including inherent pore pressure (u_f). In the case of fully consolidated soil deposits, the intercept will be the same as the original *in situ* hydrostatic pressure ($u_w = \gamma_{water}$ x depth below groundwater elevation). In Figure 2, the method of extrapolation was performed based on a linear regression analysis, using the data belonging to the last straight segment of the inverse plotting line. The coefficient of correlation (R^2) of the linear regression analysis may be used to indicate the reliability of the linearity of the extrapolated line. Using this extrapolation method, the soil deposit's consolidation state can be estimated (i.e. underconsolidated or completely consolidated). A total of ten (10) dissipation tests from field piezocone tests were analyzed by the above-described methodology to verify the effectiveness of the 1/t extrapolation method in estimating the status of consolidation. The dissipation data was obtained from tests at the site of a soil improvement project being undertaken in the south-west coastal area of South Korea.

FIELD TESTS

All dissipation tests were performed with a 15 cm^2 cross-sectional area piezocone in field verification of the 1/t extrapolation method. The dissipation tests were undertaken in accordance with ASTM D 3441. The studied dredge disposal area is located on the south-west coast of South Korea, was to be improved to support foundations for bridges and structural facilities as part of an integrated terminal to transfer cargo from ships to land transportation. The site covers a total area of 1,400,000 m^2, and was filled by dredged material during a time period of 1970 to 1997. To verify the above described technique, soil parameters such as OCR and Liquidity Index (LI) were analyzed from soil specimens sampled from borings at the site.

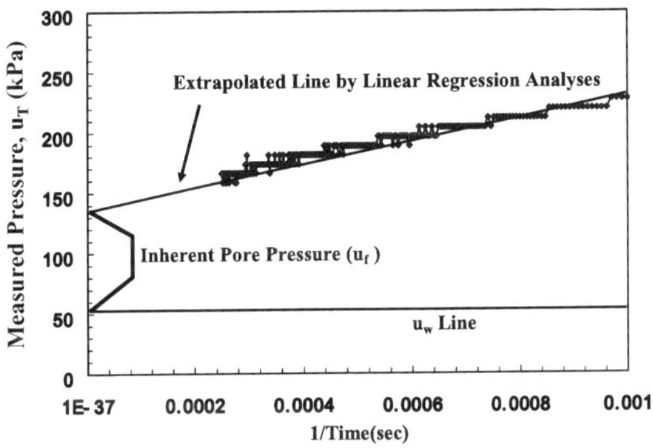

FIG. 2. Plotting inverse time scale

The site was originally a part of Gwangyang Bay. Water depths extended from 0 to -5 meters mean sea level (MSL) based on the current top surface of the natural materials. During construction of the existing Gwangyang Port, the east side hinterland area was used as the disposal site for natural clays dredged from the port. This material brought the hinterland area up to its current surface elevation of around +5 meters MSL. The dredge spoil material was contained by a sand and gravel levee which surrounds the hinterland. Based on the soil boring results, the soil conditions at the site were dominated by three zones: the six to nine meters of weak, compressible dredge spoil, which was surrounded by sand and gravel dikes; the underlying eight to fourteen meters of soft to very soft natural clay; and the more competent materials below. The major geotechnical issues in this project were stability and the settlement of the soft and compressible dredge spoil. Next in importance was the natural soft to very soft clay/silty clay, which is also relatively weak. Soil profiles from the borings indicated that the site area has basically six (6) strata, as shown in Table 1. Numerous soil specimens were tested to obtain representative engineering parameters for the design. Those parameters were obtained from historical and new consolidation tests, unconfined compression tests, triaxial tests, and water content tests. A summary of laboratory tests is provided in Table 2, using average values. Figure 3 also shows the plot of the LI and the OCR along the depth resulting from the laboratory tests. The east hinterland area was filled by 1997. There was no available data on the length of time that filling was in progress or on the elevation to which this site was originally filled. However, Figure 3 depicts that there appears to be a trend indicating the difference of stress history between the upper dredged spoil layer and the natural clay layer.

Table 1. General soil stratification for field site

Stratum No.	Stratum	Average levels	
		Top Elevation (in MSL)	Thickness (m)
1	Very Soft clay - Dredge spoil	+5	8
2	Very soft to soft clay -	-3	11
3	Natural	-14	5
4	Medium to Stiff clay	-19	2
5	Firm sand & gravel	-21	3
6	Weathered rock	-24	ND
	Rock		

ND: Bottom not determined

Table 2. Summary of soil parameters for field site

Item	Dredge spoil layer	Natural clay layer
Saturated unit weight, γ_{sat} (t/m^3)	1.5	1.6
Initial void ratio e_0	2.23	1.95
Compression index, C_c	0.78	0.78
Coefficient of consolidation, C_v (cm^2/sec)	V : 7.0 x 10^{-4} H : 1.0 x 10^{-3}	V : 7.0 x 10^{-4} H : 2.0 x 10^{-3}
Permeability, k (cm/sec)	V : 7.0 x 10^{-8} H : 1.0 x 10^{-7}	V : 7.0 x 10^{-8} H : 2.0 x 10^{-7}
Undrained shear strength, S_u (t/m^3)	0.055 x depth +0.148	0.056 x depth +0.989

Usually, the Liquidity Index (LI) provides the general trend of degree of consolidation without the impact of sampling disturbance. Generally, a LI lower than 1 means that the soil is overconsolidated, an LI equal to 1 indicates that the soil is normally consolidated, and an LI higher than 1 indicates that the soil is underconsolidated (Simons & Menzies, 1975). According to the plots, the dredged layer is between six and nine meters, followed by the natural clay layer. Figure 3 depicts that regardless of the scatter, the overall value of LI is greater than 1 in the upper dredged layer and less than 1 in the natural clay layer. Based on the LI and the OCR along the depth, the stress history of the upper dredged layer and the natural clay layer appear generally under- to normally- consolidated and slightly overconsolidated, respectively.

ANALYSES

Ten (10) dissipation tests at the field site were analyzed by the above-described technique (1/t extrapolation method). In each tested location, the dissipation test was undertaken at two different depths, which were representative for the dredge spoil layer and the natural clay layer. The groundwater level was assumed to be at the ground surface. Due to allowable page limitations for this publication, only two (2) extrapolated plots of the ten dissipation tests with u_T against 1/t are shown in Figure 4 and Figure 5. The results of all extrapolations of the ten dissipation tests are tabulated in Table 3.

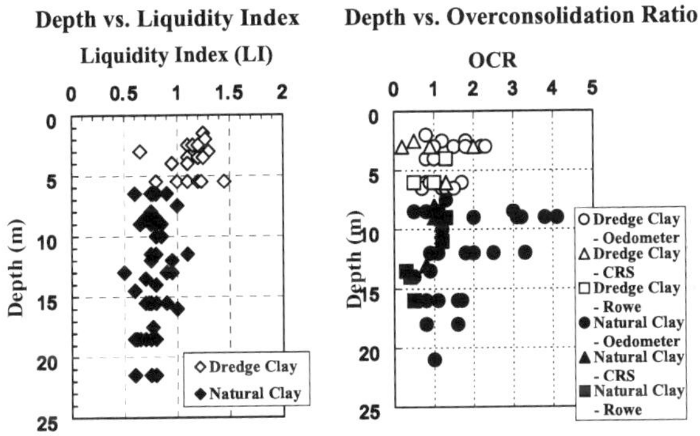

FIG. 3. LI and OCR distribution along the depth

In Figure 4 and Figure 5, dissipating u_T data points were extrapolated using linear regression to obtain the u_T intercept, u_{100}. R^2 is the coefficient of correlation for the linear regression analyses. The hydrostatic pore pressure is the equilibrium water pressure (u_0) when the dissipation of Δu approached 100 percent dissipation degree, u_{100}. In these field tests as shown Table 3, the hydrostatic pore pressures varied 19.6 to 157 kPa because the dissipation tests were performed at depths of 2.0 to 16.0 m from the ground surface. The estimated inherent pore pressures from the plots ranged from 1.4 kPa to 10.8 kPa in the dredged layer, and from -32.1 kPa to 53.9 kPa in the natural clay layer. The negative inherent pore pressures may indicate complete consolidation or possibly the overconsolidated status of the test locations.

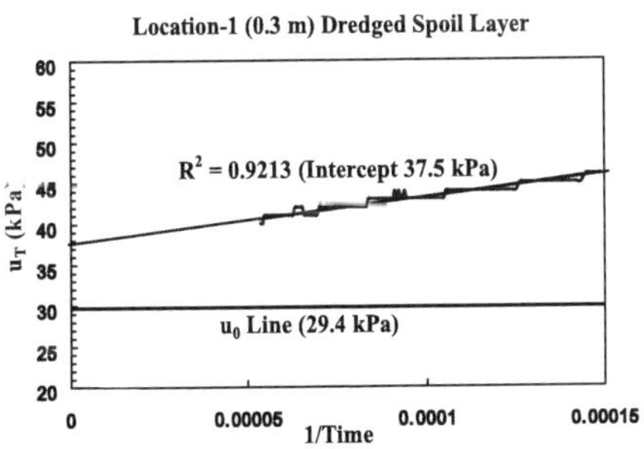

FIG. 4. Location 1 in dredged layer

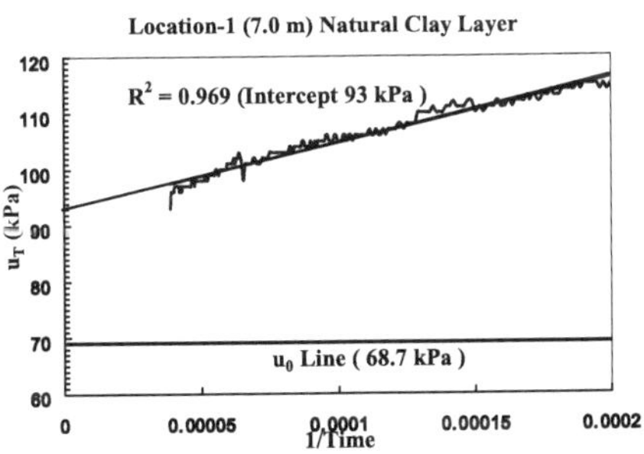

FIG. 5. Location 1 in Natural clay layer

By manipulating the u_T intercept, u_{100}, the hydrostatic pore pressures u_0, and the initial excess pore pressure at the beginning of the dissipation test u_i, the estimation of the status of consolidation may be feasible, based on the following equation:

$$U(\%) = \left\{\frac{u_i - u_{100}}{u_i - u_0}\right\} \times 100$$

where U (%) is the status of consolidation as a percentage, u_{100} is the equilibrium water pressure when 100 percent dissipation of Δu is reached, u_i is the initial excess pore pressure at the beginning of the dissipation test, and u_0 is the sum of the original hydrostatic water pressure (u_w) and inherent pore pressure (u_f). Table 4 shows a summary of the status of consolidation for the five locations where the ten dissipation tests were made.

Table 3. Summary of analyses using the 1/t extrapolation method for field test

Location	Dredged spoil layer					Natural clay layer				
	*Intercept	Hydro-static pore Pressure*	Inherent pore pressure*	R^2	Depth (M)	Intercept	Hydro-static pore Pressure*	Inherent pore pressure*	R^2	Depth (M)
1	37.5	29.4	8.1	0.92	3	93	68.7	24.3	0.97	7
2	39	28.4	10.6	0.81	2.9	203	149.1	53.9	0.99	15.2
3	50	39.2	10.8	0.99	4	66	98.1	-32.1	0.99	10
4	50	39.2	10.8	0.87	4	160	157.0	3.0	0.95	16
5	21	19.6	1.4	0.90	2	74	78.5	-4.5	0.99	8

Units = kPa.

Table 4. Summary of status of consolidation for field test

Location	Dredged spoil layer				Natural clay layer			
	U (%)	u_{100} (kPa)	u_i (kPa)	u_0 (kPa)	U (%)	u_{100} (kPa)	u_i (kPa)	u_0 (kPa)
1	70	37.5	56	29.4	74	93	163	68.7
2	62	39	56	28.4	73	203	349	149.1
3	92	50	171	39.2	110	66	411	98.1
4	72	50	78	39.2	99	160	401	157.0
5	95	21	51	19.6	103	74	211	78.5
Ave.	78	-	-	-	92	-	-	-

Based on the analyses, the dredge spoil layer was slightly underconsolidated, and the natural clay layer appeared normally consolidated or slightly overconsolidated. These trends qualitatively matched well with Figure 3. The status of consolidation is also plotted along the dissipation depth, as shown in Figure 6.

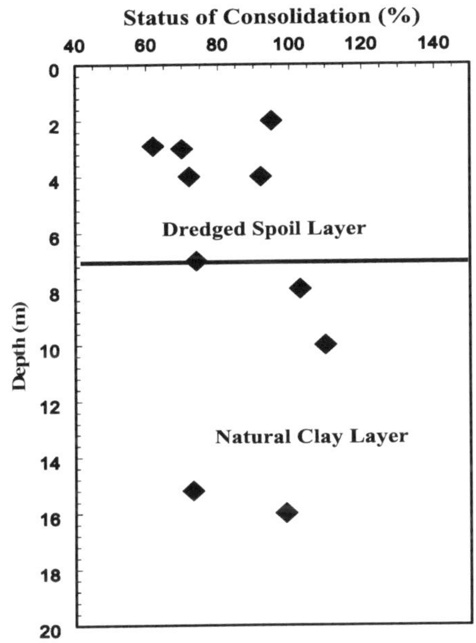

FIG. 6. Status of consolidation along the depth of dissipation test

CONCLUSIONS

In this paper, a case study to evaluate the status of consolidation, using partial piezocone dissipation tests is presented. Based on the analyses, the dredge spoil layer was underconsolidated to slightly underconsolidated; the lower natural clay layer appeared normally consolidated or slightly overconsolidated. These trends qualitatively matched well with the analysis of the LI and the OCR. These results indicate that even with partial piezocone dissipation tests, one can use the 1/t extrapolation method to reasonably estimate the status of the consolidation of the soil deposit at a site. Further verification research is recommended using calibration chamber studies and extended field tests on highly overconsolidated and underconsolidated soils.

REFERENCES

Simons, N.E.. and Menzies, B.K. (1975)."A short Course in Foundation Engineering." *Butterworth & Co.* pp 63

Whittle, A.J. and et al., 2000, "Prediction and interpretation of pore pressure dissipation for a tapered piezoprobe," *Geotechnique* Vol.51. No.7 pp 601~617.

METAL SPECIATION IN PHOSPHATE AND THERMAL STABILIZATION OF CONTAMINATED DREDGED SEDIMENTS

Peter K. Ndiba[1] and Dr. Lisa Axe[2]

(1) Civil and Environmental Engineering, New Jersey Institute of Technology, University Heights, Newark, NJ 07102, Tel.: (973) 596 6077, Fax (973) 596 5790, E-mail: pkn4@njit.edu. (2) Civil and Environmental Engineering, New Jersey Institute of Technology, University Heights, Newark, NJ 07102, Tel.: (973) 596 2477, Fax (973) 596 5790, E-mail: Axe@ADM.njit.edu.

ABSTRACT

Sediments dredged from harbors and waterways are often contaminated with metal and organic toxins, therefore, their disposal is not only expensive but unsustainable. Treatment and reuse in construction as an alternative to disposal would reduce costs and conserve resources. While thermal treatment followed by carbon adsorption theoretically addresses organics, the fate of metals is of concern. Phosphate addition followed by thermal treatment at 700 °C was investigated for immobilization of metal contaminants with X-ray absorption spectroscopy (XAS). In sediments from Belgium where Zn concentrations were elevated, Zn was precipitated as $ZnCO_3$ and adsorbed to hydrous manganese oxide and hydrous iron oxide. Phosphate and thermal treatment resulted in sparingly soluble minerals, hopeite ($Zn_3(PO_4)_2{\cdot}4H_2O$), and spinels, gahnite ($ZnAl_2O_4$) and franklinite ($ZnFe_2O_4$). Leaching assessments with the U.S. EPA toxicity characteristic leaching procedure (TCLP) confirmed Zn and other heavy metals were immobilized. Results of synthetic precipitation leaching procedure (SPLP) showed compliance with the New Jersey State impact to groundwater criteria.

INTRODUCTION

Harbor construction and maintenance requires dredging of large amounts of sediments, for example, for New York/New Jersey Harbor, 3 to 5 million m^3 annually in maintenance and improvements (Douglas et al. 2003) and 55.5 million m^3 in a harbor deepening project (Yozzo et al. 2004). Because these sediments are often contaminated with metals and organic toxins from oil spills, leaks of industrial products, and discharge of wastewater, their disposal in the ocean is restricted (Guevara-Riba et al. 2003, U.S. EPA 2004, MPRSA 2000). The alternative upland disposal is faced with increasing costs from limited placement sites (Douglas et al. 2003). Therefore, treatment techniques for reuse of the sediments, for example, in road construction and brick production, are needed.

Sediment treatment needs to address both organic and metal contaminants through verifiable physical and chemical amendments. The resulting materials should be shown to be non-hazardous; additionally, for reuse the material should comply with risk to groundwater criteria. Phosphate addition followed by thermal treatment at 700 °C and carbon adsorption theoretically addresses organics and reduces leachability of metal contaminants (Kribi at al. 2004). Phosphates minerals are used widely for immobilization of metal contaminants through precipitation of sparingly soluble phosphate phases (e.g. Crannell et al. 2000, Geysen et al. 2004). On the other hand, thermal treatment was applied to Ni-laden sludge (Shih et al. 2006) and copper contaminated surface finishing industry sludge (Lu et al. 2008). Validation of the combined phosphate addition and calcination in sediment treatment requires an understanding of metal speciation to elucidate the stabilization mechanisms. For heterogeneous systems including sediments spectroscopic techniques such as X-ray absorption spectroscopy (XAS) in combination with principal component analysis (PCA), target transformation (TT), and linear combination fit (LCF) are useful methods for assessment of trace metal speciation (Isaure et al. 2006). Although reduction in heavy metals leaching has been demonstrated with the U.S. EPA toxicity characteristic leaching procedure (TCLP) (U.S. EPA 1992) (e.g. Kribi et al. 2004), leaching from long-term weathering in beneficial reuse has not been addressed. Additionally, compliance with risk to groundwater criteria for beneficial reuse of wastes needs to be determined with the U.S. EPA synthetic precipitation leaching procedure (SPLP) (U.S. EPA 1994).

In this paper, speciation of Zn, which is abundant in sediments from Belgium, is addressed with XAS to elucidate stabilization mechanisms in phosphate and thermal treatment. Long-term leaching is evaluated with extended TCLP for Zn and other metal contaminants, Ni, Mn, Pb, and Cd, while compliance with beneficial reuse criteria is investigated with SPLP.

METHODOLOGY

Materials and Characterization

Dredged sediments were obtained from New York/New Jersey Harbor at the Passaic River Estuary and from Dampremy, Belgium. Moisture content was determined by drying at 105 °C to constant weight while organic matter was estimated by loss on ignition at 445 ± 10 °C for six hours. Total metal concentrations were obtained by HNO_3 and HCl (aqua regia) digestion. Because Zn was the most abundant heavy metal in the sediments from the two locations it was selected for analysis with XAS.

Sediment treatment was carried out with H_3PO_4 at 3% by dry weight addition under completely mixed conditions. The mixture was then dried and cured at 40 °C for 14 days. Subsequently, the cured sediments were treated by calcination at 700 °C for 3 hours. Kribi et al. (2004) found calcination at 700 °C temperature improved mechanical properties through sintering besides mineralization of organics.

XAS Data Collection and Analyses

X-ray absorption spectroscopy was conducted at the Zn K-edge at beamline X11A of the National Synchrotron Light Source (NSLS), Brookhaven National Laboratory. Spectra for sediments samples and 13 reference compounds were collected in transmission and fluorescence modes. The spectra were analyzed with WinXAS (version 3.1) (Ressler, 1998) to subtract background and convert to k (wavelength^{-1}) space. The sediment spectra were then decomposed into orthogonal components using PCA. Subsequently, the decomposed components were used with TT to assess likelihood of the reference spectra being end members of the sediment spectra. Proportions of identified reference spectra in sediments were quantified by LCF.

Leaching Tests

The TCLP (U.S. EPA 1992) was extended to 28 days to determine the efficacy of treatment, investigate whether equilibrium or pseudo equilibrium was achieved, and evaluate whether the leachate would be considered hazardous. The procedure was performed in triplicate on both untreated and treated sediments. Extractions were carried out at 20:1 liquid to solid ratio in an acetate solution at an initial pH 2.88 ± 0.05, based on the sediment pH, in 2 L HDPE bottles tumbled at 30 rpm. After the standard 18 hr TCLP leaching period, the pH was measured and 10 mL sample withdrawn with 5 mL syringes, filtered, acidified, and refrigerated at 4 °C until analysis. Extraction was continued with pH maintained at 5.0 ± 0.1 by HNO_3 and NaOH addition as necessary over sampling intervals not exceeding 4 days for up to 28 days. Analysis of the extract was carried out with flame atomic absorption spectrometry (FAA).

Risk to groundwater contamination was evaluated with SPLP (U.S. EPA 1994) which is similar to TCLP with exception of the extraction fluid, a 60/40 by weight mixture of H_2SO_4 and HNO_3 at an initial pH of 4.20.

RESULTS AND DISCUSSIONS

Sediments from both the U.S. and Belgium exhibited similar characteristics (Table 1) with heavy metal concentrations mostly within the same order of magnitude. This result implies that treatment method developed for one region would potentially be applicable to the other. Zn was the most abundant heavy metal at 1,067 and 3,400 mg kg^{-1} dry sediments for Passaic River and Dampremy, Belgium respectively.

In the XAS speciation study, the Zn K-edge spectra for treated sediments (Figure 1) were different from that of the untreated sediments indicating significant change in sediment structure. Analysis of the spectra with PCA, TT and LCF (Figure 1) revealed Zn in untreated sediments was precipitated as $ZnCO_3$ (67%) and adsorbed to hydrous manganese oxide (Zn-HMO) (18%) and hydrous iron oxide (Zn-HFO) (15%). For the treated sediments, phosphate addition resulted in hopeite $Zn_3(PO_4)_2{\cdot}4H_2O$ (22%), while calcinations induced formation of spinels, gahnite ($ZnAl_2O_4$) (44%) and franklinite ($ZnFe_2O_4$) (34%). Shih et al. (2006) similarly observed formation of nickel aluminate spinel ($NiAl_2O_4$) in nickel-laden sludge sintered with kaolinite. Thermodynamic analysis showed that these resulting minerals are sparingly soluble; therefore, reduced leaching would be expected.

Extended TCLP extractions (Figure 2) showed leaching appeared to approach equilibrium after 28 days. For Zn, leaching from the treated sediments was reduced by over 80% as compared to the untreated system. Similarly, the concentration of leachable Ni was reduced by more than 50%. Based on results of the speciation study,

TABLE 1. Sediments Characteristics: Moisture Content, Organics and Total Metal Concentrations Determined with EPA Method 3050B (U.S. EPA 1996).

	Sediments	
	Passaic River Estuary, U.S.	**Dampremy, Belgium**
Moisture content (%)	56.5	(Dry sample)
Organics (mg/kg dry weight)	132	161
Metals (mg/kg)		
Zn	1067 ± 12	3400 ± 130
Co	33 ± 0.1	30 ± 0.1
Ni	971 ± 3.5	105 ± 1.3
Mn	519 ± 4.1	1467 ± 47
Cu	520 ± 0.1	120 ± 4.5
Pb	662 ± 7.2	878 ± 9.8
Cd	18 ± 0.01	11 ± 2.3

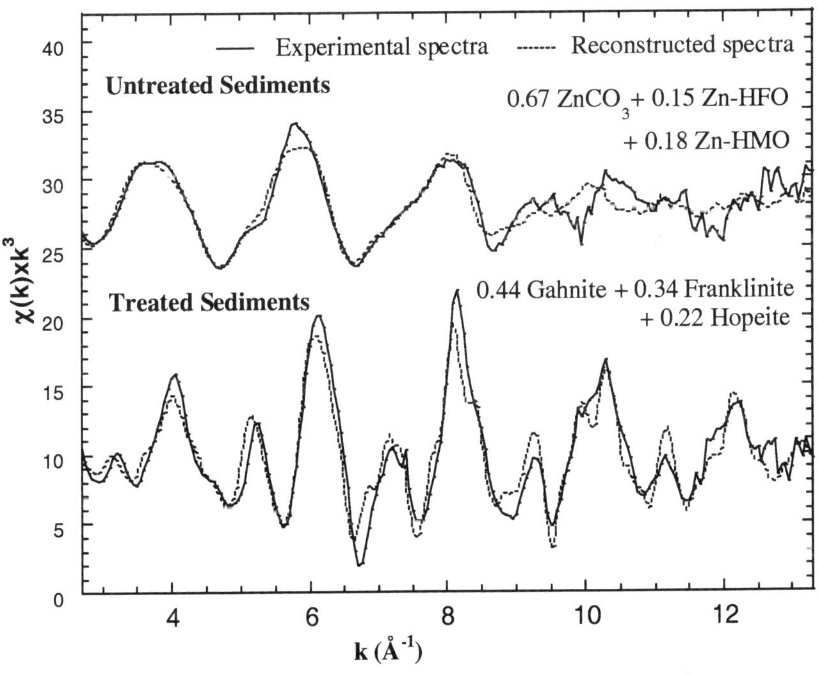

FIG. 1. Experimental and LCF reconstructed Zn-K edge $\chi(k)\cdot k^3$ spectra for untreated and treated sediments. Zn-HFO and Zn-MnO refer to Zn sorbed to hydrous iron oxide and hydrous manganese oxide respectively.

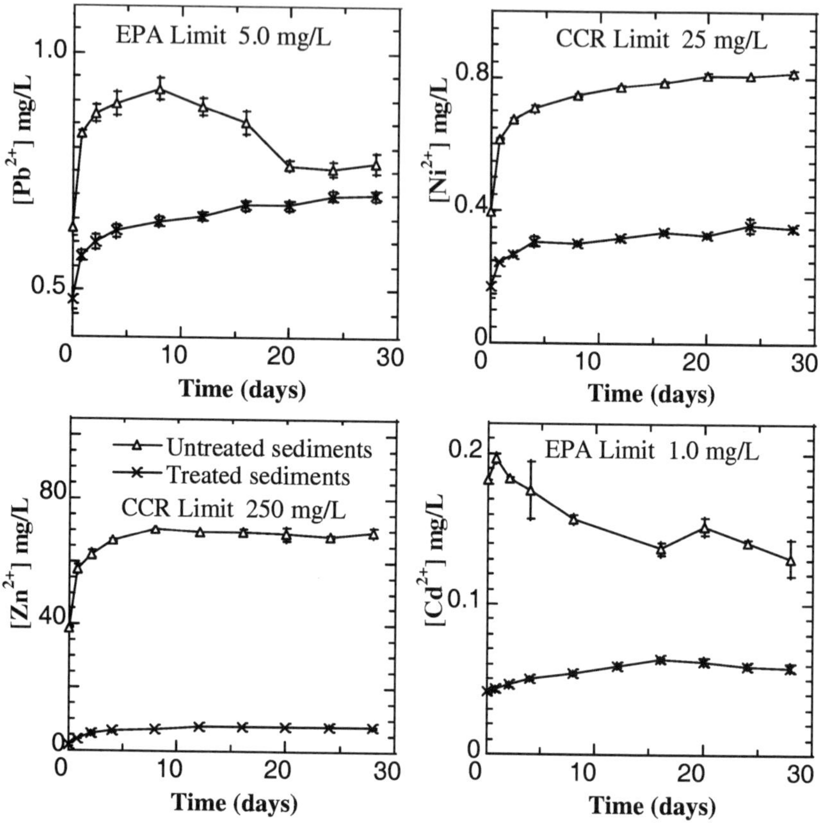

FIG. 2. TCLP leaching of Zn, Ni, Pb, and Cd from untreated and treated sediments as a function of time. CCR refers to California Code of Regulations

the observed reduction in leaching could be a result of immobilization of Zn and Ni in sparingly soluble spinels such a $ZnAl_2O_4$, $ZnFe_2O_4$, and $NiAl_2O_4$ and in the phosphate mineral $Zn_3(PO_4)_2{\cdot}4H_2O$. For Pb, initially a significant reduction in the leachable concentrations is observed for the treated systems. However, after about 10 days, Pb concentration in the aqueous phase for the untreated sediments began to decrease. Halim et al. (2005) observed a similar behavior with Pb and attributed it to initial complexation with dissolved organic matter followed by kinetically slower hydroxide precipitation. Interestingly, a similar phenomenon is observed for Cd, where after 18 h the treated sediments have leached about 25% of that from untreated sediments, yet, after the first day, Cd undergoes redistribution, either precipitating or adsorbing to other phases.

Leaching of the metals from both the treated and untreated sediments were below Resource Conservation and Recovery Act (RCRA) (40 CFR 2003) and California

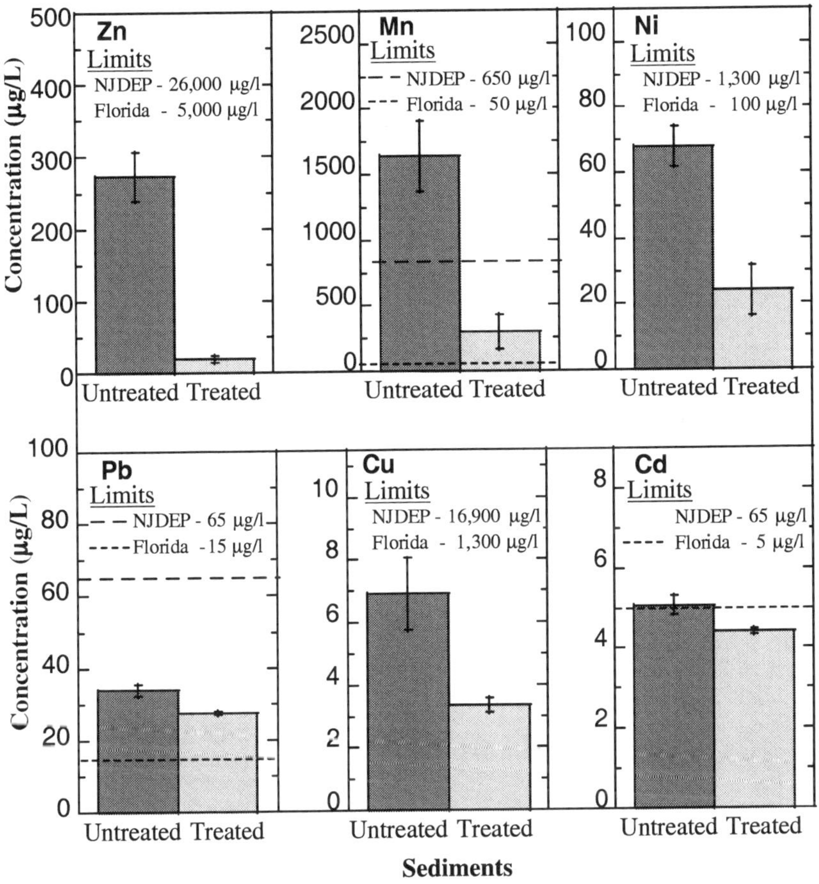

FIG. 3. SPLP leaching of metals from untreated sediments and treated 5% by dry weight H_3PO_4 addition and calcination, and states risks to groundwater limits. NJDEP refers to New Jersey Department of Environmental Protection.

Code of Regulations (CCR 1991) limits for hazardous materials even with the extended leaching period. This result indicates that the sediments are not considered hazardous for the metals analyzed. For beneficial reuse, however, further testing with SPLP is required by several states including New Jersey, Florida, and Wisconsin.

Results of SPLP (Figure 3) show similar percentage reduction in leaching as observed with the TCLP with a 93% reduction for Zn. The leaching levels were however up to two orders of magnitude less those observed for the TCLP, which is attributed to lack of buffering and complexation capacities in the SPLP extraction fluid.

In assessing the risk to groundwater New Jersey Department of Environmental protection (NJDEP) applies a dilution attenuation factor (DAF) of 13 on groundwater quality limits (GQL) to develop impact on groundwater quality criteria for comparison with SPLP results (NJDEP 2008). Other states including Florida and Wisconsin assume SPLP has an inherent dilution of 20 and apply a DAF of 1 (Townsend et al. 2006). For untreated sediments, only Mn concentrations exceeded the NJDEP groundwater quality criteria. However, in the treated sediments, the criteria for Mn were met. On the other hand, applying the state of Florida criteria with a DAF of 1 resulted in exceedances for Mn, Pb, and most treatments for Cd. Thermodynamics analysis showed that the aqueous phase was undersaturated with respect to the sparingly soluble pyromophite ($Pb_5(PO_4)_3OH$) which is assumed to be present. Therefore leaching appears to be controlled by dissolution kinetics. At the DAF of 13 applied by NJDEP, the 65 μg/L groundwater quality criteria for Pb is above pyromorphite equilibrium concentration in the pH range 7 – 8.2. On the other hand, the 15 μg/L criteria applied in Florida and Wisconsin is lower than pyromorphite solubility. Therefore, thermodynamically, the criteria are not achievable; however, slow dissolution kinetics may prevent exceedances.

CONCLUSIONS

Phosphate addition followed by calcination was investigated for immobilization of heavy metals in sediments for beneficial reuse in construction. Zinc speciation showed that the treatment resulted in more thermodynamically stable crystalline phases, the phosphate mineral, hopeite, and heat induced spinels, gahnite, and franklinite. Stability of these phases was confirmed by substantial reduction in metal leaching. SPLP leaching showed treated sediments met the requirements of the State of New Jersey impact to groundwater criteria for the metals analyzed. Thermodynamic analyses showed the Florida and Wisconsin limit for Pb may not be feasible for its stabilization in sediments.

Cited Work

40 CFR (2003). Code of Federal Regulations. *Title 40: Protection of Environment, Part 261.24 Toxicity Characteristic* 23, 59–60. U.S. Government Printing Office.

CCR (1991). California Code of Regulations. *Characteristic of Toxicity,* Title 22, Div. 4.5, Ch. 11, Art. 3, Sect. 66261.24, Register 91, No. 22, 1991.

Crannell, B.S., Eighmy, T.T., Krzanowski, J.E., Eusden Jr., J. D., Shaw, E.L., Francis, C.A., (2000). Heavy metal stabilization in municipal solid waste combustion bottom ash using soluble phosphate. *Waste Manage*, 20 (2–3), 135–148.

Douglas, W.S., Baier, L.J., Gimello, R.J., and Lodge, J. (2003). "A comprehensive strategy for managing contaminated dredged material in the Port of New York and New Jersey." *J. Dredging Engrg.* 5, 1–12.

Geysen, D., Imbrechts, K., Vandecasteele, C., Jaspers, M., Wauters, G. (2004). "Immobilization of lead and zinc in scrubber residues from MSW combustion using soluble phosphates." *Waste Manage.* 24 (5), 471–481.

Guevara-Riba, A., Sahuquillo, A., Rubio, R., and Rauret, G. (2004). "Assessment of metal mobility in dredged harbor sediments from Barcelona, Spain." *Sci. Total Environ.*, 321, 1–3, 241–255.

Halim, C.E., Scott, J.A., Amal, R., Short, S.A., Beydoun, D., Low, G., Cattle, J. (2005). Evaluating the applicability of regulatory leaching tests for assessing the hazards of Pb-contaminated soils *J. Hazard. Mater.* 120 (1-3), 101-111.

Isaure, M.-P, Laboudigue, A., Manceau, A., Sarret, G., Tiffreau, C., Trocellier, P., Lamble, G., Hazemann, J.-L., and Chateigner, D. (2002). "Quantitative Zn speciation in a contaminated dredged sediment by μ–PIXE, μ–SXRF, EXAFS spectroscopy and principal component analysis." *Geochim. Cosmochim. Acta* 66 (9), 1549–1567.

Kribi, S., Nzihou, A., and Sharrock, P. (2004). "Stabilization of heavy metals from dredged sediment. Tailoring of residue properties." *Proc. pro40: the use of recycled materials in buildings and structures*, Vázquez, E., Hendriks, Ch. F., and Janssen, G.M.T. Eds., ISBN: 2-912143-52-7. Vol. 2, 824–832.

Lu, H.-C., Chang, J.-E., Shih, P.-H., Chiang, L.-C., 2008. "Stabilization of copper sludge by high-temperature $CuFe_2O_4$ synthesis process." Journal of Hazardous Materials 150 (3), 504–509.

MPRSA (2000). "Marine Protection, Research and Sanctuaries Act of 1972." Public Law 92-532, Oct. 23, 1972, 86 Stat. 1052 and 1061, as amended through PL 106–580, Dec. 29, 2000 U.S. Government.

NJDEP (2008). "Remediation Standards, Impact to Groundwater Guidance: Synthetic Precipitation Leachate Procedure." New Jersey Department of Environmental Protection. Available at http://www.state.nj.us/dep/srp/guidance/rs/, 10/2008.

Ressler, T. (1998). WinXAS: A program for x-ray absorption spectroscopy data analysis under MS-Windows. *J. Synchr. Rad.*, 5 (2), 118–122.

Shih, K., White, T., and Leckie, J. O. (2006). "Spinel formation for stabilizing simulated nickel–laden sludge with aluminum-rich ceramic precursors." *Environ. Sci. Technol.* 40 (16), 5077–5083.

Townsend, T., Dubey, B., and Tolaymat, T. (2006). "Interpretation of Synthetic Precipitation Leaching Procedure (SPLP) results for assessing risk to groundwater from land-applied granular waste." *Environ. Engrg. Sci.* 23 (1), 239–251.

U.S. EPA 1992. SW–846 Method 1311 Update 1: *Toxicity Characteristic Leaching Procedure*, EPA: Washington, DC, 1992.

U.S. EPA 1994. SW–846 Method 1312 Update 1: *Synthetic Precipitation Leaching Procedure,* EPA: Washington, DC, 1995.

U.S. EPA 2004. "*Updated Report on the Incidence and Severity of Sediment Contamination in Surface Waters of the United States, National Sediment Quality Survey.*" EPA-823-R-04-007, Office of Water, EPA: Washington, DC, 2004.

Yozzo, D.J., Wilber, P., Will, R.J. (2004). "Beneficial use of dredged material for habitat creation, enhancement, and restoration in New York-New Jersey Harbor." *J. Environ. Manage.* 73 (1), 39–52.

Use of Geogrids to Enhance Stability of Slope in Bioreactor Landfills: A Conceptual Method

Hiroshan Hettiarachchi[1], PhD, A.M. ASCE and Louis Ge[2], PhD, A.M. ASCE

[1] Assistant Professor, Department of Civil Engineering, Lawrence Technological University, 21000 West Ten Mile Road, Southfield MI 48075; hiroshan@ltu.edu

[2] Assistant Professor, Department of Civil, Architectural, and Environmental Engineering, Missouri University of Science and Technology, 1870 Miner Circle, Rolla, MO 65409; geyun@mst.edu

ABSTRACT: In biroreactor landfills, biodegradation of municipal solid waste (MSW) is accelerated by leachate recirculation. However, the addition of moisture raises stability concerns. Reduced interlocking can also cause stability issues when MSW is shredded before disposing in bioreactor landfills. Engineers usually select safe slopes such as 3H:1V to design above ground MSW slopes in traditional dry landfills. But these slopes may not be safe for bioreactor landfills due to the stability concerns. Use of shallower slopes is not attractive because of the reduction in effective landfill capacity. This paper describes how geogrids may be used to enhance the stability of MSW slopes in bioreactor landfills. A preliminary stability analysis of un-reinforced and geogrid-reinforced MWS slopes was conducted using PLAXIS. The results indicated that the geogrids are capable of improving the factor of safety of MSW slopes. In geogrid-reinforced MSW failure surface developed at the ends of embedded geogrids separating reinforced MSW from un-reinforced MSW. This failure surface was larger than the failure surface developed for un-reinforced MSW. A different failure mechanism triggered when various lengths of geogrid were used. Although large displacements were found near the toe of the slope, a greater factor of safety was obtained.

INTRODUCTION

With only 5% of the world's population, North America produces more municipal solid waste (MSW) than any other region in the world (UNEP 2002). According to USEPA (2005), the United States generated a total of 121 million tons of MSW in 1970 and by 2005 the total increased to 245 million tons. In 2005, about 32% of MSW was recycled and about 14% was incinerated (USEPA, 2005). The rest was disposed in landfills. Is it possible to accommodate this ever increasing volume of MSW in the existing landfills? More waste is generated in the urban areas and obviously the land value is also high in such areas. This situation urges waste management industry to find more efficient and sustainable solutions to manage MSW.

Bioreactor landfills are a sustainable solution to the ever increasing demand for landfill space. Bioreactor landfills use enhanced microbiological activities to transform and stabilize the biodegradable fractions of MSW at a faster rate (Sharma and Reddy, 2004). In the bioreactor approach, biodegradation of MSW is usually accelerated by leachate recirculation. Recirculation of leachate helps the landfill to maintain a wet environment in addition to supplying nutrients required for the biodegradation. However, additional moisture introduced by leachate recirculation raises stability concerns (Koerner and Soong, 2000) and engineers are often skeptic about the stability of large bioreactor landfills. Some recommend shredding of MSW before disposing them in bioreactor landfills as it enhances the process of biodegradation (Sponza and Agdag, 2005). However, shredding decreases the interlocking characteristics that un-shredded MSW exhibits.

Landfill designers usually select safe slopes such as 3H:1V (regulatory maximum set by the USEPA, 1995) and 4H:1V (regulatory maximum set by the Ontario Ministry of Environment, 1998) to design above ground slopes in traditional dry landfills. The increased moisture content and decreased interlocking properties due to shredding suggest that the MSW slopes used in traditional dry landfills may not be appropriate to design bioreactor landfills. These stability issues may be overcome by using shallower slopes. However, use of shallower slopes is not an attractive solution due to the reduction in landfill capacity. Therefore, it is necessary to explore alternative methods to improve stability of MWS slopes without compromising the landfill capacity. The objective of this paper is to propose one such potential solution: a mechanical stabilization method to improve the stability of MSW slopes in a bioreactor landfill

PROPOSED METHOD: USE OF GEOGRID TO REINFORCE MSW SLOPES

The slope stability concerns in bioreactor landfills may be addressed by reinforcing above ground MSW slopes with layers of geogrids. Figure 1 shows how this can be practically achieved by introducing few geogrid layers into the MSW. Systematic implementation of this method may allow landfill designers to use moderate to steep but safe above ground MSW slopes in bioreactor landfills. Geogrid layers may be placed with the daily cover and installation of geogrids at the daily cover will interfere less with filling and other construction activities. It may not be necessary to install geogrids at each day cover and required vertical spacing can be decided based on the stability requirements.

For years geogrids have been used as a soil reinforcement material. Use of geogrids to reinforce soil slopes has already evolved as a standard design method. Even though geogrids are sometimes used in vertical and lateral expansions of landfills and in veneer reinforcement of landfill cover soils (Koerner, 2005), geogrids have been rarely used in MSW as a slope reinforcing material. ADD3™ system introduced by Tensar Earth Technologies Inc. (2008) is one example of how geogrids can be used to reinforce landfill slopes. However, in ADD3 system geogrids were fully embedded in a soil berm and not in the MSW as proposed in this paper. The technical background necessary for the proposed geogrid-reinforced MSW slopes can be developed based on the current knowledge of the geosynthetic-reinforced soils. The following sections briefly discuss the existing slope stability analysis techniques using finite element method (FEM) and how it can be adopted to evaluate the slope stability of geogrid-reinforced MSW.

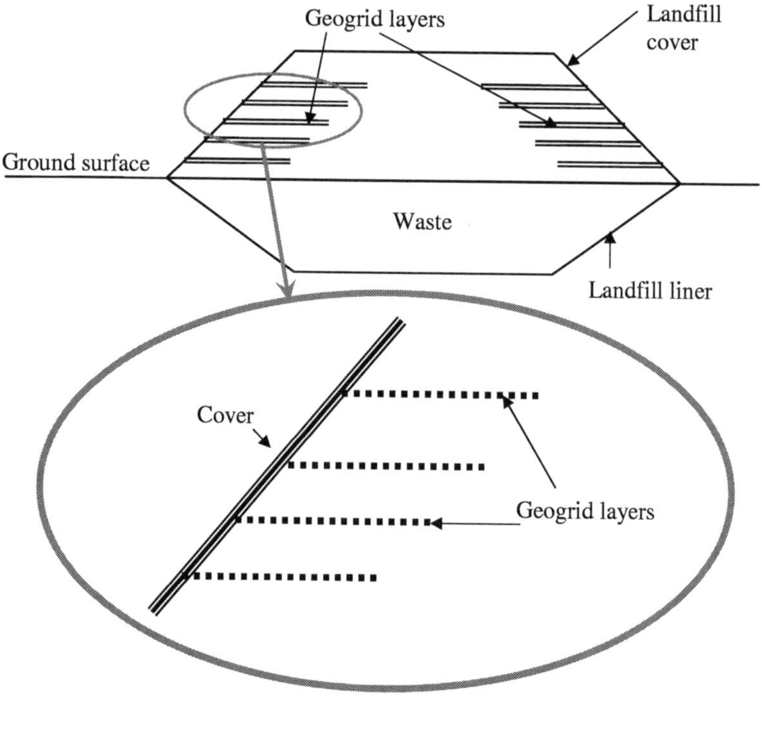

FIG. 1: Schematic diagram of the proposed geogrid-reinforced MSW slopes

SLOPE STABILITY ANALYSIS USING FEM

There are two basic approaches to conduct a slope stability analysis using FEM. The first approach is the conventional deformation analysis, where the deformation within a slope due to given boundary conditions such as loading, excavation and/or prescribed displacements, is to be determined (Duncan, 1996). This approach requires an advanced constitutive model to accurately describe the material characteristics such as hardening, softening, and nonlinear dilantancy. However, this method does not allow for a quantitative determination of a factor of safety (FS).

The second approach, shear strength reduction method, is an iterative process of weakening strength properties of soil (cohesion, c' and the tangent of the friction angle, $tan\phi'$) until a convergence could not be established within the soil slope, which defines the failure of a slope (Zienkiewicz, et al., 1975; Griffths and Lane, 1999). In general, a linearly elastic, perfectly plastic Mohr-Coulomb criterion is used, since the deformation at the ultimate failure condition within a soil slope is not the primary concern. Instead, reduced shear strength properties that cause the slope failure are determined. This reduction factor is the FS of the slope. This approach has gained much attention since it agrees with the conventional definition of FS of a slope in limit equilibrium analysis.

In the finite element shear strength reduction method, the shear strength parameters c' and ϕ' used in the slope analysis are factorized as follows, where $c'_{factored}$ and $(tan\phi')_{factored}$ are the factored shear strength parameters and SRF is the strength reduction factor.

$$c'_{\text{factored}} = \frac{c'}{SRF} \tag{1}$$

$$(\tan\phi')_{\text{factored}} = \frac{\tan\phi'}{SRF} \tag{2}$$

The SRF value is gradually increased until slope failure occurs. The SRF that corresponds to the failure is the FS of the slope. This approach leads to the same definition of the factor of safety (Eq. 3) as in the conventional limit equilibrium and method of slices approaches. A detailed discussion on these methods can be found in Griffiths and Lane (1999).

$$FS = \frac{\sum\left(c'\Delta L + \left(W\cos\alpha - u\Delta L\right)\tan\phi'\right)}{\sum W\sin\alpha} \tag{3}$$

GEOGRID-REINFORCED MSW SLOPES: EXAMPLE

To demonstrate the effectiveness of the proposed method of geogrid-reinforced MSW slopes, a preliminary slope stability analysis was conducted using PLAXIS, a computer software typically used for stability analysis. Figure 2(a) shows a cross-section of a MSW landfill with a 3H:1V slope. The material properties of MSW were assumed as follows: bulk unit weight = 12 kN/m^3; Poisson's ratio = 0.3; cohesion = 5 kN/m^2; and friction angle = 20°. The geogrid is assumed to have an elastic normal axial stiffness of 10,000 kN/m. As shown in Equations 1 and 2, MSW-geogrid interface strength properties were defined by a reduction factor R ($R \leq 1$) times the shear strength of MSW. Four cases were analyzed: (a) un-reinforced MSW slope; (b) geogrid-reinforced MSW slope with equal embedment length and R = 1 (rough MSW-geogrid interface); (c) geogrid-reinforced MSW slope with various embedment lengths and R = 1; and (d) geogrid-reinforced MSW slope with various embedment lengths and R = 0.75 (slightly smooth MSW-geogrid interface).

Displacement vector field for each case at failure condition is illustrated in Figure 2(a)-(d) and the results are summarized in Table 1. The results indicate that the geogrid is capable of increasing the factor of safety of the MSW slope. In case (b), the failure surface developed at the end of embedded geogrids separating reinforced MSW from un-reinforced MSW. This failure surface was larger than the failure surface developed for the un-reinforced MSW in case (a). A different failure mechanism triggered when various lengths of geogrid were used in case (c). In case (c) a global failure zone was observed. Case (d) shows the effect of using interface element (*R*) to simulate the MSW-geogrid interface shear strength. This resulted in FS=2.15 which is still much greater than the un-reinforced MSW.

It is noted in case (d) that the failure surface did not go through the entire geogrid reinforced layers. The failure surface was found governed by the MSW geogrid interface. Authors recommend determination of the interface strength reduction factor R through a large scale MSW-geogrid interface shear tests.

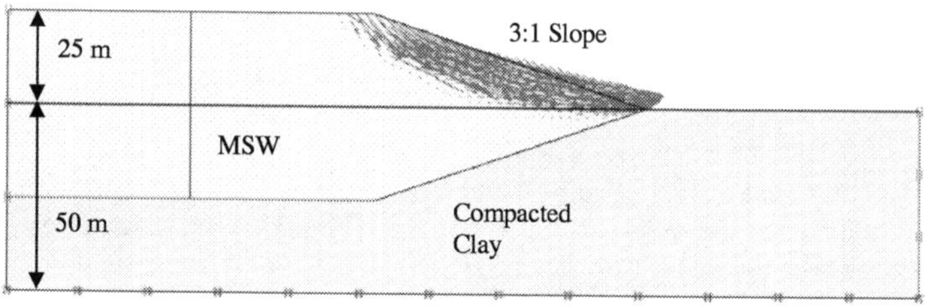

FIG. 2(a): Failure surface of un-reinforced MSW slope, FS = 1.40

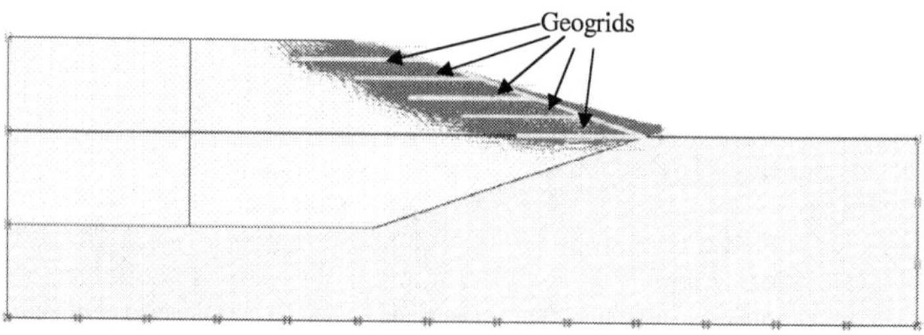

FIG. 2(b): Failure surface of geogrid-reinforced MSW slope, FS = 1.61

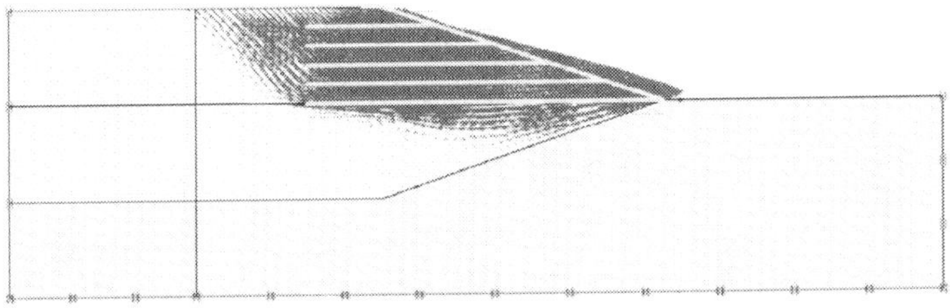

FIG. 2(c): Failure surface of geogrid-reinforced MSW slope, FS = 2.28

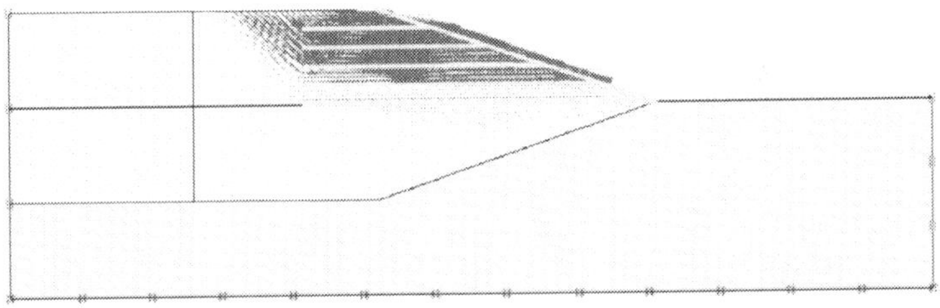

FIG. 2(d): Failure surface of geogrid-reinforced MSW slope, FS = 2.15

Table No. 1: Summary of the example stability analysis results

Cases	Factor of Safety (FS)
(a) Un-reinforced MSW slope	1.40
(b) Geogrid-reinforced MSW slope with equal embedment length and $R = 1$	1.61
(c) Geogrid-reinforced MSW slope with various embedment lengths and $R = 1$	2.28
(d) Geogrid-reinforced MSW slope with various embedment lengths and $R = 0.75$	2.15

CONCLUDING REMARKS

Installation of geogrid layers at the daily covers could be a potential solution for the slope stability issues associated with bioreactor landfills. The proposed method of geogrid-reinforced MSW may allow bioreactor landfill designers to use steep but stable slopes. However, a detailed investigation should be conducted to determine what lengths and spacing of geogrids are needed to improve the stability for a given factor of safety against failure.

Finite element strength reduction method could be a powerful tool for stability analysis of MSW slopes. Strength reduction method is attractive as the method leads to the same definition of factor of safety as in the conventional limit equilibrium methods, but unlike in the conventional limit equilibrium methods, the critical slip surface of a slope needs not to be predefined.

REFERENCES

Duncan, J.M. (1996). "State of the Art: Limit Equilibrium and Finite-Element Analysis of Slopes." *J. Geotechnical Engrg.* 122(7): 577-596.

Griffiths, D.V. and P.A. Lane (1999). "Slope Stability Analysis by Finite Elements." *Geotechnique,* 49(3): 387-403.

Koerner, R.M. and T-Y. Soong (2000). "Leachate in landfills: the stability issues." *Geotextiles and Geomembranes*, 18: 293-309.

Koerner R.M. (2005). "Designing with geosynthetics." Fifth edition, Pearson-Prentice Hall, Upper Saddle River, NJ 07458.

Ministry of the Environment (1998). "Landfill Standards - A Guideline on the Regulatory and Approval Requirements for New or Expanding Landfilling Sites." ISBN 0-7778-7568-3.

Sharma, H.D. and K.R. Reddy (2004). "Geoenvironmental Engineering: Site Remediation, Waste Containment, and Emerging Waste Management Technologies," John Wiley & Sons, NJ.

Sponza, D. T. and O.N. Agdag (2005). "Effects of shredding of wastes on the treatment of municipal solid wastes (MSWs) in simulated anaerobic recycled reactors." *Enzyme and Microbial Technology*, 36(1): 25-33.

Tensar Earth Technologies Inc. (2008). 'Increasing cost efficiency and minimizing environmental risk in municipal solid waste containment," <www.tensarcorp.com> (October 31, 2008).

UNEP 2002. "Global Environment Outlook 3 (GEO-3): Chapter 2 -Ecological footprint." ISBN 9280720872.

US EPA (2005). "Municipal Solid Waste Generation, Recycling, and Disposal in the United States: Facts and Figures."

US EPA (1995). "Decision Maker's Guide to Solid Waste Management - Vol. II, Land Disposal." EPA 530-R-95-023.

Zienkiewicz, O.C., C. Humpheson, and R.W. Lewis (1975). "Associated and Non-associated Visco-plasticity and Plasticity in Soil Mechanics." *Geotechnique*, 25(4): 671-689.

Vapor Barriers for Deep Foundation Supported Structures

Gregory T. Corcoran, P.E. and Steven M. Fitzwilliam, P.E.

Gregory T. Corcoran, P.E. Principal, Geosyntec Consultants, 10875 Rancho Bernardo Road, Ste 200, San Diego, CA 92127 – gcorcoran@geosyntec.com
Steven M Fitzwilliam, P.E., G.E. Associate, Geosyntec Consultants, 10875 Rancho Bernardo Road, Ste 200, San Diego, CA 92127 – sfitzwilliam@geosyntec.com

ABSTRACT

Building construction on Brownfield sites, especially landfill sites, often requires the design and construction of vapor control systems for the purpose of minimizing the impact of subsurface vapors on building inhabitants. Vapor control systems are typically comprised of active or passive vapor extraction systems, a vapor barrier, and vapor monitoring systems and are integrated with the building foundation system. Integration of vapor control systems with deep foundation systems on settlement-prone subsoils and municipal solid waste (MSW) poses challenges not encountered on typical projects. Specifically, vapor control system elements require detailed design to account for potentially large amounts of settlement. Potential remedies typically involve structurally supporting vapor extraction piping, geomembrane vapor barriers, and subsurface vapor monitoring probes. However, structurally supporting vapor control system elements can be challenging and result in potential post-construction problems. Case histories are presented along with design considerations to avoid damage to the vapor extraction, monitoring, and barrier systems.

BACKGROUND

Development over Brownfield sites and landfills is becoming more common as the availability of virgin or "greenfield" properties are becoming scarce in desirable areas. With the ever increasing cost of land, more and more developers are developing Brownfield sites and landfills with retail, commercial, and residential developments. Practices and regulations have been promulgated to allow this development to occur in a manner that minimizes the potential impacts to inhabitants of the buildings overlying the impacted property. In many states, specific regulations and ordinances dictate the use of vapor barriers, vapor extraction systems, and/or vapor monitoring systems when developing a property containing waste material generating soil vapors that may impact the inhabitants of the development. Waste materials in the subsurface can consist of volatile organic compounds (VOCs) from past site use or methane gas from anaerobic degradation organic materials or from natural origins. Specifically, landfilled waste materials produce methane and other vapors as the organic components in the waste materials degrade. Vapor control systems can be designed to control methane and VOCs, and to minimize these vapors from accumulating within inhabitable structures.

VAPOR CONTROL SYSTEM

A vapor control system consists of three distinct elements. These elements include: the vapor extraction, vapor barrier, and vapor monitoring systems.

The vapor extraction system consists of a continuous layer of pneumatically-permeable material that allows vapors to be collected within the layer and extracted through a network of perforated pipes. The vapor extraction layer typically consists of a 100 to 400 mm (4- to 12-inch) thick layer of clean aggregate or sand. In lieu of the aggregate of sand, a geocomposite material consisting of filter geotextile bonded to a geonet core can be employed to collect the subsurface vapors and provide a transmissive layer for the vapor to migrate to the perforated piping. Vapor extraction piping is connected to solid wall vertical riser pipes that convey the collected vapor to treatment and/or the atmosphere, typically above the roofline. The vapor extraction piping may be designed as a passive system for expected minor vapor impacts of as an active system for relatively high vapor impacts. The passive systems are often designed to be upgraded to an active system if conditions warrant such an upgrade as a result of follow-up monitoring. Passive systems are designed to use the diurnal changes in temperature and barometric changes to create air flow from under the structure and out through the vertical riser pipes. Active systems incorporate the use of a blower to create air flow.

For a deep foundation supported building, the vapor barrier consists of the concrete structural slab underlain or overlain by a geomembrane. The geomembrane can consist of spray-applied materials or sheet materials that are seamed in the field. The geomembrane is typically between 1.5 to 2.5 mm (60 to 100 mils) thick.

The vapor monitoring system consists of vapor probes located beneath the vapor barrier and/or vapor monitors within the building. The vapor probes or monitors provide a means to sample and analyze for the constituents of concern.

Granular trench backfill for utilities that enter or exit the building footprint provide a path for vapors to accumulate and migrate beneath the building from the surrounding subsurface soils. To minimize this potential path for vapor flow, trench dams are constructed at the building perimeters and/or within the utility trenches. Trench dams are zones of a utility trench that are backfilled with a low permeability material, such as a 2 percent bentonite-cement slurry. In addition, dry utilities and non-pressurized wet utilities may also provide a migration pathway and require seals and wet p-traps to minimize the potential for subsurface vapors to enter the building.

If the subsoils settle away from buildings supported on deep foundations, the vapor extraction system and vapor barrier may also settle away from the bottom of the building slab. Perforated pipes that are part of the gas extraction system are buried in an aggregate or sand material, which has been experienced to add a significant load to the piping with no support once the subsoils have settled. The granular backfill materials have been observed to remain on top of the pipes once the pipes have settled with the subsoils. The geomembrane and associated geosynthetics that are used to construct and protect the geomembrane may pull away from the bottom of the building slab and sag. To minimize the movement of these vapor control system elements and/or eliminate the damage to these system elements as the subsoils settle, engineering controls should be implemented.

ENGINEERING CONTROLS

Engineering controls to protect the vapor control system include designing and installing appropriate pipe hangers beneath the concrete structural slab and specifying materials for the vapor control system that can withstand the forces applied as a result of settlement. Voids beneath the building resulting from subsoil settlement and pipe connections at the building perimeter should also be addressed in design.

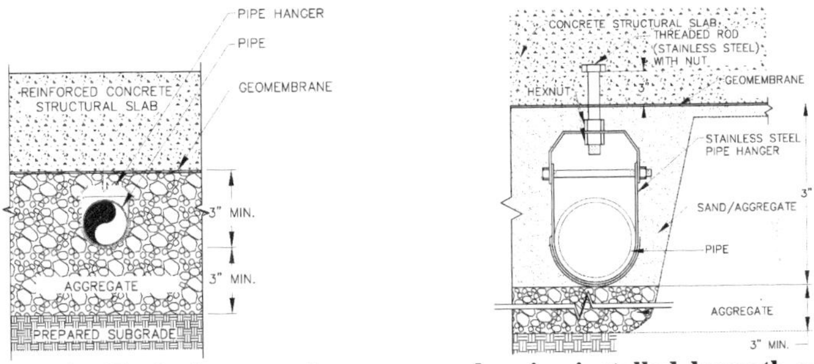

Figure 1: Typical geomembrane vapor barrier installed beneath a structural slab with a vapor extraction pipe embedded within vapor extraction aggregate. (left). Typical pipe hanger detail to support pipe (right); as settlement of soil occurs pipe is supported by pipe hanger embedded in structural slab.

Pipe hangers that are attached to the concrete structural slab will help to keep the perforated pipes within the gas extraction layer at the design grade. Water within the sagging pipes may accumulate in low spots in the pipe from condensation within the pipes as the air and soil vapors pass through the pipes, which may block the flow of air and soil vapors if the pipes are allowed to settle away from the building slab as the subsoils settle. Pipe hangers should be designed to support the selected pipe material at a regular interval along the pipe length such that the pipes are not damaged or clogged with fluids at low spots where the pipe sags. The pipe material should be selected to withstand the loading of the aggregate and backfill soils, if any, that will remain in place resting on top of the pipe as the subsoils settle away from the piping. The span between pipe hangers should be calculated based on the pipe size and diameter and material type. Pipe hangers for use in supporting the pipe can be specified from existing pre-manufactured pipe hangers or may be manufactured in the field from steel reinforcing bars. Both of these pipe hangers should be embedded in the concrete slab during slab construction. Caution should be exercised in the selection of the pipe hangar and pipe material, as aggressive subsoils may impact the pipe hangar material (cathodic protection may be required in some environments) or create cemented materials that "hang" on the pipe and increase the loading. Some designers elect to use a clean "pea gravel" consisting of rounded aggregate that is intended to "roll off" of the pipe as the settlement occurs, thereby keeping the loading to a minimum. However, strict construction quality assurance should be implemented to verify use of the correct materials. Also, consideration should be given to the large quantity of geomembrane vapor barrier penetrations that will occur, which will require the installation of boots and seals at a cost not always accounted for during the design of these systems.

The geomembrane component of the vapor barrier generally consists of a carrier geosynthetic (geotextile or thin geomembrane per the geomembrane manufacturer's requirements) for spray-applied geomembranes overlain by a spray-applied geomembrane or a sheet geomembrane (e.g. high-density polyethylene (HDPE)) seamed in the field. Either of these geomembrane materials would be overlain by a geosynthetic or sand layer to protect the geomembrane from damage during construction of the overlying slab-on-grade or structural slab. Once subsoil settlement occurs, the geomembrane vapor barrier will sag and pull away from the concrete slab as the subsoils settle. This is exacerbated by the use of sand overlying the geomembrane, which can produce a significant loading on the geomembrane, which will result in tears and other significant damage to the geomembrane vapor barrier. Pipe hangers used to support utilities and vapor extraction piping can assist in attaching the geomembrane to the bottom of the structural slab. In addition, special geomembrane hangers can be used to support the geomembrane barrier at a spacing related to the strength of the geomembrane and other overlying materials.

Figure 2: Pipe hangers (rods wrapped for protection during geomembrane spray application, removed prior to concrete installation) penetrating geocomposite vapor extraction layer (black colored material) along pipe.

The materials specified for the geomembrane layer need to be selected to be able to withstand the stresses from the weight of the geomembrane layer, including sand or geotextile cushion installed above the geomembrane layer. Generally, the geomembrane will have sufficient strength to support the geomembrane and cushion geotextile loads between the pile caps, grade beams, and pipe hanger locations, however, the tensile strength of the specified geomembrane and supporting layers should be evaluated for this condition. Specialized HDPE geoemembrane materials, consisting of small spikes on one side, can be cast into the bottom of the structural slab, thereby becoming a part of the structure and no longer susceptible to settlement. A tear or hole in the geomembrane will compromise the system, and recognition and repair for damage to the system is costly and difficult after the building slab is constructed.

Figure 3: Spray application of geomembrane vapor barrier (left), note protected reinforcing bar extending from pile caps. Completed pile cap and perimeter grade beam geomembrane application (right), note re-bar protection not yet removed.

Trench dams within the utility trenches are generally constructed from a bentonite augmented cement slurry. As differential settlement between the trench backfill and the trench dam may occur, the trench dam should be sized to extend beyond the trench perimeter. The trench dam should extend at least 150 mm (6 inches) beyond the sides of the trench and at least 150 mm (6 inches) above the pipe bedding material. This will allow the trench dam to continue to act as a barrier to air and soil vapors passing through the generally more permeable pipe bedding.

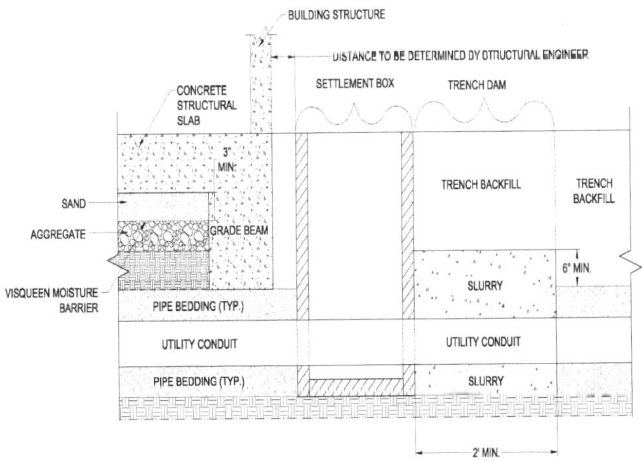

Figure 4: Typical trench dam detail to minimize vapor migration along trench backfill (minimal differential settlement anticipated).

The design of the type and quantity of the backfill materials in the gas extraction layer above the pipes or above the geomembrane layer should be considered carefully when settlement-prone subsoils are suspected at the site. The selection of a geosynthetic protection layer rather than a sand cover to provide protection of the geomembrane will apply less loading to the geomembrane as settlement occurs. Similarly, the installation of the piping as shallow as possible will reduce the loading on the pipes. The cost of installation as well as the material costs for installation of a stronger pipe or geomembrane should be considered when selecting vapor control system components.

CASE HISTORIES

Experience with building construction over landfills and settlement-sensitive soils provide evidence of the effectiveness of these engineering controls. Three case histories will be presented to indicate the effectiveness of these engineering controls. These case histories include a retail store constructed over a landfill, a commercial building complex over settlement-sensitive soils, and a commercial building constructed over a landfill.

Case I

The retail store constructed over a landfill was constructed using a driven-pile foundation system consisting of cathodically protected steel pipe piles through the unlined closed landfill. Stainless steel pipe hangers were installed every 1.5 to 3.0 meters (5 to 10 feet) along perforated pipe depending upon the depth of the bottom of the pipe beneath the bottom of the concrete slab to support the vapor extraction system piping, which was backfilled with clean pea gravel. The piping was installed with an overburden of less than 100 mm (4 inches) of pea gravel. The geomembrane vapor barrier consisted of a spray-applied geomembrane that was attached to closely spaced pile caps and grade beams. Along with the numerous pipe hangers, the pile caps and grade beams provided ample support for the geomembrane vapor barrier, which was installed on a stronger carrier geotextile to provide additional wide width strength to the geomembrane vapor barrier. The geomembrane vapor barrier was protected from concrete structural slab construction damage by a nonwoven geotextile that was adhered to the top of the geomembrane vapor barrier with an adhesive. The concrete was poured directly on the nonwoven geotextile, which embedded into the fresh concrete and when complete, was bonded to the bottom of the concrete structural slab. Trench dams were installed in all on-site utility trenches to minimize the potential for gas migration within the granular backfill. Monitoring of the vapor control system has been performed for 6 years and the system is performing as designed.

Case II

A commercial building complex consisting of four commercial buildings and a parking structure was supported on pile foundations over the settlement-sensitive soils. Settlement predictions were less than 50 mm (2 inches) for the majority of the site. A vapor control system was installed to protect the building occupants from the accumulation of naturally-occurring methane gas in the subsurface. Buildings and occupied areas of the parking structure (such as the fitness center and security guard station) were underlain by vapor extraction piping embedded in an aggregate and a spray-applied geomembrane vapor barrier. The pipes for the vapor extraction system were supported by pipe hangers every 2.4 meters (8 feet) along solid wall pipe and 3.0 meters (10 feet) along perforated pipe. The geomembrane vapor barrier was supported by the pipe hangers and closely spaced grade beams. Trench dams were installed at the building perimeter and extending 6 inches horizontally beyond the pipe trench and 150 mm (6 inches) above the top of the pipe bedding material. Monitoring of the vapor control system has been performed for over a year and the system is performing as designed.

Case III

A commercial development was constructed with some structures over a landfill and some structures within close proximity of the limits of the landfill, thus requiring a vapor control system for landfill gas. Due to the settlement-prone nature of the landfill debris, the building over the landfill was supported on drilled pier foundations through the unlined landfill. Because of the small building footprint, no gas extraction system was installed. However, a geomembrane vapor barrier was installed beneath the building and along the backside of the retaining wall. The retaining wall was incorporated into the building design with one wall of the building partially buried. As the retaining wall backfill supported the landfill debris, which is sensitive to settlement, the geomembrane material specified for the back of the retaining wall had to be selected to resist the frictional loads between the wall backfill and the geomembrane as the backfill settles. The geomembrane vapor barrier beneath the structure is supported by the pile caps and grade beams.

CONCLUSIONS

Development on Brownfield and landfill sites with vapor control systems can be successfully accomplished with structures supported on deep foundations over settlement-sensitive soils. However, the designer needs to account for elements related to the nature of the site and subsoil; environmental conditions; and impact on durability of materials of construction of the vapor control system. Vapor control system components should be designed to withstand the forces that may occur as a result of settlement of the subsoils by adding supports or using the appropriate building materials.

Foundation Replacement With No Disturbance

W. Tom Witherspoon[1], PhD, PE and Jason E. Taylor[2], Eng.

[1]Founder, S & W Foundation Contractors, Dallas, Texas, (972)235-3267, Email: tomw5@ix.netcom.com

[2]Engineer, S & W Foundation Contractors, Dallas, Texas, (972)235-3256, Email: swfoundation@sbcglobal.net

ABSTRACT

A three story hotel in Irving, Texas, home to some of the worst soil conditions in the nation, was underpinned with piers that were belled in the active zone. As a result of errors by the engineer and/or contractor, the piers uplifted and moved the building to extreme deflections that exceeded 8". To mitigate the problem, 360 – 18" diameter x 35' deep drilled piers were installed under the building in five foot of headroom. Despite the low headroom, the new piers were drilled into gray shale a minimum penetration of 10'. This process was completed while keeping the hotel in full operation with no interruption in services and minimal effects to the landscaping and appearance.

INTRODUCTION

Irving, Texas is commonly known as the home of the Dallas Cowboys, but to the engineering community it is better known for the highly active clays. This engineer has witnessed potential vertical movement in excess of 18" and an active zone that may extend 20' below the surface has made it a nightmare for geotechnical engineers. Even the best engineers have been hauled into court and learn that standard precautions for clay soil are not enough and that in Irving there is a difference between good engineering and good legal engineering.

BACKGROUND

The Staybridge hotel in Irving, Texas was designed with some pretty good engineering techniques. It was designed to be supported by 198-18" diameter drilled shafts, terminated 15' below the surface with 30" bells/underreams at the bottom for additional axial capacity and to provide resistance to upheaval. On top of the piers were to be concrete grade beams with 24" void boxes at the bottom so that when the clay swelled there would be no pressure on the underside of these beams that would act against the building. As with many projects, however, construction did not match the design.

What resulted were concrete grade beams that because of sloppy forming and over excavation resulted in an over-pour of the concrete grade beams, creating lugs that provided a conduit for heaving clays to lift against the beams and push up the three story building causing differential movement patterns and function problems in response to the floor deflection (Figure 3).

Heaving of the flatwork (drives and parking) at the perimeter reversed the positive drainage and caused water to stand along the side of the building, which exacerbated the problem by over saturating subsurface soil. This upheaval put pressure upon the pier/grade beam connection, which damage so bad in some locations that the grade beams tore away from the piers.

Figure 1 - Front View of Hotel

Figure 2 - Rear View of Hotel

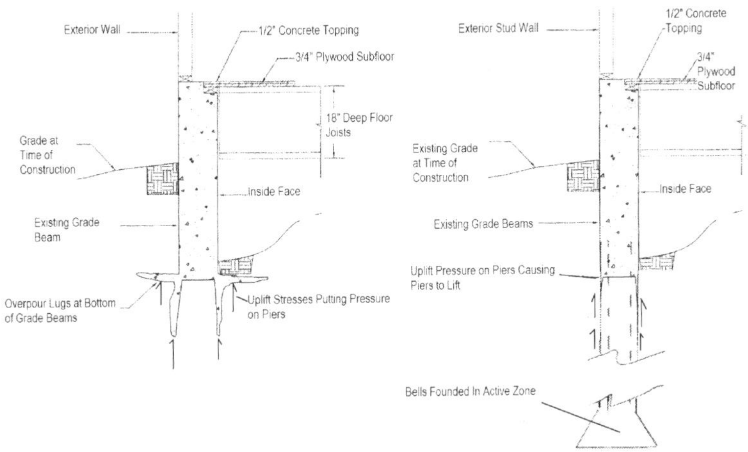

Figure 3 - Sketch of Post Construction Lugs

CORRECTION OF GRADE BEAM PROBLEMS

Realizing there was a problem, the owner required the general contractor who built the building to mitigate the situation. Thinking the only problem was drainage and over poured grade beams, Phase I of remedial repairs was to excavate along the grade beam removing the concrete lugs, excavating a void under the beams, installing fiberglass retainers at the sides of the beams, repairing some connections between piers and beams and installing a capillary drain system that would capture excess water and discharge into the city drain system. Surface drainage was separately installed in solid lines to mitigate negative drainage at the buildings caused by the parking and drives lifting so excessively.

Phase I was completed in late 2001 by S & W Foundation while keeping the hotel in full operation. This required segmental work to avoid disruption of hotel activities and honoring work hours that would be acceptable to hotel patrons.

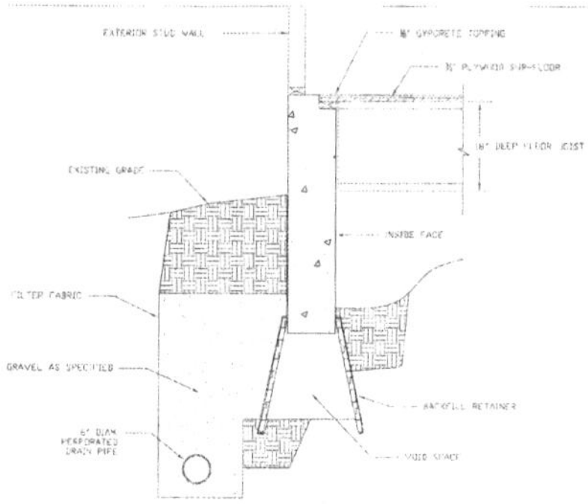

Figure 4 - Installation of Retainers and French Drain System

CORRECTION OF PIER PROBLEM

For two years after completion of phase I there were regular surveys of the interior floor and documentation of movement problems. There was also monitoring of drain performance with precise locating of movement problems. The result was that over 50% of the piers were unstable, the building was an approximate 8" out of level, functioning problems with interior doors, complaints from occupants and a floor system that exceeded tolerance for localized deflection. This led to a decision that the drain system was functioning as intended but that all the piers should be replaced. The owner hired S & W Foundation to do the pier replacement and Tom Witherspoon to design the remedial foundation system.

New soil borings were ordered by S & W, which showed there was an active depth of 15' with uplift pressures of 2,000 psf and moisture contents that ranged from 11% to 22%. The zone from 15' to 25' showed to be a non-contributing section with allowable skin friction of 650 psf in compression and resistance against uplift of 550 psf. At 25' there is gray shale that was stable and provided not only anchorage against uplift but also significant bearing capacity.

Phase II was implemented by S & W Foundation, which consisted of replacing all piers with one to two piers per replaced pier, terminate the piers into the gray shale, access the pier replacement area by excavating under the building with small rigs, relocate utilities where required and keep the building operational at all times. The piers were 18" diameter, 35' deep from paving surface and penetrated 10' into gray shale with an average SPT of 100/1".

Additional considerations were that there would be: no disruption to hotel operations, a carpenter on site during floor elevation adjustment, lawn sprinklers kept in operation, electrical system installed in crawl space, small rigs with catalytic converters to lessen emissions, ventilation system to provide safe air, starting at 8:00 am with no noise until 9:00 am and quit work at 5:30 pm, work commitment for 6 days per week, floor topographic surveys under the floor and inside the building before and during elevation adjustment, steel reinforcing drilled into existing beams connected to pier steel, sono-tube forms from pier tops to floor beams and/or grade beams, independent inspection of all pier drilling, steel and concrete placement, concrete cylinder tests of each pour, coordination with hotel staff and owner representative during all operations, installation of under floor sumps and drain lines and replacement of soil under building when complete to reduce the chance of water standing.

With the number of people using this hotel it was important to install security fencing with ramps at two sides of the building for access to the work area (figure 5). It was also mandatory that pier spoil and removed soil be hauled off so that there would be no dirt on the paving.

Figure 5 - Ramp at North End of Building

Four walk behind Bobcat MT's were utilized on the job for excavating in preparation for drilling operations. The idea was to access through a center excavated hallway and the branch off between the lines of piers to provide ground clearance for drilling. Upon completion of the piers and adjustment of the floor elevation, dirt was returned to the excavated space under the building to prevent creating a bathtub that would hold water that migrated into the crawl space.

As shown in figure 6, these piers were drilled under grade beams with as little as 5' clearance. All piers were, however, 18" diameter and over 30' deep including the 10' of shale penetration.

Figure 6 - Drilling under concrete beams

Steel diameter was kept to ¾" in order to reduce the lap, which in these tight quarters would have greatly increased steel cost because of the laps necessitated by setting steel in short sections and increased fabrication time prior to pouring concrete (figure 7).

Figure 7 - Steel extended up from piers

Concrete was gathered from outside of the building with motorized buggies and transported to each pier for placement (figure 8). This required sequencing of all movement under building since there were several operations going on during pouring of pier steel. With proper lighting and directions by the jobsite foreman this went with no safety problems.

Figure 8 - Concrete placement

Figure 9 - Main access route

As would be expected, power, sewer, water and communication utilities were in large supply under the building, which required our excavation and drilling operations to be mindful not only because of safety but because any break in these lines might shut down the entire hotel. The only damage to piping that occurred in the almost 20 months of operation was an impact with a bobcat that ruptured a sanitary sewer line. Since flushing of toilets could not be stopped, the workmen repaired this line quickly and it never happened again.

Figure 10 - Concentration of Utilities

When a section of piers were completed, it was necessary to cut away the existing piers that the remedial piers were replacing, and smooth the top of the remedial piers so that a screw jack could be installed between existing pier and beam (figure 12). Then using the new piers for jack support, the building in each section was adjusted vertically to meet the topography goal and then the screw jack was adjusted on the old piers so that jacks on the remedial piers could be removed and the formed section poured between new piers and building beams.

The method that worked best for cutting away the existing piers was drilling a pilot hole into the pier, setting a spreader bar and expanding to rupture the concrete, which saved a considerable amount of time. The figure 11 shows how the spreader broke the pier.

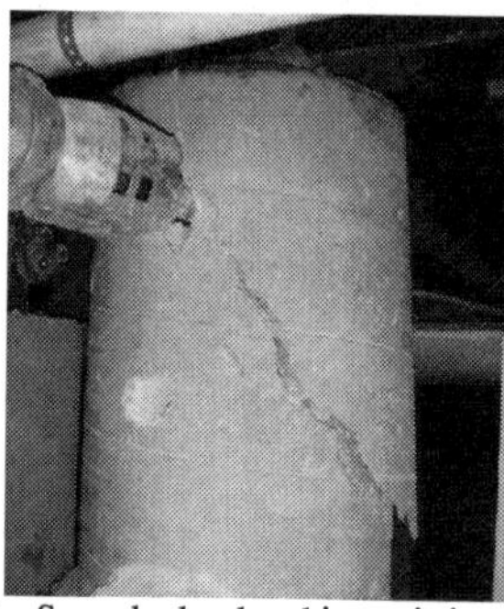

Figure 11 - Spreader bar breaking existing pier for removal

Figure 12 - Screw jack set in place on broken existing pier

Connection from the new piers to existing beams consisted of drilling into the beams, installing dowels and secured with epoxy in the drilled holes at the concrete beams. The section from new pier to existing beam was then formed and poured with Siki-Quick 1000, which had a 15 minute set time. At the wood and steel beams it was necessary to install "J" bolts to attach to the beam and secure it not only laterally but also vertically (figure 13 and figure 14).

Figure 13 - Pier cap poured under beam

Figure 14 - Finished pier columns up under foundation

Adjustment of the building had to be scheduled with building management to not only have a carpenter on site to adjust doors to each room but also provide necessary security personnel to secure valuables of each occupant's room. It was also important to produce on-going surveys of the floor so that elevation targets could be revised in line with feasibility assessments. Elevation surveys from the interior floor and below the floor in the crawl space were compared to find the actual structural elevation and restore the building to its original position. Reaction of the building structure to lifting or lowering of an area required several amendments in topography goals but since building management was directly involved in each adjustment there was agreement of final positioning.

It was decided to not lift at the elevators because any change in elevation might cause alignment problems that would compromise hotel operations. There was also one area where the concentration of utilities was so great that we could not install piers without complete rerouting of these lines, which would have shut down hotel operations. When presented with the circumstances and with management's on site inspection of actual conditions, the decision to omit remediation of these areas became an economic determination that was embraced by all.

As a result of copious planning, flexibility during the project and a partnering attitude with management, the project produced a building where localized deflection was reduced to a functional level and motel operations were never compromised. After completion of remedial underpinning the perimeter French drain was restored to capture lateral moisture migration. As an added protection, drainage was installed in the crawl space even though dirt was returned to prevent ponding.

SUMMARY AND CONCLUSIONS

Although piers at this building were not designed properly to prevent uplift, the construction anomalies (over-pour of the grade beams) induced another problem that exacerbated vertical movement and increased damage to the structure. This is another lesson that quality control in construction is critical to producing a product that meets the engineer's design.

This repair also provided a method of repair to address one of the worst and most complex remediation situations that an engineer may face when trying to put a building back together in lieu of total replacement.

Column Expansion Testing of Chromium Tailings Subgrade Fills

Charles W. Schwartz[1], Ann G. Wylie[2], Allen P. Davis[3], and Bruce R. James[4]

[1]Associate Professor, Dept. of Civil and Environmental Engineering, University of Maryland, 0147C Martin Hall, College Park, MD 20742; schwartz@umd.edu
[2]Professor, Department of Geology, University of Maryland, 1100 Main Administration Building, College Park, MD 20742; awylie@umd.edu
[3]Professor, Dept. of Civil and Environmental Engineering, University of Maryland, 1151 Martin Hall, College Park, MD 20742; apdavis@umd.edu
[4]Professor, Dept. of Environmental Science and Technology and Director, Environmental Science and Program, University of Maryland, 0220 Symons Hall, College Park, MD 20742; brjames@umd.edu

ABSTRACT: Chromium ore tailings used as fill material beneath pavements can expand significantly under specific site conditions. However, it has been difficult in the past to simulate this behavior reliably in the laboratory. Column expansion tests in an enriched carbon dioxide environment are found to reproduce expansion and induration of the chromium tailings material similar to that observed in the field. Results demonstrate that expansion pressure and degree of induration as quantified by unconfined compressive strength vary significantly with groundwater pH.

INTRODUCTION

The Dundalk Marine Terminal (DMT) operated by the Maryland Port Administration is located on fills consisting of a mixture of chromium ore tailings and dredge spoil. The tailings are waste material from past chromium refining operations in Baltimore. The chromium tailings fill was initially considered stable, but severe pavement heaving, as well as cracking and underground utility damage, has occurred episodically since the mid 1970's. The worst distresses tend to occur in the more recent fill areas where the chromium tailings are close to the paved surface and thus above the groundwater table and under only minimal overburden pressures.

Previous studies implicated hydration of calcium and magnesium oxides augmented by carbonate reactions fed by subsurface water flows as the most likely underlying causes of the continuing expansion of the chromium tailings and their concomitant hardening or induration into a rock-like material. However, there is little evidence to support these hypotheses, as few quantitative investigations have been conducted to evaluate the chemical and mechanical processes of the expansive material. An extensive laboratory exploration was therefore conducted into the chemical, mineralogical, and mechanical processes underlying the expansive behavior (Schwartz, *et al.*, 1997; 2000). Investigative activities included: collection of groundwater, soil, and rock samples from trenches; chemical analyses of groundwater and samples of both the unexpanded and expanded material; mineralogical analyses of unexpanded and expanded material; physical property testing; free swell expansion testing; and column expansion tests to reproduce the expansive behavior in the laboratory and to examine the influence of site environmental conditions on expansion. Because of length limitations, the present paper will focus only on the expansion testing.

POTENTIAL EXPANSION MECHANISMS

Viewed strictly in conventional terms, the chromium tailings would be judged a good pavement subgrade material. The unconsolidated (i.e., unexpanded) material is a medium dense to dense granular soil with generally less than 10% fines passing the No. 200 sieve (USCS category SP-SM). SPT N values range between 20 to 60, the friction angle is between 40 and 45 degrees, the cohesion is from about 12 to 29 kPa (240-600 psf), and compaction properties are good (Century, 1980; EA, 1987).

There are distinct and striking differences in the appearance and composition of the chrome tailings samples taken from unexpanded and expanded zones. The unexpanded tailings are a black, cohesionless, granular soil, while the expanded material is a tan, hard, indurated, rock-like material. Expansion/induration of the tailings material appears to occur only after burial; none was observed in the surface stockpiles prior to burial. Expansion is most severe at locations where the tailings burial is most recent and/or at the shallowest depths. Older locations with more deeply buried tailings have exhibited less or no surface heave.

Most of the prior studies at DMT have focused on an oxide hydration mechanism (e.g., EA, 1989). Calcium oxide is the largest constituent chemical in the tailings (30-49% by weight), followed by ferric oxide (15-30%) and magnesium oxide (7-10%). Although ferric oxide is relatively stable material that does not spontaneously hydrate, volumetric expansion of calcium and magnesium compounds upon hydration is well known and documented (e.g., Ramachandran *et al.*, 1964). Hydration of calcium oxides is reported to cause an 8% volume expansion under water saturated conditions and over 100% expansion under vapor saturated conditions. Weidner and Wylie (1980) found evidence of calcium oxide hydration in an earlier limited mineralogical study of chromium tailings materials from the DMT. Expansion of slag aggregate from open hearth steel furnaces, which has a chemical composition similar to that of the chromium tailings at the DMT, is well documented and generally attributed to the expansion of the calcium and magnesium oxides (e.g., Crawford and Burn, 1969).

While oxide hydration may be responsible for the initial expansion, the hypothesized mechanism for continuing expansion is ongoing carbonation reactions and precipitation of calcite and calcium-rich hydrated carbonates fed by subsurface flows of alkaline, calcium-laden groundwater. On a per-reaction basis, the expected expansion from the hydration reaction is much larger than that from the carbonation reaction. However, if the carbonation reaction is continually fed by subsurface groundwater flows, over the long term it could contribute substantial expansion.

Weidner and Wylie (1980) found extensive evidence of calcite and hydrated carbonates in all of the samples in their earlier mineralogical study of chromium tailings materials from the DMT. The extensive presence of calcite and hydrated carbonates in the indurated, expanded material is also supported by direct visual observation. Carbonate cementation is responsible for the striking difference in appearance between the blackish, granular, unconsolidated tailings soil and the tan, rock-like, indurated, expanded material.

Although there is consensus that the surface heaving is being caused by processes within the chrome tailings rather than related to the chromium itself, Raghu and Hsieh (1989) did speculate that similar expansion observed in Hudson County, New Jersey was due to the chromium. However, they posit no chemical mechanism or evidence to support this speculation. Dermatas *et al.* (2006) report significant expansion of chromite ore residue during attempts to remediate the chromium(VI), a known carcinogen, via *in situ* reduction using ferrous sulfate.

EXPANSION TESTING

One of the puzzling aspects of past investigations into the chromium tailings problem was the inability to simulate reliably the expansion behavior under laboratory conditions. Expansion testing consequently was the major focus of the present investigation.

Free Swell Expansion Tests

Free swell expansion tests were conducted using a modified soil consolidation test apparatus. Since anecdotal evidence suggested that expansion at DMT seemed to occur most during periods of fluctuating groundwater conditions, the samples were subjected to cyclical wetting and drying conditions consisting of a half-hour of full saturation followed by four hours of drainage. Three different saturants were formulated to reproduce the high pH 12 conditions measured in the groundwater at the site (see Table 1): a sodium hydroxide solution; a bicarbonate solution; and a saturant recipe that closely matched the site water chemistry. Free swell expansion tests were performed on a total of eight samples.

Overall, the free swell expansion testing was unable to achieve consistent and significant expansion. Expansion occurred in only three of the eight samples, and the maximum observed expansion levels were less than 1%, which would qualitatively appear to be too low and inconsistent with the magnitude of heaving observed at the DMT. These results are consistent with findings from past free swell expansion testing of this material.

Column Expansion Tests

Fluor-Daniel in an unpublished earlier study at another chromium tailings site reported success in obtaining expansion of the chromium tailings using a laboratory column expansion apparatus and an enriched CO_2 environment to accelerate the carbonation reactions. Consequently, the UMD team performed column expansion tests using two sets of apparatus. The first attempted to reproduce Fluor-Daniel's test conditions as closely as possible. The second, although similar to the Fluor-Daniel apparatus, attempted to model the groundwater table and the partially saturated capillary fringe zone immediately above the groundwater table.

TABLE 1: Summary of Free Swell Expansion Test Results

Sample	Saturant		
	Hydroxide	Bicarbonate	Synthetic Groundwater
1	None		
2	0.18% @ 80 hrs		
3	0.10% @ 80 hrs		
4	None		
5		0.80% @ 1400 hrs	
6		None	
7			None
8			None

Replicate of Fluor-Daniel Column Apparatus

The UMD replicate of the Fluor-Daniel column expansion apparatus is shown in Figure 1a. Tailings sand was placed at ambient moisture conditions as loosely as possible into the 64 mm (2.5 in) diameter acrylic column on top of a porous stone base. A perforated aluminum disk was placed on top of the sample as a telltale for measuring axial expansion of the material. Gaseous CO_2 was humidified by bubbling through a water-filled flask, similar to the Fluor-Daniel procedure, and then piped to the porous stone inlet at the base of the column from where it flowed vertically upward through the tailings to the porous stone outlet at the top of the column. This produced an essentially pure CO_2 pore vapor environment in the column.

Principal observations after 34 days of testing were as follows:

- There was no color change observed during the experiment. This is in contrast to the Fluor-Daniel test, where the tailings changed color early in the test.
- There was no visible expansion of the material in the column; the perforated aluminum plate at the top of the column of black sand did not move.
- At the end of the test, it was discovered that the tailings had hardened and wedged into the column so tightly that they had to be chiseled out. Qualitatively, hardening was nearly uniform over the height of the column.

Overall, the results from this test generally replicated and confirmed those from Fluor-Daniel's original tests.

Simulation of the Groundwater Table and Capillary Fringe Zone

Anecdotal evidence at DMT suggested that most of the expansion occurs in the partially saturated transition zone immediately above the groundwater table and/or within the capillary fringe zone affected by seasonal groundwater and saturation fluctuations. The second experimental apparatus illustrated in Figure 1b was designed to simulate the capillary fringe zone in an approximate way. The 865 mm column height allows for zones modeling conditions below the groundwater table, within the capillary fringe (estimated at 300 mm based on capillary rise tests), and above the capillary fringe. Small perforations in the lower 150 mm of the column permit flow of saturant through the sample material below the simulated groundwater table. Two CO_2 inlets were provided, one at the base of the column and the other above the simulated groundwater table, to ensure adequate CO_2 concentration throughout the column.

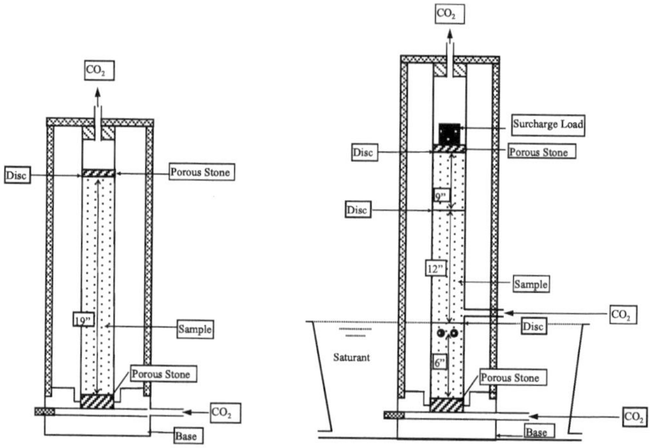

(a) Fluor-Daniel apparatus (a) UMD apparatus

FIGURE 1: Column expansion test apparatus.

Prior to testing, the tailings sand was placed in a 300°C oven for 48 hours. The sand was compacted in the column on top of the porous stone base in 150 mm lifts. Perforated thin aluminum discs for detecting axial expansion were placed at the simulated groundwater table level, at 300 mm above the simulated groundwater table level, and at the top of the column. A porous stone and a surcharge weight (equivalent pressure of 3.5 kPa) were placed on the top of the sample.

The saturant used for the simulated groundwater was the bicarbonate recipe that was most successful in the free swell expansion tests (see Table 1). The bottom of the column below the simulated groundwater table was submerged in a plastic container filled with the saturant. The saturant was continuously recycled using a pump. The pH of the saturant was monitored regularly and maintained at or above the target value of 12. Because of the CO_2 injection, regular additions of sodium hydroxide to the saturant reservoir were required to maintain the target pH.

Observations from the test after 23 days were as follows:

- In the first 24 hours, capillary water rose to the top of the sample, higher than expected based on the capillary rise tests. This might be due to the pressure and moisture in the injected CO_2.
- In the third week, the fully saturated portion started changing color from black to tan. This is consistent with field observations of the non-indurated vs. indurated material at the DMT.
- No visible expansion was observed in anywhere in the column during the test.

- When the experiment was halted, it was discovered that the material had hardened and wedged itself into the tube. It was impossible to extrude, so the tube had to be split. Qualitatively, the hardening was greatest in the lower, fully saturated portions of the sample.

The degree of induration in all zones in the second column apparatus appeared higher than in the Fluor-Daniel replicate.

Both columns in the first series of column expansion tests showed high levels of induration by end of the test. However, the only indication that expansion accompanied the induration was the fact that the indurated material wedged itself tightly into the column and had to be chiseled or cut out. A second series of column expansion tests was therefore designed to indirectly measure the expansion pressures by measuring the circumferential strains induced in the column walls. This second test series also investigated the influence of pH on the expansion phenomenon.

A total of four columns were tested. Each was similar to the apparatus shown in Figure 1b, but the columns were slightly shorter. The lower few centimeters of each column were immersed in the high pH saturant simulating natural groundwater conditions at the site. The upper portions of the columns were allowed to saturate to equilibrium conditions via capillary action. Four pH values were studied in this test series: 12, 11.3, 10.6, and 9.9. These values were selected to span the range of interest for possible field pH mitigation (The carbonate/bicarbonate equivalence point occurs at an approximate pH of 10.3.) The pH levels were monitored and adjusted on a daily basis and were generally maintained within ± 0.5 of the target values.

Expansion pressures were measured by temperature compensated circumferential strain gauges at three different heights on each column. Figure 2 summarizes the average equivalent radial expansion pressure in each column as a function of time. The equivalent radial expansion pressure is computed from the average circumferential strain:

$$p = \frac{\sigma_\theta t}{2R} = \frac{(E\varepsilon_\theta)t}{2R}$$

in which p is the equivalent radial expansion pressure, σ_θ is circumferential stress, t is the wall thickness of column, R is the column radius, E is the elastic modulus of cylinder (taken as the manufacturer's recommended mean value of 2.9 GPa), and ε_θ is the temperature-compensated average circumferential strain.

All tubes displayed continuing expansion vs. time over the entire 3.5 week duration of the tests. As shown in Figure 2, the columns having the highest saturant pH values of 12 and 11.3 exhibited the largest average radial expansion pressures, reaching maximum values of approximately 495 kPa (70 psi) or more at the end of the test. The pH=10.6 column exhibited an intermediate amount of expansion, with a final radial expansion pressure of approximately 380 kPa (55 psi), while the pH=9.9 column exhibited the smallest, but still significant, expansion of approximately 275 kPa (40 psi) at the end of the test. The pH level at which the expansion process may stop appears to be considerably lower that originally estimated. The results suggest that the pH may need to be reduced to approximately 8 or less in order to arrest the expansion process.

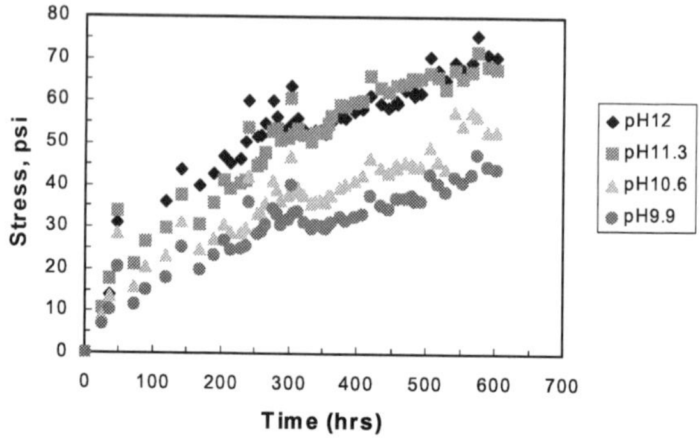

Figure 2. Radial expansion stress vs. time and pH (1 psi = 6.9 kPa).

All four columns were carefully split at the end of the testing period and the cores removed for post-mortem examination. The cores from all four columns had indurated to a substantial degree, and there was no observable qualitative difference in the degree of induration between the various columns.

Unconfined compression strength tests were performed to determine if there are any quantitative mechanical differences among the cores from the various columns and, if so, whether these differences mirror the trend of radial expansion stress vs. pH shown in Figure 2. Results from the unconfined compressive strength tests are summarized in Figure 3. Even though great care was taken in removing the indurated material from the split columns, there was still much breakage and it was thus difficult to find pieces of sufficient size for testing. Consequently, it was impossible to assess variations of compressive strength along the length of the columns. However, the data in Figure 3 clearly show the influence of pH on the average compressive strength of the post-test indurated material. The unconfined compressive strength of the pH=12 material was approximately twice as large as the unconfined compressive strength of the pH=9.9 material. The secant elastic modulus at failure for the pH=12 material was approximately four times larger than the secant elastic modulus at failure for the pH=9.9 material. Trends for the intermediate pH materials fall logically between these limits.

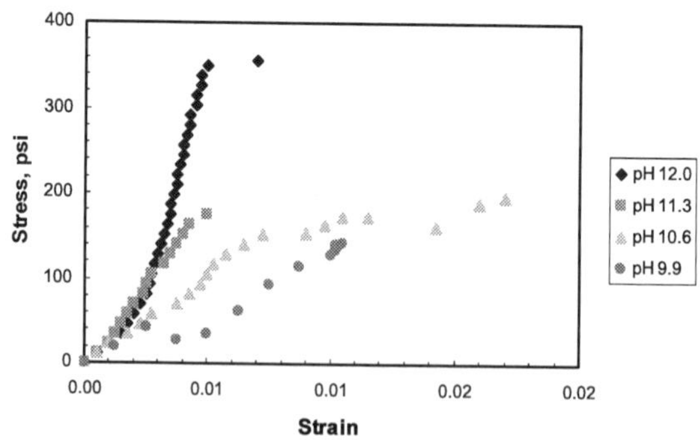

Figure 3. Unconfined compression of indurated material (1 psi = 6.9 kPa).

SUMMARY AND CONCLUSIONS

The column expansion tests reliably reproduced expansion and induration of the chromium tailings material underlying the pavements at the DMT. The tests results demonstrated that expansion pressure and degree of induration as quantified by uncompressed compressive strength decrease with pH but that an impractical reduction of pH may be required for complete mitigation. The CO_2 enrichment employed to accelerate the expansion process in the laboratory clearly does not realistically represent *in situ* conditions. Nonetheless, this testing technique can be employed to explore further the influences of site factors and possible remediation approaches on the chromium tailings expansion.

ACKNOWLEGMENTS

This work was performed with funding from the Maryland Port Administration Michael Hild and Douglas Matzke are the past and present chief engineers at the Port. All views expressed in this paper are those of the authors alone and do not necessarily represent those of the Maryland Port Administration.

REFERENCES

Century Engineering, Inc. (1980). *Geotechnical Investigation for Pavement Distress Evaluation, Dundalk Marine Terminal*, unpublished report prepared for the Maryland Port Administration.

Crawford, C.B., and Burn, K.N. (1969). "Building Damage from Expansive Steel Slag Fill." *Journal of the Soil Mechanics and Foundations Division*, ASCE, Vol. 95 (SM6): 1325-1334.

Dermatas, D., Chrysochoou, M, Moon, D.Y., Grubb, D.G., Wazne, M., and Christodoulatos, C. (2006). "Ettringite-Induced Heave in Chromite Ore Processing Residue (COPR) upon Ferrous Sulfate Treatment." *Environmental Science and Technology*, Vol. 40: 5786-5792.

EA Engineering, Science, and Technology, Inc. (1987). *Final Report: Tasks III-A and III-B Preliminary Petrographic and Mineralogic Studies, Engineering Geology of Chromium Tailings Fill at Dundalk Marine Terminal*, unpublished report prepared for the Maryland Port Administration.

EA Engineering, Science, and Technology, Inc. (1987a). *Dundalk Marine Terminal Site Contamination Assessment*, unpublished report prepared for the Maryland Port Administration.

EA Engineering, Science, and Technology, Inc. (1989). *Final Report: Task III-C Chemical and Geotechnical Tests for Engineering Geology of Chromium Tailings Fill at Dundalk Marine Terminal*, unpublished report prepared for the Maryland Port Administration.

Raghu, D., and Hseih, H.N. (1989). "Performance of Some Structures Constructed on Chromium Ore Fills." *Journal of Performance of Constructed Facilities*, ASCE, Vol. 3 (2): 113-120.

Ramchandran, V.S., Sereda, P.J., and Feldman, R.F. (1964). "Mechanism of Hydration of Calcium Oxide." *Nature* (United Kingdom), Vol. 201 (4916): 288-289.

Schwartz, C.W., James, B., Wylie, A., Davis, A., and Baecher, G. (1997). *Investigation of the Expansive Behavior of Chromium Tailings: Phase I Report*, unpublished report prepared for the Maryland Port Administration.

Schwartz, C.W., Wylie, A., Davis, A., and James, B. (2000). *Investigation of the Expansive Behavior of Chromium Tailings: Phase II Final Report*, unpublished report prepared for the Maryland Port Administration.

Weidner, J., and Wylie, A. (1980). *Preliminary Mineralogical Characterization of Chrome Tailings Material*, unpublished report prepared for the Maryland Port Administration.

Prediction of Axial Capacity of Straight Shafts and Augercast Piles in Clay Soil

W. Tom Witherspoon[1], P.E, Ph.D. and Anand J. Puppala[2], P.E., Ph.D

[1]President, S & W Foundation Contractors, Dallas, Texas, (972) 235-3267, Email: tomw5@ix.netcom.com

[2]Professor, Department of Civil Engineering, Box. 19308, The University of Texas at Arlington, Arlington, Texas 76019, 817-272-5821, Email: anand@uta.edu

ABSTRACT: A side by side comparison of straight shafts and augercast piles in expansive clay conditions was evaluated in a research project studied at the University of Texas at Arlington and these results showed the performance of both foundations was similar. There was little difference in axial capacity measurements when drilled shafts and augercast piles were constructed with the same diameter and the length. Capacity predictions with theoretical models showed good agreements with the measured results. Also, these analyses results showed that when augercast piles or drilled shafts were constructed in a clay environment, there is no perceptible difference in the skin friction and end bearing resistance factors when predicting ultimate capacity.

INTRODUCTION

Drilled shafts have been studied for a long time and these foundations were tested in various environments and soil conditions (O'Neill and Reese 1992 & 1999). Much of this literature work on drilled shafts has been sponsored by the Federal Highway Administration (FHWA), States' Departments of Transportation (DOTs) and the International Association of Foundation Drilling (ADSC) and this research has resulted in various research reports, design guides and standard practices (O'Neill and Reese 1992 and 1999).

Augercast piles were originally developed to provide an economical method of casting a foundation in subsurface conditions where the water table is high in granular soil conditions (ADSC 2005). A considerable number of published articles on augercast piles are available, but estimating their capacity has been more of a site specific testing than predictive calculations that are common with drilled shafts (ADSC 2005). Since the augercast pile is mostly used in granular conditions, there is not a plethora of research available in expansive clay soil that demonstrates how the piles transfers loads to foundation subsoils.

A research study was conducted to install various foundation systems in a high plastic expansive clay environment and these foundation systems were subjected to axial load tests at various seasonal conditions, dry conditions during summer and wet season like spring conditions. This paper presents a summary of axial ultimate loads of drilled shaft and auger cast pile foundations in both seasonal conditions. Measured results were also evaluated by comparing theoretical predictions from the available models for clayey soil conditions.

CAPACITY PREDICTION MODELS

The following Equation provides the calculation of ultimate capacity of drilled shaft foundations:

$$R_{TN} = R_{SN} + R_{BN} \qquad (1)$$

Where:

R_{TN} = total nominal ultimate capacity of the shaft (kN)
R_{SN} = nominal ultimate side resistance of the shaft (kN)
R_{BN} = nominal ultimate base resistance of the shaft (kN)

For cohesive soils similar to the ones on which this research is being conducted, where s_u is greater than 95.7 kPa and the length of the foundation is greater than three times the diameter of the foundation, the base resistance (R_{NB}) in terms of q_{max} can be expressed as:

$$q_{max} = 9 \times s_u \qquad (2)$$

For the same cohesive soils, side resistance in terms of f_{maxi} of each soil layer can be calculated based on each strata/layer that the drilled shafts penetrates is:

$$f_{maxi} = \alpha \times s_u \qquad (3)$$

where:

f_{maxi} = side resistance for compressive loading in layer i (kPa)
α = a dimensionless correlation coefficient defined as follows:
α = 0 between the surface and 1.5 m deep or the depth of seasonal moisture content change, whichever is deeper
α = 0 for a distance of the bell diameter up from the bottom of the shaft.
α = 0.55 elsewhere for $s_u / p \leq 1.5$ and varying between 0.55 and 0.45 for s_u/p between 1.5 and 2.5 as graphically depicted by Figure B-9 (FHWA 1999) and p is the atmospheric pressure, which is 14.64 psi (101 kPa).
s_u = undrained shear strength for the layer in kPa

As explained above, each layer must be considered individually and the total side resistance is calculated as the sum of resistance offered by each layer.

Augercast piles were developed in the 1940's as a result of pressure grouting open holes and holes backfilled with coarse aggregates (Neate 1988). The first patent for augercast piles was issued in 1956 with a process of pumping grout down a hollow-stem continuous flight auger as the auger is gradually extracted. Since the expiration of this patent in 1973, the augercast pile has increased in usage around the World (Neely 1989). There are two primary underpinning techniques used in Europe, the helical/screw pile and the augercast pile or continuous flight auger (CFA) piles (O'Neill 1994).

The equipment requirement for augercast piles is similar to the drilled shafts, except that the drilling tool is a continuous auger stem with the appropriate bits at the bottom. When the auger reaches the targeted depth, grout is pumped down the hollow auger stem to push soil up as the bit is extracted out of the ground (Cernica 1995). The contractor will calculate grout volume and make sure the grout stays ahead of the extraction so that no voids will be created as a result soil caving in what is normally a wet environment prone to collapse.

The engineer must evaluate the subsurface to determine which system is most favorable for a specific site condition. Calculations to estimate axial load capacity include the side friction capacity for the augercast piles but this allowance is less or eliminated for the same diameter drilled shafts in the U.S. To verify the axial loads estimated by the calculations, site specific axial load tests are many times performed prior to a construction project. Kitchen et al. (1995 and 2005) showed the significance of field testing that helped in the final foundation selection for a construction project in Georgia. This testing lowered the project cost by redesigning for drilled shafts in lieu of the original augercast design.

It should be noted that in European codes, the skin friction allowances for both augercast and drilled shafts are the same (De Cock 1997). Hence, axial load estimation and comparison between drilled shafts and augercast piles should provide valuable information that could be used to design the appropriate foundation elements for the construction projects.

Though augercast piles have been predominantly used in non-cohesive soils, there is a growing trend to use them in cohesive soils. One of the problems encountered in using augercast piles in cohesive soil is that soils block the auger, which increases resistance and many times prevents further penetration (Neely 1990). In this research, relatively shallow depths were used and hence the problem of spoil resistance to drilling process was not significant. Comparative axial load testing was performed to evaluate skin friction and end bearing components, which was examined to address the influence of skin friction component to the ultimate axial loads of these foundations.

A majority of the augercast piles are reinforced with a single bar through the center. In this research, for practical and fair comparisons, the same reinforcement as that of the drilled shaft was inserted into the grouted auger cast shaft immediately after completion of the augercast pile process.

GEOTECHNICAL INVESTIGATIONS

The site was in Irving, Texas where no rock layer was encountered for a depth of 18.3 m or higher. Soil in this region is highly expansive and active depths where the moistures fluctuate are around 3.05 m. A total of four soil borings were made in and around the site for soil property evaluation. Atterberg limits including liquid limit (LL), plastic limit (PL) and plasticity index (PI) values were first determined on soils at different depth zones and a summary of these properties from boreholes 1 and 2 are given in Table 1.

Table 1: Atterberg Limits of Soil Samples

Depth Range	LL	PL	PI
0 - 1.5 m	40	15	35
1.5 - 3 m	52	21	31
3 - 4.5 m	49	21	28
4.5 - 6.1m	39	12	27
6.1 - 9.1m	49	19	30
9.1- 12.2m	45	20	25

Also, the moisture contents in the field varied from 10% in summer conditions to above 20% in spring conditions. Based on the moisture content changes monitored at the field, the active depth of the site was taken as 3 m. Undrained shear strength (Su) values for the site were determined using UU triaixial tests on soil samples collected from the field for

Spring and Summer season. These laboratory results were then correlated with CPT results for summer conditions with very accurate corroboration. These results are reported in Table 2. These results show that these properties correspond to medium stiff nature of the present clayey subsoils.

Table 2: Undrained Soil Strength Parameters determined by CPT

	Spring 2005	Summer 2005
Depth of Section	**S_u in kPa**	**S_u in kPa**
0 - 1.5 m	69**	200
1.5 - 3 m	124.1**	193
3 - 4 m	289.5	289.6
4.5 m	117.2	117.2
4.9 - 8.5 m	200	200
8.5 - 11.9 m	255.1	255.1
11.9 -15 m	137.9	137.9
15 - 16.7 m	110.3	110.3
16.7 - 18.3 m	82.7	82.7
18.3 – 22.9 m	1323.8*	1323.8*

*Estimated from SPT measurements using accepted In Situ conversions;
**Samples tested using UU Triaxial Tests, correlated to CPT results;
1 psi – 6.9 kPa

FOUNDATION DETAILS AND ANALYSIS

To better monitor performance of foundation elements installed at different seasons, half of the piers and augercast piles were installed in August and tested in April, while the other half were installed in April and tested in August. All piers and piles were tested in accordance with ASTM 1143 using the quick test.

Design dimensions chosen for this research were considered based on those that were typically considered for installation in the design for residential building foundation applications in expansive clay environment. The drilled shafts and augercast piles were installed in the same clay environment, at the same location and with the same dimensions and reinforcement as depicted in the Figure 1. Both foundations are 0.3 m. in diameter and 0.38 m in length with 4 - #5's set vertically and #3 stirrups set at a spacing of 0.3 m o.c.

Load tests were conducted as per the standard procedures (ASTM 1143). Figures 2 and 3 present load versus deflection profiles from the axial load tests conducted on both straight shafts and augercast piles for the dry to wet season. The wet to dry season elements (piers 39, 40, 41 and piles 45, 46 & 48 were similar). Table 3 presents a summary of load test results of all straight shafts and augercast piles constructed at the present site.

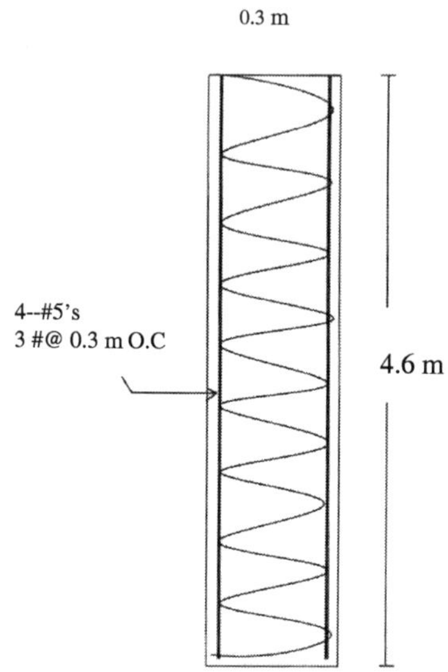

Figure 1: Pile and Pier Dimensions Used in this Research

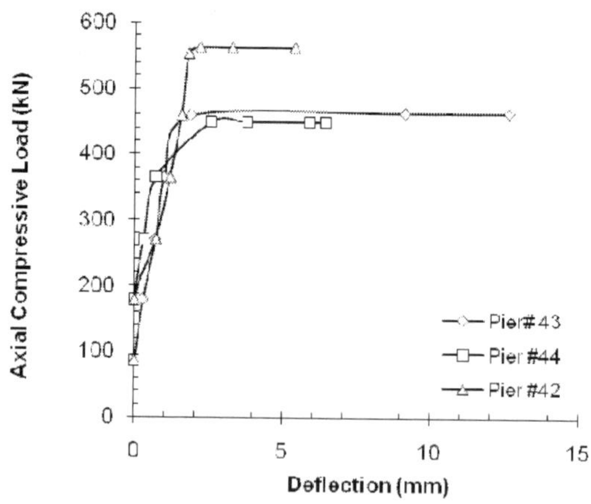

Figure 2: Load vs. Deflection Plots for Drilled Shafts Installed in August and Tested in April, 2005

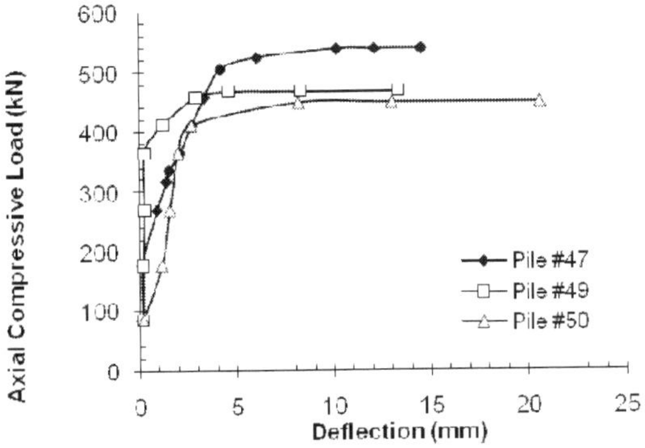

Figure 3: Load - Deflection Profiles for Augercast Piles Installed in August and Tested in April

Table 3 Load Test Results from Tests on Drilled Shafts and Augercast Piles

Foundation Type	Designation	Foundation #	Depth (m)	Measured Ultimate Load Capacity (kN)
Straight Drilled Shaft	Installed in dry season and tested in wet season	42	4.6	560.4
		43	4.6	463.5
		44	4.6	450
Straight Drilled Shaft	Installed in wet season and tested in dry season	39	4.6	474.2
		40	4.6	547.1
		41	4.6	456
Augercast Pile	Installed in dry season and tested in wet season	47	4.6	538.6
		49	4.6	468
		50	4.6	450
Augercast Pile	Installed in wet season and tested in dry season	45	4.6	489
		46	4.6	475
		48	4.6	475

Figure 4 presents a graphical comparison of the axial load capacities of both foundations. It is interesting to note that both foundations yielded almost same capacities for the similar dimensions and reinforcement details considered in this research. This indicates that both these foundations though different in the construction process yielded similar capacities under axial load testing.

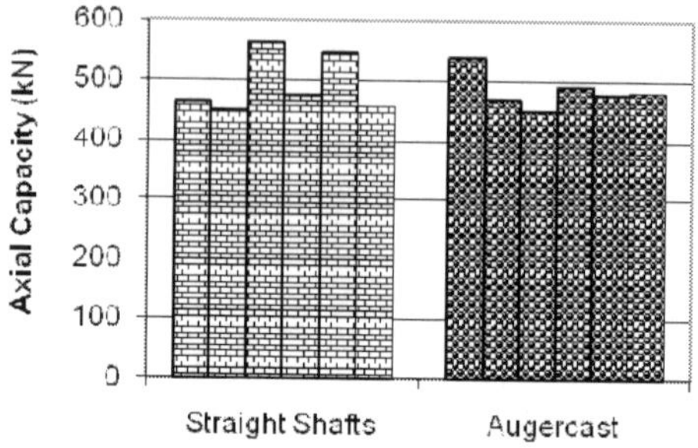

Figure 4: Comparison of Axial Ultimate Load Capacities of Straight Shafts and Augercast Piles for Both Seasons

Table 4 presents a summary of the average ultimate loads and deflections at which peak loads were recorded. Though loads are similar, the deflections at peak loads are different. Augercast piles needed higher deflections to achieve the ultimate loads and this is attributed to the bottom of hole conditions prior to concrete placement. Construction procedures for augercast pile normally do not produce a clean bottom surface whereas the drilled shaft, with a visual inspection prior to pouring of concrete, will produce higher end bearing. Inspection is typically done either by visual examination in the hole or through a camera. These variations in deflection prior to achieving ultimate capacity are presented in Table 4 below.

Table 4: Average Ultimate Loads and Deflections

Installation and Testing Condition	Foundation type	Average Load at Failure in kN	Average Deflection at failure in mm
Installed in dry season and tested in wet season	Straight shafts	491.5	4.60
	Augercast piles	485.3	7.60
Installed in wet season and tested in dry season	Straight shafts	492.4	3.05
	Augercast piles	480.5	5.56

While predicting loads using the FHWA (1999) procedure, the same axial loads were estimated as those measured in the field for both drilled shafts and augercast piles. These results can be seen in Witherspoon (2006). One important implication from this analysis is that both foundation types yielded similar skin friction estimation in the design calculations. This indicates that side friction does contribute to overall ultimate loads of both foundations, in particular drilled shafts. Also, the season of testing did not show major influence of the axial load capacities of both foundations since both foundations of 4.6 m in length are longer than active depth at the test site and there was an increase in shear strength in the dry period that compensates for shortness in contact area along the shaft.

SUMMARY and CONCLUSIONS

There was a negligible difference between ultimate axial compression capacity in the straight drilled shafts and augercast piles installed with same dimensions in similar soil conditions. The deflection readings at ultimate capacity were slightly greater for the augercast pile than for the straight shaft. This difference may be attributed to construction procedures in that the augercast pile, which normally does not produce a clean bottom surface whereas in an open hole drilled shaft, the bottom surface is inspected before placing concrete. Inspection is typically done either by visual examination in the hole or through a camera. Based on the load test results and analyses, an allowance for skin friction is warranted for the drilled shafts in expansive clay environment and this is similar to augercast piles.

REFERENCES

ADSC, "The Future of ACIP/CFA Piling in the North American Transportation Market", Findings Report, October 14, 2005, Emerging Technologies Committee, ACIP/CFA Task Force.

Black, J. P., "A Primer on Augercast Piles", (December 2005). Structure Magazine, pg. 14-16,

Cernica, J. H., *Foundation Design*, (1995), John Wiley and Sons, USA and Canada pg. 394, pg 66.

De Cock, F., Legrand, C., Lehane, B. and Mandolini, A., (1997), International Seminar on European Practice for Designing Axially Loaded Piles, Brussels, pp. 313-323.

Federal Highway Administration, Publication No. FHWA-HI-88-042 (August 1988), DRILLED SHAFTS: Construction Procedures and Design Methods,.

Federal Highway Administration, (1999) DRILLED SHAFTS: Construction Procedures and Design Methods, FHWA Publication No. FHWA-IF-99-025.

Kitchens, L., Thompson, W. R., and Housley, S., "Foundation Drilling", (March/April 2005), page 12-17.

Neate, J.J., "Augered cast-in-place piles", (1988) 13th Annual Meeting, Deep Foundation Institute, Atlanta, Georgia., pp. 167-175.

Neely, W. J., (October 1990)"Installation, Design and Quality Control of Augercast Piles." presented at the 15th Annual Members' DFI meeting and seminar, page 1 and 3.

O'Neill, M. W., Reese, L., Barnes, R., Wang, S., Morvant, M., Ochoa, M., (1992) "Effects of Stratagraphic and Construction Details on the Load Transfer Behavior of Drilled Shafts", Transportation Research Record No. 1336, Transportation Research Board,.

O'Neill, M. W., Hassan, K. M., (1993), "Perimeter Load Transfer in Drilled Shafts in the Eagle Ford Formation", Design and Performance of Deep Foundations: Piles and Piers in Soil and Soft Rock, Geotechnical Special Publication No. 38, ASCE.

O'Neill, Michael W., (1994), "Review of Augered Pile Practices Outside the United States," Paper Presented at Transportation Research Board (TRB).

SITE-SPECIFIC RISK ASSESSMENT OF SULFATE INDUCED HEAVE IN LIME-STABILIZED CLAY SOILS

Bruce Herbert[1], Dallas Little F.ASCE [2], Syam Nair[3], Chris Markley[4].

[1]Professor, Department of Geology and Geophysics, 309 M.T. Halbouty, Texas A&M University, College Station, Texas 77843-3135, E-mail: herbert@geo.tamu.edu
[2]Professor, Department of Civil Engineering, 603E CE/TTI Building, Texas A&M University, College Station, Texas 77843-3135, Ph: 979-845-9963, Fax: 979-845-0278, E-mail: d-little@tamu.edu
[3]Graduate Assistant Research, Texas Transportation Institute, 601A CE/TTI Building, Texas A&M University, College Station, Texas 77843-3135, E-mail: syam@tamu.edu
[4]Graduate Assistant Research, Department of Geology and Geophysics, 212 M.T. Halbouty, Texas A&M University, College Station, Texas 77843-3115, E-mail: cmarkley@neo.tamu.edu

ABSTRACT: Practicing engineers are at times required to ameliorate or stabilize sulfate-bearing soils that exhibit sulfate-induced heave when treated with lime. Cost effectiveness of stabilization and amelioration techniques can be improved by using geologic models that identify the factors and processes that control the distribution of sulfate through the landscape. The paper describes the use of geospatial data sets, direct measurements of soil conductivity, and geochemical modeling to develop a geologic model of sulfate distribution along the SH-130 corridor in Texas. These geologic models can effectively be used to define the risk associated with sulfate-induced heave in soils that are encountered during construction. The technique can also be used to focus sampling and testing to the high-risk locations and thereby contributing to the economic success of the project. The sensitivity of soils to form ettringite/thaumasite with increasing sulfate concentration can also be evaluated using these geologic models.

INTRODUCTION

In the past two decades, the number of reported cases where lime-stabilized soils are damaged due to formation of the expansive minerals ettringite and thaumasite has increased (Hunter 1988; Mitchell 1986). Ettringite precipitates in environments with high pH and high activities of Ca^{2+}, Al_{aq}, and SO_4^{2-} ions, whereas thaumasite, a silicon-bearing member of the ettringite mineral group, forms in similar environments at low temperature conditions when concentrations of silica and carbonates are high (Moore and Taylor 1970; Petry and Little 1992; Kohler et al. 2006). Sulfate concentrations are generally assumed to be the most important factor controlling the formation of ettringite or thaumasite in lime-amended soils (Berger et al., 2001). Sulfates in soil also have the highest spatial variability among all components needed to form ettringite/thaumasite mostly due to heterogeneity in source rocks and its hydrologic mobility as a dissolved ion. Redistribution of sulfate within the landscape is influenced by hydrologic and aoelian processes, including unsaturated flow, runoff, evaporation, aeolian deposition and erosion (Czerewko et al. 2003). The hydrologic and aeolian processes are themselves a function of environmental factors, including topography and climate. Prediction or evaluation of the distribution of sulfate across a landscape often requires extensive field sampling and laboratory testing which can be laborious, costly and time consuming. This presents a major problem to the practicing engineer in evaluating the risk of sulfate-induced heave in lime-stabilized soils or in any sulfate-bearing soil treated with a calcium-based stabilizer that induces a high pH.

The paper is based on a case study conducted to address soil stabilization issues along the southern sections of the State Highway 130 (SH-130), a 49-mile toll road intended to relieve traffic congestion along Interstate 35 between Austin and San Antonio. The paper describes the use of existing geospatial data sets as well as direct measurements of soil conductivity, a proxy for direct measurement of sulfate concentrations, to create a geologic model of sulfate distribution in the SH130 corridor in order to evaluate the risk associated with sulfate-induced heave in lime-stabilized soils.

METHODS

Identifying Sources of Sulfate at the Landscape Scale

State Soil Geographic Database (STATSGO) was used to identify the geologic sources of sulfate at the kilometer scale through an analysis of soil characteristics, including sulfate concentrations, along the SH-130 corridor (SCS 1975). In the STATSGO database, soils are represented by soil map units, each consisting of a set of soils with similar physiographic characteristics. Gross topography of the location was identified using a digital elevation model (DEM) with 1 arc-second or 30 m resolution.

Field Mapping of Sulfate Levels at the Meter Scale

Sulfate distribution at the meter scale was mapped using an EM 38 Magnetometer that measures bulk electrical conductivity of soils. Soil electrical conductivity, which is dependent on salt, water, and clay concentrations, is a reasonable proxy for direct measurement of sulfate concentrations (Little et al., 2006). Conductivity measurements were made along multiple 500m transects that are approximately 100 to 200 meters apart. Measured conductivities along selected transects were assumed to be representative of the entire soil group.

The conductivity readings were then correlated with sulfate concentrations by measuring sulfate levels in cores collected from these locations. The observations indicated a "background" EM 38 reading of approximately 45 mS/m before changes in sulfate levels could be detected in soil cores. From the regression analysis between conductivity readings and sulfate content, a conductivity reading of 50 mS/m was (conservatively) associated with a sulfate concentration of approximately 2000 ppm. Therefore, a conductivity reading of 50 mS/m was selected to distinguish between soils with low and moderate risk categories. Sulfate concentrations measured from soil cores were compared with conductivity readings in the development of the sulfate concentration - conductivity correlation.

Modeling the Behavior of Soils from Problematic Locations

Geochemical modeling of the Heiden soil, a soil from the SH 130 corridor with high sulfate concentrations, was performed to establish the sensitivity of the soil to precipitate ettringite or thaumasite as a function of sulfate concentrations. Chemical characteristics of the soil, as determined by elemental extraction, X-ray diffraction (XRD), scanning electron microscope (SEM), and Fourier-transform infrared spectroscopy, were used as input parameters in the geochemical speciation model (Table 1). Modeling was based on the Predominance approach where the soil components are allowed to react to completion. Assumptions and steps established by Little et al. (2005) were followed in developing phase diagrams that are used to identify stable mineral phases in soil after stabilization. Uncertainties and limitations in modeling were also discussed in detail by Little et al. (2005).

Table 1. Input parameters used in React program for modeling Heiden soil.

Variable	React Input	Units	Mineral	React Input (gm)
[a]Water	0.3	Free kg	Goethite	20
pH	12	-	Calcite	125
Temperature	25	^0C	[b]Quartz	100
Pressure	1.013	bar	[b]Kaolinite	100
K^+	13.9	mg / kg	Muscovite	25
Ca^{2+}	125	mg / kg	[b]Smectite (low Fe, Mg)	225
Fe^{3+}	1.58	mg / kg	Albite	35
Al^{3+}	0.54	mg / kg	Potassium Feldspar	35
Mg^{2+}	773	mg / kg	Lime	35
[b]SO_4^{2-}	11.8	mg / kg		
$SiO_2(aq)$	37.2	mg / kg		

[a]Activity held constant at 1.0
[b]Input varied for sensitivity analysis

Predictions of ettringite or thaumasite formation in the Heiden soil were validated using lab experiments conducted on samples collected from the test location. One control sample and three sulfate bearing samples were prepared for experimental study. Sodium sulfate was added to spike sulfate concentration in three sulfate-bearing samples to 5,000, 10,000 and 30,000 ppm. Twenty-five (25) grams of soil with varying concentrations of sulfates were combined with five percent calcium hydroxide and 10 ml water and subjected to accelerated cure at 40^0 C for 75 days. After the first 4 days of cure, samples were inspected periodically and the required amount of water was added to keep the samples moist during curing periods. X-ray diffraction (XRD) and scanning electron microscopy (SEM) were used to verify the presence (or absence) of ettringite or thaumasite.

Mapping of Risk Categories

Conductivity readings, measured along different transects in the SH 130 corridor, were translated in to sulfate contents in the soils and compared with empirically established sulfate threshold levels (Berger et al., 2001) that define the three risk categories for sulfate-induced heave i.e., low, medium and high. The distances along each transect that fell within each risk category was then converted to a percentage of total distance. Finally, the percentages estimated on different transects were averaged and assumed to be representative of spatial areas within in each soil class of the STATSGO database.

Results and Discussions

Risk Associated with Sulfate at the Landscape Scale

Soil characteristics extracted from the STATSGO database and digital elevation models were used to identify sites for targeted field sampling and soil characterization and to classify areas in the corridor that are at risk for sulfate-induced heave (Figure 1).

Both sulfate concentrations and conductivities appear to be the highest in TX 235 soils formed from the Navarro group of late Cretaceous age that are composed of a marine marl and carbonaceous shale. But the soils also showed scattered conductivity readings indicating a strong spatial variability in sulfate concentrations at meter scales, as discussed below.

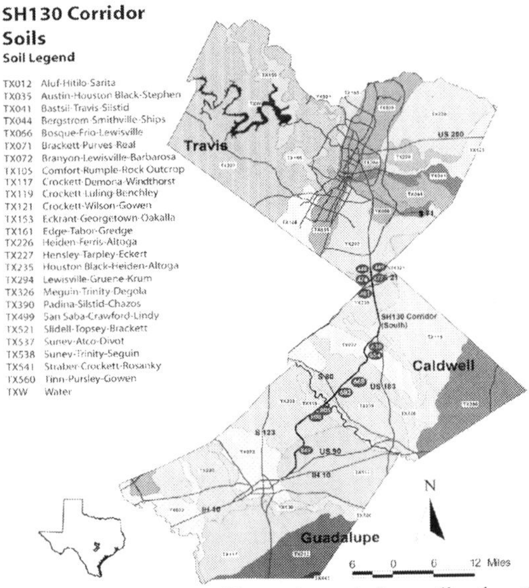

FIG. 1 General soil map for SH 130 corridor and targeted sampling locations (black ovals) identified using STATSGO database.

Field sites identified for targeted sampling were chosen to represent the different chemical characteristics, texture and topography of the soils in the construction corridor. Based on sulfate concentrations in these soils, twelve soils at risk of sulfate heave along the alignment were identified. These soils typically have high concentrations of gypsum, ranging from 0-12 percent (Table 2). Among the high sulfate soils a subset of potentially problematic locations was selected in order to consider a reasonable range in soil texture and mineralogy. These soils consist of soil groups TX 235, TX 121 and TX 119. These three soils were dominated by clays with high shrink swell properties and typically fall under AASHTO classifications A-4, A-6, A-7 or A-7-6. These clay rich soils have high salinity, higher carbonate and sulfate contents and a high water holding capacity (Table 2). The mineralogy of these soils is dominated by high concentrations of amorphous silica, smectitic clays and carbonates that may provide the major source for soluble silica, aluminum and carbonate ions needed for ettringite/thaumasite precipitation.

Characterization of Sulfate Distribution at the Meter Scale

At the meter scale, conductivity measurements indicate that topography is critical in controlling the distribution of sulfate ions (Figure 2). The measured conductivity was found to be a function of elevation and showed an inverse trend between level of conductivity and elevation (lower conductivities at higher elevations). The observed dependency of elevation and conductivity is probably due to the transfer of salts by hydrologic processes and the accumulation of salts upon evaporation. The Digital Elevation Model also showed slopes varying from 0-10 percent in the region indicating a dissected terrain shaped by stream erosion. A higher conductivity obtained along dry stream channels supports the hypothesis stated previously.

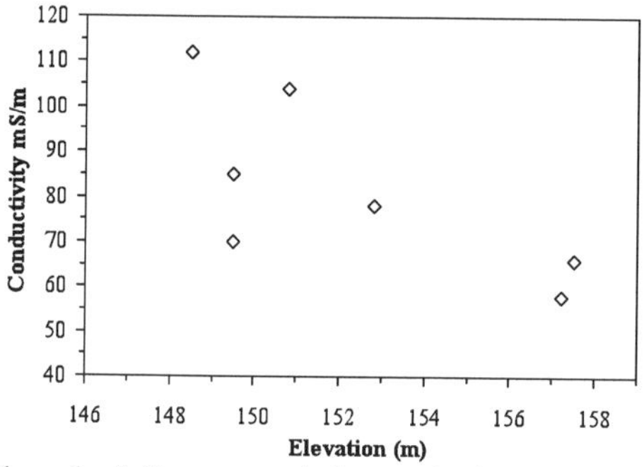

FIG. 2 Soil conductivity measured along a single transect at site 668.

Modeling Ettringite Formation in Sulfate-bearing Soils

Sulfate-induced heave is caused by the precipitation of ettringite or thaumasite at high pH conditions (Little et al., 2005). Geochemical modeling can be used to predict whether the formation of ettringite or thaumasite is thermodynamically-favorable in lime-amended soils. The Heiden soil was identified to be the major soil type in TX 235 soil group and therefore was selected for lab investigation to determine their sensitivity to ettringite formation. Characteristics of the Heiden soils, collected at the intersection of the SH 130 corridor and U.S. highway 290 was used in modeling soil behavior (Little et al. 2005). The major mineral fractions identified in the soil include goethite, calcite, quartz, feldspar, kaolinite, muscovite and smectite. These constituents were used as input to the geochemical model (Table 1). Thermodynamic modeling predicted thaumasite formation to be likely in Heiden soils (Figure 3a).

The red and black stars marked as 500 and 7,500 ppm in Figure 3a represent the locus of pH 12 and activity of sulfate ions corresponding to concentrations of 500 and 7,500 ppm, respectively. Thermodynamically, thaumasite is predicted to precipitate due to the presence of high concentrations of silica and carbonates (Hunter 1988). This prediction, though, ignores kinetic controls on ettringite/thaumasite formation. Most field observations have concluded that ettringite generally forms first and then transforms to thaumasite under specific conditions of low temperature and high silica content (Kohler et al. 2006). A sensitivity analysis of the influence of sulfate content on thaumasite formation showed thaumasite concentration to increase linearly with increase in sulfate content in the soil (Figure 3b).

(a) 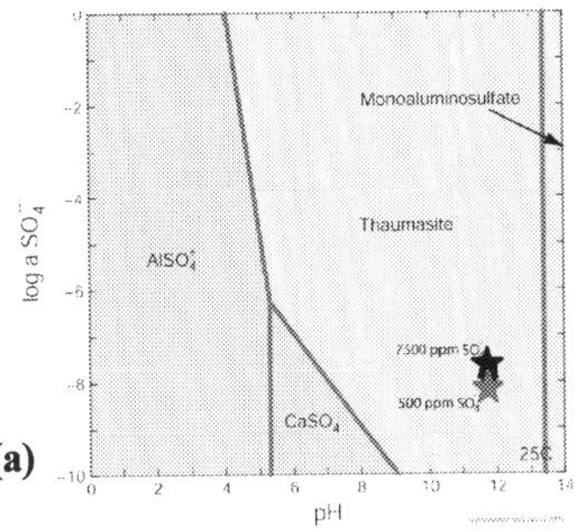

(b) 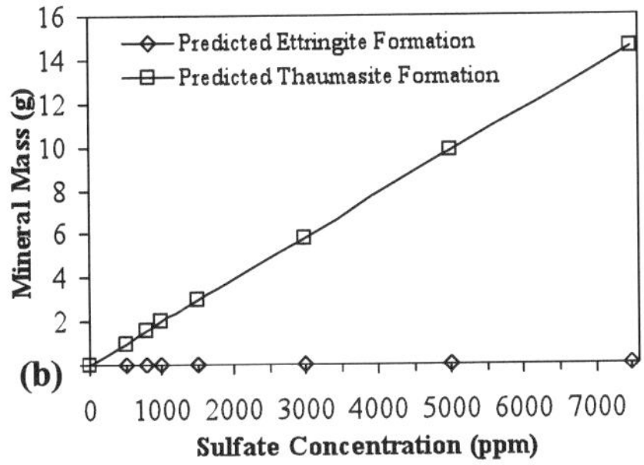

FIG. 3 (a) Geochemical models for Heiden soil with 500 ppm (red star) and 7500 ppm (black star) and (b) predicted thaumasite concentration with increasing sulfate contents.

Model results were validated based on experimental studies where soils with varying sulfate concentrations were treated with lime and cured for 75 days under accelerated curing conditions. SEM images of soil samples with 5,000 ppm sulfate (sample 2) and 15,000 ppm sulfate (sample 3) are presented in Figure 4.

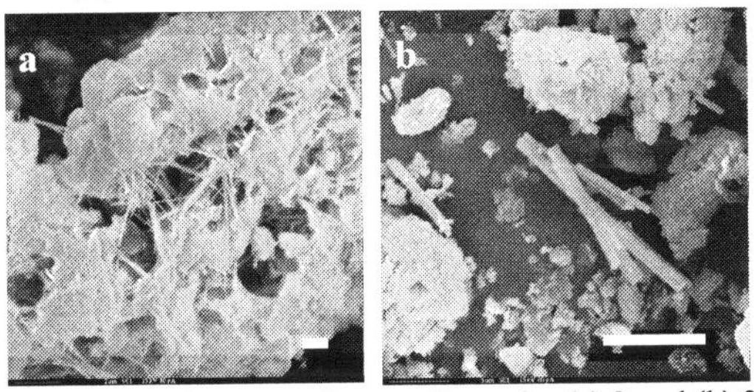

FIG. 4 Scanning electron microscopy image of samples (a) 2 and (b) 3. Bar is 1 μm.

The experimental study showed ettringite to be present in the soil and not thaumasite (Figure 4). For sample 2, the crystal formations were irregular and short and were mostly less than 1 micron in length (Figure 4a). Lime treated soils containing 15,000 ppm (sample 3) and 30,000 ppm sulfates also showed a greater number of ettringite clusters. In these samples, well-formed crystals, larger than 2 micron in length, were scattered throughout the matrix (Figure 4b). For thaumasite to precipitate and remain stable in soils, a sustained temperature below approximately 15^0 C is required (Hunter 1988). Since the laboratory samples were cured at 40°C, the formation of thaumasite would not be expected to occur.

Mapping Risk Categories of Sulfate Heave in Soils and Their Significance in Engineering Applications

The measured conductivities of soils in this corridor were translated into spatial area estimates to identify the soil with the highest potential to cause sulfate heave when treated with lime (Table 2). Spatial estimates of sulfate concentrations can be used to target sampling locations and help limit the overall cost associated with material sampling for sulfate testing. Moreover, amelioration techniques to be used during construction at high sulfate locations can be recommended using these spatial estimates, which can also be incorporated in to the design and estimation phase of the project. The spatial estimates are based on established sulfate threshold levels in soil (Berger et al., 2001). But the reactivity of soils may vary with differences in soil mineralogy (Little and Nair 2007). Geochemical models can combine threshold levels and soil mineralogy characteristics to define the sensitivity of soil to increasing concentrations of sulfates.

Table 2. Soil characteristics and risk assessment based on conductivity measurements.

Stations[a] (feet x 10^2)	**STATSGO Soil Group**	**Soil Series[b]**	**SO_4^{2-} levels (%)**	**CO_3^{2-}levels (%)**	**EM 38 Reading (mS/m)**		**Spatial Estimate for Sulfate Heave (% Area)**		
					Ave:	**Range**	**Low**	**Med:**	**High**
2938-2980	TX 235	HnC2	0-5	0-55	64.1	41.0-85.7	24	76	0
2980-2990	TX 121	BsB, BtC	nd	0-20	69.4	38.0-97.0	17	81	0
	TX 235	HeB, HoC2	0-5	10-40	79.4	52.0-133.0	0	79	21
2990-3300	TX 235	HeB	0-5	0-55	90.0	60.7-118.5	0	83	17
	TX 235	HeB, HeC2	0-5	0-55	88.0	70.0 - 114.5	0	88	12
3300-3740	TX 119	nd	nd	nd	nd	nd	nd	nd	nd
3740-3840	TX 235	nd	nd	nd	nd	nd	nd	nd	nd
3840-4100	TX 072	BrA, BrB, HhF3	0-5	0-35	20.9	4.6-39.0	100	0	0
	TX 072	BrA, BrB	0-5	0-35	28.2	21.7-35.4	100	0	0
4100-4480	TX 235	HeB, HoB, HhF3, Ts	0-5	10-40	72.0	30.1-111.4	10	82	7
	TX 235	HeB, HoB, WgC	0-5, 0-5, 0-12	10-40, 10-40, 0-20	24.7	10.0-34.0	100	0	0
4480-4560	TX 326		nd	nd	nd	nd	nd	nd	nd
4560-4750	TX 119	CgC	0-2	0-30	32.9	23.0-41.0	100	0	0
	TX 119	BrA, ByA	0-5	0-35	45.7	26.8-86.4	58	43	0
4750-4950	TX 235	nd	nd	nd	nd	nd	nd	nd	nd
4950-end	TX 849	DgE, CgC, CsC3	0-2	0-30	23.6	13.0-66.0	99	1	0

[a]Distance along the corridor in feet x 10^2

[b]Heiden soil (HeB, HeC2), Burleson clay (BsB, BtC, ByA), Houston Black clay (HoB, HnC2, HoC2), Branyon Clay (BrA, BrB), Heiden-Ferris complex (HhF3), Trinity Soils, frequently flooded (Ts), Crockett loam (CgC, CsC3), Wilson loam (WgC), Darst very gravelly sandy loam (DgE)

"nd" Data Not Available.

CONCLUSIONS

1. Spatial estimates of the potential for sulfate heave can effectively be used as a risk assessment tool during lime stabilization of soils to reduce costs and increase effectiveness. Decisions to use site-specific amelioration techniques during construction can be supported using targeted characterization in areas identified to have high risk for ettringite/thaumasite formation.
2. Techniques used to locate sampling sites rely on published sulfate threshold levels in the soil. Soil mineralogy data can be combined with the threshold sulfate levels to determine the sensitivity of soils to form ettringite and thaumasite when soils are treated with lime using thermodynamic modeling of chemical processes.
3. Field observations suggest that sulfate distributions in soil are likely to be affected by the topographic features of the region, which affect water flow. The potential of sulfate movement into pavement layers after stabilization needs to be accounted for during preliminary soil investigations.
4. Environmental conditions influence the formation of ettringite or thaumasite in lime stabilized soils, ultimately impacting the risk associated with sulfate heave.

REFERENCES

Berger, E., D.N. Little, and R. Graves, *Technical Memorandum: Guidelines for Stabilization of Soils Containing Sulfates*. 2001, National Lime Association: Arlington, VA. p. 16.

Czerewko, M.A., Cripps, J.C., Reid, J.M., and Duffell, C.G. (2003). "Sulfur species in geological materials-sources and quantification." *Cement and Concrete Composites,* 25(7), 657-671.

Hunter, D. (1988). "Lime induced heave in sulfate-bearing clay soils." *Journal of Geotechnical Engineering*, 114(2), 150-167.

Kohler, S., Heinz, D., and Urbonas, L., (2006). "Effect of ettringite on thaumasite formation." *Cement and Concrete Research*, 36(4), 697-706.

Little, D.N., B.E. Herbert, and S.N. Kunagalli. 2006. Ettringite formation in lime-treated soils: Establishing thermodynamic foundations for engineering practice. Transportation Research Board. 1936:51-59.

Little, D. N., and Nair, S. (2007). "Sensitivity of selected Colorado soils to form ettringite/thaumasite when treated with calcium-based stabilizers and when soluble sulfates are available." *Colorado Department of Transportation-Research*, Report No. CDOT-2007-14, Denver, CO.

Mitchell, J. K. (1986). "Practical Problems from Surprising Soil Behavior." *Journal of Geotechnical Engineering*, 112(3), 259-289.

Moore, A. E., and Taylor, H. F. W. (1970). "Crystal Structure of Ettringite." *Acta Crystallographica* B26, 387-393.

Petry, T.M., and Little, D.N. (1992). "Update on Sulfate-Induced Heave in Treated Clays: Problematic Sulfate Levels." *Journal of Transportation Research Record*, 1362, 51-55.

SCS. (1975). "Soil Taxonomy: A Basic System of Soil Classification for Making and Interpreting Soil Surveys." *U.S. Dept. of Agriculture,* U.S. Govt. Print. Office: Washington, DC.

STABILIZER DIFFUSION IN SWELLING SOILS

P. Harris,[1] and T. Scullion[2]

[1]Texas Transportation Institute, Materials and Pavements Division, Texas A&M University, 405 Spence, College Station, TX 77843-3135; PH (979) 845-5845; FAX (979) 845-1701; e-mail: pat-harris@tamu.edu

[2]Texas Transportation Institute, Materials and Pavements Division, Texas A&M University, 405 Spence, College Station, TX 77843-3135; PH (979) 845-9913; FAX (979) 845-1701; e-mail: t-scullion@tamu.edu

ABSTRACT

There has been much speculation about the ability of lime and cement to diffuse into highly plastic soil and chemically alter the clay minerals thereby reducing the shrink/swell capacity of the soil. This research looks at hydrated lime and type I Portland cement in a manufactured soil to obtain direct evidence of the ability of both lime and cement to diffuse into and chemically alter a clay soil. Natural soils are very complex and stabilizer/soil reaction products are difficult to identify so we chose to manufacture a soil. We manufactured five soil samples composed of pure quartz sand (70 wt. %) and pure smectite clay (30 wt. %). Electron probe microanalyses of thin-sections from the manufactured soil samples show high concentrations of calcium throughout the lime stabilized clay aggregates, but very little calcium penetrated the cement stabilized clay. X-ray diffraction confirmed the presence of smectite (clay) in all samples and suggests more calcium substitution in lime treated samples. Calcium from the lime diffuses into the clay to create a chemically altered clay that is less susceptible to shrinking and swelling. The cement forms a protective coating around the clay aggregates and very little calcium actually diffuses into the clay. Thin-section analysis of samples from an expansive, natural soil stabilized with lime and cement shows identical textures to the manufactured soil samples. This research provides direct evidence of lime and cement stabilization mechanisms in clay-rich soils which will allow engineers to make informed decisions about stabilizer options in soils that exhibit shrinking and swelling.

INTRODUCTION

Uniform mixing of chemical stabilizers (lime/cement) in highly plastic soils can be problematic if the soil is wet. The soil will often contain large clay aggregates and not meet specified gradation requirements. The ability of chemical stabilizers to effectively treat the soil in these situations is critical for constructing a durable working platform.

Extensive physical and chemical testing by Herzog and Mitchell (1963) determined that lime liberated from Portland cement hydration participates in pozzolanic reactions with clay minerals. They speculate a clay-cement structure where there is a clay-cement skeleton with a clay matrix interspersed throughout. The skeleton contains units with a core of hydrated cement gel that have layers of altered clay aggregates attached. In the vicinity of the cement grains, less altered clay aggregates could be trapped in the intercellular space.

Stocker (1963) evaluated the effects of cement stabilization on an Australian black soil with a plasticity index of 37. He concluded that relatively large clay lumps were hardened by moderate concentrations of cement by diffusion of calcium into the clay lumps.

Croft (1967a) made the following observations from thin-sections made with cement stabilized soil: (1) Deposition of cementitious products among soil particles, and aggregation of particles around cores of hydrated cement. (2) Migration of chemically saturated and "cement charged" water into cracks between aggregates and through interconnected voids into large lumps. The precipitation of cementitious products modifies soil components, and "stringers" physically support the compacted mass. (3) "Waterproofing" of large soil lumps by surface deposition and shallow penetration of hydration products. (4) Reactions between clay minerals and lime liberated during hydration produce secondary cementitious products.

Chew et al. (2004) evaluated the cement treatment of a marine clay. They concluded that illite is encapsulated by the early cementitious products and is protected from further attack by the lime so late stage pozzolanic reactions do not occur with the illite.

Previous studies give mixed results when diffusion of cement is concerned. Early studies conclude that lime liberated from cement diffuses into the clay lumps to form pozzolanic reaction products which chemically alter the clay. Other studies report that cement forms a skeletal framework in the soil and "waterproofs" or encapsulates large clay aggregates. This study uses advanced analytical equipment and sample preparation techniques, not available when the initial studies were done, to directly measure the diffusion of calcium from the chemical stabilizer through clay aggregates.

Research previously done at Texas Transportation Institute (TTI) has shown that natural soils are very complex and stabilizer/soil reaction products tend to be difficult to impossible to identify. They contain a whole range of clay minerals, as well as organics which makes mapping the fundamental changes extremely difficult. Since natural soils are so complex, we fabricated clay samples which contain a single clay mineral, smectite that is mixed with pure quartz sand to form an experimental soil. Smectite was chosen as the clay mineral because Eades et al. (1962) and Eades and Grim (1960) have shown that clay mineralogy is important in chemical stabilization: they identified Smectite as a very deleterious clay mineral for engineering purposes that responds favorably to chemical stabilization with lime (CaO). Smectite is also widely present in expansive soils in Texas and its damaging effects to highways are well known.

Samples were cured for one year to give the stabilizers sufficient time to diffuse into the expansive clay aggregates and alter the physical properties. We wanted to test the hypothesis that lime from hydrated lime as well as lime from cement hydration diffuses into expansive clay lumps over time to alter the engineering properties of the clay.

BACKGROUND

Lime/Soil Reactions

There have been numerous investigations into the chemical and mineralogical changes that occur when hydrated lime is added to a clay soil. Eades and Grim (1960) recognized that the clay minerals in a soil control the soil's physical properties. They made the first detailed study of hydrated lime reacting with eight different reference clay standards. For smectite, Eades and Grim (1960) determined that the exchangeable interlayer cations are immediately replaced by Ca^{2+}; secondly, a series of new minerals are formed (calcium silicate hydrates, CSH); and finally, the residual lime is carbonated (turned to $CaCO_3$) from atmospheric CO_2.

Hilt and Davidson (1961) reported that the reaction of lime and montmorillonite (smectite) formed a product that was isostructural with tetracalcium aluminate hydrate, but contained much more silica than alumina.

These laboratory studies were criticized for not being representative of the field. Thus Eades, Nichols, and Grim (1962) evaluated three field projects in Virginia that were stabilized with hydrated lime. They concluded that new minerals (calcium-silicate-hydrates and calcium carbonate) were indeed forming in the field similar to the minerals in earlier laboratory experiments.

Croft (1967a) differentiated between early reaction products formed by rapid reactions after short curing (i.e. CSH_{gel}) and those crystallizing after longer periods (CSH I). Short-term strength is attributed to formation of gelatinous and poorly "structurally ordered" phases. Long-term strength is due to gradual crystallization of new mineral species (CSH I) from the gelatinous reaction products (Croft, 1967a). He also stated that most of the strength gain in lime treated soils is due to the long term development of well crystallized reaction products.

Cement/Soil Reactions

Herzog and Mitchell (1963) studied two pure clay minerals (kaolinite and montmorillonite) reacting with Portland cement and C_3S. They determined that calcium hydroxide was formed during hydration of Portland cement and it participated in pozzolanic reactions with the clay minerals. The hydration of C_3S was modified in the presence of the clay minerals. They further concluded that clay-cement cannot be regarded as a simple mixture of hydrated cement particles bonding together unaltered clay aggregates, but must be considered a system where the clay and cement form secondary reaction products.

A detailed study on the influence of soil mineralogy with respect to cement stabilization was undertaken by Croft (1967b). He determined that kaolinite and illite had little to no effect on normal cement hydration (formation of CSH and CAH) since they were not actively participating in reactions with the cement. Croft (1967b) noted that strength was retarded on the montmorillonite due to reaction with the $Ca(OH)_2$

which dropped the pH and inhibited the strength development of the cement. He also made the point that cement stabilized soil strength is predominantly due to early hydration reactions (CSH and CAH) and very little strength gain can be attributed to late pozzolanic reactions between the lime liberated from the cement hydration and clay minerals in the soil.

METHODS

Five experimental samples were molded: there were 2 each, of lime and cement and only one sample was molded without stabilizer. They were composed of pure quartz sand (70 wt. %) and smectite clay (30 wt. %). The smectite was added in a limited size fraction: passing the ⅜ in sieve and retained on the ¼ in sieve. We mixed 10% dry wt. Stabilizer (hydrated lime and Type I Portland cement) and molded 10.16 cm X 11.43 cm samples in a gyratory compactor to the optimum moisture/density determined by standard Proctor compaction.

The samples were allowed to moist cure for a period of one year at 25°C prior to analysis. We sent portions of each sample to a commercial laboratory for thin-section preparation. The samples were impregnated with a blue-dyed epoxy, ground to a thickness of ~30 μm, and polished in oil to a 0.3 μm finish. We analyzed the finished thin-sections with a Zeiss polarizing microscope prior to carbon coating for electron probe microanalysis (EPMA). To construct the X-ray elemental distribution maps, the highly polished thin-sections are coated with an approximately 150-Å layer of high purity carbon. The samples are now ready for EPMA analysis.

We used a Cameca SX50 electron microprobe at Texas A&M University to construct the X-ray elemental distribution maps. We selected clay aggregates with a good polish and relatively flat surface which are required to minimize elemental counting errors (Guillemette, 2008). The microprobe can be programmed to analyze for specific elements. In our samples, four elements were analyzed: calcium, silicon, aluminum, and iron. Each element is measured at points on a grid pattern specified by the researcher. We used grids ranging from 4 X 4 mm, 5 X 5 mm, to 7 X 3 mm. The distance between each point in the grid ranged from 8 to 10 μm. The operating conditions we used for the microprobe were an accelerating voltage of 15 kV and a 20 nA beam current.

To validate results obtained with the electron microprobe, we separated clay aggregates from each sample using a binocular microscope to ensure a relatively pure clay sample for X-ray analysis. We analyzed oriented mounts of the lime, cement, and raw clay samples with a Rigaku X-ray diffractometer. The samples were scanned with copper kα radiation at 40 kV and 20 nA at a speed of 0.75°/min. from 2.1-32° two-theta with a step of 0.05 degrees.

Following the analysis of the experimental samples, we analyzed thin-sections of natural soils stabilized with lime and cement using a Zeiss polarizing microscope.

RESULTS

Figs. 1(a) and 1(b) are thin-section photomicrographs of representative clay aggregates from the lime-treated (Fig. 1(a)) and the cement-treated (Fig. 1(b)) samples taken with plane polarized light. The clay particle in each image is approximately four to five mm across and is semicircular. Note how each clay particle appears darker around the outside and gets lighter in color toward the center. The small (<1mm) white grains are quartz sand grains that were mixed with the smectite. The dark coloration indicated that the clay aggregates were possibly altered (Figs. 1(a), 1(b)) but thinning of the clays due to sample preparation will give a similar appearance.

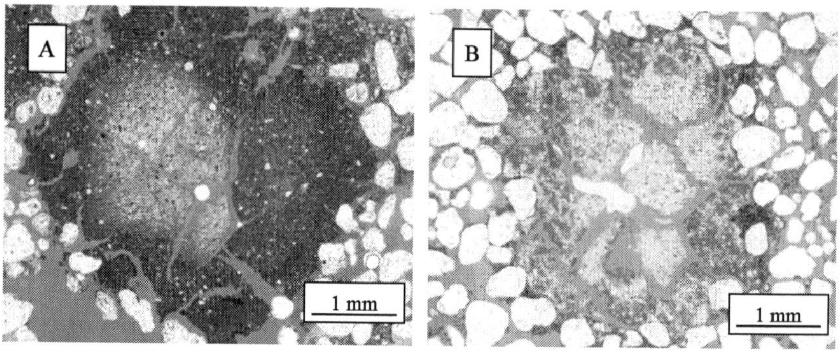

FIG. 1. Thin-section photomicrographs of lime-treated (A) and cement-treated (B) samples.

Fig. 2 is an X-ray map of a smectite aggregate from one of the lime-treated samples. The particle is approximately 5 mm across. The brighter areas represent higher calcium concentrations. The calcium concentration for the rectangular area outlined by the dashed white line across the center of the image is plotted in the graph superimposed on the image. The calcium concentration gradually decreases toward the center of the particle. At the lowest concentration, there is still an average of 25 X-ray counts of calcium. The cracks (black areas) in the smectite aggregates may be primary or they may be artifacts of sample preparation.

Fig. 3 shows the calcium concentration for a representative cement-treated smectite aggregate. As in Fig. 2 higher calcium is denoted by brighter areas in the image. As one can see, the edges of the smectite contain high concentrations of calcium (~40 ave. X-ray counts), but the interior of the smectite aggregate is extremely low in calcium: there are only an average of approximately 5 X-ray counts across the interior of the sample which contrasts with the lime treated smectite averaging 25 X-ray counts in the center of the smectite aggregate.

To measure the influence of lime and cement on the structure of the smectite aggregates, we separated smectite pieces from each core and did X-ray diffraction analyses of the smectite from lime, cement, and untreated (raw) samples. Fig. 4 shows the basal reflection of the *d*(001) smectite peak for the lime, cement, and raw samples. Note how the peak shifts from 15.11 Å for the raw soil, to 15.24 Å for cement, and 15.37 Å for the lime treated sample. As the calcium concentration in the interlayer of smectite increases, the basal spacing of the smectite will increase as well.

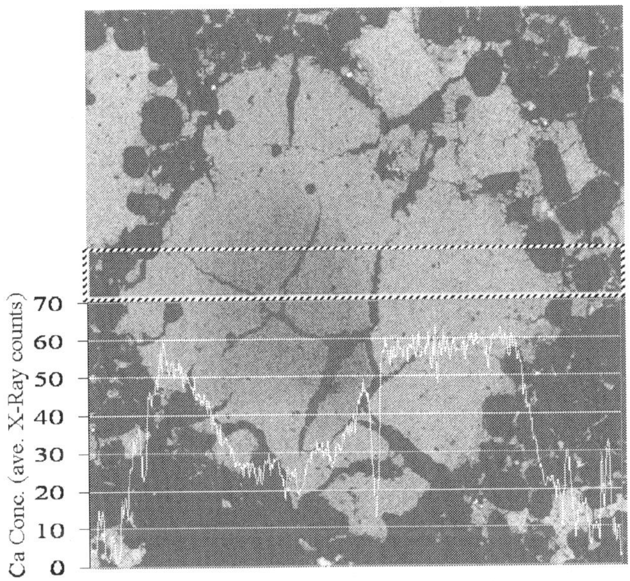

FIG. 2. Calcium concentration across the lime-treated smectite aggregate.

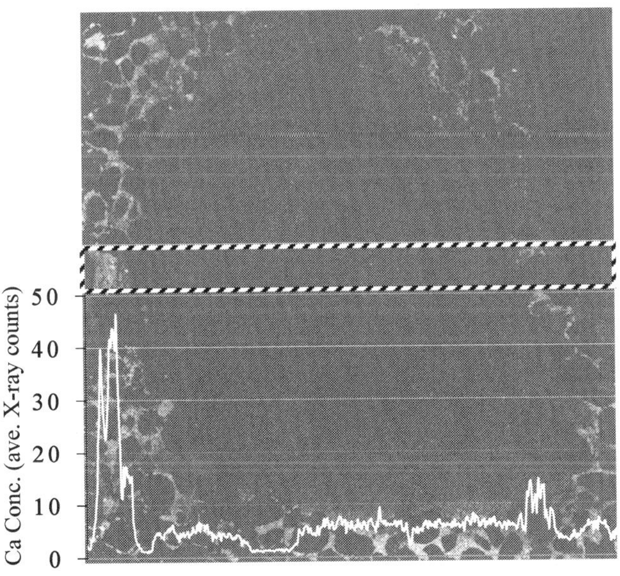

FIG. 3. Calcium concentration across the cement-treated smectite aggregate.

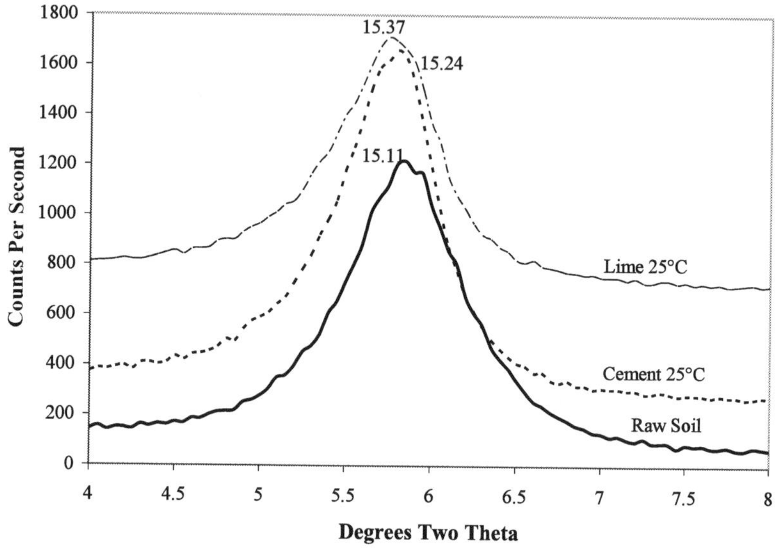

FIG. 4. Basal spacing *d*(001) of smectite aggregates showing stabilizer effects.

DISCUSSION

These data provide direct evidence of calcium diffusion into the smectite aggregates. The samples were cured for a period of 365 days: that should be sufficient time to allow the calcium-based additives (hydrated lime and type I Portland cement) to fully react. As seen in Fig. 2 the calcium from lime penetrates into the clay and alters the structure of the smectite clay aggregates. However, the cement (Fig. 3) only develops a reaction with the surface of the smectite clay aggregates: it forms more of a protective coating around the clay aggregates with less alteration of the structure of the smectite clay aggregates (Fig. 4).

MacEwan (1961) stated that a hydrated calcium smectite yields a basal spacing of 15.4 Å. The basal spacing of the X-ray patterns shown in Fig. 4 increases from 15.11 for the unstabilized soil to 15.37 Å for the lime stabilized soil. The lime more closely approaches a calcium saturated smectite than the raw or cement stabilized samples. If the relative humidity and temperature vary between samples, then the basal spacing may change, but MacEwan and Wilson (1980) state that calcium saturated smectite provides a uniform basal spacing over a large range of relative humidities. Therefore, the X-ray patterns in Fig. 4 should correlate with the calcium concentration of the sample since they were run under the same temperature and relative humidity conditions.

CONCLUSION

Based on Petrographic, EPMA, and X-ray diffraction analyses of thin-sections from the manufactured soils, one can make the following conclusions:

- Calcium from the lime diffuses deeply into the clay aggregates.
- Calcium from the cement does not appear to diffuse deeply into clay aggregates, rather the cement forms a hard shell around the clay aggregates but provides much less alteration to the structure of the clay.

REFERENCES

Chew, S.H., Kamruzzaman, A.H.M., and Lee, F.H. (2004). "Physicochemical and Engineering Behavior of Cement Treated Clays." *Journal of Geotechnical and Geoenvironmental Engineering*, 130 (7): 696-706.

Croft, J.B. (1967a). "The Structures of Soils Stabilized With Cementitious Agents." *Engineering Geology*, 2 (2): 63-80.

Croft, J.B. (1967b). "The Influence of Soil Mineralogical Composition on Cement Stabilization." *Géotechnique*, 17 (2): 119-135.

Eades, J.L., and R.E. Grim. (1960). "Reaction of Hydrated Lime with Pure Clay Minerals in Soil Stabilization." *HRB Bulletin 262*, National Research Council, Washington D.C.: 51-63.

Eades, J.L., F.P. Nichols, Jr., and R.E. Grim. (1962). "Formation of New Minerals with Lime Stabilization as Proven by Field Experiments in Virginia." *HRB Bulletin 335*, National Research Council, Washington D.C.: 31-39.

Herzog, A., and J.K. Mitchell. (1963). "Reactions Accompanying Stabilization of Clay with Cement." *Highway Research Record 36*, HRB, National Research Council, Washington D.C.: 146-171.

Hilt, H.G., and D.T. Davidson. (1961). "Isolation and Investigation of a Lime-Montmorillonite Crystalline Reaction Product." *HRB Bulletin 304*, National Research Council, Washington D.C.: 51-64.

MacEwan, D. M. C. (1961). "Montmorillonite Minerals." *The X-ray Identification and Crystal Structures of Clay Minerals*, G. Brown, ed., Mineralogical Society, London: 143-207.

MacEwan, D. M. C., and Wilson, M. J. (1980). "Interlayer and Intercalation Complexes of Clay Minerals." *Crystal Structures of Clay Minerals and Their X-ray Identification*, G. W. Brindley, and G. Brown, eds., London: 197-248.

Stocker, P.T. (1963). "A Research Project on the Mechanisms of Cement-Stabilization of Black Soil." *Australian Road Research*, 1 (1): 13-28.

Wetting induced changes in elastic modulus of a compacted low plasticity clay

Wan Soo Kim[1], Ph.D. and Roy H. Borden[2], Ph.D., P.E., M.ASCE.

[1] Geotechnical Engineer, Virginia Department of Transportation, Materials Division, 1401 East Broad Street, Richmond, VA 23219; email:wskimgeo@gmail.com
[2] Professor, Department of Civil, Construction and Environmental Engineering, North Carolina State University, Campus Box 7908, Raleigh, NC 27695-7908; email:borden@eos.ncsu.edu

ABSTRACT

The elastic moduli of specimens prepared under various compaction conditions were investigated for soaked specimens as well as specimens tested at as-compacted moisture contents. In this paper, the linearly increasing elastic modulus that was defined using a modulus number, K, was determined from the results of oedometer test. The ratios of modulus numbers of unsoaked and soaked specimens, K_{DU}/K_{DS}, were seen to vary from about 0.5 to 3. The maximum value for these ratios was observed for the specimen compacted at the lowest water content (4.7%) and high density (1.79 g/cm^3).

INTRODUCTION

The mechanical behavior of unsaturated soil is controlled by two stress state variables, while one stress state variable, the effective stress, (σ-u_w) is associated with saturated soil response (Fredlund and Rahardjo, 1993). These two stress state variables are net normal stress, (σ-u_a), and matric suction, (u_a–u_w), where σ is total normal stress, u_a is pore-air pressure and u_w is pore-water pressure. The modified oedometer or modified triaxial device, which can control pore-water pressure and pore-air pressure separately, have been used to determine stress state variables (net normal stress or matric suction) -strain relationships for unsaturated soils. However, these experimental studies are generally expensive, time-consuming, and difficult to perform.

In this paper, the stress-strain relations obtained from the conventional oedometer test are used to estimate the elastic modulus of compacted soils. The elastic modulus is evaluated for soaked specimens as well as specimens at as-compacted moisture contents. The effects of various compaction conditions (dry density and as-compacted water content) on the volume change behavior of compacted soils are highlighted. The change of elastic modulus, E, of compacted soil due to wetting (or matric suction decrease) is also discussed.

BACKGROUND

Assuming the soil behaves as an incrementally isotropic and linear-elastic material, Fredlund and Rahardjo (1993) presented the total volumetric deformation of an unsaturated soil, $d\varepsilon_v$, in an elasticity form for general three-dimensional loading (Eq. 1). This formulation (Eq. 1) can be applied to various loading conditions such as uniaxial, K_o, plane strain, plane stress, and triaxial loading conditions.

$$d\varepsilon_v = 3(\frac{1-2\mu}{E})d(\sigma_{mean} - u_a) + \frac{3}{H}d(u_a - u_w) \tag{1}$$

where E and H = elastic modulus for the soil structure with respect to a changes in net normal stress and matric suction, respectively;

σ_{mean} = average total normal stress (= $(\sigma_x+\sigma_y+\sigma_z)/3$); and,

μ = Poisson's ratio.

The volume change for the soil structure can also be presented in compressibility form, as shown in Eq. 2, and can be presented graphically in the form of a constitutive surface (Fig. 1). The coefficient of volume change, m_1^s is commonly called m_v when the soil matric suction is equal to 0 kPa (saturated soil).

$$d\varepsilon_v = \frac{\Delta V_v}{V_0} = m_1^s d(\sigma_{mean} - u_a) + m_2^s d(u_a - u_w) \tag{2}$$

where m_1^s and m_2^s = coefficient of volume change with respect to a change in net normal stress and matric suction, respectively.

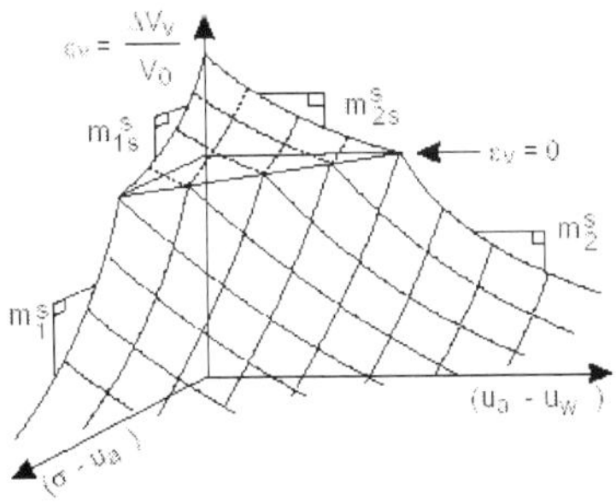

FIG. 1. Three-dimensional constitutive surface for soil structure of an unsaturated soil (Fredlund and Rahardjo, 1993)

The elasticity moduli, E and H, can be calculated from the constitutive surface by computing the coefficients of volume change, m_1^s and m_2^s, which are obtained by differentiating the constitutive surface for soil structure (Fig. 1). For K_o loading, the constrained modulus, M, is calculated from the coefficient of volume change from a one-dimensional compression test. The Elastic modulus is then computed from the relation with M as follows:

$$E = M\frac{(1+\mu)(1-2\mu)}{(1-\mu)} \tag{3}$$

where M is equal to $1/m_1^s$.

EXPERIMENTAL DATA USED IN THE ANALYSIS

In order to investigate soil suction in a low plasticity clay compacted at various water contents and with varying compactive efforts, Rorie (2006) performed collapse potential tests (ASTM D 5333-03), filter paper tests (ASTM D 5298-03), and pressure plate tests (ASTM D 6836-02). The liquid and plastic limits of the material tested are 29% and 19%, respectively. The clay content (% < 2μ) and specific gravity of the soil are 12% and 2.64, respectively. The index properties of the residual soil tested are presented in Table 1. The soil was compacted using three different energies: 10%, 30% and 100% of the standard energy (Table 2). The compaction curves developed with the three different energies are shown in Figure 2. Figure 2 also provides the average matric suctions of specimens measured at as compacted moisture contents of 4.7%, 11.7%, 16.9%, and 22%. From the figure it can clearly be seen that the matric suction of the specimens tested decreases with increasing as-compacted water content (Rorie, 2006).

Table 1. Index properties and grain sizes of residual soil (After Rorie, 2006)

Liquid limit, LL (%)	29
Plastic limit, PL (%)	19
Plasticity index, I_p (%)	10
Sand (%)	39
Silt (%)	46
Clay (%)	12
Specified gravity, Gs	2.64
USCS classification	CL
Maximum dry density (g/cm^3)	1.90
Optimum Moisture Content (%)	11.7

Table 2. Compaction methodology (After Rorie, 2006)

No. of Layers	Hammer wt. (Newton)	Drop Height (mm)	Blows/Layer	Energy (kN-m/m^3)	% of Standard Energy
3	24.4	305	25	600	100
3	24.4	152	15	180	30
3	24.4	76	10	60	10

In this research, the results of double oedometer tests were used to estimate elastic modulus of compacted soils. Typical results of a double oedometer test are shown in Fig. 3. The method requires testing two nominally identical samples (Lawton et al., 1991). One sample was initially inundated with water under a small seating load (5kPa)

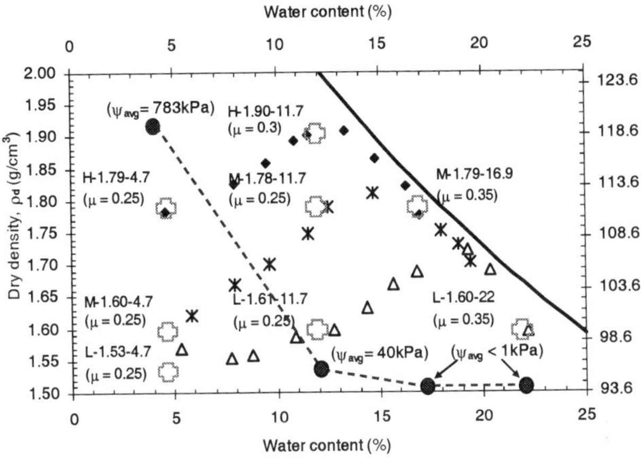

FIG. 2. Compaction curves for three different energies, average measured matric suctions(ψ_{avg}), and assumed Poisson's ratios(μ) of specimens (after Rorie, 2006)

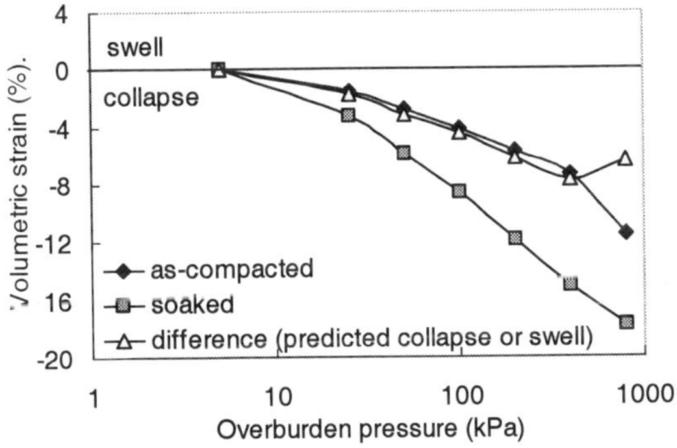

FIG. 3. Typical result of double oedometer test for the specimen

and allowed to swell until primary swelling or collapse was completed. The sample was then loaded with incrementally increasing overburden pressures (25, 50, 100, 200, 400, and 800kPa). The other sample was tested at the as-compacted moisture content using the same incremental loading procedure. The vertical strain difference between soaked (or inundated) and unsoaked (as-compacted) samples was assumed to be the collapse or swell potential. The relationships of volumetric strain versus overburden pressure for soaked and as-compacted specimens were used to determine the elastic modulus of soaked specimens and as-compacted specimens, respectively.

A total of eight specimens (noted in Fig. 2) prepared under various compaction conditions were used to develop data from which the elastic modulus could be determined. Each specimen was designated using a code to enable easy identification. The first letter in the symbol, H, M, or L indicates either High, Medium, or Low compaction energy. The following two numbers indicates the dry density (g/cm^3) and the initial as-compacted water content (%). Therefore, H-1.79-4.7 stands for a specimen compacted with the highest energy to a dry density of 1.79g/cm^3 at a moisture content of 4.7%.

ANALYSIS OF DATA

As water and air were allowed to drain through the top and bottom of the specimens during testing and sufficient time was allowed for 100% consolidation (according to the Taylor's construction), the elastic modulus estimated from the measured curves was a drained elastic modulus. Under the seating load of 5kPa, the as-compacted and subsequently soaked specimens were found to be essentially identical, as there were only small (on the order of +/- 1%) changes in density by either collapse or swell behavior during wetting.

In this research, the elastic modulus under overburden pressures ranging from 25kPa to 200kPa is determined. Figures 4(a) and (c) show the data and the best fit log-linear relationships between overburden pressure and measured volumetric strain for the soaked and unsoaked specimens, respectively. It can be seen that in this range the measured volumetric strain increases linearly with increasing overburden pressure in the semi-logarithm scale. The coefficient of volume change was obtained by differentiating the log-linear constitutive relation and the tangent elasticity modulus, *E*, was then calculated using Eq. 3 with the computed coefficient of volume change and the assumed Poisson's ratio. The different values of Poisson's ratio were considered based on the as-compacted conditions as presented in Fig. 2 (Kulhawy et al., 1969). For

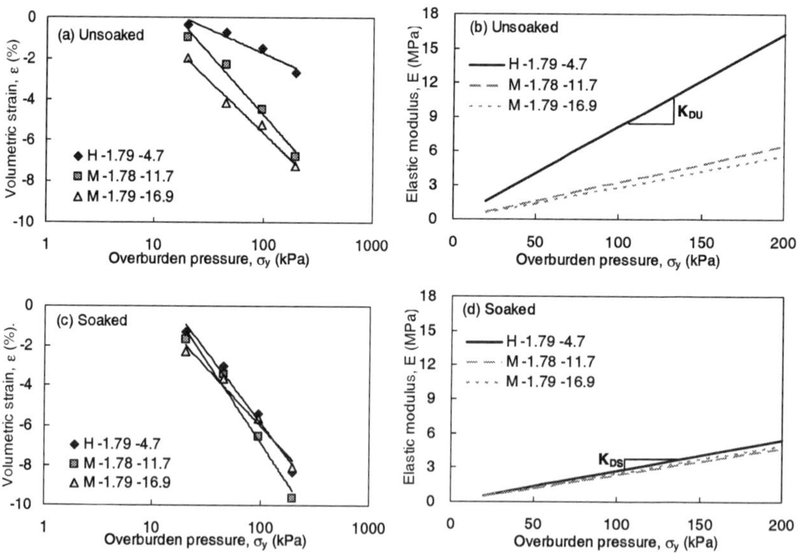

FIG. 4. Volumetric strain and elastic modulus vs. overburden pressure

example, a higher Poisson's ratio of 0.35 was selected for the specimen with higher degree of saturation and a lower Poisson's ratio of 0.25 was assigned for dryer soils.

The stress dependent modulus, which increases linearly with increasing overburden pressure can be defined using a modulus number, K, as follows:

$$E = K \cdot \sigma_y \tag{4}$$

where K = dimensionless modulus number.

The modulus numbers for the specimens tested in this research were computed from the slope of the relation between E and σ_y as shown in Figs. 4b and d. Figure 5 shows the computed modulus numbers, K_{DU} and K_{DS}, of unsoaked and soaked specimens, respectively. For the specimens compacted at a dry density of 1.79 g/cm^3, it can be clearly seen that for unsoaked specimens, the elastic modulus (or modulus number, K_{DU}) increases from 28 to 81 as the as-compacted water content decreases from 16.9% to 4.7%. A similar trend is displayed for the specimens compacted at 1.60 g/cm^3 showing that K_{DU} changes from 20 to 46. However, the elastic modulus of soaked specimens (or modulus number, K_{DS}), which were compacted at the same dry densities (1.79 g/cm^3 or 1.60 g/cm^3), were not sensitive to the magnitudes of as-compacted water content showing almost same elastic modulus.

Influence of dry density on the elastic modulus of compacted soil is also observed from Figure 5. For the values of elastic modulus numbers for both soaked and unsoaked specimens with as-compacted water contents of 4.7% and three different dry densities (1.79 g/cm^3, 1.60 g/cm^3, and 1.53 g/cm^3), as dry density increases from 1.53 g/cm^3 to 1.79 g/cm^3, the elastic modulus number of unsoaked specimen increases from 14 to 81 while that of the soaked specimen increases very little (from 21 to 27). Therefore, it can be concluded that at low as-compacted water contents (w = 4.7%), the difference between elastic moduli of soaked and unsoaked specimens increases significantly with increasing dry density but the difference may be negligible if the dry

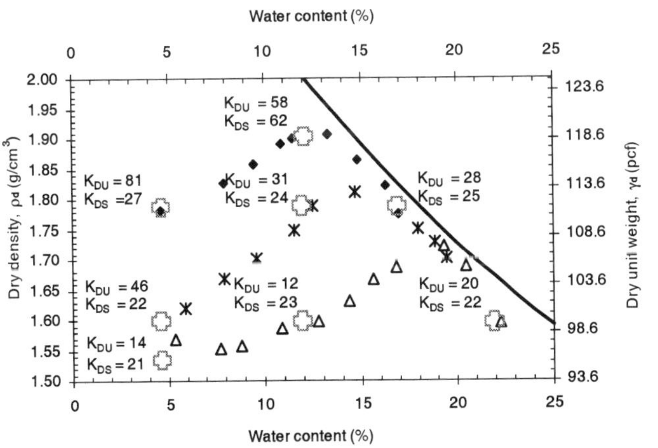

FIG. 5. Computed modulus numbers, K_{DU} and K_{DS}

density of the specimen is relatively very low (i.e. 1.53 g/cm^3). For specimens which have as-compacted water contents on the wet side of optimum (H-1.90-11.7, M-1.79-16.9, and L-1.60-22), there was little difference in soaked and unsoaked moduli because of the initial high degrees of saturation in the as-compacted state.

In general, the modulus number (or modulus) of unsoaked specimens is equal to or greater than that of soaked specimens, because we know that the soil matric suction contributes to the stiffness of unsaturated soils (i.e. unsoaked specimens) (Rahardjo et al., 2004; Thu et al., 2006). However, four pairs of specimens tested in this research (L-1.61-11.7, L-1.60-22, L-1.53-4.7, and H-1.90-11.7) show somewhat conflicting results, in that the modulus number of the unsoaked specimen is lower than that of the corresponding soaked. Among these values of modulus numbers, those of specimens, L-1.60-22 and H-1.90-11.7, which show small differences between the values of K_{DU} and K_{DS} (6% and 9%, respectively), could be considered similar, as those values are quite close. Furthermore, theses specimens were compacted to high degrees of saturation, therefore the values of modulus numbers would be expected to be similar. It is interesting that all of the "L" specimens which were compacted to the low densities using low compactive effort showed the trend of higher modulus for soaked specimens as compared to the corresponding unsoaked specimens. Possibly, for the "L" specimen the soil structure is dominant relative to the soil suction effect- i.e. changes in soil structure due to wetting and changes in volume due to initial loading from 5 to 25kPa may influence subsequent volume change behavior (25-200kPa).

The modulus numbers for both soaked and unsoaked specimens and their ratios are summarized in Table 3. The ratio K_{DU}/K_{DS} varies from less than 1.0 for the "L" specimens up to 3 for the specimen compacted at the lowest water content (4.7%) and high density (1.79 g/cm^3). The high ratios of K_{DU}/K_{DS} imply that there would be significant decreases in the elastic modulus of compacted soils when the soils are compacted very dry of optimum and then wetted by rain infiltration.

CONCLUSIONS

The drained elastic moduli of as-compacted and soaked specimens prepared under various initial conditions (under overburden pressures ranging from 25kPa to 200kPa with dry densities ranged from 1.53g/cm^3 to 1.93 g/cm^3 and as-compacted water contents ranged from 4.7% to 22%) were estimated based on assumed Poisson's ratios

Table 3. Ratios of modulus numbers (K_{DU} / K_{DS})

Sample ID	K_{DS}	K_{DU} (times K_{DS})
H-1.79-4.7	27	81 (3.0 K_{DS})
M-1.60-4.7	22	46 (2.1 K_{DS})
L-1.53-4.7	21	14 (0.7 K_{DS})
H-1.90-11.7	62	58 (0.9 K_{DS})
M-1.78-11.7	24	31 (1.3 K_{DS})
L-1.61-11.7	23	12 (0.5 K_{DS})
M-1.79-16.9	25	28 (1.1 K_{DS})
L-1.60-22	22	20 (0.9 K_{DS})

and the relationships between volumetric stress and strain, which were measured from as-compacted and soaked specimens in double oedometer tests. From the analyses of computed moduli presented in this paper, the following conclusions can be drawn:

1. At the dry densities studied (1.6 g/cm^3 or 1.79 g/cm^3), the elastic modulus of a series of as-compacted specimens increased significantly (over 130% and 190%, respectively) as the as-compacted water content decreased to 4.7% while the elastic modulus of soaked specimens remained almost constant.
2. For specimens with water contents of 4.7% and tested in their as-compacted state, the elastic modulus number increased from 14 to 81 as dry density increased from 1.53 g/cm^3 to 1.79 g/cm^3. The elastic modulus of soaked specimens increased very little, from 21 to 27. The influence of soaking (or matric suction loss) on the elastic modulus for specimens compacted at relatively high dry densities of 1.6 g/cm^3 (Relative Compaction, R.C. = 84%) and 1.79 g/cm^3 (R.C. = 94%) was seen to be significant.
3. Specimens compacted at high degrees of saturation and tested in either soaked or unsoaked conditions showed little difference in elastic moduli.
4. The linearly increasing elastic modulus was defined using a modulus number, K (Eq. 4). The ratios, K_{DU}/K_{DS} were seen to vary from about 0.5 to 3. The maximum value for these ratios was observed for the specimen compacted at the lowest water content (4.7%) and high density (1.79 g/cm^3). The ratios less than 1.0 were observed for the low density specimens compacted using low compactive effort.

REFERENCES

Fredlund, D. G., and Rahardjo, H. (1993). Soil mechanics for unsaturated soils, New York: Wiley.

Kulhawy, F. H., Duncan, J. M., and Seed, H. B. (1969). Finite element analyses of stresses and movements in embankment during construction: *Report TE-69-4*, Department of Civil Engineering, University of California at Berkeley.

Lawton, E. C., Fragaszy, R. J., and Hetherington, M. D. (1991). "View of wetting-induced collapse in compacted soil", *Journal of Geotechnical Engineering, 118(9)*, pp. 1376-1394.

Rahardjo, H., Heng, O. M., and Leong, E. C. (2004). "Shear strength of a compacted residual soil from consolidated drained and constant water content triaxial tests", *Canadian Geotechnical Journal*, 41, pp 421-436.

Rorie, D. (2006). Investigation of soil suction in a compacted low-plasticity clay. *MS thesis*, North Carolina State University, Raleigh. 246 pp.

Thu, T. M., Rahardjo, H., and Leong, E. C. (2006). "Shear strength and pore-water pressure characteristics during constant water content triaxial tests." *Journal of Geotechnical and Geoenvironmental Engineering*, 132(3), pp. 411-419.

IDENTIFICATION AND REPAIR OF KARST SUBSURFACE VOIDS

Douglas L. Gilmore, P.E.[1] and Dennis Boley, P.E.[2]

[1] Principal: EarthMetrics, Inc. Adairsville, GA, USA. douggilmore@bellsouth.net
[2] Chief Engineer: Gibson's Pressure Grouting Service, Inc. Smyrna, GA boleyd@gibsonsgrouting.com

ABSTRACT

Construction in karst areas includes the risk of encountering voids in the underlying material. Structures built above voids can experience significant settlement and extensive structural damage. In the most severe cases, the voids may open to the surface, cause catastrophic failure of the structure, and endanger the public. For commercial projects with tight deadlines, identifying potential subsurface voids early in the pre-development stage is critical in being able to provide time for further assessment and for remedial activities to be employed.

This paper will be centered on a case study of subsurface voids were identified and remediated on a site planned for the construction of a "Big Box" retail store in Northwest Georgia. The paper will present selected methods for identifying voids and the method selection decision process. It will also discuss the subsequent selection of deep soil compaction grouting to fill the voids, to limit the potential for future karstic effects.

BACKGROUND

In 2003, a major retailer planned to construct a "Big Box" retail store in northwest Georgia, USA. The site had been previously developed for another business. It was relatively level, with less than 6 feet (1.9 meters) of elevation difference across the planned development area. Minimal earthwork was anticipated for the new construction. The store was planned to be a single-story structure with a footprint of about 105,000 square feet (~9,800 square meters). The development team preferred shallow spread footing foundations with an unreinforced concrete slab-on-grade floor. The building was planned with concrete tilt-up walls and a conventional metal roof. Limiting settlement was a specific requirement concern for the retailer due to high visibility retail usage and the use of hard-tired forklifts to access high-bay storage racks.

An initial geotechnical exploration was performed for the site in February 2002. The initial study included a review of published geological mapping with a site and area reconnaissance. The geotechnical engineer also obtained, and reviewed, historical and then-current aerial photographs of the site and its surrounding area.

The initial field subsurface exploration included 44 soil test borings for the building and parking/drive areas. Twenty-seven hollow-stem augered standard penetration soil test borings were performed for the planned building. These borings were spaced on an approximate 50 foot (~15 meter) grid pattern and were planned to be extended to depths of 20 to 30 feet (~6 to 9 meters) below the then existing ground surface (bgs).

INITIAL ASSESSMENT

The literature review did not indicate the presence of obvious surface voids within the reviewed area. The site reconnaissance did not indicate evidence of open voids on the site, or subsidence that could be a hallmark of near-surface void collapse. Published geologic mapping indicated that the site was within the Valley and Ridge Physiographic Region of Georgia, an area that includes carbonate rock formations. The parent rock in the vicinity of the site had been mapped as being part of the Conasauga Group and near the contact with the Maynardville Limestone, a sub-unit of the Conasauga. The Maynardville Limestone is comprised of dolomite and limestone with intermixed siltstone and is considered a highly permeable rock unit. The Conasauga formation in the vicinity of the site is predominately shale with a secondary unit of limestone. The limestone in the Conasauga in the vicinity of the site is typically very dense with bulk densities in the range of 2.8 to 2.85.

The initial borings encountered a thin surface layer of less than 6 inches (150 mm) that consisted, depending on the boring, of topsoil, disturbed residuum or crushed stone left from a previous site development. The bulk of the borings then encountered an area-typical residual profile of firm to very stiff sandy clays, transitioning to very soft to firm sandy clays. Auger refusal was encountered in 22 of the 27 borings at depths from 3 feet to 28 feet bgs (~1 to 8.5 meters), indicating a highly variable upper rock profile. Groundwater was encountered during drilling at depths ranging from about 3 feet bgs to 11 feet (~ 1 to 3 meters), for a site with less than 6 feet (~2 meters) of elevation difference.

One boring encountered a zone of soft (4, 2, 3 and 4 blows per foot) clayey soil from 15 feet bgs (4.5 meters) to auger refusal at 32 feet bgs (~10 meters). Observation of the auger advancement suggested there were voids present. The recovery in the split-spoon samples corroborated the observations, with recoveries of less than 50%.

SUPPLEMENTAL EXPLORATION

Additional assessment was recommended to evaluate the area where voids were suspected and to obtain more data on shallow rock and soft soils present in other areas of the site. The additional field work was performed in April and May 2003. The additional exploration included soil test borings, rock coring, and the use of the geophysical test method, Electrical Resistivity testing.

Resistivity testing was selected due to the clayey nature of the soil and the engineer's prior experience. Other geophysical methods, such as Ground Penetrating Radar and Seismic Refraction, have limited effectiveness, in the geotechnical engineer's experience, in detecting voids in thick (greater than 20 feet (~6 meters)) clayey soils.

The electrical resistivity method of subsurface investigation is a portable, non-invasive ground surface testing technique. Direct current is applied at the ground surface at two specified points and the electrical potential drop is measured between another pair of electrodes. Variation in electrical resistances at depth causes variations in the potential drop. The depth of the investigation is proportional to the current electrode separation.

For the purposes of this exploration, the electrodes were set with 10 meter spacing. Near saturated to saturated soils or fractured rock with water filled voids typically yield lower resistivity values. Massive rock will yield higher values. Larger voids may be exhibited by discontinuities exhibited as shown below in Figure 1. The interpretation of resistivity data can be problematic due to variable conductive levels of the subsurface conditions, particularly in this case where layers of iron-rich shales and magnesium-rich dolomites, and their derived soils, were present.

The electrical resistivity testing was performed along six traverse lines with effective depths ranging between 60 and 100 feet (~19 and 30 meters) bgs. The traverse lines were arranged to bisect the initial soil test borings that exhibited the range of subsurface conditions. Three lines were set in the area of suspected voids.

Figure 1: Electrical Resistivity Profile

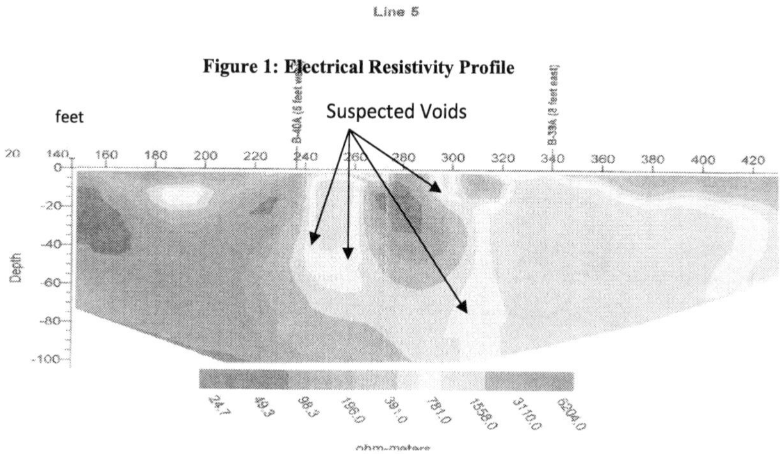

Forty-seven additional soil test borings were performed along the electrical resistivity traverse lines to help define the results of that testing, as well as to provide additional subsurface information. The additional borings were performed using mud rotary techniques, which can help identify subsurface voids through the observed loss of drilling fluid.

Undisturbed soil samples were obtained from selected boring locations where softer soils were encountered, with selected samples tested for shear strength using the triaxial shear test method (CU/PP) and for consolidation characteristics. The results of these tests indicated that the softer soils encountered above the rock layer were capable of supporting the planned building loads using shallow foundations, with acceptable predicated settlements.

Ten borings included rock coring to help characterize the physical appearance, relative rock hardness and core recovery rate. These borings were chosen after careful consideration of the results of the data obtained during the initial borings.

Figure 2 shows the boring locations and geophysical traverse lines in the suspected void area:

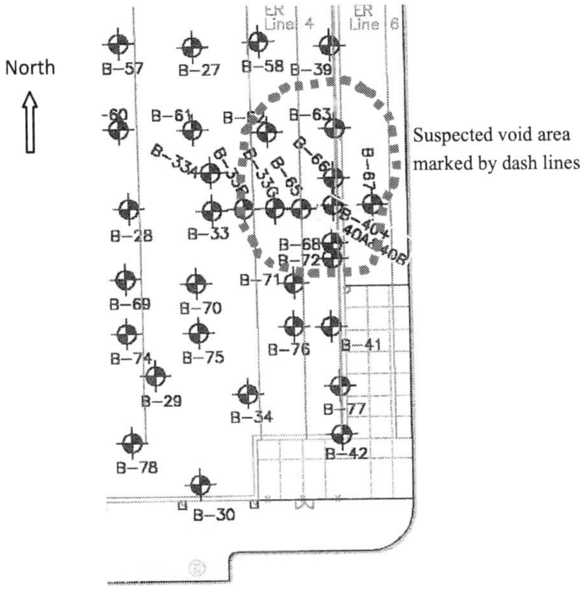

Figure 2: Soil Boring and Geophysical Traverse Line Locations

The additional borings encountered conditions similar to those in the initial borings. The boring data supported the engineer's original concerns regarding potential voids near the planned eastern wall of the building. The borings helped define the suspected void areas found by the geophysical testing.

Drilling fluid was lost in three of the borings performed in this area at depths ranging from about 20 to 60 feet (~6 to 19 meters) bgs, as typified in Figure 3.

An extensive area of voids in the underlying rock was indicated in one of the borings in this area that extended from 23 feet to 61 feet (~7 to 18.5 meters) bgs, with full loss of drilling fluid through this zon. Figure 3 depicts the idealized subsurface profile in this area. The area included relatively thin, or smaller, voids sandwiched between largely intact rock layers at depth, with typical recoveries of 95 to 100 % (for intact incremental sections of the 10-foot core runs) and RQDs of 100%. This "seamy" limestone was typical in several of the cored borings.

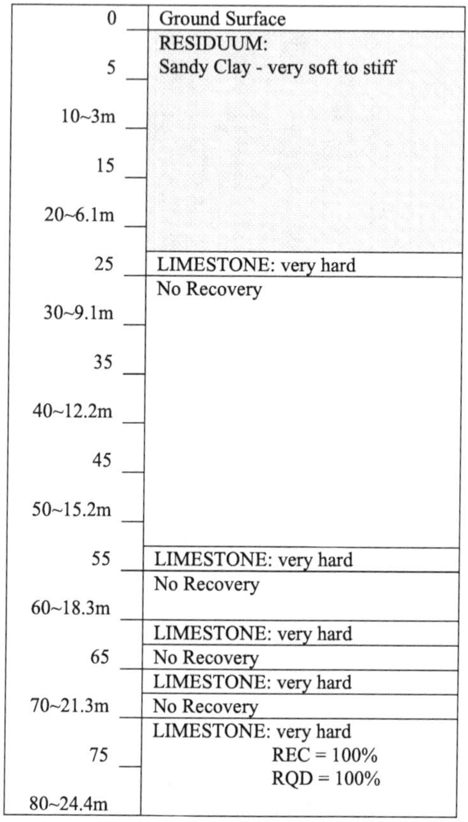

Figure 3: Idealized Subsurface Profile in area where void was detected

DECISION MAKING PROCESS FOR SITE REMEDIATION

The geotechnical engineer kept the client/owner informed of the exploration's progress and results during both phases of the exploration. Several meetings were held to discuss the findings and different options.

Alternatives were considered including moving the building outside the affected area (dismissed due to site constraints and building size requirements), the use of deep foundations (dismissed due to cost), excavation (dismissed due to the suspected depth and extent of the subsurface voids) and compaction grouting.

Compaction grouting was chosen as the most cost- and time-effective solution. Compaction grouting provided a means to fill subsurface voids and to help consolidate low shear strength soils commonly found in limestone areas, due to the chemical weathering of the rock into soil.

Compaction grouting was practical for this site due to the nature of voids in area of the site. The limestone and dolomite in the area of the site, due to its high density and mineralogical make-up, tends to be resistant to weathering, when compared to softer carbonate rocks. They do not typically exhibit very large voids as can be common elsewhere. Voids in the site's general area are rarely larger than 25 feet (~7.6 meters) in average diameter. In other areas of the country where younger, softer limestone are present, much larger voids may be present which could cause grouting to be far too expensive due to the volume of grout required to fill voids. In such areas, deep foundations could be a more suitable alternative.

COMPACTION GROUTING REPAIRS

The geotechnical engineer worked with the client in helping the client select a qualified specialty contractor. The task of developing a remediation plan was delegated to the specialty grouting contractor. The geotechnical engineer reviewed the plan and made recommendations, as judged necessary.

The grouting was planned and performed using deep soil compaction grouting techniques, with a blended pumpable cementitious grout with a required minimum 28-day compressive strength (f'_{cr}) of 1,000 psi (6.9MPa), placed at a maximum pressure of 150 psi (1MPa). The grouting was planned for a 60-foot (~18 meter) diameter area centered in an areas where borings indicated subsurface voids as depicted in Figure 3, with grout points set on approximate 16-foot (~5 meter) spacing. An initial 22 grouting points were planned.

The grouting was facilitated by drilling 2-inch (58mm) diameter (nominal) borings in which to set the grout pipes. The drilling was performed using an air-track drill. The materials noted during drilling were observed and logged by evaluating relative resistance against the drilling advancement, including monitoring pressure gauges on the drill rig, communicating with the drill operator, and by sight and sound.

The results of the air-track drilling were compared to the data from the soil test borings and rock cores and were used to help define the area of grouting. Three of the original planned grouting points were abandoned due to the air-track boring data indicating the subsurface material at those points was relatively competent. Sixteen grouting points were added, based on the air-track borings indicating soft soil-filled rock seams and/or voids.

Grouting was performed in a total of 35 locations to depths ranging from 22 to 98 feet (~6.7 to 30 meters). Four of the grouting points were performed near the center of the area depicted in Figure 3, with these points set on 8-foot (~2.4 meter) centers.

The original plan included 2100 linear feet (640 meters) of grout injection borings, with 1500 cubic yards (~1147 cubic meters) of grout. The actual installation included 1500 linear feet (458 meters) of grout injection borings and 964 cubic yards (737 cubic meters) of grout. The principal cause for the difference between the planned quantities and the as-built quantities was attributed to the encountered voids being more confined than originally estimated.

CONCLUSIONS AND SUMMARY

There was no observed evidence of subsidence during the site development or subsequent building construction. As of the date of this paper, the project has been in service for more than five years with no observed or reported subsidence incidents. The project was successful in part because of the early initial geotechnical exploration that allowed time, within the client's schedule, for subsequent testing to be performed to better define suspected areas. There are several significant keys to assessing and remediating subsurface voids in karst areas.

1. **Know the area and its history.**

   It's important that the engineer understand the geology and soils of the area of the site. What might work in one part of the world may not be effective in another part, due to soil and rock types. While the historical research on this site did not reveal the presence of surface voids within the area of the site, the authors of this paper have used aerial photographs and historical topographic maps on other projects to help define lineations of voids and to identify relic subsidence on sites that may not be observable while standing on the ground;

2. **Use a variety of testing methods that complement each other.**

   Soil test borings are necessary to obtain subsurface information, but are restricted because of their relatively small diameter and typical wide spacing on a site. Soil test borings can miss voids, even when closely spaced. Geophysical testing is more effective at providing a wider view of the site, but the results can be vague without additional invasive testing to help define what the geophysical results mean;

3. **An experienced geotechnical should be on the site during the drilling operations.**

   It is too common, in today's world, for geotechnical engineers to lay-out borings and leave the drillers to record the drilling results. This short-changes the exploration by reducing it the boring logs and jarred samples. It's important to observe the drilling operations, particularly in karst areas. Changes in resistances to drilling, recording soil sample recoveries, and judging soil moisture, color, and consistency changes in the auger cuttings are invaluable in assessing the subsurface conditions, including void presence. The on-site engineer can make decisions about additional footage or to move planned borings to help define the limits of suspected voids, without delays or additional mobilizations.

4. **The specialty contractor experience is critical.**

   The contractor's experience and ability can complement that of the geotechnical engineer. The grouting process itself can be used as an additional exploration method to help further assess the extent of subsurface anomalies. Careful observation of grout takes and changes in pressure measured at the pump can provide data that can supplement and augment the geotechnical conclusions and plan of remediation.

5. **Keep the client and the development team informed.**

Consulting with the client early, before borings are performed, can help them understand the risks associated with karst and risk mitigation. The geotechnical engineer should explain to the client and design team regarding the typical nature of karst activity in the area of the project. Simply telling a client that their site may have voids or sinkholes can provoke images of catastrophic collapse. The geotechnical engineer can help assuage the fears of the owner and design team by helping them understand viable options.

References

Sowers, G.F. (1996) *Building on Sinkholes: Design and Construction of Foundations in Karst Terrain,* ASCE

Georgia Department of Natural Resources (1976) *Geologic Map of Georgia*

C.W. Cressler, M.A. Franklin and W. G. Hester (1976) "Availability of Water Supplies in Northwest Georgia", Georgia Department of Natural Resources Bulletin 91,

Vulcan Materials Company (2008) "Materials Data – Vulcan Materials Southeast Region"

K.I. McConnell and C.E. Abrams (1984) "Geology of the Greater Atlanta Region", *Georgia Department of Natural Resources Bulletin 96*

W.L. Wilson and B.F. Beck (1988) "Evaluating Sinkhole Hazards in Mantled Karst Terrane" , *Geotechnical Aspects of Karst Terrains, Geotechnical Special Publication No.14*, ASCE

J.P. Welsh, P.E., M.ASCE (1996) "Sinkhole Rectification by Compaction Grouting," *Geotechnical Aspects of Karst Terrains, Geotechnical Special Publication No.14*, ASCE

Influence of Swell Pressure From Expansive Fill on Retaining Wall Stability

Mark G. Thomas[1], A.M. ASCE, P.G., Anand J. Puppala[2], M.ASCE, P.E. and Laureano R. Hoyos[3], M. ASCE, P.E.

[1] Senior Geologist, Parsons Brinckerhoff Americas, Inc., 2777 N. Stemmons Freeway, Ste. 1333, Dallas, Texas 75207; thomasma@pbworld.com

[2] Professor, The University of Texas at Arlington Department of Civil Engineering, 416 Yates Street, Arlington, Texas 76019; anand@uta.edu

[3] Associate Professor, The University of Texas at Arlington, Department of Civil Engineering, 416 Yates Street, Arlington, Texas 76019; lhoyos@uta.edu

ABSTRACT

Retaining wall designs most always specify use of drainable granular backfill. However, in actual practice, onsite soils are sometimes used as a cost-saving measure. In North Texas these onsite soils are often expansive cohesive materials. Traditional design using non-expansive cohesive backfill is well established. However, these methods do not consider the lateral swelling pressures that expansive soils exert upon wetting. This paper offers a method for estimating the magnitude of lateral swell pressure that may be exerted on such structures when these materials are used as backfills. In addition, stability analyses results depicting the impact of swell pressures of these materials on calculated Factors of Safety (FS) values for external stability are presented.

INTRODUCTION

Most retaining wall designs specify the use of drainable granular materials as the backfill behind those structures, primarily because the methods for calculating earth pressures on the walls are relatively simple and well established. This simplicity is due, in part, to the relatively inert nature of granular materials, i.e., the particles do not appreciably interact chemically with each other or with water.

While a majority of retaining wall systems designed and constructed in the USA and around the world perform satisfactorily, a significant number of retaining wall failures occur each year. In the Dallas-Fort Worth metroplex, recent retaining wall failures have occurred along the State Highways 360 and 161. Some retaining wall failures can be attributed to construction outside of the original design criteria, particularly the use of cohesive backfill as a substitute for granular backfill materials

(Marsh and Walsh, 1996). The main reason that cohesive materials are considered for use in these situations is almost always economic. When onsite cohesive materials are used for retaining structure backfill, not only do the potential hydrostatic buildup pressures need to be considered in stability calculations, but the lateral swell pressures that could potentially develop need to be considered as well.

This paper describes a method for estimating the magnitude of lateral swell pressure that may be exerted on retaining structures when expansive cohesive materials are used as backfills. Analyses of the impact on external Factors of Safety (FS) values by lateral swell pressures are included.

LATERAL SWELL PRESSURE MEASUREMENTS

Since it was first recognized that expansive cohesive soils constitute a potential threat to engineered structures, a number of methods have been introduced to measure the magnitude of swell in the laboratory. However, most of these methods have focused on the vertical component of swell only (Yesil, et al., 1993). To this day, one-dimensional tests are by far the most widely used means to estimate expansive soil swell pressures worldwide, largely due to the simplicity of the procedures and the ready availability of the testing equipment.

Little published research is available pertaining to the lateral components of swell pressure development in expansive soils associated with earth retaining structures. This fact is presumed to be due, at least in part, to the difficulty in measuring lateral swell pressure in the laboratory; and to the cost in both financial and time resources in instrumenting and monitoring in situ structures, through and beyond construction. In this section, the published research deemed most pertinent to the task of estimating the magnitude of potential lateral swell pressure development is reviewed. This information forms the basis of subsequent analyses.

Because of the rigid nature of the confinement ring in a one-dimensional swell test, (zero lateral deformation), all of the swell strain in this type of test is forced to occur in the vertical direction. However, in many situations this condition is not representative of conditions in nature, where the vertical swell strain is only a portion of the total volume change (Al-Shamrani and Dhowian, 2003). A number of researchers have recognized this limitation and have devised methods to simulate more realistic natural conditions. Thomas (2008) discusses the developments and improvements in methods for lateral swell parameter measurements published by several researchers. These methods generally fall into two categories: modified oedometer tests or modified triaxial cell tests, and appear capable of providing quality lateral swell data for use in retaining structure design.

Ofer (1981) determined that significant errors in lateral pressure measurement can result when even small lateral strains are allowed to occur in the test system. Building on the work of Komornik and Zeitlen (1965), he goes on to describe a similar modified oedometer setup using a thin-walled confining ring, dubbed the Lateral Soil Pressure ring MKII (LSP MK II) in a conventional oedometer base and loading frame apparatus, together with a method that measures lateral swell pressure by compensating for lateral swell strain. The difference between the two methods is that air pressure is introduced into the modified ring to counterbalance any lateral swell strain (null test), though laboratory tests can be conducted varying the compensating air pressure, thereby allowing different measures of lateral ring strain.

Of the various characteristics noted during this and a subsequent investigation (Ofer and Komornik, 1983), the authors observed that vertical swell strain and lateral pressure development is time-dependent, and that the response times of these parameters depend on the amount of applied vertical pressure. Further, they determined that the maximum lateral swell pressures were developed in specimens compacted at optimum moisture content to maximum density, and that either a decrease or increase in moisture content with an associated decrease in density results in a decrease in developed lateral swelling pressure. (Though not mentioned in this study, it would seem that modified compaction or over compaction at dry of optimum conditions would produce at least equally large lateral swell pressures.) Additionally, they confirmed that even small amounts of lateral strain reduce the lateral swell pressure considerably. Finally, they determined that clays of similar mineralogy and physical properties exhibit similar lateral swelling characteristics when tested under similar conditions, despite the presumed differences in location, origin and history of those materials.

Katti et al. (1983) conducted large-scale experiments investigating lateral swell pressure development on retaining structures with and without use of cohesive non-swelling materials (CNS) both atop high-plasticity expansive clay fill, and between the structure and the clay fill. The experiments were conducted on sand, CNS and expansive clay soils (Liquid Limit = 71, Plasticity Index = 29, 55 percent ≤ 2 μm, and a clay fraction base exchange capacity of 124.0 meq/100g). The soils were layer-compacted into a reinforced frame with dimensions adjustable up to about 1.2 m wide, 2.4 m deep and 3.7 m tall. One of the four vertical walls of the experiment frame was instrumented and equipped to restrain horizontal deflections. By maintaining zero deflections along the vertical length of wall after fill compaction was complete, the lateral swelling pressures generated after both compaction and saturation could be measured. Saturation for each test increment was conducted for a period of 60 days, though the means of saturation were not explicit.

Among the types of tests conducted, the authors evaluated the lateral pressures developed with depth for granular materials (sand), cohesive non-swelling clay and expansive clay soils in loose dry, compacted dry and compacted saturated conditions. In this case, the authors observed a linear relationship of lateral pressure with depth for the loose dry conditions of all three materials in close agreement with K_o values obtained from Jaky's equation:

$$K_o = 1 - \text{Sin}\phi \tag{1}$$

For the compacted dry condition, the observed lateral pressure relationships were also linear with depth, but the calculated values for K_o were all in excess of 1, with the sand exhibiting the highest values (e.g. 2.33 at 1 m depth). These high values are attributed to the addition of impact loads during compaction imparted to the self-weight of the fill at a given depth. Compacted saturated expansive clay materials, however, exhibited a completely different behavior in this study; though it should be noted that these soils had been compacted in an air-dried condition prior to saturation.

For the compacted saturated condition of the expansive clay, the lateral pressures against the wall increased rapidly with depth to about 1.5 m, then increased at a lesser rate below that depth. At 1.5 m depth the developed lateral pressures were measured to be about 230 kPa (~4800 psf). Calculations of the lateral pressures generated by the buoyant weight of soil, the water and the impact loads during compaction only amounted to about 19 kPa (~400 psf) at that depth. The difference is taken to be the magnitude of lateral swell pressure generated by the absorption/adsorption of water into the crystal structure of the clay minerals (predominately a smectite type clay). Fig. 1 is a graphic illustration of the measured data.

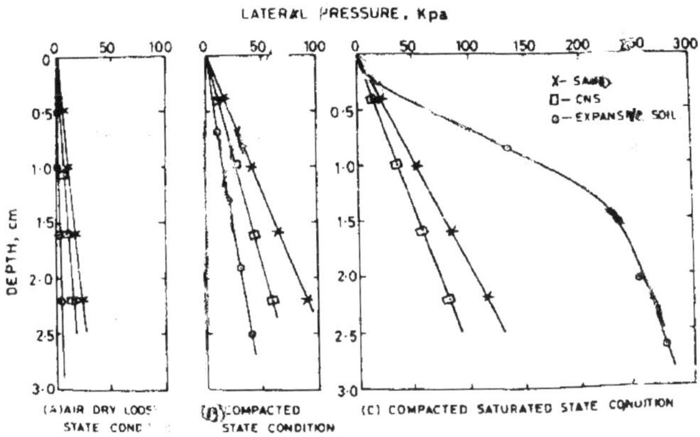

FIG. 1. Lateral Pressure Development versus Depth (from: Katti et al., 1983)

Clayton et al. (1991) investigated lateral swell pressure development in high plasticity London Clay used as backfill against retaining structures in an instrumented pilot-scale study at the Transport and Roadway Research Laboratory in England. Material properties of the London Clay studied are Liquid Limit = 78, Plasticity Index = 48 and 60 percent ≤ 2 μm. The observed magnitude of lateral swell pressure development was generally in agreement with the results of the Katti, et al, study. Additionally, a lateral displacement of 0.5 mm of one of the instrumented sections was required to relieve damaging lateral swell pressure buildup, indicating that very small strains are required to reduce sustained lateral swell pressures. A more in-depth discussion of lateral pressure measurement studies, the resultant conclusions and recommendation for further study in this field can be found in Thomas (2008).

LATERAL EARTH PRESSURES

Conventionally assumed lateral earth pressure distributions of non-cohesive and non-expansive cohesive materials, including the presence of hydrostatic and surcharge loading, are well established and can be found in any number of textbooks. Hydrostatic water pressure distribution is taken as triangular, whereas surcharge loads are often assumed to be constant with depth (i.e. a rectangular pressure distribution).

In areas of expansive soils, where these materials may be retained or used as backfill, prudent engineering design would seem to require that the magnitude of potential lateral swell pressure be estimated and that these pressures be added to the conventionally derived lateral earth pressures. It is argued that, since soil strains follow the path of least resistance, strains at and near unburdened surfaces are considered to be mostly vertical. For design calculations, and in the absence of a widely accepted method to estimate the lateral swell pressure component of lateral earth pressure, it is proposed that a composite triangular pressure distribution for lateral swell pressure development be considered as shown in Fig. 2, based on the available published research (primarily Katti et al. and Clayton et al. noted above). Subsequent analyses in this paper employ this distribution, althoughit should be noted that future studies investigating the effects of lateral wall displacement may result in a raising or lowering of the 1.5 m transition depth for sustained lateral swell pressure.

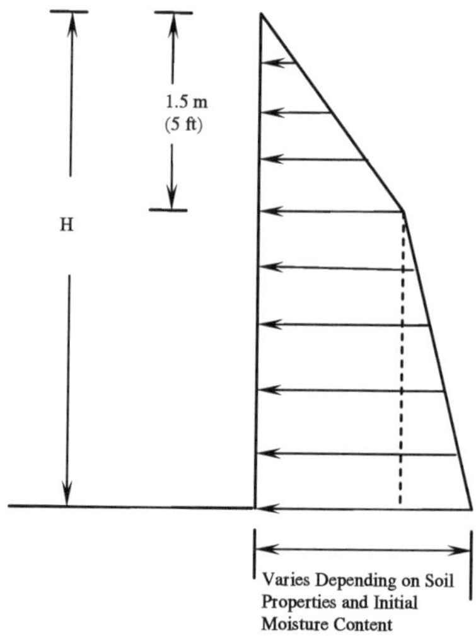

FIG. 2. Interpreted Lateral Swell Pressure Distribution

INFLUENCE OF LATERAL SWELL PRESSURE ON FS VALUES

For conventional analyses where expansive soils are not involved, it has been determined that, except for rare cases, the Factor of Safety with respect to sliding (FS_{SL}) is more critical than overturning (FS_{OT}) in external stability calculations. Efforts to improve FS_{SL} include increasing the footing width, adding a keyway and reducing the water level within the backfill and retained soil. Subgrade modifications may also improve FS_{SL}.

Data developed from previous studies indicates that very large lateral swell pressures are generated when compacted expansive soils hydrate. Depending on the moisture condition and compaction of backfill or other retained soils, the lateral swell pressures developed by hydrating expansive soils can vary greatly. Close scrutiny of the large scale and pilot scale studies cited above reveals that the moisture conditions of the backfill and retained soils were either dry of, or near the standard Proctor optimum moisture content. These pressures, as presented and considered in lateral earth pressure distribution calculations, exceed the combined pressures exerted by the active soil and the hydrostatic pressure. This condition worsens with increasing depth and surcharge pressure.

Despite these effects, knowing that increased moisture conditions and decreased compaction density will decrease the magnitude of potential lateral swell pressure, and that even very small structure deflections will result in a greatly reduced sustained applied lateral swell pressure, it is conceivable that prudently conservative retaining structure designs may be developed using expansive cohesive backfill materials, provided that these materials are placed at sufficiently high moisture contents (assuming the soil is completely wetted) and at proper compaction levels.

It is assumed that lateral soil pressure development would need to be limited to about 15.7 kPa per meter to 1.5 meter depth and an additional 6.3 kPa per meter below 1.5 meter depth (100 psf per foot to 5-foot depth and an additional 40 psf per foot of depth below 5-foot depth) to develop a prudently conservative design, since the magnitude of this parameter for walls ranging in height from 1.5 to about 13.5 m (5 to 45 feet)is about 1.0 to 2.0 times the active pressure of retained soils with an effective friction angle greater than 22° when c = 0, and about 1.0 to 4.0 times for retained soils with an effective friction angle greater than 15° when c = 9.6 kPa (200 psf). However, a discussion of the cost/benefit aspects of this solution is beyond the scope of this paper.

Clayton et al. (1991) research suggests that moisture contents from 2 to 10% above optimum moisture content (depending on bulk soil characteristics) and compaction levels between 93 and 98 percent of Standard Proctor optimum dry unit weight will limit lateral swell pressure development to less than 15.7 kPa per meter to 1.5 meter depth and an additional 6.3 kPa per meter below 1.5 meter depth (100 psf per foot to 5-foot depth and an additional 40 psf per foot of depth below 5-foot depth). However, such high moisture contents may reduce trafficability of the construction process such that increased moisture content alone may be impractical. In such cases, allowing for small wall strains with somewhat lower moisture contents may prove an acceptable solution for reducing sustained lateral swell pressures, though further research into the various contributing factors is necessary.

Even with proper soil preparation measures, a considerable reduction in FS for both overturning and sliding occurs for the same calculation when lateral swell pressures are considered as noted. The reduction for sliding is illustrated in Fig. 3 using magnitudes of lateral swell pressures equal to the 15.7 kPa / 6.3 kPa (100 psf / 40 psf) pressure distribution noted above.

The Factor of Safety for sliding in each case incorporates the effects of a 1.2 m (4-foot) deep keyway below the base of the 0.61 m (2-foot) thick footing and a lowered permanent groundwater level.

Further experiments to measure lateral swell pressure development involving instrumented large-scale, pilot-scale or in situ construction should allow refinement of the pressures and pressure distribution used in this analysis, especially the effects of small walls yielding deflections. Such research may indicate that prudent construction installation will result in lower potential lateral swell pressure development. The effects of lateral swell pressure reduction are shown in Fig. 4.

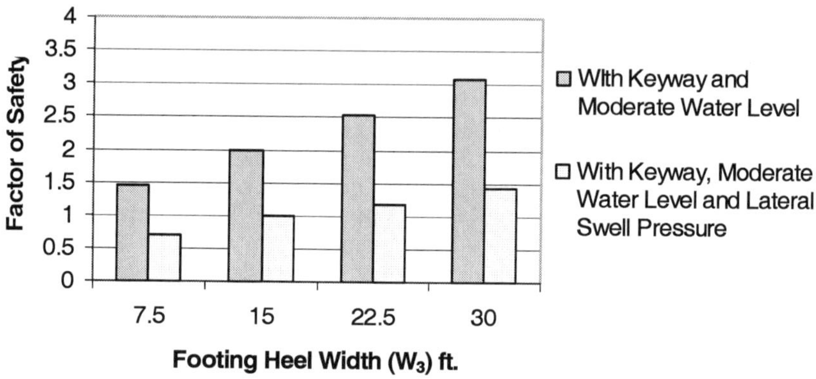

FIG. 3. Comparison of Factors of Safety With Respect to Sliding for a 7.6 m (25-foot) Tall Wall Using 15.7 kPa / 6.3 kPa (100 psf / 40 psf) Composite Triangular Lateral Swell Pressure Distribution

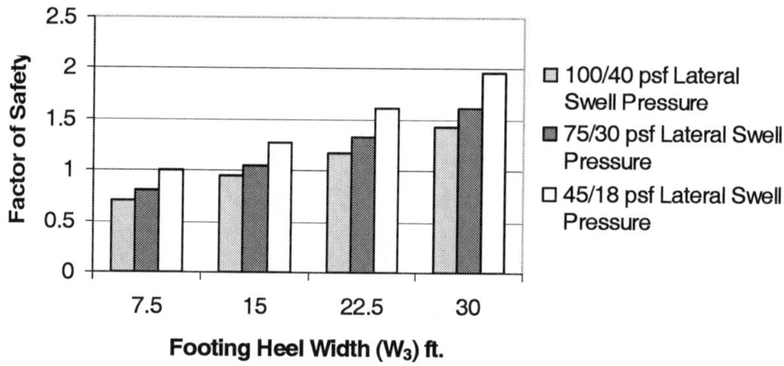

FIG. 4. Comparison of Factors of Safety With Respect to Sliding for a 7.6 m (25-foot) Tall Wall With Differing Magnitudes of Lateral Swell Pressure

CONCLUSIONS

It is evident that when considering retaining structures in expansive soil environments, or if the possibility exists for using expansive soils as retaining structure backfill, the magnitude of potential lateral swell pressure development must be incorporated into the design. At present a reliable and widely accepted method for estimating lateral swell pressure induced by use of cohesive backfill materials behind retaining structures is not yet available. However, it is clear that in order produce a economical wall design when using these backfill materials, the potential lateral swell pressure generation should be limited to less than 15.7 kPa per meter to 1.5 meter depth (100 psf per foot to 5-foot depth) with less than 6.3 kPa per meter (40 psf per foot) increase below that depth.

The analyses of the previous research utilized in this work suggests that the above lateral swell pressure values may be achievable using some combination of increased placement moisture content, reduced placement density and permitting small outward structure deflections to occur. The analyses summarized in Thomas (2008) further suggest that the relatively new, but sophisticated, 3-dimensional laboratory testing techniques now available are quite capable of providing quality lateral swell pressure data for use in future prediction research. These testing protocols should be able to be appropriately modified to investigate the effects of surcharge loading and changing soil suction potential.

Future correlation of lateral swell pressure generation with certain soil parameters should greatly facilitate the development of a practical method for estimating lateral swell pressure potential for cohesive soils based on readily obtainable bulk soil properties including plasticity, clay fraction, moisture content, and matric suction.

REFERENCES

Al-Shamrani, M. A. and Dhowian, A. W. (2003). "Experimental Study of Lateral Restraint Effect on The Potential Heave of Expansive Soils," *Engineering Geology,* Vol. 69: 63-81.

Clayton, C. R. I., Symons, I. F., and Hiedra-Coco, J. C. (1991). *"Pressure of Clay Backfill against Retaining Structures,"* Canadian Geotechnical Journal, Vol. 28, (2): 282-297.

Katti, R. K., Bhangale, E. S., and Moza, K. K. (1983). "Lateral pressure in Expansive soil with and without a cohesive non-swelling soil layer, Application to earth pressures on cross drainage structures in canals and key walls in dams", *Final Report Part-I, Submitted to Central Board of Irrigation and Power,* New Delhi, Vol. 110 (21): 1-303.

Kormornik, A. and Zeitlen J. G. (1965). "An Apparatus for Measuring Lateral Soil Swelling Pressure in the Laboratory," *Proceedings of the 6th International Conference on Soil Mechanics and Foundation Engineering*, Vol. 1: 278-281.

Marsh, E. T. and Walsh, R. K. (1996). "Common Causes of Retaining-Wall Distress: Case Study," *Journal of Performance of Constructed Facilities*: 35-38.

Ofer, Z. and Komornik, A. (1983). "Lateral Swelling Pressure of Compacted Clay", *Proceedings, 7th Asian Regional Conference on Soil Mechanics and Foundation Engineering,* Haifa, Israel: 56-63.

Ofer, Z. (1981). "Laboratory Instrument for Measuring Lateral Soil Pressure and Swelling Pressure," *Geotechnical Testing Journal,* Vol. 4, (4): 177-182.

Thomas, M. G. (2008). "Impact of Lateral Swell Pressure on Retaining Structure Design Using Expansive Cohesive Backfill", *Masters Thesis*, University of Texas at Arlington, 144 pp.

Top-Down Construction of a Bridge in Clay Shale

By: Ronald E. Smith[1], M. ASCE Ph.D., P.E.,
Doyle L. Smith[2], Jr. M. ASCE P.E.,
and Julie A. Griffin[3], M. ASCE E.I.T.

[1]Consulting Geotechnical Engineer, 1605 Benchley Ct., Henderson NV 89052, ronsmith113@yahoo.com
[2]Geotechnical Engineer/Transportation, Kleinfelder, 6850 Manhattan Blvd., Suite 300, Fort Worth TX 76120, dsmith@kleinfelder.com
[3]Geotechnical Engineer, Kleinfelder, 6850 Manhattan Blvd., Suite 300, Fort Worth TX 76120, jgriffin@kleinfelder.com

ABSTRACT: Published case histories of expansive soil/rock materials have historically reported the effects of vertical expansion, and the lessons learned from vertical expansion of soil/rock materials are numerous. In this case the abutment walls of the Las Colinas Bridge were constructed using the increasingly popular, top-down construction method which, while facilitating the construction process, restricts lateral earth movement. As a result of restraining the lateral earth movement, high horizontal stresses in the clay shale were transferred to the bridge resulting in significant structural problems in the bridge deck and guard rails.

This is a case study of a bridge built over and contemporaneous to the development of the President George Bush Turnpike (PGBT) northwest of Dallas, Texas. The geologic formation that impacted this bridge is the Eagle Ford Group (Eagle Ford) a highly plastic, residual clay soil grading to clay shale at relatively shallow depths. This paper covers the studies of the in-situ lateral pressures using the Marchetti flat plate dilatometer test (DMT) in conjunction with monitoring soil mass movement using inclinometers. The paper compares its findings with information in the literature from studies and research that have experienced comparable lateral earth pressures.

INTRODUCTION

The Las Colinas Bridge spans over the depressed roadway of the PGBT in Irving, Texas. The bridge, built in 2000, was constructed to facilitate continued use of Las Colinas Boulevard at its existing grade during and after construction of PGBT. The PGBT was to be constructed as a depressed roadway crossing some 7.62 meters below the bridge. In an effort to optimize construction schedule, it was decided to build the bridge using the top-down construction approach.

Construction of the bridge started with an excavation of about 1.5 meters followed by installation of three rows of drilled shafts to serve as the bridge foundation and support for the abutment walls at either end of the bridge. With the foundation in place, bridge construction continued with installation of the transverse beams over the three rows of foundation shafts. This was followed by placement of precast longitudinal girders to rest on and span between the transverse beams, providing support for the cast-in-place bridge deck.

With the completion of the bridge deck, excavation for the roadway began beneath the structure. Before initiation of the first excavation lift, a top row of earth anchors was installed to restrain the abutments. A subsequent row of anchors was installed after each approximately 1.5 meter excavation lift. See Figure 1. It should be noted that the drilled shafts that constituted the abutment structure were spaced approximately 3 meters apart as "soldier" piles with intervening lagging elements. The earth anchors were installed to provide horizontal restraint to the exposed drilled shafts that constituted the primary element of the bridge abutments. The excavation proceeded to the subgrade level of the PGBT roadway. To finish the abutment walls, facial elements were placed to cover the exposed drilled shafts. The net effect of the top-down installation was to significantly restrict lateral movements of the abutment walls.

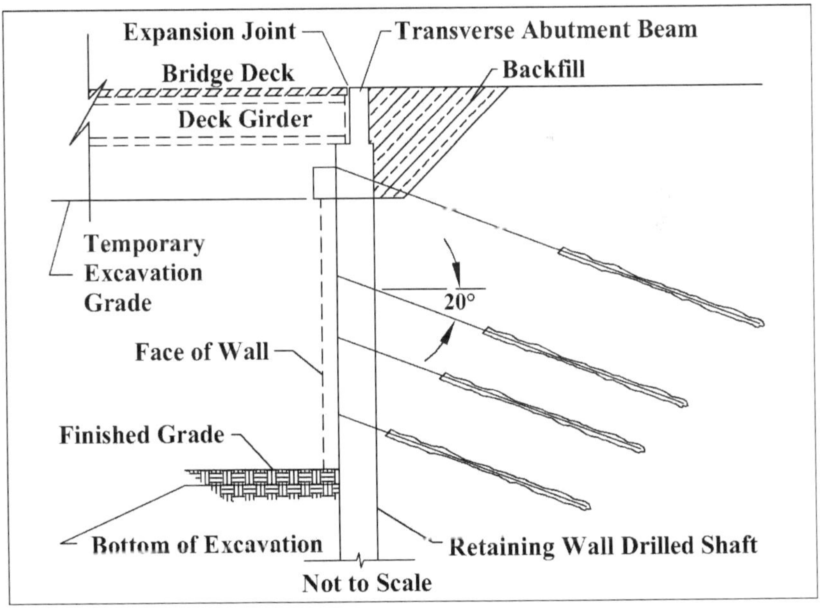

FIG. 1: Tie Back Abutment Wall Configuration

OBSERVATIONS OF BRIDGE DISTRESS

Kleinfelder's (2006) initial assignment was to conduct a forensic assessment of retaining walls along a long stretch of the PGBT roadway which included the Las Colinas Bridge. In the course of assessing the retaining walls, we became immediately aware of the condition of the bridge. It seemed evident that a root cause of the retaining wall problems along the roadway and the Las Colinas Bridge distress was the subsurface material that the walls were retaining, namely the Eagle Ford a known problem material in the area west of Dallas (Allen and Flanigan, 1986). Very early in our forensic assessment, it became obvious that the Las Colinas Bridge had been subjected to significant horizontal stresses. Evidence of the stress related distress can be seen in Figures 2 and 3.

FIG. 2: Crushed Bridge Deck Slab (2004)

FIG. 3: Patched Crushed Bridge Deck Slab (2006)

EAGLE FORD GROUP

The Eagle Ford in the project vicinity grades from a residual, highly plastic clay soil near the ground surface to a clay shale within depths of 1.5 to 2.5 meters. This formation contains numerous bentonite seams that include a high percentage of smectite. This clay mineral is among the most expansive of all soil/rock materials. In addition, there are fine sand seams and ancient fractures within the soil/rock mass that allow water to flow into the mass. In fact the Las Colinas area of Irving is well known locally for its propensity for expansive clay problems, typically expressed as either heave or slope stability problems.

In its transformation from a sedimentary soil to clay shale this material has been heavily overconsolidated. As noted by Bjerrum (1967), the high horizontal stresses that accompany overconsolidation are locked into these clay based materials by a process called diagenetic bonding. With weathering and degradation of these bonds, the materials release their recoverable strain energy which, if the material is laterally restrained, results in high in-situ horizontal stresses. These inherent lateral stresses together with the natural affinity of the smectite mineral to expand present two sources of high lateral earth pressures that when combined with the top-down construction process, provide powerful conditions for compressing the bridge. The following paragraphs explain this phenomenon.

- Overconsolidation of the Eagle Ford – The shale began as soft sedimentary soil layers that were successively placed on top of each other. In time, the weight of each succeeding layer compressed the underlying layers. With enough load, the soft soil was eventually consolidated into shale. During the process of loading the clay minerals, sheet-like in shape, were deformed and subsequently bonded at points of contact between the sheets resulting for the potential for "recoverable strain energy."

  Typical of many types of shale like the Eagle Ford, the rock forming process eventually reverses itself, as materials near the surface of the formation are weathered. This begins with the erosion of the uppermost layers that result in the removal overburden on the underlying layers and reduction of the high vertical pressures. As the overlying materials are eroded away, the newly exposed, uppermost layers are exposed to atmospheric conditions, principally changing moisture and temperature. This exposure results in weathering, and causes the degradation of the shale into a residual soil. Advancing degradation of the shale leads to breaking of the diagenetic bonds between particles and release of the recoverable strain energy. Because the mass is laterally confined by adjacent soil/rock, potential strain release is not fully recovered and as a result, higher than normal lateral earth pressures are developed within the mass.

  The effect of the overconsolidated Eagle Ford on the roadway excavation for the Las Colinas Bridge was that the highly restrained, vertical excavation into the residual soil/rock mass provided the potential for lateral stress relief. But by installing the bridge abutments, beams and decks before completing with the top-down construction process, the clay shale was structurally restrained from lateral movement and the clay shale maintained much of the high lateral stress inherent to the Eagle Ford. This phenomenon partially accounts for the higher than anticipated lateral forces acting on the bridge abutments.

- Affinity of the Smectite for Water – When exposed to moisture (either free water or humidity), smectite minerals are strongly attracted to water molecules. As water is drawn into the clay minerals, the sheets are separated farther apart until they satisfy their affinity for moisture due to the double clay layer phenomenon. This constitutes the definition of expansive clay minerals and swelling soils. It is likely that expansion of the Eagle Ford soil/rock mass also contributed to the higher than anticipated lateral earth pressures on the bridge abutments.

SUBSURFACE INVESTIGATION

Faced with the "evidence" of high lateral pressures acting on the bridge, as shown in Figures 2 and 3, a subsurface testing and monitoring program was undertaken. This program involved routine testing as well as earth anchor lift-off tests, slope inclinometer installations, and dilatometer tests that specifically addressed the lateral pressure issue. In the vicinity of the bridge, the Eagle Ford was initially encountered in borings at depths between 0 and 2.5 meters, and continued for the full depths of the borings to depths of about 15 to 17 meters. Liquid limits for this material ranged from 44% to 84%, with plasticity indices ranging from 30% to 50%. Vertical swell tests in the vicinity of the bridge, from the initial geotechnical investigation (Terra-Mar 1999), ranged from 125 to 455 kilo Newton per square meters (kN/m^2). Allen and Flanigan (1986) reported swell pressures in the upper Eagle Ford to be high as 240 to 1200 kN/m^2 and free swell up to 15%.

Lift-Off Tests – Four earth anchors of the top row (the lower anchors were covered and unavailable for testing) were tested to evaluate their existing loads. When installed, specifications called for these anchors to be initially loaded to 125% of their design load (DL) and held for a period of time and subsequently the loads were reduced to 75% DL. The top row of anchors was installed prior to excavation for the roadway, and each succeeding row of anchors were installed at the completion of the successive excavation lifts. The recommended design load for the tied-back wall was 3.8 kN/m^2 of depth (Terra-Mar 1999) or equivalent to an at rest lateral earth pressure coefficient, K_o, of 0.64.

Loads measured at lift-off of the four anchors ranged from 10% to 50% DL with an average of 34%DL. While these results are subject to alternate interpretations, the most likely explanation for these measured loads is that the anchors began to creep due to the higher than anticipated lateral earth pressure from the clay shale and the anchors subsequently lost load capacity. The lost capacity of the anchors likely resulted from the large relative movements between the anchors and the surrounding shale. This is the most likely interpretation in view of the assessment by the structural engineer that the abutment wall had moved approximately 4 inches laterally. This interpretation is based on the tightly closed expansion joints in the bridge deck and displaced position of the longitudinal deck girders relative to the transverse abutment beam seats.

Inclinometers – Four inclinometers were placed directly behind the abutment walls, two on each side of the bridge. The movements of all four were small, on the order of .006 to .013 meters, but they moved inward and outward relative to the bridge largely consistent with seasonal temperature changes. Other inclinometers on the project, behind the roadway retaining walls, showed consistent inward movements toward the depressed roadway.

Dilatometer – Eight test locations, four on each side of the bridge, were conducted using the DMT to assess the in-situ lateral stress conditions in the vicinity of the bridge. The results of these tests were interpreted using empirical correlations developed for overconsolidated clays presented by Lunne, et. al. (1990). The data is presented in Figure 4 as a plot of in-situ horizontal stress versus depth. All data points plot between K_o values of 2 and 4, with an average value of approximately 2.6.

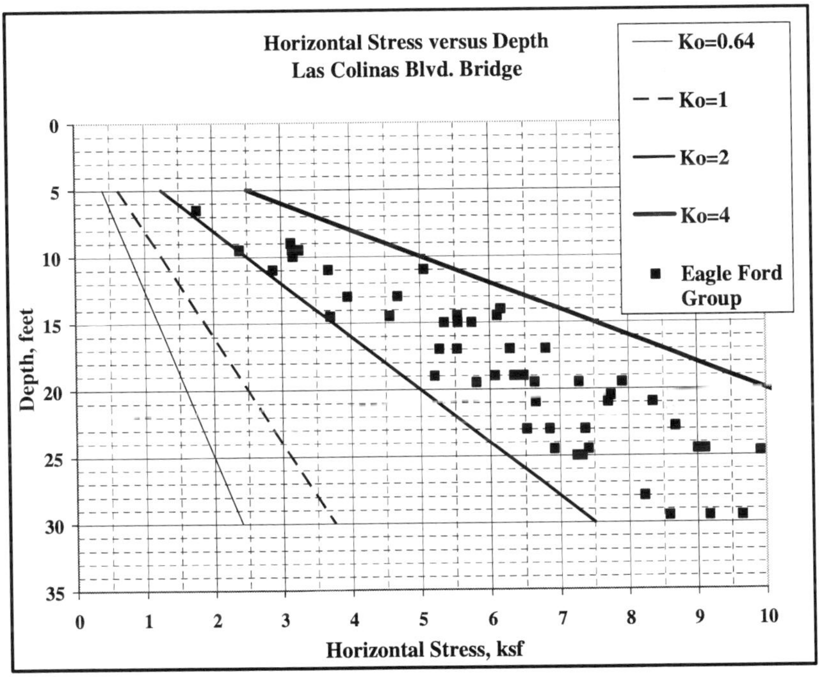

FIG.4: Las Colinas Bridge DMT Test Results

These high horizontal stress test results compare well with measured values in similar soil conditions as presented in the literature. Literature papers, which mostly focused on issues of the strength of overconsolidated clay, citied K_o values similar to those measured at the Las Colinas Bridge. These references include: Duncan and Dunlop (1969), Katti et. al. (1983), Neyer (1963) and Strazer, Bestwick, and Wilson (1974).

CONCLUSIONS

Top-Down construction in transportation projects has significant advantages relative to cost and schedule, and as such is likely to become more common place. As in the case of the PGBT, it allows for construction to proceed along parallel tracks thus reducing the critical path of a project's construction. Not only is it beneficial from a schedule perspective, it also has a significant safety aspect in that it does not require the high lifting common with much bridge construction. In summary, Top-Down construction has many positive attributes, but as we learned from the PGBT project, those who propose to use this method must understand the ramifications of the process on the earth mass it intends to retain.

In the Las Colinas Bridge case the material that required retention was predominantly clay shale, which by the nature of its origin and its constituents, had potential to develop high horizontal insitu stresses. The measured horizontal stresses were more than 4 times those assumed from the quasi-Rankine approach used for design (K_o assumed = 0.64 versus K_o measured in-situ = 2.6). The result was that excessive loads were imposed on the bridge deck, resulting in significant damage to the deck and concrete guard rail features.

The lessons learned from this case history are:

1. When involved with a new, especially a first-time approach, considerable care and research should be applied in developing a geotechnical engineering design;
2. Lateral earth pressures developed on walls restraining overconsolidated clays and clay shales must consider the potential for development of high horizontal in-situ stresses and,
3. Lateral earth pressures developed in overconsolidated clays and clay shales can not be computed using classic Rankine methods.

ACKNOWLEDGEMENTS

The authors would like to acknowledge their client the North Texas Tollway Authority in the support of this investigation and for granting permission to publish these findings. Additionally the authors express their appreciation for the work of Mr. Tony Kimmey and Mr. Mir Ali of Carter-Burgess and their Kleinfelder colleagues, Dr. Scott M. Mackiewicz, and Mr. Steve Saye.

REFERENCES

Allen, P.M., and Flanigan, W.D. (1986). "Geology of Dallas, Texas United States of America." *Bulletin of the Association of Engineering Geologist*, Vol. XXIII, No.4:363-418.

Bjerrum, L. (1967). "Progressive Failure in Slopes of Overconsolidated Plastic Clay and Clay Shales." *J. Soil Mechanics and Foundation Division*, ASCE Vol. 98, SM5: 1-49.

Duncan, J.M. and Dunlop, P. (1969). "Slopes in Stiff-Fissured Clays and Shales." *J. Soil Mechanics and Foundations Division*, ASCE, Vol. 95, SM 2.

Katti, R.K., Bhangale, E.S., and Moza, K.K. (1983). "Lateral Pressure in Expansive Soil with and without a Cohesive Non-Swelling Soil Layer – Application to Earth Pressures on Cross Drainage Structures in Canals and Key Walls in Dams (Studies on K_o Condition)." *Central Board of Irrigation and Power, New Delhi, India*, Technical Report No. 32.

Kleinfelder (2006). "Forensic Evaluation of VERT Wall System and Bridges at Los Colinas and MacArthur Boulevards along President George Bush Turnpike." *Kleinfelder,* Project 52136.

Lunne, T., Powell, J.J.M., Hauge, E.A., Mokkelbost, K.H., and Uglow, I.M. (1990). "Correlation of Dilatometer Readings with Lateral Stress in Clay." *Transportation Research Record* 1278: 183-193.

Neyer, J.C. (1963). Lateral Pressures in Preconsolidated Clay. *Thesis*, Master of Science in Civil Engineering, University of Washington.

Strazer, R.J., Bestwick, K. and Wilson, S.D. (1974). "Design Considerations for Deep Retained Excavations in Over-Consolidated Clays." *Bulletin of the Association of Engineering Geologists*, Vol. XI, No.4.

Terra-Mar (1999). "Geotechnical Report, President George Bush Turnpike, Segment V, Sections B & C From SH 114 to IH 635." TMI Report No. 98-099.

Unsaturated Soil Moisture Diffusivity Measurements in Laboratory Using Thermocouple Psychrometers

Daniel Mabirizi[1] and Rifat Bulut[2], M. ASCE

[1]Graduate Student, School of Civil and Environmental Engineering, Oklahoma State University, 207 Engineering South, Stillwater, Oklahoma 74078 USA

[2]Assistant Professor, School of Civil and Environmental Engineering, Oklahoma State University, 207 Engineering South, Stillwater, Oklahoma 74078 USA, rifat.bulut@okstate.edu

ABSTRACT: The performance of civil structures constructed on unstable or expansive unsaturated soils can be critically affected by moisture movement through these soils. The unsaturated diffusivity coefficient controls transient moisture flow conditions within a soil in response to total suctions or fluxes imposed at the boundaries of the soil profile. This study adopts and analyzes an approach originally developed by Peter Mitchell for laboratory unsaturated soil diffusion coefficient measurements. The diffusivity coefficients are determined by measuring total suction profiles over time in cylindrical soil specimens. Mitchell's methodology for evaluating the soil moisture diffusion characteristics is simple, economical, and accurate hence it can be used on a routine basis in a geotechnical laboratory. The paper presents diffusivity measurements of high plasticity clays exposed to a relatively constant low relative humidity environment in laboratory. The possible effect of soil sample lengths in determining accurate estimates of moisture diffusivity is also presented.

INTRODUCTION

The development of satisfactory designs and accurate analysis of the performance of civil structures built upon or into unsaturated soils depends on accurate predictions of the rate of diffusion of moisture through the soil. In saturated soils, hydraulic conductivity is constant; however, moisture flow in unsaturated soils depends on suction. This presents difficulties in characterizing moisture movement, predicting the resulting soil movement, and designing civil structures to accommodate the soil movement without distress to these structures. The suction technique proposed by Mitchell (1979) has the advantage of using moisture energy as a means of predicting moisture changes in unsaturated soils.

Soil moisture diffusivity properties can be either back-calculated from field measurements, determined from laboratory tests or calculated using empirical correlations involving soil index properties. Field and laboratory methods are based on suction measurements recorded over time. The laboratory methods for evaluating unsaturated soil moisture diffusivity will be discussed in this paper. The solution to the diffusion problem proposed by Mitchell (1979) and adopted in this paper has mainly two advantages. It provides a linear solution to the nonlinear flow of moisture in unsaturated soils and also provides a procedure for estimating soil diffusivity properties in the laboratory that is simple, economical and can be used routinely in a geotechnical laboratory (Aubeny et al., 2003).

Variations in soil moisture diffusivity determined on the same soil may be attributed to soil stratification and especially cracks and root-holes in the soil, which ought to be considered in the characterization of moisture flow (Lytton et al., 2004). Variations in the soil diffusion properties could also be caused by use of an inadequate sample length as will briefly be discussed in this paper. The theoretical justification of the effect of sample length on the diffusion coefficient measurements will be presented.

BACKGROUND

Mitchell (1979) developed an equation that governs moisture diffusion through an unsaturated soil mass (Equation 1), with the rate of diffusion controlled by a single diffusivity coefficient, α:

$$\frac{\partial^2 u}{\partial x^2} = \frac{1}{\alpha}\frac{\partial u}{\partial t} \tag{1}$$

where u = logarithmic measure of total suction, t = time, x = distance, and α = diffusion coefficient.

Equation 1 represents the movement of moisture through unsaturated soils. Mitchell (1979) proposed two laboratory tests for determining the diffusion coefficient in unsaturated soils namely the soaking test and the evaporating test. This paper is focused on the evaporating test. In this test, the perimeter and one end of a cylindrical soil column of specified geometry are sealed to prevent loss or gain of moisture and the other end of the sample is left open (i.e., exposed to air) to allow flow of moisture out of the sample. Mitchell (1979) described suction as a function of position and time as:

$$u(x,t) = u_a + \sum_{n=1}^{\infty} \frac{2(u_o - u_a)\sin z_n}{z_n + \sin z_n \cos z_n} \exp\left(\frac{z_n^2 \alpha t}{L^2}\right) \cos\left(\frac{z_n x}{L}\right) \tag{2}$$

where x = distance from closed end, t = time, L = sample length, u_a = atmospheric suction in laboratory, u_o = initial suction in soil sample, α = diffusion coefficient, z_n = coefficient obtained from the solution of cot $z_n = z_n/h_eL$, and h_e = the evaporation coefficient, which is a constant value equal to 0.54 cm^{-1}, based on Mitchell (1979) recommendation. The rationale for the evaporation coefficient is that the rate of evaporation of moisture from the surface of a soil mass is considered as being linearly proportional to the difference in suction at the soil surface and suction in the atmosphere by a constant parameter, h_e. Equation 2 defines the distribution of suction throughout the soil body as a function of distance and time.

The moisture diffusion coefficient can be determined indirectly by measuring suction changes in a soil column at various locations. Conversely, satisfactory estimates of α can be based upon empirical correlations to soil index properties such as Equation 3 given by Aubeny and Lytton (2003):

$$\alpha = \frac{h_o k_o S}{0.434} \frac{\gamma_w}{\gamma_d} \tag{3}$$

where k_o = saturated permeability of the soil, h_o = suction at which the soil saturates, S = slope of soil moisture characteristic curve (logarithm of suction versus gravimetric water content), γ_w = unit weight of water, and γ_d = dry unit weight of soil.

The parameter S can be obtained from the soil moisture characteristic curve, which is commonly measured with a pressure plate apparatus while the parameter k_o can be measured directly in the laboratory using permeability or consolidation tests (Aubeny and Lytton, 2003). If such data are not available, parameter S can also be obtained from the following empirical relationship (Aubeny and Lytton, 2003):

$$S = -20.29 + 0.155\,(LL) - 0.117\,(PI) + 0.0684\,(\text{percent Fines}) \tag{4}$$

where LL = liquid limit, and PI = plasticity index.

As previously mentioned, soil diffusivity can be determined using field or laboratory suction methods. Aubeny et al. (2005) findings suggest that field values of moisture diffusion are much greater than laboratory values.

LABORATORY TESTS

The laboratory procedure for the evaporating test originally proposed by Mitchell (1979) was used to determine the diffusion coefficient (α) in this study. Soil suction was measured using thermocouple psychrometers. The initial suction was determined using the filter paper test while relative humidity data was employed to calculate the atmospheric suction in the laboratory.

Initial and Atmospheric Suction Measurements

The filter paper test was used to determine the initial total suction of the soil. The testing procedure as given in Lytton et al. (2006) was adopted for the laboratory soil suction measurements. Schleicher & Schuell No. 589 – White Hard (WH) filter papers were used in this study. The tests were performed in a temperature controlled room and at least one week was allowed for vapor equilibration between the soil and filter paper to occur. Prior to testing, filter papers are calibrated to determine the relationship between equilibrium water content and relative humidity using salt solutions of known concentration. The calibration curve developed by Bulut et al. (2001) for the brand of filter paper used was employed in this study.

The atmospheric suction at the open end of the soil specimen was obtained by measuring the relative humidity of the air in the laboratory using a digital thermo-hygrometer and relating it to suction using the Kelvin's equation given by Fredlund and Rahardjo (1993):

$$h = \left(\frac{RT}{V}\right)\ln(RH) \tag{5}$$

where h = total suction, R = universal gas constant, T = absolute temperature, V = molecular volume of water, and RH = relative humidity.

The relative humidity was measured several times in a day for the duration of the diffusion test and an average of the values was obtained and using Equation 5 the atmospheric suction was determined. The relative humidity was relatively constant (a few percent change within the accuracy of the measurement) during a particular drying test.

Soil Diffusion Coefficient Measurements

The laboratory procedure for the evaporating test originally proposed by Mitchell (1979) was utilized in this study. Soil suction was measured with thermocouple psychrometers having stainless screen shields, which measure the relative humidity in the air phase of the soil pores or the region near the soil (Fredlund and Rahardjo, 1993) and a datalogger from Wescor Inc/Campbell Scientific Inc. Before using the psychrometers, the microvolt output was calibrated to suction using sodium chloride (NaCl) solution of log kPa 2.66, 3.14, 3.36, 3.51 and 3.67 by suspending the psychrometers in these solutions under carefully monitored and controlled temperature conditions of 25±0.1°C because water potential of the salt solutions are highly temperature dependent. A relationship was determined between thermocouple microvolt outputs and known total suction values of the standard NaCl solution.

For the diffusion test, small holes are drilled at various locations into the side of a cylindrical soil specimen of specified geometry and psychrometers are placed into the holes to measure total suction over a given period of time. These holes and the whole soil column are sealed in order to prevent loss or gain of moisture except for one end of the sample that is left open (i.e., exposed to air) to allow water evaporation from the soil column (Figure 1). The testing procedure for diffusion coefficient measurements followed that outlined in Lytton et al. (2004).

The duration of the test is typically four days to one week. It is important that this test is performed in a constant temperature and relative humidity environment, such as one depicted in Figure 2, which was also used in this study, in order to obtain reliable estimates of the diffusion coefficient. Having determined the initial total suction, atmospheric suction, length of soil specimen, and distance of psychrometer from closed end of sample, theoretical suction values corresponding to suction and time measurements recorded for each psychrometer can be computed using Equation 2. To solve Equation 2, α is initially estimated and then later optimized to minimize the error between the measured and theoretical suction values. Using the data interpretation procedure given by Aubeny and Lytton (2003), a typical theoretical versus measured suction values with time as presented in Figure 3 was developed.

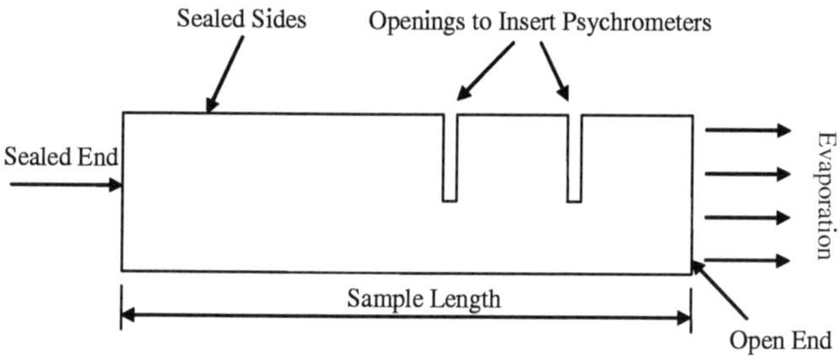

FIG. 1. Typical Cylindrical Soil Specimen Used in the Evaporation Test.

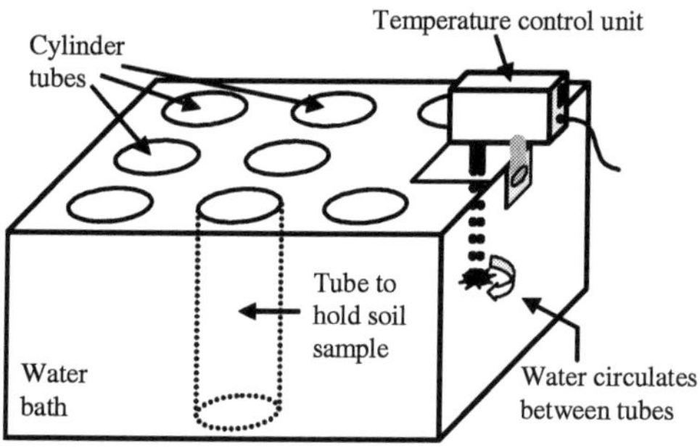

FIG. 2. Water Bath Used to Maintain Soil Specimen at Constant Temperature (Lytton et al., 2004).

SOILS TESTED

The abovementioned testing procedure was followed to determine the diffusivity coefficients of high plasticity clay soils from Texas. Laboratory tests were performed at Oklahoma State University in a controlled humidity and temperature environment. The samples as received from the field had been extruded from the sampling tubes and were wrapped in three to four layers of clear flexible plastic and aluminum foil respectively to prevent loss or gain of moisture. They were stored in a temperature controlled environment before testing. All soil tested consisted of 7.62 cm diameter tube samples obtained at depths of 0.00 to 3.66 m from compacted clay highway embankments. The length of the sample used in the diffusion tests varied from 9.60 to

18.00 cm, depending on the length of the soil column provided. Flexible plastic wraps and plastic electrical tapes provide a very good contact between the soil and the sides of the column during the testing, preventing an alternative drainage path during the test.

In this experiment, two psychrometers were used for each diffusion test. The results from the psychrometer nearest to the open end were generally used for determination of α. For the samples where this psychrometer gave insufficient suction values, α was calculated using suction values given by the second psychrometer. Mitchell (1979) recommended h_e equal to 0.54 cm^{-1} was used in data interpretation. A short summary of the results is shown in Table 1. In a few soil specimens, very small hairline cracks have formed during the drying process. The results in Table 1 do not account for the effects of the hairline cracks on the measured diffusion coefficients. The diffusion coefficients obtained in this study are comparable with some previous results in the literature (i.e., Richards, 1965 and Bulut et al., 2005).

Table 1. Diffusion Coefficient Test Results

Boring No.	Sample Depth	Sample Length	Psychrometer Distance from Closed End	Initial Total Suction	Atmospheric Suction	Diffusion Coefficient
	(m)	(cm)	(cm)	(log kPa)	(log kPa)	($x10^{-3}$ cm^2/min)
1	0.61-1.22	18.00	15.00	3.01	5.27	1.18
	1.22-1.83	14.20	11.20	3.34	5.27	0.79
	3.05-3.66	15.90	12.80	3.29	5.27	0.86
2	0.00-0.61	16.60	14.80	2.54	5.27	1.99
	1.83-2.44	16.40	13.70	2.80	5.27	1.80
3	0.00-0.61	12.00	9.00	1.75	5.11	3.29
	2.44-3.05	10.25	7.75	2.73	4.89	2.84
4	0.30-0.61	9.80	7.30	3.05	4.74	10.5
	2.44-3.05	9.60	7.10	2.53	4.74	6.22

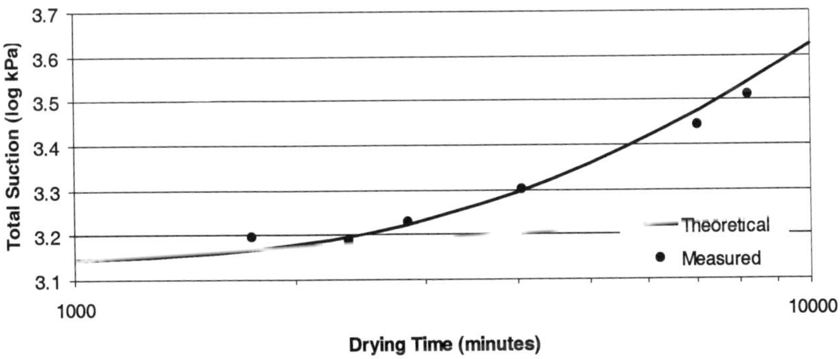

FIG. 3. Theoretical verses Measured Total Suction Values with Time.

ANALYSIS

The estimated diffusion measurements varied from 0.79×10^{-3} to 10.5×10^{-3} cm^2/min. In determination of the diffusion coefficients using Equation 2, parameters such as atmospheric suction (u_a), initial suction (u_o), evaporation coefficient ($h_e = 0.54$ cm^{-1}) are constant. A controlled temperature and relative humidity environment with good workmanship would produce reliable suction values. On the other hand, the sample length (L) to be used in the diffusion test can be varied depending on the length of the soil column available as shown in Table 1. However, Figure 4 shows that variations in sample length can cause discrepancies in z_n. The coefficient z_n is determined by solving the expression $\cot z_n = z_n/h_eL$, which is part of the solution of Equation 2. Because z_n varies with L, reliable alpha coefficient estimates may not be obtained if the appropriate sample length is not used in the diffusion test. The sample length could be optimized using the expression $\cot z_n = z_n/h_eL$ or Figure 4.

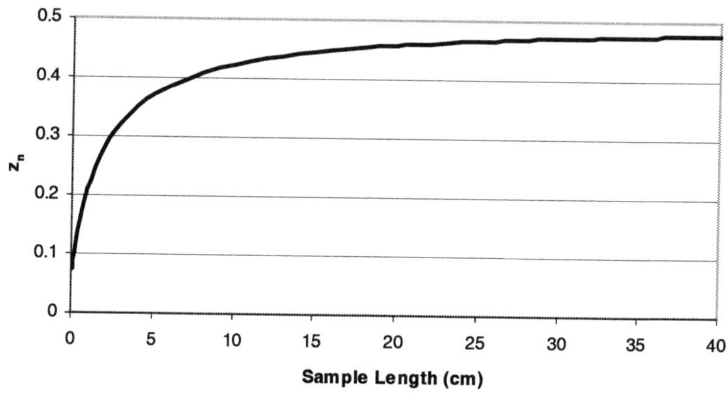

FIG. 4. Sample Length versus Optimum z_n Value.

From Figure 4 it can be seen that variations in z_n decrease with increase in sample length. Sample lengths below 10 cm may cause drastic variations in z_n while those below 5 cm are considered impractical. There are relatively low variations in z_n values for lengths of 15 cm and above. Samples of length 20 cm and above would be ideal for the diffusion test.

SUMMARY

Mitchell (1979) method for predicting moisture movement is relatively simple, economical, and α estimates obtained using this method are relatively accurate. However, an inadequate sample length may affect the α values. A sample length between 10 and 20 cm provide low variations in z_n and appropriate for determination of α coefficient. Further research in the effect of z_n to the diffusion coefficient is required.

REFERENCES

1. Aubeny, C.P., and Lytton, R.L. (2003). "Estimating Strength versus Location and Time in High Plasticity Clays." *Research Report No. FHWA/TX-03/2100-PI*, Texas Transportation Institute, College Station, Texas.
2. Aubeny, C.P., Bulut, R., and Lytton, R.L. (2005). "Laboratory versus Field Values of Unsaturated Soil Diffusivity." *Advanced Experimental Unsaturated Soil Mechanics*, Proceedings of the International Symposium on Advanced Experimental Unsaturated Soil Mechanics, Trento, Italy, pp. 495-499.
3. Aubeny, C.P., Lytton, R.L., and Tang, D., (2003). "Simplified Analysis of Unsteady Moisture Flow through Unsaturated Soil." *Geology and Properties of Earth Materials*, Transportation Research Record, Washington D.C., Publication No. 1821, pp. 75-82 .
4. Bulut, R., Aubeny, C.P., and Lytton, R.L. (2005). "Unsaturated Soil Diffusivity Measurements." *Advanced Experimental Unsaturated Soil Mechanics*, Proceedings of the International Symposium on Advanced Experimental Unsaturated Soil Mechanics, Trento, Italy, pp. 281-286.
5. Bulut, R., Lytton, R.L., and Wray, W.K. (2001). "Soil Suction Measurements by Filter Paper." *Expansive Clay Soils and Vegetative Influence on Shallow Foundations*, ASCE Geotechnical Special Publication No. 115, Houston, Texas, pp. 243-261.
6. Fredlund, D.G., and Rahardjo, H. (1993). "Soil Mechanics for Unsaturated Soils." John Wiley and Sons, New York, NY.
7. Lytton, R.L., Aubeny, C.P., and Bulut, R. (2004). "Design Procedure for Pavements on Expansive Soils." *Research Report No. FHWA/TX-05/0-4518-1*, Texas Transportation Institute, College Station.
8. Mitchell, P.W. (1979). "The Structural Analysis of Footings on Expansive Soils." *Research Report No. 1*, K. W. G. Smith & Assoc. Pty. Lty., Newton, South Australia.
9. Richards, B.G. (1965). "Determination of the Unsaturated Permeability and Diffusivity Functions from Pressure Plate Outflow Data with Non-Negligible Membrane Impedence." *Moisture Equilibria and Moisture Changes in Soils Beneath Covered Areas*, Butterworths, Sydney, pp. 47-54.

IMPACT OF A TREE ON A RESIDENTIAL BUILDING ON EXPANSIVE SOILS: A NUMERICAL APPROACH

Xiong Zhang[1], Lin Li[2], and Liangbiao Chen[2]

1. Assistant Professor, Department of Civil and Environmental Engineering, University of Alaska Fairbanks, Fairbanks, AK99775. ffxz3@uaf.edu

2. Graduate Student, Department of Civil and Environmental Engineering, University of Alaska Fairbanks, Fairbanks, AK99775.

ABSTRACT

This paper presented the analysis of a residential building built on expansive soils influenced by the local daily weather conditions and a tree over a period of two years. A recently-developed simulation system was used for the simulation in which the influence of weather conditions on evapotranspiration was evaluated by the FAO 56 PM method, the volume change of expansive soil was simulated by fully-coupled consolidation theory for saturated-unsaturated soils, the soil-structure interaction was simulated by coupled contact elements, and the slab foundation and superstructures were simulated by general shell elements. Different simulation techniques were used to combine all these components into one single system. Simulation results indicated that tree had significant influence on behaviour of residential building built on expansive soils.

INTRODUCTION

Expansive soil undergoes appreciable volume change in response to soil moisture changes. In arid and semiarid climate areas, climatic conditions changes from wet to dry and vice versa every year. These changes in the climatic conditions cause moisture change within unsaturated expansive soils near the surface and consequent vertical surface movements. When a building is constructed on expansive soils, the soil moisture underneath the building is not uniform under most situations, which can cause differential movements of soil and subsequent distresses in the building. Krohn and Slosson estimated that the annual cost of expansive soil damage in the US to be $7.0 billion in 1980 and damages to single-family and commercial buildings accounted for nearly one-third of the total amount of damage resulting from expansive soils.

In order to prevent the residential building from damages caused by differential movements, slabs-on-ground are often constructed on expansive soils to resist excess differential movements and distortion. Although numerous studies have been conducted in the past to investigate the behaviour of slabs-on-ground under the influence of weather conditions (e.g. Wray 1978; Gay 1993; Lytton 1994; Post Tensioning Institute (PTI) 2004; Bulut 2001; Wray et al. 2005; Li 2006; El-Garhy and Wray 2006), there are some limitations in the existing research such as: (1.) the influence of the daily weather variations was not reasonably evaluated (usually either constant, or sinusoidal suction, or flux boundary conditions were used to simulate the weather variations); (2.) the theory used to simulate the volume change of expansive soils only represented some aspects of the soil behavior (e.g.: the foundation loading cannot change the pore pressure or suction); (3.) the material properties were simplified (linear or log-linear elasticity, or most soil properties dependent only on soil suction); (4.) the simulation of the soil-structure interaction was either not included or did not reflect the real situation very well (the soil was considered as a solid continuum without water in it and the water flow is not simulated); and (5.) the influence of the superstructure was not considered (the load distribution was idealized). Zhang (2004) presented a fully coupled, non linear, elastic, three dimensional numerical simulation system for the analysis of slabs-on-ground which attempted to overcome the shortcomings in the existing research. In this paper, a three dimensional simulation for a virtual residential building built on expansive soils to demonstrate the application of the proposed system.

A COMPLETE SYSTEM FOR SIMULATING RESIDENTIAL BUILDINGS ON EXPANSIVE SOILS

It was proposed in Zhang (2004) that evapotranspiration at the ground surface be estimated using the FAO 56 (Food and Agriculture Organization of the United Nations) Penman-Montieth method, which has been well established as the most accurate and robust method and the international standard to estimate the reference evapotranspiration (Allen et al. 1998). The weather data includes rainfall, solar radiation, air temperature, relative humidity and wind speed, all of which are readily available from a local weather station or the internet. These data were used to determine the flux boundary conditions for the simulation. The following equation is used to estimate the evapotranspiration rate in the FAO 56 PM method for a reference grass surface:

$$ET_0 = \frac{0.408\Delta\left(R_n - G\right) + \gamma \frac{900}{T+273} u_2 \left(e_s - e_a\right)}{\Delta + \gamma\left(1 + 0.34u_2\right)} \quad (1)$$

Where ET_0 =reference evapotranspiration (mm day^{-1}); Rn =net radiation at the crop surface (MJ m^{-2} day^{-1});G =soil heat flux density (MJ m^{-2} day^{-1});T= air temperature at a 2 m height (°C);u_2 =wind speed at a height of 2 m (m s^{-1}); e_0 =saturation vapour pressure (kPa); e_a =actual vapour pressure (kPa); Δ =slope of saturation vapour pressure versus temperature curve (kPa $°C^{-1}$); λ= latent heat of vaporization, 2.45 MJ/kg, and γ =psychrometric constant (kPa $°C^{-1}$).

If the real conditions such as ground cover, canopy, aerodynamic resistance, are different from the reference grass conditions, the evapotranspiration will be different. The actual evapotranspiration $ET_{c\ adjusted}$ in the field may be lower if the vegetation or trees are under non-optimal conditions or the total suction in the soil is too high. The following equation was recommended to adjust the reference evapotranspiration in to actual evapotranspiration (Allen et al. 1998):

$$ET_{c\ adjusted} = K_s \times K_c \times ET_0 \quad (2)$$

Where $ET_{c\ adjusted}$ = crop evapotranspiration (mm/day); K_c = crop coefficient (dimensionless); K_s = water stress coefficient; and ET0 = reference grass evapotranspiration (mm/day). Recommendations for K_c for different vegetations and different seasons and K_s can be found in Allen et al. (1998).

The volume change of an unsaturated expansive soil was described by the following governing differential equations (Zhang et al. 2005):

$$(\lambda+G)\frac{\partial\varepsilon_v}{\partial x}+G\nabla^2 u-(3\lambda+2G)\alpha\frac{\partial(-u_w)}{\partial x}+X=0 \tag{3a}$$

$$(\lambda+G)\frac{\partial\varepsilon_v}{\partial y}+G\nabla^2 v-(3\lambda+2G)\alpha\frac{\partial(-u_w)}{\partial y}+Y=0 \tag{3b}$$

$$(\lambda+G)\frac{\partial\varepsilon_v}{\partial z}+G\nabla^2 w-(3\lambda+2G)\alpha\frac{\partial(-u_w)}{\partial z}+Z=0 \tag{3c}$$

$$\frac{1}{\rho_w g}\left(\frac{\partial}{\partial x}\left(k\frac{\partial(-u_w)}{\partial x}\right)+\frac{\partial}{\partial y}\left(k\frac{\partial(-u_w)}{\partial y}\right)+\frac{\partial}{\partial z}\left(k\frac{\partial(-u_w)}{\partial z}+1\right)\right) = \rho_d C_w\frac{\partial(-u_w)}{\partial t}+m_1^w\frac{\partial(\sigma_m)}{\partial t}+S' \tag{3d}$$

Where u, v, and w = displacements in x-, y-, and z-directions, respectively; X, Y, and Z = Body forces in x-, y-, and z-directions, respectively; $\lambda = E\nu/((1+\nu)(1-2\nu))$; $G = E/(2(1+\nu))$; E = Young's modulus of the soil, ν = Poisson's ratio of the soil, α = coefficient of expansion due to matric suction variation; ε_v = volumetric strain; u_w=pore water pressure; k = hydraulic conductivity; ρ_d = dry mass density of the soil; C_w = the specific water capacity of the soil (the volume of water expelled from a unit mass of soil by one kPa of matric suction); σ_m = mean net normal stress; and S'= water source term.

The general shell elements are used to simulate the behaviour of the superstructure since the thicknesses of slabs and walls are much smaller than their lengths and breadths. Coupled contact elements were used to simulate the soil-structure interaction (Zhang 2004; Zhang and Briaud 2006) and 'pseudo-moisture variation" simulation technique (Zhang 2004) were used to combine the proposed system into a unified system. The thermodynamic analogue was to implement the system in ABAQUS (2002). More detailed information can be found in Zhang (2004).

SIMULATION OF A RESIDENTIAL BUILDING BUILT ON EXPANSIVE SOILS

The proposed system was used to simulate the behaviour of residential buildings on expansive soils Fig. 1a shows the three slabs numbered as I, II and III, respectively. This EI values of the slabs were approximately 8.5×10^5 kN·m^2. Fig.1b shows the walls used in the simulation; the wall thickness and height were 0.1m and 3.5m, respectively. Doors and windows were presented by different sizes of openings in the walls. Fig. 1c shows the dimension for an idealized tree root zone. Fig. 1d shows a representative residential building formed by the slabs and walls of Figs. 1a and 1b. More complicated and realistic structure can be modelled using the same system. Fig. 1e shows the soil configuration used in the simulation. The soil was assumed to have dimensions of 48m × 48 m × 6m and include the two soil layers as those at a site in Arlington, Texas (Briaud et al. 2003). The first layer was a 1.8m dark gray silty clay layer with a grass root zone in the top 0.47 m. The second layer was a 4.2 m brown silty clay layer. All the material properties in Eqs. 3a through 3d were measured in the laboratory and can be found in Zhang (2004). The hourly weather data were collected near the site at the Arlington Municipal Airport

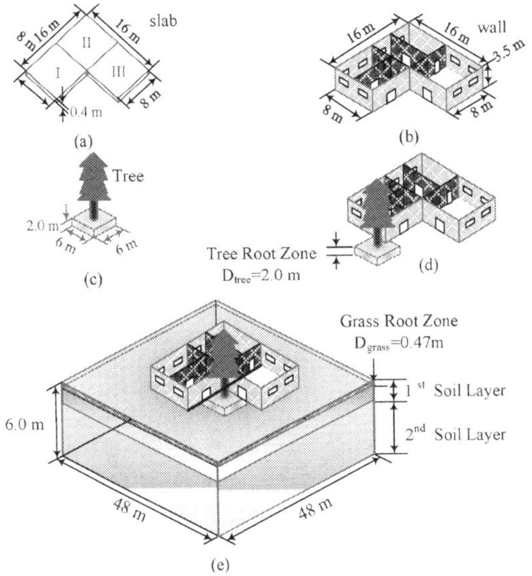

FIG. 1 . Simulation Domain. (a.) Slab; (b.)Walls; (c.) House Formed by the Slabs and Walls; (d.) Tree Root Zone; (e.) Soil Domain and the Building

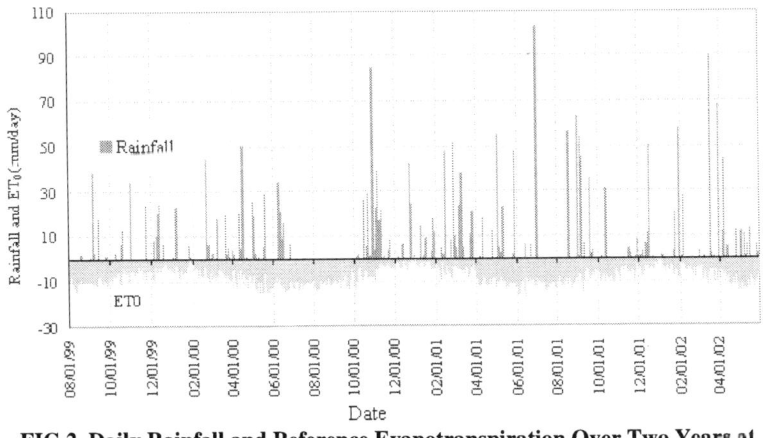

FIG.2. Daily Rainfall and Reference Evapotranspiration Over Two Years at Arlington, TX.

between August 1, 1999 and August 1, 2001(Fig.2). Fig.2 shows daily rainfall and reference evapotanspiration calculated from the daily mean temperature, relative humidity, and wind speed. The FAO-56 PM method was used to determine the evapotranspiration at the ground soil surface. The single crop coefficient for Johnsongrass was assumed to be 0.6, which was the same as that at the Arlington site. A tree with a 6m × 6 m × 2.0 m root zone as shown in Fig. 1c was planted close to a corner of the building. It was considered that rain can not infiltrate directly into the deep tree root zone. The single crop coefficient for the tree was assumed to be K_c= 0.3. Both the slabs and walls were assumed to be made of plain concrete. The mass density for concrete was 2.4 $\times 10^3$ kg/m3. The Young's Modulus of the concrete footing was E=2 $\times 10^7$ kPa. Considering that the strain in the slabs and the walls were small while the rotation was large, the section Poisson's ratio was defined as zero ν=0.0. Shell elements were used for the slab and walls and the shell thickness remained constant.

Coupled thermal-mechanical jointed (contact) elements in ABAQUS/Standard (2002) were used to simulate the interaction at the soil-concrete footing interface (Zhang and Briaud 2006). The "hard" contact relationship in ABAQUS was used to simulate the normal behaviour at the soil-slab interface. During the simulation, the program computed the thickness of the contact elements in the direction normal to the soil-structure interface. When the soil and the slab foundation were in contact (the thickness of the contact element is zero), any compressive load can be transferred from the slab to the soil. When the soil and the foundation were not in contact (the thickness of the contact element is greater than zero), no load can be transferred from the slab to the soil. The basic Coulomb friction model was used to simulate the tangential behaviour in the soil structure interaction in which the two contacting surfaces can carry shear stresses up to a certain magnitude across their interfaces before they started sliding relative to one another. The critical shear stress, τ_{crit} , at which sliding of the surfaces starts, was defined as a fraction of the contact pressure, P, between the surfaces (τ_{crit} =μP, where μ is known as the coefficient of friction). For a three-dimensional simulation, it was assumed that μ is the same in all directions (isotropic friction) at the first approximation. The static friction coefficient when there was no relative slip was different from the kinetic friction coefficient when there was relative slip. It was assumed that the friction coefficient decays exponentially from the static value to the kinetic value according to the formula (Oden and Martins 1985):

$$\mu = \mu_k + (\mu_s - \mu_k) e^{-d_c \dot{\gamma}_{eq}} = 0.1 + 0.4 e^{-4\dot{\gamma}_{eq}} \quad (4)$$

where μ_k is the kinetic friction coefficient, taken as 0.1 in the following example; μ_s is the static friction coefficient, taken as 0.5 in the following example; d_c is a user-defined decay coefficient, taken as 4 in the following examples; and $\dot{\gamma}_{eq}$ = the slip rate.

It was also assumed that no water was allowed to flow through the soil-slab interface. This condition was realized by defining a very low "gap conductance" to the jointed elements. The gap conductance of the contact elements was assumed to be 10^{-30} s^{-1} when the slab and the soil were in contact with each other. The gap conductance of the contact elements was assumed to be 0 when the slab and the soil were separated. All the other information was the same as those used in Zhang (2004).

RESULTS AND DISCUSSION

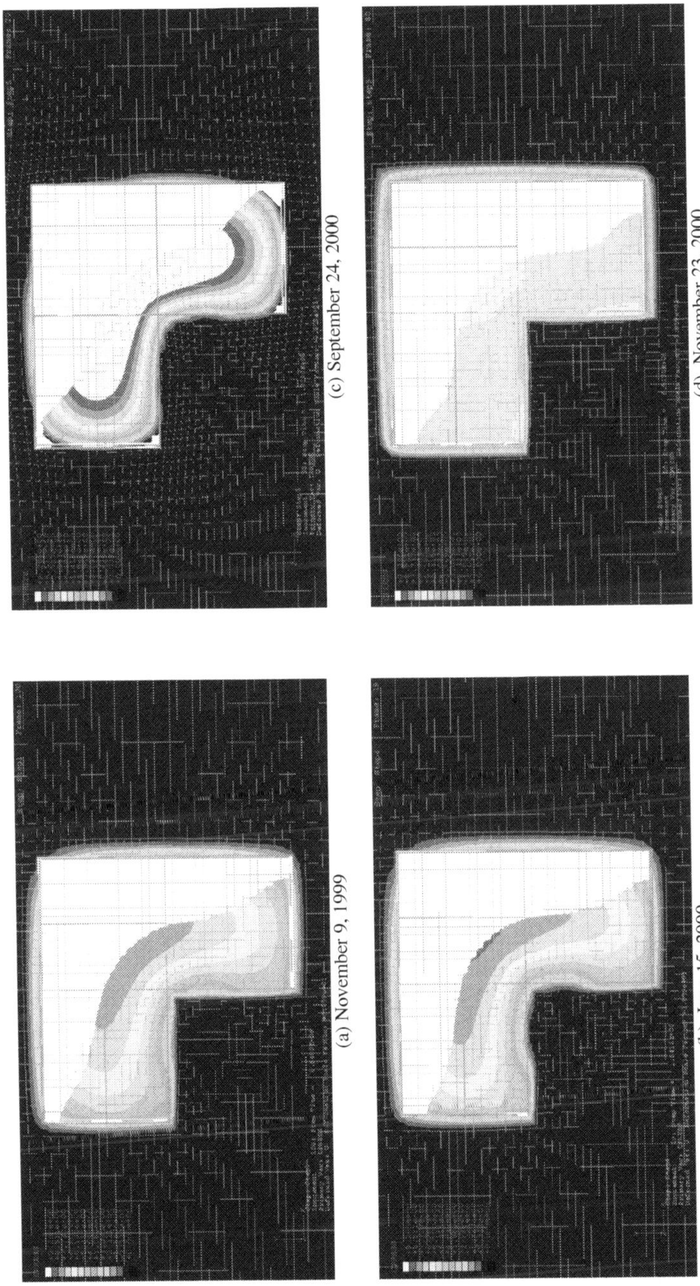

(a) November 9, 1999

(b). June 15, 2000

(c) September 24, 2000

(d). November 23, 2000

Fig.3. Changes of Gap between Soil and Slab in the Simulation Domain at Different times

A transient analysis was performed over a period of two years and the time step was chosen as one day. The simulation results have been visualized into many movies that separately showed the changing patterns of soil and slab movement, the suction in the supporting soil, and the slab directional moment, shear, stress, strain, and liftoff patterns that occurred over two years. Due to the limited space, only contact pressures between the soil and the slab were discussed for four different times as shown in Fig.3: November 9, 1999, June 15, 2000, September 24, 2000, and November 23, 2000. As shown in Fig. 2d from August 1, 1999 to June 18, 2000, rainfall and reference evapotranspiration were approximately evenly distributed, which was followed by a long dry season with no rain till October 1, 2000. After that, there was a rainy season and the rainfall and evapotranspiration were approximately evenly distributed. As shown in Fig. 3, from November 9, 1999 to June 15, 2000, there was contact pressure between the soil and the slab, indicating the soil and the slabs were in good contact. Outside of the slab, the contact pressure was zero. However, the contact pressure distribution was not uniform. At the edge of the building, the contact pressure was lower that that underneath the centre of the slab, which was mainly attributed to the shrinkage of the soil at the edges. At the corners of the slabs, the contact pressures were zero, indicating there was separation between the soil and the slab. This was due to the fact that the soils at the corners of the building were influenced by weather from two directions. There were also slight separation between the soil and the slab at the double entrance of the building, which was due to the existence of the tree. On September 24, 2000, after a long period of drying season without rain as shown in Fig.2d, there was significant shrinkage for soils at the edge of the slabs and there were separations between the soil and the slab, which was indicated by the zero contact pressure. There were separation between the soil and the slab at the double entrance of the building, which was caused by the evapotranspiration from the tree. From September 24, 2000 to November 23, 2000, there was a long rainy season and the soil outside of the building gained water through rainfall infiltration and heaved, which caused a more uniform contact pressure distribution as shown in Fig.3d.

CONCLUSIONS

In this paper, a complete system was used to simulate the behaviour of a residential building on expansive soils with slabs-on-ground foundation. Simulation results indicated that the proposed system can reasonably simulate many features of residential building built on expansive soils in a consistent way. Simulation results indicated that climatic conditions and tree had pronounced effect on residential building built on expansive soils.

REFERENCES

ABAQUS/Standard User's Manual (2002). Vol. I, II, and III, Version 6.3, Hibbit, Karlsson and Sorenson Inc., 1080 Main Street, Pawtucket, RI02860-4847, U.S.A.

Allen, R.G., Periera, L.S., Raes, D., and Smith, M. (1998). *Crop Evapotranspiration: Guidelines for Computing Crop Requirements.* Irrigation and Drainage Paper No. 56, FAO, Rome, Italy

Briaud, J.L., Zhang, Xiong and Moon, S. (2003). "The shrink test – water content method for shrink and swell predictions." *Journal of Geotechnical and Geoenvironmental Engineering, ASCE* 129(7), 590-600.

Bulut, R. (2001). "Finite Element Method Analysis of Slabs on Elastic Half Space Shrink-swell Soil Foundations." *Ph.D. Dissertation*, Department of Civil Engineering, Texas A&M University, College Station, TX.

El-Garhy, B.M. and Wray, W.K. (2006) "Investigation of a Leaking Water Line Beneath Residential Foundation on Expansive Soils." *4th International Conference on Unsaturated Soils*, April 2006, Carefree, AZ. Geotechnical Special Publication NO.147, Volume 1, Pp.367-378.

Gay, D. A. (1993). "Development of a Predictive Model for Pavement Roughness on Shrink-swell Clay." *Ph.D. Dissertation*, Department of Civil Engineering, Texas A&M University, College Station, TX.

Li, J. (2006). "Two Dimensional Simulation of a Stiffened Slab on Expansive Soil subject to a Leaking Underground Water Pipe." *4th International Conference on Unsaturated Soils,* April 2006, Carefree, AZ. Geotechnical Special Publication NO.147, Volume II, Pp.2098-2120.

Lytton, R.L. (1994). " Prediction of movements in expansive clay." In Vertical and horizontal deformations of foundations and embankments: *Proceedings of Settlement '94,* College Station, Texas, 16-18 Juan 1994. Edit by A.T. Yeung and G.Y. Feaalio. ASCE, Geotechnical Special Publication 40, pp. 1827-1845.

Oden, J. T., and Martins, J. A. C.(1985) "Models and computational methods for dynamic friction phenomena," *Computer Methods in Applied Mechanics and Engineering*, 52, 527–634.

Post-Tensioning Institute (2004). *Design and Construction of Post-Tensioned Slabs-on-Ground*, Phoenix, Arizona, USA, Third Edition.

Wray, W.K. (1978). "Development of a design procedure for residential and light commercial slabs-on-ground constructed over shrink-swell soils." *Ph.D. Dissertation*, Department of Civil Engineering, Texas A&M University, College Station, TX, December, 1978.

Wray, W.K., El-Garhy, B.M., and Youssef, A.A. (2005). "Three-dimensional model for moisture and volume changes prediction in shrink-swell soils." *Journal of Geotechnical and Geoenvironmental Engineering, ASCE* 131(3), 311-324.

Zhang, X. (2004) Consolidation theories for saturated-unsaturated soils and numerical simulations of residential buildings on shrink-swell soils. *Ph.D. Dissertation*, Department of Civil Engineering, Texas A&M University, College Station, TX.

Zhang, X. and Briaud, J. L. (2008). "Improved Approach to Construct Constitutive Surfaces for Stable-Structured Soils Covering Both Saturated and Unsaturated Conditions." *Journal of Geotechnical and Geoenvironmental Engineering, ASCE*, Volume 134, Issue 7, pp. 876-882.

Zhang, X., Lytton, R. L., and Briaud, J. L. (2005) "Coupled Consolidation Theory For Saturated-Unsaturated Soils" in the *Biot 2005 Centennial Conference* at the University of Oklahoma, Norman, USA

Author Index

Page number refers to first page of paper

Subject Index

Page number refers to first page of paper